Studies in Systems, Decision and Control

Volume 362

The series "Studies in Systems, Decision and Control" (SSDC) covers both new developments and advances, as well as the state of the art, in the various areas of broadly perceived systems, decision making and control–quickly, up to date and with a high quality. The intent is to cover the theory, applications, and perspectives on the state of the art and future developments relevant to systems, decision making, control, complex processes and related areas, as embedded in the fields of engineering, computer science, physics, economics, social and life sciences, as well as the paradigms and methodologies behind them. The series contains monographs, textbooks, lecture notes and edited volumes in systems, decision making and control spanning the areas of Cyber-Physical Systems, Autonomous Systems, Sensor Networks, Control Systems, Energy Systems, Automotive Systems, Biological Systems, Vehicular Networking and Connected Vehicles, Aerospace Systems, Automation, Manufacturing, Smart Grids, Nonlinear Systems, Power Systems, Robotics, Social Systems, Economic Systems and other. Of particular value to both the contributors and the readership are the short publication timeframe and the world-wide distribution and exposure which enable both a wide and rapid dissemination of research output.

Indexed by SCOPUS, DBLP, WTI Frankfurt eG, zbMATH, SCImago.

All books published in the series are submitted for consideration in Web of Science.

More information about this series at http://www.springer.com/series/13304

George A. Anastassiou

Constructive Fractional Analysis with Applications

Springer

George A. Anastassiou
Department of Mathematical Sciences
University of Memphis
Memphis, TN, USA

ISSN 2198-4182 ISSN 2198-4190 (electronic)
Studies in Systems, Decision and Control
ISBN 978-3-030-71480-2 ISBN 978-3-030-71481-9 (eBook)
https://doi.org/10.1007/978-3-030-71481-9

This Springer imprint is published by the registered company Springer Nature Switzerland AG
The registered company address is: Gewerbestrasse 11, 6330 Cham, Switzerland

Dedicated to my family, and to all front line medical workers fighting Covid—19

Preface

Constructive and Computational Fractional Analysis recently are more and more in the center of mathematics and of other related sciences either by themselves because of their rapid development, which is based on very old foundations, or because they cover a great variety of applications in the real world. In this monograph all presented is original work by the author given at a very general level to cover a maximum number of cases in various applications. We apply generalized fractional differentiation techniques of Riemann-Liouville, Caputo and Canavati types and of fractional variable order to various kinds of inequalities such as of Opial, Hardy, Hilbert-Pachpatte and on the spherical shell. We continue with E. R. Love left and right sides fractional integral inequalities. The fractional Landau inequalities, of left and right sides, univariate and multivariate, including ones for Semigroups occupy fourteen chapters of this monograph. They are developed to all possible directions and right side multivariate fractional Taylor formulae are proven for the purpose. We continue with several Gronwall fractional inequalities of variable order. The last part of book belongs to constructive approximation theory, we present: ordinary and fractional approximations by positive sublinear operators, and high order approximation by multivariate generalized Picard, Gauss-Weierstrass, Poisson-Cauchy and Trigonometric singular integrals which are not positive operators.

This monograph is the natural and expected evolution of recent author's research work put in a book form for the first time. The presented approaches are original, and chapters are self-contained and can be read independently. This monograph is suitable to be used in related graduate classes and research projects. We exhibit to the maximum our constructive, computational, estimating and approximation methods to all possible directions.

The motivation to write this monograph came by the following: various issues related to the modeling and analysis of ordinary and fractional order systems have gained an increased popularity, as witnessed by many books and volumes in Springer's program: http://www.springer.com/gp/search?query=fractional&submit=Prze%C5%9Blij and the purpose of our book is to capture at a very general level a deeper formal analysis on some issues that are relevant to many areas for instance: decision-making, complex processes, systems modeling and control and related areas.

The above are deeply embedded in the fields of mathematics, engineering, computer science, physics, economics, social and life sciences.

The complete list of presented topics follows:

- Abstract fractional integral inequalities of variable order
- Fractional integral inequalities of variable order on spherical shell
- E.R. Love type left fractional integral Inequalities
- E.R. Love type right side fractional integral Inequalities
- Abstract fractional Landau inequalities
- Abstract generalized fractional Landau inequalities over R
- Riemann-Liouville fractional Landau type inequalities
- Generalized Canavati fractional Landau type inequalities
- Iterated left abstract fractional Landau inequalities
- Sequential left abstract generalized fractional Landau inequalities
- Iterated abstract right side fractional Landau inequalities
- Sequential abstract generalized right side fractional Landau inequalities
- High order abstract generalized Landau Inequalities
- Multivariate Caputo left fractional Landau inequalities
- Multivariate left side Canavati fractional Landau inequalities
- Multivariate right side Caputo fractional Taylor formula and Landau inequalities
- Multivariate generalized right side fractional Taylor formula and Landau inequalities
- Landau inequality for semigroups revisited
- Gronwall Inequality of fractional variable order
- A variety of Gronwall inequalities of fractional variable order
- Advanced ordinary and fractional approximation by positive sublinear operators
- High order approximation by multivariate generalized Picard singular integrals
- Complete approximations by multivariate generalized Gauss-Weierstrass singular integrals
- Approximation by multivariate generalized Poisson-Cauchy type singular integral operators
- Approximations by multivariate generalized trigonometric type singular integral operators.

The book's results are expected to find applications in many areas of pure and applied mathematics, especially in approximation theory, inequalities and differential equations as they connect to fractionality. Thus this monograph is suitable for researchers, graduate students and seminars of the above disciplines, also to be in all science and engineering libraries.

The preparation of the book took place during 2020–2021 at the University of Memphis, at stay home of author during the Covid-19 outbreak keeping him alive, happy and in control!

The author likes to thank Prof. Alina Lupas of University of Oradea, Romania, for checking and reading the manuscript.

Memphis, TN, USA
February 2021

George A. Anastassiou

Contents

Chapter 1
Variable Order General Fractional Integral Inequalities

Here are introduced abstract left and right Riemann–Liouville generalized fractional Bochner integral operators of variable order acting over Banach space valued functions. Also are introduced abstract weighted Caputo type left and right fractional derivatives of variable order. Then, we derive abstract left and right fractional integral inequalities of variable orders of Opial, Hardy and Hilbert-Pachpatte types. We give applications to simple Riemann–Liouville fractional integrals of variable order for real valued functions. It follows [3].

1.1 Background

We are motivated by [1]. Next we introduce several new concepts.

Definition 1.1 Let $[a, b] \subset \mathbb{R}$, $(X, \|\cdot\|)$ a Banach space, $g \in C^1([a, b])$ and increasing, $f \in C([a, b], X)$, $\nu \in C([a, b])$, $\nu > 1$. We define the left Riemann-Liouville generalized fractional Bochner integral operator of variable order $\nu(\cdot)$

$$\left(I_{a+;g}^{\nu(\cdot)} f\right)(x) := \int_a^x \frac{(g(x) - g(z))^{\nu(z)-1}}{\Gamma(\nu(z))} g'(z) f(z) \, dz, \tag{1.1}$$

$\forall\, x \in [a, b]$, where Γ is the gamma function. We assume $I_{a+;g}^{\nu(\cdot)} f \in C([a, b], X)$.

We need

Definition 1.2 Let $[a, b] \subset \mathbb{R}$, $(X, \|\cdot\|)$ a Banach space, $g \in C^1([a, b])$ and increasing, $f \in C([a, b], X)$, $\nu \in C([a, b])$, $\nu > 1$. We define the right Riemann-Liouville generalized fractional Bochner integral operator of variable order $\nu(\cdot)$

G. A. Anastassiou, *Constructive Fractional Analysis with Applications*,
Studies in Systems, Decision and Control 362,
https://doi.org/10.1007/978-3-030-71481-9_1

$$\left(I_{b-;g}^{\nu(\cdot)} f\right)(x) := \int_x^b \frac{(g(z) - g(x))^{\nu(z)-1}}{\Gamma(\nu(z))} g'(z) f(z) dz, \tag{1.2}$$

$\forall\, x \in [a, b]$. We assume $I_{b-;g}^{\nu(\cdot)} f \in C([a, b], X)$.

The special case of $g = identity$ and $X = \mathbb{R}$ follows:

Definition 1.3 (*see also* [4]it, p. 235) Let $[a, b] \subset \mathbb{R}$ and $\alpha, f \in C([a, b])$, $\alpha > 1$. The left-sided Riemann-Liouville type fractional integral of variable order α is defined as

$$\left({}_{RL}I_{a+}^{\alpha(\cdot)} f\right)(t) = \int_a^t \frac{1}{\Gamma(\alpha(\tau))} \frac{f(\tau)}{(t - \tau)^{1-\alpha(\tau)}} d\tau, \tag{1.3}$$

and the right-sided Riemann-Liouville type fractional integral of variable order α as

$$\left({}_{RL}I_{b-}^{\alpha(\cdot)} f\right)(t) = \int_t^b \frac{1}{\Gamma(\alpha(\tau))} \frac{f(\tau)}{(\tau - t)^{1-\alpha(\tau)}} d\tau. \tag{1.4}$$

We need

Definition 1.4 Let $[a, b] \subset \mathbb{R}, \alpha \in C([a, b])$, such that $n - 2 < \alpha(t) < n - 1, n \in \mathbb{N}, n \neq 1, \forall\, t \in [a, b]$. Let $f \in C^n([a, b], X)$, where $(X, \|\cdot\|)$ is a Banach space. Let $\psi \in C^n([a, b])$ with ψ being increasing and $\psi'(x) \neq 0$, for all $x \in [a, b]$. The left ψ-Caputo fractional derivative of f of variable order $\alpha(\cdot)$ is given by the Bochner integral

$${}^C D_{a+}^{\alpha(\cdot),\psi} f(x) = \int_a^x \frac{(\psi(x) - \psi(z))^{n-\alpha(z)-1}}{\Gamma(n - \alpha(z))} \psi'(z) f_{\psi}^{[n]}(z) dz, \tag{1.5}$$

$\forall\, x \in [a, b]$, where

$$f_{\psi}^{[n]}(x) := \left(\frac{1}{\psi'(x)} \frac{d}{dx}\right)^n f(x), \tag{1.6}$$

with $f_{\psi}^{[0]}(x) = f(x)$.

Assume that ${}^C D_{a+}^{\alpha(\cdot),\psi} f \in C([a, b], X)$. Notice that

$${}^C D_{a+}^{\alpha(\cdot),\psi} f(x) = I_{a+;\psi}^{n-\alpha(\cdot)} f_{\psi}^{[n]}(x), \tag{1.7}$$

$\forall\, x \in [a, b]$, and $f_{\psi}^{[n]} \in C([a, b], X)$.

We also need

Definition 1.5 Let $[a, b] \subset \mathbb{R}, \alpha \in C([a, b])$, such that $n - 2 < \alpha(t) < n - 1, n \in \mathbb{N}, n \neq 1, \forall\, t \in [a, b]$. Let $f \in C^n([a, b], X)$, where $(X, \|\cdot\|)$ is a Banach space. Let $\psi \in C^n([a, b])$ with ψ being increasing and $\psi'(x) \neq 0$, for all $x \in [a, b]$. The right

ψ-Caputo fractional derivative of f of variable order $\alpha(\cdot)$ is given by the Bochner integral

$$^{C}D_{b-}^{\alpha(\cdot),\psi} f(x) = (-1)^n \int_x^b \frac{(\psi(z) - \psi(x))^{n-\alpha(z)-1}}{\Gamma(n-\alpha(z))} \psi'(z) f_{\psi}^{[n]}(z)\,dz, \tag{1.8}$$

$\forall\, x \in [a, b]$.

Assume that $^{C}D_{b-}^{\alpha(\cdot),\psi} f \in C([a, b], X)$. Notice that

$$^{C}D_{b-}^{\alpha(\cdot),\psi} f(x) = (-1)^n I_{b-;\psi}^{n-\alpha(\cdot)} f_{\psi}^{[n]}(x), \tag{1.9}$$

$\forall\, x \in [a, b]$.

Next, based on the above new generalized fractional Bochner integral operators of variable order, we derive several new interesting fractional integral inequalities.

1.2 Main Results

We present the following left fractional Opial type inequality of variable order (see also [2]).

Theorem 1.6 *All as in Definition 1.1. Let* $p, q > 1 : \frac{1}{p} + \frac{1}{q} = 1$. *Then*

$$\int_a^x \left\| \left(I_{a+;g}^{\nu(\cdot)} f \right)(w) \right\| \|f(w)\| g'(w)\,dw \le$$

$$2^{-\frac{1}{q}} \left(\int_a^x \left(\int_a^w \left(\frac{(g(w) - g(z))^{\nu(z)-1}}{\Gamma(\nu(z))} \right)^p dz \right) dw \right)^{\frac{1}{p}} \left(\int_a^x \|f(w)\|^q (g'(w))^q\,dw \right)^{\frac{2}{q}}, \tag{1.10}$$

$\forall\, x \in [a, b]$.

Proof By Hölder's inequality we obtain

$$\left\| \left(I_{a+;g}^{\nu(\cdot)} f \right)(x) \right\| = \left\| \int_a^x \frac{(g(x) - g(z))^{\nu(z)-1}}{\Gamma(\nu(z))} g'(z) f(z)\,dz \right\| \le$$

$$\int_a^x \frac{(g(x) - g(z))^{\nu(z)-1}}{\Gamma(\nu(z))} g'(z) \|f(z)\|\,dz \le$$

$$\left(\int_a^x \left(\frac{(g(x) - g(z))^{\nu(z)-1}}{\Gamma(\nu(z))} \right)^p dz \right)^{\frac{1}{p}} \left(\int_a^x (g'(z) \|f(z)\|)^q\,dz \right)^{\frac{1}{q}}, \tag{1.11}$$

$\forall\, x \in [a, b]$.

Call

$$\rho(x) := \int_a^x \left(g'(z)\,\|f(z)\|\right)^q dz, \;\; \rho(a) = 0. \tag{1.12}$$

Thus

$$\rho'(x) = \left(g'(x)\,\|f(x)\|\right)^q \geq 0, \tag{1.13}$$

and

$$\left(\rho'(x)\right)^{\frac{1}{q}} = g'(x)\,\|f(x)\| \geq 0, \;\; \forall x \in [a, b]. \tag{1.14}$$

Consequently, we get

$$\left\|\left(I_{a+;g}^{\nu(\cdot)} f\right)(w)\right\| g'(w)\,\|f(w)\| \leq \tag{1.15}$$

$$\left(\int_a^w \left(\frac{(g(w) - g(z))^{\nu(z)-1}}{\Gamma(\nu(z))}\right)^p dz\right)^{\frac{1}{p}} \left(\rho(w)\,\rho'(w)\right)^{\frac{1}{q}},$$

$\forall\, w \in [a, b]$.

Then, by applying again Hölder's inequality:

$$\int_a^x \left\|\left(I_{a+;g}^{\nu(\cdot)} f\right)(w)\right\| \|f(w)\|\, g'(w)\, dw \leq$$

$$\int_a^x \left(\int_a^w \left(\frac{(g(w) - g(z))^{\nu(z)-1}}{\Gamma(\nu(z))}\right)^p dz\right)^{\frac{1}{p}} \left(\rho(w)\,\rho'(w)\right)^{\frac{1}{q}} dw \leq$$

$$\left(\int_a^x \left(\int_a^w \left(\frac{(g(w) - g(z))^{\nu(z)-1}}{\Gamma(\nu(z))}\right)^p dz\right) dw\right)^{\frac{1}{p}} \left(\int_a^x \rho(w)\,\rho'(w)\, dw\right)^{\frac{1}{q}} =$$

$$\left(\int_a^x \left(\int_a^w \left(\frac{(g(w) - g(z))^{\nu(z)-1}}{\Gamma(\nu(z))}\right)^p dz\right) dw\right)^{\frac{1}{p}} \left(\frac{\rho^2(x)}{2}\right)^{\frac{1}{q}} =$$

$$2^{-\frac{1}{q}} \left(\int_a^x \left(\int_a^w \left(\frac{(g(w) - g(z))^{\nu(z)-1}}{\Gamma(\nu(z))}\right)^p dz\right) dw\right)^{\frac{1}{p}} \left(\int_a^x \left(g'(z)\,\|f(z)\|\right)^q\right)^{\frac{2}{q}}. \tag{1.16}$$

The theorem is proved. ■

We have the special case:

Corollary 1.7 (to Theorem 1.6) *All as in Definition 1.3. Let* $p, q > 1 : \frac{1}{p} + \frac{1}{q} = 1$. *Then*

$$\int_a^x \left| \left({}_{RL}I_{a+}^{\alpha(\cdot)} f \right)(t) \, f(t) \right| dt \leq$$

$$2^{-\frac{1}{q}} \left(\int_a^x \left(\int_a^t \left(\frac{(t-\tau)^{\alpha(\tau)-1}}{\Gamma(\alpha(\tau))} \right)^p d\tau \right) dt \right)^{\frac{1}{p}} \left(\int_a^x (f(t))^q \, dt \right)^{\frac{2}{q}}, \tag{1.17}$$

$\forall \, x \in [a, b]$.

Next comes a right fractional Opial type inequality of variable order.

Theorem 1.8 *All as in Definition 1.2. Let* $p, q > 1 : \frac{1}{p} + \frac{1}{q} = 1$. *Then*

$$\int_x^b \left\| \left(I_{b-;g}^{\nu(\cdot)} f \right)(w) \right\| \|f(w)\| \, g'(w) \, dw \leq$$

$$2^{-\frac{1}{q}} \left(\int_x^b \left(\int_w^b \left(\frac{(g(z)-g(w))^{\nu(z)-1}}{\Gamma(\nu(z))} \right)^p dz \right) dw \right)^{\frac{1}{p}} \left(\int_x^b \|f(w)\|^q \left(g'(w)\right)^q dw \right)^{\frac{2}{q}}, \tag{1.18}$$

$\forall \, x \in [a, b]$.

Proof We have that

$$\left\| \left(I_{b-;g}^{\nu(\cdot)} f \right)(x) \right\| = \left\| \int_x^b \frac{(g(z)-g(x))^{\nu(z)-1}}{\Gamma(\nu(z))} g'(z) \, f(z) \, dz \right\| \leq$$

$$\int_x^b \frac{(g(z)-g(x))^{\nu(z)-1}}{\Gamma(\nu(z))} g'(z) \, \|f(z)\| \, dz \leq$$

$$\left(\int_x^b \left(\frac{(g(z)-g(x))^{\nu(z)-1}}{\Gamma(\nu(z))} \right)^p dz \right)^{\frac{1}{p}} \left(\int_x^b \left(g'(z) \, \|f(z)\|\right)^q dz \right)^{\frac{1}{q}}, \tag{1.19}$$

$\forall \, x \in [a, b]$.

Call

$$\lambda(x) := \int_x^b \left(g'(z) \, \|f(z)\|\right)^q dz, \ \ \lambda(b) = 0. \tag{1.20}$$

Hence

$$\lambda'(x) = -\left(g'(x) \, \|f(x)\|\right)^q \leq 0, \tag{1.21}$$

and

$$-\lambda'(x) = \left(g'(x)\|f(x)\|\right)^q \geq 0, \tag{1.22}$$

and

$$\left(-\lambda'(x)\right)^{\frac{1}{q}} = g'(x)\|f(x)\| \geq 0, \ \forall x \in [a,b]. \tag{1.23}$$

Consequently, we get

$$\left\|\left(I_{b-;g}^{\nu(\cdot)} f\right)(w)\right\| g'(w)\|f(w)\| \leq$$

$$\left(\int_w^b \left(\frac{(g(z)-g(w))^{\nu(z)-1}}{\Gamma(\nu(z))}\right)^p dz\right)^{\frac{1}{p}} \left(\lambda(w)\left(-\lambda'(w)\right)\right)^{\frac{1}{q}}, \tag{1.24}$$

$\forall\, w \in [a,b]$.

Thus

$$\int_x^b \left\|\left(I_{b-;g}^{\nu(\cdot)} f\right)(w)\right\| \|f(w)\| g'(w)\, dw \leq$$

$$\int_x^b \left(\int_w^b \left(\frac{(g(z)-g(w))^{\nu(z)-1}}{\Gamma(\nu(z))}\right)^p dz\right)^{\frac{1}{p}} \left(\lambda(w)\left(-\lambda'(w)\right)\right)^{\frac{1}{q}} dw \leq$$

$$\left(\int_x^b \left(\int_w^b \left(\frac{(g(z)-g(w))^{\nu(z)-1}}{\Gamma(\nu(z))}\right)^p dz\right) dw\right)^{\frac{1}{p}} \left(-\int_x^b \lambda(w)\lambda'(w)\, dw\right)^{\frac{1}{q}} = \tag{1.25}$$

$$\left(\int_x^b \left(\int_w^b \left(\frac{(g(z)-g(w))^{\nu(z)-1}}{\Gamma(\nu(z))}\right)^p dz\right) dw\right)^{\frac{1}{p}} \left(\frac{\lambda^2(x)}{2}\right)^{\frac{1}{q}} =$$

$$2^{-\frac{1}{q}} \left(\int_x^b \left(\int_w^b \left(\frac{(g(z)-g(w))^{\nu(z)-1}}{\Gamma(\nu(z))}\right)^p dz\right) dw\right)^{\frac{1}{p}} \left(\int_x^b \left(g'(z)\|f(z)\|\right)^q\right)^{\frac{2}{q}}, \tag{1.26}$$

$\forall\, x \in [a,b]$, proving the claim. ∎

We have the special case:

Corollary 1.9 (to Theorem 1.8) *All as in Definition 1.3. Let $p, q > 1 : \frac{1}{p} + \frac{1}{q} = 1$. Then*

$$\int_x^b \left|\left({}_{RL}I_{b-}^{\alpha(\cdot)} f\right)(w)\, f(w)\right| dw \leq$$

$$2^{-\frac{1}{q}} \left(\int_x^b \left(\int_w^b \left(\frac{(z-w)^{\alpha(z)-1}}{\Gamma(\alpha(z))}\right)^p dz\right) dw\right)^{\frac{1}{p}} \left(\int_x^b |f(w)|^q\, dw\right)^{\frac{2}{q}}, \tag{1.27}$$

$\forall\, x \in [a,b]$.

We give a left fractional Opial type inequality of variable order.

Theorem 1.10 *All as in Definition 1.4. Let* $p, q > 1 : \frac{1}{p} + \frac{1}{q} = 1$. *Then*

$$\int_a^x \left\| {}^C D_{a+}^{\alpha(\cdot),\psi} f(w) \right\| \left\| f_\psi^{[n]}(w) \right\| \psi'(w)\, dw \leq$$

$$2^{-\frac{1}{q}} \left(\int_a^x \left(\int_a^w \left(\frac{(\psi(w) - \psi(z))^{n-\alpha(z)-1}}{\Gamma(n-\alpha(z))} \right)^p dz \right) dw \right)^{\frac{1}{p}}$$

$$\left(\int_a^x \left\| f_\psi^{[n]}(w) \right\|^q (\psi'(w))^q\, dw \right)^{\frac{2}{q}}, \tag{1.28}$$

$\forall\, x \in [a, b]$.

Proof By (1.7) and Theorem 1.6. ■

We present a right fractional Opial type inequality of variable order.

Theorem 1.11 *All as in Definition 1.5. Let* $p, q > 1 : \frac{1}{p} + \frac{1}{q} = 1$. *Then*

$$\int_x^b \left\| {}^C D_{b-}^{\alpha(\cdot),\psi} f(w) \right\| \left\| f_\psi^{[n]}(w) \right\| \psi'(w)\, dw \leq$$

$$2^{-\frac{1}{q}} \left(\int_x^b \left(\int_w^b \left(\frac{(\psi(z) - \psi(w))^{n-\alpha(z)-1}}{\Gamma(n-\alpha(z))} \right)^p dz \right) dw \right)^{\frac{1}{p}}$$

$$\left(\int_x^b \left\| f_\psi^{[n]}(w) \right\|^q (\psi'(w))^q\, dw \right)^{\frac{2}{q}}, \tag{1.29}$$

$\forall\, x \in [a, b]$.

Proof By (1.9) and Theorem 1.8. ■

We give two extreme Opial inequalities cases ($p = 1, q = \infty$).

Theorem 1.12 *All as in Definition 1.1. Then*

$$\int_a^x \left\| \left(I_{a+;g}^{\nu(\cdot)} f \right)(w) \right\| \| f(w) \|\, dw \leq$$

$$\left(\int_a^x \left(\int_a^w \frac{(g(w) - g(z))^{\nu(z)-1}}{\Gamma(\nu(z))} g'(z)\, dz \right) dw \right) \| f \|_\infty^2, \tag{1.30}$$

$\forall\, x \in [a, b]$.

Proof We have (by (1.1))

$$\left\| \left(I_{a+;g}^{\nu(\cdot)} f \right)(x) \right\| \leq \int_a^x \frac{(g(x)-g(z))^{\nu(z)-1}}{\Gamma(\nu(z))} g'(z) \|f(z)\| \, dz \leq$$

$$\left(\int_a^x \frac{(g(x)-g(z))^{\nu(z)-1}}{\Gamma(\nu(z))} g'(z) \, dz \right) \|f\|_\infty . \tag{1.31}$$

Hence

$$\left\| \left(I_{a+;g}^{\nu(\cdot)} f \right)(w) \right\| \|f(w)\| \leq \left(\int_a^w \frac{(g(w)-g(z))^{\nu(z)-1}}{\Gamma(\nu(z))} g'(z) \, dz \right) \|f\|_\infty^2 . \tag{1.32}$$

Now integrate (1.32) over $[a, x]$ to get (1.30). ■

Theorem 1.13 *All as in Definition 1.2. Then*

$$\int_x^b \left\| \left(I_{b-;g}^{\nu(\cdot)} f \right)(w) \right\| \|f(w)\| \, dw \leq$$

$$\left(\int_x^b \left(\int_w^b \frac{(g(z)-g(w))^{\nu(z)-1}}{\Gamma(\nu(z))} g'(z) \, dz \right) dw \right) \|f\|_\infty^2 , \tag{1.33}$$

$\forall \, x \in [a, b]$.

Proof We have (by (1.2))

$$\left\| \left(I_{b-;g}^{\nu(\cdot)} f \right)(x) \right\| \leq \int_x^b \frac{(g(z)-g(x))^{\nu(z)-1}}{\Gamma(\nu(z))} g'(z) \|f(z)\| \, dz \leq$$

$$\left(\int_x^b \frac{(g(z)-g(x))^{\nu(z)-1}}{\Gamma(\nu(z))} g'(z) \, dz \right) \|f\|_\infty . \tag{1.34}$$

Hence

$$\left\| \left(I_{b-;g}^{\nu(\cdot)} f \right)(w) \right\| \|f(w)\| \leq \left(\int_w^b \frac{(g(z)-g(w))^{\nu(z)-1}}{\Gamma(\nu(z))} g'(z) \, dz \right) \|f\|_\infty^2 . \tag{1.35}$$

Now integrate (1.35) over $[x, b]$ to get (1.33). ■

Corollary 1.14 *All as in Definition 1.4. Then*

$$\int_a^x \left\| {}^C D_{a+}^{\alpha(\cdot),\psi} f(w) \right\| \left\| f_\psi^{[n]}(w) \right\| dw \leq$$

$$\left(\int_a^x \left(\int_a^w \frac{(\psi(w)-\psi(z))^{n-\alpha(z)-1}}{\Gamma(n-\alpha(z))}\psi'(z)\,dz\right)dw\right)\left\|f_\psi^{[n]}\right\|_\infty^2. \tag{1.36}$$

$\forall\, x \in [a,b]$.

Proof Use of Theorem 1.12, (1.7). ■

Corollary 1.15 *All as in Definition 1.5. Then*

$$\int_x^b \left\|{}^C D_{b-}^{\alpha(\cdot),\psi} f(w)\right\| \left\|f_\psi^{[n]}(w)\right\| dw \le$$

$$\left(\int_x^b \left(\int_w^b \frac{(\psi(z)-\psi(w))^{n-\alpha(z)-1}}{\Gamma(n-\alpha(z))}\psi'(z)\,dz\right)dw\right)\left\|f_\psi^{[n]}\right\|_\infty^2. \tag{1.37}$$

$\forall\, x \in [a,b]$.

Proof Use of Theorem 1.13, (1.9). ■

Next we give some fractional Hardy's type inequalities of variable order.

Theorem 1.16 *All as in Definition 1.1. Let $p,q>1: \frac{1}{p}+\frac{1}{q}=1$. Then*

$$\left\|\left\|I_{a+;g}^{\nu(\cdot)} f\right\|\right\|_q \le \left(\int_a^b \left(\int_a^x \left(\frac{(g(x)-g(z))^{\nu(z)-1}}{\Gamma(\nu(z))}\right)^p dz\right)^{\frac{q}{p}} dx\right)^{\frac{1}{q}} \left\|g'\,\|f\|\right\|_q. \tag{1.38}$$

Proof By (1.11) we get ($\forall\, x \in [a,b]$)

$$\left\|\left(I_{a+;g}^{\nu(\cdot)} f\right)(x)\right\| \le$$

$$\left(\int_a^x \left(\frac{(g(x)-g(z))^{\nu(z)-1}}{\Gamma(\nu(z))}\right)^p dz\right)^{\frac{1}{p}} \left(\int_a^b \left(g'(z)\,\|f(z)\|\right)^q dz\right)^{\frac{1}{q}}. \tag{1.39}$$

Hence it holds

$$\left\|\left(I_{a+;g}^{\nu(\cdot)} f\right)(x)\right\|^q \le \left(\int_a^x \left(\frac{(g(x)-g(z))^{\nu(z)-1}}{\Gamma(\nu(z))}\right)^p dz\right)^{\frac{q}{p}} \int_a^b \left(g'(z)\,\|f(z)\|\right)^q dz, \tag{1.40}$$

$\forall\, x \in [a,b]$.

Therefore we get

$$\int_a^b \left\|\left(I_{a+;g}^{\nu(\cdot)} f\right)(x)\right\|^q dx \le$$

$$\left(\int_a^b\left(\int_a^x\left(\frac{(g(x)-g(z))^{\nu(z)-1}}{\Gamma(\nu(z))}\right)^p dz\right)^{\frac{q}{p}} dx\right)\left(\int_a^b\left(g'(z)\|f(z)\|\right)^q dz\right), \tag{1.41}$$

proving the claim. ■

Theorem 1.17 *All as in Definition 1.2. Let* $p, q > 1 : \frac{1}{p} + \frac{1}{q} = 1$. *Then*

$$\left\|\left\|I_{b-;g}^{\nu(\cdot)} f\right\|\right\|_q \leq \left(\int_a^b\left(\int_x^b\left(\frac{(g(z)-g(x))^{\nu(z)-1}}{\Gamma(\nu(z))}\right)^p dz\right)^{\frac{q}{p}} dx\right)^{\frac{1}{q}} \left\|g'\|f\|\right\|_q. \tag{1.42}$$

Proof By (1.19) ($\forall\, x \in [a, b]$) we get

$$\left\|\left(I_{b-;g}^{\nu(\cdot)} f\right)(x)\right\| \leq$$

$$\left(\int_x^b\left(\frac{(g(z)-g(x))^{\nu(z)-1}}{\Gamma(\nu(z))}\right)^p dz\right)^{\frac{1}{p}}\left(\int_a^b\left(g'(z)\|f(z)\|\right)^q dz\right)^{\frac{1}{q}}. \tag{1.43}$$

Hence

$$\left\|\left(I_{b-;g}^{\nu(\cdot)} f\right)(x)\right\|^q \leq \left(\int_x^b\left(\frac{(g(z)-g(x))^{\nu(z)-1}}{\Gamma(\nu(z))}\right)^p dz\right)^{\frac{q}{p}}\left(\int_a^b\left(g'(z)\|f(z)\|\right)^q dz\right), \tag{1.44}$$

$\forall\, x \in [a, b]$.

Therefore it holds

$$\int_a^b\left\|\left(I_{b-;g}^{\nu(\cdot)} f\right)(x)\right\|^q dx \leq$$

$$\left(\int_a^b\left(\int_x^b\left(\frac{(g(z)-g(x))^{\nu(z)-1}}{\Gamma(\nu(z))}\right)^p dz\right)^{\frac{q}{p}} dx\right)\left(\int_a^b\left(g'(z)\|f(z)\|\right)^q dz\right), \tag{1.45}$$

proving the claim. ■

We present also:

Theorem 1.18 *All as in Definition 1.4. Let* $p, q > 1 : \frac{1}{p} + \frac{1}{q} = 1$. *Then*

$$\left\|\left\|{}^C D_{a+}^{\alpha(\cdot),\psi}\right\|\right\|_q \leq$$

$$\left(\int_a^b \left(\int_a^x \left(\frac{(\psi(x) - \psi(z))^{n-\alpha(z)-1}}{\Gamma(n - \alpha(z))} \right)^p dz \right)^{\frac{q}{p}} dx \right)^{\frac{1}{q}} \left\| \psi' \left\| f_\psi^{[n]} \right\| \right\|_q . \tag{1.46}$$

Proof By Theorem 1.16 and (1.7). ■

Theorem 1.19 *All as in Definition 1.5. Let* $p, q > 1 : \frac{1}{p} + \frac{1}{q} = 1$. *Then*

$$\left\| \left\| {}^C D_{b-}^{\alpha(\cdot),\psi} \right\| \right\|_q \leq$$

$$\left(\int_a^b \left(\int_x^b \left(\frac{(\psi(z) - \psi(x))^{n-\alpha(z)-1}}{\Gamma(n - \alpha(z))} \right)^p dz \right)^{\frac{q}{p}} dx \right)^{\frac{1}{q}} \left\| \psi' \left\| f_\psi^{[n]} \right\| \right\|_q . \tag{1.47}$$

Proof By Theorem 1.17 and (1.9). ■

Next we give a Hilbert-Pachpatte left fractional inequality of variable orders.

Theorem 1.20 *Let* $p, q > 1 : \frac{1}{p} + \frac{1}{q} = 1$. *Here* $i = 1, 2$. *Let* $[a_i, b_i] \subset \mathbb{R}$, $(X, \|\cdot\|)$ *a Banach space,* $g_i \in C^1([a_i, b_i])$ *and strictly increasing,* $f_i \in C([a_i, b_i], X)$, $\nu_i \in C([a_i, b_i])$, $\nu_i > 1$. *Assume* $I_{a_i+;g_i}^{\nu_i(\cdot)} f_i \in C([a_i, b_i], X)$ *.Then*

$$\int_{a_1}^{b_1} \int_{a_2}^{b_2} \frac{\left\| \left(I_{a_1+;g_1}^{\nu_1(\cdot)} f_1 \right)(x_1) \right\| \left\| \left(I_{a_2+;g_2}^{\nu_2(\cdot)} f_2 \right)(x_2) \right\|}{\left[\frac{\left(\int_{a_1}^{x_1} \left(\frac{(g_1(x_1) - g_1(z_1))^{\nu_1(z_1)-1}}{\Gamma(\nu_1(z_1))} \right)^p dz_1 \right)}{p} + \frac{\left(\int_{a_2}^{x_2} \left(\frac{(g_2(x_2) - g_2(z_2))^{\nu_2(z_2)-1}}{\Gamma(\nu_2(z_2))} \right)^q dz_2 \right)}{q} \right]} dx_1 dx_2$$

$$\leq (b_1 - a_1)(b_2 - a_2) \left\| g_1' \|f_1\| \right\|_q \left\| g_2' \|f_2\| \right\|_p . \tag{1.48}$$

Proof We have that $(i = 1, 2)$

$$\left(I_{a_i+;g_i}^{\nu_i(\cdot)} f_i \right)(x_i) \overset{(1.1)}{=} \int_{a_i}^{x_i} \frac{(g_i(x_i) - g_i(z_i))^{\nu_i(z_i)-1}}{\Gamma(\nu_i(z_i))} g_i'(z_i) f_i(z_i) dz_i, \tag{1.49}$$

$\forall x_i \in [a_i, b_i]$.

Then

$$\left\| \left(I_{a_i+;g_i}^{\nu_i(\cdot)} f_i \right)(x_i) \right\| \leq \int_{a_i}^{x_i} \frac{(g_i(x_i) - g_i(z_i))^{\nu_i(z_i)-1}}{\Gamma(\nu_i(z_i))} g_i'(z_i) \|f_i(z_i)\| dz_i, \tag{1.50}$$

$i = 1, 2, \forall x_i \in [a_i, b_i]$.

By applying Hölder's inequality twice we get:

$$\left\| \left(I_{a_1+;g_1}^{\nu_1(\cdot)} f_1 \right) (x_1) \right\| \leq$$

$$\left(\int_{a_1}^{x_1} \left(\frac{(g_1(x_1) - g_1(z_1))^{\nu_1(z_1)-1}}{\Gamma(\nu_1(z_1))} \right)^p dz_1 \right)^{\frac{1}{p}} \left(\int_{a_1}^{x_1} (g_1'(z_1) \|f_1(z_1)\|)^q dz_1 \right)^{\frac{1}{q}}, \tag{1.51}$$

$\forall\, x_1 \in [a_1, b_1]$,

and

$$\left\| \left(I_{a_2+;g_2}^{\nu_2(\cdot)} f_2 \right) (x_2) \right\| \leq$$

$$\left(\int_{a_2}^{x_2} \left(\frac{(g_2(x_2) - g_2(z_2))^{\nu_2(z_2)-1}}{\Gamma(\nu_2(z_2))} \right)^q dz_2 \right)^{\frac{1}{q}} \left(\int_{a_2}^{x_2} (g_2'(z_2) \|f_2(z_2)\|)^p dz_2 \right)^{\frac{1}{p}}, \tag{1.52}$$

$\forall\, x_2 \in [a_2, b_2]$.

Hence we have (by (1.51), (1.52))

$$\left\| \left(I_{a_1+;g_1}^{\nu_1(\cdot)} f_1 \right) (x_1) \right\| \left\| \left(I_{a_2+;g_2}^{\nu_2(\cdot)} f_2 \right) (x_2) \right\| \leq$$

$$\left(\int_{a_1}^{x_1} \left(\frac{(g_1(x_1) - g_1(z_1))^{\nu_1(z_1)-1}}{\Gamma(\nu_1(z_1))} \right)^p dz_1 \right)^{\frac{1}{p}} \tag{1.53}$$

$$\left(\int_{a_2}^{x_2} \left(\frac{(g_2(x_2) - g_2(z_2))^{\nu_2(z_2)-1}}{\Gamma(\nu_2(z_2))} \right)^q dz_2 \right)^{\frac{1}{q}} \left\| g_1' \|f_1\| \right\|_q \left\| g_2' \|f_2\| \right\|_p \leq$$

(using Yang's inequality for $a, b \geq 0$, $a^{\frac{1}{p}} b^{\frac{1}{q}} \leq \frac{a}{p} + \frac{b}{q}$)

$$\left[\frac{\left(\int_{a_1}^{x_1} \left(\frac{(g_1(x_1) - g_1(z_1))^{\nu_1(z_1)-1}}{\Gamma(\nu_1(z_1))} \right)^p dz_1 \right)}{p} + \frac{\left(\int_{a_2}^{x_2} \left(\frac{(g_2(x_2) - g_2(z_2))^{\nu_2(z_2)-1}}{\Gamma(\nu_2(z_2))} \right)^q dz_2 \right)}{q} \right]$$

$$\left\| g_1' \|f_1\| \right\|_q \left\| g_2' \|f_2\| \right\|_p, \tag{1.54}$$

$\forall\, x_i \in [a_i, b_i]$, $i = 1, 2$.

So far we have

$$\frac{\left\| \left(I_{a_1+;g_1}^{\nu_1(\cdot)} f_1 \right) (x_1) \right\| \left\| \left(I_{a_2+;g_2}^{\nu_2(\cdot)} f_2 \right) (x_2) \right\|}{\left[\frac{\left(\int_{a_1}^{x_1} \left(\frac{(g_1(x_1) - g_1(z_1))^{\nu_1(z_1)-1}}{\Gamma(\nu_1(z_1))} \right)^p dz_1 \right)}{p} + \frac{\left(\int_{a_2}^{x_2} \left(\frac{(g_2(x_2) - g_2(z_2))^{\nu_2(z_2)-1}}{\Gamma(\nu_2(z_2))} \right)^q dz_2 \right)}{q} \right]}$$

$$\leq \left\| g_1' \left\| f_1 \right\| \right\|_q \left\| g_2' \left\| f_2 \right\| \right\|_p , \tag{1.55}$$

$\forall\, x_i \in [a_i, b_i]$, $i = 1, 2$.

The denominator in (1.55) can be zero only when $x_1 = a_1$ and $x_2 = a_2$. Therefore we obtain (1.48) by integrating (1.55) over $[a_1, b_1] \times [a_2, b_2]$. ■

It follows a Hilbert-Pachpatte right fractional inequality of variable orders.

Theorem 1.21 *All as in Theorem 1.20. Assume that* $I_{b_i-;g_i}^{\nu_i(\cdot)} f_i \in C([a_i, b_i], X)$, $i = 1, 2$. *Then*

$$\int_{a_1}^{b_1}\int_{a_2}^{b_2} \frac{\left\| \left(I_{b_1-;g_1}^{\nu_1(\cdot)} f_1 \right)(x_1) \right\| \left\| \left(I_{b_2-;g_2}^{\nu_2(\cdot)} f_2 \right)(x_2) \right\|}{\left[\frac{\left(\int_{x_1}^{b_1} \left(\frac{(g_1(z_1)-g_1(x_1))^{\nu_1(z_1)-1}}{\Gamma(\nu_1(z_1))} \right)^p dz_1 \right)}{p} + \frac{\left(\int_{x_2}^{b_2} \left(\frac{(g_2(z_2)-g_2(x_2))^{\nu_2(z_2)-1}}{\Gamma(\nu_2(z_2))} \right)^q dz_2 \right)}{q} \right]} dx_1 dx_2$$

$$\leq (b_1 - a_1)(b_2 - a_2) \left\| g_1' \left\| f_1 \right\| \right\|_q \left\| g_2' \left\| f_2 \right\| \right\|_p . \tag{1.56}$$

Proof As similar to the proof of Theorem 1.20 is omitted. ■

We also give (use of Theorem 1.20 and (1.7)):

Theorem 1.22 *Let* $p, q > 1 : \frac{1}{p} + \frac{1}{q} = 1$. *Here* $i = 1, 2$. *Let* $[a_i, b_i] \subset \mathbb{R}$, $\alpha_i \in C([a_i, b_i])$, *such that* $n_i - 2 < \alpha_i(t_i) < n_i - 1$, $n_i \in \mathbb{N}$, $n_i \neq 1$, $\forall\, t_i \in [a_i, b_i]$. *Let* $f_i \in C^{n_i}([a_i, b_i], X)$, *where* $(X, \|\cdot\|)$ *is a Banach space. Let* $\psi_i \in C^{n_i}([a_i, b_i])$ *with* ψ_i *being stricly increasing and* $\psi_i'(x) \neq 0$, *for all* $x_i \in [a_i, b_i]$. *Assume that* ${}^C D_{a_i+}^{\alpha_i(\cdot),\psi_i} f_i \in C([a_i, b_i], X)$. *Then*

$$\int_{a_1}^{b_1}\int_{a_2}^{b_2} \frac{\left\| {}^C D_{a_1+}^{\alpha_1(\cdot),\psi_1} f_1(x_1) \right\| \left\| {}^C D_{a_2+}^{\alpha_2(\cdot),\psi} f_2(x_2) \right\|}{\left[\frac{\left(\int_{a_1}^{x_1} \left(\frac{(\psi_1(x_1)-\psi_1(z_1))^{n_1-\alpha_1(z_1)-1}}{\Gamma(n_1-\alpha_1(z_1))} \right)^p dz_1 \right)}{p} + \frac{\left(\int_{a_2}^{x_2} \left(\frac{(\psi_2(x_2)-\psi_2(z_2))^{n_2-\alpha_2(z_2)-1}}{\Gamma(n_2-\alpha_2(z_2))} \right)^q dz_2 \right)}{q} \right]} dx_1 dx_2$$

$$\leq (b_1 - a_1)(b_2 - a_2) \left\| \psi_1' \left\| f_{1\psi_1}^{[n_1]} \right\| \right\|_q \left\| \psi_2' \left\| f_{2\psi_2}^{[n_2]} \right\| \right\|_p . \tag{1.57}$$

We finish with (use of Theorem 1.21 and (1.9))

Theorem 1.23 *All as in Theorem 1.22. Assume that* ${}^C D_{b_i-}^{\alpha_i(\cdot),\psi_i} f_i \in C([a_i, b_i], X)$, $i = 1, 2$. *Then*

$$\int_{a_1}^{b_1}\int_{a_2}^{b_2} \frac{\left\| {}^C D_{b_1-}^{\alpha_1(\cdot),\psi_1} f_1(x_1) \right\| \left\| {}^C D_{b_2-}^{\alpha_2(\cdot),\psi_2} f_2(x_2) \right\|}{\left[\frac{\left(\int_{x_1}^{b_1} \left(\frac{(\psi_1(z_1)-\psi_1(x_1))^{n_1-\alpha_1(z_1)-1}}{\Gamma(n_1-\alpha_1(z_1))} \right)^p dz_1 \right)}{p} + \frac{\left(\int_{x_2}^{b_2} \left(\frac{(\psi_2(z_2)-\psi_2(x_2))^{n_2-\alpha_2(z_2)-1}}{\Gamma(n_2-\alpha_2(z_2))} \right)^q dz_2 \right)}{q} \right]} dx_1 dx_2$$

$$\leq (b_1 - a_1)(b_2 - a_2) \left\| \psi_1' \left\| f_{1\psi_1}^{[n_1]} \right\| \right\|_q \left\| \psi_2' \left\| f_{2\psi_2}^{[n_2]} \right\| \right\|_p . \tag{1.58}$$

References

1. Anastassiou, G.A.: Fractional Differentiation Inequalities. Springer, Heidelberg, New York (2009)
2. Anastassiou, G.A.: Intelligent Computations: Abstract Fractional Calculus, Inequalities, Approximations. Springer, Heidelberg, New York (2018)
3. Anastassiou, G.A.: Abstract fractional integral inequalities of variable order. Indian J. Math. (2020)
4. Yang, X.-J.: General Fractional Derivatives. CRC Press-Taylor & Francis, Boca Raton, London, New York (2019)

Chapter 2
Variable Order Fractional Integral Inequalities for Spherical Shell

Here are introduced left and right Riemann–Liouville generalized fractional radial integral operators of variable order over a spherical shell. Also are introduced left and right weighted Caputo type generalized fractional radial derivatives of variable order over a spherical shell. After proving continuity of these operators, we establish a series of left and right fractional integral inequalities of variable order over the spherical shell of Opial and Hardy types. Extreme cases are met. It follows [5].

2.1 Background

W are inspired by [1–3].

Let $N \geq 2$, $S^{N-1} := \left\{x \in \mathbb{R}^N : |x| = 1\right\}$ the unit sphere on $\mathbb{R}^N$, where $|\cdot|$ stands for the Euclidean norm in $\mathbb{R}^N$. Also denote the ball

$$B(0, R) := \left\{x \in \mathbb{R}^N : |x| < R\right\} \subseteq \mathbb{R}^N,\ R > 0,$$

and the spherical shell

$$A := B(0, R_2) - \overline{B(0, R_1)},\ 0 < R_1 < R_2.$$

For the following see [6, pp. 149–150] and [7, pp. 87–88].

For $x \in \mathbb{R}^N - \{0\}$ we can write uniquely $x = r\omega$, where $r = |x| > 0$, and $\omega = \frac{x}{r} \in S^{N-1}$, $|\omega| = 1$. Clearly here

$$\mathbb{R}^N - \{0\} = (0, \infty) \times S^{N-1},$$

G. A. Anastassiou, *Constructive Fractional Analysis with Applications*,
Studies in Systems, Decision and Control 362,
https://doi.org/10.1007/978-3-030-71481-9_2

and

$$\overline{A} = [R_1, R_2] \times S^{N-1}.$$

In the sequel the following related theorem will be used:

Theorem 2.1 ([1, p. 458]) *Let $f : A \to \mathbb{R}$ be a Lebesgue integrable function. Then*

$$\int_A f(x)\,dx = \int_{S^{N-1}} \left(\int_{R_1}^{R_2} f(r\omega)\, r^{N-1} dr \right) d\omega. \tag{2.1}$$

So we are able to write an integral in polar form using the polar coordinates (r, ω).

The area of S^{N-1} is $\omega_N = \int_{S^{N-1}} d\omega = \frac{2\pi^{\frac{N}{2}}}{\Gamma\left(\frac{N}{2}\right)}$.

The volume of $B(0, R)$, $Vol(B(0, R)) = \int_{B(0,R)} dy = \frac{\pi^{\frac{N}{2}} R^N}{\Gamma\left(\frac{N}{2}+1\right)}$.

And the volume of A, $Vol(A) = \frac{\pi^{\frac{N}{2}}\left(R_2^N - R_1^N\right)}{\Gamma\left(\frac{N}{2}+1\right)}$.

We need

Definition 2.2 Let $[R_1, R_2] \subset \mathbb{R}$, $R_2 > R_1 > 0$, and $\alpha \in C([R_1, R_2])$, $\alpha > 1$. Let $\psi \in C^1([R_1, R_2])$ being increasing and $f \in C\left(\overline{A}\right)$. We define the left Riemann–Liouville generalized fractional radial integral operator for $f : A \to \mathbb{R}$ of variable order $\alpha(\cdot)$:

$$\left(I_{R_1+;\psi}^{\alpha(\cdot)} f\right)(x) = \left(I_{R_1+;\psi}^{\alpha(\cdot)} f\right)(r\omega) := \int_{R_1}^{r} \frac{(\psi(r) - \psi(z))^{\alpha(z)-1}}{\Gamma(\alpha(z))} \psi'(z)\, f(z\omega)\, dz, \tag{2.2}$$

$\forall\, x \in A$, where Γ is the gamma function.

We also define the right Riemann–Liouville generalized farctional radial integral operator for $f : A \to \mathbb{R}$ of variable order $\alpha(\cdot)$:

$$\left(I_{R_2-;\psi}^{\alpha(\cdot)} f\right)(x) = \left(I_{R_2-;\psi}^{\alpha(\cdot)} f\right)(r\omega) := \int_{r}^{R_2} \frac{(\psi(z) - \psi(r))^{\alpha(z)-1}}{\Gamma(\alpha(z))} \psi'(z)\, f(z\omega)\, dz, \tag{2.3}$$

$\forall\, x \in A$.

We give

Theorem 2.3 (related to Definition 2.2) *We have that $I_{R_1+;\psi}^{\alpha(\cdot)} f$, $I_{R_2-;\psi}^{\alpha(\cdot)} f \in C\left(\overline{A}\right)$.*

Proof We will prove in detail only that $I_{R_1+;\psi}^{\alpha(\cdot)} f \in C\left(\overline{A}\right)$, the proof for $I_{R_2-;\psi}^{\alpha(\cdot)} f \in C\left(\overline{A}\right)$ as similar is omitted. Indeed here $f(r\omega) \in C([R_1, R_2])$, $\forall\, \omega \in S^{N-1}$. We can write

$$\left(I_{R_1+;\psi}^{\alpha(\cdot)} f\right)(r\omega) = \int_{R_1}^{R_2} \chi_{[R_1,r]}(z) \frac{(\psi(r) - \psi(z))^{\alpha(z)-1}}{\Gamma(\alpha(z))} \psi'(z)\, f(z\omega)\, dz, \tag{2.4}$$

where χ is the characteristic function, $\forall\, \omega \in S^{N-1}$.

Let $r_n \to r$, $\omega_n \to \omega$; then $\chi_{[R_1,r_n]}(z) \to \chi_{[R_1,r]}(z)$, a.e., $\frac{(\psi(r_n)-\psi(z))^{\alpha(z)-1}}{\Gamma(\alpha(z))}\psi'(z)$ $\to \frac{(\psi(r)-\psi(z))^{\alpha(z)-1}}{\Gamma(\alpha(z))}\psi'(z)$, and $f(z\omega_n) \to f(z\omega)$.

Furthermore it holds that

$$\chi_{[R_1,r_n]}(z)\frac{(\psi(r_n)-\psi(z))^{\alpha(z)-1}}{\Gamma(\alpha(z))}\psi'(z) f(z\omega_n) \to$$

$$\chi_{[R_1,r]}(z)\frac{(\psi(r)-\psi(z))^{\alpha(z)-1}}{\Gamma(\alpha(z))}\psi'(z) f(z\omega), \tag{2.5}$$

a.e. on $[R_1, R_2]$.

Here it is $\alpha(z)-1>0$ and $\alpha-1\in C([R_1,R_2])$. Thus, $\inf(\alpha-1)\le\alpha(z)-1\le\sup(\alpha-1)$, $\forall z\in[R_1,R_2]$ and $\inf(\alpha-1)$, $\sup(\alpha-1)>0$.

If $|\psi(r_n)-\psi(z)|\le 1$, then

$$|\psi(r_n)-\psi(z)|^{\alpha(z)-1}\le|\psi(r_n)-\psi(z)|^{\inf(\alpha-1)}, \tag{2.6}$$

and if $|\psi(r_n)-\psi(z)|>1$, then

$$|\psi(r_n)-\psi(z)|^{\alpha(z)-1}\le|\psi(r_n)-\psi(z)|^{\sup(\alpha-1)}. \tag{2.7}$$

That is

$$|\psi(r_n)-\psi(z)|^{\alpha(z)-1}\le|\psi(r_n)-\psi(z)|^{\inf(\alpha-1)}+|\psi(r_n)-\psi(z)|^{\sup(\alpha-1)}$$

$$\le(\psi(R_2)-\psi(R_1))^{\inf(\alpha-1)}+(\psi(R_2)-\psi(R_1))^{\sup(\alpha-1)}=:\lambda<\infty, \tag{2.8}$$

true for all $r_n, z\in[R_1,R_2]$.

Therefore it holds

$$\chi_{[R_1,r_n]}(z)\frac{|\psi(r_n)-\psi(z)|^{\alpha(z)-1}}{\Gamma(\alpha(z))}\psi'(z)|f(z\omega_n)|\le$$

$$\frac{\lambda}{\inf_{[R_1,R_2]}\Gamma(\alpha(z))}\left\|\psi'\right\|_{\infty,[R_1,R_2]}\|f\|_{\infty,\overline{A}}<\infty. \tag{2.9}$$

Thus, by the denominated convergence theorem we obtain

$$\int_{R_1}^{R_2}\chi_{[R_1,r_n]}(z)\frac{(\psi(r_n)-\psi(z))^{\alpha(z)-1}}{\Gamma(\alpha(z))}\psi'(z) f(z\omega_n)\,dz \to$$

$$\int_{R_1}^{R_2}\chi_{[R_1,r]}(z)\frac{(\psi(r)-\psi(z))^{\alpha(z)-1}}{\Gamma(\alpha(z))}\psi'(z) f(z\omega)\,dz, \tag{2.10}$$

proving the claim. ■

We need

Definition 2.4 Let $[R_1, R_2] \subset \mathbb{R}$, $R_2 > R_1 > 0$, and $\alpha \in C([R_1, R_2])$, such that $n-2 < \alpha(t) < n-1$, $n \in \mathbb{N} - \{1\}$, $\forall\, t \in [R_1, R_2]$. Let $f \in C^n\left(\overline{A}\right)$ and $\psi \in C^n([R_1, R_2])$ with ψ being increasing and $\psi'(t) \neq 0$, for all $t \in [R_1, R_2]$. The left ψ-Caputo generalized fractional radial derivative of $f : A \to \mathbb{R}$ of variable order $\alpha(\cdot)$ is given by:

$$ {}^{C}D_{R_1+}^{\alpha(\cdot),\psi} f(x) = {}^{C}D_{R_1+}^{\alpha(\cdot),\psi} f(r\omega) := \int_{R_1}^{r} \frac{(\psi(r) - \psi(t))^{n-\alpha(t)-1}}{\Gamma(n-\alpha(t))} \psi'(t) f_{\psi}^{[n]}(t\omega)\, dt, \tag{2.11} $$

$\forall\, x \in A$, where

$$ f_{\psi}^{[n]}(x) = f_{\psi}^{[n]}(r\omega) := \left(\frac{1}{\psi'(r)} \frac{d}{dr}\right)^{n} f(r\omega), \tag{2.12} $$

with $f_{\psi}^{[0]}(r\omega) = f(r\omega)$, $\forall\, \omega \in S^{N-1}$, $\forall\, r \in [R_1, R_2]$.

And, the right ψ-Caputo generalized fractional radial derivative of $f : A \to \mathbb{R}$ of variable order $\alpha(\cdot)$ is given by:

$$ {}^{C}D_{R_2-}^{\alpha(\cdot),\psi} f(x) = {}^{C}D_{R_2-}^{\alpha(\cdot),\psi} f(r\omega) := $$

$$ (-1)^n \int_{r}^{R_2} \frac{(\psi(t) - \psi(r))^{n-\alpha(t)-1}}{\Gamma(n-\alpha(t))} \psi'(t) f_{\psi}^{[n]}(t\omega)\, dt, \tag{2.13} $$

$\forall\, x \in A$.

We make

Remark 2.5 (*to Definition* 2.4) We observe that

$$ {}^{C}D_{R_1+}^{\alpha(\cdot),\psi} f(x) = I_{R_1+;\psi}^{n-\alpha(\cdot)} f_{\psi}^{[n]}(x), \tag{2.14} $$

and

$$ {}^{C}D_{R_2-}^{\alpha(\cdot),\psi} f(x) = (-1)^n I_{R_2-;\psi}^{n-\alpha(\cdot)} f_{\psi}^{[n]}(x), \tag{2.15} $$

$\forall\, x \in A$.

By Theorem 2.3, both ${}^{C}D_{R_1+}^{\alpha(\cdot),\psi} f(x)$, ${}^{C}D_{R_2-}^{\alpha(\cdot),\psi} f \in C(A)$.

All the next in this section come from [4] and they will be used in the main results.

Definition 2.6 Let $[a, b] \subset \mathbb{R}$, $(X, \|\cdot\|)$ a Banach space, $g \in C^1([a, b])$ and increasing, $f \in C([a, b], X)$, $\nu \in C([a, b])$, $\nu > 1$. We define the left Riemann–Liouville generalized fractional Bochner integral operator of variable order $\nu(\cdot)$

$$ \left({}^{B}I_{a+;g}^{\nu(\cdot)} f\right)(x) := \int_{a}^{x} \frac{(g(x) - g(z))^{\nu(z)-1}}{\Gamma(\nu(z))} g'(z) f(z)\, dz, \tag{2.16} $$

$\forall\, x \in [a,b]$. We assume ${}^{B}I^{\nu(\cdot)}_{a+;g} f \in C([a,b], X)$.

Definition 2.7 Let $[a,b] \subset \mathbb{R}$, $(X, \|\cdot\|)$ a Banach space, $g \in C^1([a,b])$ and increasing, $f \in C([a,b], X)$, $\nu \in C([a,b])$, $\nu > 1$. We define the right Riemann–Liouville generalized fractional Bochner integral operator of variable order $\nu(\cdot)$

$$\left({}^{B}I^{\nu(\cdot)}_{b-;g} f\right)(x) := \int_x^b \frac{(g(z) - g(x))^{\nu(z)-1}}{\Gamma(\nu(z))} g'(z) f(z)\, dz, \tag{2.17}$$

$\forall\, x \in [a,b]$. We assume ${}^{B}I^{\nu(\cdot)}_{b-;g} f \in C([a,b], X)$.

We mention the following left fractional Opial type inequality of variable order (see also [1]).

Theorem 2.8 *All as in Definition 2.6. Let* $p, q > 1 : \frac{1}{p} + \frac{1}{q} = 1$. *Then*

$$\int_a^x \left\| \left({}^{B}I^{\nu(\cdot)}_{a+;g} f\right)(w) \right\| \|f(w)\| g'(w)\, dw \leq$$

$$2^{-\frac{1}{q}} \left(\int_a^x \left(\int_a^w \left(\frac{(g(w) - g(z))^{\nu(z)-1}}{\Gamma(\nu(z))} \right)^p dz \right) dw \right)^{\frac{1}{p}} \left(\int_a^x \|f(w)\|^q (g'(w))^q\, dw \right)^{\frac{2}{q}}, \tag{2.18}$$

$\forall\, x \in [a,b]$.

Next follows a right fractional Opial type inequality of variable order.

Theorem 2.9 *All as in Definition 2.7. Let* $p, q > 1 : \frac{1}{p} + \frac{1}{q} = 1$. *Then*

$$\int_x^b \left\| \left({}^{B}I^{\nu(\cdot)}_{b-;g} f\right)(w) \right\| \|f(w)\| g'(w)\, dw \leq$$

$$2^{-\frac{1}{q}} \left(\int_x^b \left(\int_w^b \left(\frac{(g(z) - g(w))^{\nu(z)-1}}{\Gamma(\nu(z))} \right)^p dz \right) dw \right)^{\frac{1}{p}} \left(\int_x^b \|f(w)\|^q (g'(w))^q\, dw \right)^{\frac{2}{q}}, \tag{2.19}$$

$\forall\, x \in [a,b]$.

We mention two extreme Opial inequalities ($p = 1$, $q = \infty$ case).

Theorem 2.10 *All as in Definition 2.6. Then*

$$\int_a^x \left\| \left({}^{B}I^{\nu(\cdot)}_{a+;g} f\right)(w) \right\| \|f(w)\|\, dw \leq$$

$$\left(\int_a^x \left(\int_a^w \frac{(g(w) - g(z))^{\nu(z)-1}}{\Gamma(\nu(z))} g'(z)\, dz \right) dw \right) \|f\|_\infty^2, \tag{2.20}$$

$\forall\, x \in [a,b]$.

Theorem 2.11 *All as in Definition 2.7. Then*

$$\int_x^b \left\| \left({}^B I_{b-;g}^{\nu(\cdot)} f \right)(w) \right\| \|f(w)\| \, dw \le$$

$$\left(\int_x^b \left(\int_w^b \frac{(g(z) - g(w))^{\nu(z)-1}}{\Gamma(\nu(z))} g'(z) \, dz \right) dw \right) \|f\|_\infty^2, \tag{2.21}$$

$\forall \, x \in [a, b]$.

Next we mention two fractional Hardy's type inequalities of variable order.

Theorem 2.12 *All as in Definition 2.6. Let* $p, q > 1 : \frac{1}{p} + \frac{1}{q} = 1$. *Then*

$$\left\| \left\| {}^B I_{a+;g}^{\nu(\cdot)} f \right\| \right\|_q \le \left(\int_a^b \left(\int_a^x \left(\frac{(g(x) - g(z))^{\nu(z)-1}}{\Gamma(\nu(z))} \right)^p dz \right)^{\frac{q}{p}} dx \right)^{\frac{1}{q}} \left\| g' \|f\| \right\|_q. \tag{2.22}$$

Theorem 2.13 *All as in Definition 2.7. Let* $p, q > 1 : \frac{1}{p} + \frac{1}{q} = 1$. *Then*

$$\left\| \left\| {}^B I_{b-;g}^{\nu(\cdot)} f \right\| \right\|_q \le \left(\int_a^b \left(\int_x^b \left(\frac{(g(z) - g(x))^{\nu(z)-1}}{\Gamma(\nu(z))} \right)^p dz \right)^{\frac{q}{p}} dx \right)^{\frac{1}{q}} \left\| g' \|f\| \right\|_q. \tag{2.23}$$

The above mentioned results will be applied for $(X, \|\cdot\|) = (\mathbb{R}, |\cdot|)$.

The purpose of this chapter is to derive interesting fractional radial integral inequalities of variable order.

2.2 Main Results

We present the following left fractional Opial type inequality of variable order over the spherical shell.

Theorem 2.14 *All as in Definition 2.2 with* ψ *strictly increasing. Then*

$$\int_A \left| \left(I_{R_1+;\psi}^{\alpha(\cdot)} f \right)(x) \right| |f(x)| \, dx \le 2^{-\frac{1}{2}} \left(\frac{R_2}{R_1} \right)^{N-1} \left(\frac{\|\psi'\|_\infty^2}{\inf_{[R_1,R_2]} (\psi')} \right)$$

$$\left(\int_{R_1}^{R_2} \left(\int_{R_1}^r \left(\frac{(\psi(r) - \psi(z))^{\alpha(z)-1}}{\Gamma(\alpha(z))} \right)^2 dz \right) dr \right)^{\frac{1}{2}} \left(\int_A f^2(x) \, dx \right). \tag{2.24}$$

Proof Here we apply (2.18) for $q = 2$ and for $f(\cdot\omega) \in C([R_1, R_2])$. We have that

$$\int_{R_1}^{R_2} \left|\left(I_{R_1+;\psi}^{\alpha(\cdot)} f\right)(r\omega)\right| |f(r\omega)| \psi'(r)\, dr \le$$

$$2^{-\frac{1}{2}} \left(\int_{R_1}^{R_2} \left(\int_{R_1}^{r} \left(\frac{(\psi(r) - \psi(z))^{\alpha(z)-1}}{\Gamma(\alpha(z))}\right)^2 dz\right) dr\right)^{\frac{1}{2}} \tag{2.25}$$

$$\left(\int_{R_1}^{R_2} (f(r\omega))^2 \left(\psi'(r)\right)^2 dr\right), \quad \forall\, \omega \in S^{N-1}.$$

Hence it holds

$$\inf_{[R_1,R_2]} (\psi') \int_{R_1}^{R_2} \left|\left(I_{R_1+;\psi}^{\alpha(\cdot)} f\right)(r\omega)\right| |f(r\omega)|\, dr \le 2^{-\frac{1}{2}} \left\|\psi'\right\|_\infty^2$$

$$\left(\int_{R_1}^{R_2} \left(\int_{R_1}^{r} \left(\frac{(\psi(r) - \psi(z))^{\alpha(z)-1}}{\Gamma(\alpha(z))}\right)^2 dz\right) dr\right)^{\frac{1}{2}} \left(\int_{R_1}^{R_2} (f(r\omega))^2\, dr\right), \tag{2.26}$$

$\forall\, \omega \in S^{N-1}$.

By Theorem 2.3 we have that $\left(I_{R_1+;\psi}^{\alpha(\cdot)} f\right)(\cdot\omega) \in C([R_1, R_2])$, $\forall\ \omega \in S^{N-1}$, where $\overline{A} = [R_1, R_2] \times S^{N-1}$. By $r^{N-1} r^{1-N} = 1$, where $R_1 \le r \le R_2$, and $R_2^{1-N} \le r^{1-N} \le R_1^{1-N}$, we obtain:

$$R_2^{1-N} \inf_{[R_1,R_2]} (\psi') \int_{R_1}^{R_2} r^{N-1} \left|\left(I_{R_1+;\psi}^{\alpha(\cdot)} f\right)(r\omega)\right| |f(r\omega)|\, dr \le$$

$$\inf_{[R_1,R_2]} (\psi') \int_{R_1}^{R_2} r^{N-1} r^{1-N} \left|\left(I_{R_1+;\psi}^{\alpha(\cdot)} f\right)(r\omega)\right| |f(r\omega)|\, dr =$$

$$\inf_{[R_1,R_2]} (\psi') \int_{R_1}^{R_2} \left|\left(I_{R_1+;\psi}^{\alpha(\cdot)} f\right)(r\omega)\right| |f(r\omega)|\, dr \overset{(2.26)}{\le}$$

$$2^{-\frac{1}{2}} \left\|\psi'\right\|_\infty^2 \left(\int_{R_1}^{R_2} \left(\int_{R_1}^{r} \left(\frac{(\psi(r) - \psi(z))^{\alpha(z)-1}}{\Gamma(\alpha(z))}\right)^2 dz\right) dr\right)^{\frac{1}{2}} \tag{2.27}$$

$$\left(\int_{R_1}^{R_2} r^{N-1} r^{1-N} (f(r\omega))^2\, dr\right) \le$$

$$2^{-\frac{1}{2}} R_1^{1-N} \left\|\psi'\right\|_\infty^2 \left(\int_{R_1}^{R_2} \left(\int_{R_1}^{r} \left(\frac{(\psi(r) - \psi(z))^{\alpha(z)-1}}{\Gamma(\alpha(z))} \right)^2 dz \right) dr \right)^{\frac{1}{2}} \tag{2.28}$$

$$\left(\int_{R_1}^{R_2} r^{N-1} (f(r\omega))^2 dr \right), \quad \forall \omega \in S^{N-1}.$$

So far we have

$$\int_{R_1}^{R_2} r^{N-1} \left| \left(I_{R_1+;\psi}^{\alpha(\cdot)} f \right)(r\omega) \right| |f(r\omega)| \, dr \leq$$

$$2^{-\frac{1}{2}} \left(\frac{R_2}{R_1} \right)^{N-1} \left(\frac{\left\|\psi'\right\|_\infty^2}{\inf\limits_{[R_1,R_2]} (\psi')} \right) \left(\int_{R_1}^{R_2} \left(\int_{R_1}^{r} \left(\frac{(\psi(r) - \psi(z))^{\alpha(z)-1}}{\Gamma(\alpha(z))} \right)^2 dz \right) dr \right)^{\frac{1}{2}}$$

$$\left(\int_{R_1}^{R_2} r^{N-1} (f(r\omega))^2 dr \right), \quad \forall \omega \in S^{N-1}. \tag{2.29}$$

Hence it holds

$$\int_A \left| \left(I_{R_1+;\psi}^{\alpha(\cdot)} f \right)(x) \right| |f(x)| \, dx =$$

$$\int_{S^{N-1}} \left(\int_{R_1}^{R_2} r^{N-1} \left| \left(I_{R_1+;\psi}^{\alpha(\cdot)} f \right)(r\omega) \right| |f(r\omega)| \, dr \right) d\omega \leq$$

$$2^{-\frac{1}{2}} \left(\frac{R_2}{R_1} \right)^{N-1} \left(\frac{\left\|\psi'\right\|_\infty^2}{\inf\limits_{[R_1,R_2]} (\psi')} \right) \left(\int_{R_1}^{R_2} \left(\int_{R_1}^{r} \left(\frac{(\psi(r) - \psi(z))^{\alpha(z)-1}}{\Gamma(\alpha(z))} \right)^2 dz \right) dr \right)^{\frac{1}{2}}$$

$$\left(\int_{S^{N-1}} \left(\int_{R_1}^{R_2} r^{N-1} (f(r\omega))^2 dr \right) d\omega \right) =$$

$$2^{-\frac{1}{2}} \left(\frac{R_2}{R_1} \right)^{N-1} \left(\frac{\left\|\psi'\right\|_\infty^2}{\inf\limits_{[R_1,R_2]} (\psi')} \right) \left(\int_{R_1}^{R_2} \left(\int_{R_1}^{r} \left(\frac{(\psi(r) - \psi(z))^{\alpha(z)-1}}{\Gamma(\alpha(z))} \right)^2 dz \right) dr \right)^{\frac{1}{2}}$$

$$\left(\int_A (f(x))^2 dx \right), \tag{2.30}$$

proving the claim. ■

We give a right fractional Opial type inequality of variable order over the spherical shell.

Theorem 2.15 *All as in Definition 2.2 with ψ strictly increasing. Then*

$$\int_A \left|\left(I^{\alpha(\cdot)}_{R_2-;\psi} f\right)(x)\right| |f(x)|\, dx \le 2^{-\frac{1}{2}} \left(\frac{R_2}{R_1}\right)^{N-1} \left(\frac{\|\psi'\|_\infty^2}{\inf\limits_{[R_1,R_2]} (\psi')}\right)$$

$$\left(\int_{R_1}^{R_2} \left(\int_r^{R_2} \left(\frac{(\psi(z)-\psi(r))^{\alpha(z)-1}}{\Gamma(\alpha(z))}\right)^2 dz\right) dr\right)^{\frac{1}{2}} \left(\int_A f^2(x)\, dx\right). \quad (2.31)$$

Proof Here we apply (2.19) for $q = 2$ and for $f(\cdot\omega) \in C([R_1, R_2])$. We have that

$$\int_{R_1}^{R_2} \left|\left(I^{\alpha(\cdot)}_{R_2-;\psi} f\right)(r\omega)\right| |f(r\omega)|\, \psi'(r)\, dr \le$$

$$2^{-\frac{1}{2}} \left(\int_{R_1}^{R_2} \left(\int_r^{R_2} \left(\frac{(\psi(z)-\psi(r))^{\alpha(z)-1}}{\Gamma(\alpha(z))}\right)^2 dz\right) dr\right)^{\frac{1}{2}} \quad (2.32)$$

$$\left(\int_{R_1}^{R_2} (f(r\omega))^2 \left(\psi'(r)\right)^2 dr\right), \quad \forall \omega \in S^{N-1}.$$

Hence it holds

$$\inf_{[R_1,R_2]} (\psi') \int_{R_1}^{R_2} \left|\left(I^{\alpha(\cdot)}_{R_2-;\psi} f\right)(r\omega)\right| |f(r\omega)|\, dr \le 2^{-\frac{1}{2}} \|\psi'\|_\infty^2$$

$$\left(\int_{R_1}^{R_2} \left(\int_r^{R_2} \left(\frac{(\psi(z)-\psi(r))^{\alpha(z)-1}}{\Gamma(\alpha(z))}\right)^2 dz\right) dr\right)^{\frac{1}{2}} \left(\int_{R_1}^{R_2} (f(r\omega))^2\, dr\right), \quad (2.33)$$

$\forall\, \omega \in S^{N-1}$.

By Theorem 2.3 we have that $\left(I^{\alpha(\cdot)}_{R_2-;\psi} f\right)(\cdot\omega) \in C([R_1, R_2])$, $\forall\, \omega \in S^{N-1}$. We obtain

$$R_2^{1-N} \inf_{[R_1,R_2]} (\psi') \int_{R_1}^{R_2} r^{N-1} \left|\left(I^{\alpha(\cdot)}_{R_2-;\psi} f\right)(r\omega)\right| |f(r\omega)|\, dr \le$$

$$\inf_{[R_1,R_2]} (\psi') \int_{R_1}^{R_2} r^{N-1} r^{1-N} \left|\left(I^{\alpha(\cdot)}_{R_2-;\psi} f\right)(r\omega)\right| |f(r\omega)|\, dr =$$

$$\inf_{[R_1,R_2]} (\psi') \int_{R_1}^{R_2} \left|\left(I^{\alpha(\cdot)}_{R_2-;\psi} f\right)(r\omega)\right| |f(r\omega)|\, dr \overset{(2.33)}{\le}$$

$$2^{-\frac{1}{2}} \left\|\psi'\right\|_{\infty}^{2} \left(\int_{R_1}^{R_2} \left(\int_{r}^{R_2} \left(\frac{(\psi(z)-\psi(r))^{\alpha(z)-1}}{\Gamma(\alpha(z))}\right)^2 dz\right) dr\right)^{\frac{1}{2}} \tag{2.34}$$

$$\left(\int_{R_1}^{R_2} r^{N-1} r^{1-N} (f(r\omega))^2 dr\right) \leq$$

$$2^{-\frac{1}{2}} R_1^{1-N} \left\|\psi'\right\|_{\infty}^{2} \left(\int_{R_1}^{R_2} \left(\int_{r}^{R_2} \left(\frac{(\psi(z)-\psi(r))^{\alpha(z)-1}}{\Gamma(\alpha(z))}\right)^2 dz\right) dr\right)^{\frac{1}{2}}$$

$$\left(\int_{R_1}^{R_2} r^{N-1} (f(r\omega))^2 dr\right), \quad \forall \omega \in S^{N-1}. \tag{2.35}$$

So far we have

$$\int_{R_1}^{R_2} r^{N-1} \left|\left(I_{R_2-;\psi}^{\alpha(\cdot)} f\right)(r\omega)\right| |f(r\omega)| dr \leq$$

$$2^{-\frac{1}{2}} \left(\frac{R_2}{R_1}\right)^{N-1} \left(\frac{\left\|\psi'\right\|_{\infty}^{2}}{\inf\limits_{[R_1,R_2]} (\psi')}\right) \left(\int_{R_1}^{R_2} \left(\int_{r}^{R_2} \left(\frac{(\psi(z)-\psi(r))^{\alpha(z)-1}}{\Gamma(\alpha(z))}\right)^2 dz\right) dr\right)^{\frac{1}{2}}$$

$$\left(\int_{R_1}^{R_2} r^{N-1} (f(r\omega))^2 dr\right), \quad \forall \omega \in S^{N-1}. \tag{2.36}$$

Hence it holds

$$\int_A \left|\left(I_{R_2-;\psi}^{\alpha(\cdot)} f\right)(x)\right| |f(x)| dx =$$

$$\int_{S^{N-1}} \left(\int_{R_1}^{R_2} r^{N-1} \left|\left(I_{R_2-;\psi}^{\alpha(\cdot)} f\right)(r\omega)\right| |f(r\omega)| dr\right) d\omega \leq$$

$$2^{-\frac{1}{2}} \left(\frac{R_2}{R_1}\right)^{N-1} \left(\frac{\left\|\psi'\right\|_{\infty}^{2}}{\inf\limits_{[R_1,R_2]} (\psi')}\right) \left(\int_{R_1}^{R_2} \left(\int_{r}^{R_2} \left(\frac{(\psi(z)-\psi(r))^{\alpha(z)-1}}{\Gamma(\alpha(z))}\right)^2 dz\right) dr\right)^{\frac{1}{2}}$$

$$\left(\int_{S^{N-1}} \left(\int_{R_1}^{R_2} r^{N-1} (f(r\omega))^2 dr\right) d\omega\right) =$$

$$2^{-\frac{1}{2}} \left(\frac{R_2}{R_1}\right)^{N-1} \left(\frac{\left\|\psi'\right\|_{\infty}^{2}}{\inf\limits_{[R_1,R_2]} (\psi')}\right) \left(\int_{R_1}^{R_2} \left(\int_{r}^{R_2} \left(\frac{(\psi(z)-\psi(r))^{\alpha(z)-1}}{\Gamma(\alpha(z))}\right)^2 dz\right) dr\right)^{\frac{1}{2}}$$

$$\left(\int_A (f(x))^2 \, dx\right), \tag{2.37}$$

proving the claim. ■

We give two extreme Opial inequalities.

Theorem 2.16 *All as in Definition 2.2. Then*

$$\int_A \left| \left(I^{\alpha(\cdot)}_{R_1+;\psi} f \right)(x) \right| |f(x)| \, dx \leq \frac{2\pi^{\frac{N}{2}} R_2^{N-1}}{\Gamma\left(\frac{N}{2}\right)}$$

$$\left(\int_{R_1}^{R_2} \left(\int_{R_1}^{r} \frac{(\psi(r) - \psi(z))^{\alpha(z)-1}}{\Gamma(\alpha(z))} \psi'(z) \, dz \right) dr \right) \|f\|^2_{\infty,\overline{A}}. \tag{2.38}$$

Proof Here we apply (2.20). We have that

$$R_2^{1-N} \int_{R_1}^{R_2} r^{N-1} \left| \left(I^{\alpha(\cdot)}_{R_1+;\psi} f \right)(r\omega) \right| |f(r\omega)| \, dr \leq$$

$$\int_{R_1}^{R_2} r^{N-1} r^{1-N} \left| \left(I^{\alpha(\cdot)}_{R_1+;\psi} f \right)(r\omega) \right| |f(r\omega)| \, dr =$$

$$\int_{R_1}^{R_2} \left| \left(I^{\alpha(\cdot)}_{R_1+;\psi} f \right)(r\omega) \right| |f(r\omega)| \, dr \overset{(2.20)}{\leq}$$

$$\left(\int_{R_1}^{R_2} \left(\int_{R_1}^{r} \frac{(\psi(r) - \psi(z))^{\alpha(z)-1}}{\Gamma(\alpha(z))} \psi'(z) \, dz \right) dr \right) \|f(\cdot\omega)\|^2_{\infty,[R_1,R_2]} \leq \tag{2.39}$$

$$\left(\int_{R_1}^{R_2} \left(\int_{R_1}^{r} \frac{(\psi(r) - \psi(z))^{\alpha(z)-1}}{\Gamma(\alpha(z))} \psi'(z) \, dz \right) dr \right) \|f\|^2_{\infty,\overline{A}},$$

$\forall \omega \in S^{N-1}$.

Hence it holds

$$\int_A \left| \left(I^{\alpha(\cdot)}_{R_1+;\psi} f \right)(x) \right| |f(x)| \, dx =$$

$$\int_{S^{N-1}} \left(\int_{R_1}^{R_2} r^{N-1} \left| \left(I^{\alpha(\cdot)}_{R_1+;\psi} f \right)(r\omega) \right| |f(r\omega)| \, dr \right) d\omega \leq \tag{2.40}$$

$$\frac{2R_2^{N-1}\pi^{\frac{N}{2}}}{\Gamma\left(\frac{N}{2}\right)} \left(\int_{R_1}^{R_2} \left(\int_{R_1}^{r} \frac{(\psi(r) - \psi(z))^{\alpha(z)-1}}{\Gamma(\alpha(z))} \psi'(z) \, dz \right) dr \right) \|f\|^2_{\infty,\overline{A}},$$

proving the claim. ■

Theorem 2.17 *All as in Definition 2.2. Then*

$$\int_A \left| \left(I^{\alpha(\cdot)}_{R_2-;\psi} f \right)(x) \right| |f(x)| \, dx \leq \frac{2\pi^{\frac{N}{2}} R_2^{N-1}}{\Gamma\left(\frac{N}{2}\right)}$$

$$\left(\int_{R_1}^{R_2} \left(\int_r^{R_2} \frac{(\psi(z) - \psi(r))^{\alpha(z)-1}}{\Gamma(\alpha(z))} \psi'(z) \, dz \right) dr \right) \|f\|^2_{\infty,\overline{A}}. \tag{2.41}$$

Proof Here we apply (2.21). We have that

$$R_2^{1-N} \int_{R_1}^{R_2} r^{N-1} \left| \left(I^{\alpha(\cdot)}_{R_2-;\psi} f \right)(r\omega) \right| |f(r\omega)| \, dr \leq$$

$$\int_{R_1}^{R_2} r^{N-1} r^{1-N} \left| \left(I^{\alpha(\cdot)}_{R_2-;\psi} f \right)(r\omega) \right| |f(r\omega)| \, dr =$$

$$\int_{R_1}^{R_2} \left| \left(I^{\alpha(\cdot)}_{R_2-;\psi} f \right)(r\omega) \right| |f(r\omega)| \, dr \overset{(2.21)}{\leq}$$

$$\left(\int_{R_1}^{R_2} \left(\int_r^{R_2} \frac{(\psi(z) - \psi(r))^{\alpha(z)-1}}{\Gamma(\alpha(z))} \psi'(z) \, dz \right) dr \right) \|f(\cdot\omega)\|^2_{\infty,[R_1,R_2]} \leq \tag{2.42}$$

$$\left(\int_{R_1}^{R_2} \left(\int_r^{R_2} \frac{(\psi(z) - \psi(r))^{\alpha(z)-1}}{\Gamma(\alpha(z))} \psi'(z) \, dz \right) dr \right) \|f\|^2_{\infty,\overline{A}},$$

$\forall \omega \in S^{N-1}$.

Hence it holds

$$\int_A \left| \left(I^{\alpha(\cdot)}_{R_2-;\psi} f \right)(x) \right| |f(x)| \, dx =$$

$$\int_{S^{N-1}} \left(\int_{R_1}^{R_2} r^{N-1} \left| \left(I^{\alpha(\cdot)}_{R_2-;\psi} f \right)(r\omega) \right| |f(r\omega)| \, dr \right) d\omega \leq$$

$$\frac{2R_2^{N-1}\pi^{\frac{N}{2}}}{\Gamma\left(\frac{N}{2}\right)} \left(\int_{R_1}^{R_2} \left(\int_r^{R_2} \frac{(\psi(z) - \psi(r))^{\alpha(z)-1}}{\Gamma(\alpha(z))} \psi'(z) \, dz \right) dr \right) \|f\|^2_{\infty,\overline{A}}, \tag{2.43}$$

proving the claim. ■

Two Hardy's type fractional inequalities of variable order on the spherical shell follow.

Theorem 2.18 *All as in Definition 2.2. Then*

$$\int_A \left(I^{\alpha(\cdot)}_{R_1+;\psi} f(x)\right)^2 dx \leq \left(\frac{R_2}{R_1}\right)^{N-1} \left\|\psi'\right\|^2_{\infty,[R_1,R_2]}$$

$$\left(\int_{R_1}^{R_2}\left(\int_{R_1}^{r}\left(\frac{(\psi(r)-\psi(z))^{\alpha(z)-1}}{\Gamma(\alpha(z))}\right)^2 dz\right)dr\right)\left(\int_A f^2(x)\,dx\right). \tag{2.44}$$

Proof By (2.22) ($p = q = 2$ case) we have

$$R_2^{1-N}\int_{R_1}^{R_2} r^{N-1}\left(I^{\alpha(\cdot)}_{R_1+;\psi} f(r\omega)\right)^2 dr \leq$$

$$\int_{R_1}^{R_2} r^{N-1}r^{1-N}\left(I^{\alpha(\cdot)}_{R_1+;\psi} f(r\omega)\right)^2 dr = \int_{R_1}^{R_2}\left(I^{\alpha(\cdot)}_{R_1+;\psi} f(r\omega)\right)^2 dr \overset{(2.22)}{\leq}$$

$$\left(\int_{R_1}^{R_2}\left(\int_{R_1}^{r}\left(\frac{(\psi(r)-\psi(z))^{\alpha(z)-1}}{\Gamma(\alpha(z))}\right)^2 dz\right)dr\right) \tag{2.45}$$

$$\left(\int_{R_1}^{R_2}\left(\psi'(r)\right)^2 (f(r\omega))^2\,dr\right) \leq$$

$$\left\|\psi'\right\|^2_{\infty,[R_1,R_2]}\left(\int_{R_1}^{R_2}\left(\int_{R_1}^{r}\left(\frac{(\psi(r)-\psi(z))^{\alpha(z)-1}}{\Gamma(\alpha(z))}\right)^2 dz\right)dr\right)$$

$$\left(\int_{R_1}^{R_2} r^{N-1}r^{1-N}(f(r\omega))^2\,dr\right) \leq$$

$$R_1^{1-N}\left\|\psi'\right\|^2_{\infty,[R_1,R_2]}\left(\int_{R_1}^{R_2}\left(\int_{R_1}^{r}\left(\frac{(\psi(r)-\psi(z))^{\alpha(z)-1}}{\Gamma(\alpha(z))}\right)^2 dz\right)dr\right)$$

$$\left(\int_{R_1}^{R_2} r^{N-1}(f(r\omega))^2\,dr\right), \ \forall\omega\in S^{N-1}. \tag{2.46}$$

So far we have proved that

$$\int_{R_1}^{R_2} r^{N-1}\left(I^{\alpha(\cdot)}_{R_1+;\psi} f(r\omega)\right)^2 dr \leq$$

$$\left(\frac{R_2}{R_1}\right)^{N-1} \|\psi'\|^2_{\infty,[R_1,R_2]} \left(\int_{R_1}^{R_2} \left(\int_{R_1}^{r} \left(\frac{(\psi(r)-\psi(z))^{\alpha(z)-1}}{\Gamma(\alpha(z))}\right)^2 dz\right) dr\right)$$

$$\left(\int_{R_1}^{R_2} r^{N-1} (f(r\omega))^2 dr\right), \quad \forall \omega \in S^{N-1}. \tag{2.47}$$

Hence it holds

$$\int_A \left(I^{\alpha(\cdot)}_{R_1+;\psi} f(x)\right)^2 dx = \int_{S^{N-1}} \left(\int_{R_1}^{R_2} r^{N-1} \left(I^{\alpha(\cdot)}_{R_1+;\psi} f(r\omega)\right)^2 dr\right) d\omega \le$$

$$\left(\frac{R_2}{R_1}\right)^{N-1} \|\psi'\|^2_{\infty,[R_1,R_2]} \left(\int_{R_1}^{R_2} \left(\int_{R_1}^{r} \left(\frac{(\psi(r)-\psi(z))^{\alpha(z)-1}}{\Gamma(\alpha(z))}\right)^2 dz\right) dr\right)$$

$$\left(\int_{S^{N-1}} \left(\int_{R_1}^{R_2} r^{N-1} (f(r\omega))^2 dr\right) d\omega\right) =$$

$$\left(\frac{R_2}{R_1}\right)^{N-1} \|\psi'\|^2_{\infty,[R_1,R_2]} \left(\int_{R_1}^{R_2} \left(\int_{R_1}^{r} \left(\frac{(\psi(r)-\psi(z))^{\alpha(z)-1}}{\Gamma(\alpha(z))}\right)^2 dz\right) dr\right)$$

$$\left(\int_A (f(x))^2 dx\right), \tag{2.48}$$

proving the claim. ■

We continue with

Theorem 2.19 *All as in Definition 2.2. Then*

$$\int_A \left(I^{\alpha(\cdot)}_{R_2-;\psi} f(x)\right)^2 dx \le \left(\frac{R_2}{R_1}\right)^{N-1} \|\psi'\|^2_{\infty,[R_1,R_2]}$$

$$\left(\int_{R_1}^{R_2} \left(\int_{r}^{R_2} \left(\frac{(\psi(z)-\psi(r))^{\alpha(z)-1}}{\Gamma(\alpha(z))}\right)^2 dz\right) dr\right) \left(\int_A f^2(x)\, dx\right). \tag{2.49}$$

Proof By (2.23) ($p = q = 2$ case) we have

$$R_2^{1-N} \int_{R_1}^{R_2} r^{N-1} \left(I^{\alpha(\cdot)}_{R_2-;\psi} f(r\omega)\right)^2 dr \le$$

$$\int_{R_1}^{R_2} r^{N-1} r^{1-N} \left(I_{R_2-;\psi}^{\alpha(\cdot)} f(r\omega) \right)^2 dr = \int_{R_1}^{R_2} \left(I_{R_2-;\psi}^{\alpha(\cdot)} f(r\omega) \right)^2 dr \overset{(2.23)}{\leq} \tag{2.50}$$

$$\left(\int_{R_1}^{R_2} \left(\int_r^{R_2} \left(\frac{(\psi(z) - \psi(r))^{\alpha(z)-1}}{\Gamma(\alpha(z))} \right)^2 dz \right) dr \right)$$

$$\left(\int_{R_1}^{R_2} (\psi'(r))^2 (f(r\omega))^2 dr \right) \leq$$

$$\|\psi'\|_{\infty,[R_1,R_2]}^2 \left(\int_{R_1}^{R_2} \left(\int_r^{R_2} \left(\frac{(\psi(z) - \psi(r))^{\alpha(z)-1}}{\Gamma(\alpha(z))} \right)^2 dz \right) dr \right)$$

$$\left(\int_{R_1}^{R_2} r^{N-1} r^{1-N} (f(r\omega))^2 dr \right) \leq$$

$$R_1^{1-N} \|\psi'\|_{\infty,[R_1,R_2]}^2 \left(\int_{R_1}^{R_2} \left(\int_r^{R_2} \left(\frac{(\psi(z) - \psi(r))^{\alpha(z)-1}}{\Gamma(\alpha(z))} \right)^2 dz \right) dr \right)$$

$$\left(\int_{R_1}^{R_2} r^{N-1} (f(r\omega))^2 dr \right), \ \forall \omega \in S^{N-1}. \tag{2.51}$$

So far we have proved that

$$\int_{R_1}^{R_2} r^{N-1} \left(I_{R_2-;\psi}^{\alpha(\cdot)} f(r\omega) \right)^2 dr \leq$$

$$\left(\frac{R_2}{R_1} \right)^{N-1} \|\psi'\|_{\infty,[R_1,R_2]}^2 \left(\int_{R_1}^{R_2} \left(\int_r^{R_2} \left(\frac{(\psi(z) - \psi(r))^{\alpha(z)-1}}{\Gamma(\alpha(z))} \right)^2 dz \right) dr \right)$$

$$\left(\int_{R_1}^{R_2} r^{N-1} (f(r\omega))^2 dr \right), \ \forall \omega \in S^{N-1}. \tag{2.52}$$

Hence it holds

$$\int_A \left(I_{R_2-;\psi}^{\alpha(\cdot)} f(x) \right)^2 dx = \int_{S^{N-1}} \left(\int_{R_1}^{R_2} r^{N-1} \left(I_{R_2-;\psi}^{\alpha(\cdot)} f(r\omega) \right)^2 dr \right) d\omega \leq$$

$$\left(\frac{R_2}{R_1}\right)^{N-1} \left\|\psi'\right\|_{\infty,[R_1,R_2]}^2 \left(\int_{R_1}^{R_2}\left(\int_r^{R_2}\left(\frac{(\psi(z)-\psi(r))^{\alpha(z)-1}}{\Gamma(\alpha(z))}\right)^2 dz\right)dr\right)$$

$$\left(\int_{S^{N-1}}\left(\int_{R_1}^{R_2} r^{N-1}(f(r\omega))^2 dr\right)d\omega\right) =$$

$$\left(\frac{R_2}{R_1}\right)^{N-1} \left\|\psi'\right\|_{\infty,[R_1,R_2]}^2 \left(\int_{R_1}^{R_2}\left(\int_r^{R_2}\left(\frac{(\psi(z)-\psi(r))^{\alpha(z)-1}}{\Gamma(\alpha(z))}\right)^2 dz\right)dr\right)$$

$$\left(\int_A (f(x))^2 dx\right), \tag{2.53}$$

proving the claim. ■

Fractional inequalities for the generalized fractional radial derivatives of variable order follow.

First Opial type inequalities:

Theorem 2.20 *All as in Definition 2.4. Then*

$$\int_A \left|\left({}^C D_{R_1+}^{\alpha(\cdot),\psi} f\right)(x)\right| \left|f_\psi^{[n]}(x)\right| dx \leq 2^{-\frac{1}{2}} \left(\frac{R_2}{R_1}\right)^{N-1} \left(\frac{\left\|\psi'\right\|_\infty^2}{\inf\limits_{[R_1,R_2]}(\psi')}\right)$$

$$\left(\int_{R_1}^{R_2}\left(\int_{R_1}^{r}\left(\frac{(\psi(r)-\psi(z))^{n-\alpha(z)-1}}{\Gamma(n-\alpha(z))}\right)^2 dz\right)dr\right)^{\frac{1}{2}} \left(\int_A \left(f_\psi^{[n]}(x)\right)^2 dx\right). \tag{2.54}$$

Proof As similar to the proof of Theorem 2.14 is omitted, see also (2.14). ■

Theorem 2.21 *All as in Definition 2.4. Then*

$$\int_A \left|\left({}^C D_{R_2-}^{\alpha(\cdot),\psi} f\right)(x)\right| \left|f_\psi^{[n]}(x)\right| dx \leq 2^{-\frac{1}{2}} \left(\frac{R_2}{R_1}\right)^{N-1} \left(\frac{\left\|\psi'\right\|_\infty^2}{\inf\limits_{[R_1,R_2]}(\psi')}\right)$$

$$\left(\int_{R_1}^{R_2}\left(\int_{r}^{R_2}\left(\frac{(\psi(z)-\psi(r))^{n-\alpha(z)-1}}{\Gamma(n-\alpha(z))}\right)^2 dz\right)dr\right)^{\frac{1}{2}} \left(\int_A \left(f_\psi^{[n]}(x)\right)^2 dx\right). \tag{2.55}$$

Proof As similar to the proof of Theorem 2.15 is omitted, see also (2.15). ■

We continue with extreme Opial type inequalities:

Theorem 2.22 *All as in Definition 2.4. Then*

$$\int_A \left| \left({}^C D_{R_1+}^{\alpha(\cdot),\psi} f \right)(x) \right| \left| f_\psi^{[n]}(x) \right| dx \leq \frac{2\pi^{\frac{N}{2}} R_2^{N-1}}{\Gamma\left(\frac{N}{2}\right)}$$

$$\left(\int_{R_1}^{R_2} \left(\int_{R_1}^{r} \frac{(\psi(r) - \psi(z))^{n-\alpha(z)-1}}{\Gamma(n-\alpha(z))} \psi'(z)\, dz \right) dr \right) \left\| f_\psi^{[n]} \right\|_{\infty,\overline{A}}^2 . \tag{2.56}$$

Proof As in Theorem 2.16, use of (2.14). ■

Theorem 2.23 *All as in Definition 2.4. Then*

$$\int_A \left| \left({}^C D_{R_2-}^{\alpha(\cdot),\psi} f \right)(x) \right| \left| f_\psi^{[n]}(x) \right| dx \leq \frac{2\pi^{\frac{N}{2}} R_2^{N-1}}{\Gamma\left(\frac{N}{2}\right)}$$

$$\left(\int_{R_1}^{R_2} \left(\int_{r}^{R_2} \frac{(\psi(z) - \psi(r))^{n-\alpha(z)-1}}{\Gamma(n-\alpha(z))} \psi'(z)\, dz \right) dr \right) \left\| f_\psi^{[n]} \right\|_{\infty,\overline{A}}^2 . \tag{2.57}$$

Proof As in Theorem 2.17, use of (2.15). ■

We finish with Hardy's type inequalities:

Theorem 2.24 *All as in Definition 2.4. Then*

$$\int_A \left({}^C D_{R_1+}^{\alpha(\cdot),\psi} f(x) \right)^2 dx \leq \left(\frac{R_2}{R_1} \right)^{N-1} \left\| \psi' \right\|_{\infty,[R_1,R_2]}^2$$

$$\left(\int_{R_1}^{R_2} \left(\int_{R_1}^{r} \left(\frac{(\psi(r) - \psi(z))^{n-\alpha(z)-1}}{\Gamma(n-\alpha(z))} \right)^2 dz \right) dr \right) \left(\int_A \left(f_\psi^{[n]}(x) \right)^2 dx \right). \tag{2.58}$$

Proof As in Theorem 2.18, use of (2.14). ■

Theorem 2.25 *All as in Definition 2.4. Then*

$$\int_A \left({}^C D_{R_2-}^{\alpha(\cdot),\psi} f(x) \right)^2 dx \leq \left(\frac{R_2}{R_1} \right)^{N-1} \left\| \psi' \right\|_{\infty,[R_1,R_2]}^2$$

$$\left(\int_{R_1}^{R_2} \left(\int_{r}^{R_2} \left(\frac{(\psi(z) - \psi(r))^{n-\alpha(z)-1}}{\Gamma(n-\alpha(z))} \right)^2 dz \right) dr \right) \left(\int_A \left(f_\psi^{[n]}(x) \right)^2 dx \right). \tag{2.59}$$

Proof As in Theorem 2.19, use of (2.15). ■

References

1. Anastassiou, G.A.: Fractional Differentiation Inequalities. Springer, Heidelberg, New York (2009)
2. Anastassiou, G.A.: Intelligent: Fractional Inequalities and Approximations Expanded. Springer, Heidelberg, New York (2020)
3. Anastassiou, G.A.: Generalized Fractional Calculus: New Advancements and Applications (accepted for publication). Springer, Heidelberg, New York (2020)
4. Anastassiou, G.A.: Abstract Fractional Integral Inequalities of Variable Order (2020)
5. Anastassiou, G.A.: Fractional Integral Inequalities of Variable Order on Spherical Shell (2020)
6. Rudin, W.: Real and Complex Analysis, International Student Edition. Mc Graw Hill, London, New York (1970)
7. Stroock, D.: A Concise Introduction to the Theory of Integration, 3rd edn. Birkäuser, Boston, Basel, Berlin (1999)

Chapter 3
Left Fractional Integral Inequalities of E.R. Love Type

Here first we derive a general reverse Minkowski integral inequality. Then motivated by the work of E.R. Love ([5], 1985) on integral inequalities we produce general reverse and direct integral inequalities. We apply these to ordinary and left fractional integral inequalities. The last involve ordinary derivatives, left Riemann–Liouville fractional integrals, left Caputo fractional derivatives, and left generalized fractional derivatives. These inequalities are of Opial type. It follows [2].

3.1 Introduction

This chapter deals with ordinary and left fractional integral inequalities. We are motivated by the following results:

Theorem 3.1 (Hardy's Inequality, integral version [4, Theorem 327]) *If f is a complex-valued function in $L^r(0,\infty)$, $\|\cdot\|$ is the $L^r(0,\infty)$ norm and $r>1$, then*

$$\left\| \frac{1}{x} \int_0^x f(t)\, dt \right\| \leq \frac{r}{r-1} \|f\| . \tag{3.1}$$

Theorem 3.2 (E.R. Love, [5]) *If $s \geq r \geq 1$, $0 \leq a < b \leq \infty$, γ is real, $\omega(x)$ is decreasing and positive in (a,b), $f(x)$ and $H(x,y)$ are measurable and non-negative on (a,b), $H(x,y)$ is homogeneous of degree -1,*

$$(Hf)(x) = \int_a^x H(x,y)\, f(y)\, dy \tag{3.2}$$

G. A. Anastassiou, *Constructive Fractional Analysis with Applications*,
Studies in Systems, Decision and Control 362,
https://doi.org/10.1007/978-3-030-71481-9_3

and

$$\|f\|_r = \left(\int_a^b f(x)^r x^{\gamma-1} \omega(x) \, dx \right)^{\frac{1}{r}}, \tag{3.3}$$

then

$$\|Hf\|_r \leq C \|f\|_s, \tag{3.4}$$

where

$$C = \int_{\frac{a}{b}}^1 H(1,t) t^{-\frac{\gamma}{r}} \left(\int_a^{bt} x^{\gamma-1} \omega(x) \, dx \right)^{\frac{1}{r}-\frac{1}{s}} dt. \tag{3.5}$$

Here $\frac{a}{b}$ is to mean 0 if $a = 0$ or $b = \infty$ or both; and bt is to mean ∞ if $b = \infty$.

An application of Theorem 3.2 follows:

Theorem 3.3 (E.R. Love, [5]) *If $p > 0$, $q > 0$, $p + q = r \geq 1$, $0 \leq a < b \leq \infty$, $\gamma < r$, $\omega(x)$ is decreasing and positive in (a, b), $f(x)$ is measurable and non-negative on (a, b), I^α is the left Riemann–Liouville operator of fractional integration defined by*

$$(I^\alpha f)(x) = \int_a^x \frac{(x-t)^{\alpha-1}}{\Gamma(\alpha)} f(t) \, dt \quad \text{for } \alpha > 0, \tag{3.6}$$

$I^0 f(x) = f(x)$, where Γ is the gamma function, and $I^\beta f$ is defined similarly for $\beta \geq 0$, then

$$\int_a^b \left[(I^\alpha f)(x) \right]^p \left[\left(I^\beta f \right)(x) \right]^q x^{\gamma-\alpha p-\beta q-1} \omega(x) \, dx \leq C \int_a^b f(x)^r x^{\gamma-1} \omega(x) \, dx, \tag{3.7}$$

where

$$C = \left(\frac{\Gamma\left(1-\frac{\gamma}{r}\right)}{\Gamma\left(\alpha+1-\frac{\gamma}{r}\right)} \right)^p \left(\frac{\Gamma\left(1-\frac{\gamma}{r}\right)}{\Gamma\left(\beta+1-\frac{\gamma}{r}\right)} \right)^q. \tag{3.8}$$

Also Theorem 3.2 implies Theorem 3.1 (by 3.4), just take $a = 0$, $b = \infty$, $\gamma = 1$, $s = r > 1$, $\omega(x) = 1$ and $H(x, y) = \frac{1}{x}$.

3.2 Main Results

We start with a general result, see also [3].

Theorem 3.4 (Reverse Minkowski integral inequality) *Let $(X, \mathcal{A}, \mu)$ and $(Y, \mathcal{B}, \nu)$ be σ-finite measure spaces and let $0 < p < 1$. Here f is a nonnegative function on $X \times Y$ with $f(x, y) > 0$ for almost all $x \in X$, almost all $y \in Y$ and $\int_Y (f(x, y))^p \, d\nu(y) < \infty$ for almost all $x \in X$.*

Then

$$\left(\int_Y\left(\int_X f(x,y)\,d\mu(x)\right)^p d\nu(y)\right)^{\frac{1}{p}} \geq \int_X\left(\int_Y (f(x,y))^p\,d\nu(y)\right)^{\frac{1}{p}} d\mu(x), \tag{3.9}$$

if left-hand side is finite.

Proof Notice that $\int_X f(x,y)\,d\mu(x) > 0$, for almost all $y \in Y$ and

$$\int_Y\left(\int_X f(x,y)\,d\mu(x)\right)^p d\nu(y) > 0.$$

We observe that

$$\int_Y\left(\int_X f(x,y)\,d\mu(x)\right)^p d\nu(y) =$$

$$\int_Y\left[\left(\int_X f(x,y)\,d\mu(x)\right)\left(\int_X f(x',y)\,d\mu(x')\right)^{p-1}\right]d\nu(y) =$$

$$\int_Y\left[\int_X f(x,y)\left(\int_X f(x',y)\,d\mu(x')\right)^{p-1} d\mu(x)\right]d\nu(y) =$$

(by Tonelli's theorem)

$$\int_X\left[\int_Y f(x,y)\left(\int_X f(x',y)\,d\mu(x')\right)^{p-1} d\nu(y)\right]d\mu(x)$$

(by applying the reverse Hölder's inequality in the inside we get)

$$\geq \int_X\left[\left(\int_Y (f(x,y))^p\,d\nu(y)\right)^{\frac{1}{p}}\left(\int_Y\left(\int_X f(x',y)\,d\mu(x')\right)^p d\nu(y)\right)^{\frac{p-1}{p}}\right]d\mu(x) \tag{3.10}$$

$$= \left[\int_X\left(\int_Y (f(x,y))^p\,d\nu(y)\right)^{\frac{1}{p}} d\mu(x)\right]\left[\int_Y\left(\int_X f(x,y)\,d\mu(x)\right)^p d\nu(y)\right]^{\frac{p-1}{p}}.$$

Finally, divide both ends of (3.10) by $\left[\int_Y\left(\int_X f(x,y)\,d\mu(x)\right)^p d\nu(y)\right]^{\frac{p-1}{p}} > 0$ to obtain (3.9). ■

We continue with a reverse analog of Theorem 3.2. The proof involves a special kind of variation of reverse Minkowski integral inequality which we establish completely.

We present

Theorem 3.5 *Let* $0 < r < 1$, $0 < a < b < \infty$, $f(x)$ *and* $H(x, y)$ *are measurable and non-negative on* (a, b), $(a, b)^2$, *respectively,* $H(x, y)$ *is homogeneous of degree* -1,

$$(Hf)(x) = \int_a^x H(x, y) f(y) dy, \tag{3.11}$$

and suppose that

$$\|Hf\|_{r,[a,b]} = \left(\int_a^b (Hf)(x)^r dx \right)^{\frac{1}{r}} < \infty, \tag{3.12}$$

and $\|f\|_{r,[a,b]}$ *is defined similarly.*

We assume that $H(1, t) f(x, t) > 0$, *for almost all* $t \in \left[\frac{a}{x}, 1\right]$, *for almost all* $x \in [a, b]$, *and* $\frac{H(1,t)}{t^{\frac{1}{r}}} \|f\|_{r,[a,bt]} < \infty$, *for almost all* $t \in \left[\frac{a}{b}, 1\right]$.

Then

$$\|Hf\|_{r,[a,b]} \geq \int_{\frac{a}{b}}^1 \frac{H(1, t)}{t^{\frac{1}{r}}} \|f\|_{r,[a,bt]} dt. \tag{3.13}$$

Proof For $a < x < b$ the homogeneity of degree -1 of H gives

$$(Hf)(x) = \int_{\frac{a}{x}}^1 H(x, xt) f(xt) x dt = \int_{\frac{a}{x}}^1 H(1, t) f(xt) dt,$$

where $t = \frac{y}{x}$. As $a \leq y \leq x$, then $0 < t \leq 1$. We will prove first $(0 < r < 1)$

$$\|Hf\|_{r,[a,b]} = \left(\int_a^b \left(\int_{\frac{a}{x}}^1 H(1, t) f(xt) dt \right)^r dx \right)^{\frac{1}{r}} \geq \tag{3.14}$$

$$\int_{\frac{a}{b}}^1 \left(\int_{\frac{a}{t}}^b H(1, t)^r f(xt)^r dx \right)^{\frac{1}{r}} dt = (*).$$

Indeed we observe that

$$\int_a^b \left(\int_{\frac{a}{x}}^1 H(1, t) f(xt) dt \right)^r dx = \int_a^b \left(\int_0^1 \chi_{\left[\frac{a}{x},1\right]}(t) H(1, t) f(xt) dt \right)^r dx =$$

(where χ is the characteristic function)

$$\int_a^b \left\{ \left(\int_0^1 \chi_{\left[\frac{a}{x},1\right]}(t) H(1, t) f(xt) dt \right) \left(\int_0^1 \chi_{\left[\frac{a}{x},1\right]}(t') H(1, t') f(xt') dt' \right)^{r-1} \right\} dx$$

$$= \int_a^b \left\{ \int_0^1 \chi_{\left[\frac{a}{x},1\right]}(t)\, H(1,t)\, f(xt) \left(\int_0^1 \chi_{\left[\frac{a}{x},1\right]}(t')\, H(1,t')\, f(xt')\, dt' \right)^{r-1} dt \right\} dx$$

$$\overset{\text{(by Tonelli's theorem)}}{=}$$

$$\int_0^1 \left\{ \int_{\frac{a}{t}}^b \chi_{\left[\frac{a}{b},1\right]}(t)\, H(1,t)\, f(xt) \left(\int_0^1 \chi_{\left[\frac{a}{x},1\right]}(t')\, H(1,t')\, f(xt')\, dt' \right)^{r-1} dx \right\} dt$$

$$= \int_{\frac{a}{b}}^1 \left[\int_{\frac{a}{t}}^b H(1,t)\, f(xt) \left(\int_{\frac{a}{x}}^1 H(1,t')\, f(xt')\, dt' \right)^{r-1} dx \right] dt$$

(by reverse Hölder's inequality applied inside)

$$\geq \int_{\frac{a}{b}}^1 \left[\left(\int_{\frac{a}{t}}^b H(1,t)\, f(xt)^r\, dx \right)^{\frac{1}{r}} \left(\int_a^b \left(\int_{\frac{a}{x}}^1 H(1,t')\, f(xt')\, dt' \right)^r dx \right)^{\frac{r-1}{r}} \right] dt =$$

$$\left(\int_{\frac{a}{b}}^1 \left(\int_{\frac{a}{t}}^b (H(1,t)\, f(xt))^r\, dx \right)^{\frac{1}{r}} dt \right) \left(\int_a^b \left(\int_{\frac{a}{x}}^1 H(1,t)\, f(xt)\, dt \right)^r dx \right)^{\frac{r-1}{r}}. \tag{3.15}$$

Clearly here, by the assumptions, it holds $\int_a^b \left(\int_{\frac{a}{x}}^1 H(1,t)\, f(xt)\, dt \right)^r dx > 0$ and all we did they make sense.

Finally, we divide both ends of (3.15) by $\left(\int_a^b \left(\int_{\frac{a}{x}}^1 H(1,t)\, f(xt)\, dt \right)^r dx \right)^{\frac{r-1}{r}} > 0$, to validate (3.14), which is a particular case of a reverse Minkowski type integral inequality.

By (3.14) we continue

$$(*) = \int_{\frac{a}{b}}^1 \frac{H(1,t)}{t^{\frac{1}{r}}} \left(\int_{\frac{a}{t}}^b f(xt)^r\, t dx \right)^{\frac{1}{r}} dt = \int_{\frac{a}{b}}^1 \frac{H(1,t)}{t^{\frac{1}{r}}} \left(\int_a^{bt} f(y)^r\, dy \right)^{\frac{1}{r}} dt \tag{3.16}$$

$$= \int_{\frac{a}{b}}^1 \frac{H(1,t)}{t^{\frac{1}{r}}} \|f\|_{r,[a,bt]}\, dt,$$

proving (3.13). ■

We give a reverse left fractional inequality.

Corollary 3.6 (to Theorem 3.5) *Let* $0 < r < p$ *with* $r < 1$, $0 < a < b < \infty$, f *is measurable and non-negative on* (a,b) *such that* $f(x) > 0$ *almost everywhere on* $[a,b]$. *Here* I^{α} *is the left fractional Riemann–Liouville integral operator of order*

$\alpha > 0$, *defined by*

$$\left(I^{\alpha} f\right)(x) = \int_a^x \frac{(x-t)^{\alpha-1}}{\Gamma(\alpha)} f(t)\,dt, \quad I^0 f(x) = f(x), \tag{3.17}$$

$\forall x \in [a, b]$, *and suppose that* $\left\| x^{-\alpha} I^{\alpha} f(x) \right\|_{r,[a,b]} < \infty$. *We assume that* $\| f \|_{r,[a,bt]} < \infty$, *for almost all* $t \in \left[\frac{a}{b}, 1\right]$. *Then*

$$\int_a^b \left(I^{\alpha} f(x)\right)^p (f(x))^{r-p} x^{-\alpha p} dx \geq \frac{1}{\Gamma(\alpha)^p} \left(\int_{\frac{a}{b}}^1 \frac{(1-t)^{\alpha-1}}{t^{\frac{1}{r}}} \| f \|_{r,[a,bt]}\, dt \right)^p \| f \|_r^{r-p}. \tag{3.18}$$

Proof In Theorem 3.5, let $H(x, y) = \frac{(x-y)^{\alpha-1}}{x^{\alpha}\Gamma(\alpha)}$ for $x > y > a$ and $\alpha > 0$, which is homogeneous of degree -1. Then $(Hf)(x) = x^{-\alpha} I^{\alpha} f(x)$, and so by (3.13)

$$\| Hf \|_{r,[a,b]} \geq \int_{\frac{a}{b}}^1 \frac{H(1,t)}{t^{\frac{1}{r}}} \| f \|_{r,[a,bt]}\, dt = \frac{1}{\Gamma(\alpha)} \int_{\frac{a}{b}}^1 \frac{(1-t)^{\alpha-1}}{t^{\frac{1}{r}}} \| f \|_{r,[a,bt]}\, dt. \tag{3.19}$$

Here $0 < r < p$, hence $0 < \frac{r}{p} < 1$, also $\frac{r}{r-p} < 0$, and $f(x) > 0$ almost everywhere in $[a, b]$. Next we apply the reverse Hölder's inequality:

$$\int_a^b \left(x^{-\alpha} \left(I^{\alpha} f\right)(x)\right)^p (f(x))^{r-p} dx \geq$$

$$\left(\int_a^b \left(x^{-\alpha} \left(I^{\alpha} f\right)(x)\right)^r dx \right)^{\frac{p}{r}} \left(\int_a^b (f(x))^r dx \right)^{\frac{r-p}{r}} =$$

$$\| Hf \|_{r,[a,b]}^p \| f \|_r^{r-p} \overset{(3.19)}{\geq} \frac{1}{\Gamma(\alpha)^p} \left(\int_{\frac{a}{b}}^1 \frac{(1-t)^{\alpha-1}}{t^{\frac{1}{r}}} \| f \|_{r,[a,bt]}\, dt \right)^p \| f \|_r^{r-p}, \tag{3.20}$$

proving the claim. ■

Next we present a reverse Opial type [6] inequality.

Corollary 3.7 (to Corollary 3.6) *Let* $0 < r < p$ *with* $r < 1$, $m \in \mathbb{N}$, $0 < a < b < \infty$, $f : [a, b] \to \mathbb{R}$ *such that* $f^{(m-1)}$ *is an absolutely continuous function over* $[a, b]$, *where* $f(a) = f'(a) = \ldots = f^{(m-1)}(a) = 0$, *and* $f^{(m)}$ *is non-negative, with* $f^{(m)}(x) > 0$ *almost everywhere over* $[a, b]$. *We assume that* $\left\| x^{-m} f(x) \right\|_{r,[a,b]} < \infty$, *and* $\left\| f^{(m)} \right\|_{r,[a,bt]} < \infty$, *a.e. for* $t \in \left[\frac{a}{b}, 1\right]$. *Then*

$$\int_a^b (f(x))^p \left(f^{(m)}(x)\right)^{r-p} x^{-mp} dx \geq$$

$$\frac{1}{((m-1)!)^p}\left(\int_{\frac{a}{b}}^{1}\frac{(1-t)^{m-1}}{t^{\frac{1}{r}}}\left\|f^{(m)}\right\|_{r,[a,bt]}dt\right)^p\left\|f^{(m)}\right\|_r^{r-p}. \tag{3.21}$$

Proof By Taylor's formula with integral remainder we have

$$f(x)=\int_a^x\frac{(x-t)^{m-1}}{(m-1)!}f^{(m)}(t)\,dt=\left(I^m f^{(m)}\right)(x),\ \forall\, x\in[a,b] \tag{3.22}$$

then apply Corollary 3.6 for $f^{(m)}$. ■

We need

Definition 3.8 Let $\alpha>0$, $n=\lceil\alpha\rceil$ ($\lceil\cdot\rceil$ is the ceiling), $f\in AC^n([a,b])$ (i.e. $f^{(n-1)}$ is absolutely continuous function). The left Caputo fractional derivative is given by

$$D_{*a}^{\alpha}f(x)=\frac{1}{\Gamma(n-\alpha)}\int_a^x(x-t)^{n-\alpha-1}f^{(n)}(t)\,dt \tag{3.23}$$

and exists almost everywhere for x in $[a,b]$, $D_{*a}^{0}f=f$, see [1, p. 394].

We mention

Corollary 3.9 ([1, p. 395]) *Let* $\alpha>0$, $n=\lceil\alpha\rceil$, $f\in AC^n([a,b])$, *and* $f^{(k)}(a)=0$, $k=0,1,\ldots,n-1$. *Then*

$$f(x)=\frac{1}{\Gamma(\alpha)}\int_a^x(x-t)^{\alpha-1}D_{*a}^{\alpha}f(t)\,dt=I^{\alpha}D_{*a}^{\alpha}f(x),\ \ \forall\, x\in[a,b]. \tag{3.24}$$

We give a reverse left fractional Opial type [1] inequality.

Corollary 3.10 (to Corollary 3.6) *Let* $0<r<p$ *with* $r<1$, $0<a<b<\infty$, $\alpha>0$, $n=\lceil\alpha\rceil$, $f\in AC^n([a,b])$, *and* $f^{(k)}(a)=0$, $k=0,1,\ldots,n-1$. *Assume here that* $D_{*a}^{\alpha}f$ *is non-negative over* (a,b) *such that* $D_{*a}^{\alpha}f>0$ *almost everywhere on* $[a,b]$. *Suppose that* $\left\|x^{-\alpha}f(x)\right\|_{r,[a,b]}<\infty$ *and* $\left\|D_{*a}^{\alpha}f\right\|_{r,[a,bt]}<\infty$, *for almost all* $t\in\left[\frac{a}{b},1\right]$. *Then*

$$\int_a^b(f(x))^p\left(D_{*a}^{\alpha}f(x)\right)^{r-p}x^{-\alpha p}dx\geq$$

$$\frac{1}{\Gamma(\alpha)^p}\left(\int_{\frac{a}{b}}^{1}\frac{(1-t)^{\alpha-1}}{t^{\frac{1}{r}}}\left\|D_{*a}^{\alpha}f\right\|_{r,[a,bt]}dt\right)^p\left\|D_{*a}^{\alpha}f\right\|_r^{r-p}. \tag{3.25}$$

Proof By Corollaries 3.6, 3.9, see also (3.23). ■

We need

Definition 3.11 ([1, pp. 7–8]) Let $\nu > 0$, $n := [\nu]$ $[\cdot]$ the integral part, and $\alpha := \nu - n$ $(0 < \alpha < 1)$; $x, x_0 \in [a, b] \subset \mathbb{R}$ such that $x \geq x_0$, x_0 is fixed. Let $f \in C([a, b])$ and define

$$\left(J_{\nu}^{x_0} f\right)(x) := \frac{1}{\Gamma(\nu)} \int_{x_0}^{x} (x - t)^{\nu-1} f(t)\, dt, \quad x_0 \leq x \leq b, \tag{3.26}$$

the left Riemann–Liouville integral. We define the subspace $C_{x_0}^{\nu}([a, b])$ of $C^n([a, b])$:

$$C_{x_0}^{\nu}([a, b]) := \left\{ f \in C^n([a, b]) : J_{1-\alpha}^{x_0} D^n f \in C^1([x_0, b]) \right\}. \tag{3.27}$$

For $f \in C_{x_0}^{\nu}([a, b])$ we define the left generalized ν-fractional derivative of f over $[x_0, b]$ as

$$D_{x_0+}^{\nu} f := D J_{1-\alpha}^{x_0} f^{(n)} \quad \left(f^{(n)} := D^n f\right). \tag{3.28}$$

Notice $D_{x_0+}^{\nu} f \in C([x_0, b])$.

We also need

Theorem 3.12 (from Theorem 2.1, p. 8 of [1]) *Let $f \in C_{x_0}^{\nu}([a, b])$, $x_0 \in [a, b]$ fixed.*

(i) If $\nu \geq 1$, and $f^{(i)}(x_0) = 0$, $i = 0, 1, \dots, n-1$, then $f(x) = \left(J_{\nu}^{x_0} D_{x_0+}^{\nu} f\right)(x)$, all $x \in [a, b] : x \geq x_0$.

(ii) If $0 < \nu < 1$, then $f(x) = \left(J_{\nu}^{x_0} D_{x_0+}^{\nu} f\right)(x)$, all $x \in [a, b] : x \geq x_0$.

That is, in both cases we have

$$f(x) = \frac{1}{\Gamma(\nu)} \int_{x_0}^{x} (x - t)^{\nu-1} \left(D_{x_0+}^{\nu} f\right)(t)\, dt, \quad x_0 \leq x \leq b. \tag{3.29}$$

If $x_0 = a$, we get

$$f(x) = \frac{1}{\Gamma(\nu)} \int_{a}^{x} (x - t)^{\nu-1} \left(D_{a+}^{\nu} f\right)(t)\, dt = \left(J_{\nu}^{a} D_{a+}^{\nu} f\right)(x), \quad \text{all } a \leq x \leq b. \tag{3.30}$$

We give another reverse left fractional Opial type inequality.

Corollary 3.13 (to Corollary 3.6) *Let $0 < r < p$ with $r < 1$, $0 < a < b < \infty$, $\nu > 0$, $n = [\nu]$; $f \in C_a^{\nu}([a, b])$, such that $f^{(i)}(a) = 0$, $i = 0, 1, \dots, n-1$ for only the case of $\nu \geq 1$. Assume here that $D_{a+}^{\nu} f$ is non-negative over (a, b) such that $D_{a+}^{\nu} f > 0$ almost everywhere on $[a, b]$. Suppose that $\left\| x^{-\nu} f(x) \right\|_{r,[a,b]} < \infty$ and $\left\| D_{a+}^{\nu} f \right\|_{r,[a,bt]} < \infty$, for almost all $t \in \left[\frac{a}{b}, 1\right]$. Then*

$$\int_{a}^{b} (f(x))^{p} \left(D_{a+}^{\nu} f(x)\right)^{r-p} x^{-\nu p} dx \geq$$

$$\frac{1}{\Gamma(\nu)^p}\left(\int_{\frac{a}{b}}^{1}\frac{(1-t)^{\nu-1}}{t^{\frac{1}{r}}}\left\|D_{a+}^{\nu}f\right\|_{r,[a,bt]}dt\right)^p\left\|D_{a+}^{\nu}f\right\|_r^{r-p}. \tag{3.31}$$

Proof By Corollary 3.6, Theorem 3.12, see also (3.28). ■

We need the following representation result.

Theorem 3.14 ([1, p. 395]) *Let $\nu \geq \overline{\gamma}+1$, $\overline{\gamma} \geq 0$. Call $n = \lceil \nu \rceil$, $m := \lceil \overline{\gamma} \rceil$. Assume $f \in AC^n([a,b])$, such that $f^{(k)}(a) = 0$, $k = 0, 1, \ldots, n-1$, and $D_{*a}^{\nu}f \in L_{\infty}(a,b)$. Then $D_{*a}^{\overline{\gamma}}f \in C([a,b])$, $D_{*a}^{\overline{\gamma}}f(x) = I^{m-\overline{\gamma}}f^{(m)}(x)$, and*

$$D_{*a}^{\overline{\gamma}}f(x) = \frac{1}{\Gamma(\nu-\overline{\gamma})}\int_a^x (x-t)^{\nu-\overline{\gamma}-1}D_{*a}^{\nu}f(t)\,dt = \left(I^{\nu-\overline{\gamma}}D_{*a}^{\nu}f\right)(x), \tag{3.32}$$

$\forall\, x \in [a,b]$.

Remark 3.15 (*to Theorem* 3.14) By Corollary 3.9 we also have

$$f(x) = \frac{1}{\Gamma(\nu)}\int_a^x (x-t)^{\nu-1}D_{*a}^{\nu}f(t)\,dt = \left(I^{\nu}D_{*a}^{\nu}f\right)(x), \tag{3.33}$$

$\forall\, x \in [a,b]$.

It follows left fractional direct Opial type integral inequalities.

Theorem 3.16 *If $p > 0$, $q > 0$, $p+q = r \geq 1$, $0 \leq a < b < \infty$, $\gamma < r$, $\omega(x)$ is decreasing and positive in (a,b). Let $\nu \geq \overline{\gamma}+1$, $\overline{\gamma} \geq 0$, call $n = \lceil \nu \rceil$, $f \in AC^n([a,b]) : f^{(k)}(a) = 0$, $k = 0, 1, \ldots, n-1$; $D_{*a}^{\nu}f \in L_{\infty}(a,b)$, with $D_{*a}^{\nu}f \geq 0$ over (a,b). Then*

$$\int_a^b (f(x))^p\left(D_{*a}^{\overline{\gamma}}f(x)\right)^q x^{\gamma-\nu p-(\nu-\overline{\gamma})q-1}\omega(x)\,dx \leq$$

$$C\int_a^b \left(\left(D_{*a}^{\nu}f\right)(x)\right)^r x^{\gamma-1}\omega(x)\,dx, \tag{3.34}$$

where

$$C = \left(\frac{\Gamma\left(1-\frac{\gamma}{r}\right)}{\Gamma\left(\nu+1-\frac{\gamma}{r}\right)}\right)^p\left(\frac{\Gamma\left(1-\frac{\gamma}{r}\right)}{\Gamma\left(\nu-\overline{\gamma}+1-\frac{\gamma}{r}\right)}\right)^q. \tag{3.35}$$

Proof Directly from Theorem 3.3. Notice that

$$\int_a^b \left[\left(I^{\nu}D_{*a}^{\nu}f\right)(x)\right]^p\left[\left(I^{\nu-\overline{\gamma}}D_{*a}^{\nu}f\right)(x)\right]^q x^{\gamma-\nu p-(\nu-\overline{\gamma})q-1}\omega(x)\,dx \overset{\text{(by (3.32), (3.33))}}{=}$$

$$\int_a^b (f(x))^p\left(D_{*a}^{\overline{\gamma}}f(x)\right)^q x^{\gamma-\nu p-(\nu-\overline{\gamma})q-1}\omega(x)\,dx. \tag{3.36}$$

So, in applying (3.7), now instead of f we take $D_{*a}^{\nu} f$. ∎

Theorem 3.17 *If $p > 0$, $q > 0$, $p + q = r \geq 1$, $0 \leq a < b < \infty$, $\gamma < r$, $\omega(x)$ is decreasing and positive in (a, b). Let $\nu \geq \overline{\gamma_i} + 1$, $\overline{\gamma_i} \geq 0$, $i = 1, 2$, call $n = \lceil \nu \rceil$, $f \in AC^n([a, b]) : f^{(k)}(a) = 0$, $k = 0, 1, \ldots, n-1$; $D_{*a}^{\nu} f \in L_\infty(a, b)$, with $D_{*a}^{\nu} f \geq 0$ over (a, b). Then*

$$\int_a^b \left(D_{*a}^{\overline{\gamma_1}} f(x)\right)^p \left(D_{*a}^{\overline{\gamma_2}} f(x)\right)^q x^{\gamma-(\nu-\overline{\gamma_1})p-(\nu-\overline{\gamma_2})q-1} \omega(x)\, dx \leq$$

$$C^* \int_a^b \left(\left(D_{*a}^{\nu} f\right)(x)\right)^r x^{\gamma-1} \omega(x)\, dx, \tag{3.37}$$

where

$$C^* = \left(\frac{\Gamma\left(1-\frac{\gamma}{r}\right)}{\Gamma\left(\nu-\overline{\gamma_1}+1-\frac{\gamma}{r}\right)}\right)^p \left(\frac{\Gamma\left(1-\frac{\gamma}{r}\right)}{\Gamma\left(\nu-\overline{\gamma_2}+1-\frac{\gamma}{r}\right)}\right)^q. \tag{3.38}$$

Proof Use of Theorem 3.14 and similar to Theorem 3.16. ∎

Corollary 3.18 *All as in Theorem 3.16. Then*

$$\int_a^b \left(D_{*a}^{\overline{\gamma}} f(x)\right)^p \left(\left(D_{*a}^{\nu} f\right)(x)\right)^q x^{\gamma-(\nu-\overline{\gamma})p-1} \omega(x)\, dx \leq$$

$$\overline{C} \int_a^b \left(\left(D_{*a}^{\nu} f\right)(x)\right)^r x^{\gamma-1} \omega(x)\, dx, \tag{3.39}$$

where

$$\overline{C} = \left(\frac{\Gamma\left(1-\frac{\gamma}{r}\right)}{\Gamma\left(\nu-\overline{\gamma}+1-\frac{\gamma}{r}\right)}\right)^p.$$

Proof By Theorems 3.3, 3.14. ∎

We need

Remark 3.19 (*see* [1, p. 26]) Let $\nu \geq \overline{\gamma} + 1$, $\overline{\gamma} \geq 0$, $n = [\nu]$, $x_0 \in [a, b]$ fixed, $f \in C_{x_0}^{\nu}([a, b]) : f^{(i)}(x_0) = 0$, $i = 0, 1, \ldots, n-1$. Then

$$\left(D_{x_0+}^{\overline{\gamma}} f\right)(x) = \frac{1}{\Gamma(\nu-\overline{\gamma})} \int_{x_0}^x (x-t)^{(\nu-\overline{\gamma})-1} \left(D_{x_0+}^{\nu} f\right)(t)\, dt, \tag{3.40}$$

which is continuous in x on $[x_0, b]$.

We continue with

Theorem 3.20 *If $p > 0$, $q > 0$, $p + q = r \geq 1$, $0 \leq a < b < \infty$, $\gamma < r$, $\omega(x)$ is decreasing and positive in (a, b). Let $\nu \geq \overline{\gamma} + 1$, $\overline{\gamma} \geq 0$, $n = [\nu]$, $f \in C_a^\nu([a, b])$: $f^{(i)}(a) = 0$, $i = 0, 1, \ldots, n-1$. Assume that $D_{a+}^\nu f \geq 0$ on (a, b). Then*

$$\int_a^b \left(D_{a+}^{\overline{\gamma}} f(x)\right)^p \left(D_{a+}^\nu f(x)\right)^q x^{\gamma-(\nu-\overline{\gamma})p-1} \omega(x)\, dx \leq$$

$$C_1 \int_a^b \left(D_{a+}^\nu f(x)\right)^r x^{\gamma-1} \omega(x)\, dx, \tag{3.41}$$

where

$$C_1 = \left(\frac{\Gamma\left(1 - \frac{\gamma}{r}\right)}{\Gamma\left(\nu - \overline{\gamma} + 1 - \frac{\gamma}{r}\right)}\right)^p.$$

Proof By Theorem 3.3 and see Remark 3.19. ■

We make

Remark 3.21 By [5], see Theorem 3.2, let $s = r \geq 1$, $0 \leq a < b \leq \infty$, γ is real, $\omega(x)$ is decreasing and positive in (a, b), $f(x)$ and $H_k(x, y)$ $(k = 1, \ldots, n)$ are measurable and non-negative on (a, b), $H_k(x, y)$ is homogeneous of degree -1,

$$(H_k f)(x) = \int_a^b H_k(x, y) f(y)\, dy, \; k = 1, \ldots, n, \tag{3.42}$$

and

$$\|f\|_r = \left(\int_a^b f(x)^r x^{\gamma-1} \omega(x)\, dx\right)^{\frac{1}{r}}, \tag{3.43}$$

then

$$\|H_k f\|_r \leq C_k \|f\|_r, \tag{3.44}$$

where

$$C_k = \int_{\frac{a}{b}}^1 H_k(1, t)\, t^{-\frac{\gamma}{r}} dt, \quad k = 1, \ldots, n. \tag{3.45}$$

Here $\frac{a}{b}$ means 0 if $a = 0$ or $b = \infty$ or both; and bt means ∞ if $b = \infty$.

Let now $p_k > 0$ such that $\sum_{k=1}^n p_k = r$.

We notice the following (apply generalized Hölder's inequality)

$$\int_a^b \prod_{k=1}^n (H_k f(x))^{p_k} x^{\gamma-1} \omega(x)\, dx \leq \prod_{k=1}^n \left(\int_a^b (H_k f(x))^r x^{\gamma-1} \omega(x)\, dx\right)^{\frac{p_k}{r}} =$$

$$\prod_{k=1}^{n} \|H_k f\|_r^{p_k} \overset{(3.44)}{\leq} \prod_{k=1}^{n} C_k^{p_k} \|f\|_r^{p_k} = \left(\prod_{k=1}^{n} C_k^{p_k}\right) \|f\|_r^{\sum_{k=1}^{n} p_k} = \tag{3.46}$$

$$\left(\prod_{k=1}^{n} C_k^{p_k}\right) \|f\|_r^r = \widehat{C} \left(\int_a^b f(x)^r x^{\gamma-1} \omega(x)\, dx\right),$$

where

$$\widehat{C} := \prod_{k=1}^{n} C_k^{p_k}. \tag{3.47}$$

We have proved

Theorem 3.22 *Let* $0 \leq a < b \leq \infty$, γ *is real,* $\omega(x)$ *is decreasing and positive in* (a, b), $f(x)$ *and* $H_k(x, y)$ $(k = 1, \ldots, n)$ *are measurable and non-negative on* (a, b), $H_k(x, y)$ *is homogeneous of degree* -1,

$$(H_k f)(x) = \int_a^x H_k(x, y) f(y)\, dy, \; k = 1, \ldots, n. \tag{3.48}$$

Let $p_k > 0 : \sum_{k=1}^{n} p_k = r \geq 1$. *Then*

$$\int_a^b \prod_{k=1}^{n} (H_k f(x))^{p_k} x^{\gamma-1} \omega(x)\, dx \leq \widehat{C} \left(\int_a^b f(x)^r x^{\gamma-1} \omega(x)\, dx\right), \tag{3.49}$$

where

$$\widehat{C} = \prod_{k=1}^{n} \left(\int_{\frac{a}{b}}^{1} H_k(1, t)\, t^{-\frac{\gamma}{r}} dt\right)^{p_k}. \tag{3.50}$$

Next we give an application.

Theorem 3.23 *Here* $p_k > 0 : \sum_{k=1}^{n} p_k = r \geq 1$. *Let* $0 \leq a < b \leq \infty$, $\gamma < r$, $\omega(x)$ *is decreasing and positive in* (a, b), $f(x)$ *is measurable and non-negative on* (a, b), I^{α_k} *is the left Riemann–Liouville operator of fractional integration defined by*

$$(I^{\alpha_k} f)(x) = \int_a^x \frac{(x-t)^{\alpha_k - 1}}{\Gamma(\alpha_k)} f(t)\, dt, \;\; \textit{for}\; \alpha_k > 0, \tag{3.51}$$

and

$$I^{\alpha_k} f(x) = f(x), \; for \alpha_k = 0; k = 1, \ldots, n.$$

Then

$$\int_a^b \prod_{k=1}^n \left(\left(I^{\alpha_k} f\right)(x)\right)^{p_k} x^{\gamma - \sum_{k=1}^n \alpha_k p_k - 1} \omega(x)\, dx \leq \widetilde{C} \left(\int_a^b f(x)^r x^{\gamma-1} \omega(x)\, dx \right), \tag{3.52}$$

where

$$\widetilde{C} = \prod_{k=1}^n \left(\frac{\Gamma\left(1 - \frac{\gamma}{r}\right)}{\Gamma\left(\alpha_k + 1 - \frac{\gamma}{r}\right)} \right)^{p_k}. \tag{3.53}$$

Proof Here we apply Theorem 3.22. Inequality (3.52) derives from (3.49) directly. We let

$$H_k(x, y) = \frac{(x-y)^{\alpha_k - 1}}{x^{\alpha_k} \Gamma(\alpha_k)} \text{ for } x > y > a \text{ and } \alpha_k > 0.$$

Then $H_k f(x) = x^{-\alpha_k} I^{\alpha_k} f(x)$, $k \in \{1, \ldots, n\}$. Notice that

$$\int_{\frac{a}{b}}^1 H_k(1, t)\, t^{-\frac{\gamma}{r}} dt = \int_{\frac{a}{b}}^1 \frac{(1-t)^{\alpha_k - 1}}{\Gamma(\alpha_k)} t^{-\frac{\gamma}{r}} dt \leq$$

$$\int_0^1 \frac{(1-t)^{\alpha_k - 1}}{\Gamma(\alpha_k)} t^{-\frac{\gamma}{r}} dt = \frac{\Gamma\left(1 - \frac{\gamma}{r}\right)}{\Gamma\left(\alpha_k + 1 - \frac{\gamma}{r}\right)}, \tag{3.54}$$

for $k \in \{1, \ldots, n\}$.

Therefore

$$\widetilde{C} = \prod_{k=1}^n \left(\frac{\Gamma\left(1 - \frac{\gamma}{r}\right)}{\Gamma\left(\alpha_k + 1 - \frac{\gamma}{r}\right)} \right)^{p_k}.$$

■

Next we give general left fractional direct Opial type integral inequalities.

Theorem 3.24 *Here* $p_j > 0 : \sum_{j=1}^N p_j = r \geq 1$. *Let* $0 \leq a < b < \infty$, $\gamma < r$, $\omega(x)$ *is decreasing and positive in* (a, b). *Let* $\nu \geq \overline{\gamma_j} + 1$, $\overline{\gamma_j} \geq 0$, $j = 2, \ldots, N$, $n = \lceil \nu \rceil$, $f \in AC^n([a, b]) : f^{(k)}(a) = 0$, $k = 0, 1, \ldots, n-1$, *and* $D_{*a}^{\nu} f \in L_\infty(a, b)$, *with* $D_{*a}^{\nu} f \geq 0$ *over* (a, b). *Then*

$$\int_a^b (f(x))^{p_1} \prod_{j=2}^N \left(D_{*a}^{\overline{\gamma_j}} f(x) \right)^{p_j} x^{\gamma - \nu p_1 - \sum_{j=2}^N (\nu - \overline{\gamma_j}) p_j - 1} \omega(x)\, dx \leq$$

$$\left[\left(\frac{\Gamma\left(1 - \frac{\gamma}{r}\right)}{\Gamma\left(\nu + 1 - \frac{\gamma}{r}\right)} \right)^{p_1} \prod_{j=2}^N \left(\frac{\Gamma\left(1 - \frac{\gamma}{r}\right)}{\Gamma\left(\nu - \overline{\gamma_j} + 1 - \frac{\gamma}{r}\right)} \right)^{p_j} \right]$$

$$\left(\int_a^b \left(D_{*a}^{\nu} f(x) \right)^r x^{\gamma-1} \omega(x) \, dx \right). \tag{3.55}$$

Proof By Theorem 3.23, use of Theorem 3.14 and (3.33). ■

Theorem 3.25 *All as in Theorem 3.24. Then*

$$\int_a^b \left(D_{*a}^{\nu} f(x) \right)^{p_1} \prod_{j=2}^{N} \left(D_{*a}^{\overline{\gamma_j}} f(x) \right)^{p_j} x^{\gamma - \sum_{j=2}^{N} (\nu - \overline{\gamma_j}) p_j - 1} \omega(x) \, dx \leq$$

$$\left(\prod_{j=2}^{N} \left(\frac{\Gamma\left(1 - \frac{\gamma}{r}\right)}{\Gamma\left(\nu - \overline{\gamma_j} + 1 - \frac{\gamma}{r}\right)} \right)^{p_j} \right) \left(\int_a^b \left(D_{*a}^{\nu} f(x) \right)^r x^{\gamma-1} \omega(x) \, dx \right). \tag{3.56}$$

Proof By Theorem 3.23, use of Theorem 3.14. ■

Corollary 3.26 *All as in Theorem 3.24, and* $\overline{\gamma_2} = \overline{\gamma_3} = \ldots = \overline{\gamma_\lambda}$, $\overline{\gamma_{\lambda+1}} = \overline{\gamma_{\lambda+2}} = \ldots = \overline{\gamma_\mu}$, *and* $\overline{\gamma_{\mu+1}} = \overline{\gamma_{\mu+2}} = \ldots = \overline{\gamma_N}$. *Then*

$$\int_a^b \left(D_{*a}^{\nu} f(x) \right)^{p_1} \left(D_{*a}^{\overline{\gamma_\lambda}} f(x) \right)^{\sum_{j=2}^{\lambda} p_j} \left(D_{*a}^{\overline{\gamma_\mu}} f(x) \right)^{\sum_{j=\lambda+1}^{\mu} p_j} \left(D_{*a}^{\overline{\gamma_N}} f(x) \right)^{\sum_{j=\mu+1}^{N} p_j}$$

$$x^{\gamma - (\nu - \overline{\gamma_\lambda}) \left(\sum_{j=2}^{\lambda} p_j \right) - (\nu - \overline{\gamma_\mu}) \left(\sum_{j=\lambda+1}^{\mu} p_j \right) - (\nu - \overline{\gamma_N}) \left(\sum_{j=\mu+1}^{N} p_j \right) - 1} \omega(x) \, dx \leq$$

$$\left(\frac{\Gamma\left(1 - \frac{\gamma}{r}\right)}{\Gamma\left(\nu - \overline{\gamma_\lambda} + 1 - \frac{\gamma}{r}\right)} \right)^{\sum_{j=2}^{\lambda} p_j} \left(\frac{\Gamma\left(1 - \frac{\gamma}{r}\right)}{\Gamma\left(\nu - \overline{\gamma_\mu} + 1 - \frac{\gamma}{r}\right)} \right)^{\sum_{j=\lambda+1}^{\mu} p_j}$$

$$\left(\frac{\Gamma\left(1 - \frac{\gamma}{r}\right)}{\Gamma\left(\nu - \overline{\gamma_N} + 1 - \frac{\gamma}{r}\right)} \right)^{\sum_{j=\mu+1}^{N} p_j} \left(\int_a^b \left(D_{*a}^{\nu} f(x) \right)^r x^{\gamma-1} \omega(x) \, dx \right). \tag{3.57}$$

Proof By Theorem 3.25. ■

We finish with

Theorem 3.27 *Here* $p_j > 0 : \sum_{j=1}^{N} p_j = r \geq 1$. *Let* $0 \leq a < b < \infty$, $\gamma < r$, $\omega(x)$ *is decreasing and positive in* (a, b). *Let* $\nu \geq \overline{\gamma_j} + 1$, $\overline{\gamma_j} \geq 0$, $j = 2, \ldots, N$, $n = [\nu]$, $f \in C_a^{\nu}([a, b]) : f^{(k)}(a) = 0$, $k = 0, 1, \ldots, n-1$. *Assume that* $D_{a+}^{\nu} f \geq 0$ *on* (a, b). *Then*

$$\int_a^b \left(D_{a+}^{\nu} f(x) \right)^{p_1} \prod_{j=2}^{N} \left(D_{a+}^{\overline{\gamma_j}} f(x) \right)^{p_j} x^{\gamma - \sum_{j=2}^{N} (\nu - \overline{\gamma_j}) p_j - 1} \omega(x) \, dx \leq$$

$$\left(\prod_{j=2}^{N}\left(\frac{\Gamma\left(1-\frac{\gamma}{r}\right)}{\Gamma\left(\nu-\overline{\gamma_j}+1-\frac{\gamma}{r}\right)}\right)^{p_j}\right)\left(\int_a^b \left(D_{a+}^{\nu} f(x)\right)^r x^{\gamma-1}\omega(x)\,dx\right). \quad (3.58)$$

Proof By Theorem 3.23, use of (3.40). ∎

Comment 3.28 *With the exhibited methods above one can derive all kinds of variation of left fractional Opial type integral inequalities, as well as of ordinary differentiation ones, due to lack of space we omit this task.*

References

1. Anastassiou, G.A.: Fractional Differentiation Inequalities. Springer, Heidelberg, New York (2009)
2. Anastassiou, G.A.: E.R. Love type left fractional integral inequalities. Issues Anal. **9**(27) No. 3, 14–30 (2020)
3. Benaissa, B.: On the reverse Minkowski's integral inequality. Kragujev. J. Math. **46**(3), 407–416 (2022)
4. Hardy, G.H., Littlewood, J.E., Pólya, G.: Inequalities. Cambridge University Press, Cambridge, UK (1934)
5. Love, E.R.: Inequalities like Opial's inequality, Rocznik naukowo dydaktyczny WSP w Krakowie. Pr. Mat. **97**, 109–118 (1985)
6. Opial, Z.: Sur une inégalité. Ann. Polon. Math. **8**, 29–32 (1960)

Chapter 4
Right Side Fractional Integral Inequalities of E.R. Love Type

Motivated for the work of E.R. Love ([5], 1985) on integral inequalities we produce general right side direct and reverse integral inequalities. We apply these to ordinary and right side fractional integral inequalities. The last involves ordinary derivatives, right side Riemann–Liouville fractional integrals, right side Caputo fractional derivatives, and right side generalized fractional derivatives. These inequalities are of Opial type ([6]). It follows [3].

4.1 Introduction

This chapter deals with ordinary and right side fractional integral inequalities. We are motivated by the following left side results:

Theorem 4.1 (Hardy's Inequality, integral version [4, Theorem 327]) *If f is a complex-valued function in $L^r(0,\infty)$, $\|\cdot\|$ is the $L^r(0,\infty)$ norm and $r > 1$, then*

$$\left\| \frac{1}{x} \int_0^x f(t)\,dt \right\| \leq \frac{r}{r-1} \|f\|. \tag{4.1}$$

Theorem 4.2 (E.R. Love, [5]) *If $s \geq r \geq 1$, $0 \leq a < b \leq \infty$, γ is real, $\omega(x)$ is decreasing and positive in (a,b), $f(x)$ and $H(x,y)$ are measurable and non-negative on (a,b), $H(x,y)$ is homogeneous of degree -1,*

$$(Hf)(x) = \int_a^x H(x,y) f(y)\,dy \tag{4.2}$$

G. A. Anastassiou, *Constructive Fractional Analysis with Applications*,
Studies in Systems, Decision and Control 362,
https://doi.org/10.1007/978-3-030-71481-9_4

and

$$\|f\|_r = \left(\int_a^b f(x)^r x^{\gamma-1} \omega(x) \, dx \right)^{\frac{1}{r}}, \tag{4.3}$$

then

$$\|Hf\|_r \leq C \|f\|_s, \tag{4.4}$$

where

$$C = \int_{\frac{a}{b}}^{1} H(1,t) \, t^{-\frac{\gamma}{r}} \left(\int_a^{bt} x^{\gamma-1} \omega(x) \, dx \right)^{\frac{1}{r}-\frac{1}{s}} dt. \tag{4.5}$$

Here $\frac{a}{b}$ *is to mean* 0 *if* $a = 0$ *or* $b = \infty$ *or both; and* bt *is to mean* ∞ *if* $b = \infty$.

We present direct and reverse integral inequalities.

4.2 Main Results

4.2.1 *Part I*

First we give direct right side results. We present the following direct general right side result:

Theorem 4.3 *If* $s \geq r \geq 1,$, $0 \leq a < b \leq \infty$, $\gamma \in \mathbb{R}$, $\omega(x)$ *is increasing and positive in* (a, b), $f(x)$ *and* $H(z, x)$ *are measurable and non-negative on* (a, b), *and* $(a, b)^2$, *respectively,* $H(z, x)$ *is homogeneous of degree* -1,

$$(Hf)(x) = \int_x^b H(z,x) f(z) \, dz, \tag{4.6}$$

and

$$\|f\|_r = \left(\int_a^b f^r(x) \, x^{\gamma-1} \omega(x) \, dx \right)^{\frac{1}{r}}, \tag{4.7}$$

then

$$\|Hf\|_r \leq C \|f\|_s, \tag{4.8}$$

where

$$C := \int_1^{\frac{b}{a}} H(t,1) \, t^{-\frac{\gamma}{r}} \left(\int_{at}^{b} x^{\gamma-1} \omega(x) \, dx \right)^{\frac{1}{r}-\frac{1}{s}} dt. \tag{4.9}$$

Here $\frac{b}{a}$ *is to mean* ∞, *if* $a = 0$ *or* $b = \infty$ *or both; and* at *is to mean* 0 *if* $a = 0$.

Proof (i) Let $x \in (a, b)$.

Here $x \leq z \leq b$ and $1 \leq \frac{z}{x} \leq \frac{b}{x}$. set $t = \frac{z}{x}$, that is $z = tx$, and $1 \leq t \leq \frac{b}{x}$. Infact $1 \leq t < \frac{b}{a}$. By degree -1 homogenity of H we have

$$(Hf)(x) = \int_1^{\frac{b}{x}} H(tx, x) f(tx) x dt =$$

$$\int_1^{\frac{b}{x}} x^{-1} H(t, 1) f(tx) x dt = \int_1^{\frac{b}{x}} H(t, 1) f(tx) dt.$$

That is

$$(Hf)(x) = \int_1^{\frac{b}{x}} H(t, 1) f(tx) dt. \tag{4.10}$$

Using Minkowski's inequality at (4.11), and the increasing property of ω at (4.12), we get

$$\|Hf\|_r = \left(\int_a^b \left(\int_1^{\frac{b}{x}} H(t, 1) f(tx) dt \right)^r x^{\gamma-1} \omega(x) dx \right)^{\frac{1}{r}} \leq$$

$$\int_1^{\frac{b}{a}} \left(\int_a^{\frac{b}{t}} H(t, 1)^r f(tx)^r x^{\gamma-1} \omega(x) dx \right)^{\frac{1}{r}} dt \leq \tag{4.11}$$

$$\int_1^{\frac{b}{a}} H(t, 1) t^{-\frac{\gamma}{r}} \left(\int_a^{\frac{b}{t}} f(tx)^r (tx)^{\gamma-1} \omega(xt) t dx \right)^{\frac{1}{r}} dt = \tag{4.12}$$

$$\int_1^{\frac{b}{a}} H(t, 1) t^{-\frac{\gamma}{r}} \left(\int_{at}^b f(z)^r z^{\gamma-1} \omega(z) dz \right)^{\frac{1}{r}} dt \leq \tag{4.13}$$

$$\left(\int_1^{\frac{b}{a}} H(t, 1) t^{-\frac{\gamma}{r}} dt \right) \left(\int_{at}^b f(z)^r z^{\gamma-1} \omega(z) dz \right)^{\frac{1}{r}}. \tag{4.14}$$

If $s = r$, this is the required inequality.

(ii) Assume that $s > r$. Applying Hölder's inequality with indices $\frac{s}{r}$ and $\left(\frac{s}{s-r}\right)$ in the inside integral of (4.13) we get

$$\int_{at}^b f(z)^r z^{\gamma-1} \omega(z) dz = \int_{at}^b f(z)^r \left(z^{\gamma-1} \omega(z) \right)^{\frac{r}{s}} \left(z^{\gamma-1} \omega(z) \right)^{1-\frac{r}{s}} dz \leq$$

$$\left(\int_{at}^b f(z)^s z^{\gamma-1} \omega(z) dz \right)^{\frac{r}{s}} \left(\int_{at}^b z^{\gamma-1} \omega(z) dz \right)^{1-\frac{r}{s}} \leq$$

$$\left(\int_a^b f(z)^s z^{\gamma-1} \omega(z) dz \right)^{\frac{r}{s}} \left(\int_{at}^b z^{\gamma-1} \omega(z) dz \right)^{1-\frac{r}{s}} = \|f\|_s^r \left(\int_{at}^b z^{\gamma-1} \omega(z) dz \right)^{1-\frac{r}{s}}. \tag{4.15}$$

So we have (by use of (4.14) and (4.15))

$$\|Hf\|_r \leq \int_1^{\frac{b}{a}} H(t,1)\, t^{-\frac{\gamma}{r}} \left(\int_{at}^{b} f(z)^r z^{\gamma-1} \omega(z)\, dz \right)^{\frac{1}{r}} dt \leq$$

$$\|f\|_s \int_1^{\frac{b}{a}} H(t,1)\, t^{-\frac{\gamma}{r}} \left(\int_{at}^{b} z^{\gamma-1} \omega(z)\, dz \right)^{\frac{1}{r}-\frac{1}{s}} dt = C\, \|f\|_s , \tag{4.16}$$

proving the claim. ■

Remark 4.4 If $r = s$, then (see (4.9))

$$C := \int_1^{\frac{b}{a}} H(t,1)\, t^{-\frac{\gamma}{r}} dt, \tag{4.17}$$

independent of ω.

We give

Corollary 4.5 *Let* $r > 1$ *and* $f \in L^r(0,\infty)$. *Then*

$$\left\| \int_x^{\infty} \frac{f(t)}{t} dt \right\|_r \leq r\, \|f\|_r . \tag{4.18}$$

Proof Apply (4.8), with $a = 0$, $b = \infty$, $x \in \mathbb{R}_+$, $\gamma = 1$, $s = r > 1$, $\omega(x) = 1$, and $H(z,x) = \frac{1}{z}$. ■

We continue with a right side fractional result.

Theorem 4.6 *If* $p > 0$, $q > 0$, $p + q = r \geq 1$, $0 < a < b < \infty$, $\gamma > 0$, $\omega(x)$ *is increasing and positive in* (a,b), $f(x)$ *is measurable and non-negative on* (a,b), I_{b-}^{α} *is the right Riemann–Liouville operator of fractional integration defined by*

$$I_{b-}^{\alpha} f(x) = \int_x^b \frac{(z-x)^{\alpha-1}}{\Gamma(\alpha)} f(z)\, dz, \tag{4.19}$$

where Γ *is the gamma function,* $\alpha > 0$,

$$I_{b-}^{0} f(x) = f(x),$$

and $I_{b-}^{\beta} f$ *is defined similarly for* $\beta \geq 0$, *then*

$$\int_a^b \left(I_{b-}^{\alpha} f(x)\right)^p \left(I_{b-}^{\beta} f(x)\right)^q x^{\gamma-\alpha p-\beta q-1} \omega(x)\, dx \leq C^* \int_a^b f^r(x)\, x^{\gamma-1} \omega(x)\, dx, \tag{4.20}$$

where

$$C^* = \left(\frac{b-a}{a}\right)^{(\alpha p+\beta q)} \frac{1}{\Gamma^r(\alpha+1)}. \quad (4.21)$$

Proof Consider $H(z,x) = \frac{(z-x)^{\alpha-1}}{x^{\alpha}\Gamma(\alpha)}, a < x < z < b$. Notice $H(tz,tx) = t^{-1}H(z,x)$, so that H is homogeneous of degree -1. Here we consider

$$(Hf)(x) = \frac{1}{x^{\alpha}} \int_x^b \frac{(z-x)^{\alpha-1}}{\Gamma(\alpha)} f(z)\,dz = x^{-\alpha} I_{b-}^{\alpha} f(x),$$

and by (4.8) we have

$$\left\| x^{-\alpha} I_{b-}^{\alpha} f(x) \right\|_r \leq A \|f\|_r, \quad (4.22)$$

where

$$A = \frac{1}{\Gamma(\alpha)} \int_1^{\frac{b}{a}} (t-1)^{\alpha-1} t^{-\frac{\gamma}{r}}\,dt.$$

We notice that

$$\int_1^{\frac{b}{a}} (t-1)^{\alpha-1} t^{-\frac{\gamma}{r}}\,dt \leq \int_1^{\frac{b}{a}} (t-1)^{\alpha-1}\,dt = \frac{(b-a)^{\alpha}}{a^{\alpha}\alpha}. \quad (4.23)$$

That is

$$A \leq \frac{(b-a)^{\alpha}}{a^{\alpha}\Gamma(\alpha+1)}. \quad (4.24)$$

Therefore it holds

$$\left\| x^{-\alpha} I_{b-}^{\alpha} f(x) \right\|_r \leq \frac{(b-a)^{\alpha}}{a^{\alpha}\Gamma(\alpha+1)} \|f\|_r, \quad (4.25)$$

and

$$\left\| x^{-\beta} I_{b-}^{\beta} f(x) \right\|_r \leq \frac{(b-a)^{\beta}}{a^{\beta}\Gamma(\beta+1)} \|f\|_r. \quad (4.26)$$

Using Hölder's inequality with indices $\frac{r}{p}$ and $\frac{r}{q}$, we get

$$\int_a^b \left(x^{-\alpha} I_{b-}^{\alpha} f(x)\right)^p \left(x^{-\beta} I_{b-}^{\beta} f(x)\right)^q x^{\gamma-1}\omega(x)\,dx \leq$$

$$\left(\int_a^b \left(x^{-\alpha} I_{b-}^{\alpha} f(x)\right)^r x^{\gamma-1}\omega(x)\,dx\right)^{\frac{p}{r}} \left(\int_a^b \left(x^{-\beta} I_{b-}^{\beta} f(x)\right)^r x^{\gamma-1}\omega(x)\,dx\right)^{\frac{q}{r}} =$$

$$\left\| x^{-\alpha} I_{b-}^{\alpha} f(x) \right\|_r^p \left\| x^{-\beta} I_{b-}^{\beta} f(x) \right\|_r^q \leq \left(\frac{(b-a)^{\alpha}}{a^{\alpha}\Gamma(\alpha+1)}\right)^p \left(\frac{(b-a)^{\beta}}{a^{\beta}\Gamma(\beta+1)}\right)^q \|f\|_r^p \|f\|_r^q$$

$$= \frac{(b-a)^{\alpha p+\beta q}}{a^{\alpha p+\beta q}\Gamma(\alpha+1)^r}\|f\|_r^r, \tag{4.27}$$

proving the claim. ■

We continue with

Corollary 4.7 *If $p > 0$, $q > 0$, $p + q = r \geq 1$, $0 < a < b < \infty$, $\gamma > 0$, $\omega(x)$ is increasing and positive in (a, b), $f(x)$ is measurable and non-negative on (a, b), I_{b-}^{α} is the right Riemann–Liouville fractional integral, $\alpha > 0$, then*

$$\int_a^b \left(I_{b-}^{\alpha} f(x)\right)^p f(x)^q x^{\gamma-\alpha p-1}\omega(x)\,dx \leq$$

$$\left(\frac{b-a}{a}\right)^{\alpha p}\frac{1}{\Gamma^r(\alpha+1)}\int_a^b f^r(x)\,x^{\gamma-1}\omega(x)\,dx. \tag{4.28}$$

Proof By Theorem 4.6 for $\beta = 0$. ■

We also give

Corollary 4.8 *Here $p, q > 0$, $p + q = r \geq 1$, $0 < a < b < \infty$, $\gamma > 0$, $\omega(x)$ is increasing and positive in (a, b). Let $f \in AC^n([a, b])$ (i.e. $f^{(n-1)}$ is absolutely continuous), n is even and $f^{(n)} \geq 0$. Assume further that $f^{(k)}(b) = 0$, $k = 0, 1, \ldots, n-1$. Then*

$$\int_a^b f^p(x)\left(f^{(n)}(x)\right)^q x^{\gamma-np-1}\omega(x)\,dx \leq$$

$$\left(\frac{b-a}{a}\right)^{np}\frac{1}{(n!)^r}\int_a^b \left(f^{(n)}(x)\right)^r x^{\gamma-1}\omega(x)\,dx. \tag{4.29}$$

If $\gamma = 1$, $\omega(x) = 1$, we have

$$\int_a^b f^p(x)\left(f^{(n)}(x)\right)^q x^{-np}dx \leq \left(\frac{b-a}{a}\right)^{np}\frac{1}{(n!)^r}\int_a^b \left(f^{(n)}(x)\right)^r dx. \tag{4.30}$$

Proof Since $f \in AC^n([a, b])$, $x \in [a, b]$, we have that (Taylor's formula)

$$f(x) = \sum_{k=0}^{n-1}\frac{f^{(k)}(b)}{k!}(x-b)^k + \frac{1}{(n-1)!}\int_b^x (x-t)^{n-1} f^{(n)}(t)\,dt. \tag{4.31}$$

By the assumptions we get that

$$f(x) = \frac{1}{(n-1)!}\int_x^b (t-x)^{n-1} f^{(n)}(t)\,dt = I_{b-}^{n} f^{(n)}(x). \tag{4.32}$$

Here we have that there exists $f^{(n)}$ almost everywhere and $f^{(n)} \in L_1([a,b])$. Direct application of (4.28) produces (4.29). ■

Remark 4.9 Since $a \leq x \leq b$, by (4.30) we easily get that

$$\int_a^b f^p(x)\left(f^{(n)}(x)\right)^q dx \leq \left(\frac{b(b-a)}{a}\right)^{np} \frac{1}{(n!)^r} \int_a^b \left(f^{(n)}(x)\right)^r dx. \tag{4.33}$$

Setting $p = q = 1$, we have

$$\int_a^b f(x) f^{(n)}(x) dx \leq \left(\frac{b(b-a)}{a}\right)^n \frac{1}{(n!)^2} \int_a^b \left(f^{(n)}(x)\right)^2 dx, \tag{4.34}$$

which is a ride side Opial's type inequality.

We would like to formalize the last result.

Corollary 4.10 *Let* $0 < a < b < \infty$, $f \in AC^n([a,b])$, *n is even and* $f^{(n)} \geq 0$. *Assume that* $f^{(k)}(b) = 0$, $k = 0, 1, \ldots, n-1$. *Then*

$$\int_a^b f(x) f^{(n)}(x) dx \leq \left(\frac{b(b-a)}{a}\right)^n \frac{1}{(n!)^2} \int_a^b \left(f^{(n)}(x)\right)^2 dx. \tag{4.35}$$

We specialize (4.35) as follows:

Corollary 4.11 *Let* $f \in AC^2([1,2])$, *with* $f^{(2)} \geq 0$. *Assume that* $f(2) = f'(2) = 0$. *Then*

$$\int_1^2 f(x) f^{(2)}(x) dx \leq \int_1^2 \left(f^{(2)}(x)\right)^2 dx. \tag{4.36}$$

We need

Definition 4.12 ([1, p. 336]) Let $f \in AC^m([a,b])$, $m \in \mathbb{N}$, where $m = \lceil \alpha \rceil$, $\alpha > 0$ ($\lceil \cdot \rceil$ the ceiling of number). We define the right Caputo fractional derivative of order α, by

$$D_{b-}^{\alpha} f(x) := (-1)^m I_{b-}^{m-\alpha} f^{(m)}(x), \ \forall\, x \in [a,b], \tag{4.37}$$

that is

$$D_{b-}^{\alpha} f(x) = \frac{(-1)^m}{\Gamma(m-\alpha)} \int_x^b (z-x)^{m-\alpha-1} f^{(m)}(z) dz. \tag{4.38}$$

We mention the right Caputo fractional Taylor formula:

Theorem 4.13 ([1, p. 341]) *Let* $f \in AC^m([a,b])$, $x \in [a,b]$, $\alpha > 0$, $m = \lceil \alpha \rceil$. *Then*

(1)

$$f(x) = \sum_{k=0}^{m-1} \frac{f^{(k)}(b)}{k!}(x-b)^k + \frac{1}{\Gamma(\alpha)} \int_x^b (z-x)^{\alpha-1} D_{b-}^{\alpha} f(z)\, dz, \tag{4.39}$$

(2) in case of $f^{(k)}(b) = 0$, $k = 0, 1, \dots, m-1$, *we have*

$$f(x) = \frac{1}{\Gamma(\alpha)} \int_x^b (z-x)^{\alpha-1} D_{b-}^{\alpha} f(z)\, dz = \left(I_{b-}^{\alpha}\left(D_{b-}^{\alpha} f\right)\right)(x), \tag{4.40}$$

$\forall\, x \in [a, b]$.

We present

Corollary 4.14 *Let* $p, q > 0$, $p + q = r \geq 1$, $0 < a < b < \infty$, $\gamma > 0$, $\omega(x)$ *is increasing and positive in* (a, b), $\alpha > 0$, $m = \lceil \alpha \rceil$. *Assume that* $f \in AC^m([a, b])$ *such that* $f^{(k)}(b) = 0$, $k = 0, 1, \dots, m-1$ *and* $D_{b-}^{\alpha} f \geq 0$ *over* (a, b). *Then*

$$\int_a^b (f(x))^p \left(\left(D_{b-}^{\alpha} f\right)(x)\right)^q x^{\gamma-\alpha p-1} \omega(x)\, dx \leq$$

$$\left(\frac{b-a}{a}\right)^{\alpha p} \frac{1}{\Gamma^r(\alpha+1)} \int_a^b \left(\left(D_{b-}^{\alpha} f\right)(x)\right)^r x^{\gamma-1} \omega(x)\, dx. \tag{4.41}$$

Proof Apply (4.28) and (4.40). ∎

We continue with

Definition 4.15 ([1, p. 345]) Let $\nu > 0$, $n := [\nu]$ ($[\cdot]$ integral part), $\alpha = \nu - n$, $0 < \alpha < 1$, $f \in C([a, b])$. Consider the subspace of functions

$$C_{b-}^{\nu}([a, b]) := \left\{ f \in C^n([a, b]) : I_{b-}^{1-\alpha} f^{(n)} \in C^1([a, b]) \right\}. \tag{4.42}$$

We define the right generalized ν-fractional derivative of f over $[a, b]$ as

$$\overline{D}_{b-}^{\nu} f = (-1)^{n-1} \left(I_{b-}^{1-\alpha} f^{(n)}\right)'. \tag{4.43}$$

That is

$$\left(\overline{D}_{b-}^{\nu} f\right)(x) = \frac{(-1)^{n-1}}{\Gamma(n-\nu+1)} \frac{d}{dx} \int_x^b (z-x)^{n-\nu} f^{(n)}(z)\, dz, \tag{4.44}$$

$\forall\, x \in [a, b]$.

We need the following right fractional Taylor formula.

Theorem 4.16 ([1, p. 348]) *Let* $f \in C_{b-}^{\nu}([a,b])$, $\nu > 0$, $n = [\nu]$.
(1) If $\nu \geq 1$, *then*

$$f(x) = \sum_{k=0}^{n-1} \frac{f^{(k)}(b)}{k!}(x-b)^k + \left(I_{b-}^{\nu}\overline{D}_{b-}^{\nu}f\right)(x), \tag{4.45}$$

$\forall\, x \in [a,b]$.
(2) If $0 < \nu < 1$, *then*

$$f(x) = I_{b-}^{\nu}\overline{D}_{b-}^{\nu}f(x), \tag{4.46}$$

$\forall\, x \in [a,b]$.

We make

Remark 4.17 By Theorem 4.16 we have: for $f \in C_{b-}^{\nu}([a,b])$, $\nu > 0$, $n = [\nu]$, we get that $f(x) = I_{b-}^{\nu}\overline{D}_{b-}^{\nu}f(x)$, $\forall\, x \in [a,b]$, given that $0 < \nu < 1$, or $\nu \geq 1$ and $f^{(k)}(b) = 0$, $k = 0, 1, \ldots, n-1$.

We present

Corollary 4.18 *Let* $p, q > 0$, $p + q = r \geq 1$, $0 < a < b < \infty$, $\gamma > 0$, $\omega(x)$ *is increasing and positive in* (a,b), $\nu > 0$, $n = [\nu]$. *Assume that* $f \in C_{b-}^{\nu}([a,b])$, *in case of* $\nu \geq 1$ *we have that* $f^{(k)}(b) = 0$, $k = 0, 1, \ldots, n-1$ *and* $\overline{D}_{b-}^{\nu}f \geq 0$ *over* (a,b). *Then*

$$\int_a^b (f(x))^p \left(\left(\overline{D}_{b-}^{\nu}f\right)(x)\right)^q x^{\gamma-\nu p-1}\omega(x)\,dx \leq$$

$$\left(\frac{b-a}{a}\right)^{\nu p} \frac{1}{\Gamma^r(\nu+1)} \int_a^b \left(\left(\overline{D}_{b-}^{\nu}f\right)(x)\right)^r x^{\gamma-1}\omega(x)\,dx. \tag{4.47}$$

Proof Apply (4.28) and Remark 4.17. ■

4.2.2 Part II

Next we get reverse right side results.

We present

Theorem 4.19 *Let* $0 < r < 1$, $0 < a < b < \infty$, $f(x)$ *and* $H(z,x)$ *are measurable and non-negative on* (a,b), $(a,b)^2$, *respectively,* $H(z,x)$ *is homogeneous of degree* -1,

$$(Hf)(x) = \int_x^b H(z,x)\,f(z)\,dz, \tag{4.48}$$

and

$$\|f\|_{r,[a,b]}=\left(\int_a^b f^r(x)\,dx\right)^{\frac{1}{r}}. \tag{4.49}$$

We suppose that $\|Hf\|_{r,[a,b]}<\infty$ *and* $H(t,1)f(tx)>0$, *for almost all* $t\in\left[1,\frac{b}{x}\right]$, *for almost all* $x\in[a,b]$, *and* $\frac{H(t,1)}{t^{\frac{1}{r}}}\|f\|_{r,[at,b]}<\infty$, *for almost all* $t\in\left[1,\frac{b}{a}\right]$.

Then

$$\|Hf\|_{r,[a,b]}\geq\int_1^{\frac{b}{a}}\frac{H(t,1)}{t^{\frac{1}{r}}}\|f\|_{r,[at,b]}\,dt. \tag{4.50}$$

Proof For $a<x<b$, the homogeneity of degree -1 of H gives

$$(Hf)(x)=\int_1^{\frac{b}{x}}H(t,1)f(tx)\,dt,$$

where $x\leq z\leq b$, $1\leq\frac{z}{x}\leq\frac{b}{x}$, setting $t:=\frac{z}{x}$ then $z=tx$ and $1\leq t\leq\frac{b}{x}$. Infact $1\leq t<\frac{b}{a}$. Using the reverse Minkowski integral inequality [2] we obtain

$$\|Hf\|_{r,[a,b]}=\left(\int_a^b\left(\int_1^{\frac{b}{x}}H(t,1)f(tx)\,dt\right)^r dx\right)^{\frac{1}{r}}\geq$$

$$\int_1^{\frac{b}{a}}\left(\int_a^{\frac{b}{t}}H(t,1)^r f(tx)^r\,dx\right)^{\frac{1}{r}}dt=\int_1^{\frac{b}{a}}\frac{H(t,1)}{t^{\frac{1}{r}}}\left(\int_a^{\frac{b}{t}}f(tx)^r\,tdx\right)^{\frac{1}{r}}dt=$$

$$\int_1^{\frac{b}{a}}\frac{H(t,1)}{t^{\frac{1}{r}}}\left(\int_{at}^b f(y)^r\,dy\right)^{\frac{1}{r}}dt=\int_1^{\frac{b}{a}}\frac{H(t,1)}{t^{\frac{1}{r}}}\|f\|_{r,[at,b]}\,dt, \tag{4.51}$$

proving (4.50). ∎

Next we apply Theorem 4.19.

We give

Proposition 4.20 *Let* $0<r<1$, $0<a<b<\infty$, $\alpha>0$, $f(x)$ *is measurable and non-negative on* (a,b). *Assume that* $\left\|x^{-\alpha}I_{b-}^{\alpha}f(x)\right\|_{r,[a,b]}<\infty$; $(t-1)^{\alpha-1}f(tx)>0$, *for almost all* $t\in\left[1,\frac{b}{x}\right]$, *for almost all* $x\in[a,b]$ *and* $\frac{(t-1)^{\alpha-1}}{t^{\frac{1}{r}}}\|f\|_{r,[at,b]}<\infty$, *for almost all* $t\in\left[1,\frac{b}{a}\right]$. *Then*

$$\left\|x^{-\alpha}I_{b-}^{\alpha}f(x)\right\|_{r,[a,b]}\geq\frac{1}{\Gamma(\alpha)}\int_1^{\frac{b}{a}}\frac{(t-1)^{\alpha-1}}{t^{\frac{1}{r}}}\|f\|_{r,[at,b]}\,dt. \tag{4.52}$$

Proof Consider here $H(z,x)=\frac{(z-x)^{\alpha-1}}{x^{\alpha}\Gamma(\alpha)}$, $a<x<z<b$, then $(Hf)(x)=x^{-\alpha}I_{b-}^{\alpha}f(x)$. Finally we apply Theorem 4.19. ∎

We give a reverse right side fractional integral inequality.

Theorem 4.21 *Let* $0 < r < p$, *with* $r < 1$, $0 < a < b < \infty$, f *is measurable and non-negative on* (a, b) *such that* $f(x) > 0$ *almost everywhere on* $[a, b]$, $\alpha > 0$. *Assume that* $\left\|x^{-\alpha} I_{b-}^{\alpha} f(x)\right\|_{r,[a,b]} < \infty$, *and* $\|f\|_{r,[at,b]} < \infty$, *for almost all* $t \in \left[1, \frac{b}{a}\right]$. *Then*

$$\int_a^b \left(I_{b-}^{\alpha} f(x)\right)^p (f(x))^{r-p} x^{-\alpha p} dx \geq$$

$$\frac{1}{\Gamma^p(\alpha)} \left(\int_1^{\frac{b}{a}} \frac{(t-1)^{\alpha-1}}{t^{\frac{1}{r}}} \|f\|_{r,[at,b]} dt\right)^p \|f\|_{r,[a,b]}^{r-p}. \tag{4.53}$$

Proof We will use Proposition 4.20.

Here $0 < r < p$, hence $0 < \frac{r}{p} < 1$, also $\frac{r}{r-p} < 0$, and $f(x) > 0$ almost everywhere in $[a, b]$. Next we apply the reverse Hölder's inequality:

$$\int_a^b \left(x^{-\alpha}\left(I_{b-}^{\alpha} f\right)(x)\right)^p (f(x))^{r-p} dx \geq$$

$$\left(\int_a^b \left(x^{-\alpha}\left(I_{b-}^{\alpha} f\right)(x)\right)^r dx\right)^{\frac{p}{r}} \left(\int_a^b (f(x))^r dx\right)^{\frac{r-p}{r}} =$$

$$\left\|x^{-\alpha}\left(I_{b-}^{\alpha} f\right)(x)\right\|_{r,[a,b]}^p \|f\|_{r,[a,b]}^{r-p} \geq \tag{4.54}$$

$$\frac{1}{\Gamma^p(\alpha)} \left(\int_1^{\frac{b}{a}} \frac{(t-1)^{\alpha-1}}{t^{\frac{1}{r}}} \|f\|_{r,[at,b]} dt\right)^p \|f\|_{r,[a,b]}^{r-p},$$

proving the claim. ■

Next we give some reverse right side Opial type inequalities.

We start with an ordinary derivatives result.

Corollary 4.22 *Let* $0 < r < p$, *with* $r < 1$, $0 < a < b < \infty$, $f \in AC^n([a, b])$, n *is even and* $f^{(n)} \geq 0$ *on* (a, b) *with* $f^{(n)} > 0$ *almost everywhere on* $[a, b]$. *Assume that* $f^{(k)}(b) = 0$, $k = 0, 1, \ldots, n-1$, *and* $\left\|x^{-\alpha} f(x)\right\|_{r,[a,b]} < \infty$, *and* $\left\|f^{(n)}\right\|_{r,[at,b]} < \infty$, *for almost all* $t \in \left[1, \frac{b}{a}\right]$. *Then*

$$\int_a^b (f(x))^p \left(f^{(n)}(x)\right)^{r-p} x^{-np} dx \geq$$

$$\frac{1}{((n-1)!)^p} \left(\int_1^{\frac{b}{a}} \frac{(t-1)^{\alpha-1}}{t^{\frac{1}{r}}} \left\|f^{(n)}\right\|_{r,[at,b]} dt\right)^p \left\|f^{(n)}\right\|_{r,[a,b]}^{r-p}. \tag{4.55}$$

Proof Apply Theorem 4.21 and see the proof Corollary 4.8. ■

We continue with fractional reverse results.

Corollary 4.23 *Let* $0 < r < p$, *with* $r < 1$, $0 < a < b < \infty$, $\alpha > 0$, $m = \lceil \alpha \rceil$, $f \in AC^m([a,b])$, *such that* $f^{(k)}(b) = 0$, $k = 0, 1, \ldots, m-1$, *and* $D_{b-}^{\alpha} f \geq 0$ *over* (a,b) *with* $D_{b-}^{\alpha} f > 0$ *almost everywhere on* $[a,b]$. *Assume that* $\left\| x^{-\alpha} f(x) \right\|_{r,[a,b]} < \infty$, *and* $\left\| D_{b-}^{\alpha} f \right\|_{r,[at,b]} < \infty$, *for almost all* $t \in \left[1, \frac{b}{a}\right]$. *Then*

$$\int_a^b (f(x))^p \left(D_{b-}^{\alpha} f(x)\right)^{r-p} x^{-\alpha p} dx \geq$$

$$\frac{1}{\Gamma^p(\alpha)} \left(\int_1^{\frac{b}{a}} \frac{(t-1)^{\alpha-1}}{t^{\frac{1}{r}}} \left\| D_{b-}^{\alpha} f \right\|_{r,[at,b]} dt \right)^p \left\| D_{b-}^{\alpha} f \right\|_{r,[a,b]}^{r-p}. \tag{4.56}$$

Proof Use of Theorems 4.13 and 4.21. ■

We finish with

Corollary 4.24 *Let* $0 < r < p$, *with* $r < 1$, $0 < a < b < \infty$, $\nu > 0$, $n = [\nu]$, $f \in C_{b-}^{\nu}([a,b])$, *for either* $0 < \nu < 1$, *or* $\nu \geq 1$ *and* $f^{(k)}(b) = 0$, $k = 0, 1, \ldots, n-1$. *Suppose that* $\overline{D}_{b-}^{\nu} f \geq 0$ *over* (a,b) *with* $\overline{D}_{b-}^{\nu} f > 0$ *almost everywhere on* $[a,b]$. *Assume further that* $\left\| x^{-\nu} f(x) \right\|_{r,[a,b]} < \infty$, *and* $\left\| \overline{D}_{b-}^{\nu} f \right\|_{r,[at,b]} < \infty$, *for almost all* $t \in \left[1, \frac{b}{a}\right]$. *Then*

$$\int_a^b (f(x))^p \left(\overline{D}_{b-}^{\nu} f(x)\right)^{r-p} x^{-\nu p} dx \geq$$

$$\frac{1}{\Gamma^p(\nu)} \left(\int_1^{\frac{b}{a}} \frac{(t-1)^{\nu-1}}{t^{\frac{1}{r}}} \left\| \overline{D}_{b-}^{\nu} f \right\|_{r,[at,b]} dt \right)^p \left\| \overline{D}_{b-}^{\nu} f \right\|_{r,[a,b]}^{r-p}. \tag{4.57}$$

Proof Use of Remark 4.17 and Theorem 4.21. ■

References

1. Anastassiou, G.A.: Intelligent Mathematics: Computational Analysis. Springer, Heidelberg, New York (2011)
2. Anastassiou, G.A.: E.R. Love type left fractional integral inequalities. Issues Anal. **9**(27)(3), 1–17 (2020)
3. Anastassiou, G.A.: E.R. Love type right side fractional integral inequalities (2020)
4. Hardy, G.H., Littlewood, J.E., Pólya, G.: Inequalities, Cambridge, UK (1934)
5. Love, E.R.: Inequalities like Opial's inequality, Rocznik naukowo dydaktyczny WSP w Krakowie. Pr. Mat. **97**, 109–118 (1985)
6. Opial, Z.: Sur une inégalité. Ann. Polon. Math. **8**, 29–32 (1960)

Chapter 5
General Fractional Landau Inequalities

We present uniform and L_p mixed Caputo–Bochner abstract fractional Landau inequalities over $\mathbb{R}$ of fractional orders $2 < \nu \leq 3$. These estimate the size of first and second derivatives of a Banach space valued function over $\mathbb{R}$. We give applications when $\nu = 2.5$. It follows [5].

5.1 Introduction

Let $p \in [1, \infty]$, $I = \mathbb{R}_+$ or $I = \mathbb{R}$ and $f : I \to \mathbb{R}$ is twice differentiable with $f, f'' \in L_p(I)$, then $f' \in L_p(I)$. Moreover, there exists a constant $C_p(I) > 0$ independent of f, such that

$$\left\| f' \right\|_{p,I} \leq C_p(I) \left\| f \right\|_{p,I}^{\frac{1}{2}} \left\| f'' \right\|_{p,I}^{\frac{1}{2}}, \tag{5.1}$$

where $\|\cdot\|_{p,I}$ is the p-norm on the interval I, see [1, 6].

The research on these inequalities started by Landau [12] in 1913. For the case of $p = \infty$ he proved that

$$C_\infty(\mathbb{R}_+) = 2 \text{ and } C_\infty(\mathbb{R}) = \sqrt{2}, \tag{5.2}$$

are the best constants in (5.1).

In 1932, Hardy and Littlewood [8] proved (5.1) for $p = 2$, with the best constants

$$C_2(\mathbb{R}_+) = \sqrt{2} \text{ and } C_2(\mathbb{R}) = 1. \tag{5.3}$$

In 1935, Hardy et al. [9] showed that the best constants $C_p(\mathbb{R}_+)$ in (5.1) satisfies the estimate

$$C_p(\mathbb{R}_+) \leq 2, \text{ for } p \in [1, \infty), \tag{5.4}$$

G. A. Anastassiou, *Constructive Fractional Analysis with Applications*,
Studies in Systems, Decision and Control 362,
https://doi.org/10.1007/978-3-030-71481-9_5

which yields $C_p(\mathbb{R}) \leq 2$ for $p \in [1, \infty)$.

In fact, in [7, 10] was shown that $C_p(\mathbb{R}) \leq \sqrt{2}$.

In this chapter we are greatly inspired by interesting articles [11, 15].

We need the following concepts from abstract fractional calculus.

Our integrals next are of Bochner type [13].

We need

Definition 5.1 ([4, p. 150]) Let $[a, b] \subset \mathbb{R}$, X be a Banach space, $\alpha > 0$; $m = \lceil \alpha \rceil \in \mathbb{N}$, ($\lceil \cdot \rceil$ is the ceiling of the number), $f : [a, b] \to X$. We assume that $f^{(m)} \in L_1([a, b], X)$. We call the Caputo–Bochner left fractional derivative of order α:

$$\left(D_{*a}^{\alpha} f\right)(x) := \frac{1}{\Gamma(m-\alpha)} \int_a^x (x-t)^{m-\alpha-1} f^{(m)}(t)\, dt, \ \forall\, x \in [a, b], \tag{5.5}$$

where Γ is the gamma function.

If $\alpha \in \mathbb{N}$, we set $D_{*a}^{\alpha} f := f^{(m)}$ the ordinary X-valued derivative (defined similar to numerical one, see [14], p. 83), and also set $D_{*a}^{0} f := f$.

By ([4, p. 2]), $\left(D_{*a}^{\alpha} f\right)(x)$ exists almost everywhere in $x \in [a, b]$ and $D_{*a}^{\alpha} f \in L_1([a, b], X)$.

If $\left\| f^{(m)} \right\|_{L_\infty([a,b],X)} < \infty$, and $\alpha \notin \mathbb{N}$, then by ([4, p. 3]), $D_{*a}^{\alpha} f \in C([a, b], X)$, hence $\left\| D_{*a}^{\alpha} f \right\| \in C([a, b])$.

We also need

Definition 5.2 ([4, p. 150]) Let $[a, b] \subset \mathbb{R}$, X be a Banach space, $\alpha > 0$, $m := \lceil \alpha \rceil$. We assume that $f^{(m)} \in L_1([a, b], X)$, where $f : [a, b] \to X$. We call the Caputo–Bochner right fractional derivative of order α:

$$\left(D_{b-}^{\alpha} f\right)(x) := \frac{(-1)^m}{\Gamma(m-\alpha)} \int_x^b (z-x)^{m-\alpha-1} f^{(m)}(z)\, dz, \ \forall\, x \in [a, b]. \tag{5.6}$$

We observe that $\left(D_{b-}^{m} f\right)(x) = (-1)^m f^{(m)}(x)$, for $m \in \mathbb{N}$, and $\left(D_{b-}^{0} f\right)(x) = f(x)$.

By ([4, p. 34]), $\left(D_{b-}^{\alpha} f\right)(x)$ exists almost everywhere on $[a, b]$ and $\left(D_{b-}^{\alpha} f\right) \in L_1([a, b], X)$.

If $\left\| f^{(m)} \right\|_{L_\infty([a,b],X)} < \infty$, and $\alpha \notin \mathbb{N}$, by ([4, p. 37]), $D_{b-}^{\alpha} f \in C([a, b], X)$, hence $\left\| D_{b-}^{\alpha} f \right\| \in C([a, b])$.

We mention the important:

Corollary 5.3 ([4, p. 157]) *Let $f \in C^m([a, b], X)$, $m = \lceil \alpha \rceil$, $\alpha > 0$, $x, x_0 \in [a, b]$. Then $D_{*x_0}^{a} f(x)$, $D_{x_0-}^{a} f(x)$ are jointly continuous functions in (x, x_0) from $[a, b]^2$ into X, X is a Banach space.*

By convention we suppose that

$$D_{*x_0}^{\alpha} f(x) = 0, \text{ for } x < x_0,$$

and

$$D_{x_0-}^{\alpha} f(x) = 0, \text{ for } x > x_0,$$

for all $x, x_0 \in [a, b]$.

The author has already done an extensive body of work on fractional Landau inequalities, see [3], and on abstract fractional Landau inequalities, see [4]. However there the proving methods came out of applications of fractional Ostrowski inequalities ([2, 4]) and the derived inequalities were for small fractional orders, i.e. $\alpha \in (0, 1)$. Usually there the domains where $[A, +\infty)$ or $(-\infty, B]$, with $A, B \in \mathbb{R}$ and in one mixed case the domain was all of $\mathbb{R}$.

In this chapter with less assumptions we establish uniform and L_p type mixed Caputo–Bochner abstract fractional Landau inequalities over $\mathbb{R}$ for fractional orders $2 < \nu \leq 3$. The method of proving is based on left and right Caputo–Bochner fractional Taylor's formulae with integral remainders, see [4], pp. 151–152.

We give also applications when $\nu = 2.5$. Regardless to say that we are also inspired by [3, 4].

5.2 Main Results

We present the following abstract mixed fractional Landau inequalities over $\mathbb{R}$.

Theorem 5.4 *Let $2 < \nu \leq 3$ and $f \in C^3(\mathbb{R}, X)$, where $(X, \|\cdot\|)$ is a Banach space. We assume that $\|\|f\|\|_{\infty,\mathbb{R}} < \infty$, and that*

$$K := \max\left\{\left\|\left\|D_{*a}^{\nu} f(z)\right\|\right\|_{\infty,\mathbb{R}^2}, \left\|\left\|D_{a-}^{\nu} f(z)\right\|\right\|_{\infty,\mathbb{R}^2}\right\} < \infty, \tag{5.7}$$

where $(a, z) \in \mathbb{R}^2$.

Then

$$\left\|\left\|f'\right\|\right\|_{\infty,\mathbb{R}} \leq \nu\left(\frac{K}{\Gamma(\nu+1)}\right)^{\frac{1}{\nu}}\left(\frac{\|\|f\|\|_{\infty,\mathbb{R}}}{\nu-1}\right)^{\frac{\nu-1}{\nu}}, \tag{5.8}$$

and

$$\left\|\left\|f''\right\|\right\|_{\infty,\mathbb{R}} \leq \nu\left(\frac{K}{\Gamma(\nu+1)}\right)^{\frac{2}{\nu}}\left(\frac{4\|\|f\|\|_{\infty,\mathbb{R}}}{\nu-2}\right)^{\frac{\nu-2}{\nu}}. \tag{5.9}$$

That is $\left\|\left\|f'\right\|\right\|_{\infty,\mathbb{R}}, \left\|\left\|f''\right\|\right\|_{\infty,\mathbb{R}} < \infty$.

Proof Here $2 < \nu \leq 3$, i.e. $\lceil \nu \rceil = 3$. Let $f \in C^3(\mathbb{R}, X)$, where $(X, \|\cdot\|)$ is a Banach space, $a \in \mathbb{R}$.

We need the following abstract fractional Taylor formulae.
By Theorem 5.8, p. 151 ([4]) we get,

$$f(x) - f(a) - (x-a) f'(a) - \frac{(x-a)^2}{2} f''(a) = \tag{5.10}$$

$$\frac{1}{\Gamma(\nu)} \int_a^x (x-z)^{\nu-1} \left(D_{*a}^{\nu} f\right)(z)\, dz, \ \forall\, x \geq a.$$

And by Theorem 5.9, p. 152 ([4]), see also Definition 5.10, p. 152 ([4]):

$$f(x) - f(a) - (x-a) f'(a) - \frac{(x-a)^2}{2} f''(a) =$$

$$\frac{1}{\Gamma(\nu)} \int_x^a (z-x)^{\nu-1} \left(D_{a-}^{\nu} f\right)(z)\, dz, \ \forall\, x \leq a. \tag{5.11}$$

The remainders are continuous functions.

Let $x_1 > a$, then

$$f(x_1) - f(a) - (x_1-a) f'(a) - \frac{(x_1-a)^2}{2} f''(a) =$$

$$\frac{1}{\Gamma(\nu)} \int_a^{x_1} (x_1-z)^{\nu-1} \left(D_{*a}^{\nu} f\right)(z)\, dz, \tag{5.12}$$

and for $x_2 < a$, we get

$$f(x_2) - f(a) - (x_2-a) f'(a) - \frac{(x_2-a)^2}{2} f''(a) =$$

$$\frac{1}{\Gamma(\nu)} \int_{x_2}^a (z-x_2)^{\nu-1} \left(D_{a-}^{\nu} f\right)(z)\, dz. \tag{5.13}$$

That is, we have the system of two equations with unknowns $f'(a)$, $f''(a)$:

$$(x_1-a) f'(a) + \frac{(x_1-a)^2}{2} f''(a) =$$

$$f(x_1) - f(a) - \frac{1}{\Gamma(\nu)} \int_a^{x_1} (x_1-z)^{\nu-1} \left(D_{*a}^{\nu} f\right)(z)\, dz =: A, \tag{5.14}$$

and

$$(x_2-a) f'(a) + \frac{(x_2-a)^2}{2} f''(a) =$$

$$f(x_2) - f(a) - \frac{1}{\Gamma(\nu)} \int_{x_2}^{a} (z - x_2)^{\nu - 1} \left(D_{a-}^{\nu} f \right)(z)\, dz =: B. \tag{5.15}$$

The coefficient determinant is $D = \frac{1}{2}(x_1 - a)(x_2 - a)(x_2 - x_1)$. It is $D \neq 0$.
We have the system:

$$\begin{aligned} (x_1 - a) f'(a) + \tfrac{(x_1 - a)^2}{2} f''(a) &= A, \\ (x_2 - a) f'(a) + \tfrac{(x_2 - a)^2}{2} f''(a) &= B. \end{aligned} \tag{5.16}$$

Solving the system (5.16), it has a unique non-trivial solution, we get:

$$f'(a) = \frac{\begin{vmatrix} A & \frac{(x_1 - a)^2}{2} \\ B & \frac{(x_2 - a)^2}{2} \end{vmatrix}}{\frac{1}{2}(x_1 - a)(x_2 - a)(x_2 - x_1)} = \frac{A(x_2 - a)^2 - B(x_1 - a)^2}{(x_1 - a)(x_2 - a)(x_2 - x_1)}, \tag{5.17}$$

and

$$f''(a) = \frac{\begin{vmatrix} x_1 - a & A \\ x_2 - a & B \end{vmatrix}}{\frac{1}{2}(x_1 - a)(x_2 - a)(x_2 - x_1)} = \frac{2\left[B(x_1 - a) - A(x_2 - a)\right]}{(x_1 - a)(x_2 - a)(x_2 - x_1)}. \tag{5.18}$$

We derive that

$$f'(a) = \left[\left[f(x_1) - f(a) - \frac{1}{\Gamma(\nu)} \int_a^{x_1} (x_1 - z)^{\nu - 1} \left(D_{*a}^{\nu} f \right)(z)\, dz \right] (x_2 - a)^2 - \right.$$

$$\left. \left[f(x_2) - f(a) - \frac{1}{\Gamma(\nu)} \int_{x_2}^{a} (z - x_2)^{\nu - 1} \left(D_{a-}^{\nu} f \right)(z)\, dz \right] (x_1 - a)^2 \right] \cdot$$

$$\frac{1}{(x_1 - a)(x_2 - a)(x_2 - x_1)}, \tag{5.19}$$

and

$$f''(a) = 2 \left[\left[f(x_2) - f(a) - \frac{1}{\Gamma(\nu)} \int_{x_2}^{a} (z - x_2)^{\nu - 1} \left(D_{a-}^{\nu} f \right)(z)\, dz \right] (x_1 - a) - \right.$$

$$\left. \left[f(x_1) - f(a) - \frac{1}{\Gamma(\nu)} \int_a^{x_1} (x_1 - z)^{\nu - 1} \left(D_{*a}^{\nu} f \right)(z)\, dz \right] (x_2 - a) \right] \cdot$$

$$\frac{1}{(x_1 - a)(x_2 - a)(x_2 - x_1)}. \tag{5.20}$$

Simplifying things we have

$$f'(a) = \frac{\left[f(x_1) - f(a) - \frac{1}{\Gamma(\nu)}\int_a^{x_1}(x_1 - z)^{\nu-1}\left(D_{*a}^{\nu}f\right)(z)\,dz\right](x_2 - a)}{(x_1 - a)(x_2 - x_1)} -$$

$$\frac{\left[f(x_2) - f(a) - \frac{1}{\Gamma(\nu)}\int_{x_2}^{a}(z - x_2)^{\nu-1}\left(D_{a-}^{\nu}f\right)(z)\,dz\right](x_1 - a)}{(x_2 - a)(x_2 - x_1)}, \tag{5.21}$$

and

$$f''(a) = \frac{2\left[f(x_2) - f(a) - \frac{1}{\Gamma(\nu)}\int_{x_2}^{a}(z - x_2)^{\nu-1}\left(D_{a-}^{\nu}f\right)(z)\,dz\right]}{(x_2 - a)(x_2 - x_1)} -$$

$$\frac{2\left[f(x_1) - f(a) - \frac{1}{\Gamma(\nu)}\int_a^{x_1}(x_1 - z)^{\nu-1}\left(D_{*a}^{\nu}f\right)(z)\,dz\right]}{(x_1 - a)(x_2 - x_1)}. \tag{5.22}$$

Rewriting above solutions we have

$$f'(a) = \frac{(f(x_1) - f(a))(x_2 - a)}{(x_1 - a)(x_2 - x_1)} - \frac{(f(x_2) - f(a))(x_1 - a)}{(x_2 - a)(x_2 - x_1)} -$$

$$\frac{\left(\int_a^{x_1}(x_1 - z)^{\nu-1}\left(D_{*a}^{\nu}f\right)(z)\,dz\right)(x_2 - a)}{\Gamma(\nu)(x_1 - a)(x_2 - x_1)} + \frac{\left(\int_{x_2}^{a}(z - x_2)^{\nu-1}\left(D_{a-}^{\nu}f\right)(z)\,dz\right)(x_1 - a)}{\Gamma(\nu)(x_2 - a)(x_2 - x_1)}, \tag{5.23}$$

and

$$f''(a) = \frac{2(f(x_2) - f(a))}{(x_2 - a)(x_2 - x_1)} - \frac{2(f(x_1) - f(a))}{(x_1 - a)(x_2 - x_1)} -$$

$$\frac{2\left(\int_{x_2}^{a}(z - x_2)^{\nu-1}\left(D_{a-}^{\nu}f\right)(z)\,dz\right)}{\Gamma(\nu)(x_2 - a)(x_2 - x_1)} + \frac{2\left(\int_a^{x_1}(x_1 - z)^{\nu-1}\left(D_{*a}^{\nu}f\right)(z)\,dz\right)}{\Gamma(\nu)(x_1 - a)(x_2 - x_1)}. \tag{5.24}$$

Let now $h > 0$, and choose $x_1 = a + h$ and $x_2 = a - h$. That is $x_1 - a = h$, $x_2 - a = -h$, and $x_2 - x_1 = -2h$.

We derive the following:

$$f'(a) = \frac{(f(a+h) - f(a))}{2h} - \frac{(f(a-h) - f(a))}{2h} -$$

$$\frac{\left(\int_a^{a+h}(a + h - z)^{\nu-1}\left(D_{*a}^{\nu}f\right)(z)\,dz\right)}{2h\Gamma(\nu)} + \frac{\left(\int_{a-h}^{a}(z - a + h)^{\nu-1}\left(D_{a-}^{\nu}f\right)(z)\,dz\right)}{2h\Gamma(\nu)}, \tag{5.25}$$

and similarly,

$$f''(a) = \frac{(f(a-h) - f(a))}{h^2} - \frac{(f(a+h) - f(a))}{h^2} -$$

$$\frac{\left(\int_{a-h}^{a} (z-a+h)^{\nu-1} \left(D_{a-}^{\nu} f\right)(z)\, dz\right)}{h^2 \Gamma(\nu)} - \frac{\left(\int_{a}^{a+h} (a+h-z)^{\nu-1} \left(D_{*a}^{\nu} f\right)(z)\, dz\right)}{h^2 \Gamma(\nu)}. \tag{5.26}$$

Putting things together we derive

$$f'(a) = \frac{(f(a+h) - f(a-h))}{2h} - \frac{1}{2h\Gamma(\nu)}$$

$$\left[\int_{a}^{a+h} (a+h-z)^{\nu-1} \left(D_{*a}^{\nu} f\right)(z)\, dz - \int_{a-h}^{a} (z-(a-h))^{\nu-1} \left(D_{a-}^{\nu} f\right)(z)\, dz\right], \tag{5.27}$$

and

$$f''(a) = \frac{(f(a+h) + f(a-h) - 2f(a))}{h^2} - \frac{1}{h^2\Gamma(\nu)}$$

$$\left[\int_{a}^{a+h} (a+h-z)^{\nu-1} \left(D_{*a}^{\nu} f\right)(z)\, dz + \int_{a-h}^{a} (z-(a-h))^{\nu-1} \left(D_{a-}^{\nu} f\right)(z)\, dz\right]. \tag{5.28}$$

By Corollary 5.3 we obtain that $\left(D_{*a}^{\nu} f\right)(z)$, $\left(D_{a-}^{\nu} f\right)(z)$ are jointly continuous functions in (z, a) from $\mathbb{R}^2$ into X, where X is a Banach space. We assumed here that $\left\| \left\| D_{*a}^{\nu} f(z) \right\| \right\|_{\infty,\mathbb{R}^2}$, $\left\| \left\| D_{a-}^{\nu} f(z) \right\| \right\|_{\infty,\mathbb{R}^2} < \infty$.

By (5.27) we get that

$$\left\| f'(a) \right\| \leq \frac{\| f(a+h) - f(a-h) \|}{2h} +$$

$$\frac{1}{2h\Gamma(\nu)} \left[\int_{a}^{a+h} (a+h-z)^{\nu-1} \left\| \left(D_{*a}^{\nu} f\right)(z) \right\| dz + \right.$$

$$\left. \int_{a-h}^{a} (z-(a-h))^{\nu-1} \left\| \left(D_{a-}^{\nu} f\right)(z) \right\| dz \right] \leq \tag{5.29}$$

$$\frac{\| \| f \| \|_{\infty,\mathbb{R}}}{h} + \frac{K}{2h\Gamma(\nu)} \left[\int_{a}^{a+h} (a+h-z)^{\nu-1} dz + \int_{a-h}^{a} (z-(a-h))^{\nu-1} dz\right] =$$

$$\frac{\| \| f \| \|_{\infty,\mathbb{R}}}{h} + \frac{K}{2h\Gamma(\nu)} \left[\frac{2h^{\nu}}{\nu}\right] = \frac{\| \| f \| \|_{\infty,\mathbb{R}}}{h} + \frac{K h^{\nu-1}}{\Gamma(\nu+1)}.$$

That is

$$\left\| f'(a) \right\| \leq \frac{\| \| f \| \|_{\infty,\mathbb{R}}}{h} + \frac{K h^{\nu-1}}{\Gamma(\nu+1)}, \tag{5.30}$$

$\forall\, a \in \mathbb{R}, \forall\, h > 0.$

I.e. it holds

$$\left\|\left\|f'\right\|\right\|_{\infty,\mathbb{R}} \leq \frac{\left\|\left\|f\right\|\right\|_{\infty,\mathbb{R}}}{h} + \frac{Kh^{\nu-1}}{\Gamma(\nu+1)}, \tag{5.31}$$

$\forall\, h > 0, 2 < \nu \leq 3.$

Call

$$\mu := \left\|\left\|f\right\|\right\|_{\infty,\mathbb{R}}, \qquad \theta = \frac{K}{\Gamma(\nu+1)}, \tag{5.32}$$

both are greater than zero.

Set also $\rho := \nu - 1 > 1$.

We consider the function

$$y(h) := \frac{\mu}{h} + \theta h^{\rho},\ \forall\, h > 0. \tag{5.33}$$

We have

$$y'(h) = -\frac{\mu}{h^2} + \rho\theta h^{\rho-1} = 0, \tag{5.34}$$

then

$$\rho\theta h^{\rho-1} = \frac{\mu}{h^2},$$

and

$$\rho\theta h^{\rho+1} = \mu,$$

with a unique solution

$$h_0 := h_{crit.no} = \left(\frac{\mu}{\rho\theta}\right)^{\frac{1}{\rho+1}}. \tag{5.35}$$

We have that

$$y'(h) = -\mu h^{-2} + \rho\theta h^{\rho-1},$$

and

$$y''(h) = 2\mu h^{-3} + \rho(\rho-1)\theta h^{\rho-2}. \tag{5.36}$$

We see that

$$y''(h_0) = 2\mu\left(\frac{\mu}{\rho\theta}\right)^{-\frac{3}{\rho+1}} + \rho(\rho-1)\theta\left(\frac{\mu}{\rho\theta}\right)^{\frac{((\rho+1)-3)}{\rho+1}} =$$

$$\left(\frac{\mu}{\rho\theta}\right)^{-\frac{3}{\rho+1}}\left[2\mu + \rho(\rho-1)\theta\left(\frac{\mu}{\rho\theta}\right)\right] =$$

$$\left(\frac{\mu}{\rho\theta}\right)^{-\frac{3}{\rho+1}}[2\mu+\mu(\rho-1)]=\mu\left(\frac{\mu}{\rho\theta}\right)^{-\frac{3}{\rho+1}}(\rho+1)>0. \tag{5.37}$$

Therefore y has a global minimum at $h_0=\left(\frac{\mu}{\rho\theta}\right)^{\frac{1}{\rho+1}}$, which is

$$y(h_0)=\frac{\mu}{\left(\frac{\mu}{\rho\theta}\right)^{\frac{1}{\rho+1}}}+\theta\left(\frac{\mu}{\rho\theta}\right)^{\frac{\rho}{\rho+1}}=$$

$$(\rho\theta)^{\frac{1}{\rho+1}}\frac{\mu}{\mu^{\frac{1}{\rho+1}}}+\frac{\theta\mu^{\frac{\rho}{\rho+1}}}{\rho^{\frac{\rho}{\rho+1}}\theta^{\frac{\rho}{\rho+1}}}=(\rho\theta)^{\frac{1}{\rho+1}}\mu^{1-\frac{1}{\rho+1}}+\frac{\theta^{1-\frac{\rho}{\rho+1}}\mu^{\frac{\rho}{\rho+1}}}{\rho^{\frac{\rho}{\rho+1}}}=$$

$$\rho^{\frac{1}{\rho+1}}\theta^{\frac{1}{\rho+1}}\mu^{\frac{\rho}{\rho+1}}+\theta^{\frac{1}{\rho+1}}\mu^{\frac{\rho}{\rho+1}}\rho^{-\frac{\rho}{\rho+1}}=\theta^{\frac{1}{\rho+1}}\mu^{\frac{\rho}{\rho+1}}\left[\rho^{\frac{1}{\rho+1}}+\rho^{-\frac{\rho}{\rho+1}}\right]= \tag{5.38}$$

$$(\theta\mu^{\rho})^{\frac{1}{\rho+1}}\left(\rho^{\frac{1}{\rho+1}}+\frac{1}{\rho^{\frac{\rho}{\rho+1}}}\right)=(\theta\mu^{\rho})^{\frac{1}{\rho+1}}\left(\frac{\rho+1}{\rho^{\frac{\rho}{\rho+1}}}\right)=$$

$$(\theta\mu^{\rho})^{\frac{1}{\rho+1}}(\rho+1)\,\rho^{-\frac{\rho}{\rho+1}}.$$

That is

$$y(h_0)=(\theta\mu^{\rho})^{\frac{1}{\rho+1}}(\rho+1)\,\rho^{-\frac{\rho}{\rho+1}}. \tag{5.39}$$

Consequently

$$y(h_0)=\left(\frac{K}{\Gamma(\nu+1)}\left\|\|f\|\right\|_{\infty,\mathbb{R}}^{\nu-1}\right)^{\frac{1}{\nu}}\nu(\nu-1)^{-\left(\frac{\nu-1}{\nu}\right)}. \tag{5.40}$$

We have proved that

$$\left\|\|f'\|\right\|_{\infty,\mathbb{R}}\le\left(\frac{K}{\Gamma(\nu+1)}\left\|\|f\|\right\|_{\infty,\mathbb{R}}^{\nu-1}\right)^{\frac{1}{\nu}}\nu(\nu-1)^{-\left(\frac{\nu-1}{\nu}\right)}. \tag{5.41}$$

From (5.28) we derive

$$\left\|f''(a)\right\|\le\frac{4\left\|\|f\|\right\|_{\infty,\mathbb{R}}}{h^2}+$$

$$\frac{K}{h^2\Gamma(\nu)}\left[\int_a^{a+h}(a+h-z)^{\nu-1}dz+\int_{a-h}^{a}(z-(a-h))^{\nu-1}dz\right]$$

$$=\frac{4\left\|\|f\|\right\|_{\infty,\mathbb{R}}}{h^2}+\frac{K}{h^2\Gamma(\nu)}\left[\frac{2h^{\nu}}{\nu}\right]=\frac{4\left\|\|f\|\right\|_{\infty,\mathbb{R}}}{h^2}+\frac{2Kh^{\nu-2}}{\Gamma(\nu+1)}. \tag{5.42}$$

That is

$$\left\| f''(a) \right\| \leq \frac{4 \left\| \| f \| \right\|_{\infty,\mathbb{R}}}{h^2} + \frac{2Kh^{\nu-2}}{\Gamma(\nu+1)}, \tag{5.43}$$

$\forall\, a \in \mathbb{R}, \forall\, h > 0$.

Hence it holds

$$\left\| \| f'' \| \right\|_{\infty,\mathbb{R}} \leq \frac{4 \left\| \| f \| \right\|_{\infty,\mathbb{R}}}{h^2} + \frac{2Kh^{\nu-2}}{\Gamma(\nu+1)}, \tag{5.44}$$

$\forall\, h > 0,\ 2 < \nu \leq 3$.

Call

$$\begin{gathered} \xi := 4 \left\| \| f \| \right\|_{\infty,\mathbb{R}}, \\ \psi = \tfrac{2K}{\Gamma(\nu+1)}, \end{gathered} \tag{5.45}$$

both are greater than zero.

Set also $\varphi := \nu - 2 > 0$.

We consider the function

$$\gamma(h) := \frac{\xi}{h^2} + \psi h^{\varphi} = \xi h^{-2} + \psi h^{\varphi},\ \forall\, h > 0. \tag{5.46}$$

We have

$$\gamma'(h) = -2\xi h^{-3} + \varphi\psi h^{\varphi-1} = 0, \tag{5.47}$$

then

$$\varphi\psi h^{\varphi-1} = 2\xi h^{-3},$$

and

$$\varphi\psi h^{\varphi+2} = 2\xi,$$

with a unique solution

$$h_0 := h_{crit.no} = \left(\frac{2\xi}{\varphi\psi}\right)^{\frac{1}{\varphi+2}}. \tag{5.48}$$

We have that

$$\gamma''(h) = 6\xi h^{-4} + \varphi(\varphi-1)\psi h^{\varphi-2}. \tag{5.49}$$

We see that

$$\gamma''(h_0) = 6\xi \left(\frac{2\xi}{\varphi\psi}\right)^{-\frac{4}{\varphi+2}} + \varphi(\varphi-1)\psi\left(\frac{2\xi}{\varphi\psi}\right)^{\frac{(\varphi+2)-4}{\varphi+2}} =$$

$$\left(\frac{2\xi}{\varphi\psi}\right)^{-\frac{4}{\varphi+2}}\left[6\xi+\varphi\left(\varphi-1\right)\psi\frac{2\xi}{\varphi\psi}\right]=$$

$$\left(\frac{2\xi}{\varphi\psi}\right)^{-\frac{4}{\varphi+2}}\left[6\xi+\left(\varphi-1\right)2\xi\right]=2\xi\left(\frac{2\xi}{\varphi\psi}\right)^{-\frac{4}{\varphi+2}}\left(\varphi+2\right)>0. \tag{5.50}$$

Therefore γ has a global minimum at $h_0=\left(\frac{2\xi}{\varphi\psi}\right)^{\frac{1}{\varphi+2}}$, which is

$$\gamma\left(h_0\right)=\xi\left(\frac{2\xi}{\varphi\psi}\right)^{-\frac{2}{\varphi+2}}+\psi\left(\frac{2\xi}{\varphi\psi}\right)^{\frac{\varphi+2-2}{\varphi+2}}=$$

$$\left(\frac{2\xi}{\varphi\psi}\right)^{-\frac{2}{\varphi+2}}\left[\xi+\psi\frac{2\xi}{\varphi\psi}\right]=\xi\left(\frac{2\xi}{\varphi\psi}\right)^{-\frac{2}{\varphi+2}}\left[1+\frac{2}{\varphi}\right]=$$

$$\frac{\xi}{\varphi}\left(\frac{2\xi}{\varphi\psi}\right)^{-\frac{2}{\varphi+2}}\left(\varphi+2\right)=\frac{\xi}{\varphi}\left(\frac{\xi}{\varphi}\right)^{-\frac{2}{\varphi+2}}\left(\frac{2}{\psi}\right)^{-\frac{2}{\varphi+2}}\left(\varphi+2\right)= \tag{5.51}$$

$$\left(\frac{\xi}{\varphi}\right)^{\frac{\varphi}{\varphi+2}}\left(\frac{\psi}{2}\right)^{\frac{2}{\varphi+2}}\left(\varphi+2\right).$$

That is

$$\gamma\left(h_0\right)=\left(\frac{\xi}{\varphi}\right)^{\frac{\varphi}{\varphi+2}}\left(\frac{\psi}{2}\right)^{\frac{2}{\varphi+2}}\left(\varphi+2\right). \tag{5.52}$$

Consequently

$$\gamma\left(h_0\right)=\left(\frac{4\left\|\|f\|\right\|_{\infty,\mathbb{R}}}{\nu-2}\right)^{\frac{\nu-2}{\nu}}\left(\frac{K}{\Gamma\left(\nu+1\right)}\right)^{\frac{2}{\nu}}\nu. \tag{5.53}$$

We have proved that

$$\left\|\|f''\|\right\|_{\infty,\mathbb{R}}\leq\left(\frac{4\left\|\|f\|\right\|_{\infty,\mathbb{R}}}{\nu-2}\right)^{\frac{\nu-2}{\nu}}\left(\frac{K}{\Gamma\left(\nu+1\right)}\right)^{\frac{2}{\nu}}\nu. \tag{5.54}$$

The theorem is proved. ■

We also give an L_p analog of a fractional Landau inequality

Theorem 5.5 *Let* $p,q>1:\frac{1}{p}+\frac{1}{q}=1,\ \ 2<\nu\leq3$ *and* $f\in C^3\left(\mathbb{R},X\right)$, *where* $\left(X,\|\cdot\|\right)$ *is a Banach space. We assume that* $\left\|\|f\|\right\|_{\infty,\mathbb{R}}<\infty$, *and that*

$$M:=\max\left\{\sup_{a\in\mathbb{R}}\left\|\left\|D_{*a}^{\nu}f\right\|\right\|_{p,\mathbb{R}},\ \sup_{a\in\mathbb{R}}\left\|\left\|D_{a-}^{\nu}f\right\|\right\|_{p,\mathbb{R}}\right\}<\infty. \tag{5.55}$$

Then

(1)

$$\left\| \left\| f' \right\| \right\|_{\infty,\mathbb{R}} \leq \left(\nu - \frac{1}{p} \right) \left(\frac{M}{\Gamma(\nu) \left(q(\nu-1)+1 \right)^{\frac{1}{q}}} \right)^{\frac{1}{\left(\nu-\frac{1}{p}\right)}} \left(\frac{\left\| \left\| f \right\| \right\|_{\infty,\mathbb{R}}}{\nu - 1 - \frac{1}{p}} \right)^{\left(\frac{\nu-1-\frac{1}{p}}{\nu-\frac{1}{p}} \right)}, \tag{5.56}$$

and

(2) under the additional assumption $2 + \frac{1}{p} < \nu \leq 3$, *we have*

$$\left\| \left\| f'' \right\| \right\|_{\infty,\mathbb{R}} \leq \left(\nu - \frac{1}{p} \right) \left(\frac{M}{\Gamma(\nu) \left(q(\nu-1)+1 \right)^{\frac{1}{q}}} \right)^{\left(\frac{2}{\nu-\frac{1}{p}}\right)} \left(\frac{4 \left\| \left\| f \right\| \right\|_{\infty,\mathbb{R}}}{\nu - 2 - \frac{1}{p}} \right)^{\left(\frac{\nu-2-\frac{1}{p}}{\nu-\frac{1}{p}} \right)}. \tag{5.57}$$

That is $\left\| \left\| f' \right\| \right\|_{\infty,\mathbb{R}}, \left\| \left\| f'' \right\| \right\|_{\infty,\mathbb{R}} < \infty$.

Proof From (5.27) we get

$$\left\| f'(a) \right\| \leq \frac{\left\| f(a+h) - f(a-h) \right\|}{2h} +$$

$$\frac{1}{2h\Gamma(\nu)} \left[\int_a^{a+h} (a+h-z)^{\nu-1} \left\| \left(D_{*a}^{\nu} f \right)(z) \right\| dz + \right.$$

$$\left. \int_{a-h}^{a} (z-(a-h))^{\nu-1} \left\| \left(D_{a-}^{\nu} f \right)(z) \right\| dz \right] \leq$$

$$\frac{\left\| \left\| f \right\| \right\|_{\infty,\mathbb{R}}}{h} + \frac{1}{2h\Gamma(\nu)} \left[\left(\int_a^{a+h} (a+h-z)^{q(\nu-1)} dz \right)^{\frac{1}{q}} \left(\int_a^{a+h} \left\| \left(D_{*a}^{\nu} f \right)(z) \right\|^p dz \right)^{\frac{1}{p}} \right.$$

$$\left. + \left(\int_{a-h}^{a} (z-(a-h))^{q(\nu-1)} dz \right)^{\frac{1}{q}} \left(\int_{a-h}^{a} \left\| \left(D_{a-}^{\nu} f \right)(z) \right\|^p dz \right)^{\frac{1}{p}} \right] \leq \tag{5.58}$$

$$\frac{\left\| \left\| f \right\| \right\|_{\infty,\mathbb{R}}}{h} + \frac{1}{2h\Gamma(\nu)} \frac{h^{\nu-\frac{1}{p}}}{\left(q(\nu-1)+1 \right)^{\frac{1}{q}}}$$

$$\left(\sup_{a\in\mathbb{R}} \left\| \left\| D_{*a}^{\nu} f \right\| \right\|_{p,\mathbb{R}} + \sup_{a\in\mathbb{R}} \left\| \left\| D_{a-}^{\nu} f \right\| \right\|_{p,\mathbb{R}} \right) \leq$$

$$\frac{\left\| \left\| f \right\| \right\|_{\infty,\mathbb{R}}}{h} + \frac{M h^{\left(\nu-1-\frac{1}{p}\right)}}{\Gamma(\nu) \left(q(\nu-1)+1 \right)^{\frac{1}{q}}}. \tag{5.59}$$

That is

$$\left\| f'(a) \right\| \leq \frac{\left\| \| f \| \right\|_{\infty,\mathbb{R}}}{h} + \frac{M}{\Gamma(\nu)(q(\nu-1)+1)^{\frac{1}{q}}} h^{\left(\nu-1-\frac{1}{p}\right)}, \tag{5.60}$$

$\forall\, a \in \mathbb{R}, \forall\, h > 0.$

I.e. it holds

$$\left\| \| f' \| \right\|_{\infty,\mathbb{R}} \leq \frac{\left\| \| f \| \right\|_{\infty,\mathbb{R}}}{h} + \frac{M}{\Gamma(\nu)(q(\nu-1)+1)^{\frac{1}{q}}} h^{\left(\nu-1-\frac{1}{p}\right)}, \tag{5.61}$$

$\forall\, h > 0, 2 < \nu \leq 3.$

Call

$$\begin{aligned} \mu &:= \left\| \| f \| \right\|_{\infty,\mathbb{R}}, \\ \theta &= \frac{M}{\Gamma(\nu)(q(\nu-1)+1)^{\frac{1}{q}}}, \end{aligned} \tag{5.62}$$

both are greater than zero.

Set also $\rho := \nu - 1 - \frac{1}{p} > \frac{1}{q} > 0.$

We consider the function

$$y(h) := \frac{\mu}{h} + \theta h^{\rho}, \ \forall\, h > 0. \tag{5.63}$$

As in the proof of Theorem 5.4 it has only one critical number

$$h_0 := h_{crit.no} = \left(\frac{\mu}{\rho\theta} \right)^{\frac{1}{\rho+1}}, \tag{5.64}$$

and a global minimum

$$y(h_0) = \theta^{\frac{1}{\rho+1}} \mu^{\frac{\rho}{\rho+1}} (\rho+1) \rho^{-\frac{\rho}{\rho+1}}. \tag{5.65}$$

Consequently

$$y(h_0) = \left(\frac{M}{\Gamma(\nu)(q(\nu-1)+1)^{\frac{1}{q}}} \right)^{\frac{1}{\left(\nu-\frac{1}{p}\right)}} \left(\left\| \| f \| \right\|_{\infty,\mathbb{R}} \right)^{\left(\frac{\nu-1-\frac{1}{p}}{\nu-\frac{1}{p}} \right)}$$

$$\left(\nu - \frac{1}{p} \right) \left(\nu - 1 - \frac{1}{p} \right)^{-\left(\frac{\nu-1-\frac{1}{p}}{\nu-\frac{1}{p}} \right)}. \tag{5.66}$$

We have proved that

$$\left\| \left\| f' \right\| \right\|_{\infty,\mathbb{R}} \leq \left(\frac{M}{\Gamma(\nu)(q(\nu-1)+1)^{\frac{1}{q}}} \right)^{\frac{1}{\left(\nu-\frac{1}{p}\right)}} \left(\frac{\| \| f \| \|_{\infty,\mathbb{R}}}{\nu-1-\frac{1}{p}} \right)^{\left(\frac{\nu-1-\frac{1}{p}}{\nu-\frac{1}{p}}\right)} \left(\nu - \frac{1}{p} \right). \tag{5.67}$$

From (5.28) we derive

$$\left\| f''(a) \right\| \leq \frac{4 \| \| f \| \|_{\infty,\mathbb{R}}}{h^2} + \frac{1}{h^2 \Gamma(\nu)} \left[\int_a^{a+h} (a+h-z)^{\nu-1} \left\| \left(D_{*a}^{\nu} f \right)(z) \right\| dz \right. \tag{5.68}$$

$$\left. + \int_{a-h}^{a} (z-(a-h))^{\nu-1} \left\| \left(D_{a-}^{\nu} f \right)(z) \right\| dz \right] \leq$$

$$\frac{4 \| \| f \| \|_{\infty,\mathbb{R}}}{h^2} + \frac{1}{h^2 \Gamma(\nu)} \left[\left(\int_a^{a+h} (a+h-z)^{q(\nu-1)} dz \right)^{\frac{1}{q}} \sup_{a \in \mathbb{R}} \left\| \left\| D_{*a}^{\nu} f \right\| \right\|_{p,\mathbb{R}} \right.$$

$$\left. + \left(\int_{a-h}^{a} (z-(a-h))^{q(\nu-1)} dz \right)^{\frac{1}{q}} \sup_{a \in \mathbb{R}} \left\| \left\| D_{a-}^{\nu} f \right\| \right\|_{p,\mathbb{R}} \right] \leq$$

$$\frac{4 \| \| f \| \|_{\infty,\mathbb{R}}}{h^2} + \frac{M}{h^2 \Gamma(\nu)} \left(\frac{2h^{\nu-\frac{1}{p}}}{(q(\nu-1)+1)^{\frac{1}{q}}} \right) =$$

$$\frac{4 \| \| f \| \|_{\infty,\mathbb{R}}}{h^2} + \frac{2Mh^{\nu-2-\frac{1}{p}}}{\Gamma(\nu)(q(\nu-1)+1)^{\frac{1}{q}}}. \tag{5.69}$$

That is

$$\left\| f''(a) \right\| \leq \frac{4 \| \| f \| \|_{\infty,\mathbb{R}}}{h^2} + \left(\frac{2M}{\Gamma(\nu)(q(\nu-1)+1)^{\frac{1}{q}}} \right) h^{\nu-2-\frac{1}{p}}, \tag{5.70}$$

$\forall\, a \in \mathbb{R}$, $\forall\, h > 0$.

I.e. it holds

$$\left\| \left\| f'' \right\| \right\|_{\infty,\mathbb{R}} \leq \frac{4 \| \| f \| \|_{\infty,\mathbb{R}}}{h^2} + \left(\frac{2M}{\Gamma(\nu)(q(\nu-1)+1)^{\frac{1}{q}}} \right) h^{\nu-2-\frac{1}{p}}, \tag{5.71}$$

$\forall\, h > 0$, under $2 + \frac{1}{p} < \nu \leq 3$.

Call

$$\xi := 4 \| \| f \| \|_{\infty,\mathbb{R}},$$

$$\psi := \frac{2M}{\Gamma(\nu)(q(\nu-1)+1)^{\frac{1}{q}}}, \tag{5.72}$$

both are greater than zero.

Set also $\varphi := \nu - 2 - \frac{1}{p} > 0$.

We consider the function

$$\gamma(h) := \frac{\xi}{h^2} + \psi h^{\varphi}, \ \forall\, h > 0. \tag{5.73}$$

We have as in the proof of Theorem 5.4 that γ has a global minimum at

$$h_0 = \left(\frac{2\xi}{\varphi\psi}\right)^{\frac{1}{\varphi+2}}, \tag{5.74}$$

which is

$$\gamma(h_0) = \left(\frac{\xi}{\varphi}\right)^{\frac{\varphi}{\varphi+2}} \left(\frac{\psi}{2}\right)^{\frac{2}{\varphi+2}} (\varphi + 2). \tag{5.75}$$

Consequently

$$\gamma(h_0) = \left(\frac{4\,\|\|f\|\|_{\infty,\mathbb{R}}}{\nu - 2 - \frac{1}{p}}\right)^{\left(\frac{\nu-2-\frac{1}{p}}{\nu-\frac{1}{p}}\right)} \left(\frac{M}{\Gamma(\nu)\,(q(\nu-1)+1)^{\frac{1}{q}}}\right)^{\left(\frac{2}{\nu-\frac{1}{p}}\right)} \left(\nu - \frac{1}{p}\right). \tag{5.76}$$

We have proved that

$$\left\|\left\|f''\right\|\right\|_{\infty,\mathbb{R}} \le \left(\frac{4\,\|\|f\|\|_{\infty,\mathbb{R}}}{\nu - 2 - \frac{1}{p}}\right)^{\left(\frac{\nu-2-\frac{1}{p}}{\nu-\frac{1}{p}}\right)} \left(\frac{M}{\Gamma(\nu)\,(q(\nu-1)+1)^{\frac{1}{q}}}\right)^{\left(\frac{2}{\nu-\frac{1}{p}}\right)} \left(\nu - \frac{1}{p}\right). \tag{5.77}$$

The theorem is proved. ■

We give

Corollary 5.6 (case of $\nu = 2.5$) *Let $f \in C^3(\mathbb{R}, X)$, where $(X, \|\cdot\|)$ is a Banach space. We assume that $\|\|f\|\|_{\infty,\mathbb{R}} < \infty$, and that*

$$K_{2.5} := \max\left\{\left\|\left\|D_{*a}^{2.5} f(z)\right\|\right\|_{\infty,\mathbb{R}^2}, \left\|\left\|D_{a-}^{2.5} f(z)\right\|\right\|_{\infty,\mathbb{R}^2}\right\} < \infty, \tag{5.78}$$

where $(a, z) \in \mathbb{R}^2$.

Then

$$\left\|\left\|f'\right\|\right\|_{\infty,\mathbb{R}} \le 1.21136\,(K_{2.5})^{0.4}\left(\|\|f\|\|_{\infty,\mathbb{R}}\right)^{0.6}, \tag{5.79}$$

and

$$\left\|\left\|f''\right\|\right\|_{\infty,\mathbb{R}} \le 1.44713\,(K_{2.5})^{0.8}\left(\|\|f\|\|_{\infty,\mathbb{R}}\right)^{0.2}. \tag{5.80}$$

That is $\left\|\left\|f'\right\|\right\|_{\infty,\mathbb{R}}, \left\|\left\|f''\right\|\right\|_{\infty,\mathbb{R}} < \infty$.

Proof By Theorem 5.4. ∎

We finish with

Corollary 5.7 (case of $\nu = 2.5$, $p = q = 2$) *Let* $f \in C^3(\mathbb{R}, X)$, *where* $(X, \|\cdot\|)$ *is a Banach space. We assume that* $\|\|f\|\|_{\infty,\mathbb{R}} < \infty$, *and that*

$$M_{2.5} := \max\left\{\sup_{a\in\mathbb{R}} \left\|\left\|D_{*a}^{2.5} f\right\|\right\|_{2,\mathbb{R}}, \sup_{a\in\mathbb{R}} \left\|\left\|D_{a-}^{2.5} f\right\|\right\|_{2,\mathbb{R}}\right\} < \infty. \tag{5.81}$$

Then

$$\left\|\left\|f'\right\|\right\|_{\infty,\mathbb{R}} \leq 1.226583057\sqrt{M_{2.5} \, \|\|f\|\|_{\infty,\mathbb{R}}}. \tag{5.82}$$

That is $\left\|\left\|f'\right\|\right\|_{\infty,\mathbb{R}} < \infty$.

Proof By Theorem 5.5. ∎

References

1. Aglic Aljinovic, A., Marangunic, LJ., Pecaric, J.: On Landau type inequalities via Ostrowski inequalities. Nonlinear Funct. Anal. Appl. **10**(4), pp. 565–579 (2005)
2. Anastassiou, G.: Fractional Differentiation Inequalities. Research Monograph. Springer, New York (2009)
3. Anastassiou, G.A.: Advances on Fractional Inequalities. Springer, New York (2011)
4. Anastassiou, G.A.: Intelligent Computations: Abstract Fractional Calculus, Inequalities, Approximations. Springer, Heidelberg, New York (2018)
5. Anastassiou, G.A.: Abstract Fractional Landau Inequalities. Ann. Univ. Sci. Budapest, Sect. Comp. (2020)
6. Barnett, N.S., Dragomir, S.S.: Some Landau type inequalities for functions whose derivatives are of locally bounded variation. Tamkang J. Math. **37**(4), 301–308 (2006)
7. Ditzian, Z.: Remarks, questions and conjectures on Landau–Kolmogorov-type inequalities. Math. Inequal. Appl. **3**, 15–24 (2000)
8. Hardy, G.H., Littlewood, J.E.: Some integral inequalities connected with the calculus of variations. Quart. J. Math. Oxford Ser. **3**, 241–252 (1932)
9. Hardy, G.H., Landau, E., Littlewood, J.E.: Some inequalities satisfied by the integrals or derivatives of real or analytic functions. Math. Z. **39**, 677–695 (1935)
10. Kallman R.R., Rota, G.C.: On the inequality $\|f'\|^2 \leq 4\|f\| \cdot \|f''\|$. In: O. Shisha, (ed.) Inequalities, vol. II, pp. 187–192. Academic Press, New York (1970)
11. Kraljevic, H., Pecaric, J.: Some Landau's type inequalities for infinitesimal generators. Aequationes Mathematicae **40**, 147–153 (1990)
12. Landau, E.: Einige Ungleichungen für zweimal differentzierban funktionen. Proc. London Math. Soc. **13**, 43–49 (1913)
13. Mikusinski, J.: The Bochner Integral. Academic Press, New York (1978)
14. Shilov, G.E.: Elementary Functional Analysis. Dover Publications Inc., New York (1996)
15. Xiao, Y.: Landau type inequalities for Banach space valued functions. J. Math. Inequal. **7**(1), 103–114 (2013)

Chapter 6
Abstract Fractional Landau Inequalities

We present uniform and L_p mixed Caputo-Bochner abstract generalized fractional Landau inequalities over $\mathbb{R}$ of fractional orders $2 < \alpha \leq 3$. These estimate the size of first and second derivatives of a composition with a Banach space valued function over $\mathbb{R}$. We give applications when $\alpha = 2.5$. It follows [5].

6.1 Introduction

Let $p \in [1, \infty]$, $I = \mathbb{R}_+$ or $I = \mathbb{R}$ and $f : I \to \mathbb{R}$ is twice differentiable with $f, f'' \in L_p(I)$, then $f' \in L_p(I)$. Moreover, there exists a constant $C_p(I) > 0$ independent of f, such that

$$\left\| f' \right\|_{p,I} \leq C_p(I) \left\| f \right\|_{p,I}^{\frac{1}{2}} \left\| f'' \right\|_{p,I}^{\frac{1}{2}}, \tag{6.1}$$

where $\|\cdot\|_{p,I}$ is the p-norm on the interval I, see [1, 6].

The research on these inequalities started by Landau [11] in 1913. For the case of $p = \infty$ he proved that

$$C_\infty(\mathbb{R}_+) = 2 \text{ and } C_\infty(\mathbb{R}) = \sqrt{2}, \tag{6.2}$$

are the best constants in (6.1).

In 1932, Hardy and Littlewood [8] proved (6.1) for $p = 2$, with the best constants

$$C_2(\mathbb{R}_+) = \sqrt{2} \text{ and } C_2(\mathbb{R}) = 1. \tag{6.3}$$

In 1935, Hardy, Landau and Littlewood [9] showed that the best constants $C_p(\mathbb{R}_+)$ in (6.1) satisfies the estimate

G. A. Anastassiou, *Constructive Fractional Analysis with Applications*,
Studies in Systems, Decision and Control 362,
https://doi.org/10.1007/978-3-030-71481-9_6

$$C_p(\mathbb{R}_+) \leq 2, \text{ for } p \in [1, \infty), \tag{6.4}$$

which yields $C_p(\mathbb{R}) \leq 2$ for $p \in [1, \infty)$.

In fact, in [7, 10] was shown that $C_p(\mathbb{R}) \leq \sqrt{2}$.

We need the following concepts from abstract generalized fractional calculus.

Our integrals next are of Bochner type [12].

We need

Definition 6.1 ([4], *p. 104*) Let $[a, b] \subset \mathbb{R}$, $(X, \|\cdot\|)$ a Banach space, $g \in C^1([a, b])$ and increasing, $f \in C([a, b], X)$, $\nu > 0$.

We define the left Riemann-Liouville generalized fractional Bochner integral operator

$$\left(J_{a;g}^{\nu} f\right)(x) := \frac{1}{\Gamma(\nu)} \int_a^x (g(x) - g(z))^{\nu-1} g'(z) f(z) \, dz, \tag{6.5}$$

$\forall\, x \in [a, b]$, where Γ is the gamma function.

The last integral is of Bochner type. Since $f \in C([a, b], X)$, then $f \in L_{\infty}([a, b], X)$. By Theorem 4.10, p. 98, [4], we get that $\left(J_{a;g}^{\nu} f\right) \in C([a, b], X)$. Above we set $J_{a;g}^{0} f := f$ and see that $\left(J_{a;g}^{\nu} f\right)(a) = 0$.

We need

Definition 6.2 ([4], *p. 105*) Let $[a, b] \subset \mathbb{R}$, $(X, \|\cdot\|)$ a Banach space, $g \in C^1([a, b])$ and increasing, $f \in C([a, b], X)$, $\nu > 0$.

We define the right Riemann-Liouville generalized fractional Bochner integral operator

$$\left(J_{b-;g}^{\nu} f\right)(x) := \frac{1}{\Gamma(\nu)} \int_x^b (g(z) - g(x))^{\nu-1} g'(z) f(z) \, dz, \tag{6.6}$$

$\forall\, x \in [a, b]$, where Γ is the gamma function.

The last integral is of Bochner type. Since $f \in C([a, b], X)$, then $f \in L_{\infty}([a, b], X)$. By Theorem 4.11, p. 101, [4], we get that $\left(J_{b-;g}^{\nu} f\right) \in C([a, b], X)$. Above we set $J_{b-;g}^{0} f := f$ and see that $\left(J_{b-;g}^{\nu} f\right)(b) = 0$.

We also need

Definition 6.3 ([4], *p. 106*) Let $\alpha > 0$, $\lceil \alpha \rceil = n$, $\lceil \cdot \rceil$ the ceiling of the number. Let $f \in C^n([a, b], X)$, where $[a, b] \subset \mathbb{R}$, and $(X, \|\cdot\|)$ is a Banach space. Let $g \in C^1([a, b])$, strictly increasing, such that $g^{-1} \in C^n([g(a), g(b)])$.

We define the left generalized g-fractional derivative X-valued of f of order α as follows:

$$\left(D_{a+;g}^{\alpha}f\right)(x) := \frac{1}{\Gamma(n-\alpha)} \int_{a}^{x} (g(x)-g(t))^{n-\alpha-1} g'(t) \left(f \circ g^{-1}\right)^{(n)} (g(t))\, dt, \tag{6.7}$$

$\forall\, x \in [a,b]$. The last integral is of Bochner type.

Ordinary vector valued derivative is as in [13], similar to numerical one.

If $\alpha \notin \mathbb{N}$, by Theorem 4.10, p. 98, [4], we have that $\left(D_{a+;g}^{\alpha}f\right) \in C([a,b], X)$.

We see that

$$\left(J_{a;g}^{n-\alpha}\left(\left(f \circ g^{-1}\right)^{(n)} \circ g\right)\right)(x) = \left(D_{a+;g}^{\alpha}f\right)(x), \quad \forall\, x \in [a,b]. \tag{6.8}$$

We set

$$D_{a+;g}^{n}f(x) := \left(\left(f \circ g^{-1}\right)^{n} \circ g\right)(x) \in C([a,b], X), \quad n \in \mathbb{N}, \tag{6.9}$$

$$D_{a+;g}^{0}f(x) = f(x), \quad \forall\, x \in [a,b].$$

When $g = id$, then

$$D_{a+;g}^{\alpha}f = D_{a+;id}^{\alpha}f = D_{*a}^{\alpha}f, \tag{6.10}$$

the usual left X-valued Caputo fractional derivative, see [4], Chap. 1

We mention

Definition 6.4 ([4], *p. 107*) Let $\alpha > 0$, $\lceil \alpha \rceil = n$, $\lceil \cdot \rceil$ the ceiling of the number. Let $f \in C^{n}([a,b], X)$, where $[a,b] \subset \mathbb{R}$, and $(X, \|\cdot\|)$ is a Banach space. Let $g \in C^{1}([a,b])$, strictly increasing, such that $g^{-1} \in C^{n}([g(a), g(b)])$.

We define the right generalized g-fractional derivative X-valued of f of order α as follows:

$$\left(D_{b-;g}^{\alpha}f\right)(x) := \frac{(-1)^{n}}{\Gamma(n-\alpha)} \int_{x}^{b} (g(t)-g(x))^{n-\alpha-1} g'(t) \left(f \circ g^{-1}\right)^{(n)} (g(t))\, dt, \tag{6.11}$$

$\forall\, x \in [a,b]$. The last integral is of Bochner type.

If $\alpha \notin \mathbb{N}$, by Theorem 4.11, p. 101, [4], we have that $\left(D_{b-;g}^{\alpha}f\right) \in C([a,b], X)$.

We see that

$$J_{b-;g}^{n-\alpha}\left((-1)^{n}\left(f \circ g^{-1}\right)^{(n)} \circ g\right)(x) = \left(D_{b-;g}^{\alpha}f\right)(x), \quad a \leq x \leq b. \tag{6.12}$$

We set

$$D_{b-;g}^{n}f(x) := (-1)^{n}\left(\left(f \circ g^{-1}\right)^{n} \circ g\right)(x) \in C([a,b], X), \quad n \in \mathbb{N}, \tag{6.13}$$

$$D_{b-;g}^{0}f(x) := f(x), \quad \forall\, x \in [a,b].$$

When $g = id$, then

$$D_{b-;g}^{\alpha} f(x) = D_{b-;id}^{\alpha} f(x) = D_{b-}^{\alpha} f, \tag{6.14}$$

the usual right X-valued Caputo fractional derivative, see [4], Chap. 2.

We mention the generalized left fractional Taylor formula:

Theorem 6.5 ([4], p. 107) *Let $\alpha > 0$, $n = \lceil \alpha \rceil$, and $f \in C^n([a,b], X)$, where $[a,b] \subset \mathbb{R}$ and $(X, \|\cdot\|)$ is a Banach space. Let $g \in C^1([a,b])$, strictly increasing, such that $g^{-1} \in C^n([g(a), g(b)])$, $a \leq x \leq b$. Then*

$$f(x) = f(a) + \sum_{i=1}^{n-1} \frac{(g(x) - g(a))^i}{i!} \left(f \circ g^{-1}\right)^{(i)} (g(a)) +$$

$$\frac{1}{\Gamma(\alpha)} \int_a^x (g(x) - g(t))^{\alpha-1} g'(t) \left(D_{a+;g}^{\alpha} f\right)(t)\, dt =$$

$$f(a) + \sum_{i=1}^{n-1} \frac{(g(x) - g(a))^i}{i!} \left(f \circ g^{-1}\right)^{(i)} (g(a)) +$$

$$\frac{1}{\Gamma(\alpha)} \int_{g(a)}^{g(x)} (g(x) - z)^{\alpha-1} \left(\left(D_{a+;g}^{\alpha} f\right) \circ g^{-1}\right)(z)\, dz. \tag{6.15}$$

We also mention the generalized right fractional Taylor formula:

Theorem 6.6 ([4], p. 108) *Let $\alpha > 0$, $n = \lceil \alpha \rceil$, and $f \in C^n([a,b], X)$, where $[a,b] \subset \mathbb{R}$ and $(X, \|\cdot\|)$ is a Banach space. Let $g \in C^1([a,b])$, strictly increasing, such that $g^{-1} \in C^n([g(a), g(b)])$, $a \leq x \leq b$. Then*

$$f(x) = f(b) + \sum_{i=1}^{n-1} \frac{(g(x) - g(b))^i}{i!} \left(f \circ g^{-1}\right)^{(i)} (g(b)) +$$

$$\frac{1}{\Gamma(\alpha)} \int_x^b (g(t) - g(x))^{\alpha-1} g'(t) \left(D_{b-;g}^{\alpha} f\right)(t)\, dt =$$

$$f(b) + \sum_{i=1}^{n-1} \frac{(g(x) - g(b))^i}{i!} \left(f \circ g^{-1}\right)^{(i)} (g(b)) +$$

$$\frac{1}{\Gamma(\alpha)} \int_{g(x)}^{g(b)} (z - g(x))^{\alpha-1} \left(\left(D_{b-;g}^{\alpha} f\right) \circ g^{-1}\right)(z)\, dz. \tag{6.16}$$

By convention we suppose that

$$\left(D^{\alpha}_{x_0+;g}f\right)(x)=0,\ \text{for } x<x_0,$$
$$\left(D^{\alpha}_{x_0-;g}f\right)(x)=0,\ \text{for } x>x_0, \tag{6.17}$$

for any $x, x_0 \in [a, b]$.

The author has already done an extensive amount of work on fractional Landau inequalities, see [3], and on abstract fractional Landau inequalities, see [4]. However there the proving methods came out of applications of fractional Ostrowski inequalities [2, 4] and the derived inequalities were for small fractional orders, i.e. $\alpha \in (0, 1)$. Usually there the domains where $[A, +\infty)$ or $(-\infty, B]$, with $A, B \in \mathbb{R}$ and in one mixed case the domain was all of $\mathbb{R}$.

In this chapter with less assumptions we establish uniform and L_p type mixed Caputo–Bochner abstract generalized fractional Landau inequalities over $\mathbb{R}$ for fractional orders $2 < \alpha \leq 3$. The method of proving is based on left and right Caputo–Bochner generalized fractional Taylor's formulae with integral remainder, see Theorems 6.5, 6.6.

We give also an applications for $\alpha = 2.5$. Certainly we are also inspired by [3, 4].

6.2 Main Results

We give the following abstract mixed generalized fractional Landau inequalities over $\mathbb{R}$.

Theorem 6.7 *Let $2 < \alpha \leq 3$ and $f \in C^3(\mathbb{R}, X)$, where $(X, \|\cdot\|)$ is a Banach space. Let $g \in C^1(\mathbb{R})$, strictly increasing, such that $g^{-1} \in C^3(g(\mathbb{R}))$. We assume that $\|\|f\|\|_{\infty,\mathbb{R}} < \infty$ and that*

$$K := \max\left\{\left\|\left\|\left(\left(D^{\alpha}_{a+;g}f\right)\circ g^{-1}\right)(z)\right\|\right\|_{\infty,\mathbb{R}\times g(\mathbb{R})},\right.$$
$$\left.\left\|\left\|\left(\left(D^{\alpha}_{a-;g}f\right)\circ g^{-1}\right)(z)\right\|\right\|_{\infty,\mathbb{R}\times g(\mathbb{R})}\right\} < \infty, \tag{6.18}$$

where $(a, z) \in \mathbb{R} \times g(\mathbb{R})$.

Then

$$\left\|\left\|\left(f\circ g^{-1}\right)'\circ g\right\|\right\|_{\infty,\mathbb{R}} \leq \alpha\left(\frac{K}{\Gamma(\alpha+1)}\right)^{\frac{1}{\alpha}}\left(\frac{\|\|f\|\|_{\infty,\mathbb{R}}}{\alpha-1}\right)^{\frac{\alpha-1}{\alpha}}, \tag{6.19}$$

and

$$\left\|\left\|\left(f\circ g^{-1}\right)''\circ g\right\|\right\|_{\infty,\mathbb{R}} \leq \alpha\left(\frac{K}{\Gamma(\alpha+1)}\right)^{\frac{2}{\alpha}}\left(\frac{4\,\|\|f\|\|_{\infty,\mathbb{R}}}{\alpha-2}\right)^{\frac{\alpha-2}{\alpha}}. \tag{6.20}$$

That is $\left\| \left\| \left(f \circ g^{-1} \right)' \circ g \right\| \right\|_{\infty,\mathbb{R}}, \left\| \left\| \left(f \circ g^{-1} \right)'' \circ g \right\| \right\|_{\infty,\mathbb{R}} < \infty.$

Proof Here $2 < \alpha \leq 3$, i.e. $\lceil \alpha \rceil = 3$. Let $f \in C^3(\mathbb{R}, X)$, where $(X, \|\cdot\|)$ is a Banach space, $a \in \mathbb{R}$ is fixed momentarily. We need the following abstract generalized fractional Taylor formulae for $n = 3$.

By Theorem 6.5 we get

$$f(x) - f(a) = (g(x) - g(a)) \left(f \circ g^{-1} \right)' (g(a)) + \frac{(g(x) - g(a))^2}{2} \left(f \circ g^{-1} \right)'' (g(a)) \tag{6.21}$$
$$+ \frac{1}{\Gamma(\alpha)} \int_{g(a)}^{g(x)} (g(x) - z)^{\alpha-1} \left(\left(D^{\alpha}_{a+;g} f \right) \circ g^{-1} \right)(z)\, dz, \quad \forall\, x \geq a.$$

And by Theorem 6.6 we get

$$f(x) - f(a) = (g(x) - g(a)) \left(f \circ g^{-1} \right)' (g(a)) + \frac{(g(x) - g(a))^2}{2} \left(f \circ g^{-1} \right)'' (g(a)) \tag{6.22}$$
$$+ \frac{1}{\Gamma(\alpha)} \int_{g(x)}^{g(a)} (z - g(x))^{\alpha-1} \left(\left(D^{\alpha}_{a-;g} f \right) \circ g^{-1} \right)(z)\, dz, \quad \forall\, x \leq a.$$

Let $x_1 > a$, then

$$(g(x_1) - g(a)) \left(f \circ g^{-1} \right)' (g(a)) + \frac{(g(x_1) - g(a))^2}{2} \left(f \circ g^{-1} \right)'' (g(a)) =$$

$$(f(x_1) - f(a)) - \frac{1}{\Gamma(\alpha)} \int_{g(a)}^{g(x_1)} (g(x_1) - z)^{\alpha-1} \left(\left(D^{\alpha}_{a+;g} f \right) \circ g^{-1} \right)(z)\, dz =: A, \tag{6.23}$$

and let $x_2 < a$, then

$$(g(x_2) - g(a)) \left(f \circ g^{-1} \right)' (g(a)) + \frac{(g(x_2) - g(a))^2}{2} \left(f \circ g^{-1} \right)'' (g(a))$$

$$= (f(x_2) - f(a)) - \frac{1}{\Gamma(\alpha)} \int_{g(x_2)}^{g(a)} (z - g(x_2))^{\alpha-1} \left(\left(D^{\alpha}_{a-;g} f \right) \circ g^{-1} \right)(z)\, dz =: B. \tag{6.24}$$

Let $h > 0$, we can choose x_1 such that $g(x_1) - g(a) = h$ and we can choose x_2 such that $g(a) - g(x_2) = h$. That is $g(x_1) = g(a) + h$ and $g(x_2) = g(a) - h$, and $g(x_2) - g(a) = -h$. Furthermore it holds $g(x_2) - g(x_1) = -2h$.

We can rewrite (6.23) as

$$h \left(f \circ g^{-1} \right)' (g(a)) + \frac{h^2}{2} \left(f \circ g^{-1} \right)'' (g(a)) = A, \tag{6.25}$$

and we can rewrite (6.24) as

$$-h\left(f\circ g^{-1}\right)'(g(a))+\frac{h^{2}}{2}\left(f\circ g^{-1}\right)''(g(a))=B. \tag{6.26}$$

Solving the system of (6.25) and (6.26) we find

$$\begin{array}{l}\left(f\circ g^{-1}\right)'(g(a))=\frac{A-B}{2h},\\ \text{and}\\ \left(f\circ g^{-1}\right)''(g(a))=\frac{A+B}{h^{2}}.\end{array} \tag{6.27}$$

We assumed that $\left\|\left\|\left(\left(D^{\alpha}_{a+;g}f\right)\circ g^{-1}\right)(z)\right\|\right\|_{\infty,\mathbb{R}\times g(\mathbb{R})}$, $\left\|\left\|\left(\left(D^{\alpha}_{a-;g}f\right)\circ g^{-1}\right)(z)\right\|\right\|_{\infty,\mathbb{R}\times g(\mathbb{R})}<\infty$.

We obtain

$$\begin{array}{l}\left\|\left(f\circ g^{-1}\right)'(g(a))\right\|=\frac{1}{2h}\left\|A-B\right\|,\\ \text{and}\\ \left\|\left(f\circ g^{-1}\right)''(g(a))\right\|\leq\frac{1}{h^{2}}\left(\left\|A\right\|+\left\|B\right\|\right).\end{array} \tag{6.28}$$

We get

$$\left\|A\right\|=\left\|\left(f(x_1)-f(a)\right)-\frac{1}{\Gamma(\alpha)}\int_{g(a)}^{g(x_1)}\left(g(x_1)-z\right)^{\alpha-1}\left(\left(D^{\alpha}_{a+;g}f\right)\circ g^{-1}\right)(z)\,dz\right\|$$

$$\leq 2\left\|\left\|f\right\|\right\|_{\infty,\mathbb{R}}+\frac{1}{\Gamma(\alpha)}\int_{g(a)}^{g(x_1)}\left(g(x_1)-z\right)^{\alpha-1}\left\|\left(\left(D^{\alpha}_{a+;g}f\right)\circ g^{-1}\right)(z)\right\|dz\leq$$

$$2\left\|\left\|f\right\|\right\|_{\infty,\mathbb{R}}+\frac{K}{\Gamma(\alpha)}\left(\int_{g(a)}^{g(x_1)}\left(g(x_1)-z\right)^{\alpha-1}dz\right)=$$

$$2\left\|\left\|f\right\|\right\|_{\infty,\mathbb{R}}+\frac{K}{\Gamma(\alpha+1)}\left(g(x_1)-g(a)\right)^{\alpha}=2\left\|\left\|f\right\|\right\|_{\infty,\mathbb{R}}+\frac{K}{\Gamma(\alpha+1)}h^{\alpha}. \tag{6.29}$$

That is

$$\left\|A\right\|\leq 2\left\|\left\|f\right\|\right\|_{\infty,\mathbb{R}}+\frac{K}{\Gamma(\alpha+1)}h^{\alpha},\quad h>0. \tag{6.30}$$

Similarly, it holds

$$\left\|B\right\|=\left\|\left(f(x_2)-f(a)\right)-\frac{1}{\Gamma(\alpha)}\int_{g(x_2)}^{g(a)}\left(z-g(x_2)\right)^{\alpha-1}\left(\left(D^{\alpha}_{a-;g}f\right)\circ g^{-1}\right)(z)\,dz\right\|$$

$$\leq 2 \|\|f\|\|_{\infty,\mathbb{R}} + \frac{1}{\Gamma(\alpha)} \int_{g(x_2)}^{g(a)} (z - g(x_2))^{\alpha-1} \left\| \left(\left(D_{a-;g}^{\alpha} f \right) \circ g^{-1} \right)(z) \right\| dz \leq$$

$$2 \|\|f\|\|_{\infty,\mathbb{R}} + \frac{K}{\Gamma(\alpha)} \left(\int_{g(x_2)}^{g(a)} (z - g(x_2))^{\alpha-1} dz \right) =$$

$$2 \|\|f\|\|_{\infty,\mathbb{R}} + \frac{K}{\Gamma(\alpha+1)} (g(a) - g(x_2))^{\alpha} = 2 \|\|f\|\|_{\infty,\mathbb{R}} + \frac{K}{\Gamma(\alpha+1)} h^{\alpha}. \tag{6.31}$$

That is

$$\|B\| \leq 2 \|\|f\|\|_{\infty,\mathbb{R}} + \frac{K}{\Gamma(\alpha+1)} h^{\alpha}, \quad h > 0. \tag{6.32}$$

Furthermore we have

$$\|A\| + \|B\| \leq 4 \|\|f\|\|_{\infty,\mathbb{R}} + \frac{2K}{\Gamma(\alpha+1)} h^{\alpha}, \quad h > 0. \tag{6.33}$$

We also notice that

$$\|A - B\| =$$

$$\left\| f(x_1) - f(a) - \frac{1}{\Gamma(\alpha)} \int_{g(a)}^{g(x_1)} (g(x_1) - z)^{\alpha-1} \left(\left(D_{a+;g}^{\alpha} f \right) \circ g^{-1} \right)(z) dz \right.$$

$$\left. - f(x_2) + f(a) + \frac{1}{\Gamma(\alpha)} \int_{g(x_2)}^{g(a)} (z - g(x_2))^{\alpha-1} \left(\left(D_{a-;g}^{\alpha} f \right) \circ g^{-1} \right)(z) dz \right\| \leq$$

$$\|f(x_1) - f(x_2)\| + \frac{1}{\Gamma(\alpha)} \left[\int_{g(a)}^{g(x_1)} (g(x_1) - z)^{\alpha-1} \left\| \left(\left(D_{a+;g}^{\alpha} f \right) \circ g^{-1} \right)(z) \right\| dz \right. \tag{6.34}$$

$$\left. + \int_{g(x_2)}^{g(a)} (z - g(x_2))^{\alpha-1} \left\| \left(\left(D_{a-;g}^{\alpha} f \right) \circ g^{-1} \right)(z) \right\| dz \right] \leq$$

$$2 \|\|f\|\|_{\infty,\mathbb{R}} + \frac{K}{\Gamma(\alpha)} \left[\int_{g(a)}^{g(x_1)} (g(x_1) - z)^{\alpha-1} dz + \int_{g(x_2)}^{g(a)} (z - g(x_2))^{\alpha-1} dz \right] =$$

$$2 \|\|f\|\|_{\infty,\mathbb{R}} + \frac{K}{\Gamma(\alpha+1)} \left[(g(x_1) - g(a))^{\alpha} + (g(a) - g(x_2))^{\alpha} \right] =$$

$$2 \|\|f\|\|_{\infty,\mathbb{R}} + \frac{2Kh^{\alpha}}{\Gamma(\alpha+1)}.$$

That is

$$\frac{\|A - B\|}{2} \leq \|\|f\|\|_{\infty,\mathbb{R}} + \frac{Kh^{\alpha}}{\Gamma(\alpha+1)}, \quad h > 0. \tag{6.35}$$

Consequently we obtain

$$\begin{aligned} &\left\| \left(f \circ g^{-1}\right)' (g(a)) \right\| \overset{((6.28.),\,(6.35.))}{\leq} \frac{\||f\||_{\infty,\mathbb{R}}}{h} + \frac{K h^{\alpha-1}}{\Gamma(\alpha+1)}, \\ &\text{and} \\ &\left\| \left(f \circ g^{-1}\right)'' (g(a)) \right\| \overset{((6.28.),\,(6.33.))}{\leq} \frac{4\||f\||_{\infty,\mathbb{R}}}{h^2} + \frac{2K h^{\alpha-2}}{\Gamma(\alpha+1)}, \end{aligned} \tag{6.36}$$

$h > 0$, for any $a \in \mathbb{R}$.

Hence

$$\begin{aligned} &\left\| \left\| \left(f \circ g^{-1}\right)' \circ g \right\| \right\|_{\infty,\mathbb{R}} \leq \frac{\||f\||_{\infty,\mathbb{R}}}{h} + \frac{K h^{\alpha-1}}{\Gamma(\alpha+1)}, \\ &\text{and} \\ &\left\| \left\| \left(f \circ g^{-1}\right)'' \circ g \right\| \right\|_{\infty,\mathbb{R}} \leq \frac{4\||f\||_{\infty,\mathbb{R}}}{h^2} + \frac{2K h^{\alpha-2}}{\Gamma(\alpha+1)}, \end{aligned} \tag{6.37}$$

true $\forall\, h > 0$, $2 < \alpha \leq 3$.

Call

$$\begin{aligned} \mu &:= \||f\||_{\infty,\mathbb{R}}, \\ \theta &= \frac{K}{\Gamma(\alpha+1)}, \end{aligned} \tag{6.38}$$

both are greater than zero.

Set also $\rho := \alpha - 1 > 1$.

We consider the function

$$y(h) := \frac{\mu}{h} + \theta h^{\rho}, \quad \forall\, h > 0. \tag{6.39}$$

We have

$$y'(h) = -\frac{\mu}{h^2} + \rho\theta h^{\rho-1} = 0,$$

then

$$\rho\theta h^{\rho+1} = \mu,$$

with a unique solution

$$h_0 := h_{crit.no} = \left(\frac{\mu}{\rho\theta}\right)^{\frac{1}{\rho+1}}. \tag{6.40}$$

We have that

$$y''(h) = 2\mu h^{-3} + \rho(\rho-1)\theta h^{\rho-2}. \tag{6.41}$$

We observe that

$$y''(h_0) = 2\mu \left(\frac{\mu}{\rho\theta}\right)^{-\frac{3}{\rho+1}} + \rho(\rho-1)\theta \left(\frac{\mu}{\rho\theta}\right)^{\frac{((\rho+1)-3)}{\rho+1}} =$$

$$\left(\frac{\mu}{\rho\theta}\right)^{-\frac{3}{\rho+1}}\left[2\mu+\mu\left(\rho-1\right)\right]=\mu\left(\frac{\mu}{\rho\theta}\right)^{-\frac{3}{\rho+1}}\left(\rho+1\right)>0. \tag{6.42}$$

Therefore y has a global minimum at $h_0=\left(\frac{\mu}{\rho\theta}\right)^{\frac{1}{\rho+1}}$, which is

$$y\left(h_0\right)=\frac{\mu}{\left(\frac{\mu}{\rho\theta}\right)^{\frac{1}{\rho+1}}}+\theta\left(\frac{\mu}{\rho\theta}\right)^{\frac{\rho}{\rho+1}}=$$

$$\left(\rho\theta\right)^{\frac{1}{\rho+1}}\frac{\mu}{\mu^{\frac{1}{\rho+1}}}+\frac{\theta\mu^{\frac{\rho}{\rho+1}}}{\rho^{\frac{\rho}{\rho+1}}\theta^{\frac{\rho}{\rho+1}}}=$$

$$\left(\theta\mu^{\rho}\right)^{\frac{1}{\rho+1}}\left(\rho^{\frac{1}{\rho+1}}+\frac{1}{\rho^{\frac{\rho}{\rho+1}}}\right)=\left(\theta\mu^{\rho}\right)^{\frac{1}{\rho+1}}\left(\frac{\rho+1}{\rho^{\frac{\rho}{\rho+1}}}\right)= \tag{6.43}$$

$$\left(\theta\mu^{\rho}\right)^{\frac{1}{\rho+1}}\left(\rho+1\right)\rho^{-\frac{\rho}{\rho+1}}.$$

That is

$$y\left(h_0\right)=\left(\theta\mu^{\rho}\right)^{\frac{1}{\rho+1}}\left(\rho+1\right)\rho^{-\frac{\rho}{\rho+1}}. \tag{6.44}$$

Consequently

$$y\left(h_0\right)=\left(\frac{K}{\Gamma\left(\alpha+1\right)}\left\|\left\|f\right\|\right\|_{\infty,\mathbb{R}}^{\alpha-1}\right)^{\frac{1}{\alpha}}\alpha\left(\alpha-1\right)^{-\left(\frac{\alpha-1}{\alpha}\right)}. \tag{6.45}$$

We have proved that

$$\left\|\left\|\left(f\circ g^{-1}\right)'\circ g\right\|\right\|_{\infty,\mathbb{R}}\leq\left(\frac{K}{\Gamma\left(\alpha+1\right)}\left\|\left\|f\right\|\right\|_{\infty,\mathbb{R}}^{\alpha-1}\right)^{\frac{1}{\alpha}}\alpha\left(\alpha-1\right)^{-\left(\frac{\alpha-1}{\alpha}\right)}. \tag{6.46}$$

Next call

$$\begin{aligned}\xi&:=4\left\|\left\|f\right\|\right\|_{\infty,\mathbb{R}},\\ \psi&=\tfrac{2K}{\Gamma(\alpha+1)},\end{aligned} \tag{6.47}$$

both are greater than zero.

Set also $\varphi:=\alpha-2>0$.

We consider the function

$$\gamma\left(h\right):=\frac{\xi}{h^2}+\psi h^{\varphi}=\xi h^{-2}+\psi h^{\varphi},\quad\forall\, h>0. \tag{6.48}$$

We have

$$\gamma'(h) = -2\xi h^{-3} + \varphi\psi h^{\varphi-1} = 0,$$

then

$$\varphi\psi h^{\varphi+2} = 2\xi,$$

with a unique solution

$$h_0 := h_{crit.no} = \left(\frac{2\xi}{\varphi\psi}\right)^{\frac{1}{\varphi+2}}. \tag{6.49}$$

We have that

$$\gamma''(h) = 6\xi h^{-4} + \varphi(\varphi-1)\psi h^{\varphi-2}. \tag{6.50}$$

We see that

$$\gamma''(h_0) = 6\xi\left(\frac{2\xi}{\varphi\psi}\right)^{-\frac{4}{\varphi+2}} + \varphi(\varphi-1)\psi\left(\frac{2\xi}{\varphi\psi}\right)^{\frac{(\varphi+2)-4}{\varphi+2}} =$$

$$\left(\frac{2\xi}{\varphi\psi}\right)^{-\frac{4}{\varphi+2}}[6\xi + (\varphi-1)2\xi] = 2\xi\left(\frac{2\xi}{\varphi\psi}\right)^{-\frac{4}{\varphi+2}}(\varphi+2) > 0. \tag{6.51}$$

Therefore γ has a global minimum at $h_0 = \left(\frac{2\xi}{\varphi\psi}\right)^{\frac{1}{\varphi+2}}$, which is

$$\gamma(h_0) = \xi\left(\frac{2\xi}{\varphi\psi}\right)^{-\frac{2}{\varphi+2}} + \psi\left(\frac{2\xi}{\varphi\psi}\right)^{\frac{\varphi+2-2}{\varphi+2}} =$$

$$\left(\frac{2\xi}{\varphi\psi}\right)^{-\frac{2}{\varphi+2}}\left[\xi + \psi\frac{2\xi}{\varphi\psi}\right] = \frac{\xi}{\varphi}\left(\frac{\xi}{\varphi}\right)^{-\frac{2}{\varphi+2}}\left(\frac{2}{\psi}\right)^{-\frac{2}{\varphi+2}}(\varphi+2) = \tag{6.52}$$

$$\left(\frac{\xi}{\varphi}\right)^{\frac{\varphi}{\varphi+2}}\left(\frac{\psi}{2}\right)^{\frac{2}{\varphi+2}}(\varphi+2).$$

That is

$$\gamma(h_0) = \left(\frac{\xi}{\varphi}\right)^{\frac{\varphi}{\varphi+2}}\left(\frac{\psi}{2}\right)^{\frac{2}{\varphi+2}}(\varphi+2). \tag{6.53}$$

Consequently

$$\gamma(h_0) = \left(\frac{4\,\|\|f\|\|_{\infty,\mathbb{R}}}{\alpha-2}\right)^{\frac{\alpha-2}{\alpha}}\left(\frac{K}{\Gamma(\alpha+1)}\right)^{\frac{2}{\alpha}}\alpha. \tag{6.54}$$

We have proved that

$$\left\| \left\| \left(f \circ g^{-1} \right)'' \circ g \right\| \right\|_{\infty,\mathbb{R}} \leq \left(\frac{4 \| \| f \| \|_{\infty,\mathbb{R}}}{\alpha - 2} \right)^{\frac{\alpha-2}{\alpha}} \left(\frac{K}{\Gamma(\alpha+1)} \right)^{\frac{2}{\alpha}} \alpha. \tag{6.55}$$

The theorem is established. ■

We also give an L_p analog of a generalized fractional Landau inequality

Theorem 6.8 *Let $p, q > 1 : \frac{1}{p} + \frac{1}{q} = 1$, $2 < \alpha \leq 3$ and $f \in C^3(\mathbb{R}, X)$, where $(X, \|\cdot\|)$ is a Banach space. Let $g \in C^1(\mathbb{R})$, strictly increasing, such that $g^{-1} \in C^3(g(\mathbb{R}))$. We assume that $\| \| f \| \|_{\infty,\mathbb{R}} < \infty$, and that*

$$M := \max \left\{ \sup_{a \in \mathbb{R}} \left\| \left\| \left(\left(D_{a+;g}^{\alpha} f \right) \circ g^{-1} \right)(z) \right\| \right\|_{p,g(\mathbb{R})}, \right.$$

$$\left. \sup_{a \in \mathbb{R}} \left\| \left\| \left(\left(D_{a-;g}^{\alpha} f \right) \circ g^{-1} \right)(z) \right\| \right\|_{p,g(\mathbb{R})} \right\} < \infty. \tag{6.56}$$

Then

(1)

$$\left\| \left\| \left(f \circ g^{-1} \right)' \circ g \right\| \right\|_{\infty,\mathbb{R}} \leq$$

$$\left(\alpha - \frac{1}{p} \right) \left(\frac{M}{\Gamma(\alpha) \left(q(\alpha-1)+1 \right)^{\frac{1}{q}}} \right)^{\left(\frac{1}{\alpha - \frac{1}{p}} \right)} \left(\frac{\| \| f \| \|_{\infty,\mathbb{R}}}{\alpha - 1 - \frac{1}{p}} \right)^{\left(\frac{\alpha - 1 - \frac{1}{p}}{\alpha - \frac{1}{p}} \right)}, \tag{6.57}$$

and

(2) under the additional assumption $2 + \frac{1}{p} < \alpha \leq 3$, we have

$$\left\| \left\| \left(f \circ g^{-1} \right)'' \circ g \right\| \right\|_{\infty,\mathbb{R}} \leq$$

$$\left(\alpha - \frac{1}{p} \right) \left(\frac{M}{\Gamma(\alpha) \left(q(\alpha-1)+1 \right)^{\frac{1}{q}}} \right)^{\left(\frac{2}{\alpha - \frac{1}{p}} \right)} \left(\frac{4 \| \| f \| \|_{\infty,\mathbb{R}}}{\alpha - 2 - \frac{1}{p}} \right)^{\left(\frac{\alpha - 2 - \frac{1}{p}}{\alpha - \frac{1}{p}} \right)}. \tag{6.58}$$

That is $\left\| \left\| \left(f \circ g^{-1} \right)' \circ g \right\| \right\|_{\infty,\mathbb{R}}, \left\| \left\| \left(f \circ g^{-1} \right)'' \circ g \right\| \right\|_{\infty,\mathbb{R}} < \infty.$

Proof We continue with the proof of Theorem 6.7.

By (6.23) we have

$$\|A\| = \left\| \left(f(x_1) - f(a) \right) - \frac{1}{\Gamma(\alpha)} \int_{g(a)}^{g(x_1)} \left(g(x_1) - z \right)^{\alpha-1} \left(\left(D_{a+;g}^{\alpha} f \right) \circ g^{-1} \right)(z)\, dz \right\|$$

$$\leq 2\,\||f\||_{\infty,\mathbb{R}} + \frac{1}{\Gamma(\alpha)} \int_{g(a)}^{g(x_1)} (g(x_1) - z)^{\alpha-1} \left\| \left(\left(D_{a+;g}^{\alpha} f \right) \circ g^{-1} \right)(z) \right\| dz$$

$$\leq 2\,\||f\||_{\infty,\mathbb{R}} + \frac{1}{\Gamma(\alpha)} \left(\int_{g(a)}^{g(x_1)} (g(x_1) - z)^{q(\alpha-1)} dz \right)^{\frac{1}{q}}$$

$$\left(\int_{g(a)}^{g(x_1)} \left\| \left(\left(D_{a+;g}^{\alpha} f \right) \circ g^{-1} \right)(z) \right\|^p dz \right)^{\frac{1}{p}} \leq \tag{6.59}$$

$$2\,\||f\||_{\infty,\mathbb{R}} + \frac{1}{\Gamma(\alpha)} \frac{(g(x_1) - g(a))^{\frac{(q(\alpha-1)+1)}{q}}}{(q(\alpha-1)+1)^{\frac{1}{q}}} \left\| \left\| \left(\left(D_{a+;g}^{\alpha} f \right) \circ g^{-1} \right)(z) \right\| \right\|_{p,g(\mathbb{R})} \leq$$

$$2\,\||f\||_{\infty,\mathbb{R}} + \frac{1}{\Gamma(\alpha)} \frac{h^{\alpha-\frac{1}{p}}}{(q(\alpha-1)+1)^{\frac{1}{q}}} \left(\sup_{a\in\mathbb{R}} \left\| \left\| \left(\left(D_{a+;g}^{\alpha} f \right) \circ g^{-1} \right)(z) \right\| \right\|_{p,g(\mathbb{R})} \right)$$

$$\leq 2\,\||f\||_{\infty,\mathbb{R}} + \frac{h^{\alpha-\frac{1}{p}}}{\Gamma(\alpha)(q(\alpha-1)+1)^{\frac{1}{q}}} M.$$

That is

$$\|A\| \leq 2\,\||f\||_{\infty,\mathbb{R}} + \frac{M}{\Gamma(\alpha)(q(\alpha-1)+1)^{\frac{1}{q}}} h^{\alpha-\frac{1}{p}}, \quad h > 0. \tag{6.60}$$

Similarly, from (6.24) we get

$$\|B\| = \left\| (f(x_2) - f(a)) - \frac{1}{\Gamma(\alpha)} \int_{g(x_2)}^{g(a)} (z - g(x_2))^{\alpha-1} \left(\left(D_{a-;g}^{\alpha} f \right) \circ g^{-1} \right)(z)\, dz \right\|$$

$$\leq 2\,\||f\||_{\infty,\mathbb{R}} + \frac{1}{\Gamma(\alpha)} \int_{g(x_2)}^{g(a)} (z - g(x_2))^{\alpha-1} \left\| \left(\left(D_{a-;g}^{\alpha} f \right) \circ g^{-1} \right)(z) \right\| dz$$

$$\leq 2\,\||f\||_{\infty,\mathbb{R}} + \frac{1}{\Gamma(\alpha)} \left(\int_{g(x_2)}^{g(a)} (z - g(x_2))^{q(\alpha-1)} dz \right)^{\frac{1}{q}}$$

$$\left(\int_{g(x_2)}^{g(a)} \left\| \left(\left(D_{a-;g}^{\alpha} f \right) \circ g^{-1} \right)(z) \right\|^p dz \right)^{\frac{1}{p}} \leq$$

$$2\,\||f\||_{\infty,\mathbb{R}} + \frac{1}{\Gamma(\alpha)} \frac{(g(a) - g(x_2))^{\alpha-\frac{1}{p}}}{(q(\alpha-1)+1)^{\frac{1}{q}}} \left(\sup_{a\in\mathbb{R}} \left\| \left\| \left(\left(D_{a-;g}^{\alpha} f \right) \circ g^{-1} \right)(z) \right\| \right\|_{p,g(\mathbb{R})} \right) \tag{6.61}$$

$$\leq 2 \|\|f\|\|_{\infty,\mathbb{R}} + \frac{h^{\alpha-\frac{1}{p}}}{\Gamma(\alpha)(q(\alpha-1)+1)^{\frac{1}{q}}} M.$$

That is

$$\|B\| \leq 2 \|\|f\|\|_{\infty,\mathbb{R}} + \frac{M}{\Gamma(\alpha)(q(\alpha-1)+1)^{\frac{1}{q}}} h^{\alpha-\frac{1}{p}}, \quad h > 0. \tag{6.62}$$

Hence it holds

$$\|A+B\| \leq \|A\| + \|B\| \overset{\text{(by (6.60.), (6.62.))}}{\leq}$$

$$4 \|\|f\|\|_{\infty,\mathbb{R}} + \frac{2M}{\Gamma(\alpha)(q(\alpha-1)+1)^{\frac{1}{q}}} h^{\alpha-\frac{1}{p}}, \quad h > 0. \tag{6.63}$$

Furthermore we have

$$\|A-B\| \overset{(6.34.)}{\leq} 2 \|\|f\|\|_{\infty,\mathbb{R}} +$$

$$\frac{1}{\Gamma(\alpha)} \left[\left(\int_{g(a)}^{g(x_1)} (g(x_1)-z)^{q(\alpha-1)} dz \right)^{\frac{1}{q}} \left\| \left\| \left(\left(D_{a+;g}^{\alpha} f \right) \circ g^{-1} \right)(z) \right\| \right\|_{p,g(\mathbb{R})} \right.$$

$$\left. + \left(\int_{g(x_2)}^{g(a)} (z-g(x_2))^{q(\alpha-1)} dz \right)^{\frac{1}{q}} \left\| \left\| \left(\left(D_{a-;g}^{\alpha} f \right) \circ g^{-1} \right)(z) \right\| \right\|_{p,g(\mathbb{R})} \right] \leq$$

$$2 \|\|f\|\|_{\infty,\mathbb{R}} + \frac{M}{\Gamma(\alpha)} \left[\frac{2h^{\alpha-\frac{1}{p}}}{(q(\alpha-1)+1)^{\frac{1}{q}}} \right]. \tag{6.64}$$

We have proved that

$$\frac{\|A-B\|}{2} \leq \|\|f\|\|_{\infty,\mathbb{R}} + \frac{M}{\Gamma(\alpha)(q(\alpha-1)+1)^{\frac{1}{q}}} h^{\alpha-\frac{1}{p}}, \quad h > 0. \tag{6.65}$$

From (6.27), (6.65), we have

$$\left\| \left(f \circ g^{-1} \right)' (g(a)) \right\| \leq \frac{\|\|f\|\|_{\infty,\mathbb{R}}}{h} + \frac{M}{\Gamma(\alpha)(q(\alpha-1)+1)^{\frac{1}{q}}} h^{\alpha-\frac{1}{p}-1}, \tag{6.66}$$

$h > 0$, any $a \in \mathbb{R}$.

And from (6.27), (6.63), we get that

$$\left\| \left(f \circ g^{-1} \right)'' (g(a)) \right\| \leq \frac{4 \|\|f\|\|_{\infty,\mathbb{R}}}{h^2} + \frac{2M}{\Gamma(\alpha)(q(\alpha-1)+1)^{\frac{1}{q}}} h^{\alpha-\frac{1}{p}-2}, \tag{6.67}$$

$h > 0$, any $a \in \mathbb{R}$.

Hence

$$\left\| \left\| \left(f \circ g^{-1}\right)' \circ g \right\| \right\|_{\infty,\mathbb{R}} \leq \frac{\| \| f \| \|_{\infty,\mathbb{R}}}{h} + \left(\frac{M}{\Gamma(\alpha)\left(q(\alpha-1)+1\right)^{\frac{1}{q}}} \right) h^{\left(\alpha-\frac{1}{p}-1\right)}, \tag{6.68}$$

and

$$\left\| \left\| \left(f \circ g^{-1}\right)'' \circ g \right\| \right\|_{\infty,\mathbb{R}} \leq \frac{4 \| \| f \| \|_{\infty,\mathbb{R}}}{h^2} + \left(\frac{2M}{\Gamma(\alpha)\left(q(\alpha-1)+1\right)^{\frac{1}{q}}} \right) h^{\left(\alpha-\frac{1}{p}-2\right)}, \tag{6.69}$$

true $\forall\, h > 0$, $2 < \alpha \leq 3$, $p, q > 1 : \frac{1}{p} + \frac{1}{q} = 1$.

In (6.69) we restrict ourselves to $2 + \frac{1}{p} < \alpha \leq 3$.

Call

$$\mu := \| \| f \| \|_{\infty,\mathbb{R}}, \quad \theta = \frac{M}{\Gamma(\alpha)(q(\alpha-1)+1)^{\frac{1}{q}}}, \tag{6.70}$$

both are greater than zero.

Set also $\rho := \alpha - 1 - \frac{1}{p} > \frac{1}{q} > 0$.

We consider the function

$$y(h) := \frac{\mu}{h} + \theta h^{\rho}, \quad \forall\, h > 0. \tag{6.71}$$

As in the proof of Theorem 6.7 it has only one critical number

$$h_0 := h_{crit.no} = \left(\frac{\mu}{\rho\theta} \right)^{\frac{1}{\rho+1}}, \tag{6.72}$$

and a global minimum

$$y(h_0) = \theta^{\frac{1}{\rho+1}} \mu^{\frac{\rho}{\rho+1}} (\rho+1) \rho^{-\frac{\rho}{\rho+1}}. \tag{6.73}$$

Consequently

$$y(h_0) = \left(\frac{M}{\Gamma(\alpha)\left(q(\alpha-1)+1\right)^{\frac{1}{q}}} \right)^{\frac{1}{\left(\alpha-\frac{1}{p}\right)}} \left(\| \| f \| \|_{\infty,\mathbb{R}} \right)^{\left(\frac{\alpha-1-\frac{1}{p}}{\alpha-\frac{1}{p}} \right)}$$

$$\left(\alpha - \frac{1}{p} \right) \left(\alpha - 1 - \frac{1}{p} \right)^{-\left(\frac{\alpha-1-\frac{1}{p}}{\alpha-\frac{1}{p}} \right)}. \tag{6.74}$$

We have proved that (see (6.68))

$$\left\| \left\| \left(f \circ g^{-1} \right)' \circ g \right\| \right\|_{\infty,\mathbb{R}} \leq$$

$$\left(\frac{M}{\Gamma(\alpha)\left(q(\alpha-1)+1\right)^{\frac{1}{q}}} \right)^{\frac{1}{\left(\alpha-\frac{1}{p}\right)}} \left(\frac{\| \| f \| \|_{\infty,\mathbb{R}}}{\alpha-1-\frac{1}{p}} \right)^{\left(\frac{\alpha-1-\frac{1}{p}}{\alpha-\frac{1}{p}}\right)} \left(\alpha - \frac{1}{p} \right). \tag{6.75}$$

We also call

$$\xi := 4 \| \| f \| \|_{\infty,\mathbb{R}},$$
$$\psi = \frac{2M}{\Gamma(\alpha)(q(\alpha-1)+1)^{\frac{1}{q}}}, \tag{6.76}$$

both are greater than zero.

Set also $\varphi := \alpha - 2 - \frac{1}{p} > 0$.

We consider the function

$$\gamma(h) := \frac{\xi}{h^2} + \psi h^{\varphi}, \quad \forall\, h > 0. \tag{6.77}$$

As in the proof of Theorem 6.7, γ has a global minimum at

$$h_0 = \left(\frac{2\xi}{\varphi\psi} \right)^{\frac{1}{\varphi+2}}, \tag{6.78}$$

which is

$$\gamma(h_0) = \left(\frac{\xi}{\varphi} \right)^{\frac{\varphi}{\varphi+2}} \left(\frac{\psi}{2} \right)^{\frac{2}{\varphi+2}} (\varphi + 2). \tag{6.79}$$

Consequently

$$\gamma(h_0) = \left(\frac{4 \| \| f \| \|_{\infty,\mathbb{R}}}{\alpha-2-\frac{1}{p}} \right)^{\left(\frac{\alpha-2-\frac{1}{p}}{\alpha-\frac{1}{p}}\right)} \left(\frac{M}{\Gamma(\alpha)\left(q(\alpha-1)+1\right)^{\frac{1}{q}}} \right)^{\left(\frac{2}{\alpha-\frac{1}{p}}\right)} \left(\alpha - \frac{1}{p} \right). \tag{6.80}$$

We have proved that (see (6.69))

$$\left\| \left\| \left(f \circ g^{-1} \right)'' \circ g \right\| \right\|_{\infty,\mathbb{R}} \leq$$

$$\left(\frac{4 \| \| f \| \|_{\infty,\mathbb{R}}}{\alpha-2-\frac{1}{p}} \right)^{\left(\frac{\alpha-2-\frac{1}{p}}{\alpha-\frac{1}{p}}\right)} \left(\frac{M}{\Gamma(\alpha)\left(q(\alpha-1)+1\right)^{\frac{1}{q}}} \right)^{\left(\frac{2}{\alpha-\frac{1}{p}}\right)} \left(\alpha - \frac{1}{p} \right). \tag{6.81}$$

The theorem is established. ■

Next we apply Theorems 6.7, 6.8 for $g(t) = e^t$, $t \in \mathbb{R}$ and $\alpha = 2.5$.

Corollary 6.9 *Let $f \in C^3(\mathbb{R}, X)$, where $(X, \|\cdot\|)$ is a Banach space. We assume that $\|\|f\|\|_{\infty,\mathbb{R}} < \infty$, and that*

$$K_{2.5} := \max\left\{\left\|\left\|\left(\left(D^{2.5}_{a+;e^t} f\right) \circ \ln\right)(z)\right\|\right\|_{\infty,\mathbb{R}\times(0,\infty)},\right.$$

$$\left.\left\|\left\|\left(\left(D^{2.5}_{a-;e^t} f\right) \circ \ln\right)(z)\right\|\right\|_{\infty,\mathbb{R}\times(0,\infty)}\right\} < \infty, \quad (6.82)$$

where $(a, z) \in \mathbb{R} \times (0, \infty)$.

Then

$$\left\|\left\|(f \circ \ln)' \circ e^t\right\|\right\|_{\infty,\mathbb{R}} \leq 1.21136\,(K_{2.5})^{0.4}\left(\|\|f\|\|_{\infty,\mathbb{R}}\right)^{0.6}, \quad (6.83)$$

and

$$\left\|\left\|(f \circ \ln)'' \circ e^t\right\|\right\|_{\infty,\mathbb{R}} \leq 1.44713\,(K_{2.5})^{0.8}\left(\|\|f\|\|_{\infty,\mathbb{R}}\right)^{0.2}. \quad (6.84)$$

That is $\left\|\left\|(f \circ \ln)' \circ e^t\right\|\right\|_{\infty,\mathbb{R}}, \left\|\left\|(f \circ \ln)'' \circ e^t\right\|\right\|_{\infty,\mathbb{R}} < \infty$.

Proof By Theorem 6.7. ■

We finish with

Corollary 6.10 *(case of $g(t) = e^t$, $\alpha = 2.5$, $p = q = 2$) Let $f \in C^3(\mathbb{R}, X)$, where $(X, \|\cdot\|)$ is a Banach space. We assume that $\|\|f\|\|_{\infty,\mathbb{R}} < \infty$, and that*

$$M_{2.5} := \max\left\{\sup_{a\in\mathbb{R}}\left\|\left\|\left(\left(D^{2.5}_{a+;e^t} f\right) \circ \ln\right)(z)\right\|\right\|_{2,(0,\infty)},\right.$$

$$\left.\sup_{a\in\mathbb{R}}\left\|\left\|\left(\left(D^{2.5}_{a-;e^t} f\right) \circ \ln\right)(z)\right\|\right\|_{2,(0,\infty)}\right\} < \infty. \quad (6.85)$$

Then

$$\left\|\left\|(f \circ \ln)' \circ e^t\right\|\right\|_{\infty,\mathbb{R}} \leq 1.226583057\sqrt{M_{2.5}\,\|\|f\|\|_{\infty,\mathbb{R}}} < \infty. \quad (6.86)$$

Proof By Theorem 6.8, (6.57). ■

References

1. A. Aglic Aljinovic, L. Marangunic, J. Pecaric, On Landau type inequalities via Ostrowski inequalities. Nonlinear Funct. Anal. Appl. **10**(4) 565–579 (2005)
2. Anastassiou, G.: Fractional Differentiation Inequalities. Research Monograph. Springer, New York (2009)

3. Anastassiou, G.A.: Advances on Fractional Inequalities. Springer, New York (2011)
4. Anastassiou, G.A.: Intelligent Computations: Abstract Fractional Calculus, Inequalities Approximations. Springer, Heidelberg (2018)
5. G.A. Anastassiou, G.A.: Abstract Generalized Fractional Landau inequalities over $\mathbb{R}$, Constructive Mathematical Analysis, accepted (2020)
6. Barnett, N.S., Dragomir, S.S.: Some Landau type inequalities for functions whose derivatives are of locally bounded variation. Tamkang J. Math. **37**(4), 301–308 (Winter 2006)
7. Ditzian, Z.: Remarks, questions and conjectures on Landau-Kolmogorov-type inequalities. Math. Inequal. Appl. **3**, 15–24 (2000)
8. Hardy, G.H., Littlewood, J.E.: Some integral inequalities connected with the calculus of variations. Quart. J. Math. Oxford Ser. **3**, 241–252 (1932)
9. Hardy, G.H., Landau, E., Littlewood, J.E.: Some inequalities satisfied by the integrals or derivatives of real or analytic functions. Math. Z. **39**, 677–695 (1935)
10. Kallman, R.R., Rota, G.C.: On theinequality $\|'\|^2 \leq 4\|\| \cdot \|''\|$. In: Shisha, O., (ed.), Inequalities , vol. II, pp. 187–192. Academic, New York (1970)
11. Landau, E.: Einige Ungleichungen für zweimal differentzierban funktionen. Proc. Lond. Math. Soc. **13**, 43–49 (1913)
12. Mikusinski, J.: The Bochner Integral. Academic, New York (1978)
13. Shilov, G.E.: Elementary Functional Analysis. Dover Publications Inc., New York (1996)

Chapter 7
Fractional Landau Inequalities of Riemann–Liouville Type

We present uniform Riemann–Liouville left and right fractional Landau inequalities over $\mathbb{R}_+$ and $\mathbb{R}_-$, respectively, of fractional orders $1 < \nu < 2$ and $2 < \nu < 3$ and we estimate lower order fractional derivatives. These inequalities are sharp or nearly sharp with completely determined constants. We give applications when $\nu = 1.5$. We finish with a related new Ostrowski like inequality for $\nu > 0$, $\nu \notin \mathbb{N}$. It follows [1].

7.1 Introduction

Landau in 1913 [7] proved if $f \in C^2([0,1])$, $\|f\|_\infty = 1$, $\|f''\|_\infty = 4$, then $\|f'\|_\infty \leq 4$, with 4 the best constant and the result is not necessarily true for an interval of length < 1. Now Landau's inequality has taken the form [2]

$$\|f'\|_{p,I} \leq C_p(I) \|f\|_{p,I}^{\frac{1}{2}} \|f''\|_{p,I}^{\frac{1}{2}}, \tag{7.1}$$

where $\|\cdot\|_{p,I}$ is the p-norm on the interval I; $p \in [1, \infty]$, $I = \mathbb{R}_+$ or $I = \mathbb{R}$, and $f : I \to \mathbb{R}$ is twice differentiable with $f, f'' \in L_p(I)$. $C_p(I) > 0$ is independent of f. And the best constants are $C_\infty(\mathbb{R}_+) = 2$ and $C_\infty(\mathbb{R}) = \sqrt{2}$.

Research about Landau type inequalities has expanded to many different directions and it is a very active topic of mathematical activity. Here we are concerned about fractional Landau inequalities of Riemann–Liouville type.

The main motivation to write this chapter is [8], and especially the result that follows:

Let α be a complex number with $Re\alpha > 0$. We define I_L^α and D_L^α to be the classical Riemann–Liouville fractional integration and differentiation operators of order α,

G. A. Anastassiou, *Constructive Fractional Analysis with Applications*,
Studies in Systems, Decision and Control 362,
https://doi.org/10.1007/978-3-030-71481-9_7

$$I_L^{\alpha} f(x) = \frac{1}{\Gamma(\alpha)} \int_a^x (x-t)^{\alpha-1} f(t)\, dt,$$

$$D_L^{\alpha} f(x) = \frac{d^{n+1}}{dx^{n+1}} \left(I_L^{n+1-\alpha} f \right)(x),$$

where Γ is the gamma function and $n = [Re\alpha]$ (integral part).

The above operators are defined in the natural domain of appropriate function spaces.

We state

Theorem 7.1 ([8]) *Let I_L^{α} and D_L^{α} be the fractional Riemann–Liouville operators in the Fréchet space $L_{loc}^1([0,+\infty))$ with the usual topology. If α, β and γ are complex numbers for which $0 < Re\alpha < Re\beta < Re\gamma$, for every seminorm $\|\cdot\|_p$ there exist constant $F_1(p,\alpha,\beta,\gamma)$ and $F_2(p,\alpha,\beta,\gamma)$ that depend on p, α, β, and γ, with the properties that*

(a)

$$\left\| I_L^{\beta} f \right\|_p \leq F_1(p,\alpha,\beta,\gamma) \left\| I_L^{\alpha} f \right\|_p^{\frac{Re(\gamma-\beta)}{Re(\gamma-\alpha)}} \left\| I_L^{\gamma} f \right\|_p^{\frac{Re(\beta-\alpha)}{Re(\gamma-\alpha)}}, \quad f \in L_{loc}^1([0,+\infty)), \tag{7.2}$$

and

$$\left\| D_L^{\beta} f \right\|_p \leq F_2(p,\alpha,\beta,\gamma) \left\| D_L^{\alpha} f \right\|_p^{\frac{Re(\gamma-\beta)}{Re(\gamma-\alpha)}} \left\| D_L^{\gamma} f \right\|_p^{\frac{Re(\beta-\alpha)}{Re(\gamma-\alpha)}}, \tag{7.3}$$

$f \in \mathcal{D}\left(D_L^{\gamma}\right)$ (domain of D_L^{γ}).

If $Im\alpha = Im\beta = Im\gamma$, the constants $F_1(p,\alpha,\beta,\gamma)$ and $F_2(p,\alpha,\beta,\gamma)$ depend only on p and on an arbitrary integer $n > Re\gamma$.

For more see [8]. Other inspirations are the papers [4, 5].

The above motivating results and references are existential without giving exact constants which are not the best possible. Their methods are based on functional analysis and semigroup theory, in particular dealing with moment inequalities for fractional powers of operators in Banach spaces.

As a last motivation we also mention [3]. Our method is classical analytic and we will produce fractional Landau type inequalities, involving the Riemann–Liouville left and right fractional derivatives, that are sharp or nearly sharp and all constant are exactly calculated.

We need the following background (based on [9], Sect. 7.2, and [6], Chap. 2).

Let $-\infty < a < b < \infty$, the left and right Riemann–Liouville fractional integrals of order $\alpha \in \mathbb{C}$ ($\mathcal{R}(\alpha) > 0$) are defined by

$$\left(I_a^{\alpha} f \right)(x) := \frac{1}{\Gamma(\alpha)} \int_a^x (x-t)^{\alpha-1} f(t)\, dt \tag{7.4}$$

$(x > a; \mathcal{R}(\alpha) > 0)$

and

$$\left(I_{b-}^{\alpha}f\right)(x) := \frac{1}{\Gamma(\alpha)}\int_x^b (t-x)^{\alpha-1} f(t)\,dt \tag{7.5}$$

$(x < b; \mathcal{R}(\alpha) > 0)$, respectively.

The Riemann–Liouville left and right fractional derivatives of order $\alpha \in \mathbb{C}$ $(\mathcal{R}(\alpha) \geq 0)$ are defined by

$$\left(D_a^{\alpha}y\right)(x) := \left(\frac{d}{dx}\right)^n \left(I_a^{n-\alpha}y\right)(x) = \frac{1}{\Gamma(n-\alpha)}\left(\frac{d}{dx}\right)^n \int_a^x (x-t)^{n-\alpha-1} y(t)\,dt \tag{7.6}$$

$(n = [\mathcal{R}(\alpha)] + 1; x > a)$

and

$$\left(D_{b-}^{\alpha}y\right)(x) := (-1)^n \left(\frac{d}{dx}\right)^n \left(I_{b-}^{n-\alpha}y\right)(x)$$

$$= \frac{(-1)^n}{\Gamma(n-\alpha)}\left(\frac{d}{dx}\right)^n \int_x^b (t-x)^{n-\alpha-1} y(t)\,dt \tag{7.7}$$

$(n = [\mathcal{R}(\alpha)] + 1; x < b)$,

respectively, where $\mathcal{R}(\alpha)$ is the real part of α.

In particular, when $\alpha = n \in \mathbb{Z}_+$, then

$$\left(D_a^0 y\right)(x) = \left(D_{b-}^0 y\right)(x) = y(x);$$

$$\left(D_a^n y\right)(x) = y^{(n)}(x), \text{ and } \left(D_{b-}^n y\right)(x) = (-1)^n y^{(n)}(x) \quad (n \in \mathbb{N}).$$

We denote by

$$\left(D_a^{-\alpha}f\right)(x) := \left(I_a^{\alpha}f\right)(x) = \frac{1}{\Gamma(\alpha)}\int_a^x (x-t)^{\alpha-1} f(t)\,dt \tag{7.8}$$

$(x > a; \mathcal{R}(\alpha) > 0)$,

and

$$\left(D_{b-}^{-\alpha}f\right)(x) := \left(I_{b-}^{\alpha}f\right)(x) = \frac{1}{\Gamma(\alpha)}\int_x^b (t-x)^{\alpha-1} f(t)\,dt \tag{7.9}$$

$(x < b; \mathcal{R}(\alpha) > 0)$.

Based on [9], pp. 44–45, and [6], pp. 74–76 we will use the following two fractional Taylor theorems. The left one first:

Theorem 7.2 *Let $\nu > 0$, $\nu \notin \mathbb{N}$ and $m = \lceil \nu \rceil$ (ceiling of ν); φ, $D_a^{\nu}\varphi \in C([a,b])$. Then*

$$\varphi(b) = \sum_{j=1}^{m} \frac{D_a^{\nu-j}\varphi(a)}{\Gamma(\nu-j+1)}(b-a)^{\nu-j} + \frac{1}{\Gamma(\nu)}\int_a^b (b-t)^{\nu-1} D_a^{\nu}\varphi(t)\,dt. \tag{7.10}$$

It follows the right one:

Theorem 7.3 *Let* $\nu > 0$, $\nu \notin \mathbb{N}$ *and* $m = \lceil \nu \rceil$; φ, $D_{b-}^{\nu}\varphi \in C([a,b])$. *Then*

$$\varphi(a) = \sum_{j=1}^{m} \frac{D_{b-}^{\nu-j}\varphi(b)}{\Gamma(\nu-j+1)}(b-a)^{\nu-j} + \frac{1}{\Gamma(\nu)}\int_a^b (t-a)^{\nu-1} D_{b-}^{\nu}\varphi(t)\,dt. \tag{7.11}$$

7.2 Main Results

First we give left Riemann–Liouville fractional Landau type inequalities.

Convention 7.4 Define $D_a^{\nu}\varphi(x) := 0$, for any $x < a$, for any $\nu > 0$, $\nu \notin \mathbb{N}$.

We present

Theorem 7.5 *Here* $1 < \nu < 2$ *and* $\varphi \in C_B(\mathbb{R}_+)$ *and* $D_a^{\nu}\varphi \in C_B([a,+\infty))$, $\forall\, a \in \mathbb{R}_+$, *where* C_B *means continuous and bounded functions. We assume that* $\sup_{a\in\mathbb{R}_+} \left\| D_a^{\nu}\varphi \right\|_{\infty,\mathbb{R}_+} < +\infty$.

Then

$$\sup_{a\in\mathbb{R}_+} \left| D_a^{\nu-1}\varphi(a) \right| \le \left(\frac{\Gamma(\nu+1)}{(\nu-1)^{\nu-1}} \right)^{\frac{1}{\nu}} \|\varphi\|_{\infty,\mathbb{R}_+}^{\frac{1}{\nu}} \left(\sup_{a\in\mathbb{R}_+} \left\| D_a^{\nu}\varphi \right\|_{\infty,\mathbb{R}_+} \right)^{\frac{\nu-1}{\nu}}, \tag{7.12}$$

a sharp inequality with a precisely determined constant. That is $\sup_{a\in\mathbb{R}_+} \left| D_a^{\nu-1}\varphi(a) \right| < +\infty$.

Proof For any $b \in (a,+\infty)$ we get (by Theorem 7.2)

$$\varphi(b) = \frac{D_a^{\nu-1}\varphi(a)}{\Gamma(\nu)}(b-a)^{\nu-1} + \frac{D_a^{\nu-2}\varphi(a)}{\Gamma(\nu-1)}(b-a)^{\nu-2}$$
$$+ \frac{1}{\Gamma(\nu)}\int_a^b (b-t)^{\nu-1} D_a^{\nu}\varphi(t)\,dt. \tag{7.13}$$

Notice that ($x \in [a,+\infty)$)

$$D_a^{\nu-2}\varphi(x) = D_a^{-(2-\nu)}\varphi(x) = \frac{1}{\Gamma(2-\nu)}\int_a^x (x-t)^{1-\nu}\varphi(t)\,dt,$$

and

$$\left|D_a^{\nu-2}\varphi(x)\right| \leq \frac{1}{\Gamma(2-\nu)}\int_a^x (x-t)^{1-\nu}|\varphi(t)|\,dt \leq \frac{\|\varphi\|_{\infty,[a,+\infty)}}{\Gamma(2-\nu)}\int_a^x (x-t)^{1-\nu}\,dt$$

$$= \frac{\|\varphi\|_{\infty,[a,+\infty)}}{\Gamma(2-\nu)}\frac{(x-a)^{2-\nu}}{(2-\nu)} = \frac{\|\varphi\|_{\infty,[a,+\infty)}}{\Gamma(3-\nu)}(x-a)^{2-\nu}. \tag{7.14}$$

That is

$$\left|D_a^{\nu-2}\varphi(x)\right| \leq \frac{\|\varphi\|_{\infty,[a,+\infty)}}{\Gamma(3-\nu)}(x-a)^{2-\nu}, \tag{7.15}$$

$\forall\, x \in [a,+\infty)$.

Clearly, it is $D_a^{\nu-2}\varphi(a) = 0$.

Hence by (7.13) we get

$$\varphi(b) = \frac{D_a^{\nu-1}\varphi(a)(b-a)^{\nu-1}}{\Gamma(\nu)} + \frac{1}{\Gamma(\nu)}\int_a^b (b-t)^{\nu-1} D_a^{\nu}\varphi(t)\,dt, \tag{7.16}$$

$\forall\, b \in (a,+\infty)$.

So that

$$\frac{D_a^{\nu-1}\varphi(a)}{\Gamma(\nu)}(b-a)^{\nu-1} = \varphi(b) - \frac{1}{\Gamma(\nu)}\int_a^b (b-t)^{\nu-1} D_a^{\nu}\varphi(t)\,dt, \tag{7.17}$$

$\forall\, b \in (a,+\infty)$.

Hence (set $h := b-a > 0$)

$$\left|D_a^{\nu-1}\varphi(a)\right| h^{\nu-1} = \left|\Gamma(\nu)\varphi(b) - \int_a^b (b-t)^{\nu-1} D_a^{\nu}\varphi(t)\,dt\right| \leq$$

$$\Gamma(\nu)|\varphi(b)| + \int_a^b (b-t)^{\nu-1}\left|D_a^{\nu}\varphi(t)\right| dt \leq$$

$$\Gamma(\nu)|\varphi(b)| + \left\|D_a^{\nu}\varphi\right\|_{\infty,[a,+\infty)}\int_a^b (b-t)^{\nu-1}\,dt =$$

$$\Gamma(\nu)|\varphi(b)| + \left\|D_a^{\nu}\varphi\right\|_{\infty,[a,+\infty)}\frac{(b-a)^{\nu}}{\nu} = \Gamma(\nu)|\varphi(b)| + \left\|D_a^{\nu}\varphi\right\|_{\infty,[a,+\infty)}\frac{h^{\nu}}{\nu}. \tag{7.18}$$

That is we have

$$\left|D_a^{\nu-1}\varphi(a)\right| h^{\nu-1} \leq \Gamma(\nu)|\varphi(b)| + \left\|D_a^{\nu}\varphi\right\|_{\infty,[a,+\infty)}\frac{h^{\nu}}{\nu} \leq \tag{7.19}$$

$$\Gamma(\nu)\|\varphi\|_{\infty,\mathbb{R}_+} + \left(\sup_{a\in\mathbb{R}_+}\left\|D_a^{\nu}\varphi\right\|_{\infty,[a,+\infty)}\right)\frac{h^{\nu}}{\nu}.$$

Furthermore we see that

$$\left|D_a^{\nu-1}\varphi(a)\right| \leq \frac{\Gamma(\nu)\|\varphi\|_{\infty,\mathbb{R}_+}}{h^{\nu-1}} + \frac{\left(\sup_{a\in\mathbb{R}_+}\left\|D_a^{\nu}\varphi\right\|_{\infty,[a,+\infty)}\right)}{\nu}h, \tag{7.20}$$

$\forall\, h > 0$ and $\forall\, a \in \mathbb{R}_+$.

Consequently it holds

$$\sup_{a\in\mathbb{R}_+}\left|D_a^{\nu-1}\varphi(a)\right| \leq \frac{\left(\Gamma(\nu)\|\varphi\|_{\infty,\mathbb{R}_+}\right)}{h^{\nu-1}} + \frac{\left(\sup_{a\in\mathbb{R}_+}\left\|D_a^{\nu}\varphi\right\|_{\infty,\mathbb{R}_+}\right)}{\nu}h, \tag{7.21}$$

$\forall\, h > 0$.

Call

$$\mu := \Gamma(\nu)\|\varphi\|_{\infty,\mathbb{R}_+}, \tag{7.22}$$

$$\theta := \left(\frac{\sup_{a\in\mathbb{R}_+}\left\|D_a^{\nu}\varphi\right\|_{\infty,\mathbb{R}_+}}{\nu}\right), \tag{7.23}$$

both are greater than zero.

Here $1 < \nu < 2$ and $0 < \nu - 1 < 1$. Set also $\rho := \nu - 1$, i.e. $0 < \rho < 1$.

We consider the function

$$y(h) := \mu h^{-\rho} + \theta h, \quad \forall\, h > 0. \tag{7.24}$$

We have

$$y'(h) = -\rho\mu h^{-(\rho+1)} + \theta = 0 \tag{7.25}$$

and

$$\rho\mu h^{-(\rho+1)} = \theta,$$

then

$$h^{-(\rho+1)} = \frac{\theta}{\rho\mu},$$

with the only critical number

$$h_0 = \left(\frac{\rho\mu}{\theta}\right)^{\frac{1}{\rho+1}} > 0. \tag{7.26}$$

Also it holds

$$y''(h_0) = \rho(\rho+1)\mu h_0^{-\rho-2} > 0. \tag{7.27}$$

Thus y has a global minimum over $\mathbb{R}_+$ which is

$$y(h_0) = \mu h_0^{-\rho} + \theta h_0 = \frac{(\rho+1)}{\rho^{\frac{\rho}{\rho+1}}} \theta^{\frac{\rho}{\rho+1}} \mu^{\frac{1}{\rho+1}}. \tag{7.28}$$

Consequently we derive that

$$\sup_{a\in\mathbb{R}_+} \left| D_a^{\nu-1} \varphi(a) \right| \leq \left(\frac{\nu}{(\nu-1)^{\frac{\nu-1}{\nu}}} \right) \left(\frac{\sup_{a\in\mathbb{R}_+} \left\| D_a^{\nu} \varphi \right\|_{\infty,\mathbb{R}_+}}{\nu} \right)^{\frac{\nu-1}{\nu}} \left(\Gamma(\nu) \|\varphi\|_{\infty,\mathbb{R}_+} \right)^{\frac{1}{\nu}}$$

$$= \left(\frac{\Gamma(\nu+1)}{(\nu-1)^{\nu-1}} \right)^{\frac{1}{\nu}} \|\varphi\|_{\infty,\mathbb{R}_+}^{\frac{1}{\nu}} \left(\sup_{a\in\mathbb{R}_+} \left\| D_a^{\nu} \varphi \right\|_{\infty,\mathbb{R}_+} \right)^{\frac{\nu-1}{\nu}}, \tag{7.29}$$

for $1 < \nu < 2$.

The theorem is proved. ■

We continue with

Theorem 7.6 *Here* $2 < \nu < 3$ *and* $\varphi \in C_B(\mathbb{R}_+)$ *and* $D_a^{\nu}\varphi \in C_B([a,+\infty))$, $\forall a \in \mathbb{R}_+$. *We assume that* $\sup_{a\in\mathbb{R}_+} \left\| D_a^{\nu} \varphi \right\|_{\infty,\mathbb{R}_+} < +\infty$. *Then*

(i)

$$\sup_{a\in\mathbb{R}_+} \left| D_a^{\nu-1} \varphi(a) \right| \leq$$

$$\left(\frac{5}{\nu-1} \right)^{\left(\frac{\nu-1}{\nu}\right)} \left(\nu \left(1 + 2^{2-\nu} \right) \Gamma(\nu) \right)^{\frac{1}{\nu}} \|\varphi\|_{\infty,\mathbb{R}_+}^{\frac{1}{\nu}} \left(\sup_{a\in\mathbb{R}_+} \left\| D_a^{\nu} \varphi \right\|_{\infty,\mathbb{R}_+} \right)^{\left(\frac{\nu-1}{\nu}\right)}, \tag{7.30}$$

and

(ii)

$$\sup_{a\in\mathbb{R}_+} \left| D_a^{\nu-2} \varphi(a) \right| \leq \left(\frac{6}{(\nu-1)(\nu-2)} \right)^{\left(\frac{\nu-2}{\nu}\right)}$$

$$\left(\nu \left(1 + 2^{1-\nu} \right) \Gamma(\nu-1) \right)^{\frac{2}{\nu}} \|\varphi\|_{\infty,\mathbb{R}_+}^{\frac{2}{\nu}} \left(\sup_{a\in\mathbb{R}_+} \left\| D_a^{\nu} \varphi \right\|_{\infty,\mathbb{R}_+} \right)^{\left(\frac{\nu-2}{\nu}\right)}. \tag{7.31}$$

The above inequalities are nearly sharp with precisely determined constants. That is $\sup_{a\in\mathbb{R}_+} \left| D_a^{\nu-1} \varphi(a) \right|$, $\sup_{a\in\mathbb{R}_+} \left| D_a^{\nu-2} \varphi(a) \right| < +\infty$.

Proof For any $b \in (a,+\infty)$ we get (by Theorem 7.2)

$$\varphi(b) = \frac{D_a^{\nu-1}\varphi(a)}{\Gamma(\nu)}(b-a)^{\nu-1} + \frac{D_a^{\nu-2}\varphi(a)}{\Gamma(\nu-1)}(b-a)^{\nu-2} + \tag{7.32}$$

$$\frac{D_a^{\nu-3}\varphi(a)}{\Gamma(\nu-2)}(b-a)^{\nu-3} + \frac{1}{\Gamma(\nu)}\int_a^b (b-t)^{\nu-1} D_a^{\nu}\varphi(t)\,dt.$$

Notice that ($x \in [a, +\infty)$)

$$D_a^{\nu-3}\varphi(x) = D_a^{-(3-\nu)}\varphi(x) = \frac{1}{\Gamma(3-\nu)}\int_a^x (x-t)^{2-\nu}\varphi(t)\,dt,$$

and

$$\left|D_a^{\nu-3}\varphi(x)\right| \leq \frac{1}{\Gamma(3-\nu)}\int_a^x (x-t)^{2-\nu}|\varphi(t)|\,dt \leq \frac{\|\varphi\|_{\infty,[a,+\infty)}}{\Gamma(3-\nu)}\int_a^x (x-t)^{2-\nu}\,dt$$

$$= \frac{\|\varphi\|_{\infty,[a,+\infty)}}{\Gamma(3-\nu)}\frac{(x-a)^{3-\nu}}{(3-\nu)} = \frac{\|\varphi\|_{\infty,[a,+\infty)}}{\Gamma(4-\nu)}(x-a)^{3-\nu}. \tag{7.33}$$

That is

$$\left|D_a^{\nu-3}\varphi(x)\right| \leq \frac{\|\varphi\|_{\infty,[a,+\infty)}}{\Gamma(4-\nu)}(x-a)^{3-\nu}, \tag{7.34}$$

$\forall\, x \in [a, +\infty)$.

Clearly, it is $D_a^{\nu-3}\varphi(a) = 0$.

Hence by (7.32) we get

$$\varphi(b) = \frac{D_a^{\nu-1}\varphi(a)}{\Gamma(\nu)}(b-a)^{\nu-1} + \frac{D_a^{\nu-2}\varphi(a)}{\Gamma(\nu-1)}(b-a)^{\nu-2}$$

$$+\frac{1}{\Gamma(\nu)}\int_a^b (b-t)^{\nu-1} D_a^{\nu}\varphi(t)\,dt, \tag{7.35}$$

$\forall\, b \in (a, +\infty)$.

So that

$$\frac{D_a^{\nu-1}\varphi(a)}{\Gamma(\nu)}(b-a)^{\nu-1} + \frac{D_a^{\nu-2}\varphi(a)}{\Gamma(\nu-1)}(b-a)^{\nu-2} =$$

$$\varphi(b) - \frac{1}{\Gamma(\nu)}\int_a^b (b-t)^{\nu-1} D_a^{\nu}\varphi(t)\,dt, \tag{7.36}$$

$\forall\, b \in (a, +\infty)$.

Let $b = a + h$, $h > 0$, i.e. $b - a = h$, then

$$\frac{D_a^{\nu-1}\varphi(a)}{\Gamma(\nu)}h^{\nu-1} + \frac{D_a^{\nu-2}\varphi(a)}{\Gamma(\nu-1)}h^{\nu-2} =$$

$$\varphi(a+h) - \frac{1}{\Gamma(\nu)} \int_a^{a+h} (a+h-t)^{\nu-1} D_a^{\nu}\varphi(t)\, dt =: A. \quad (7.37)$$

For $b = a + 2h$, i.e. $b - a = 2h$, we get

$$\frac{D_a^{\nu-1}\varphi(a)}{\Gamma(\nu)} 2^{\nu-1} h^{\nu-1} + \frac{D_a^{\nu-2}\varphi(a)}{\Gamma(\nu-1)} 2^{\nu-2} h^{\nu-2} =$$

$$\varphi(a+2h) - \frac{1}{\Gamma(\nu)} \int_a^{a+2h} (a+2h-t)^{\nu-1} D_a^{\nu}\varphi(t)\, dt =: B. \quad (7.38)$$

We solve the system of two equations (7.37), (7.38) for two unknowns $D_a^{\nu-1}\varphi(a)$ and $D_a^{\nu-2}\varphi(a)$. We calculate the determinants:

$$D = \begin{vmatrix} \frac{h^{\nu-1}}{\Gamma(\nu)} & \frac{h^{\nu-2}}{\Gamma(\nu-1)} \\ \frac{2^{\nu-1}h^{\nu-1}}{\Gamma(\nu)} & \frac{2^{\nu-2}h^{\nu-2}}{\Gamma(\nu-1)} \end{vmatrix} = \frac{2^{\nu-2}h^{2\nu-3} - 2^{\nu-1}h^{2\nu-3}}{\Gamma(\nu)\Gamma(\nu-1)} =$$

$$\frac{h^{2\nu-3}}{\Gamma(\nu)\Gamma(\nu-1)} \left(2^{\nu-2} - 2^{\nu-1}\right) = -\frac{2^{\nu-2}h^{2\nu-3}}{\Gamma(\nu)\Gamma(\nu-1)}. \quad (7.39)$$

Hence we get

$$D = -\frac{2^{\nu-2}h^{2\nu-3}}{\Gamma(\nu)\Gamma(\nu-1)} < 0. \quad (7.40)$$

Next we find

$$D_1 = \begin{vmatrix} A & \frac{h^{\nu-2}}{\Gamma(\nu-1)} \\ B & \frac{2^{\nu-2}h^{\nu-2}}{\Gamma(\nu-1)} \end{vmatrix} = \frac{h^{\nu-2}}{\Gamma(\nu-1)} \left(2^{\nu-2}A - B\right), \quad (7.41)$$

and

$$D_2 = \begin{vmatrix} \frac{h^{\nu-1}}{\Gamma(\nu)} & A \\ \frac{2^{\nu-1}h^{\nu-1}}{\Gamma(\nu)} & B \end{vmatrix} = \frac{h^{\nu-1}}{\Gamma(\nu)} \left(B - 2^{\nu-1}A\right). \quad (7.42)$$

Therefore we derive

$$D_a^{\nu-1}\varphi(a) = \frac{D_1}{D} = \frac{\frac{h^{\nu-2}}{\Gamma(\nu-1)}\left(2^{\nu-2}A - B\right)}{-\frac{2^{\nu-2}h^{2\nu-3}}{\Gamma(\nu)\Gamma(\nu-1)}} = \frac{\Gamma(\nu)}{2^{\nu-2}} \frac{\left(B - 2^{\nu-2}A\right)}{h^{\nu-1}}. \quad (7.43)$$

Similarly, we get

$$D_a^{\nu-2}\varphi(a)=\frac{D_2}{D}=\frac{\frac{h^{\nu-1}}{\Gamma(\nu)}\left(B-2^{\nu-1}A\right)}{-\frac{2^{\nu-2}h^{2\nu-3}}{\Gamma(\nu)\Gamma(\nu-1)}}=\frac{\Gamma(\nu-1)}{2^{\nu-2}}\frac{\left(2^{\nu-1}A-B\right)}{h^{\nu-2}}. \tag{7.44}$$

We notice that

$$|A|=\left|\varphi(a+h)-\frac{1}{\Gamma(\nu)}\int_a^{a+h}(a+h-t)^{\nu-1}D_a^{\nu}\varphi(t)\,dt\right|\leq$$

$$\|\varphi\|_{\infty,\mathbb{R}_+}+\frac{\sup_{a\in\mathbb{R}_+}\left\|D_a^{\nu}\varphi\right\|_{\infty,\mathbb{R}_+}}{\Gamma(\nu+1)}h^{\nu}, \tag{7.45}$$

and

$$|B|=\left|\varphi(a+2h)-\frac{1}{\Gamma(\nu)}\int_a^{a+2h}(a+2h-t)^{\nu-1}D_a^{\nu}\varphi(t)\,dt\right|\leq$$

$$\|\varphi\|_{\infty,\mathbb{R}_+}+\frac{\sup_{a\in\mathbb{R}_+}\left\|D_a^{\nu}\varphi\right\|_{\infty,\mathbb{R}_+}}{\Gamma(\nu+1)}2^{\nu}h^{\nu}. \tag{7.46}$$

Therefore we have

$$\left|D_a^{\nu-1}\varphi(a)\right|\leq\frac{\Gamma(\nu)}{2^{\nu-2}}\frac{\left(|B|+2^{\nu-2}|A|\right)}{h^{\nu-1}}\leq\frac{\Gamma(\nu)}{2^{\nu-2}h^{\nu-1}}$$

$$\left(\|\varphi\|_{\infty,\mathbb{R}_+}+\frac{\sup_{a\in\mathbb{R}_+}\left\|D_a^{\nu}\varphi\right\|_{\infty,\mathbb{R}_+}}{\Gamma(\nu+1)}2^{\nu}h^{\nu}+2^{\nu-2}\|\varphi\|_{\infty,\mathbb{R}_+}+2^{\nu-2}\frac{\sup_{a\in\mathbb{R}_+}\left\|D_a^{\nu}\varphi\right\|_{\infty,\mathbb{R}_+}}{\Gamma(\nu+1)}h^{\nu}\right) \tag{7.47}$$

$$=\frac{\Gamma(\nu)}{2^{\nu-2}h^{\nu-1}}\left[\left(2^{\nu-2}+1\right)\|\varphi\|_{\infty,\mathbb{R}_+}+h^{\nu}\frac{\sup_{a\in\mathbb{R}_+}\left\|D_a^{\nu}\varphi\right\|_{\infty,\mathbb{R}_+}}{\Gamma(\nu+1)}\left(2^{\nu}+2^{\nu-2}\right)\right]$$

$$=\frac{\Gamma(\nu)}{2^{\nu-2}h^{\nu-1}}\left[\left(2^{\nu-2}+1\right)\|\varphi\|_{\infty,\mathbb{R}_+}+\frac{\left(5\cdot2^{\nu-2}\sup_{a\in\mathbb{R}_+}\left\|D_a^{\nu}\varphi\right\|_{\infty,\mathbb{R}_+}\right)}{\Gamma(\nu+1)}h^{\nu}\right]$$

$$=\frac{\left(1+2^{2-\nu}\right)\Gamma(\nu)\|\varphi\|_{\infty,\mathbb{R}_+}}{h^{\nu-1}}+\frac{\left(5\sup_{a\in\mathbb{R}_+}\left\|D_a^{\nu}\varphi\right\|_{\infty,\mathbb{R}_+}\right)}{\nu}h. \tag{7.48}$$

That is

$$\sup_{a\in\mathbb{R}_+}\left|D_a^{\nu-1}\varphi(a)\right| \leq \frac{\left(1+2^{2-\nu}\right)\Gamma(\nu)\|\varphi\|_{\infty,\mathbb{R}_+}}{h^{\nu-1}} + \frac{\left(5\left(\sup_{a\in\mathbb{R}_+}\left\|D_a^{\nu}\varphi\right\|_{\infty,\mathbb{R}_+}\right)\right)}{\nu}h, \tag{7.49}$$

$\forall\, h>0,\ 2<\nu<3.$

Furthermore we have

$$\left|D_a^{\nu-2}\varphi(a)\right| \leq \frac{\Gamma(\nu-1)}{2^{\nu-2}h^{\nu-2}}\left(2^{\nu-1}|A|+|B|\right) \leq$$

$$\frac{\Gamma(\nu-1)}{2^{\nu-2}h^{\nu-2}}\left[2^{\nu-1}\left(\|\varphi\|_{\infty,\mathbb{R}_+} + \frac{\sup_{a\in\mathbb{R}_+}\left\|D_a^{\nu}\varphi\right\|_{\infty,\mathbb{R}_+}}{\Gamma(\nu+1)}h^{\nu}\right)\right.$$

$$\left.+\left(\|\varphi\|_{\infty,\mathbb{R}_+} + \frac{\sup_{a\in\mathbb{R}_+}\left\|D_a^{\nu}\varphi\right\|_{\infty,\mathbb{R}_+}}{\Gamma(\nu+1)}2^{\nu}h^{\nu}\right)\right] = \tag{7.50}$$

$$\frac{\Gamma(\nu-1)}{2^{\nu-2}h^{\nu-2}}\left[\left(2^{\nu-1}+1\right)\|\varphi\|_{\infty,\mathbb{R}_+} + h^{\nu}\left(\frac{\sup_{a\in\mathbb{R}_+}\left\|D_a^{\nu}\varphi\right\|_{\infty,\mathbb{R}_+}}{\Gamma(\nu+1)}\right)\left(2^{\nu-1}+2^{\nu}\right)\right] =$$

$$\left[\frac{\left(2^{\nu-1}+1\right)\Gamma(\nu-1)\|\varphi\|_{\infty,\mathbb{R}_+}}{2^{\nu-2}h^{\nu-2}} + \Gamma(\nu-1)\left(\frac{\sup_{a\in\mathbb{R}_+}\left\|D_a^{\nu}\varphi\right\|_{\infty,\mathbb{R}_+}}{\Gamma(\nu+1)}\right)\frac{\left(2^{\nu-1}+2^{\nu}\right)}{2^{\nu-2}}h^2\right] \tag{7.51}$$

$$= \frac{\left(2\left(1+2^{1-\nu}\right)\Gamma(\nu-1)\|\varphi\|_{\infty,\mathbb{R}_+}\right)}{h^{\nu-2}} + \frac{6\left(\sup_{a\in\mathbb{R}_+}\left\|D_a^{\nu}\varphi\right\|_{\infty,\mathbb{R}_+}\right)}{\nu(\nu-1)}h^2.$$

That is

$$\sup_{a\in\mathbb{R}_+}\left|D_a^{\nu-2}\varphi(a)\right| \leq \frac{\left(2\left(1+2^{1-\nu}\right)\Gamma(\nu-1)\|\varphi\|_{\infty,\mathbb{R}_+}\right)}{h^{\nu-2}}$$

$$+\left(\frac{6\left(\sup_{a\in\mathbb{R}_+}\left\|D_a^{\nu}\varphi\right\|_{\infty,\mathbb{R}_+}\right)}{\nu(\nu-1)}\right)h^2, \quad \forall h>0,\ 2<\nu<3. \tag{7.52}$$

Next we working on (7.49).

Call

$$\mu := \left(1 + 2^{2-\nu}\right) \Gamma(\nu) \|\varphi\|_{\infty,\mathbb{R}_+},$$

$$\theta := \left(\frac{5\left(\sup_{a\in\mathbb{R}_+} \left\| D_a^{\nu}\varphi \right\|_{\infty,\mathbb{R}_+} \right)}{\nu} \right), \tag{7.53}$$

both are greater than zero.

Here $2 < \nu < 3$ and $1 < \nu - 1 < 2$. Set also $\rho := \nu - 1$, i.e. $1 < \rho < 2$.

We consider again the function

$$y(h) := \mu h^{-\rho} + \theta h, \quad \forall\, h > 0. \tag{7.54}$$

As in the proof of Theorem 7.5 y has a global minimum at $h_0 = \left(\frac{\rho\mu}{\theta}\right)^{\frac{1}{\rho+1}} > 0$, which is

$$y(h_0) = \frac{(\rho+1)}{\rho^{\frac{\rho}{\rho+1}}} \theta^{\frac{\rho}{\rho+1}} \mu^{\frac{1}{\rho+1}}. \tag{7.55}$$

Consequently we derive

$$\sup_{a\in\mathbb{R}_+} \left| D_a^{\nu-1}\varphi(a) \right| \leq \frac{\nu}{(\nu-1)^{\left(\frac{\nu-1}{\nu}\right)}}$$

$$\left(\frac{5\left(\sup_{a\in\mathbb{R}_+} \left\| D_a^{\nu}\varphi \right\|_{\infty,\mathbb{R}_+} \right)}{\nu} \right)^{\left(\frac{\nu-1}{\nu}\right)} \left(\left(1 + 2^{2-\nu}\right) \Gamma(\nu) \|\varphi\|_{\infty,\mathbb{R}_+} \right)^{\frac{1}{\nu}} =$$

$$\left(\frac{5}{\nu-1}\right)^{\left(\frac{\nu-1}{\nu}\right)} \left(\nu\left(1 + 2^{2-\nu}\right)\Gamma(\nu)\right)^{\frac{1}{\nu}} \|\varphi\|_{\infty,\mathbb{R}_+}^{\frac{1}{\nu}} \left(\sup_{a\in\mathbb{R}_+} \left\| D_a^{\nu}\varphi \right\|_{\infty,\mathbb{R}_+} \right)^{\left(\frac{\nu-1}{\nu}\right)}, \tag{7.56}$$

$2 < \nu < 3$.

Next we are working on (7.52).

Call

$$\xi := \left(2\left(1 + 2^{1-\nu}\right) \Gamma(\nu-1) \|\varphi\|_{\infty,\mathbb{R}_+}\right), \tag{7.57}$$

$$\psi := \left(\frac{6\left(\sup_{a\in\mathbb{R}_+} \left\| D_a^{\nu}\varphi \right\|_{\infty,\mathbb{R}_+} \right)}{\nu(\nu-1)} \right), \tag{7.58}$$

both are greater than zero.

Call also $\rho := \nu - 2, 2 < \nu < 3$.

We consider the function

$$y(h) := \xi h^{-\rho} + \psi h^2, \quad \forall\, h > 0. \tag{7.59}$$

Hence

$$y'(h) = -\rho\xi h^{-(\rho+1)} + 2\psi h = 0,$$

with the only critical number

$$h_0 = h_{crit.} = \left(\frac{\rho\xi}{2\psi}\right)^{\frac{1}{\rho+2}}. \tag{7.60}$$

We need

$$y''(h) = \rho(\rho+1)\xi h^{-(\rho+2)} + 2\psi. \tag{7.61}$$

We calculate

$$y''(h_0) = \rho(\rho+1)\xi h_0^{-(\rho+2)} + 2\psi =$$

$$\rho(\rho+1)\xi\left(\left(\frac{\rho\xi}{2\psi}\right)^{\frac{1}{\rho+2}}\right)^{-(\rho+2)} + 2\psi =$$

$$\rho(\rho+1)\xi\left(\frac{2\psi}{\rho\xi}\right) + 2\psi = (\rho+1)2\psi + 2\psi = 2\psi(\rho+2) > 0. \tag{7.62}$$

Thus y has a global minimum, which is

$$y(h_0) = \xi h_0^{-\rho} + \psi h_0^2 = h_0^2\left(\xi h_0^{-(\rho+2)} + \psi\right) =$$

$$\left(\frac{\rho\xi}{2\psi}\right)^{\frac{2}{\rho+2}}\left(\xi\left(\left(\frac{\rho\xi}{2\psi}\right)^{\frac{1}{\rho+2}}\right)^{-(\rho+2)} + \psi\right) =$$

$$\left(\frac{\rho\xi}{2\psi}\right)^{\frac{2}{\rho+2}}\left(\xi\left(\frac{2\psi}{\rho\xi}\right) + \psi\right) = \left(\frac{\rho\xi}{2\psi}\right)^{\frac{2}{\rho+2}}\left(\frac{2\psi}{\rho} + \psi\right) =$$

$$\psi\left(\frac{\rho+2}{\rho}\right)\left(\frac{\rho\xi}{2\psi}\right)^{\frac{2}{\rho+2}} = (\rho+2)\left(\frac{\xi}{2}\right)^{\frac{2}{2+\rho}}\left(\frac{\psi}{\rho}\right)^{\frac{\rho}{\rho+2}}. \tag{7.63}$$

That is

$$y\left(h_0\right)=(\rho+2)\left(\frac{\xi}{2}\right)^{\frac{2}{2+\rho}}\left(\frac{\psi}{\rho}\right)^{\frac{\rho}{\rho+2}}. \tag{7.64}$$

Therefore it holds

$$\sup_{a\in\mathbb{R}_+}\left|D_a^{\nu-2}\varphi(a)\right|\leq\nu\left(\left(1+2^{1-\nu}\right)\Gamma(\nu-1)\|\varphi\|_{\infty,\mathbb{R}_+}\right)^{\frac{2}{\nu}}$$

$$\left(\frac{6\left(\sup_{a\in\mathbb{R}_+}\left\|D_a^{\nu}\varphi\right\|_{\infty,\mathbb{R}_+}\right)}{\nu(\nu-1)(\nu-2)}\right)^{\left(\frac{\nu-2}{\nu}\right)}=\left(\frac{6}{(\nu-1)(\nu-2)}\right)^{\left(\frac{\nu-2}{\nu}\right)}$$

$$\left(\nu\left(1+2^{1-\nu}\right)\Gamma(\nu-1)\right)^{\frac{2}{\nu}}\|\varphi\|_{\infty,\mathbb{R}_+}^{\frac{2}{\nu}}\left(\sup_{a\in\mathbb{R}_+}\left\|D_a^{\nu}\varphi\right\|_{\infty,\mathbb{R}_+}\right)^{\left(\frac{\nu-2}{\nu}\right)}, \tag{7.65}$$

$2<\nu<3$.

The theorem is proved. ■

Next we give right Riemann–Liouville fractional Landau type inequalities.

Convention 7.7 Define $D_{b-}^{\nu}\varphi(x):=0$, for any $x>b$, for any $\nu>0$, $\nu\notin\mathbb{N}$.

We present

Theorem 7.8 *Here* $1<\nu<2$ *and* $\varphi\in C_B(\mathbb{R}_-)$ *and* $D_{b-}^{\nu}\varphi\in C_B((-\infty,b])$, $\forall\, b\in\mathbb{R}_-$, *where* C_B *means continuous and bounded functions.*

We assume that $\sup_{b\in\mathbb{R}_-}\left\|D_{b-}^{\nu}\varphi\right\|_{\infty,\mathbb{R}_-}<+\infty$.

Then

$$\sup_{b\in\mathbb{R}_-}\left|D_{b-}^{\nu-1}\varphi(b)\right|\leq\left(\frac{\Gamma(\nu+1)}{(\nu-1)^{\nu-1}}\right)^{\frac{1}{\nu}}\|\varphi\|_{\infty,\mathbb{R}_-}^{\frac{1}{\nu}}\left(\sup_{b\in\mathbb{R}_-}\left\|D_{b-}^{\nu}\varphi\right\|_{\infty,\mathbb{R}_-}\right)^{\frac{\nu-1}{\nu}}, \tag{7.66}$$

a sharp inequality with a precisely determined constant. That is $\sup_{b\in\mathbb{R}_-}\left|D_{b-}^{\nu-1}\varphi(b)\right|<+\infty$.

Proof For any $a\in(-\infty,b)$ we get (by Theorem 7.3)

$$\varphi(a)=\frac{D_{b-}^{\nu-1}\varphi(b)}{\Gamma(\nu)}(b-a)^{\nu-1}+\frac{D_{b-}^{\nu-2}\varphi(b)}{\Gamma(\nu-1)}(b-a)^{\nu-2}$$

$$+\frac{1}{\Gamma(\nu)}\int_a^b(t-a)^{\nu-1}D_{b-}^{\nu}\varphi(t)\,dt. \tag{7.67}$$

Notice that ($x \in (-\infty, b]$)

$$D_{b-}^{\nu-2}\varphi(x) = D_{b-}^{-(2-\nu)}\varphi(x) = \frac{1}{\Gamma(2-\nu)}\int_x^b (t-x)^{1-\nu}\varphi(t)\,dt,$$

and

$$\left|D_{b-}^{\nu-2}\varphi(x)\right| \leq \frac{1}{\Gamma(2-\nu)}\int_x^b (t-x)^{1-\nu}|\varphi(t)|\,dt \leq \frac{\|\varphi\|_{\infty,(-\infty,b]}}{\Gamma(2-\nu)}\int_x^b (t-x)^{1-\nu}\,dt$$

$$= \frac{\|\varphi\|_{\infty,(-\infty,b]}}{\Gamma(3-\nu)}(b-x)^{2-\nu}. \tag{7.68}$$

That is

$$\left|D_{b-}^{\nu-2}\varphi(x)\right| \leq \frac{\|\varphi\|_{\infty,(-\infty,b]}}{\Gamma(3-\nu)}(b-x)^{2-\nu}, \tag{7.69}$$

$\forall\, x \in (-\infty, b]$.

In particular we have $D_{b-}^{\nu-2}\varphi(b) = 0$.

Hence by (7.67) we get

$$\varphi(a) = \frac{D_{b-}^{\nu-1}\varphi(b)}{\Gamma(\nu)}(b-a)^{\nu-1} + \frac{1}{\Gamma(\nu)}\int_a^b (t-a)^{\nu-1} D_{b-}^{\nu}\varphi(t)\,dt, \tag{7.70}$$

$\forall\, a \in (-\infty, b)$.

So that

$$\frac{D_{b-}^{\nu-1}\varphi(b)}{\Gamma(\nu)}(b-a)^{\nu-1} = \varphi(a) - \frac{1}{\Gamma(\nu)}\int_a^b (t-a)^{\nu-1} D_{b-}^{\nu}\varphi(t)\,dt, \tag{7.71}$$

$\forall\, a \in (-\infty, b)$.

Hence (set $h := b - a > 0$)

$$\left|D_{b-}^{\nu-1}\varphi(b)\right| h^{\nu-1} = \left|\Gamma(\nu)\varphi(a) - \int_a^b (t-a)^{\nu-1} D_{b-}^{\nu}\varphi(t)\,dt\right| \leq$$

$$\Gamma(\nu)|\varphi(a)| + \int_a^b (t-a)^{\nu-1}\left|D_{b-}^{\nu}\varphi(t)\right| dt \leq$$

$$\Gamma(\nu)\|\varphi\|_{\infty,\mathbb{R}_-} + \left\|D_{b-}^{\nu}\varphi\right\|_{\infty,(-\infty,b]}\int_a^b (t-a)^{\nu-1}\,dt =$$

$$\Gamma(\nu)\|\varphi\|_{\infty,\mathbb{R}_-} + \left\|D_{b-}^{\nu}\varphi\right\|_{\infty,(-\infty,b]}\frac{(b-a)^{\nu}}{\nu} = \Gamma(\nu)\|\varphi\|_{\infty,\mathbb{R}_-} + \left\|D_{b-}^{\nu}\varphi\right\|_{\infty,(-\infty,b]}\frac{h^{\nu}}{\nu}. \tag{7.72}$$

That is

$$\left|D_{b-}^{\nu-1}\varphi(b)\right| h^{\nu-1} \leq \Gamma(\nu)\|\varphi\|_{\infty,\mathbb{R}_-} + \left\|D_{b-}^{\nu}\varphi\right\|_{\infty,(-\infty,b]}\frac{h^{\nu}}{\nu} \leq \tag{7.73}$$

$$\Gamma(\nu)\|\varphi\|_{\infty,\mathbb{R}_-} + \left(\sup_{b\in\mathbb{R}_-}\left\|D_{b-}^{\nu}\varphi\right\|_{\infty,(-\infty,b]}\right)\frac{h^{\nu}}{\nu}.$$

Furthermore we see that

$$\left|D_{b-}^{\nu-1}\varphi(b)\right| \leq \frac{\Gamma(\nu)\|\varphi\|_{\infty,\mathbb{R}_-}}{h^{\nu-1}} + \frac{\left(\sup_{b\in\mathbb{R}_-}\left\|D_{b-}^{\nu}\varphi\right\|_{\infty,(-\infty,b]}\right)}{\nu}h, \tag{7.74}$$

$\forall\, h > 0$ and $\forall\, b \in \mathbb{R}_-$.

Consequently it holds

$$\sup_{b\in\mathbb{R}_-}\left|D_{b-}^{\nu-1}\varphi(b)\right| \leq \frac{\Gamma(\nu)\|\varphi\|_{\infty,\mathbb{R}_-}}{h^{\nu-1}} + \frac{\left(\sup_{b\in\mathbb{R}_-}\left\|D_{b-}^{\nu}\varphi\right\|_{\infty,\mathbb{R}_-}\right)}{\nu}h, \tag{7.75}$$

$\forall\, h > 0$.

Call

$$\mu := \Gamma(\nu)\|\varphi\|_{\infty,\mathbb{R}_-}, \tag{7.76}$$

$$\theta := \left(\frac{\sup_{b\in\mathbb{R}_-}\left\|D_{b-}^{\nu}\varphi\right\|_{\infty,\mathbb{R}_-}}{\nu}\right), \tag{7.77}$$

both are greater than zero.

Here $1 < \nu < 2$ and $0 < \nu - 1 < 1$. Set also $\rho := \nu - 1$, i.e. $0 < \rho < 1$.

We consider the function

$$y(h) := \mu h^{-\rho} + \theta h, \quad \forall\, h > 0. \tag{7.78}$$

As in the proof of Theorem 7.5, y has a global minimum over $\mathbb{R}_+$ which is

$$y(h_0) = \frac{(\rho+1)}{\rho^{\frac{\rho}{\rho+1}}}\theta^{\frac{\rho}{\rho+1}}\mu^{\frac{1}{\rho+1}}, \tag{7.79}$$

where the critical number

$$h_0 = \left(\frac{\rho\mu}{\theta}\right)^{\frac{1}{\rho+1}} > 0.$$

Consequently we derive that

$$\sup_{b\in\mathbb{R}_-}\left|D_{b-}^{\nu-1}\varphi(b)\right| \leq \left(\frac{\nu}{(\nu-1)^{\frac{\nu-1}{\nu}}}\right)\left(\frac{\sup_{b\in\mathbb{R}_-}\left\|D_{b-}^{\nu}\varphi\right\|_{\infty,\mathbb{R}_-}}{\nu}\right)^{\frac{\nu-1}{\nu}}\left(\Gamma(\nu)\|\varphi\|_{\infty,\mathbb{R}_-}\right)^{\frac{1}{\nu}}$$

$$= \left(\frac{\Gamma(\nu+1)}{(\nu-1)^{\nu-1}}\right)^{\frac{1}{\nu}}\|\varphi\|_{\infty,\mathbb{R}_-}^{\frac{1}{\nu}}\left(\sup_{b\in\mathbb{R}_-}\left\|D_{b-}^{\nu}\varphi\right\|_{\infty,\mathbb{R}_-}\right)^{\frac{\nu-1}{\nu}}, \tag{7.80}$$

for $1 < \nu < 2$.

The theorem is proved. ∎

We continue with

Theorem 7.9 *Here* $2 < \nu < 3$ *and* $\varphi \in C_B(\mathbb{R}_-)$ *and* $D_{b-}^{\nu}\varphi \in C_B((-\infty, b])$, $\forall\, b \in \mathbb{R}_-$. *We assume that* $\sup_{b\in\mathbb{R}_-}\left\|D_{b-}^{\nu}\varphi\right\|_{\infty,\mathbb{R}_-} < +\infty$. *Then*

(i)

$$\sup_{b\in\mathbb{R}_-}\left|D_{b-}^{\nu-1}\varphi(b)\right| \leq$$

$$\left(\frac{5}{\nu-1}\right)^{\left(\frac{\nu-1}{\nu}\right)}\left(\nu\left(1+2^{2-\nu}\right)\Gamma(\nu)\right)^{\frac{1}{\nu}}\|\varphi\|_{\infty,\mathbb{R}_-}^{\frac{1}{\nu}}\left(\sup_{b\in\mathbb{R}_-}\left\|D_{b-}^{\nu}\varphi\right\|_{\infty,\mathbb{R}_-}\right)^{\left(\frac{\nu-1}{\nu}\right)}, \tag{7.81}$$

and

(ii)

$$\sup_{b\in\mathbb{R}_-}\left|D_{b-}^{\nu-2}\varphi(b)\right| \leq \left(\frac{6}{(\nu-1)(\nu-2)}\right)^{\left(\frac{\nu-2}{\nu}\right)}$$

$$\left(\nu\left(1+2^{1-\nu}\right)\Gamma(\nu-1)\right)^{\frac{2}{\nu}}\|\varphi\|_{\infty,\mathbb{R}_-}^{\frac{2}{\nu}}\left(\sup_{b\in\mathbb{R}_-}\left\|D_{b-}^{\nu}\varphi\right\|_{\infty,\mathbb{R}_-}\right)^{\left(\frac{\nu-2}{\nu}\right)}. \tag{7.82}$$

The above inequalities are nearly sharp with precisely determined constants. That is $\sup_{b\in\mathbb{R}_-}\left|D_{b-}^{\nu-1}\varphi(b)\right|, \sup_{b\in\mathbb{R}_-}\left|D_{b-}^{\nu-2}\varphi(b)\right| < +\infty$.

Proof For any $a \in (-\infty, b)$ we get (by Theorem 7.3)

$$\varphi(a) = \frac{D_{b-}^{\nu-1}\varphi(b)}{\Gamma(\nu)}(b-a)^{\nu-1} + \frac{D_{b-}^{\nu-2}\varphi(b)}{\Gamma(\nu-1)}(b-a)^{\nu-2} + \tag{7.83}$$

$$\frac{D_{b-}^{\nu-3}\varphi(b)}{\Gamma(\nu-2)}(b-a)^{\nu-3} + \frac{1}{\Gamma(\nu)}\int_a^b (t-a)^{\nu-1} D_{b-}^{\nu}\varphi(t)\,dt.$$

Notice that ($x \in (-\infty, b]$)

$$D_{b-}^{\nu-3}\varphi(x) = D_{b-}^{-(3-\nu)}\varphi(x) = \frac{1}{\Gamma(3-\nu)} \int_x^b (t-x)^{2-\nu} \varphi(t)\, dt,$$

and

$$\left| D_{b-}^{\nu-3}\varphi(x) \right| \leq \frac{\|\varphi\|_{\infty,(-\infty,b]}}{\Gamma(4-\nu)} (b-x)^{3-\nu}, \tag{7.84}$$

$\forall\, x \in (-\infty, b]$.

That is $D_{b-}^{\nu-3}\varphi(b) = 0$.

Hence by (7.83) we get

$$\frac{D_{b-}^{\nu-1}\varphi(b)}{\Gamma(\nu)} (b-a)^{\nu-1} + \frac{D_{b-}^{\nu-2}\varphi(b)}{\Gamma(\nu-1)} (b-a)^{\nu-2} =$$

$$\varphi(a) - \frac{1}{\Gamma(\nu)} \int_a^b (t-a)^{\nu-1} D_{b-}^{\nu}\varphi(t)\, dt, \tag{7.85}$$

$\forall\, a \in (-\infty, b)$.

Let $h > 0 : b - a = h,$, i.e. $b = a + h$, then

$$\frac{D_{b-}^{\nu-1}\varphi(b)}{\Gamma(\nu)} h^{\nu-1} + \frac{D_{b-}^{\nu-2}\varphi(b)}{\Gamma(\nu-1)} h^{\nu-2} =$$

$$\varphi(a) - \frac{1}{\Gamma(\nu)} \int_a^{a+h} (t-a)^{\nu-1} D_{b-}^{\nu}\varphi(t)\, dt =: \overline{A}. \tag{7.86}$$

For $b - a = 2h$, i.e. $b = a + 2h$, we get

$$\frac{D_{b-}^{\nu-1}\varphi(b)}{\Gamma(\nu)} 2^{\nu-1} h^{\nu-1} + \frac{D_{b-}^{\nu-2}\varphi(b)}{\Gamma(\nu-1)} 2^{\nu-2} h^{\nu-2} =$$

$$\varphi(a) - \frac{1}{\Gamma(\nu)} \int_a^{a+2h} (t-a)^{\nu-1} D_{b-}^{\nu}\varphi(t)\, dt =: \overline{B}. \tag{7.87}$$

We solve the system of two equations (7.86), (7.87) for two unknowns $D_{b-}^{\nu-1}\varphi(b)$ and $D_{b-}^{\nu-2}\varphi(b)$. As in the proof of Theorem 7.6 we find

$$D_{b-}^{\nu-1}\varphi(b) = \frac{\Gamma(\nu)}{2^{\nu-2}} \frac{\left(\overline{B} - 2^{\nu-2}\overline{A}\right)}{h^{\nu-1}}, \tag{7.88}$$

and

$$D_{b-}^{\nu-2}\varphi(b) = \frac{\Gamma(\nu-1)}{2^{\nu-2}} \frac{\left(2^{\nu-1}\overline{A} - \overline{B}\right)}{h^{\nu-2}}. \tag{7.89}$$

We notice that

$$\left|\overline{A}\right| \leq \|\varphi\|_{\infty,\mathbb{R}_-} + \left(\frac{\sup\limits_{b\in\mathbb{R}_-} \left\| D_{b-}^{\nu}\varphi \right\|_{\infty,\mathbb{R}_-}}{\Gamma(\nu+1)} \right) h^{\nu}, \tag{7.90}$$

and

$$\left|\overline{B}\right| \leq \|\varphi\|_{\infty,\mathbb{R}_-} + \left(\frac{\sup\limits_{b\in\mathbb{R}_-} \left\| D_{b-}^{\nu}\varphi \right\|_{\infty,\mathbb{R}_-}}{\Gamma(\nu+1)} \right) 2^{\nu} h^{\nu}. \tag{7.91}$$

As in the proof of Theorem 7.6 we obtain

$$\sup_{b\in\mathbb{R}_-} \left| D_{b-}^{\nu-1}\varphi(b) \right| \leq \frac{\left(1+2^{2-\nu}\right)\Gamma(\nu)\|\varphi\|_{\infty,\mathbb{R}_-}}{h^{\nu-1}} + \frac{\left(5\left(\sup\limits_{b\in\mathbb{R}_-} \left\| D_{b-}^{\nu}\varphi \right\|_{\infty,\mathbb{R}_-}\right)\right)}{\nu} h, \tag{7.92}$$

and

$$\sup_{b\in\mathbb{R}_-} \left| D_{b-}^{\nu-2}\varphi(b) \right| \leq \frac{\left(2\left(1+2^{1-\nu}\right)\Gamma(\nu-1)\|\varphi\|_{\infty,\mathbb{R}_-}\right)}{h^{\nu-2}}$$

$$+ \left(\frac{6\left(\sup\limits_{b\in\mathbb{R}_-} \left\| D_{b-}^{\nu}\varphi \right\|_{\infty,\mathbb{R}_-}\right)}{\nu(\nu-1)} \right) h^2, \tag{7.93}$$

$\forall\, h > 0,\ 2 < \nu < 3.$

As in the proof of Theorem 7.6 we find similarly optimal upper bounds in (7.92) and (7.93).

We derive

$$\sup_{b\in\mathbb{R}_-} \left| D_{b-}^{\nu-1}\varphi(b) \right| \leq \frac{\nu}{(\nu-1)^{\left(\frac{\nu-1}{\nu}\right)}}$$

$$\left(\frac{5\left(\sup\limits_{b\in\mathbb{R}_-} \left\| D_{b-}^{\nu}\varphi \right\|_{\infty,\mathbb{R}_-}\right)}{\nu} \right)^{\left(\frac{\nu-1}{\nu}\right)} \left(\left(1+2^{2-\nu}\right)\Gamma(\nu)\|\varphi\|_{\infty,\mathbb{R}_-}\right)^{\frac{1}{\nu}} =$$

$$\left(\frac{5}{\nu-1}\right)^{\left(\frac{\nu-1}{\nu}\right)} \left(\nu\left(1+2^{2-\nu}\right)\Gamma(\nu)\right)^{\frac{1}{\nu}} \|\varphi\|_{\infty,\mathbb{R}_-}^{\frac{1}{\nu}} \left(\sup_{b\in\mathbb{R}_-} \left\| D_{b-}^{\nu}\varphi \right\|_{\infty,\mathbb{R}_-}\right)^{\left(\frac{\nu-1}{\nu}\right)}, \tag{7.94}$$

and

$$\sup_{b\in\mathbb{R}_-}\left|D_{b-}^{\nu-2}\varphi(b)\right| \leq \nu\left(\left(1+2^{1-\nu}\right)\Gamma(\nu-1)\|\varphi\|_{\infty,\mathbb{R}_-}\right)^{\frac{2}{\nu}}$$

$$\left(\frac{6\left(\sup_{b\in\mathbb{R}_-}\left\|D_{b-}^{\nu}\varphi\right\|_{\infty,\mathbb{R}_-}\right)}{\nu(\nu-1)(\nu-2)}\right)^{\left(\frac{\nu-2}{\nu}\right)} = \left(\frac{6}{(\nu-1)(\nu-2)}\right)^{\left(\frac{\nu-2}{\nu}\right)}$$

$$\left(\nu\left(1+2^{1-\nu}\right)\Gamma(\nu-1)\right)^{\frac{2}{\nu}}\|\varphi\|_{\infty,\mathbb{R}_-}^{\frac{2}{\nu}}\left(\sup_{b\in\mathbb{R}_-}\left\|D_{b-}^{\nu}\varphi\right\|_{\infty,\mathbb{R}_-}\right)^{\left(\frac{\nu-2}{\nu}\right)}, \tag{7.95}$$

$2<\nu<3$.

The theorem is proved. ■

An application of Theorem 7.5 follows ($\nu = 1.5$).

Corollary 7.10 *Let $\varphi \in C_B(\mathbb{R}_+)$ such that $D_a^{1.5}\varphi \in C_B([a,+\infty))$, $\forall\ a \in \mathbb{R}_+$. Assume that $\sup_{a\in\mathbb{R}_+}\left\|D_a^{1.5}\varphi\right\|_{\infty,\mathbb{R}_+} < +\infty$. Then*

$$\sup_{a\in\mathbb{R}_+}\left|D_a^{0.5}\varphi(a)\right| \leq (1.52323665)\|\varphi\|_{\infty,\mathbb{R}_+}^{0.666666667}\left(\sup_{a\in\mathbb{R}_+}\left\|D_a^{1.5}\varphi\right\|_{\infty,\mathbb{R}_+}\right)^{0.333333333}, \tag{7.96}$$

a sharp inequality. That is $\sup_{a\in\mathbb{R}_+}\left|D_a^{0.5}\varphi(a)\right| < +\infty$.

An application of Theorem 7.8 follows ($\nu = 1.5$).

Corollary 7.11 *Let $\varphi \in C_B(\mathbb{R}_-)$ such that $D_{b-}^{1.5}\varphi \in C_B((-\infty,b])$, $\forall\ b \in \mathbb{R}_-$. Assume that $\sup_{b\in\mathbb{R}_-}\left\|D_{b-}^{1.5}\varphi\right\|_{\infty,\mathbb{R}_-} < +\infty$. Then*

$$\sup_{b\in\mathbb{R}_-}\left|D_{b-}^{0.5}\varphi(b)\right| \leq (1.52323665)\|\varphi\|_{\infty,\mathbb{R}_-}^{0.666666667}\left(\sup_{b\in\mathbb{R}_-}\left\|D_{b-}^{1.5}\varphi\right\|_{\infty,\mathbb{R}_-}\right)^{0.333333333}, \tag{7.97}$$

a sharp inequality. That is $\sup_{b\in\mathbb{R}_-}\left|D_{b-}^{0.5}\varphi(b)\right| < +\infty$.

7.3 Appendix

We give the following simplified fractional Taylor formulae.

Corollary 7.12 *Let* $\nu > 0$, $\nu \notin \mathbb{N}$ *and* $m = \lceil \nu \rceil$; φ, $D_a^{\nu}\varphi \in C([a,b])$. *Then*

$$\varphi(b) = \sum_{j=1}^{m-1} \frac{D_a^{\nu-j}\varphi(a)}{\Gamma(\nu-j+1)} (b-a)^{\nu-j} + \frac{1}{\Gamma(\nu)} \int_a^b (b-t)^{\nu-1} D_a^{\nu}\varphi(t)\, dt. \quad (7.98)$$

Proof It is $m > \nu$, hence $m - \nu > 0$. We have that

$$D_a^{\nu-m}\varphi(x) = D_a^{-(m-\nu)}\varphi(x) = \frac{1}{\Gamma(m-\nu)} \int_a^x (x-t)^{m-\nu-1} \varphi(t)\, dt,$$

and

$$\left|D_a^{\nu-m}\varphi(x)\right| \leq \frac{1}{\Gamma(m-\nu)} \int_a^x (x-t)^{m-\nu-1} |\varphi(t)|\, dt \leq$$

$$\frac{\|\varphi\|_{\infty,[a,b]}}{\Gamma(m-\nu)} \int_a^x (x-t)^{m-\nu-1}\, dt = \frac{\|\varphi\|_{\infty,[a,b]}}{\Gamma(m-\nu)} \frac{(x-a)^{m-\nu}}{(m-\nu)} \quad (7.99)$$

$$= \frac{\|\varphi\|_{\infty,[a,b]}}{\Gamma(m-\nu+1)} (x-a)^{m-\nu}.$$

That is

$$\left|D_a^{\nu-m}\varphi(x)\right| \leq \frac{\|\varphi\|_{\infty,[a,b]}}{\Gamma(m-\nu+1)} (x-a)^{m-\nu}, \quad (7.100)$$

$\forall\, x \in [a,b]$.

Hence $D_a^{\nu-m}\varphi(a) = 0$.

Use also Theorem 7.2. ∎

Corollary 7.13 *Let* $\nu > 0$, $\nu \notin \mathbb{N}$ *and* $m = \lceil \nu \rceil$; φ, $D_{b-}^{\nu}\varphi \in C([a,b])$. *Then*

$$\varphi(a) = \sum_{j=1}^{m-1} \frac{D_{b-}^{\nu-j}\varphi(b)}{\Gamma(\nu-j+1)} (b-a)^{\nu-j} + \frac{1}{\Gamma(\nu)} \int_a^b (t-a)^{\nu-1} D_{b-}^{\nu}\varphi(t)\, dt. \quad (7.101)$$

Proof We have that

$$D_{b-}^{\nu-m}\varphi(x) = D_{b-}^{-(m-\nu)}\varphi(x) = \frac{1}{\Gamma(m-\nu)} \int_x^b (t-x)^{m-\nu-1} \varphi(t)\, dt,$$

and

$$\left|D_{b-}^{\nu-m}\varphi(x)\right| \leq \frac{\|\varphi\|_{\infty,[a,b]}}{\Gamma(m-\nu+1)} (b-x)^{m-\nu}, \quad (7.102)$$

$\forall\, x \in [a,b]$.

Hence $D_{b-}^{\nu-m}\varphi(b) = 0$. Use also Theorem 7.3. ∎

We give also an interesting new fractional inequality of Ostrowski like.

Theorem 7.14 *Let $\nu > 0$, $\nu \notin \mathbb{N}$ and $m = \lceil \nu \rceil$, with $\varphi \in C([a,b])$. Let also $x_0 \in [a,b]$ be fixed. Assumed that $D_{x_0}^{\varphi} \in C([x_0,b])$ and $D_{x_0-}^{\nu}\varphi \in C([a,x_0])$. Then*

$$\left| \int_a^b \varphi(x)\,dx - \sum_{j=1}^{m-1} \frac{1}{\Gamma(\nu-j+2)} \left[D_{x_0}^{\nu-j}\varphi(x_0)(b-x_0)^{\nu-j+1} \right.\right.$$

$$\left.\left. + D_{x_0-}^{\nu-j}\varphi(x_0)(x_0-a)^{\nu-j+1} \right] \right| \leq$$

$$\frac{1}{\Gamma(\nu+2)} \left[\left\| D_{x_0}^{\nu}\varphi \right\|_{\infty,[x_0,b]} (b-x_0)^{\nu+1} + \left\| D_{x_0-}^{\nu}\varphi \right\|_{\infty,[a,x_0]} (x_0-a)^{\nu+1} \right] \leq$$

$$\frac{\max\left\{ \left\| D_{x_0-}^{\nu}\varphi \right\|_{\infty,[a,x_0]}, \left\| D_{x_0}^{\nu}\varphi \right\|_{\infty,[x_0,b]} \right\}}{\Gamma(\nu+2)} \left[(b-x_0)^{\nu+1} + (x_0-a)^{\nu+1} \right]. \quad (7.103)$$

Proof We have that

$$\varphi(x) = \sum_{j=1}^{m-1} \frac{D_{x_0}^{\nu-j}\varphi(x_0)}{\Gamma(\nu-j+1)} (x-x_0)^{\nu-j} + \frac{1}{\Gamma(\nu)} \int_{x_0}^{x} (x-t)^{\nu-1} D_{x_0}^{\nu}\varphi(t)\,dt, \quad (7.104)$$

$\forall\, x \in [x_0, b]$.

Also we have

$$\varphi(x) = \sum_{j=1}^{m-1} \frac{D_{x_0-}^{\nu-j}\varphi(x_0)}{\Gamma(\nu-j+1)} (x_0-x)^{\nu-j} + \frac{1}{\Gamma(\nu)} \int_{x}^{x_0} (t-x)^{\nu-1} D_{x_0-}^{\nu}\varphi(t)\,dt, \quad (7.105)$$

$\forall\, x \in [a, x_0]$.

Thus

$$\varphi(x) - \sum_{j=1}^{m-1} \frac{D_{x_0}^{\nu-j}\varphi(x_0)}{\Gamma(\nu-j+1)} (x-x_0)^{\nu-j} = \frac{1}{\Gamma(\nu)} \int_{x_0}^{x} (x-t)^{\nu-1} D_{x_0}^{\nu}\varphi(t)\,dt, \quad (7.106)$$

$\forall\, x \in [x_0, b]$,

and

$$\varphi(x) - \sum_{j=1}^{m-1} \frac{D_{x_0-}^{\nu-j}\varphi(x_0)}{\Gamma(\nu-j+1)} (x_0-x)^{\nu-j} = \frac{1}{\Gamma(\nu)} \int_{x}^{x_0} (t-x)^{\nu-1} D_{x_0-}^{\nu}\varphi(t)\,dt, \quad (7.107)$$

$\forall\, x \in [a, x_0]$.

Therefore we get

$$\int_{x_0}^{b} \varphi(x)\,dx - \sum_{j=1}^{m-1} \frac{D_{x_0}^{\nu-j}\varphi(x_0)}{\Gamma(\nu-j+2)} (b-x_0)^{\nu-j+1} =$$

$$\frac{1}{\Gamma(\nu)} \int_{x_0}^{b} \left(\int_{x_0}^{x} (x-t)^{\nu-1} D_{x_0}^{\nu}\varphi(t)\,dt \right) dx \tag{7.108}$$

and

$$\int_{a}^{x_0} \varphi(x)\,dx - \sum_{j=1}^{m-1} \frac{D_{x_0-}^{\nu-j}\varphi(x_0)}{\Gamma(\nu-j+2)} (x_0-a)^{\nu-j+1} =$$

$$\frac{1}{\Gamma(\nu)} \int_{a}^{x_0} \left(\int_{x}^{x_0} (t-x)^{\nu-1} D_{x_0-}^{\nu}\varphi(t)\,dt \right) dx. \tag{7.109}$$

Adding (7.108) and (7.109) we derive

$$E := \int_{a}^{b} \varphi(x)\,dx -$$

$$\sum_{j=1}^{m-1} \frac{1}{\Gamma(\nu-j+2)} \left[D_{x_0}^{\nu-j}\varphi(x_0)(b-x_0)^{\nu-j+1} + D_{x_0-}^{\nu-j}\varphi(x_0)(x_0-a)^{\nu-j+1} \right]$$

$$= \frac{1}{\Gamma(\nu)} \left[\int_{x_0}^{b} \left(\int_{x_0}^{x} (x-t)^{\nu-1} D_{x_0}^{\nu}\varphi(t)\,dt \right) dx \right. \tag{7.110}$$

$$\left. + \int_{a}^{x_0} \left(\int_{x}^{x_0} (t-x)^{\nu-1} D_{x_0-}^{\nu}\varphi(t)\,dt \right) dx \right].$$

Therefore we obtain

$$|E| \leq \frac{1}{\Gamma(\nu)} \left[\int_{x_0}^{b} \left(\int_{x_0}^{x} (x-t)^{\nu-1} \left| D_{x_0}^{\nu}\varphi(t) \right| dt \right) dx + \right.$$

$$\left. \int_{a}^{x_0} \left(\int_{x}^{x_0} (t-x)^{\nu-1} \left| D_{x_0-}^{\nu}\varphi(t) \right| dt \right) dx \right] \leq$$

$$\frac{1}{\Gamma(\nu)} \left[\left\| D_{x_0}^{\nu}\varphi \right\|_{\infty,[x_0,b]} \int_{x_0}^{b} \left(\int_{x_0}^{x} (x-t)^{\nu-1} dt \right) dx \right.$$

$$\left. + \left\| D_{x_0-}^{\nu}\varphi \right\|_{\infty,[a,x_0]} \int_{a}^{x_0} \left(\int_{x}^{x_0} (t-x)^{\nu-1} dt \right) dx \right] =$$

$$\frac{1}{\Gamma(\nu+2)}\left[\left\|D_{x_0}^{\nu}\varphi\right\|_{\infty,[x_0,b]}(b-x_0)^{\nu+1}+\left\|D_{x_0-}^{\nu}\varphi\right\|_{\infty,[a,x_0]}(x_0-a)^{\nu+1}\right]\leq$$

$$\frac{\max\left\{\left\|D_{x_0-}^{\nu}\varphi\right\|_{\infty,[a,x_0]},\left\|D_{x_0}^{\nu}\varphi\right\|_{\infty,[x_0,b]}\right\}}{\Gamma(\nu+2)}\left[(b-x_0)^{\nu+1}+(x_0-a)^{\nu+1}\right]. \tag{7.111}$$

■

We finish with

Corollary 7.15 *All as in Theorem 7.14 with* $1<\nu<2$. *Then*

$$\left|\int_a^b \varphi(x)\,dx-\frac{1}{\Gamma(\nu+1)}\left[D_{x_0}^{\nu-1}\varphi(x_0)(b-x_0)^{\nu}+D_{x_0-}^{\nu-1}\varphi(x_0)(x_0-a)^{\nu}\right]\right|$$

$$\leq\frac{1}{\Gamma(\nu+2)}\left[\left\|D_{x_0}^{\nu}\varphi\right\|_{\infty,[x_0,b]}(b-x_0)^{\nu+1}+\left\|D_{x_0-}^{\nu}\varphi\right\|_{\infty,[a,x_0]}(x_0-a)^{\nu+1}\right]$$

$$\leq\max\left\{\left\|D_{x_0-}^{\nu}\varphi\right\|_{\infty,[a,x_0]},\left\|D_{x_0}^{\nu}\varphi\right\|_{\infty,[x_0,b]}\right\}\frac{\left[(b-x_0)^{\nu+1}+(x_0-a)^{\nu+1}\right]}{\Gamma(\nu+2)}. \tag{7.112}$$

References

1. Anastassiou, G.A.: Riemann-Liouville fractional Landau type inequalities. J. Nonlinear Evolution Equa. Appls., accepted (2020)
2. Barnett, N.S., Dragomir, S.S.: Some Landau type inequalities for functions whose derivatives are of locally bounded variation. Tamkang J. Math. **37**(4), 301–308 (Winter 2006)
3. Hardy, G.H., Landau, E., Littlewood, J.E.: Some inequalities satisfied by the integrals or derivatives of real or analytic functions. Math. Z. **39**, 677–695 (1935)
4. Hughes, R.J.: On fractional integrals and derivatives in L^p. Indiana Univ. Math. J. **26**(2), 325–328 (1977)
5. Hughes, R.J.: Hardy-Landau-Litlewwod inequalities for fractional derivatives in weightedL^p *spaces*. J. Lond. Math. Soc. **35**(2), 489–498 (1987)
6. Kilbas, A.A., Srivastava, H.M., Trujillo, J.J.: Theory and Applications of Fractional Differential Equations. North Holland, Amsterdam (2006)
7. Landau, E.: Einige Ungleichungen für zweimal differentzierban funktionen. Proc. Lond. Math. Soc. **13**, 43–49 (1913)
8. Martines, S., Sans, M., Martines, M.D.: Some inequalities for fractional integrals and derivatives. Soviet. Math. Dokl. **42**(3), 876–879 (1991)
9. Samko, S.G., Kilbas, A.A., Marichev, O.I.: Fractional Integrals and Derivatives: Theory and Applications. Gordon and Breach Science Publishers, Switzerland (1993)

Chapter 8
Generalized Canavati Fractional Landau Inequalities

We present uniform generalized Canavati fractional Landau inequalities over $\mathbb{R}$ of fractional orders $4 < \nu < 5$ and we estimate the size of first three ordinary derivatives of the on hand function. The inequalities are optimal by giving best upper bounds and the constants there are completely determined. We give an application when $\nu = 4.5$. It follows [3].

8.1 Introduction

Landau in 1913 ([9]) proved if $f \in C^2([0,1])$, $\|f\|_\infty = 1$, $\|f''\|_\infty = 4$, then $\|f'\|_\infty \le 4$, with 4 the best constant and the result is not necessarily true for an interval of length < 1. Now Landau's inequality has taken the form ([3])

$$\|f'\|_{p,I} \le C_p(I)\, \|f\|_{p,I}^{\frac{1}{2}} \, \|f''\|_{p,I}^{\frac{1}{2}}, \tag{8.1}$$

where $\|\cdot\|_{p,I}$ is the p-norm on the interval I; $p \in [1, \infty]$, $I = \mathbb{R}_+$ or $I = \mathbb{R}$, and $f : I \to \mathbb{R}$ is twice differentiable with $f, f'' \in L_p(I)$. $C_p(I) > 0$ is independent of f. And the best constants are $C_\infty(\mathbb{R}_+) = 2$ and $C_\infty(\mathbb{R}) = \sqrt{2}$.

Research about Landau type inequalities has expanded to many different directions and it is a very active topic of mathematical activity. Here we are concerned about fractional Landau inequalities of generalized Canavati type.

The main motivation to write this chapter is [10], and especially the result that follows:

Let α be a complex number with $Re\alpha > 0$. We define I_L^α and D_L^α to be the classical Riemann-Liouville fractional integration and differentiation operators of order α,

G. A. Anastassiou, *Constructive Fractional Analysis with Applications*,
Studies in Systems, Decision and Control 362,
https://doi.org/10.1007/978-3-030-71481-9_8

$$I_L^{\alpha} f(x) = \frac{1}{\Gamma(\alpha)} \int_a^x (x-t)^{\alpha-1} f(t)\, dt,$$

$$D_L^{\alpha} f(x) = \frac{d^{n+1}}{dx^{n+1}} \left(I_L^{n+1-\alpha} f \right)(x),$$

where Γ is the gamma function and $n = [Re\alpha]$ (integral part).

The above operators are defined in the natural domain of appropriate function spaces.

We state

Theorem 8.1 *([10]) Let I_L^{α} and D_L^{α} be the fractional Riemann-Liouville operators in the Fréchet space $L_{loc}^1([0,+\infty))$ with the usual topology. If α, β and γ are complex numbers for which $0 < Re\alpha < Re\beta < Re\gamma$, for every seminorm $\|\cdot\|_p$ there exist constant $F_1(p,\alpha,\beta,\gamma)$ and $F_2(p,\alpha,\beta,\gamma)$ that depend on p, α, β, and γ, with the properties that*

(a)

$$\left\| I_L^{\beta} f \right\|_p \le F_1(p,\alpha,\beta,\gamma) \left\| I_L^{\alpha} f \right\|_p^{\frac{Re(\gamma-\beta)}{Re(\gamma-\alpha)}} \left\| I_L^{\gamma} f \right\|_p^{\frac{Re(\beta-\alpha)}{Re(\gamma-\alpha)}}, \quad f \in L_{loc}^1([0,+\infty)), \tag{8.2}$$

and

$$\left\| D_L^{\beta} f \right\|_p \le F_2(p,\alpha,\beta,\gamma) \left\| D_L^{\alpha} f \right\|_p^{\frac{Re(\gamma-\beta)}{Re(\gamma-\alpha)}} \left\| D_L^{\gamma} f \right\|_p^{\frac{Re(\beta-\alpha)}{Re(\gamma-\alpha)}}, \tag{8.3}$$

$f \in \mathcal{D}\left(D_L^{\gamma}\right)$ (domain of D_L^{γ}).

If $Im\alpha = Im\beta = Im\gamma$, the constants $F_1(p,\alpha,\beta,\gamma)$ and $F_2(p,\alpha,\beta,\gamma)$ depend only on p and on an arbitrary integer $n > Re\gamma$.

For more see [10]. Other inspirations are the papers [7], [8].

The above motivating results and references are existential without giving exact constants which are not the best possible. Their methods are based on functional analysis and semigroup theory, in particular dealing with moment inequalities for fractional powers of operators in Banach spaces.

As a last motivation we also mention [6]. Our method is classical analytic and we will produce fractional Landau type inequalities, involving the generalized Canavati left and right fractional derivatives, that are optimal by giving best upper bounds and all constants are exactly calculated.

We need the following background:

First we follow [1], pp. 7-9.

Let $\mathbb{N} \not\ni \nu > 0$, $n := [\nu]$, $\alpha := \nu - n$ $(0 < \alpha < 1)$. Here $f \in C([a,b])$, $[a,b] \subset \mathbb{R}$ and define

$$\left(J_{\nu}^a f \right)(x) := \frac{1}{\Gamma(\nu)} \int_a^x (x-t)^{\nu-1} f(t)\, dt, \quad a \le x \le b, \tag{8.4}$$

the left Riemann-Liouville fractional integral.

We define the subspace $C_a^{\nu}([a,b])$ of $C^n([a,b])$:

$$C_a^{\nu}([a,b]) := \left\{ f \in C^n([a,b]) : J_{1-\alpha}^a f^{(n)} \in C^1([a,b]) \right\}.$$

So let $f \in C_a^{\nu}([a,b])$; we define the left generalized Canavati fractional derivative of f of order ν over $[a,b]$, as

$$D_a^{\nu} f := \left(J_{1-\alpha}^a f^{(n)} \right)', \tag{8.5}$$

(see also [4]) that is

$$\left(D_a^{\nu} f \right)(x) = \frac{1}{\Gamma(1-\alpha)} \frac{d}{dx} \int_a^x (x-t)^{-\alpha} f^{(n)}(t)\, dt, \tag{8.6}$$

all $a \leq x \leq b$, exists for $f \in C_a^{\nu}([a,b])$.

We need the left generalized fractional Taylor formula (see [1], p. 8):

Theorem 8.2 *Let $f \in C_a^{\nu}([a,b])$.*

(i) If $\nu \geq 1$, then

$$f(x) = f(a) + f'(a)(x-a) + f''(a)\frac{(x-a)^2}{2} + ... +$$

$$f^{(n-1)}(a)\frac{(x-a)^{n-1}}{(n-1)!} + \frac{1}{\Gamma(\nu)} \int_a^x (x-t)^{\nu-1} \left(D_a^{\nu} f \right)(t)\, dt, \tag{8.7}$$

all $x \in [a,b]$.

(ii) If $0 < \nu < 1$ we get

$$f(x) = \frac{1}{\Gamma(\nu)} \int_a^x (x-t)^{\nu-1} \left(D_a^{\nu} f \right)(t)\, dt, \tag{8.8}$$

all $x \in [a,b]$.

We also define the subspace

$$C_a^{\nu}([a,+\infty)) := \left\{ f \in C^n(\mathbb{R}) : f \in C_a^{\nu}([a,b]), \forall\, b > a \right\}, \tag{8.9}$$

where $a \in \mathbb{R}$.

Next we follow [2], pp. 345-348.

Let $\mathbb{N} \not\ni \nu > 0$, $n := [\nu]$, $\alpha := \nu - n$, $0 < \alpha < 1$, $f \in C([a,b])$, call the right Riemann-Liouville fractional integral operator by

$$\left(J_{b-}^{\nu} f \right)(x) := \frac{1}{\Gamma(\nu)} \int_x^b (z-x)^{\nu-1} f(z)\, dz, \tag{8.10}$$

$\forall\, x \in [a,b]$. Define the subspace of functions

$$C_{b-}^{\nu}([a,b]) := \left\{ f \in C^n([a,b]) : J_{b-}^{1-\alpha} f^{(n)} \in C^1([a,b]) \right\}.$$

Define the right generalized ν-fractional derivative of f over $[a,b]$ (called also right generalized Canavati fractional derivative) as

$$D_{b-}^{\nu} f := (-1)^{n-1} \left(J_{b-}^{1-\alpha} f^{(n)} \right)'. \tag{8.11}$$

Notice that

$$D_{b-}^{\nu} f(x) = \frac{(-1)^{n-1}}{\Gamma(1-\alpha)} \frac{d}{dx} \int_x^b (z-x)^{-\alpha} f^{(n)}(z)\, dz, \tag{8.12}$$

exists for $f \in C_{b-}^{\nu}([a,b])$, all $x \in [a,b]$.

We need also the following right fractional Taylor formula ([2], p. 348):

Theorem 8.3 *Let $f \in C_{b-}^{\nu}([a,b])$, $\nu > 0$, $n := [\nu]$. Then*

(1) If $\nu \geq 1$, we get

$$f(x) = \sum_{k=0}^{n-1} \frac{f^{(k)}(b)}{k!} (x-b)^k + \frac{1}{\Gamma(\nu)} \int_x^b (z-x)^{\nu-1} \left(D_{b-}^{\nu} f \right)(z)\, dz, \tag{8.13}$$

$\forall\, x \in [a,b]$.

(2) If $0 < \nu < 1$, we obtain

$$f(x) = \frac{1}{\Gamma(\nu)} \int_x^b (z-x)^{\nu-1} \left(D_{b-}^{\nu} f \right)(z)\, dz, \tag{8.14}$$

$\forall\, x \in [a,b]$.

We need also to define the subspace

$$C_{b-}^{\nu}((-\infty,b]) := \left\{ f \in C^n(\mathbb{R}) : f \in C_{b-}^{\nu}([a,b]),\ \forall\, a < b \right\}, \tag{8.15}$$

where $b \in \mathbb{R}$.

8.2 Main Results

We present a set of fractional generalized Canavati type Landau inequalities over $\mathbb{R}$.

Theorem 8.4 *Let $4 < \nu < 5$ and let $f \in C^4(\mathbb{R})$ with $f \in C_B(\mathbb{R})$ (continuous and bounded). We assume that $f \in C_a^{\nu}([a,+\infty))$ and $f \in C_{a-}^{\nu}((-\infty,a])$, $\forall\, a \in \mathbb{R}$. Assume also that*

$$K := \max\left\{\left\|\left(D_{a}^{\nu}f\right)(t)\right\|_{\infty,\mathbb{R}^2}, \left\|\left(D_{a-}^{\nu}f\right)(t)\right\|_{\infty,\mathbb{R}^2}\right\} < +\infty, \tag{8.16}$$

where $(a, t) \in \mathbb{R}^2$. *Let also* $0 < \lambda < 1$. *Then*

(i)

$$\left\|f'\right\|_{\infty,\mathbb{R}} \leq \frac{\nu}{\left(\Gamma(\nu+1)(\nu-1)^{(\nu-1)}\right)^{\frac{1}{\nu}}} \frac{\left[\left(\lambda^{\nu}+\lambda^{3}\right)\left(1+\lambda^{3}\right)^{(\nu-1)}\right]^{\frac{1}{\nu}}}{\lambda\left(1-\lambda^{2}\right)} \|f\|_{\infty,\mathbb{R}}^{\left(\frac{\nu-1}{\nu}\right)} K^{\frac{1}{\nu}}, \tag{8.17}$$

(ii)

$$\left\|f''\right\|_{\infty,\mathbb{R}} \leq \nu\left(\frac{4}{\nu-2}\right)^{\left(\frac{\nu-2}{\nu}\right)} (\Gamma(\nu+1))^{-\frac{2}{\nu}} \|f\|_{\infty,\mathbb{R}}^{\left(\frac{\nu-2}{\nu}\right)} K^{\frac{2}{\nu}}, \tag{8.18}$$

and

(iii)

$$\left\|f'''\right\|_{\infty,\mathbb{R}} \leq 2\nu\left(\frac{3^{(\nu-3)}}{(\nu-3)^{(\nu-3)}(\Gamma(\nu+1))^{3}}\right)^{\frac{1}{\nu}}\left[\left(\frac{\lambda+\lambda^{\nu}}{1+\lambda}\right)^{\frac{3}{\nu}} \frac{1}{\lambda(1-\lambda)}\right] \|f\|_{\infty,\mathbb{R}}^{\left(\frac{\nu-3}{\nu}\right)} K^{\frac{3}{\nu}}. \tag{8.19}$$

The above inequalities are optimal by giving best upper bounds. We also obtain $\left\|f^{(i)}\right\|_{\infty,\mathbb{R}} < +\infty$, *for* $i = 1, 2, 3$.

Proof Let any $a \in \mathbb{R}$. Here it is $n = [\nu] = 4$.

By Theorem 8.2 we have that

$$f(x) = f(a) + f'(a)(x-a) + f''(a)\frac{(x-a)^2}{2} + f'''(a)\frac{(x-a)^3}{6} + \frac{1}{\Gamma(\nu)}\int_a^x (x-t)^{\nu-1} D_a^{\nu} f(t)\,dt, \tag{8.20}$$

$\forall\, x \in [a, +\infty)$. And by Theorem 8.3 we have that

$$f(x) = f(a) + f'(a)(x-a) + f''(a)\frac{(x-a)^2}{2} + f'''(a)\frac{(x-a)^3}{6} + \frac{1}{\Gamma(\nu)}\int_x^a (t-x)^{\nu-1}\left(D_{a-}^{\nu} f\right)(t)\,dt, \tag{8.21}$$

$\forall\, x \in (-\infty, a]$.

Momentarily we fix $a \in \mathbb{R}$.

Let $x_1 > a$, then

$$f'(a)(x_1-a)+f''(a)\frac{(x_1-a)^2}{2}+f'''(a)\frac{(x_1-a)^3}{6}= \tag{8.22}$$

$$f(x_1)-f(a)-\frac{1}{\Gamma(\nu)}\int_a^{x_1}(x_1-t)^{\nu-1}D_a^{\nu}f(t)\,dt.$$

Let also $x_2<a$, then

$$f'(a)(x_2-a)+f''(a)\frac{(x_2-a)^2}{2}+f'''(a)\frac{(x_2-a)^3}{6}= \tag{8.23}$$

$$f(x_2)-f(a)-\frac{1}{\Gamma(\nu)}\int_{x_2}^{a}(t-x_2)^{\nu-1}\left(D_{a-}^{\nu}f\right)(t)\,dt.$$

Choose first $x_1-a=h>0$ and $x_2-a=-h$. Thus

$$f'(a)h+f''(a)\frac{h^2}{2}+f'''(a)\frac{h^3}{6}=$$

$$f(a+h)-f(a)-\frac{1}{\Gamma(\nu)}\int_a^{a+h}(a+h-t)^{\nu-1}D_a^{\nu}f(t)\,dt=:A_1, \tag{8.24}$$

and

$$-f'(a)h+f''(a)\frac{h^2}{2}-f'''(a)\frac{h^3}{6}=$$

$$f(a-h)-f(a)-\frac{1}{\Gamma(\nu)}\int_{a-h}^{a}(t-(a-h))^{\nu-1}\left(D_{a-}^{\nu}f\right)(t)\,dt=:A_2. \tag{8.25}$$

Let $0<\lambda<1$. Next we choose $x_1-a=\lambda h$ and $x_2-a=-\lambda h$, $h>0$. By (8.22) and (8.23) we get

$$f'(a)\lambda h+f''(a)\frac{(\lambda h)^2}{2}+f'''(a)\frac{(\lambda h)^3}{6}=$$

$$f(a+\lambda h)-f(a)-\frac{1}{\Gamma(\nu)}\int_a^{a+\lambda h}(a+\lambda h-t)^{\nu-1}D_a^{\nu}f(t)\,dt=:B_1, \tag{8.26}$$

and

$$-f'(a)\lambda h+f''(a)\frac{(\lambda h)^2}{2}-f'''(a)\frac{(\lambda h)^3}{6}=$$

$$f(a-\lambda h)-f(a)-\frac{1}{\Gamma(\nu)}\int_{a-\lambda h}^{a}(t-(a-\lambda h))^{\nu-1}\left(D_{a-}^{\nu}f\right)(t)\,dt=:B_2. \tag{8.27}$$

Adding (8.24) and (8.25) we get

$$f''(a)\,h^2 = A_1 + A_2. \tag{8.28}$$

Subtracting (8.27) from (8.26), we obtain

$$2f'(a)\,\lambda h + f'''(a)\,\frac{(\lambda h)^3}{3} = B_1 - B_2, \tag{8.29}$$

and by subtracting (8.25) from (8.24), we derive

$$2f'(a)\,h + f'''(a)\,\frac{h^3}{3} = A_1 - A_2. \tag{8.30}$$

From (8.28) we find

$$f''(a) = \frac{A_1 + A_2}{h^2}. \tag{8.31}$$

Next we solve the system of two equations in the two unknowns $f'(a)$ and $f'''(a)$. We calculate

$$D = \begin{vmatrix} 2\lambda h & \frac{(\lambda h)^3}{3} \\ 2h & \frac{h^3}{3} \end{vmatrix} = \frac{2\lambda h^4}{3}\left(1 - \lambda^2\right) > 0. \tag{8.32}$$

Furthermore we have

$$D_1 = \begin{vmatrix} B_1 - B_2 & \frac{(\lambda h)^3}{3} \\ A_1 - A_2 & \frac{h^3}{3} \end{vmatrix} = \frac{h^3}{3}\left[(B_1 - B_2) - \lambda^3 (A_1 - A_2)\right] \tag{8.33}$$

and

$$D_2 = \begin{vmatrix} 2\lambda h & B_1 - B_2 \\ 2h & A_1 - A_2 \end{vmatrix} = 2h\left[\lambda (A_1 - A_2) - (B_1 - B_2)\right]. \tag{8.34}$$

Therefore we find

$$f'(a) = \frac{\frac{h^3}{3}\left[(B_1 - B_2) - \lambda^3 (A_1 - A_2)\right]}{\frac{2\lambda h^4}{3}\left(1 - \lambda^2\right)} = \frac{\left[(B_1 - B_2) - \lambda^3 (A_1 - A_2)\right]}{2\lambda\left(1 - \lambda^2\right)h}, \tag{8.35}$$

and

$$f'''(a) = \frac{2h\left[\lambda (A_1 - A_2) - (B_1 - B_2)\right]}{\frac{2\lambda h^4}{3}\left(1 - \lambda^2\right)} = \frac{3\left[\lambda (A_1 - A_2) - (B_1 - B_2)\right]}{\lambda\left(1 - \lambda^2\right)h^3}. \tag{8.36}$$

Next we estimate $f'(a)$, $f''(a)$, $f'''(a)$.

We have that

$$|A_1 + A_2| = |f(a+h) + f(a-h) - 2f(a) -$$

$$\frac{1}{\Gamma(\nu)} \left[\int_a^{a+h} (a+h-t)^{\nu-1} D_a^\nu f(t)\, dt + \int_{a-h}^{a} (t-(a-h))^{\nu-1} \left(D_{a-}^\nu f\right)(t)\, dt \right] \Bigg|$$

$$\leq 4\|f\|_{\infty,\mathbb{R}} + \frac{2\max\left\{ \left\| \left(D_a^\nu f\right)(t) \right\|_{\infty,\mathbb{R}^2}, \left\| \left(D_{a-}^\nu f\right)(t) \right\|_{\infty,\mathbb{R}^2} \right\}}{\Gamma(\nu+1)} h^\nu. \tag{8.37}$$

That is

$$\left|f''(a)\right| \leq \frac{4\|f\|_{\infty,\mathbb{R}}}{h^2} + \frac{2K}{\Gamma(\nu+1)} h^{\nu-2}, \tag{8.38}$$

$\forall\ a \in \mathbb{R}, \forall\, h > 0$.

Consequently it holds

$$\left\|f''\right\|_{\infty,\mathbb{R}} \leq \frac{4\|f\|_{\infty,\mathbb{R}}}{h^2} + \frac{2K}{\Gamma(\nu+1)} h^{\nu-2}, \tag{8.39}$$

$\forall\, h > 0$.

We notice that

$$B_1 - B_2 = f(a+\lambda h) + f(a-\lambda h) - \frac{1}{\Gamma(\nu)} \int_a^{a+\lambda h} (a+\lambda h - t)^{\nu-1} D_a^\nu f(t)\, dt$$
$$+ \frac{1}{\Gamma(\nu)} \int_{a-\lambda h}^{a} (t-(a-\lambda h))^{\nu-1} \left(D_{a-}^\nu f\right)(t)\, dt \tag{8.40}$$

and

$$\lambda^3 (A_1 - A_2) = \lambda^3 \left[f(a+h) - f(a-h) - \frac{1}{\Gamma(\nu)} \int_a^{a+h} (a+h-t)^{\nu-1} D_a^\nu f(t)\, dt \right.$$
$$\left. + \frac{1}{\Gamma(\nu)} \int_{a-h}^{a} (t-(a-h))^{\nu-1} \left(D_{a-}^\nu f\right)(t)\, dt \right]. \tag{8.41}$$

Furthermore it holds

$$|B_1 - B_2| \leq 2\|f\|_{\infty,\mathbb{R}} + \frac{K}{\Gamma(\nu)} \left[\frac{2\lambda^\nu h^\nu}{\nu} \right] = 2\|f\|_{\infty,\mathbb{R}} + \frac{2K\lambda^\nu}{\Gamma(\nu+1)} h^\nu, \tag{8.42}$$

and

$$\lambda^3 |A_1 - A_2| \leq 2\|f\|_{\infty,\mathbb{R}} \lambda^3 + \frac{2K\lambda^3}{\Gamma(\nu+1)} h^\nu. \tag{8.43}$$

Consequently we have

$$\left|f'(a)\right| \le \frac{2\|f\|_{\infty,\mathbb{R}}\left(\lambda^3+1\right)+\frac{2K}{\Gamma(\nu+1)}\left(\lambda^\nu+\lambda^3\right)h^\nu}{2\lambda\left(1-\lambda^2\right)h} \tag{8.44}$$

$$= \frac{1}{\lambda\left(1-\lambda^2\right)}\left[\frac{\|f\|_{\infty,\mathbb{R}}\left(1+\lambda^3\right)}{h}+\frac{K}{\Gamma(\nu+1)}\left(\lambda^\nu+\lambda^3\right)h^{\nu-1}\right], \tag{8.45}$$

$\forall\, a \in \mathbb{R}, \forall\, h > 0.$

We have proved that

$$\left\|f'\right\|_{\infty,\mathbb{R}} \le \frac{1}{\lambda\left(1-\lambda^2\right)}\left[\frac{\|f\|_{\infty,\mathbb{R}}\left(1+\lambda^3\right)}{h}+\frac{K\left(\lambda^\nu+\lambda^3\right)}{\Gamma(\nu+1)}h^{\nu-1}\right], \tag{8.46}$$

$\forall\, h > 0.$

We have

$$\left|f'''(a)\right| \le \frac{3}{\lambda\left(1-\lambda^2\right)h^3}\left[\lambda\left|A_1-A_2\right|+\left|B_1-B_2\right|\right] \le$$

$$\frac{3}{\lambda\left(1-\lambda^2\right)h^3}\left[2\|f\|_{\infty,\mathbb{R}}\lambda+\frac{2K\lambda}{\Gamma(\nu+1)}h^\nu+2\|f\|_{\infty,\mathbb{R}}+\frac{2K}{\Gamma(\nu+1)}\lambda^\nu h^\nu\right]=$$

$$\frac{3}{\lambda\left(1-\lambda^2\right)h^3}\left[2\|f\|_{\infty,\mathbb{R}}(\lambda+1)+\frac{2K}{\Gamma(\nu+1)}\left(\lambda+\lambda^\nu\right)h^\nu\right]= \tag{8.47}$$

$$\frac{6}{\lambda\left(1-\lambda^2\right)}\left[\frac{\|f\|_{\infty,\mathbb{R}}(\lambda+1)}{h^3}+\frac{K}{\Gamma(\nu+1)}\left(\lambda+\lambda^\nu\right)h^{\nu-3}\right],$$

$\forall\, h > 0, \forall\, a \in \mathbb{R}.$

We have derived that

$$\left\|f'''\right\|_{\infty,\mathbb{R}} \le \frac{6}{\lambda\left(1-\lambda^2\right)}\left[\frac{\|f\|_{\infty,\mathbb{R}}(1+\lambda)}{h^3}+\frac{K\left(\lambda+\lambda^\nu\right)}{\Gamma(\nu+1)}h^{\nu-3}\right], \tag{8.48}$$

$\forall\, h > 0.$

We are working in (8.39).

Call

$$\xi := 4\|f\|_{\infty,\mathbb{R}}, \qquad \psi := \frac{2K}{\Gamma(\nu+1)}, \tag{8.49}$$

both are greater than zero.

Set also $\varphi := \nu - 2 > 0$.

We consider the function

$$\gamma(h) := \xi h^{-2} + \psi h^{\varphi}, \quad \forall\, h > 0. \tag{8.50}$$

We have

$$\gamma'(h) = -2\xi h^{-3} + \varphi\psi h^{\varphi-1} = 0,$$

with one critical number

$$h_0 := h_{crit.no.} = \left(\frac{2\xi}{\varphi\psi}\right)^{\frac{1}{\varphi+2}}. \tag{8.51}$$

We have that

$$\gamma''(h) = 6\xi h^{-4} + \varphi(\varphi-1)\psi h^{\varphi-2}. \tag{8.52}$$

We have that

$$\gamma''(h_0) = 2\xi\left(\frac{2\xi}{\varphi\psi}\right)^{-\frac{4}{\varphi+2}}(\varphi+2) > 0. \tag{8.53}$$

Therefore γ has a global minimum at $h_0 = \left(\frac{2\xi}{\varphi\psi}\right)^{\frac{1}{\varphi+2}}$, which is

$$\gamma(h_0) = \left(\frac{\xi}{\varphi}\right)^{\frac{\varphi}{\varphi+2}}\left(\frac{\psi}{2}\right)^{\frac{2}{\varphi+2}}(\varphi+2). \tag{8.54}$$

Consequently

$$\gamma(h_0) = \left(\frac{4\|f\|_{\infty,\mathbb{R}}}{\nu-2}\right)^{\frac{\nu-2}{\nu}}\left(\frac{K}{\Gamma(\nu+1)}\right)^{\frac{2}{\nu}}\nu. \tag{8.55}$$

We have proved that

$$\|f''\|_{\infty,\mathbb{R}} \leq \left(\frac{4\|f\|_{\infty,\mathbb{R}}}{\nu-2}\right)^{\frac{\nu-2}{\nu}}\left(\frac{K}{\Gamma(\nu+1)}\right)^{\frac{2}{\nu}}\nu =$$

$$\nu\left(\frac{4}{\nu-2}\right)^{\frac{\nu-2}{\nu}}(\Gamma(\nu+1))^{-\frac{2}{\nu}}\|f\|_{\infty,\mathbb{R}}^{\frac{\nu-2}{\nu}}K^{\frac{2}{\nu}}.$$

Next we work on (8.46).

Let

$$\begin{aligned}\mu &:= \|f\|_{\infty,\mathbb{R}}\left(1+\lambda^3\right),\\ \theta &:= \frac{K\left(\lambda^{\nu}+\lambda^3\right)}{\Gamma(\nu+1)},\end{aligned} \tag{8.56}$$

both are greater than zero.

Set also $\rho := \nu - 1 > 3$.

We consider the function

$$y(h) := \frac{\mu}{h} + \theta h^{\rho}, \quad \forall\, h > 0. \tag{8.57}$$

We have

$$y'(h) = -\frac{\mu}{h^2} + \rho\theta h^{\rho-1} = 0, \tag{8.58}$$

with only one critical number

$$h_0 := h_{crit.no.} = \left(\frac{\mu}{\rho\theta}\right)^{\frac{1}{\rho+1}}. \tag{8.59}$$

We have that

$$y''(h) = 2\mu h^{-3} + \rho(\rho-1)\theta h^{\rho-2}. \tag{8.60}$$

We observe that

$$y''(h_0) = \mu\left(\frac{\mu}{\rho\theta}\right)^{-\frac{3}{\rho+1}}(\rho+1) > 0. \tag{8.61}$$

Therefore y has a global minimum at $h_0 = \left(\frac{\mu}{\rho\theta}\right)^{\frac{1}{\rho+1}}$, which is

$$y(h_0) = (\theta\mu^{\rho})^{\frac{1}{\rho+1}}(\rho+1)\rho^{-\frac{\rho}{\rho+1}}. \tag{8.62}$$

That is

$$y(h_0) = \left(\frac{K(\lambda^{\nu}+\lambda^3)}{\Gamma(\nu+1)}\right)^{\frac{1}{\nu}}\left(\|f\|_{\infty,\mathbb{R}}(1+\lambda^3)\right)^{\frac{\nu-1}{\nu}}\nu(\nu-1)^{-\left(\frac{\nu-1}{\nu}\right)}. \tag{8.63}$$

We have found that

$$\|f'\|_{\infty,\mathbb{R}} \leq \frac{\nu(\nu-1)^{-\left(\frac{\nu-1}{\nu}\right)}}{(\Gamma(\nu+1))^{\frac{1}{\nu}}}\frac{(\lambda^{\nu}+\lambda^3)^{\frac{1}{\nu}}}{\lambda(1-\lambda^2)}(1+\lambda^3)^{\frac{\nu-1}{\nu}}\|f\|_{\infty,\mathbb{R}}^{\frac{\nu-1}{\nu}}K^{\frac{1}{\nu}} =$$

$$\frac{\nu}{\left(\Gamma(\nu+1)(\nu-1)^{(\nu-1)}\right)^{\frac{1}{\nu}}}\frac{\left[(\lambda^{\nu}+\lambda^3)(1+\lambda^3)^{(\nu-1)}\right]^{\frac{1}{\nu}}}{\lambda(1-\lambda^2)}\|f\|_{\infty,\mathbb{R}}^{\frac{\nu-1}{\nu}}K^{\frac{1}{\nu}}. \tag{8.64}$$

Finally we work on (8.48).

We study

$$z(h) := \frac{\|f\|_{\infty,\mathbb{R}}(1+\lambda)}{h^3} + \frac{K(\lambda+\lambda^{\nu})}{\Gamma(\nu+1)}h^{\nu-3}, \tag{8.65}$$

$\forall\, h > 0$.

Call

$$w := \|f\|_{\infty,\mathbb{R}} (1+\lambda), \quad l := \frac{K(\lambda+\lambda^{\nu})}{\Gamma(\nu+1)}, \tag{8.66}$$

both are greater than zero.

Set also $d := \nu - 3 > 1$, i.e. $\nu = d + 3$.

Thus we can write

$$z(h) := \frac{w}{h^3} + lh^d, \quad \forall\, h > 0. \tag{8.67}$$

We set

$$z'(h) = -3wh^{-4} + dlh^{d-1} = 0. \tag{8.68}$$

The only critical number here is

$$h_0 := h_{crit.no.} = \left(\frac{3w}{dl}\right)^{\frac{1}{d+3}} > 0.$$

We need

$$z''(h) = 12wh^{-5} + d(d-1)\,lh^{d-2}, \tag{8.69}$$

and we find

$$z''(h_0) = (dl)^{\frac{5}{d+3}} (3w)^{\frac{d-2}{d+3}} (d+3) > 0. \tag{8.70}$$

Hence z has a global minimum which is

$$z(h_0) = \left(\frac{d+3}{3}\right)\left(\frac{3}{d}\right)^{\left(\frac{d}{d+3}\right)} l^{\left(\frac{3}{d+3}\right)} w^{\left(\frac{d}{d+3}\right)}. \tag{8.71}$$

That is

$$z(h_0) = \left(\frac{\nu}{3}\right)\left(\frac{3}{\nu-3}\right)^{\left(\frac{\nu-3}{\nu}\right)} \left(\frac{K(\lambda+\lambda^{\nu})}{\Gamma(\nu+1)}\right)^{\frac{3}{\nu}} \left(\|f\|_{\infty,\mathbb{R}}(1+\lambda)\right)^{\left(\frac{\nu-3}{\nu}\right)} = \tag{8.72}$$

$$\frac{\left(\frac{\nu}{3}\right)\left(\frac{3}{\nu-3}\right)^{\left(\frac{\nu-3}{\nu}\right)}}{(\Gamma(\nu+1))^{\frac{3}{\nu}}} (\lambda+\lambda^{\nu})^{\frac{3}{\nu}} (1+\lambda)^{\left(\frac{\nu-3}{\nu}\right)} \|f\|_{\infty,\mathbb{R}}^{\left(\frac{\nu-3}{\nu}\right)} K^{\frac{3}{\nu}}.$$

We have proved that

$$\|f'''\|_{\infty,\mathbb{R}} \leq 2\nu \left(\frac{3^{\nu-3}}{(\nu-3)^{\nu-3}(\Gamma(\nu+1))^3}\right)^{\frac{1}{\nu}} \left(\frac{(\lambda+\lambda^{\nu})^{\frac{3}{\nu}}}{\lambda(1-\lambda)(1+\lambda)^{\frac{3}{\nu}}}\right) \|f\|_{\infty,\mathbb{R}}^{\left(\frac{\nu-3}{\nu}\right)} K^{\frac{3}{\nu}}. \tag{8.73}$$

The proof of the theorem is finished. ∎

It follows an application when $\nu = 4.5$.

Corollary 8.5 *Let* $f \in C^4(\mathbb{R})$ *with* $f \in C_B(\mathbb{R})$*. We assume that* $f \in C_a^{4.5}([a, +\infty))$ *and* $f \in C_{a-}^{4.5}((-\infty, a])$*,* $\forall\, a \in \mathbb{R}$*. Assume also that*

$$K_{4.5} := \max\left\{ \left\| \left(D_a^{4.5} f\right)(t) \right\|_{\infty,\mathbb{R}^2}, \left\| \left(D_{a-}^{4.5} f\right)(t) \right\|_{\infty,\mathbb{R}^2} \right\} < +\infty, \tag{8.74}$$

where $(a, t) \in \mathbb{R}^2$*.*

Then

$$\left\| f'' \right\|_{\infty,\mathbb{R}} \leq (1.006181121) \left\| f \right\|_{\infty,\mathbb{R}}^{(0.555555556)} K_{4.5}^{(0.444444444)}, \tag{8.75}$$

an optimal inequality giving a best upper bound with a constant almost 1*.*

Hence $\left\| f'' \right\|_{\infty,\mathbb{R}} < +\infty$*.*

Proof By Theorem 8.4 (ii). ∎

References

1. Anastassiou, G.A.: Fractional Differentiation Inequalities. Springer, Heidelberg (2009)
2. Anastassiou, G.A.: Intelligent Mathematics: Computational Analysis. Springer, Heidelberg (2011)
3. Anastassiou, G.A.: Generalized Canavati fractional Landau type inequalities. Commun. Appl. Nonlinear Anal., accepted (2020)
4. Barnett, N.S., Dragomir, S.S.: Some Landau type inequalities for functions whose derivatives are of locally bounded variation. Tamkang J. Math. **37**(4), 301–308 (Winter 2006)
5. Canavati, J.A.: The Riemann-Liouville integral. Nieuw Archief Voor Wiskunde **5**(1), 53–75 (1987)
6. Hardy, G.H., Landau, E., Littlewood, J.E.: Some inequalities satisfied by the integrals or derivatives of real or analytic functions. Math. Z. **39**, 677–695 (1935)
7. Hughes, R.J.: On fractional integrals and derivatives inL^p. Indiana Univ. Math. J. **26**(2), 325–328 (1977)
8. Hughes, R.J.: Hardy-Landau-Litlewwod inequalities for fractional derivatives in weighted L^p spaces. J. Lond. Math. Soc. **35**(2), 489–498 (1987)
9. Landau, E.: Einige Ungleichungen für zweimal differentzierban funktionen. Proc. Lond. Math. Soc. **13**, 43–49 (1913)
10. Martines, S., Sans, M., Martines, M.D.: Some inequalities for fractional integrals and derivatives. Soviet. Math. Dokl. **42**(3), 876–879 (1991)

Chapter 9
Sequential Left Abstract Fractional Landau Inequalities

We present uniform and L_p left Caputo–Bochner abstract sequential fractional Landau inequalities over $\mathbb{R}_+$. These estimate the size of second and third iterated left abstract fractional derivatives of a Banach space valued function over $\mathbb{R}_+$. We give an application when the basic fractional order is $\frac{1}{2}$. It follows [5].

9.1 Introduction

Let $p \in [1, \infty]$, $I = \mathbb{R}_+$ or $I = \mathbb{R}$ and $f : I \to \mathbb{R}$ is twice differentiable with $f, f'' \in L_p(I)$, then $f' \in L_p(I)$. Moreover, there exists a constant $C_p(I) > 0$ independent of f, such that

$$\left\| f' \right\|_{p,I} \leq C_p(I) \left\| f \right\|_{p,I}^{\frac{1}{2}} \left\| f'' \right\|_{p,I}^{\frac{1}{2}}, \tag{9.1}$$

where $\|\cdot\|_{p,I}$ is the p-norm on the interval I, see [1, 6].

The research on these inequalities started by Landau [11] in 1913. For the case of $p = \infty$ he proved that

$$C_\infty(\mathbb{R}_+) = 2 \text{ and } C_\infty(\mathbb{R}) = \sqrt{2}, \tag{9.2}$$

are the best constants in (9.1).

In 1932, Hardy and Littlewood [8] proved (9.1) for $p = 2$, with the best constants

$$C_2(\mathbb{R}_+) = \sqrt{2} \text{ and } C_2(\mathbb{R}) = 1. \tag{9.3}$$

In 1935, Hardy, Landau and Littlewood [9] showed that the best constants $C_p(\mathbb{R}_+)$ in (9.1) satisfies the estimate

G. A. Anastassiou, *Constructive Fractional Analysis with Applications*,
Studies in Systems, Decision and Control 362,
https://doi.org/10.1007/978-3-030-71481-9_9

$$C_p(\mathbb{R}_+) \leq 2, \text{ for } p \in [1, \infty), \tag{9.4}$$

which yields $C_p(\mathbb{R}) \leq 2$ for $p \in [1, \infty)$.

In fact, in [7, 10] was shown that $C_p(\mathbb{R}) \leq \sqrt{2}$.

We need the following concept from abstract fractional calculus.

Our integral next is of Bochner type [12].

We need

Definition 9.1 ([4], *p. 150*) Let $[a, b] \subset \mathbb{R}$, X be a Banach space, $\alpha > 0$; $m = \lceil \alpha \rceil \in \mathbb{N}$, ($\lceil \cdot \rceil$ is the ceiling of the number), $f : [a, b] \to X$. We assume that $f^{(m)} \in L_1([a, b], X)$. We call the Caputo–Bochner left fractional derivative of order α:

$$\left(D_{*a}^{\alpha} f\right)(x) := \frac{1}{\Gamma(m-\alpha)} \int_a^x (x-t)^{m-\alpha-1} f^{(m)}(t)\, dt, \quad \forall\, x \in [a, b], \tag{9.5}$$

where Γ is the gamma function.

If $\alpha \in \mathbb{N}$, we set $D_{*a}^{\alpha} f := f^{(m)}$ the ordinary X-valued derivative (defined similar to numerical one, see [13], p. 83), and also set $D_{*a}^{0} f := f$.

By ([4], p. 2), $\left(D_{*a}^{\alpha} f\right)(x)$ exists almost everywhere in $x \in [a, b]$ and $D_{*a}^{\alpha} f \in L_1([a, b], X)$.

If $\left\| f^{(m)} \right\|_{L_\infty([a,b],X)} < \infty$, and $\alpha \notin \mathbb{N}$, then by ([4], p. 3), $D_{*a}^{\alpha} f \in C([a, b], X)$, hence $\left\| D_{*a}^{\alpha} f \right\| \in C([a, b])$.

We mention the important:

Corollary 9.2 ([4], p. 157) *Let $f \in C^m([a, b], X)$, $m = \lceil \alpha \rceil$, $\alpha > 0$, $x, x_0 \in [a, b]$. Then $D_{*x_0}^{a} f(x)$ is jointly continuous functions in (x, x_0) from $[a, b]^2$ into X, X is a Banach space.*

By convention we suppose that

$$D_{*x_0}^{\alpha} f(x) = 0, \text{ for } x < x_0,$$

for all $x, x_0 \in [a, b]$.

The author has already done an extensive body of work on fractional Landau inequalities, see [3], and on abstract fractional Landau inequalities, see [4]. However there the proving methods came out of applications of fractional Ostrowski inequalities [2, 4]. Usually there the domains where $[A, +\infty)$ or $(-\infty, B]$, with $A, B \in \mathbb{R}$ and in one mixed case the domain was all of $\mathbb{R}$.

In this chapter with less assumptions we establish uniform and L_p type left Caputo–Bochner abstract iterated fractional Landau inequalities over $\mathbb{R}_+$. The method of proving is based on left Caputo–Bochner iterated fractional Taylor's formula with integral remainder, see [4], pp. 129.

We give also an application for $\alpha = \frac{1}{2}$. Regardless to say that we are also inspired by [3, 4].

9.2 Main Results

We consider ($\alpha > 0$) the composition $D_{*a}^{n\alpha} := D_{*a}^{\alpha} D_{*a}^{\alpha} ... D_{*a}^{\alpha}$ (n-times), $n \in \mathbb{N}$.

We mention the following left modified X-valued Taylor's formula, $(X, \|\cdot\|)$ is a Banach space.

Theorem 9.3 ([4], p. 129) *Let $0 < \alpha \leq 1$, $n \in \mathbb{N}$, $f \in C^1([a,b], X)$. For $k = 1, ..., n$, we assume that $D_{*a}^{k\alpha} f \in C^1([a,b], X)$ and $D_{*a}^{(n+1)\alpha} f \in C([a,b], X)$. Then*

$$f(x) = \sum_{i=0}^{n} \frac{(x-a)^{i\alpha}}{\Gamma(i\alpha+1)} \left(D_{*a}^{i\alpha} f\right)(a) + \tag{9.6}$$

$$\frac{1}{\Gamma((n+1)\alpha)} \int_a^x (x-t)^{(n+1)\alpha-1} \left(D_{*a}^{(n+1)\alpha} f\right)(t)\, dt,$$

$\forall\, x \in [a,b]$.

When $0 < \alpha < 1$ and $f \in C^1([a,b], X)$, by [4], p. 130, we get that $\left(D_{*a}^{\alpha} f\right)(a) = 0$.

We present the following abstract left fractional Landau inequalities over $\mathbb{R}_+$.

Theorem 9.4 *Let $0 < \alpha < 1$, $f \in C^1(\mathbb{R}_+, X)$ with $\| \|f\| \|_{\infty,\mathbb{R}_+}$, $\left\| \|f'\| \right\|_{\infty,\mathbb{R}_+} < \infty$. For $k = 1, 2, 3$, we assume that $D_{*a}^{k\alpha} f \in C^1([a,\infty), X)$ and $D_{*a}^{4\alpha} f \in C([a,\infty), X)$, $\forall\, a \in \mathbb{R}_+$. We further assume that*

$$K := \left\| \left\| D_{*a}^{4\alpha} f(t) \right\| \right\|_{\infty,\mathbb{R}_+^2} < \infty, \tag{9.7}$$

where $(a,t) \in \mathbb{R}_+^2$.

Then

$$\sup_{a \in \mathbb{R}_+} \left\| \left(D_{*a}^{2\alpha} f\right)(a) \right\| \leq \frac{\Gamma(2\alpha+1)}{2^{2\alpha-1}(2^{\alpha}-1)} \sqrt{\frac{2^{3\alpha+1}\left(2^{3\alpha}+1\right)\left(2^{\alpha}+1\right)}{\Gamma(4\alpha+1)} \| \|f\| \|_{\infty,\mathbb{R}_+} K}, \tag{9.8}$$

and

$$\sup_{a \in \mathbb{R}_+} \left\| \left(D_{*a}^{3\alpha} f\right)(a) \right\| \leq \frac{4\sqrt[4]{2}\,\Gamma(3\alpha+1)\left(\Gamma(4\alpha+1)\right)^{-\frac{3}{4}}\left(2^{2\alpha}+1\right)}{\left(\sqrt[4]{3}\right)^3 \left(\sqrt{2}\right)^{\alpha} \left(2^{\alpha}-1\right)} \| \|f\| \|_{\infty,\mathbb{R}_+}^{\frac{1}{4}} K^{\frac{3}{4}}. \tag{9.9}$$

That is $\sup_{a \in \mathbb{R}_+} \left\| \left(D_{*a}^{2\alpha} f\right)(a) \right\|$, $\sup_{a \in \mathbb{R}_+} \left\| \left(D_{*a}^{3\alpha} f\right)(a) \right\| < \infty$.

Proof We notice again here that $\left(D_{*a}^{\alpha} f\right)(a) = 0$, $\forall\, a \in \mathbb{R}_+$.

We make use of Theorem 9.3 for $0 < \alpha < 1$ and $n = 3$, applied for any $a \in \mathbb{R}_+$ and $b = \infty$.

Momentarily we fix $a \in \mathbb{R}_+$. Let $x_2 > x_1 > a$, then

$$f(x_1) - f(a) = \frac{(x_1 - a)^{2\alpha}}{\Gamma(2\alpha+1)} \left(D_{*a}^{2\alpha} f\right)(a) + \frac{(x_1 - a)^{3\alpha}}{\Gamma(3\alpha+1)} \left(D_{*a}^{3\alpha} f\right)(a) + \tag{9.10}$$

$$\frac{1}{\Gamma(4\alpha)} \int_a^{x_1} (x_1 - t)^{4\alpha-1} \left(D_{*a}^{4\alpha} f\right)(t)\, dt,$$

and

$$f(x_2) - f(a) = \frac{(x_2 - a)^{2\alpha}}{\Gamma(2\alpha+1)} \left(D_{*a}^{2\alpha} f\right)(a) + \frac{(x_2 - a)^{3\alpha}}{\Gamma(3\alpha+1)} \left(D_{*a}^{3\alpha} f\right)(a) + \tag{9.11}$$

$$\frac{1}{\Gamma(4\alpha)} \int_a^{x_2} (x_2 - t)^{4\alpha-1} \left(D_{*a}^{4\alpha} f\right)(t)\, dt.$$

That is

$$\frac{(x_1 - a)^{2\alpha}}{\Gamma(2\alpha+1)} \left(D_{*a}^{2\alpha} f\right)(a) + \frac{(x_1 - a)^{3\alpha}}{\Gamma(3\alpha+1)} \left(D_{*a}^{3\alpha} f\right)(a) = \tag{9.12}$$

$$f(x_1) - f(a) - \frac{1}{\Gamma(4\alpha)} \int_a^{x_1} (x_1 - t)^{4\alpha-1} \left(D_{*a}^{4\alpha} f\right)(t)\, dt =: A,$$

and

$$\frac{(x_2 - a)^{2\alpha}}{\Gamma(2\alpha+1)} \left(D_{*a}^{2\alpha} f\right)(a) + \frac{(x_2 - a)^{3\alpha}}{\Gamma(3\alpha+1)} \left(D_{*a}^{3\alpha} f\right)(a) = \tag{9.13}$$

$$f(x_2) - f(a) - \frac{1}{\Gamma(4\alpha)} \int_a^{x_2} (x_2 - t)^{4\alpha-1} \left(D_{*a}^{4\alpha} f\right)(t)\, dt =: B.$$

We are solving the above system of two equations with two unknowns $\left(D_{*a}^{2\alpha} f\right)(a)$, $\left(D_{*a}^{3\alpha} f\right)(a)$.

The main determinant of system is

$$D := \begin{vmatrix} \frac{(x_1-a)^{2\alpha}}{\Gamma(2\alpha+1)} & \frac{(x_1-a)^{3\alpha}}{\Gamma(3\alpha+1)} \\ \frac{(x_2-a)^{2\alpha}}{\Gamma(2\alpha+1)} & \frac{(x_2-a)^{3\alpha}}{\Gamma(3\alpha+1)} \end{vmatrix} =$$

$$\frac{1}{\Gamma(2\alpha+1)\,\Gamma(3\alpha+1)} \left[(x_1 - a)^{2\alpha} (x_2 - a)^{3\alpha} - (x_1 - a)^{3\alpha} (x_2 - a)^{2\alpha}\right] =$$

$$\frac{(x_1 - a)^{2\alpha} (x_2 - a)^{2\alpha}}{\Gamma(2\alpha+1)\,\Gamma(3\alpha+1)} \left[(x_2 - a)^{\alpha} - (x_1 - a)^{\alpha}\right] > 0.$$

I.e.

$$D=\frac{(x_1-a)^{2\alpha}(x_2-a)^{2\alpha}}{\Gamma(2\alpha+1)\Gamma(3\alpha+1)}\left[(x_2-a)^{\alpha}-(x_1-a)^{\alpha}\right]>0. \tag{9.14}$$

We obtain the unique solution

$$\left(D_{*a}^{2\alpha}f\right)(a)=\frac{\begin{vmatrix} A & \frac{(x_1-a)^{3\alpha}}{\Gamma(3\alpha+1)} \\ B & \frac{(x_2-a)^{3\alpha}}{\Gamma(3\alpha+1)}\end{vmatrix}}{D},$$

$$\left(D_{*a}^{3\alpha}f\right)(a)=\frac{\begin{vmatrix} \frac{(x_1-a)^{2\alpha}}{\Gamma(2\alpha+1)} & A \\ \frac{(x_2-a)^{2\alpha}}{\Gamma(2\alpha+1)} & B\end{vmatrix}}{D}. \tag{9.15}$$

Therefore we have

$$\left(D_{*a}^{2\alpha}f\right)(a)=\frac{\frac{(x_2-a)^{3\alpha}}{\Gamma(3\alpha+1)}A-\frac{(x_1-a)^{3\alpha}}{\Gamma(3\alpha+1)}B}{D},$$

and

$$\left(D_{*a}^{3\alpha}f\right)(a)=\frac{\frac{(x_1-a)^{2\alpha}}{\Gamma(2\alpha+1)}B-\frac{(x_2-a)^{2\alpha}}{\Gamma(2\alpha+1)}A}{D}. \tag{9.16}$$

We have the following

$$\|A\|=\left\|f(x_1)-f(a)-\frac{1}{\Gamma(4\alpha)}\int_a^{x_1}(x_1-t)^{4\alpha-1}\left(D_{*a}^{4\alpha}f\right)(t)\,dt\right\|\leq$$

$$2\,\|\|f\|\|_{\infty,\mathbb{R}_+}+\frac{\|\|D_{*a}^{4\alpha}f(t)\|\|_{\infty,\mathbb{R}_+^2}}{\Gamma(4\alpha+1)}(x_1-a)^{4\alpha}, \tag{9.17}$$

under the assumption $\|\|f\|\|_{\infty,\mathbb{R}_+}<\infty$.

That is

$$\|A\|\leq 2\,\|\|f\|\|_{\infty,\mathbb{R}_+}+\frac{K}{\Gamma(4\alpha+1)}(x_1-a)^{4\alpha}, \tag{9.18}$$

and similarly,

$$\|B\|\leq 2\,\|\|f\|\|_{\infty,\mathbb{R}_+}+\frac{K}{\Gamma(4\alpha+1)}(x_2-a)^{4\alpha}, \tag{9.19}$$

where by assumption

$$K:=\|\|D_{*a}^{4\alpha}f(t)\|\|_{\infty,\mathbb{R}_+^2}<\infty, \tag{9.20}$$

with $(a,t)\in\mathbb{R}_+^2$.

Consequently we have

$$\left\| \left(D_{*a}^{2\alpha} f\right)(a)\right\| \leq \frac{1}{\Gamma(3\alpha+1) D}\left[(x_2-a)^{3\alpha}\left(2\left\|\|f\|\right\|_{\infty,\mathbb{R}_+}+\frac{K}{\Gamma(4\alpha+1)}(x_1-a)^{4\alpha}\right)\right.$$

$$\left.+(x_1-a)^{3\alpha}\left(2\left\|\|f\|\right\|_{\infty,\mathbb{R}_+}+\frac{K}{\Gamma(4\alpha+1)}(x_2-a)^{4\alpha}\right)\right], \tag{9.21}$$

and

$$\left\| \left(D_{*a}^{3\alpha} f\right)(a)\right\| \leq \frac{1}{\Gamma(2\alpha+1) D}\left[(x_1-a)^{2\alpha}\left(2\left\|\|f\|\right\|_{\infty,\mathbb{R}_+}+\frac{K}{\Gamma(4\alpha+1)}(x_2-a)^{4\alpha}\right)\right.$$

$$\left.+(x_2-a)^{2\alpha}\left(2\left\|\|f\|\right\|_{\infty,\mathbb{R}_+}+\frac{K}{\Gamma(4\alpha+1)}(x_1-a)^{4\alpha}\right)\right]. \tag{9.22}$$

Set now $x_1 := a+h$, $x_2 := a+2h$, where $h>0$, so that $x_1-a=h$, $x_2-a=2h$.

Hence we get

$$D=\frac{2^{2\alpha}h^{5\alpha}\left(2^{\alpha}-1\right)}{\Gamma(2\alpha+1)\,\Gamma(3\alpha+1)}>0. \tag{9.23}$$

Therefore we derive (from (9.16))

$$\left\| \left(D_{*a}^{2\alpha} f\right)(a)\right\| \leq \frac{\Gamma(2\alpha+1)}{2^{2\alpha}h^{5\alpha}\left(2^{\alpha}-1\right)}\left[2^{3\alpha}h^{3\alpha}\left(2\left\|\|f\|\right\|_{\infty,\mathbb{R}_+}+\frac{K}{\Gamma(4\alpha+1)}h^{4\alpha}\right)\right.$$

$$\left.+h^{3\alpha}\left(2\left\|\|f\|\right\|_{\infty,\mathbb{R}_+}+\frac{K}{\Gamma(4\alpha+1)}2^{4\alpha}h^{4\alpha}\right)\right]= \tag{9.24}$$

$$\frac{\Gamma(2\alpha+1)}{2^{2\alpha}h^{5\alpha}\left(2^{\alpha}-1\right)}\left[2\left\|\|f\|\right\|_{\infty,\mathbb{R}_+}\left(2^{3\alpha}+1\right)h^{3\alpha}+\frac{K}{\Gamma(4\alpha+1)}\left(2^{3\alpha}+2^{4\alpha}\right)h^{7\alpha}\right]$$

$$=\left(\frac{\Gamma(2\alpha+1)}{2^{2\alpha}\left(2^{\alpha}-1\right)}\right)\left[\frac{2\left(2^{3\alpha}+1\right)\left\|\|f\|\right\|_{\infty,\mathbb{R}_+}}{h^{2\alpha}}+\frac{2^{3\alpha}\left(2^{\alpha}+1\right)}{\Gamma(4\alpha+1)}Kh^{2\alpha}\right]. \tag{9.25}$$

That is

$$\left\| \left(D_{*a}^{2\alpha} f\right)(a)\right\| \leq \left(\frac{\Gamma(2\alpha+1)}{2^{2\alpha}\left(2^{\alpha}-1\right)}\right)$$

$$\left[\frac{2\left(2^{3\alpha}+1\right)\left\|\|f\|\right\|_{\infty,\mathbb{R}_+}}{h^{2\alpha}}+\frac{2^{3\alpha}\left(2^{\alpha}+1\right)}{\Gamma(4\alpha+1)}Kh^{2\alpha}\right], \tag{9.26}$$

$\forall\, a\in\mathbb{R}_+$, $\forall\, h>0$.

I.e. it holds

$$\sup_{a\in\mathbb{R}_+}\left\| \left(D_{*a}^{2\alpha} f\right)(a)\right\| \leq \left(\frac{\Gamma(2\alpha+1)}{2^{2\alpha}\left(2^{\alpha}-1\right)}\right)$$

$$\left[\frac{2\left(2^{3\alpha}+1\right)\|\|f\|\|_{\infty,\mathbb{R}_+}}{h^{2\alpha}}+\frac{2^{3\alpha}\left(2^{\alpha}+1\right)}{\Gamma\left(4\alpha+1\right)}Kh^{2\alpha}\right]<\infty, \tag{9.27}$$

$\forall\, h>0,\, 0<\alpha<1$.

By (9.16) we derive

$$\left\|\left(D_{*a}^{3\alpha}f\right)(a)\right\|\leq\frac{\Gamma\left(3\alpha+1\right)}{2^{2\alpha}h^{5\alpha}\left(2^{\alpha}-1\right)}\left[h^{2\alpha}\left(2\,\|\|f\|\|_{\infty,\mathbb{R}_+}+\frac{K}{\Gamma\left(4\alpha+1\right)}2^{4\alpha}h^{4\alpha}\right)\right.$$

$$\left.+2^{2\alpha}h^{2\alpha}\left(2\,\|\|f\|\|_{\infty,\mathbb{R}_+}+\frac{K}{\Gamma\left(4\alpha+1\right)}h^{4\alpha}\right)\right]=$$

$$\frac{\Gamma\left(3\alpha+1\right)}{2^{2\alpha}h^{5\alpha}\left(2^{\alpha}-1\right)}\left[2\,\|\|f\|\|_{\infty,\mathbb{R}_+}\left(2^{2\alpha}+1\right)h^{2\alpha}+\frac{K}{\Gamma\left(4\alpha+1\right)}\left(2^{4\alpha}+2^{2\alpha}\right)h^{6\alpha}\right] \tag{9.28}$$

$$=\left(\frac{\Gamma\left(3\alpha+1\right)}{2^{2\alpha}\left(2^{\alpha}-1\right)}\right)\left[\frac{2\left(2^{2\alpha}+1\right)\|\|f\|\|_{\infty,\mathbb{R}_+}}{h^{3\alpha}}+\frac{2^{2\alpha}\left(2^{2\alpha}+1\right)}{\Gamma\left(4\alpha+1\right)}Kh^{\alpha}\right]$$

$$=\frac{\Gamma\left(3\alpha+1\right)\left(2^{2\alpha}+1\right)}{2^{2\alpha}\left(2^{\alpha}-1\right)}\left[\frac{2\,\|\|f\|\|_{\infty,\mathbb{R}_+}}{h^{3\alpha}}+\frac{2^{2\alpha}K}{\Gamma\left(4\alpha+1\right)}h^{\alpha}\right].$$

That is

$$\left\|\left(D_{*a}^{3\alpha}f\right)(a)\right\|\leq\frac{\Gamma\left(3\alpha+1\right)\left(2^{2\alpha}+1\right)}{2^{2\alpha}\left(2^{\alpha}-1\right)}$$

$$\left[\frac{2\,\|\|f\|\|_{\infty,\mathbb{R}_+}}{h^{3\alpha}}+\frac{2^{2\alpha}K}{\Gamma\left(4\alpha+1\right)}h^{\alpha}\right], \tag{9.29}$$

$\forall\, a\in\mathbb{R}_+,\,\forall\, h>0$.

I.e. it holds

$$\sup_{a\in\mathbb{R}_+}\left\|\left(D_{*a}^{3\alpha}f\right)(a)\right\|\leq\frac{\Gamma\left(3\alpha+1\right)\left(2^{2\alpha}+1\right)}{2^{2\alpha}\left(2^{\alpha}-1\right)}$$

$$\left[\frac{2\,\|\|f\|\|_{\infty,\mathbb{R}_+}}{h^{3\alpha}}+\frac{2^{2\alpha}K}{\Gamma\left(4\alpha+1\right)}h^{\alpha}\right]<\infty, \tag{9.30}$$

$\forall\, h>0,\, 0<\alpha<1$.

Call

$$\mu:=2\left(2^{3\alpha}+1\right)\|\|f\|\|_{\infty,\mathbb{R}_+}, \tag{9.31}$$

$$\theta=\frac{2^{3\alpha}\left(2^{\alpha}+1\right)K}{\Gamma\left(4\alpha+1\right)}, \tag{9.32}$$

both are greater than zero.

Set also $\rho := 2\alpha$; $0 < \rho < 2$. We consider the function

$$y(h) := \mu h^{-\rho} + \theta h^{\rho}, \quad \forall\, h > 0. \tag{9.33}$$

We have

$$y'(h) = -\rho\mu h^{-\rho-1} + \rho\theta h^{\rho-1} = 0, \tag{9.34}$$

then

$$\theta h^{\rho-1} = \mu h^{-\rho-1}$$

and

$$\theta h^{2\rho} = \mu,$$

with a unique solution

$$h_0 := h_{crit.no.} = \left(\frac{\mu}{\theta}\right)^{\frac{1}{2\rho}}. \tag{9.35}$$

We have that

$$y''(h) = \rho(\rho+1)\mu h^{-\rho-2} + \rho(\rho-1)\theta h^{\rho-2}. \tag{9.36}$$

We see that

$$y''(h_0) = y''\left(\left(\frac{\mu}{\theta}\right)^{\frac{1}{2\rho}}\right) = \rho(\rho+1)\mu\left(\frac{\mu}{\theta}\right)^{\frac{-\rho-2}{2\rho}} + \rho(\rho-1)\theta\left(\frac{\mu}{\theta}\right)^{\frac{\rho-2}{2\rho}} =$$

$$\rho\left(\frac{\mu}{\theta}\right)^{-\frac{1}{\rho}}\left[(\rho+1)\mu\left(\frac{\mu}{\theta}\right)^{-\frac{1}{2}} + (\rho-1)\theta\left(\frac{\mu}{\theta}\right)^{\frac{1}{2}}\right] =$$

$$\rho\left(\frac{\theta}{\mu}\right)^{\frac{1}{\rho}}\left[(\rho+1)\sqrt{\mu\theta} + (\rho-1)\sqrt{\mu\theta}\right] =$$

$$\rho\left(\frac{\theta}{\mu}\right)^{\frac{1}{\rho}}\left(2\rho\sqrt{\mu\theta}\right) = 2\rho^2\sqrt{\mu\theta}\left(\frac{\theta}{\mu}\right)^{\frac{1}{\rho}} > 0.$$

Therefore y has a global minimum at $h_0 = \left(\frac{\mu}{\theta}\right)^{\frac{1}{2\rho}}$, which is

$$y(h_0) = \mu\left(\frac{\mu}{\theta}\right)^{-\frac{1}{2}} + \theta\left(\frac{\mu}{\theta}\right)^{\frac{1}{2}} = \mu\left(\frac{\theta}{\mu}\right)^{\frac{1}{2}} + \sqrt{\theta\mu} = 2\sqrt{\theta\mu}.$$

We have proved that (see (9.27))

$$\sup_{a\in\mathbb{R}_+}\left\|\left(D_{*a}^{2\alpha}f\right)(a)\right\|\leq\frac{\Gamma(2\alpha+1)}{2^{2\alpha-1}(2^{\alpha}-1)}$$

$$\sqrt{\frac{2^{3\alpha+1}\left(2^{3\alpha}+1\right)(2^{\alpha}+1)}{\Gamma(4\alpha+1)}\left\|\|f\|\right\|_{\infty,\mathbb{R}_+}K}. \tag{9.37}$$

Call

$$\begin{aligned}\xi &:= 2\left\|\|f\|\right\|_{\infty,\mathbb{R}_+},\\ \psi &:= \tfrac{2^{2\alpha}K}{\Gamma(4\alpha+1)},\end{aligned} \tag{9.38}$$

both are greater than zero.

We consider the function

$$\gamma(h) := \xi h^{-3\alpha}+\psi h^{\alpha},\ \ \forall\, h>0. \tag{9.39}$$

We have

$$\gamma'(h) = -3\alpha\xi h^{-3\alpha-1}+\alpha\psi h^{\alpha-1}=0,$$

then

$$\psi h^{\alpha-1}=3\xi h^{-3\alpha-1}$$

and

$$\psi h^{4\alpha}=3\xi,$$

with unique solution

$$h_0 := h_{crit.no.}=\left(\frac{3\xi}{\psi}\right)^{\frac{1}{4\alpha}}. \tag{9.40}$$

We have that

$$\gamma''(h)=3\alpha(3\alpha+1)\xi h^{-3\alpha-2}+\alpha(\alpha-1)\psi h^{\alpha-2}. \tag{9.41}$$

We see

$$\gamma''(h_0)=3\alpha(3\alpha+1)\xi\left(\frac{3\xi}{\psi}\right)^{\frac{-3\alpha-2}{4\alpha}}+\alpha(\alpha-1)\psi\left(\frac{3\xi}{\psi}\right)^{\frac{\alpha-2}{4\alpha}}=$$

$$\alpha\left(\frac{3\xi}{\psi}\right)^{\frac{\alpha-2}{4\alpha}}\left[3(3\alpha+1)\xi\left(\frac{3\xi}{\psi}\right)^{-1}+(\alpha-1)\psi\right]=$$

$$\alpha\left(\frac{3\xi}{\psi}\right)^{\frac{\alpha-2}{4\alpha}}\left[3(3\alpha+1)\xi\frac{\psi}{3\xi}+(\alpha-1)\psi\right]=$$

$$\alpha\left(\frac{3\xi}{\psi}\right)^{\frac{\alpha-2}{4\alpha}}\left[(3\alpha+1)\psi+(\alpha-1)\psi\right]=$$

$$\alpha\left(\frac{3\xi}{\psi}\right)^{\frac{\alpha-2}{4\alpha}}(4\alpha\psi)=4\alpha^{2}\psi\left(\frac{3\xi}{\psi}\right)^{\frac{\alpha-2}{4\alpha}}>0. \tag{9.42}$$

Therefore γ has a global minimum at $h_0=\left(\frac{3\xi}{\psi}\right)^{\frac{1}{4\alpha}}$, which is

$$\gamma(h_0)=\xi\left(\frac{3\xi}{\psi}\right)^{-\frac{3}{4}}+\psi\left(\frac{3\xi}{\psi}\right)^{\frac{1}{4}}=$$

$$\left(\frac{3\xi}{\psi}\right)^{\frac{1}{4}}\left[\xi\left(\frac{3\xi}{\psi}\right)^{-1}+\psi\right]=\left(\frac{3\xi}{\psi}\right)^{\frac{1}{4}}\left(\xi\frac{\psi}{3\xi}+\psi\right)= \tag{9.43}$$

$$\left(\frac{3\xi}{\psi}\right)^{\frac{1}{4}}\left(\frac{4}{3}\psi\right)=\frac{4}{3}\psi\left(\frac{3\xi}{\psi}\right)^{\frac{1}{4}}.$$

Consequently,

$$\gamma(h_0)=\frac{4}{3}\psi\left(\frac{3\xi}{\psi}\right)^{\frac{1}{4}}=\frac{4}{\left(\sqrt[4]{3}\right)^{3}}\psi^{\frac{3}{4}}\xi^{\frac{1}{4}}. \tag{9.44}$$

We have proved that (see (9.30))

$$\sup_{a\in\mathbb{R}_+}\left\|\left(D_{*a}^{3\alpha}f\right)(a)\right\|\leq\frac{4\Gamma(3\alpha+1)\left(2^{2\alpha}+1\right)}{\left(\sqrt[4]{3}\right)^{3}2^{2\alpha}\left(2^{\alpha}-1\right)}$$

$$\left(2\,\|\|f\|\|_{\infty,\mathbb{R}_+}\right)^{\frac{1}{4}}\left(\frac{2^{2\alpha}K}{\Gamma(4\alpha+1)}\right)^{\frac{3}{4}}=$$

$$\frac{4\sqrt[4]{2}\,\Gamma(3\alpha+1)\,\Gamma(4\alpha+1)^{-\frac{3}{4}}\left(2^{2\alpha}+1\right)}{\left(\sqrt[4]{3}\right)^{3}2^{\frac{\alpha}{2}}\left(2^{\alpha}-1\right)}\|\|f\|\|_{\infty,\mathbb{R}_+}^{\frac{1}{4}}K^{\frac{3}{4}}. \tag{9.45}$$

The theorem is proved. ■

We continue with abstract L_p left fractional Landau inequalities over $\mathbb{R}_+$.

Theorem 9.5 *Let* $p,q>1:\frac{1}{p}+\frac{1}{q}=1$, $0<\alpha<1$. *Let* $f\in C^{1}(\mathbb{R}_+,X)$ *with* $\|\|f\|\|_{\infty,\mathbb{R}_+}$, $\left\|\|f'\|\right\|_{\infty,\mathbb{R}_+}<\infty$. *For* $k=1,2,3$, *we assume that* $D_{*a}^{k\alpha}f\in C^{1}([a,\infty),X)$ *and* $D_{*a}^{4\alpha}f\in C([a,\infty),X)$, $\forall\,a\in\mathbb{R}_+$. *We further assume that*

$$\left(\sup_{a \in \mathbb{R}_+} \left\| \left\| D_{*a}^{4\alpha} f \right\| \right\|_{p, \mathbb{R}_+} \right) < \infty. \tag{9.46}$$

Then

(1) under $\frac{1}{2p} < \alpha < 1$, *we get*

$$\sup_{a \in \mathbb{R}_+} \left\| \left(D_{*a}^{2\alpha} f \right)(a) \right\| \leq \left[\left(\frac{2^{\alpha} \Gamma(2\alpha) \left(4\alpha - \frac{1}{p} \right)}{2^{\alpha} - 1} \right) \left(\frac{4\alpha \left(1 + 2^{-3\alpha} \right)}{2\alpha - \frac{1}{p}} \right)^{\left(\frac{2\alpha - \frac{1}{p}}{4\alpha - \frac{1}{p}} \right)} \right.$$

$$\left. \left(\frac{1 + 2^{\alpha - \frac{1}{p}}}{\Gamma(4\alpha) \left(q(4\alpha - 1) + 1 \right)^{\frac{1}{q}}} \right)^{\left(\frac{2\alpha}{4\alpha - \frac{1}{p}} \right)} \right] \left\| \left\| f \right\| \right\|_{\infty, \mathbb{R}_+}^{\left(\frac{2\alpha - \frac{1}{p}}{4\alpha - \frac{1}{p}} \right)}$$

$$\left(\sup_{a \in \mathbb{R}_+} \left\| \left\| D_{*a}^{4\alpha} f \right\| \right\|_{p, \mathbb{R}_+} \right)^{\left(\frac{2\alpha}{4\alpha - \frac{1}{p}} \right)} < \infty. \tag{9.47}$$

(2) under $\frac{1}{p} < \alpha < 1$, *we get*

$$\sup_{a \in \mathbb{R}_+} \left\| \left(D_{*a}^{3\alpha} f \right)(a) \right\| \leq \left[\left(\frac{\Gamma(3\alpha) \left(4\alpha - \frac{1}{p} \right)}{2^{\alpha} - 1} \right) \left(\frac{6\alpha \left(1 + 2^{-2\alpha} \right)}{\alpha - \frac{1}{p}} \right)^{\left(\frac{\alpha - \frac{1}{p}}{4\alpha - \frac{1}{p}} \right)} \right.$$

$$\left. \left(\frac{1 + 2^{2\alpha - \frac{1}{p}}}{\Gamma(4\alpha) \left(q(4\alpha - 1) + 1 \right)^{\frac{1}{q}}} \right)^{\left(\frac{3\alpha}{4\alpha - \frac{1}{p}} \right)} \right] \left\| \left\| f \right\| \right\|_{\infty, \mathbb{R}_+}^{\left(\frac{\alpha - \frac{1}{p}}{4\alpha - \frac{1}{p}} \right)}$$

$$\left(\sup_{a \in \mathbb{R}_+} \left\| \left\| D_{*a}^{4\alpha} f \right\| \right\|_{p, \mathbb{R}_+} \right)^{\left(\frac{3\alpha}{4\alpha - \frac{1}{p}} \right)} < \infty. \tag{9.48}$$

That is $\sup_{a \in \mathbb{R}_+} \left\| \left(D_{*a}^{2\alpha} f \right)(a) \right\|, \sup_{a \in \mathbb{R}_+} \left\| \left(D_{*a}^{3\alpha} f \right)(a) \right\| < \infty.$

Proof As in the proof of Theorem 9.4 we have that

$$\|A\| \overset{(9.12)}{=} \left\| f(x_1) - f(a) - \frac{1}{\Gamma(4\alpha)} \int_a^{x_1} (x_1 - t)^{4\alpha - 1} \left(D_{*a}^{4\alpha} f \right)(t) \, dt \right\| \leq$$

$$2 \left\| \left\| f \right\| \right\|_{\infty, \mathbb{R}_+} + \frac{1}{\Gamma(4\alpha)} \int_a^{x_1} (x_1 - t)^{4\alpha - 1} \left\| \left(D_{*a}^{4\alpha} f \right)(t) \right\| dt \leq$$

$$2\,\|\|f\|\|_{\infty,\mathbb{R}_+} + \frac{1}{\Gamma(4\alpha)} \frac{(x_1-a)^{\frac{(q(4\alpha-1)+1)}{q}}}{(q(4\alpha-1)+1)^{\frac{1}{q}}} \left(\sup_{a\in\mathbb{R}_+} \left\| \left\| D_{*a}^{4\alpha} f \right\| \right\|_{p,\mathbb{R}_+} \right) \overset{(x_1-a=:h>0)}{=}$$
$$2\,\|\|f\|\|_{\infty,\mathbb{R}_+} + \frac{1}{\Gamma(4\alpha)} \frac{h^{\left(4\alpha-\frac{1}{p}\right)}}{(q(4\alpha-1)+1)^{\frac{1}{q}}} \left(\sup_{a\in\mathbb{R}_+} \left\| \left\| D_{*a}^{4\alpha} f \right\| \right\|_{p,\mathbb{R}_+} \right), \tag{9.49}$$

with $\frac{1}{4p} < \alpha < 1$.

That is

$$\|A\| \le 2\,\|\|f\|\|_{\infty,\mathbb{R}_+} + \frac{h^{4\alpha-\frac{1}{p}}}{\Gamma(4\alpha)(q(4\alpha-1)+1)^{\frac{1}{q}}} \left(\sup_{a\in\mathbb{R}_+} \left\| \left\| D_{*a}^{4\alpha} f \right\| \right\|_{p,\mathbb{R}_+} \right), \tag{9.50}$$

where $\frac{1}{4p} < \alpha < 1$.

We also have

$$\|B\| \overset{(9.13)}{=} \left\| f(x_2) - f(a) - \frac{1}{\Gamma(4\alpha)} \int_a^{x_2} (x_2-t)^{4\alpha-1} \left(D_{*a}^{4\alpha} f \right)(t)\, dt \right\| \le$$

$$2\,\|\|f\|\|_{\infty,\mathbb{R}_+} + \frac{(x_2-a)^{\frac{(q(4\alpha-1)+1)}{q}}}{\Gamma(4\alpha)(q(4\alpha-1)+1)^{\frac{1}{q}}} \left(\sup_{a\in\mathbb{R}_+} \left\| \left\| D_{*a}^{4\alpha} f \right\| \right\|_{p,\mathbb{R}_+} \right) \overset{(x_2-a=:2h)}{=}$$
$$2\,\|\|f\|\|_{\infty,\mathbb{R}_+} + \frac{2^{4\alpha-\frac{1}{p}} h^{4\alpha-\frac{1}{p}}}{\Gamma(4\alpha)(q(4\alpha-1)+1)^{\frac{1}{q}}} \left(\sup_{a\in\mathbb{R}_+} \left\| \left\| D_{*a}^{4\alpha} f \right\| \right\|_{p,\mathbb{R}_+} \right). \tag{9.51}$$

That is

$$\|B\| \le 2\,\|\|f\|\|_{\infty,\mathbb{R}_+} + \frac{2^{4\alpha-\frac{1}{p}} h^{4\alpha-\frac{1}{p}}}{\Gamma(4\alpha)(q(4\alpha-1)+1)^{\frac{1}{q}}} \left(\sup_{a\in\mathbb{R}_+} \left\| \left\| D_{*a}^{4\alpha} f \right\| \right\|_{p,\mathbb{R}_+} \right), \tag{9.52}$$

where $\frac{1}{4p} < \alpha < 1$.

We have assumed that

$$M := \left(\sup_{a\in\mathbb{R}_+} \left\| \left\| D_{*a}^{4\alpha} f \right\| \right\|_{p,\mathbb{R}_+} \right) < \infty. \tag{9.53}$$

For convenience we call

$$c := \Gamma(4\alpha)(q(4\alpha-1)+1)^{\frac{1}{q}} > 0. \tag{9.54}$$

So we have

$$\|A\| \leq 2 \, \||f\||_{\infty,\mathbb{R}_+} + \frac{h^{4\alpha-\frac{1}{p}}}{c} M,$$
and
$$\|B\| \leq 2 \, \||f\||_{\infty,\mathbb{R}_+} + \frac{2^{4\alpha-\frac{1}{p}} h^{4\alpha-\frac{1}{p}}}{c} M, \tag{9.55}$$

where $\frac{1}{4p} < \alpha < 1$.

Next we estimate the (9.16)-quantities and we have:

$$\left\| \left(D_{*a}^{2\alpha} f \right)(a) \right\| \leq \frac{1}{D\Gamma(3\alpha+1)} \left[2^{3\alpha} h^{3\alpha} \|A\| + h^{3\alpha} \|B\| \right] \overset{(9.23)}{=}$$

$$\frac{h^{3\alpha}\Gamma(2\alpha+1)}{2^{2\alpha} h^{5\alpha} (2^{\alpha}-1)} \left[2^{3\alpha} \|A\| + \|B\| \right] \overset{(9.55)}{\leq} \tag{9.56}$$

$$\frac{\Gamma(2\alpha+1)}{2^{2\alpha}(2^{\alpha}-1) h^{2\alpha}} \left[2^{3\alpha+1} \, \||f\||_{\infty,\mathbb{R}_+} + \frac{2^{3\alpha} h^{4\alpha-\frac{1}{p}}}{c} M + \right.$$

$$\left. 2 \, \||f\||_{\infty,\mathbb{R}_+} + \frac{2^{4\alpha-\frac{1}{p}} h^{4\alpha-\frac{1}{p}}}{c} M \right] =$$

$$\frac{\Gamma(2\alpha+1)}{2^{2\alpha}(2^{\alpha}-1) h^{2\alpha}} \left[\left(2^{3\alpha+1}+2 \right) \||f\||_{\infty,\mathbb{R}_+} + \frac{\left(2^{3\alpha} + 2^{4\alpha-\frac{1}{p}} \right)}{c} M h^{4\alpha-\frac{1}{p}} \right] =$$

$$\frac{\Gamma(2\alpha+1)}{2^{2\alpha}(2^{\alpha}-1)} \left[\frac{\left(2^{3\alpha+1}+2 \right) \||f\||_{\infty,\mathbb{R}_+}}{h^{2\alpha}} + \frac{\left(2^{3\alpha} + 2^{4\alpha-\frac{1}{p}} \right)}{c} M h^{2\alpha-\frac{1}{p}} \right] =$$

$$\frac{2^{\alpha}\Gamma(2\alpha+1)}{(2^{\alpha}-1)} \left[\frac{2\left(1+2^{-3\alpha} \right) \||f\||_{\infty,\mathbb{R}_+}}{h^{2\alpha}} + \frac{\left(1 + 2^{\alpha-\frac{1}{p}} \right) M}{c} h^{2\alpha-\frac{1}{p}} \right]. \tag{9.57}$$

That is

$$\left\| \left(D_{*a}^{2\alpha} f \right)(a) \right\| \leq$$

$$\left(\frac{2^{\alpha}\Gamma(2\alpha+1)}{2^{\alpha}-1} \right) \left[\frac{2\left(1+2^{-3\alpha} \right) \||f\||_{\infty,\mathbb{R}_+}}{h^{2\alpha}} + \frac{\left(1 + 2^{\alpha-\frac{1}{p}} \right) M}{c} h^{2\alpha-\frac{1}{p}} \right], \tag{9.58}$$

$\forall \, a \in \mathbb{R}_+, \forall \, h > 0$.

I.e. it holds

$$\sup_{a \in \mathbb{R}_+} \left\| \left(D_{*a}^{2\alpha} f \right)(a) \right\| \leq \left(\frac{2^{\alpha}\Gamma(2\alpha+1)}{2^{\alpha}-1} \right)$$

$$\left[\frac{2\left(1+2^{-3\alpha}\right)\|\|f\|\|_{\infty,\mathbb{R}_+}}{h^{2\alpha}}+\frac{\left(1+2^{\alpha-\frac{1}{p}}\right)M}{c}h^{2\alpha-\frac{1}{p}}\right], \tag{9.59}$$

$\forall\, h>0$, under $\frac{1}{4p}<\alpha<1$.

Again from (9.16) we get

$$\left\|\left(D_{*a}^{3\alpha}f\right)(a)\right\| \leq \frac{1}{\Gamma(2\alpha+1)D}\left[h^{2\alpha}\|B\|+2^{2\alpha}h^{2\alpha}\|A\|\right]=$$

$$\frac{h^{2\alpha}\Gamma(3\alpha+1)}{2^{2\alpha}h^{5\alpha}(2^{\alpha}-1)}\left[\|B\|+2^{2\alpha}\|A\|\right]\leq \tag{9.60}$$

$$\left(\frac{h^{2\alpha}\Gamma(3\alpha+1)}{2^{2\alpha}h^{5\alpha}(2^{\alpha}-1)}\right)\left[2\|\|f\|\|_{\infty,\mathbb{R}_+}+\frac{2^{4\alpha-\frac{1}{p}}h^{4\alpha-\frac{1}{p}}}{c}M+\right.$$

$$\left.2^{2\alpha+1}\|\|f\|\|_{\infty,\mathbb{R}_+}+\frac{2^{2\alpha}h^{4\alpha-\frac{1}{p}}}{c}M\right]=$$

$$\left(\frac{\Gamma(3\alpha+1)}{2^{2\alpha}(2^{\alpha}-1)h^{3\alpha}}\right)\left[\left(2+2^{2\alpha+1}\right)\|\|f\|\|_{\infty,\mathbb{R}_+}+\frac{\left(2^{2\alpha}+2^{4\alpha-\frac{1}{p}}\right)}{c}Mh^{4\alpha-\frac{1}{p}}\right]=$$

$$\frac{\Gamma(3\alpha+1)}{2^{2\alpha}(2^{\alpha}-1)}\left[\frac{\left(2+2^{2\alpha+1}\right)\|\|f\|\|_{\infty,\mathbb{R}_+}}{h^{3\alpha}}+\frac{\left(2^{2\alpha}+2^{4\alpha-\frac{1}{p}}\right)}{c}Mh^{\alpha-\frac{1}{p}}\right]= \tag{9.61}$$

$$\frac{\Gamma(3\alpha+1)}{(2^{\alpha}-1)}\left[\frac{2\left(1+2^{-2\alpha}\right)\|\|f\|\|_{\infty,\mathbb{R}_+}}{h^{3\alpha}}+\frac{\left(1+2^{2\alpha-\frac{1}{p}}\right)}{c}Mh^{\alpha-\frac{1}{p}}\right].$$

That is

$$\left\|\left(D_{*a}^{3\alpha}f\right)(a)\right\|\leq\left(\frac{\Gamma(3\alpha+1)}{2^{\alpha}-1}\right)\left[\frac{2\left(1+2^{-2\alpha}\right)\|\|f\|\|_{\infty,\mathbb{R}_+}}{h^{3\alpha}}+\frac{\left(1+2^{2\alpha-\frac{1}{p}}\right)}{c}Mh^{\alpha-\frac{1}{p}}\right], \tag{9.62}$$

$\forall\, a\in\mathbb{R}_+, \forall\, h>0$.

I.e. it holds

$$\sup_{a\in\mathbb{R}_+}\left\|\left(D_{*a}^{3\alpha}f\right)(a)\right\|\leq\left(\frac{\Gamma(3\alpha+1)}{2^{\alpha}-1}\right)$$

$$\left[\frac{2\left(1+2^{-2\alpha}\right)\||f\||_{\infty,\mathbb{R}_+}}{h^{3\alpha}}+\frac{\left(1+2^{2\alpha-\frac{1}{p}}\right)}{c}Mh^{\alpha-\frac{1}{p}}\right], \tag{9.63}$$

$\forall\, h>0$, $\frac{1}{4p}<\alpha<1$.

Call

$$\mu := 2\left(1+2^{-3\alpha}\right)\||f\||_{\infty,\mathbb{R}_+}, \qquad \theta := \frac{\left(1+2^{\alpha-\frac{1}{p}}\right)M}{c}, \tag{9.64}$$

both are greater than zero.

We consider the function

$$y(h)=\mu h^{-2\alpha}+\theta h^{2\alpha-\frac{1}{p}}, \quad \forall\, h>0. \tag{9.65}$$

We have

$$y'(h)=-2\alpha\mu h^{-2\alpha-1}+\left(2\alpha-\frac{1}{p}\right)\theta h^{2\alpha-\frac{1}{p}-1}=0, \tag{9.66}$$

then

$$\left(2\alpha-\frac{1}{p}\right)\theta h^{2\alpha-\frac{1}{p}-1}=2\alpha\mu h^{-2\alpha-1}$$

and

$$\left(2\alpha-\frac{1}{p}\right)\theta h^{2\alpha-\frac{1}{p}-1}h^{2\alpha+1}=2\alpha\mu,$$

i.e.

$$\left(2\alpha-\frac{1}{p}\right)\theta h^{4\alpha-\frac{1}{p}}=2\alpha\mu,$$

with a unique solution

$$h_0 := h_{crit.no.}=\left(\frac{2\alpha\mu}{\left(2\alpha-\frac{1}{p}\right)\theta}\right)^{\frac{1}{4\alpha-\frac{1}{p}}} \tag{9.67}$$

(assuming $\frac{1}{2p}<\alpha<1$).

We have that

$$y''(h)=2\alpha(2\alpha+1)\mu h^{-2\alpha-2}+\left(2\alpha-\frac{1}{p}\right)\left(2\alpha-\frac{1}{p}-1\right)\theta h^{2\alpha-\frac{1}{p}-2}. \tag{9.68}$$

We see that

$$y''(h_0) = 2\alpha(2\alpha+1)\mu\left(\frac{2\alpha\mu}{\left(2\alpha-\frac{1}{p}\right)\theta}\right)^{\frac{-2\alpha-2}{4\alpha-\frac{1}{p}}} +$$

$$\left(2\alpha-\frac{1}{p}\right)\left(2\alpha-\frac{1}{p}-1\right)\theta\left(\frac{2\alpha\mu}{\left(2\alpha-\frac{1}{p}\right)\theta}\right)^{\frac{2\alpha-\frac{1}{p}-2}{4\alpha-\frac{1}{p}}} =$$

$$\left(\frac{2\alpha\mu}{\left(2\alpha-\frac{1}{p}\right)\theta}\right)^{\frac{-2\alpha-2}{4\alpha-\frac{1}{p}}}[2\alpha(2\alpha+1)\mu+$$

$$\left.\left(2\alpha-\frac{1}{p}\right)\left(2\alpha-\frac{1}{p}-1\right)\theta\left(\frac{2\alpha\mu}{\left(2\alpha-\frac{1}{p}\right)\theta}\right)\right] = \tag{9.69}$$

$$\left(\frac{2\alpha\mu}{\left(2\alpha-\frac{1}{p}\right)\theta}\right)^{\frac{-2\alpha-2}{4\alpha-\frac{1}{p}}}\left[2\alpha(2\alpha+1)\mu+2\alpha\mu\left(2\alpha-\frac{1}{p}-1\right)\right] =$$

$$2\alpha\mu\left(\frac{\left(2\alpha-\frac{1}{p}\right)\theta}{2\alpha\mu}\right)^{\left(\frac{2(\alpha+1)}{4\alpha-\frac{1}{p}}\right)}\left(4\alpha-\frac{1}{p}\right) > 0.$$

Therefore y has a global minimum at

$$h_0 = \left(\frac{2\alpha\mu}{\left(2\alpha-\frac{1}{p}\right)\theta}\right)^{\frac{1}{4\alpha-\frac{1}{p}}},$$

which is

$$y(h_0) = \mu\left(\frac{2\alpha\mu}{\left(2\alpha-\frac{1}{p}\right)\theta}\right)^{\frac{-2\alpha}{4\alpha-\frac{1}{p}}} + \theta\left(\frac{2\alpha\mu}{\left(2\alpha-\frac{1}{p}\right)\theta}\right)^{\frac{2\alpha-\frac{1}{p}}{4\alpha-\frac{1}{p}}} = \tag{9.70}$$

$$\left(\frac{2\alpha\mu}{\left(2\alpha-\frac{1}{p}\right)\theta}\right)^{\frac{-2\alpha}{4\alpha-\frac{1}{p}}}\left[\mu+\theta\frac{2\alpha\mu}{\left(2\alpha-\frac{1}{p}\right)\theta}\right] = \mu\left(\frac{\left(2\alpha-\frac{1}{p}\right)\theta}{2\alpha\mu}\right)^{\left(\frac{2\alpha}{4\alpha-\frac{1}{p}}\right)}\left(1+\frac{2\alpha}{2\alpha-\frac{1}{p}}\right)$$

$$= \mu \left(\frac{\left(2\alpha - \frac{1}{p}\right)\theta}{2\alpha\mu} \right)^{\left(\frac{2\alpha}{4\alpha - \frac{1}{p}}\right)} \left(\frac{4\alpha - \frac{1}{p}}{2\alpha - \frac{1}{p}} \right) =$$

$$\left(\frac{4\alpha - \frac{1}{p}}{2\alpha - \frac{1}{p}} \right) \left(\frac{2\alpha - \frac{1}{p}}{2\alpha} \right)^{\left(\frac{2\alpha}{4\alpha - \frac{1}{p}}\right)} \mu^{\left(\frac{2\alpha - \frac{1}{p}}{4\alpha - \frac{1}{p}}\right)} \theta^{\left(\frac{2\alpha}{4\alpha - \frac{1}{p}}\right)} =$$

$$\frac{\left(4\alpha - \frac{1}{p}\right)}{(2\alpha)^{\left(\frac{2\alpha}{4\alpha - \frac{1}{p}}\right)}} \left(2\alpha - \frac{1}{p}\right)^{\frac{-2\alpha + \frac{1}{p}}{4\alpha - \frac{1}{p}}} \mu^{\left(\frac{2\alpha - \frac{1}{p}}{4\alpha - \frac{1}{p}}\right)} \theta^{\left(\frac{2\alpha}{4\alpha - \frac{1}{p}}\right)}. \tag{9.71}$$

That is

$$y(h_0) = \frac{\left(4\alpha - \frac{1}{p}\right) \left(2\alpha - \frac{1}{p}\right)^{\left(\frac{-2\alpha + \frac{1}{p}}{4\alpha - \frac{1}{p}}\right)}}{(2\alpha)^{\left(\frac{2\alpha}{4\alpha - \frac{1}{p}}\right)}} \mu^{\left(\frac{2\alpha - \frac{1}{p}}{4\alpha - \frac{1}{p}}\right)} \theta^{\left(\frac{2\alpha}{4\alpha - \frac{1}{p}}\right)}. \tag{9.72}$$

Therefore we derive (see (9.59))

$$\sup_{a \in \mathbb{R}_+} \left\| \left(D_{*a}^{2\alpha} f\right)(a) \right\| \leq \left(\frac{2^{\alpha} \Gamma(2\alpha)}{2^{\alpha} - 1} \right) \left(\frac{2\alpha}{2\alpha - \frac{1}{p}} \right)^{\frac{\left(2\alpha - \frac{1}{p}\right)}{\left(4\alpha - \frac{1}{p}\right)}} \left(4\alpha - \frac{1}{p}\right)$$

$$\left(2\left(1 + 2^{-3\alpha}\right)\right)^{\left(\frac{2\alpha - \frac{1}{p}}{4\alpha - \frac{1}{p}}\right)} \left(\frac{\left(1 + 2^{\alpha - \frac{1}{p}}\right)}{\Gamma(4\alpha)\left(q(4\alpha - 1) + 1\right)^{\frac{1}{q}}} \right)^{\left(\frac{2\alpha}{4\alpha - \frac{1}{p}}\right)} \left\| \| f \| \right\|_{\infty, \mathbb{R}_+}^{\left(\frac{2\alpha - \frac{1}{p}}{4\alpha - \frac{1}{p}}\right)}$$

$$\left(\sup_{a \in \mathbb{R}_+} \left\| \left\| D_{*a}^{4\alpha} f \right\| \right\|_{p, \mathbb{R}_+} \right)^{\left(\frac{2\alpha}{4\alpha - \frac{1}{p}}\right)} < \infty, \tag{9.73}$$

where $\frac{1}{2p} < \alpha < 1$.

Call

$$\xi := 2\left(1 + 2^{-2\alpha}\right) \left\| \| f \| \right\|_{\infty, \mathbb{R}_+},$$

$$\psi := \frac{\left(1 + 2^{2\alpha - \frac{1}{p}}\right)}{c} M, \tag{9.74}$$

both are greater than zero.

We consider the function

$$\gamma(h) := \xi h^{-3\alpha} + \psi h^{\alpha-\frac{1}{p}}, \quad \forall\, h > 0. \tag{9.75}$$

We have

$$\gamma'(h) = -3\alpha\xi h^{-3\alpha-1} + \left(\alpha - \frac{1}{p}\right)\psi h^{\alpha-\frac{1}{p}-1} = 0,$$

then

$$\left(\alpha - \frac{1}{p}\right)\psi h^{\alpha-\frac{1}{p}-1} = 3\alpha\xi h^{-3\alpha-1}$$

and

$$\left(\alpha - \frac{1}{p}\right)\psi h^{\alpha-\frac{1}{p}-1} h^{3\alpha+1} = 3\alpha\xi,$$

and

$$\left(\alpha - \frac{1}{p}\right)\psi h^{4\alpha-\frac{1}{p}} = 3\alpha\xi,$$

with unique solution

$$h_0 := h_{crit.no.} = \left(\frac{3\alpha\xi}{\left(\alpha - \frac{1}{p}\right)\psi}\right)^{\frac{1}{4\alpha-\frac{1}{p}}} \tag{9.76}$$

(assuming $\frac{1}{p} < \alpha < 1$).

We have that

$$\gamma''(h) = 3\alpha(3\alpha+1)\xi h^{-3\alpha-2} + \left(\alpha - \frac{1}{p}\right)\left(\alpha - \frac{1}{p} - 1\right)\psi h^{\alpha-\frac{1}{p}-2}. \tag{9.77}$$

We observe that

$$\gamma''(h_0) = 3\alpha(3\alpha+1)\xi\left(\frac{3\alpha\xi}{\left(\alpha - \frac{1}{p}\right)\psi}\right)^{\left(\frac{-3\alpha-2}{4\alpha-\frac{1}{p}}\right)} +$$

$$\left(\alpha - \frac{1}{p}\right)\left(\alpha - \frac{1}{p} - 1\right)\psi\left(\frac{3\alpha\xi}{\left(\alpha - \frac{1}{p}\right)\psi}\right)^{\left(\frac{\alpha-\frac{1}{p}-2}{4\alpha-\frac{1}{p}}\right)} =$$

$$\left(\frac{3\alpha\xi}{\left(\alpha - \frac{1}{p}\right)\psi}\right)^{\left(\frac{-3\alpha-2}{4\alpha-\frac{1}{p}}\right)}\left[3\alpha(3\alpha+1)\xi + \left(\alpha - \frac{1}{p}\right)\left(\alpha - \frac{1}{p} - 1\right)\psi\left(\frac{3\alpha\xi}{\left(\alpha - \frac{1}{p}\right)\psi}\right)\right]$$

$$= \left(\frac{3\alpha\xi}{\left(\alpha-\frac{1}{p}\right)\psi}\right)^{\left(\frac{-3\alpha-2}{4\alpha-\frac{1}{p}}\right)} \left[3\alpha\left(3\alpha+1\right)\xi + \left(\alpha-\frac{1}{p}-1\right)3\alpha\xi\right] = \tag{9.78}$$

$$3\alpha\xi\left(\frac{3\alpha\xi}{\left(\alpha-\frac{1}{p}\right)\psi}\right)^{\left(\frac{-3\alpha-2}{4\alpha-\frac{1}{p}}\right)} \left[\left(3\alpha+1\right) + \left(\alpha-\frac{1}{p}-1\right)\right] =$$

$$3\alpha\xi\left(\frac{3\alpha\xi}{\left(\alpha-\frac{1}{p}\right)\psi}\right)^{\left(\frac{-3\alpha-2}{4\alpha-\frac{1}{p}}\right)} \left(4\alpha-\frac{1}{p}\right) > 0.$$

Therefore y has a global minimum at

$$h_0 = \left(\frac{3\alpha\xi}{\left(\alpha-\frac{1}{p}\right)\psi}\right)^{\frac{1}{4\alpha-\frac{1}{p}}},$$

which is

$$\gamma\left(h_0\right) = \xi h_0^{-3\alpha} + \psi h_0^{\alpha-\frac{1}{p}} = h_0^{-3\alpha}\left(\xi + \psi h_0^{4\alpha-\frac{1}{p}}\right) = \tag{9.79}$$

$$\left(\frac{3\alpha\xi}{\left(\alpha-\frac{1}{p}\right)\psi}\right)^{\left(\frac{-3\alpha}{4\alpha-\frac{1}{p}}\right)} \left(\xi + \psi\left(\frac{3\alpha\xi}{\left(\alpha-\frac{1}{p}\right)\psi}\right)\right) =$$

$$\xi\left(\frac{3\alpha\xi}{\left(\alpha-\frac{1}{p}\right)\psi}\right)^{\left(\frac{-3\alpha}{4\alpha-\frac{1}{p}}\right)} \left(1 + \frac{3\alpha}{\alpha-\frac{1}{p}}\right) =$$

$$\xi\left(\frac{3\alpha\xi}{\left(\alpha-\frac{1}{p}\right)\psi}\right)^{\left(\frac{-3\alpha}{4\alpha-\frac{1}{p}}\right)} \left(\frac{4\alpha-\frac{1}{p}}{\alpha-\frac{1}{p}}\right).$$

That is

$$\gamma\left(h_0\right) = \xi\left(\frac{4\alpha-\frac{1}{p}}{\alpha-\frac{1}{p}}\right)\left(\frac{3\alpha\xi}{\left(\alpha-\frac{1}{p}\right)\psi}\right)^{\left(\frac{-3\alpha}{4\alpha-\frac{1}{p}}\right)} = \tag{9.80}$$

$$\left(\frac{4\alpha-\frac{1}{p}}{\alpha-\frac{1}{p}}\right)\xi\left(\frac{\left(\alpha-\frac{1}{p}\right)\psi}{3\alpha\xi}\right)^{\left(\frac{3\alpha}{4\alpha-\frac{1}{p}}\right)}=$$

$$\left(\frac{4\alpha-\frac{1}{p}}{\alpha-\frac{1}{p}}\right)\xi^{\left(\frac{\alpha-\frac{1}{p}}{4\alpha-\frac{1}{p}}\right)}\left(\frac{\left(\alpha-\frac{1}{p}\right)}{3\alpha}\right)^{\left(\frac{3\alpha}{4\alpha-\frac{1}{p}}\right)}\psi^{\left(\frac{3\alpha}{4\alpha-\frac{1}{p}}\right)}.$$

I.e. we have found

$$\gamma\left(h_0\right)=\frac{\left(4\alpha-\frac{1}{p}\right)}{\left(\alpha-\frac{1}{p}\right)^{\left(\frac{\alpha-\frac{1}{p}}{4\alpha-\frac{1}{p}}\right)}(3\alpha)^{\left(\frac{3\alpha}{4\alpha-\frac{1}{p}}\right)}}\xi^{\left(\frac{\alpha-\frac{1}{p}}{4\alpha-\frac{1}{p}}\right)}\psi^{\left(\frac{3\alpha}{4\alpha-\frac{1}{p}}\right)}. \tag{9.81}$$

We have proved that (see (9.63))

$$\sup_{a\in\mathbb{R}_+}\left\|\left(D_{*a}^{3\alpha}f\right)(a)\right\|\leq\left(\frac{\left(4\alpha-\frac{1}{p}\right)\Gamma(3\alpha)}{2^{\alpha}-1}\right)\left(\frac{3\alpha}{\alpha-\frac{1}{p}}\right)^{\left(\frac{\alpha-\frac{1}{p}}{4\alpha-\frac{1}{p}}\right)} \tag{9.82}$$

$$\left(2\left(1+2^{-2\alpha}\right)\right)^{\left(\frac{\alpha-\frac{1}{p}}{4\alpha-\frac{1}{p}}\right)}\left(\frac{\left(1+2^{2\alpha-\frac{1}{p}}\right)}{\Gamma(4\alpha)\left(q(4\alpha-1)+1\right)^{\frac{1}{q}}}\right)^{\left(\frac{3\alpha}{4\alpha-\frac{1}{p}}\right)}\|\|f\|\|_{\infty,\mathbb{R}_+}^{\left(\frac{\alpha-\frac{1}{p}}{4\alpha-\frac{1}{p}}\right)}$$

$$\left(\sup_{a\in\mathbb{R}_+}\left\|\left\|D_{*a}^{4\alpha}f\right\|\right\|_{p,\mathbb{R}_+}\right)^{\left(\frac{3\alpha}{4\alpha-\frac{1}{p}}\right)}<\infty,$$

where $\frac{1}{p}<\alpha<1$.

The theorem is proved. ■

We give an application when $\alpha=\frac{1}{2}$.

Corollary 9.6 *Let $f\in C^1(\mathbb{R}_+,X)$ with $\|\|f\|\|_{\infty,\mathbb{R}_+}$, $\left\|\left\|f'\right\|\right\|_{\infty,\mathbb{R}_+}<\infty$, where $(X,\|\cdot\|)$ is a Banach space. For $k=1,2,3$, we assume that $D_{*a}^{k\frac{1}{2}}f\in C^1([a,\infty),X)$ and $D_{*a}^{4\frac{1}{2}}f\in C([a,\infty),X)$, $\forall\, a\in\mathbb{R}_+$. We further assume that*

$$K:=\left\|\left\|D_{*a}^{4\frac{1}{2}}f(t)\right\|\right\|_{\infty,\mathbb{R}_+^2}<\infty, \tag{9.83}$$

where $(a,t)\in\mathbb{R}_+^2$.

Then

$$\sup_{a\in\mathbb{R}_+}\left\|\left(D_{*a}^{2\frac{1}{2}}f\right)(a)\right\| \leq \left(\frac{\sqrt{12+6\sqrt{2}}}{\sqrt{2}-1}\right)\|\|f\|\|_{\infty,\mathbb{R}_+}^{\frac{1}{2}}\left(\left\|\left\|D_{*a}^{4\frac{1}{2}}f(t)\right\|\right\|_{\infty,\mathbb{R}_+^2}\right)^{\frac{1}{2}} < \infty, \tag{9.84}$$

and

$$\sup_{a\in\mathbb{R}_+}\left\|\left(D_{*a}^{3\frac{1}{2}}f\right)(a)\right\| \leq$$

$$\left(\frac{9\sqrt{\pi}}{\left(2-\sqrt{2}\right)\sqrt[4]{2}\left(\sqrt[4]{3}\right)^3}\right)\|\|f\|\|_{\infty,\mathbb{R}_+}^{\frac{1}{4}}\left(\left\|\left\|D_{*a}^{4\frac{1}{2}}f(t)\right\|\right\|_{\infty,\mathbb{R}_+^2}\right)^{\frac{3}{4}} < \infty. \tag{9.85}$$

That is $\sup_{a\in\mathbb{R}_+}\left\|\left(D_{*a}^{2\frac{1}{2}}f\right)(a)\right\|$, $\sup_{a\in\mathbb{R}_+}\left\|\left(D_{*a}^{3\frac{1}{2}}f\right)(a)\right\| < \infty$.

Proof By Theorem 9.4. ■

References

1. Aglic Aljinovic, A., Marangunic, L., Pecaric, J.: On Landau type inequalities via Ostrowski inequalities. Nonlinear Funct. Anal. Appl. **10**(4), 565–579 (2005)
2. Anastassiou, G.: Fractional Differentiation Inequalities. Research Monograph. Springer, New York (2009)
3. Anastassiou, G.A.: Advances on Fractional Inequalities. Springer, New York (2011)
4. Anastassiou, G.A.: Intelligent Computations: Abstract Fractional Calculus, Inequalities, Approximations. Springer, Heidelberg (2018)
5. Anastassiou, G.A.: Iterated left abstract fractional Landau inequalities. J. App. Math. Inf., accepted, (2020)
6. Barnett, N.S., Dragomir, S.S.: Some Landau type inequalities for functions whose derivatives are of locally bounded variation. Tamkang J. Math. **37**(4), 301–308 (Winter 2006)
7. Ditzian, Z.: Remarks, questions and conjectures on Landau-Kolmogorov-type inequalities. Math. Inequal. Appl. **3**, 15–24 (2000)
8. Hardy, G.H., Littlewood, J.E.: Some integral inequalities connected with the calculus of variations. Quart. J. Math. Oxford Ser. **3**, 241–252 (1932)
9. Hardy, G.H., Landau, E., Littlewood, J.E.: Some inequalities satisfied by the integrals or derivatives of real or analytic functions. Math. Z. **39**, 677–695 (1935)
10. Kallman, R.R., Rota, G.C.: On theinequality $\|f'\|^2 \leq 4\|f\|\| \cdot \|f''\|$. In: Shisha, O., (ed.) Inequalities, vol. II, pp. 187–192. Academic, New York (1970)
11. Landau, E.: Einige Ungleichungen für zweimal differentzierban funktionen. Proc. Lond. Math. Soc. **13**, 43–49 (1913)
12. Mikusinski, J.: The Bochner Integral. Academic, New York (1978)
13. Shilov, G.E.: Elementary Functional Analysis. Dover Publications Inc., New York (1996)

Chapter 10
Iterated Left Abstract Generalized Fractional Landau Inequalities

We give uniform and L_p left Caputo–Bochner abstract iterated generalized fractional Landau inequalities over $\mathbb{R}_+$. These estimate the size of second and third sequential left abstract generalized fractional derivatives of a Banach space valued function over $\mathbb{R}_+$. We give an application when the basic fractional order is $\frac{1}{2}$. It follows [5].

10.1 Introduction

Let $p \in [1, \infty]$, $I = \mathbb{R}_+$ or $I = \mathbb{R}$ and $f : I \to \mathbb{R}$ is twice differentiable with $f, f'' \in L_p(I)$, then $f' \in L_p(I)$. Moreover, there exists a constant $C_p(I) > 0$ independent of f, such that

$$\|f'\|_{p,I} \leq C_p(I) \|f\|_{p,I}^{\frac{1}{2}} \|f''\|_{p,I}^{\frac{1}{2}}, \tag{10.1}$$

where $\|\cdot\|_{p,I}$ is the p-norm on the interval I, see [1, 6].

The research on these inequalities started by Landau [11] in 1913. For the case of $p = \infty$ he proved that

$$C_\infty(\mathbb{R}_+) = 2 \text{ and } C_\infty(\mathbb{R}) = \sqrt{2}, \tag{10.2}$$

are the best constants in (10.1).

In 1932, Hardy and Littlewood [8] proved (10.1) for $p = 2$, with the best constants

$$C_2(\mathbb{R}_+) = \sqrt{2} \text{ and } C_2(\mathbb{R}) = 1. \tag{10.3}$$

In 1935, Hardy, Landau and Littlewood [9] showed that the best constants $C_p(\mathbb{R}_+)$ in (10.1) satisfies the estimate

G. A. Anastassiou, *Constructive Fractional Analysis with Applications*,
Studies in Systems, Decision and Control 362,
https://doi.org/10.1007/978-3-030-71481-9_10

$$C_p(\mathbb{R}_+) \leq 2, \text{ for } p \in [1, \infty), \tag{10.4}$$

which yields $C_p(\mathbb{R}) \leq 2$ for $p \in [1, \infty)$.

In fact, in [7, 10] was shown that $C_p(\mathbb{R}) \leq \sqrt{2}$.

We need the following concept from abstract fractional calculus.

Our integral next is of Bochner type [12].

We need

Definition 10.1 ([4], *p. 104*) Let $[a, b] \subset \mathbb{R}$, $(X, \|\cdot\|)$ a Banach space, $g \in C^1([a, b])$ and increasing, $f \in C([a, b], X)$, $\nu > 0$.

We define the left Riemann–Liouville generalized fractional Bochner integral operator

$$\left(J_{a;g}^{\nu} f\right)(x) := \frac{1}{\Gamma(\nu)} \int_a^x (g(x) - g(z))^{\nu-1} g'(z) f(z) dz, \tag{10.5}$$

$\forall x \in [a, b]$, where Γ is the gamma function.

The last integral is of Bochner type. Since $f \in C([a, b], X)$, then $f \in L_\infty([a, b], X)$. By Theorem 4.10, p. 98, [4], we get that $\left(J_{a;g}^{\nu} f\right) \in C([a, b], X)$. Above we set $J_{a;g}^0 f := f$ and see that $\left(J_{a;g}^{\nu} f\right)(a) = 0$.

We also need

Definition 10.2 ([4], *p. 106*) Let $\alpha > 0$, $\lceil \alpha \rceil = n$, $\lceil \cdot \rceil$ the ceiling of the number. Let $f \in C^n([a, b], X)$, where $[a, b] \subset \mathbb{R}$, and $(X, \|\cdot\|)$ is a Banach space. Let $g \in C^1([a, b])$, strictly increasing, such that $g^{-1} \in C^n([g(a), g(b)])$.

We define the left generalized g-fractional derivative X-valued of f of order α as follows:

$$\left(D_{a+;g}^{\alpha} f\right)(x) := \frac{1}{\Gamma(n-\alpha)} \int_a^x (g(x) - g(t))^{n-\alpha-1} g'(t) \left(f \circ g^{-1}\right)^{(n)}(g(t)) dt, \tag{10.6}$$

$\forall x \in [a, b]$. The last integral is of Bochner type.

Ordinary vector valued derivative is as in [13], similar to numerical one.

If $\alpha \notin \mathbb{N}$, by Theorem 4.10, p. 98, [4], we have that $\left(D_{a+;g}^{\alpha} f\right) \in C([a, b], X)$.

We see that

$$\left(J_{a;g}^{n-\alpha}\left(\left(f \circ g^{-1}\right)^{(n)} \circ g\right)\right)(x) = \left(D_{a+;g}^{\alpha} f\right)(x), \quad \forall x \in [a, b]. \tag{10.7}$$

We set

$$D_{a+;g}^{n} f(x) := \left(\left(f \circ g^{-1}\right)^{n} \circ g\right)(x) \in C([a, b], X), \quad n \in \mathbb{N}, \tag{10.8}$$

$$D_{a+;g}^{0} f(x) = f(x), \quad \forall x \in [a, b].$$

When $g = id$, then

$$D^{\alpha}_{a+;g} f = D^{\alpha}_{a+;id} f = D^{\alpha}_{*a} f, \tag{10.9}$$

the usual left X-valued Caputo fractional derivative, see [4], Chap. 1.

By convention we suppose that

$$\left(D^{\alpha}_{x_0+;g} f\right)(x) = 0, \text{ for } x < x_0, \tag{10.10}$$

for any $x, x_0 \in [a, b]$.

Denote the sequential (also called iterated) generalized left fractional derivative by

$$D^{n\alpha}_{a+;g} := D^{\alpha}_{a+;g} D^{\alpha}_{a+;g} ... D^{\alpha}_{a+;g} \quad (n \text{ times}), n \in \mathbb{N}. \tag{10.11}$$

We need the following g-left generalized modified X-valued Taylor's formula.

Theorem 10.3 ([4], p. 117) *Let* $0 < \alpha \leq 1$, $n \in \mathbb{N}$, $f \in C^1([a, b], X)$, $g \in C^1([a, b])$ *strictly increasing, such that* $g^{-1} \in C^1([g(a), g(b)])$. *Let* $F_k := D^{k\alpha}_{a+;g} f$, $k = 1, ..., n$, *that fulfill* $F_k \in C^1([a, b], X)$, *and* $F_{n+1} \in C([a, b], X)$.

Then

$$f(x) = \sum_{i=0}^{n} \frac{(g(x) - g(a))^{i\alpha}}{\Gamma(i\alpha + 1)} \left(D^{i\alpha}_{a+;g} f\right)(a) + \tag{10.12}$$

$$\frac{1}{\Gamma((n+1)\alpha)} \int_a^x (g(x) - g(t))^{(n+1)\alpha - 1} g'(t) \left(D^{(n+1)\alpha}_{a+;g} f\right)(t)\, dt,$$

$\forall\, x \in [a, b]$.

We make

Remark 10.4 (*to Theorem* 10.3) When $0 < \alpha < 1$, by (10.6) we get

$$\left(D^{\alpha}_{a+;g} f\right)(x) = \frac{1}{\Gamma(1-\alpha)} \int_a^x (g(x) - g(t))^{-\alpha} g'(t) \left(f \circ g^{-1}\right)'(g(t))\, dt, \tag{10.13}$$

$\forall\, x \in [a, b]$.

Hence

$$\left\| \left(D^{\alpha}_{a+;g} f\right)(x) \right\| \leq \frac{1}{\Gamma(1-\alpha)} \int_a^x (g(x) - g(t))^{-\alpha} g'(t) \left\| \left(f \circ g^{-1}\right)'(g(t)) \right\| dt$$

$$\leq \frac{\left\| \left\| \left(f \circ g^{-1}\right)' \circ g \right\| \right\|_{\infty,[a,b]}}{\Gamma(1-\alpha)} \left(\int_a^x (g(x) - g(t))^{-\alpha} g'(t)\, dt \right) \tag{10.14}$$

$$= \frac{\left\| \left\| \left(f \circ g^{-1} \right)' \circ g \right\| \right\|_{\infty,[a,b]}}{\Gamma(2-\alpha)} (g(x) - g(a))^{1-\alpha}.$$

That is

$$\left\| \left(D_{a+;g}^{\alpha} f \right)(x) \right\| \leq \frac{\left\| \left\| \left(f \circ g^{-1} \right)' \circ g \right\| \right\|_{\infty,[a,b]}}{\Gamma(2-\alpha)} (g(x) - g(a))^{1-\alpha} < \infty, \quad (10.15)$$

$\forall\, x \in [a,b]$, $0 < \alpha < 1$.

Hence it holds

$$\left\| \left(D_{a+;g}^{\alpha} f \right)(a) \right\| = 0,$$

i.e.

$$\left(D_{a+;g}^{\alpha} f \right)(a) = 0, \quad (10.16)$$

when $0 < \alpha < 1$.

The author has already done an extensive amount of work on fractional Landau inequalities, see [3], and on abstract fractional Landau inequalities, see [4]. However there the proving methods came out of applications of fractional Ostrowski inequalities [2, 4]. Usually there the domains where $[A, +\infty)$ or $(-\infty, B]$, with $A, B \in \mathbb{R}$ and in one mixed case the domain was all of $\mathbb{R}$.

In this chapter with less assumptions we establish uniform and L_p type left Caputo–Bochner abstract sequential generalized fractional Landau inequalities over $\mathbb{R}_+$. The method of proving is based on left Caputo–Bochner sequential generalized fractional Taylor's formula with integral remainder, see Theorem 10.3.

We give also an application for $\alpha = \frac{1}{2}$. Regardless to say that we are also inspired by [3, 4].

10.2 Main Results

We present the following abstract iterated generalized fractional Landau inequalities over $\mathbb{R}_+$.

Theorem 10.5 *Let $g \in C^1(\mathbb{R}_+)$ strictly increasing, with $g^{-1} \in C^1(g(\mathbb{R}_+))$. Let $0 < \alpha < 1$, $f \in C^1(\mathbb{R}_+, X)$ with $\|\|f\|\|_{\infty,\mathbb{R}_+}$, $\left\| \left\| \left(f \circ g^{-1} \right)' \circ g \right\| \right\|_{\infty,\mathbb{R}_+} < \infty$. For $k = 1, 2, 3$, we assume that $D_{a+;g}^{k\alpha} f \in C^1([a,\infty), X)$ and $D_{a+;g}^{4\alpha} f \in C([a,\infty), X)$, $\forall\, a \in \mathbb{R}_+$. We further assume that*

$$K_g := \left\| \left\| D_{a+;g}^{4\alpha} f(t) \right\| \right\|_{\infty,\mathbb{R}_+^2} < \infty, \quad (10.17)$$

where $(a,t) \in \mathbb{R}_+^2$.

Then

$$\sup_{a\in\mathbb{R}_+}\left\|\left(D_{a+;g}^{2\alpha}f\right)(a)\right\|\leq\frac{\Gamma(2\alpha+1)}{2^{2\alpha-1}(2^{\alpha}-1)}\sqrt{\frac{2^{3\alpha+1}\left(2^{3\alpha}+1\right)(2^{\alpha}+1)}{\Gamma(4\alpha+1)}}\,\|\|f\|\|_{\infty,\mathbb{R}_+}K_g, \tag{10.18}$$

and

$$\sup_{a\in\mathbb{R}_+}\left\|\left(D_{a+;g}^{3\alpha}f\right)(a)\right\|\leq\frac{4\sqrt[4]{2}\,\Gamma(3\alpha+1)\left(\Gamma(4\alpha+1)\right)^{-\frac{3}{4}}\left(2^{2\alpha}+1\right)}{\left(\sqrt[4]{3}\right)^{3}\left(\sqrt{2}\right)^{\alpha}(2^{\alpha}-1)}\,\|\|f\|\|_{\infty,\mathbb{R}_+}^{\frac{1}{4}}K_g^{\frac{3}{4}}. \tag{10.19}$$

That is $\sup_{a\in\mathbb{R}_+}\left\|\left(D_{a+;g}^{2\alpha}f\right)(a)\right\|,\ \sup_{a\in\mathbb{R}_+}\left\|\left(D_{a+;g}^{3\alpha}f\right)(a)\right\|<\infty.$

Proof We notice easily again here that $\left(D_{a+;g}^{\alpha}f\right)(a)=0,\ \forall\, a\in\mathbb{R}_+$.

We make use of Theorem 10.3 for $0<\alpha<1$ and $n=3$, applied for any $a\in\mathbb{R}_+$ and $b=\infty$.

Momentarily we fix $a\in\mathbb{R}_+$. Let $x_2>x_1>a$, then $g(x_2)>g(x_1)>g(a)$, and

$$f(x_1)-f(a)=\frac{(g(x_1)-g(a))^{2\alpha}}{\Gamma(2\alpha+1)}\left(D_{a+;g}^{2\alpha}f\right)(a)+\frac{(g(x_1)-g(a))^{3\alpha}}{\Gamma(3\alpha+1)}\left(D_{a+;g}^{3\alpha}f\right)(a)$$

$$+\frac{1}{\Gamma(4\alpha)}\int_a^{x_1}(g(x_1)-g(t))^{4\alpha-1}g'(t)\left(D_{a+;g}^{4\alpha}f\right)(t)\,dt, \tag{10.20}$$

and

$$f(x_2)-f(a)=\frac{(g(x_2)-g(a))^{2\alpha}}{\Gamma(2\alpha+1)}\left(D_{a+;g}^{2\alpha}f\right)(a)+\frac{(g(x_2)-g(a))^{3\alpha}}{\Gamma(3\alpha+1)}\left(D_{a+;g}^{3\alpha}f\right)(a)$$

$$+\frac{1}{\Gamma(4\alpha)}\int_a^{x_2}(g(x_2)-g(t))^{4\alpha-1}g'(t)\left(D_{a+;g}^{4\alpha}f\right)(t)\,dt. \tag{10.21}$$

That is

$$\frac{(g(x_1)-g(a))^{2\alpha}}{\Gamma(2\alpha+1)}\left(D_{a+;g}^{2\alpha}f\right)(a)+\frac{(g(x_1)-g(a))^{3\alpha}}{\Gamma(3\alpha+1)}\left(D_{a+;g}^{3\alpha}f\right)(a)= \tag{10.22}$$

$$f(x_1)-f(a)-\frac{1}{\Gamma(4\alpha)}\int_a^{x_1}(g(x_1)-g(t))^{4\alpha-1}g'(t)\left(D_{a+;g}^{4\alpha}f\right)(t)\,dt=:A,$$

and

$$\frac{(g(x_2)-g(a))^{2\alpha}}{\Gamma(2\alpha+1)}\left(D_{a+;g}^{2\alpha}f\right)(a)+\frac{(g(x_2)-g(a))^{3\alpha}}{\Gamma(3\alpha+1)}\left(D_{a+;g}^{3\alpha}f\right)(a)= \tag{10.23}$$

$$f(x_2) - f(a) - \frac{1}{\Gamma(4\alpha)} \int_a^{x_2} (g(x_2) - g(t))^{4\alpha-1} g'(t) \left(D_{a+;g}^{4\alpha} f\right)(t)\, dt =: B.$$

We are solving the above system of two equations with two unknowns $\left(D_{a+;g}^{2\alpha} f\right)(a)$, $\left(D_{a+;g}^{3\alpha} f\right)(a)$.

The main determinant of system is

$$D := \begin{vmatrix} \frac{(g(x_1)-g(a))^{2\alpha}}{\Gamma(2\alpha+1)} & \frac{(g(x_1)-g(a))^{3\alpha}}{\Gamma(3\alpha+1)} \\ \frac{(g(x_2)-g(a))^{2\alpha}}{\Gamma(2\alpha+1)} & \frac{(g(x_2)-g(a))^{3\alpha}}{\Gamma(3\alpha+1)} \end{vmatrix} = \frac{1}{\Gamma(2\alpha+1)\Gamma(3\alpha+1)}$$

$$\left[(g(x_1) - g(a))^{2\alpha} (g(x_2) - g(a))^{3\alpha} - (g(x_1) - g(a))^{3\alpha} (g(x_2) - g(a))^{2\alpha}\right] =$$

$$\frac{(g(x_1) - g(a))^{2\alpha} (g(x_2) - g(a))^{2\alpha}}{\Gamma(2\alpha+1)\Gamma(3\alpha+1)} \left[(g(x_2) - g(a))^{\alpha} - (g(x_1) - g(a))^{\alpha}\right] > 0.$$

I.e.

$$D = \frac{(g(x_1) - g(a))^{2\alpha} (g(x_2) - g(a))^{2\alpha}}{\Gamma(2\alpha+1)\Gamma(3\alpha+1)} \left[(g(x_2) - g(a))^{\alpha} - (g(x_1) - g(a))^{\alpha}\right] > 0. \tag{10.24}$$

We obtain the unique solution

$$\left(D_{a+;g}^{2\alpha} f\right)(a) = \frac{\begin{vmatrix} A & \frac{(g(x_1)-g(a))^{3\alpha}}{\Gamma(3\alpha+1)} \\ B & \frac{(g(x_2)-g(a))^{3\alpha}}{\Gamma(3\alpha+1)} \end{vmatrix}}{D},$$

$$\left(D_{a+;g}^{3\alpha} f\right)(a) = \frac{\begin{vmatrix} \frac{(g(x_1)-g(a))^{2\alpha}}{\Gamma(2\alpha+1)} & A \\ \frac{(g(x_2)-g(a))^{2\alpha}}{\Gamma(2\alpha+1)} & B \end{vmatrix}}{D}. \tag{10.25}$$

Therefore we have

$$\left(D_{a+;g}^{2\alpha} f\right)(a) = \frac{\frac{(g(x_2)-g(a))^{3\alpha}}{\Gamma(3\alpha+1)} A - \frac{(g(x_1)-g(a))^{3\alpha}}{\Gamma(3\alpha+1)} B}{D},$$

and

$$\left(D_{a+;g}^{3\alpha} f\right)(a) = \frac{\frac{(g(x_1)-g(a))^{2\alpha}}{\Gamma(2\alpha+1)} B - \frac{(g(x_2)-g(a))^{2\alpha}}{\Gamma(2\alpha+1)} A}{D}. \tag{10.26}$$

We have the following

$$\|A\| = \left\| f(x_1) - f(a) - \frac{1}{\Gamma(4\alpha)} \int_a^{x_1} (g(x_1) - g(t))^{4\alpha-1} g'(t) \left(D_{a+;g}^{4\alpha} f \right)(t)\, dt \right\| \le$$

$$2\, \| \|f\| \|_{\infty,\mathbb{R}_+} + \frac{\left\| \left\| D_{a+;g}^{4\alpha} f(t) \right\| \right\|_{\infty,\mathbb{R}_+^2}}{\Gamma(4\alpha+1)} (g(x_1) - g(a))^{4\alpha}, \tag{10.27}$$

under the assumption $\| \|f\| \|_{\infty,\mathbb{R}_+} < \infty$.

That is

$$\|A\| \le 2\, \| \|f\| \|_{\infty,\mathbb{R}_+} + \frac{K_g}{\Gamma(4\alpha+1)} (g(x_1) - g(a))^{4\alpha}, \tag{10.28}$$

and similarly,

$$\|B\| \le 2\, \| \|f\| \|_{\infty,\mathbb{R}_+} + \frac{K_g}{\Gamma(4\alpha+1)} (g(x_2) - g(a))^{4\alpha}, \tag{10.29}$$

where by assumption

$$K_g := \left\| \left\| D_{a+;g}^{4\alpha} f(t) \right\| \right\|_{\infty,\mathbb{R}_+^2} < \infty, \tag{10.30}$$

with $(a,t) \in \mathbb{R}_+^2$.

Consequently we have

$$\left\| \left(D_{a+;g}^{2\alpha} f \right)(a) \right\| \le \frac{1}{\Gamma(3\alpha+1)\, D}$$

$$\left[(g(x_2) - g(a))^{3\alpha} \left(2\, \| \|f\| \|_{\infty,\mathbb{R}_+} + \frac{K_g}{\Gamma(4\alpha+1)} (g(x_1) - g(a))^{4\alpha} \right) \right.$$

$$\left. + (g(x_1) - g(a))^{3\alpha} \left(2\, \| \|f\| \|_{\infty,\mathbb{R}_+} + \frac{K_g}{\Gamma(4\alpha+1)} (g(x_2) - g(a))^{4\alpha} \right) \right], \tag{10.31}$$

and

$$\left\| \left(D_{a+;g}^{3\alpha} f \right)(a) \right\| \le \frac{1}{\Gamma(2\alpha+1)\, D}$$

$$\left[(g(x_1) - g(a))^{2\alpha} \left(2\, \| \|f\| \|_{\infty,\mathbb{R}_+} + \frac{K_g}{\Gamma(4\alpha+1)} (g(x_2) - g(a))^{4\alpha} \right) \right.$$

$$\left. + (g(x_2) - g(a))^{2\alpha} \left(2\, \| \|f\| \|_{\infty,\mathbb{R}_+} + \frac{K_g}{\Gamma(4\alpha+1)} (g(x_1) - g(a))^{4\alpha} \right) \right]. \tag{10.32}$$

Set now $g(x_1) := g(a) + h$, $g(x_2) := g(a) + 2h$, where $h > 0$, so that $g(x_1) - g(a) = h$, $g(x_2) - g(a) = 2h$.

Hence we get

$$D=\frac{2^{2\alpha}h^{5\alpha}\left(2^{\alpha}-1\right)}{\Gamma\left(2\alpha+1\right)\Gamma\left(3\alpha+1\right)}>0. \tag{10.33}$$

Therefore we derive (from (10.26))

$$\left\|\left(D_{a+;g}^{2\alpha}f\right)(a)\right\|\leq\frac{\Gamma\left(2\alpha+1\right)}{2^{2\alpha}h^{5\alpha}\left(2^{\alpha}-1\right)}\left[2^{3\alpha}h^{3\alpha}\left(2\left\|\|f\|\right\|_{\infty,\mathbb{R}_{+}}+\frac{K_{g}}{\Gamma\left(4\alpha+1\right)}h^{4\alpha}\right)\right.$$

$$\left.+h^{3\alpha}\left(2\left\|\|f\|\right\|_{\infty,\mathbb{R}_{+}}+\frac{K_{g}}{\Gamma\left(4\alpha+1\right)}2^{4\alpha}h^{4\alpha}\right)\right]= \tag{10.34}$$

$$\frac{\Gamma\left(2\alpha+1\right)}{2^{2\alpha}h^{5\alpha}\left(2^{\alpha}-1\right)}\left[2\left\|\|f\|\right\|_{\infty,\mathbb{R}_{+}}\left(2^{3\alpha}+1\right)h^{3\alpha}+\frac{K_{g}}{\Gamma\left(4\alpha+1\right)}\left(2^{3\alpha}+2^{4\alpha}\right)h^{7\alpha}\right]$$

$$=\left(\frac{\Gamma\left(2\alpha+1\right)}{2^{2\alpha}\left(2^{\alpha}-1\right)}\right)\left[\frac{2\left(2^{3\alpha}+1\right)\left\|\|f\|\right\|_{\infty,\mathbb{R}_{+}}}{h^{2\alpha}}+\frac{2^{3\alpha}\left(2^{\alpha}+1\right)}{\Gamma\left(4\alpha+1\right)}K_{g}h^{2\alpha}\right]. \tag{10.35}$$

That is

$$\left\|\left(D_{a+;g}^{2\alpha}f\right)(a)\right\|\leq\left(\frac{\Gamma\left(2\alpha+1\right)}{2^{2\alpha}\left(2^{\alpha}-1\right)}\right)$$

$$\left[\frac{2\left(2^{3\alpha}+1\right)\left\|\|f\|\right\|_{\infty,\mathbb{R}_{+}}}{h^{2\alpha}}+\frac{2^{3\alpha}\left(2^{\alpha}+1\right)}{\Gamma\left(4\alpha+1\right)}K_{g}h^{2\alpha}\right], \tag{10.36}$$

$\forall\, a\in\mathbb{R}_{+}$, $\forall\, h>0$.

I.e. it holds

$$\sup_{a\in\mathbb{R}_{+}}\left\|\left(D_{a+;g}^{2\alpha}f\right)(a)\right\|\leq\left(\frac{\Gamma\left(2\alpha+1\right)}{2^{2\alpha}\left(2^{\alpha}-1\right)}\right)$$

$$\left[\frac{2\left(2^{3\alpha}+1\right)\left\|\|f\|\right\|_{\infty,\mathbb{R}_{+}}}{h^{2\alpha}}+\frac{2^{3\alpha}\left(2^{\alpha}+1\right)}{\Gamma\left(4\alpha+1\right)}K_{g}h^{2\alpha}\right]<\infty, \tag{10.37}$$

$\forall\, h>0$, $0<\alpha<1$.

By (10.26) we derive

$$\left\|\left(D_{a+;g}^{3\alpha}f\right)(a)\right\|\leq\frac{\Gamma\left(3\alpha+1\right)}{2^{2\alpha}h^{5\alpha}\left(2^{\alpha}-1\right)}\left[h^{2\alpha}\left(2\left\|\|f\|\right\|_{\infty,\mathbb{R}_{+}}+\frac{K_{g}}{\Gamma\left(4\alpha+1\right)}2^{4\alpha}h^{4\alpha}\right)\right.$$

$$\left.+2^{2\alpha}h^{2\alpha}\left(2\left\|\|f\|\right\|_{\infty,\mathbb{R}_{+}}+\frac{K_{g}}{\Gamma\left(4\alpha+1\right)}h^{4\alpha}\right)\right]=$$

$$\frac{\Gamma(3\alpha+1)}{2^{2\alpha}h^{5\alpha}(2^{\alpha}-1)}\left[2\,\|\|f\|\|_{\infty,\mathbb{R}_+}\left(2^{2\alpha}+1\right)h^{2\alpha}+\frac{K_g}{\Gamma(4\alpha+1)}\left(2^{4\alpha}+2^{2\alpha}\right)h^{6\alpha}\right] \tag{10.38}$$

$$=\left(\frac{\Gamma(3\alpha+1)}{2^{2\alpha}(2^{\alpha}-1)}\right)\left[\frac{2\left(2^{2\alpha}+1\right)\|\|f\|\|_{\infty,\mathbb{R}_+}}{h^{3\alpha}}+\frac{2^{2\alpha}\left(2^{2\alpha}+1\right)}{\Gamma(4\alpha+1)}K_g h^{\alpha}\right]$$

$$=\frac{\Gamma(3\alpha+1)\left(2^{2\alpha}+1\right)}{2^{2\alpha}(2^{\alpha}-1)}\left[\frac{2\,\|\|f\|\|_{\infty,\mathbb{R}_+}}{h^{3\alpha}}+\frac{2^{2\alpha}K_g}{\Gamma(4\alpha+1)}h^{\alpha}\right].$$

That is

$$\left\|\left(D^{3\alpha}_{a+;g}f\right)(a)\right\|\leq\frac{\Gamma(3\alpha+1)\left(2^{2\alpha}+1\right)}{2^{2\alpha}(2^{\alpha}-1)}$$

$$\left[\frac{2\,\|\|f\|\|_{\infty,\mathbb{R}_+}}{h^{3\alpha}}+\frac{2^{2\alpha}K_g}{\Gamma(4\alpha+1)}h^{\alpha}\right], \tag{10.39}$$

$\forall\, a\in\mathbb{R}_+, \forall\, h>0$.

I.e. it holds

$$\sup_{a\in\mathbb{R}_+}\left\|\left(D^{3\alpha}_{a+;g}f\right)(a)\right\|\leq\frac{\Gamma(3\alpha+1)\left(2^{2\alpha}+1\right)}{2^{2\alpha}(2^{\alpha}-1)}$$

$$\left[\frac{2\,\|\|f\|\|_{\infty,\mathbb{R}_+}}{h^{3\alpha}}+\frac{2^{2\alpha}K_g}{\Gamma(4\alpha+1)}h^{\alpha}\right]<\infty, \tag{10.40}$$

$\forall\, h>0, 0<\alpha<1$.

Call

$$\mu:=2\left(2^{3\alpha}+1\right)\|\|f\|\|_{\infty,\mathbb{R}_+}, \tag{10.41}$$

$$\theta=\frac{2^{3\alpha}(2^{\alpha}+1)K_g}{\Gamma(4\alpha+1)}, \tag{10.42}$$

both are greater than zero.

Set also $\rho:=2\alpha$; $0<\rho<2$. We consider the function

$$y(h):=\mu h^{-\rho}+\theta h^{\rho},\quad \forall\, h>0. \tag{10.43}$$

We have

$$y'(h)=-\rho\mu h^{-\rho-1}+\rho\theta h^{\rho-1}=0, \tag{10.44}$$

then

$$\theta h^{2\rho}=\mu,$$

with a unique solution

$$h_0 := h_{crit.no.} = \left(\frac{\mu}{\theta}\right)^{\frac{1}{2\rho}}. \tag{10.45}$$

We have that

$$y''(h) = \rho(\rho+1)\mu h^{-\rho-2} + \rho(\rho-1)\theta h^{\rho-2}. \tag{10.46}$$

We see that

$$y''(h_0) = y''\left(\left(\frac{\mu}{\theta}\right)^{\frac{1}{2\rho}}\right) = \rho(\rho+1)\mu\left(\frac{\mu}{\theta}\right)^{\frac{-\rho-2}{2\rho}} + \rho(\rho-1)\theta\left(\frac{\mu}{\theta}\right)^{\frac{\rho-2}{2\rho}} =$$

$$\rho\left(\frac{\theta}{\mu}\right)^{\frac{1}{\rho}}\left[(\rho+1)\sqrt{\mu\theta} + (\rho-1)\sqrt{\mu\theta}\right] =$$

$$\rho\left(\frac{\theta}{\mu}\right)^{\frac{1}{\rho}}\left(2\rho\sqrt{\mu\theta}\right) = 2\rho^2\sqrt{\mu\theta}\left(\frac{\theta}{\mu}\right)^{\frac{1}{\rho}} > 0.$$

Therefore y has a global minimum at $h_0 = \left(\frac{\mu}{\theta}\right)^{\frac{1}{2\rho}}$, which is

$$y(h_0) = \mu\left(\frac{\mu}{\theta}\right)^{-\frac{1}{2}} + \theta\left(\frac{\mu}{\theta}\right)^{\frac{1}{2}} = \mu\left(\frac{\theta}{\mu}\right)^{\frac{1}{2}} + \sqrt{\theta\mu} = 2\sqrt{\theta\mu}.$$

We have proved that (see (10.37))

$$\sup_{a\in\mathbb{R}_+}\left\|\left(D^{2\alpha}_{a+;g}f\right)(a)\right\| \leq \frac{\Gamma(2\alpha+1)}{2^{2\alpha-1}(2^\alpha-1)}$$

$$\sqrt{\frac{2^{3\alpha+1}\left(2^{3\alpha}+1\right)(2^\alpha+1)}{\Gamma(4\alpha+1)}\,\||f\||_{\infty,\mathbb{R}_+}\,K_g}. \tag{10.47}$$

Call

$$\xi := 2\,\||f\||_{\infty,\mathbb{R}_+},$$
$$\psi := \frac{2^{2\alpha}K_g}{\Gamma(4\alpha+1)}, \tag{10.48}$$

both are greater than zero.

We consider the function

$$\gamma(h) := \xi h^{-3\alpha} + \psi h^\alpha, \quad \forall\, h > 0. \tag{10.49}$$

We have

$$\gamma'(h) = -3\alpha\xi h^{-3\alpha-1} + \alpha\psi h^{\alpha-1} = 0,$$

then

$$\psi h^{4\alpha} = 3\xi,$$

with unique solution

$$h_0 := h_{crit.no.} = \left(\frac{3\xi}{\psi}\right)^{\frac{1}{4\alpha}}. \tag{10.50}$$

We have that

$$\gamma''(h) = 3\alpha(3\alpha+1)\xi h^{-3\alpha-2} + \alpha(\alpha-1)\psi h^{\alpha-2}. \tag{10.51}$$

We see

$$\gamma''(h_0) = 3\alpha(3\alpha+1)\xi\left(\frac{3\xi}{\psi}\right)^{\frac{-3\alpha-2}{4\alpha}} + \alpha(\alpha-1)\psi\left(\frac{3\xi}{\psi}\right)^{\frac{\alpha-2}{4\alpha}} =$$

$$\alpha\left(\frac{3\xi}{\psi}\right)^{\frac{\alpha-2}{4\alpha}}\left[3(3\alpha+1)\xi\frac{\psi}{3\xi} + (\alpha-1)\psi\right] =$$

$$\alpha\left(\frac{3\xi}{\psi}\right)^{\frac{\alpha-2}{4\alpha}}(4\alpha\psi) = 4\alpha^2\psi\left(\frac{3\xi}{\psi}\right)^{\frac{\alpha-2}{4\alpha}} > 0. \tag{10.52}$$

Therefore γ has a global minimum at $h_0 = \left(\frac{3\xi}{\psi}\right)^{\frac{1}{4\alpha}}$, which is

$$\gamma(h_0) = \xi\left(\frac{3\xi}{\psi}\right)^{-\frac{3}{4}} + \psi\left(\frac{3\xi}{\psi}\right)^{\frac{1}{4}} =$$

$$\left(\frac{3\xi}{\psi}\right)^{\frac{1}{4}}\left(\xi\frac{\psi}{3\xi} + \psi\right) = \frac{4}{3}\psi\left(\frac{3\xi}{\psi}\right)^{\frac{1}{4}}. \tag{10.53}$$

Consequently,

$$\gamma(h_0) = \frac{4}{3}\psi\left(\frac{3\xi}{\psi}\right)^{\frac{1}{4}} = \frac{4}{\left(\sqrt[4]{3}\right)^3}\psi^{\frac{3}{4}}\xi^{\frac{1}{4}}. \tag{10.54}$$

We have proved that (see (10.40))

$$\sup_{a\in\mathbb{R}_+}\left\|\left(D_{a+;g}^{3\alpha}f\right)(a)\right\| \leq \frac{4\Gamma(3\alpha+1)\left(2^{2\alpha}+1\right)}{\left(\sqrt[4]{3}\right)^3 2^{2\alpha}\left(2^{\alpha}-1\right)}$$

$$\left(2\left\|\|f\|\right\|_{\infty,\mathbb{R}_+}\right)^{\frac{1}{4}}\left(\frac{2^{2\alpha}K_g}{\Gamma(4\alpha+1)}\right)^{\frac{3}{4}}=$$

$$\frac{4\sqrt[4]{2}\Gamma(3\alpha+1)\Gamma(4\alpha+1)^{-\frac{3}{4}}\left(2^{2\alpha}+1\right)}{\left(\sqrt[4]{3}\right)^{3}2^{\frac{\alpha}{2}}\left(2^{\alpha}-1\right)}\left\|\|f\|\right\|_{\infty,\mathbb{R}_+}^{\frac{1}{4}}K_g^{\frac{3}{4}}. \tag{10.55}$$

The theorem is proved. ■

We continue with abstract L_p left sequential generalized fractional Landau inequalities over $\mathbb{R}_+$.

Theorem 10.6 *Let $g\in C^1(\mathbb{R}_+)$ strictly increasing, with $g^{-1}\in C^1(g(\mathbb{R}_+))$. Let $p,q>1:\frac{1}{p}+\frac{1}{q}=1$, $0<\alpha<1$. Let $f\in C^1(\mathbb{R}_+,X)$ with $\left\|\|f\|\right\|_{\infty,\mathbb{R}_+}$, $\left\|\left\|\left(f\circ g^{-1}\right)'\circ g\right\|\right\|_{\infty,\mathbb{R}_+}<\infty$. For $k=1,2,3$, we assume that $D_{a+;g}^{k\alpha}f\in C^1([a,\infty),X)$ and $D_{a+;g}^{4\alpha}f\in C([a,\infty),X)$, $\forall\, a\in\mathbb{R}_+$. We further assume that*

$$\left(\sup_{a\in\mathbb{R}_+}\left\|\left\|D_{a+;g}^{4\alpha}f\right\|\right\|_{p,\mathbb{R}_+}\right)<\infty. \tag{10.56}$$

Then

(1) under $\frac{1}{2p}<\alpha<1$, we get

$$\sup_{a\in\mathbb{R}_+}\left\|\left(D_{a+;g}^{2\alpha}f\right)(a)\right\|\leq\left[\left(\frac{2^{\alpha}\Gamma(2\alpha)\left(4\alpha-\frac{1}{p}\right)}{2^{\alpha}-1}\right)\left(\frac{4\alpha\left(1+2^{-3\alpha}\right)}{2\alpha-\frac{1}{p}}\right)^{\left(\frac{2\alpha-\frac{1}{p}}{4\alpha-\frac{1}{p}}\right)}\right.$$

$$\left.\left(\frac{1+2^{\alpha-\frac{1}{p}}}{\Gamma(4\alpha)\left(q(4\alpha-1)+1\right)^{\frac{1}{q}}}\right)^{\left(\frac{2\alpha}{4\alpha-\frac{1}{p}}\right)}\right]\left\|\|f\|\right\|_{\infty,\mathbb{R}_+}^{\left(\frac{2\alpha-\frac{1}{p}}{4\alpha-\frac{1}{p}}\right)}$$

$$\left(\sup_{a\in\mathbb{R}_+}\left\|\left\|D_{a+;g}^{4\alpha}f\right\|\right\|_{p,\mathbb{R}_+}\right)^{\left(\frac{2\alpha}{4\alpha-\frac{1}{p}}\right)}<\infty. \tag{10.57}$$

(2) under $\frac{1}{p}<\alpha<1$, we get

$$\sup_{a\in\mathbb{R}_+}\left\|\left(D_{a+;g}^{3\alpha}f\right)(a)\right\|\leq\left[\left(\frac{\Gamma(3\alpha)\left(4\alpha-\frac{1}{p}\right)}{2^{\alpha}-1}\right)\left(\frac{6\alpha\left(1+2^{-2\alpha}\right)}{\alpha-\frac{1}{p}}\right)^{\left(\frac{\alpha-\frac{1}{p}}{4\alpha-\frac{1}{p}}\right)}\right.$$

$$\left(\frac{1+2^{2\alpha-\frac{1}{p}}}{\Gamma(4\alpha)(q(4\alpha-1)+1)^{\frac{1}{q}}}\right)^{\left(\frac{3\alpha}{4\alpha-\frac{1}{p}}\right)}\Bigg]\, \||f|\|_{\infty,\mathbb{R}_+}^{\left(\frac{\alpha-\frac{1}{p}}{4\alpha-\frac{1}{p}}\right)}$$

$$\left(\sup_{a\in\mathbb{R}_+}\left\|\left\|D_{a+;g}^{4\alpha}f\right\|\right\|_{p,\mathbb{R}_+}\right)^{\left(\frac{3\alpha}{4\alpha-\frac{1}{p}}\right)}<\infty. \tag{10.58}$$

That is $\sup_{a\in\mathbb{R}_+}\left\|\left(D_{a+;g}^{2\alpha}f\right)(a)\right\|,\ \sup_{a\in\mathbb{R}_+}\left\|\left(D_{a+;g}^{3\alpha}f\right)(a)\right\|<\infty.$

Proof As in the proof of Theorem 10.5 we have that

$$\|A\| \overset{(10.22.)}{=} \left\|f(x_1)-f(a)-\frac{1}{\Gamma(4\alpha)}\int_a^{x_1}(g(x_1)-g(t))^{4\alpha-1}g'(t)\left(D_{a+;g}^{4\alpha}f\right)(t)\,dt\right\| \leq$$

$$2\||f|\|_{\infty,\mathbb{R}_+}+\frac{1}{\Gamma(4\alpha)}\int_a^{x_1}(g(x_1)-g(t))^{4\alpha-1}g'(t)\left\|\left(D_{a+;g}^{4\alpha}f\right)(t)\right\|dt \leq$$

$$2\||f|\|_{\infty,\mathbb{R}_+}+\frac{1}{\Gamma(4\alpha)}\frac{(g(x_1)-g(a))^{\frac{(q(4\alpha-1)+1)}{q}}}{(q(4\alpha-1)+1)^{\frac{1}{q}}}\left(\sup_{a\in\mathbb{R}_+}\left\|\left\|D_{a+;g}^{4\alpha}f\right\|\right\|_{p,\mathbb{R}_+}\right)\overset{(g(x_1)-g(a)=:h>0)}{=}$$

$$2\||f|\|_{\infty,\mathbb{R}_+}+\frac{1}{\Gamma(4\alpha)}\frac{h^{\left(4\alpha-\frac{1}{p}\right)}}{(q(4\alpha-1)+1)^{\frac{1}{q}}}\left(\sup_{a\in\mathbb{R}_+}\left\|\left\|D_{a+;g}^{4\alpha}f\right\|\right\|_{p,\mathbb{R}_+}\right), \tag{10.59}$$

with $\frac{1}{4p}<\alpha<1$.

That is

$$\|A\|\leq 2\||f|\|_{\infty,\mathbb{R}_+}+\frac{h^{4\alpha-\frac{1}{p}}}{\Gamma(4\alpha)(q(4\alpha-1)+1)^{\frac{1}{q}}}\left(\sup_{a\in\mathbb{R}_+}\left\|\left\|D_{a+;g}^{4\alpha}f\right\|\right\|_{p,\mathbb{R}_+}\right), \tag{10.60}$$

where $\frac{1}{4p}<\alpha<1$.

We also have

$$\|B\| \overset{(10.23.)}{=} \left\|f(x_2)-f(a)-\frac{1}{\Gamma(4\alpha)}\int_a^{x_2}(g(x_2)-g(t))^{4\alpha-1}g'(t)\left(D_{a+;g}^{4\alpha}f\right)(t)\,dt\right\| \leq$$

$$2\||f|\|_{\infty,\mathbb{R}_+}+\frac{(g(x_2)-g(a))^{\frac{(q(4\alpha-1)+1)}{q}}}{\Gamma(4\alpha)(q(4\alpha-1)+1)^{\frac{1}{q}}}\left(\sup_{a\in\mathbb{R}_+}\left\|\left\|D_{a+;g}^{4\alpha}f\right\|\right\|_{p,\mathbb{R}_+}\right)\overset{(g(x_2)-g(a)=:2h)}{=}$$

$$2\||f|\|_{\infty,\mathbb{R}_+}+\frac{2^{4\alpha-\frac{1}{p}}h^{4\alpha-\frac{1}{p}}}{\Gamma(4\alpha)(q(4\alpha-1)+1)^{\frac{1}{q}}}\left(\sup_{a\in\mathbb{R}_+}\left\|\left\|D_{a+;g}^{4\alpha}f\right\|\right\|_{p,\mathbb{R}_+}\right). \tag{10.61}$$

That is

$$\|B\| \leq 2 \||| f \|||_{\infty,\mathbb{R}_+} + \frac{2^{4\alpha-\frac{1}{p}} h^{4\alpha-\frac{1}{p}}}{\Gamma(4\alpha)(q(4\alpha-1)+1)^{\frac{1}{q}}} \left(\sup_{a\in\mathbb{R}_+} \left\| \left\| D_{a+;g}^{4\alpha} f \right\| \right\|_{p,\mathbb{R}_+} \right), \tag{10.62}$$

where $\frac{1}{4p} < \alpha < 1$.

We have assumed that

$$M_g := \left(\sup_{a\in\mathbb{R}_+} \left\| \left\| D_{a+;g}^{4\alpha} f \right\| \right\|_{p,\mathbb{R}_+} \right) < \infty. \tag{10.63}$$

For convenience we call

$$c := \Gamma(4\alpha)(q(4\alpha-1)+1)^{\frac{1}{q}} > 0. \tag{10.64}$$

So we have

$$\begin{aligned} &\|A\| \leq 2 \||| f \|||_{\infty,\mathbb{R}_+} + \tfrac{h^{4\alpha-\frac{1}{p}}}{c} M_g, \\ &\text{and} \\ &\|B\| \leq 2 \||| f \|||_{\infty,\mathbb{R}_+} + \tfrac{2^{4\alpha-\frac{1}{p}} h^{4\alpha-\frac{1}{p}}}{c} M_g, \end{aligned} \tag{10.65}$$

where $\frac{1}{4p} < \alpha < 1$.

Next we estimate the (10.26)-quantities and we have:

$$\left\| \left(D_{a+;g}^{2\alpha} f \right)(a) \right\| \leq \frac{1}{D\Gamma(3\alpha+1)} \left[2^{3\alpha} h^{3\alpha} \|A\| + h^{3\alpha} \|B\| \right] \overset{(10.33.)}{=}$$

$$\frac{h^{3\alpha}\Gamma(2\alpha+1)}{2^{2\alpha}h^{5\alpha}(2^{\alpha}-1)} \left[2^{3\alpha} \|A\| + \|B\| \right] \overset{(10.65.)}{\leq} \tag{10.66}$$

$$\frac{\Gamma(2\alpha+1)}{2^{2\alpha}(2^{\alpha}-1)h^{2\alpha}} \left[2^{3\alpha+1} \||| f \|||_{\infty,\mathbb{R}_+} + \frac{2^{3\alpha} h^{4\alpha-\frac{1}{p}}}{c} M_g + \right.$$

$$\left. 2 \||| f \|||_{\infty,\mathbb{R}_+} + \frac{2^{4\alpha-\frac{1}{p}} h^{4\alpha-\frac{1}{p}}}{c} M_g \right] =$$

$$\frac{\Gamma(2\alpha+1)}{2^{2\alpha}(2^{\alpha}-1)} \left[\frac{\left(2^{3\alpha+1}+2\right) \||| f \|||_{\infty,\mathbb{R}_+}}{h^{2\alpha}} + \frac{\left(2^{3\alpha}+2^{4\alpha-\frac{1}{p}}\right)}{c} M_g h^{2\alpha-\frac{1}{p}} \right] =$$

$$\frac{2^{\alpha}\Gamma(2\alpha+1)}{(2^{\alpha}-1)} \left[\frac{2\left(1+2^{-3\alpha}\right) \||| f \|||_{\infty,\mathbb{R}_+}}{h^{2\alpha}} + \frac{\left(1+2^{\alpha-\frac{1}{p}}\right) M_g}{c} h^{2\alpha-\frac{1}{p}} \right]. \tag{10.67}$$

That is

$$\left\| \left(D_{a+;g}^{2\alpha} f\right)(a) \right\| \leq$$

$$\left(\frac{2^{\alpha}\Gamma(2\alpha+1)}{2^{\alpha}-1}\right)\left[\frac{2\left(1+2^{-3\alpha}\right)\||f|\|_{\infty,\mathbb{R}_{+}}}{h^{2\alpha}}+\frac{\left(1+2^{\alpha-\frac{1}{p}}\right)M_{g}}{c}h^{2\alpha-\frac{1}{p}}\right], \tag{10.68}$$

$\forall\, a \in \mathbb{R}_{+}, \forall\, h > 0.$

I.e. it holds

$$\sup_{a\in\mathbb{R}_{+}} \left\| \left(D_{a+;g}^{2\alpha} f\right)(a) \right\| \leq \left(\frac{2^{\alpha}\Gamma(2\alpha+1)}{2^{\alpha}-1}\right)$$

$$\left[\frac{2\left(1+2^{-3\alpha}\right)\||f|\|_{\infty,\mathbb{R}_{+}}}{h^{2\alpha}}+\frac{\left(1+2^{\alpha-\frac{1}{p}}\right)M_{g}}{c}h^{2\alpha-\frac{1}{p}}\right], \tag{10.69}$$

$\forall\, h > 0$, under $\frac{1}{4p} < \alpha < 1$.

Again from (10.26) we get

$$\left\| \left(D_{a+;g}^{3\alpha} f\right)(a) \right\| \leq \frac{1}{\Gamma(2\alpha+1)\, D}\left[h^{2\alpha}\|B\|+2^{2\alpha}h^{2\alpha}\|A\|\right]=$$

$$\frac{h^{2\alpha}\Gamma(3\alpha+1)}{2^{2\alpha}h^{5\alpha}\left(2^{\alpha}-1\right)}\left[\|B\|+2^{2\alpha}\|A\|\right]\leq \tag{10.70}$$

$$\left(\frac{h^{2\alpha}\Gamma(3\alpha+1)}{2^{2\alpha}h^{5\alpha}\left(2^{\alpha}-1\right)}\right)\left[2\||f|\|_{\infty,\mathbb{R}_{+}}+\frac{2^{4\alpha-\frac{1}{p}}h^{4\alpha-\frac{1}{p}}}{c}M_{g}+\right.$$

$$\left.2^{2\alpha+1}\||f|\|_{\infty,\mathbb{R}_{+}}+\frac{2^{2\alpha}h^{4\alpha-\frac{1}{p}}}{c}M_{g}\right]=$$

$$\frac{\Gamma(3\alpha+1)}{2^{2\alpha}\left(2^{\alpha}-1\right)}\left[\frac{\left(2+2^{2\alpha+1}\right)\||f|\|_{\infty,\mathbb{R}_{+}}}{h^{3\alpha}}+\frac{\left(2^{2\alpha}+2^{4\alpha-\frac{1}{p}}\right)}{c}M_{g}h^{\alpha-\frac{1}{p}}\right]= \tag{10.71}$$

$$\frac{\Gamma(3\alpha+1)}{\left(2^{\alpha}-1\right)}\left[\frac{2\left(1+2^{-2\alpha}\right)\||f|\|_{\infty,\mathbb{R}_{+}}}{h^{3\alpha}}+\frac{\left(1+2^{2\alpha-\frac{1}{p}}\right)}{c}M_{g}h^{\alpha-\frac{1}{p}}\right].$$

That is

$$\left\| \left(D_{a+;g}^{3\alpha} f\right)(a) \right\| \leq$$

$$\left(\frac{\Gamma(3\alpha+1)}{2^{\alpha}-1}\right)\left[\frac{2\left(1+2^{-2\alpha}\right)\||f|\|_{\infty,\mathbb{R}_+}}{h^{3\alpha}}+\frac{\left(1+2^{2\alpha-\frac{1}{p}}\right)}{c}M_g h^{\alpha-\frac{1}{p}}\right], \quad (10.72)$$

$\forall\, a \in \mathbb{R}_+, \forall\, h > 0.$

I.e. it holds

$$\sup_{a\in\mathbb{R}_+}\left\|\left(D^{3\alpha}_{a+;g}f\right)(a)\right\| \le \left(\frac{\Gamma(3\alpha+1)}{2^{\alpha}-1}\right)$$

$$\left[\frac{2\left(1+2^{-2\alpha}\right)\||f|\|_{\infty,\mathbb{R}_+}}{h^{3\alpha}}+\frac{\left(1+2^{2\alpha-\frac{1}{p}}\right)}{c}M_g h^{\alpha-\frac{1}{p}}\right], \quad (10.73)$$

$\forall\, h > 0, \frac{1}{4p} < \alpha < 1.$

Call

$$\mu := 2\left(1+2^{-3\alpha}\right)\||f|\|_{\infty,\mathbb{R}_+},$$

$$\theta := \frac{\left(1+2^{\alpha-\frac{1}{p}}\right)M_g}{c}, \quad (10.74)$$

both are greater than zero.

We consider the function

$$y(h) = \mu h^{-2\alpha} + \theta h^{2\alpha-\frac{1}{p}}, \quad \forall\, h > 0. \quad (10.75)$$

We have

$$y'(h) = -2\alpha\mu h^{-2\alpha-1} + \left(2\alpha - \frac{1}{p}\right)\theta h^{2\alpha-\frac{1}{p}-1} = 0, \quad (10.76)$$

then

$$\left(2\alpha - \frac{1}{p}\right)\theta h^{2\alpha-\frac{1}{p}-1} = 2\alpha\mu h^{-2\alpha-1},$$

i.e.

$$\left(2\alpha - \frac{1}{p}\right)\theta h^{4\alpha-\frac{1}{p}} = 2\alpha\mu,$$

with a unique solution

$$h_0 := h_{crit.no.} = \left(\frac{2\alpha\mu}{\left(2\alpha-\frac{1}{p}\right)\theta}\right)^{\frac{1}{4\alpha-\frac{1}{p}}} \quad (10.77)$$

(assuming $\frac{1}{2p} < \alpha < 1$).

We have that

$$y''(h) = 2\alpha(2\alpha+1)\mu h^{-2\alpha-2} + \left(2\alpha - \frac{1}{p}\right)\left(2\alpha - \frac{1}{p} - 1\right)\theta h^{2\alpha - \frac{1}{p} - 2}. \quad (10.78)$$

We see that

$$y''(h_0) = 2\alpha(2\alpha+1)\mu\left(\frac{2\alpha\mu}{\left(2\alpha-\frac{1}{p}\right)\theta}\right)^{\frac{-2\alpha-2}{4\alpha-\frac{1}{p}}} +$$

$$\left(2\alpha - \frac{1}{p}\right)\left(2\alpha - \frac{1}{p} - 1\right)\theta\left(\frac{2\alpha\mu}{\left(2\alpha-\frac{1}{p}\right)\theta}\right)^{\frac{2\alpha-\frac{1}{p}-2}{4\alpha-\frac{1}{p}}} =$$

$$\left(\frac{2\alpha\mu}{\left(2\alpha-\frac{1}{p}\right)\theta}\right)^{\frac{-2\alpha-2}{4\alpha-\frac{1}{p}}}\left[2\alpha(2\alpha+1)\mu + 2\alpha\mu\left(2\alpha - \frac{1}{p} - 1\right)\right] = \quad (10.79)$$

$$2\alpha\mu\left(\frac{\left(2\alpha-\frac{1}{p}\right)\theta}{2\alpha\mu}\right)^{\left(\frac{2(\alpha+1)}{4\alpha-\frac{1}{p}}\right)}\left(4\alpha - \frac{1}{p}\right) > 0.$$

Therefore y has a global minimum at

$$h_0 = \left(\frac{2\alpha\mu}{\left(2\alpha-\frac{1}{p}\right)\theta}\right)^{\frac{1}{4\alpha-\frac{1}{p}}},$$

which is

$$y(h_0) = \mu\left(\frac{2\alpha\mu}{\left(2\alpha-\frac{1}{p}\right)\theta}\right)^{\frac{-2\alpha}{4\alpha-\frac{1}{p}}} + \theta\left(\frac{2\alpha\mu}{\left(2\alpha-\frac{1}{p}\right)\theta}\right)^{\frac{2\alpha-\frac{1}{p}}{4\alpha-\frac{1}{p}}} = \quad (10.80)$$

$$\left(\frac{2\alpha\mu}{\left(2\alpha-\frac{1}{p}\right)\theta}\right)^{\frac{-2\alpha}{4\alpha-\frac{1}{p}}}\left[\mu + \theta\frac{2\alpha\mu}{\left(2\alpha-\frac{1}{p}\right)\theta}\right] = \mu\left(\frac{\left(2\alpha-\frac{1}{p}\right)\theta}{2\alpha\mu}\right)^{\left(\frac{2\alpha}{4\alpha-\frac{1}{p}}\right)}\left(1 + \frac{2\alpha}{2\alpha-\frac{1}{p}}\right)$$

$$= \frac{\left(4\alpha-\frac{1}{p}\right)}{(2\alpha)^{\left(\frac{2\alpha}{4\alpha-\frac{1}{p}}\right)}}\left(2\alpha - \frac{1}{p}\right)^{\frac{-2\alpha+\frac{1}{p}}{4\alpha-\frac{1}{p}}}\mu^{\left(\frac{2\alpha-\frac{1}{p}}{4\alpha-\frac{1}{p}}\right)}\theta^{\left(\frac{2\alpha}{4\alpha-\frac{1}{p}}\right)}. \quad (10.81)$$

That is

$$y(h_0) = \frac{\left(4\alpha - \frac{1}{p}\right)\left(2\alpha - \frac{1}{p}\right)^{\left(\frac{-2\alpha+\frac{1}{p}}{4\alpha-\frac{1}{p}}\right)}}{(2\alpha)^{\left(\frac{2\alpha}{4\alpha-\frac{1}{p}}\right)}} \mu^{\left(\frac{2\alpha-\frac{1}{p}}{4\alpha-\frac{1}{p}}\right)} \theta^{\left(\frac{2\alpha}{4\alpha-\frac{1}{p}}\right)}. \tag{10.82}$$

Therefore we derive (see (10.69))

$$\sup_{a\in\mathbb{R}_+} \left\| \left(D^{2\alpha}_{a+;g} f\right)(a) \right\| \leq \left(\frac{2^{\alpha}\Gamma(2\alpha)}{2^{\alpha}-1}\right)\left(\frac{2\alpha}{2\alpha-\frac{1}{p}}\right)^{\frac{\left(2\alpha-\frac{1}{p}\right)}{\left(4\alpha-\frac{1}{p}\right)}} \left(4\alpha - \frac{1}{p}\right)$$

$$\left(2\left(1+2^{-3\alpha}\right)\right)^{\left(\frac{2\alpha-\frac{1}{p}}{4\alpha-\frac{1}{p}}\right)} \left(\frac{\left(1+2^{\alpha-\frac{1}{p}}\right)}{\Gamma(4\alpha)\left(q(4\alpha-1)+1\right)^{\frac{1}{q}}}\right)^{\left(\frac{2\alpha}{4\alpha-\frac{1}{p}}\right)} \||| f |||_{\infty,\mathbb{R}_+}^{\left(\frac{2\alpha-\frac{1}{p}}{4\alpha-\frac{1}{p}}\right)}$$

$$\left(\sup_{a\in\mathbb{R}_+} \left\| \left\| D^{4\alpha}_{a+;g} f \right\| \right\|_{p,\mathbb{R}_+}\right)^{\left(\frac{2\alpha}{4\alpha-\frac{1}{p}}\right)} < \infty, \tag{10.83}$$

where $\frac{1}{2p} < \alpha < 1$.

Call

$$\xi := 2\left(1+2^{-2\alpha}\right) \||| f |||_{\infty,\mathbb{R}_+},$$

$$\psi := \frac{\left(1+2^{2\alpha-\frac{1}{p}}\right)}{c} M_g, \tag{10.84}$$

both are greater than zero.

We consider the function

$$\gamma(h) := \xi h^{-3\alpha} + \psi h^{\alpha-\frac{1}{p}}, \quad \forall\, h > 0. \tag{10.85}$$

We have

$$\gamma'(h) = -3\alpha\xi h^{-3\alpha-1} + \left(\alpha - \frac{1}{p}\right)\psi h^{\alpha-\frac{1}{p}-1} = 0,$$

then

$$\left(\alpha - \frac{1}{p}\right)\psi h^{\alpha-\frac{1}{p}-1} = 3\alpha\xi h^{-3\alpha-1},$$

and

$$\left(\alpha - \frac{1}{p}\right)\psi h^{4\alpha-\frac{1}{p}} = 3\alpha\xi,$$

with unique solution

$$h_0 := h_{crit.no.} = \left(\frac{3\alpha\xi}{\left(\alpha - \frac{1}{p}\right)\psi} \right)^{\frac{1}{4\alpha - \frac{1}{p}}} \tag{10.86}$$

(assuming $\frac{1}{p} < \alpha < 1$).

We have that

$$\gamma''(h) = 3\alpha(3\alpha+1)\xi h^{-3\alpha-2} + \left(\alpha - \frac{1}{p}\right)\left(\alpha - \frac{1}{p} - 1\right)\psi h^{\alpha - \frac{1}{p} - 2}. \tag{10.87}$$

We observe

$$\gamma''(h_0) = 3\alpha(3\alpha+1)\xi \left(\frac{3\alpha\xi}{\left(\alpha - \frac{1}{p}\right)\psi} \right)^{\left(\frac{-3\alpha-2}{4\alpha - \frac{1}{p}}\right)} +$$

$$\left(\alpha - \frac{1}{p}\right)\left(\alpha - \frac{1}{p} - 1\right)\psi \left(\frac{3\alpha\xi}{\left(\alpha - \frac{1}{p}\right)\psi} \right)^{\left(\frac{\alpha - \frac{1}{p} - 2}{4\alpha - \frac{1}{p}}\right)} =$$

$$\left(\frac{3\alpha\xi}{\left(\alpha - \frac{1}{p}\right)\psi} \right)^{\left(\frac{-3\alpha-2}{4\alpha - \frac{1}{p}}\right)} \left[3\alpha(3\alpha+1)\xi + \left(\alpha - \frac{1}{p} - 1\right)3\alpha\xi\right] = \tag{10.88}$$

$$3\alpha\xi \left(\frac{3\alpha\xi}{\left(\alpha - \frac{1}{p}\right)\psi} \right)^{\left(\frac{-3\alpha-2}{4\alpha - \frac{1}{p}}\right)} \left(4\alpha - \frac{1}{p}\right) > 0.$$

Therefore y has a global minimum at

$$h_0 = \left(\frac{3\alpha\xi}{\left(\alpha - \frac{1}{p}\right)\psi} \right)^{\frac{1}{4\alpha - \frac{1}{p}}},$$

which is

$$\gamma(h_0) = \xi h_0^{-3\alpha} + \psi h_0^{\alpha - \frac{1}{p}} = h_0^{-3\alpha}\left(\xi + \psi h_0^{4\alpha - \frac{1}{p}}\right) = \tag{10.89}$$

$$\left(\frac{3\alpha\xi}{\left(\alpha-\frac{1}{p}\right)\psi}\right)^{\left(\frac{-3\alpha}{4\alpha-\frac{1}{p}}\right)}\left(\xi+\psi\left(\frac{3\alpha\xi}{\left(\alpha-\frac{1}{p}\right)\psi}\right)\right)=$$

$$\xi\left(\frac{3\alpha\xi}{\left(\alpha-\frac{1}{p}\right)\psi}\right)^{\left(\frac{-3\alpha}{4\alpha-\frac{1}{p}}\right)}\left(\frac{4\alpha-\frac{1}{p}}{\alpha-\frac{1}{p}}\right).$$

That is

$$\gamma\left(h_0\right)=\xi\left(\frac{4\alpha-\frac{1}{p}}{\alpha-\frac{1}{p}}\right)\left(\frac{3\alpha\xi}{\left(\alpha-\frac{1}{p}\right)\psi}\right)^{\left(\frac{-3\alpha}{4\alpha-\frac{1}{p}}\right)}= \tag{10.90}$$

$$\left(\frac{4\alpha-\frac{1}{p}}{\alpha-\frac{1}{p}}\right)\xi^{\left(\frac{\alpha-\frac{1}{p}}{4\alpha-\frac{1}{p}}\right)}\left(\frac{\left(\alpha-\frac{1}{p}\right)}{3\alpha}\right)^{\left(\frac{3\alpha}{4\alpha-\frac{1}{p}}\right)}\psi^{\left(\frac{3\alpha}{4\alpha-\frac{1}{p}}\right)}.$$

I.e. we have found

$$\gamma\left(h_0\right)=\frac{\left(4\alpha-\frac{1}{p}\right)}{\left(\alpha-\frac{1}{p}\right)^{\left(\frac{\alpha-\frac{1}{p}}{4\alpha-\frac{1}{p}}\right)}(3\alpha)^{\left(\frac{3\alpha}{4\alpha-\frac{1}{p}}\right)}}\xi^{\left(\frac{\alpha-\frac{1}{p}}{4\alpha-\frac{1}{p}}\right)}\psi^{\left(\frac{3\alpha}{4\alpha-\frac{1}{p}}\right)}. \tag{10.91}$$

We have proved that (see (10.73))

$$\sup_{a\in\mathbb{R}_+}\left\|\left(D_{a+;g}^{3\alpha}f\right)(a)\right\|\leq\left(\frac{\left(4\alpha-\frac{1}{p}\right)\Gamma(3\alpha)}{2^{\alpha}-1}\right)\left(\frac{3\alpha}{\alpha-\frac{1}{p}}\right)^{\left(\frac{\alpha-\frac{1}{p}}{4\alpha-\frac{1}{p}}\right)} \tag{10.92}$$

$$\left(2\left(1+2^{-2\alpha}\right)\right)^{\left(\frac{\alpha-\frac{1}{p}}{4\alpha-\frac{1}{p}}\right)}\left(\frac{\left(1+2^{2\alpha-\frac{1}{p}}\right)}{\Gamma(4\alpha)\left(q(4\alpha-1)+1\right)^{\frac{1}{q}}}\right)^{\left(\frac{3\alpha}{4\alpha-\frac{1}{p}}\right)}\|\|f\|\|_{\infty,\mathbb{R}_+}^{\left(\frac{\alpha-\frac{1}{p}}{4\alpha-\frac{1}{p}}\right)}$$

$$\left(\sup_{a\in\mathbb{R}_+}\left\|\left\|D_{a+;g}^{4\alpha}f\right\|\right\|_{p,\mathbb{R}_+}\right)^{\left(\frac{3\alpha}{4\alpha-\frac{1}{p}}\right)}<\infty,$$

where $\frac{1}{p}<\alpha<1$.

The theorem is proved. ■

We give an application when $\alpha = \frac{1}{2}$ and $g(t) = e^t|_{\mathbb{R}_+}$.

Corollary 10.7 *Let $f \in C^1(\mathbb{R}_+, X)$ with $\| \|f\| \|_{\infty,\mathbb{R}_+}$, $\left\| \|(f \circ \ln)' \circ e^t\| \right\|_{\infty,\mathbb{R}_+} < \infty$, where $(X, \|\cdot\|)$ is a Banach space. For $k = 1, 2, 3$, we assume that $D^{k\frac{1}{2}}_{a+;e^t} f \in C^1([a, \infty), X)$ and $D^{4\frac{1}{2}}_{a+;e^t} f \in C([a, \infty), X)$, $\forall\, a \in \mathbb{R}_+$. We further assume that*

$$\left\| \left\| D^{4\frac{1}{2}}_{a+;e^t} f(t) \right\| \right\|_{\infty,\mathbb{R}^2_+} < \infty, \tag{10.93}$$

where $(a, t) \in \mathbb{R}^2_+$.

Then

$$\sup_{a\in\mathbb{R}_+} \left\| \left(D^{2\frac{1}{2}}_{a+;e^t} f \right)(a) \right\| \leq \left(\frac{\sqrt{12 + 6\sqrt{2}}}{\sqrt{2} - 1} \right) \| \|f\| \|^{\frac{1}{2}}_{\infty,\mathbb{R}_+} \left(\left\| \left\| D^{4\frac{1}{2}}_{a+;e^t} f(t) \right\| \right\|_{\infty,\mathbb{R}^2_+} \right)^{\frac{1}{2}} < \infty, \tag{10.94}$$

and

$$\sup_{a\in\mathbb{R}_+} \left\| \left(D^{3\frac{1}{2}}_{a+;e^t} f \right)(a) \right\| \leq$$

$$\left(\frac{9\sqrt{\pi}}{\left(2 - \sqrt{2}\right)\sqrt{42}\left(\sqrt[4]{3}\right)^3} \right) \| \|f\| \|^{\frac{1}{4}}_{\infty,\mathbb{R}_+} \left(\left\| \left\| D^{4\frac{1}{2}}_{a+;e^t} f(t) \right\| \right\|_{\infty,\mathbb{R}^2_+} \right)^{\frac{3}{4}} < \infty. \tag{10.95}$$

That is $\sup_{a\in\mathbb{R}_+} \left\| \left(D^{2\frac{1}{2}}_{a+;e^t} f \right)(a) \right\|$, $\sup_{a\in\mathbb{R}_+} \left\| \left(D^{3\frac{1}{2}}_{a+;e^t} f \right)(a) \right\| < \infty$.

Proof By Theorem 10.5. ∎

References

1. Aglic Aljinovic, A., Marangunic, Lj., Pecaric, J.: On Landau type inequalities via Ostrowski inequalities. Nonlinear Funct. Anal. Appl. **10**(4), 565–579 (2005)
2. Anastassiou, G.: Fractional Differentiation inequalities. Research Monograph. Springer, New York (2009)
3. Anastassiou, G.A.: Advances on Fractional Inequalities. Springer, New York (2011)
4. Anastassiou, G.A.: Intelligent Computations: Abstract Fractional Calculus, Inequalities, Approximations. Springer, Heidelberg (2018)
5. Anastassiou, G.A.: Sequential Left Abstract Generalized Fractional Landau Inequalities, submitted (2020)
6. Barnett, N.S., Dragomir, S.S.: Some Landau type inequalities for functions whose derivatives are of locally bounded variation. Tamkang J. Math. **37**(4), 301–308 (2006)
7. Ditzian, Z.: Remarks, questions and conjectures on Landau-Kolmogorov-type inequalities. Math. Inequal. Appl. **3**, 15–24 (2000)

8. Hardy, G.H., Littlewood, J.E.: Some integral inequalities connected with the calculus of variations. Quart. J. Math. Oxford Ser. **3**, 241–252 (1932)
9. Hardy, G.H., Landau, E., Littlewood, J.E.: Some inequalities satisfied by the integrals or derivatives of real or analytic functions. Math. Z. **39**, 677–695 (1935)
10. Kallman, R.R., Rota, G.C.: On theinequality $\|f'\|^2 \leq 4\|\| \cdot \|f''\|$. In: Shisha, O. (ed.) Inequalities, vol. II, pp. 187–192. Academic, New York (1970)
11. Landau, E.: Einige Ungleichungen für zweimal differentzierban funktionen. Proc. Lon. Math. Soc. **13**, 43–49 (1913)
12. Mikusinski, J.: The Bochner Integral. Academic, New York (1978)
13. Shilov, G.E.: Elementary Functional Analysis. Dover Publications Inc., New York (1996)

Chapter 11
Sequential General Right Side Fractional Landau Inequalities

We derive uniform and L_p right Caputo–Bochner abstract sequential fractional Landau inequalities over $\mathbb{R}_-$. These estimate the size of second and third iterated right abstract fractional derivatives of a Banach space valued function over $\mathbb{R}_-$. We give an application when the basic fractional order is $\frac{1}{2}$. It follows [5].

11.1 Introduction

Let $p \in [1, \infty]$, $I = \mathbb{R}_+$ or $I = \mathbb{R}$ and $f : I \to \mathbb{R}$ is twice differentiable with $f, f'' \in L_p(I)$, then $f' \in L_p(I)$. Moreover, there exists a constant $C_p(I) > 0$ independent of f, such that

$$\left\| f' \right\|_{p,I} \leq C_p(I) \left\| f \right\|_{p,I}^{\frac{1}{2}} \left\| f'' \right\|_{p,I}^{\frac{1}{2}}, \tag{11.1}$$

where $\|\cdot\|_{p,I}$ is the p-norm on the interval I, see [1, 6].

The research on these inequalities started by Landau [11] in 1913. For the case of $p = \infty$ he proved that

$$C_\infty(\mathbb{R}_+) = 2 \text{ and } C_\infty(\mathbb{R}) = \sqrt{2}, \tag{11.2}$$

are the best constants in (11.1).

In 1932, Hardy and Littlewood [8] proved (11.1) for $p = 2$, with the best constants

$$C_2(\mathbb{R}_+) = \sqrt{2} \text{ and } C_2(\mathbb{R}) = 1. \tag{11.3}$$

In 1935, Hardy, Landau and Littlewood [9] showed that the best constants $C_p(\mathbb{R}_+)$ in (11.1) satisfies the estimate

G. A. Anastassiou, *Constructive Fractional Analysis with Applications*,
Studies in Systems, Decision and Control 362,
https://doi.org/10.1007/978-3-030-71481-9_11

$$C_p(\mathbb{R}_+) \leq 2, \text{ for } p \in [1, \infty), \tag{11.4}$$

which yields $C_p(\mathbb{R}) \leq 2$ for $p \in [1, \infty)$.

In fact, in [7, 10] was shown that $C_p(\mathbb{R}) \leq \sqrt{2}$.

We need the following concept from abstract fractional calculus.

Our integral next is of Bochner type [12].

We need

Definition 11.1 ([4], *p. 150*) Let $[a, b] \subset \mathbb{R}$, X be a Banach space, $\alpha > 0$; $m = \lceil \alpha \rceil \in \mathbb{N}$, ($\lceil \cdot \rceil$ is the ceiling of the number), $f : [a, b] \to X$. We assume that $f^{(m)} \in L_1([a, b], X)$. We call the Caputo–Bochner right fractional derivative of order α:

$$\left(D_{b-}^{\alpha} f\right)(x) := \frac{(-1)^m}{\Gamma(m-\alpha)} \int_x^b (z-x)^{m-\alpha-1} f^{(m)}(z)\, dz, \quad \forall\, x \in [a, b], \tag{11.5}$$

where Γ is the gamma function.

If $\alpha = m \in \mathbb{N}$, we observe that $D_{b-}^{\alpha} f = (-1)^m f^{(m)}$ the ordinary X-valued derivative (defined similar to numerical one, see [13], p. 83), and also set $D_{b-}^{0} f := f$.

By ([4], p. 34), $\left(D_{b-}^{\alpha} f\right)(x)$ exists almost everywhere in $x \in [a, b]$ and $D_{b-}^{\alpha} f \in L_1([a, b], X)$.

If $\left\| f^{(m)} \right\|_{L_\infty([a,b],X)} < \infty$, and $\alpha \notin \mathbb{N}$, then by ([4], p. 37), $D_{b-}^{\alpha} f \in C([a, b], X)$, hence $\left\| D_{b-}^{\alpha} f \right\| \in C([a, b])$.

We mention the important:

Corollary 11.2 ([4], p. 157) *Let $f \in C^m([a, b], X)$, $m = \lceil \alpha \rceil$, $\alpha > 0$, $x, x_0 \in [a, b]$. Then $D_{x_0-}^{a} f(x)$ is jointly continuous functions in (x, x_0) from $[a, b]^2$ into X, X is a Banach space.*

By convention we suppose that

$$D_{x_0-}^{\alpha} f(x) = 0, \text{ for } x > x_0,$$

for all $x, x_0 \in [a, b]$.

The author has already done an extensive body of work on fractional Landau inequalities, see [3], and on abstract fractional Landau inequalities, see [4]. However there the proving methods came out of applications of fractional Ostrowski inequalities [2, 4]. Usually there the domains where $[A, +\infty)$ or $(-\infty, B]$, with $A, B \in \mathbb{R}$ and in one mixed case the domain was all of $\mathbb{R}$.

In this chapter with less assumptions we establish uniform and L_p type right Caputo–Bochner abstract iterated fractional Landau inequalities over $\mathbb{R}_-$. The method of proving is based on right Caputo–Bochner iterated fractional Taylor's formula with integral remainder, see [4], pp. 129.

We give also an application for $\alpha = \frac{1}{2}$. Regardless to say that we are also inspired by [3, 4].

11.2 Main Results

We consider ($\alpha > 0$) the composition $D_{b-}^{n\alpha} := D_{b-}^{\alpha} D_{b-}^{\alpha} ... D_{b-}^{\alpha}$ (n-times), $n \in \mathbb{N}$.

We mention the following right modified X-valued Taylor's formula, $(X, \|\cdot\|)$ is a Banach space.

Theorem 11.3 ([4], p. 129) *Let $0 < \alpha \leq 1$, $n \in \mathbb{N}$, $f \in C^1([a,b], X)$. For $k = 1, ..., n$, we assume that $D_{b-}^{k\alpha} f \in C^1([a,b], X)$ and $D_{b-}^{(n+1)\alpha} f \in C([a,b], X)$. Then*

$$f(x) = \sum_{i=0}^{n} \frac{(b-x)^{i\alpha}}{\Gamma(i\alpha+1)} \left(D_{b-}^{i\alpha} f\right)(b) + \tag{11.6}$$

$$\frac{1}{\Gamma((n+1)\alpha)} \int_x^b (t-x)^{(n+1)\alpha-1} \left(D_{b-}^{(n+1)\alpha} f\right)(t)\, dt,$$

$\forall\, x \in [a,b]$.

When $0 < \alpha < 1$ and $f \in C^1([a,b], X)$, by [4], p. 135, we get that $\left(D_{b-}^{\alpha} f\right)(b) = 0$.

We present the following abstract right fractional Landau inequalities over $\mathbb{R}_-$.

Theorem 11.4 *Let $0 < \alpha < 1$, $f \in C^1(\mathbb{R}_-, X)$ with $\|\|f\|\|_{\infty,\mathbb{R}_-}$, $\left\|\|f'\|\right\|_{\infty,\mathbb{R}_-} < \infty$. For $k = 1, 2, 3$, we assume that $D_{b-}^{k\alpha} f \in C^1((-\infty, b], X)$ and $D_{b-}^{4\alpha} f \in C((-\infty, b], X)$, $\forall\, b \in \mathbb{R}_-$. We further assume that*

$$K := \left\|\left\|D_{b-}^{4\alpha} f(t)\right\|\right\|_{\infty,\mathbb{R}_-^2} < \infty, \tag{11.7}$$

where $(b,t) \in \mathbb{R}_-^2$.

Then

$$\sup_{b\in\mathbb{R}_-} \left\|\left(D_{b-}^{2\alpha} f\right)(b)\right\| \leq \frac{\Gamma(2\alpha+1)}{2^{2\alpha-1}(2^{\alpha}-1)} \sqrt{\frac{2^{3\alpha+1}\left(2^{3\alpha}+1\right)\left(2^{\alpha}+1\right)}{\Gamma(4\alpha+1)}\, \|\|f\|\|_{\infty,\mathbb{R}_-} K}, \tag{11.8}$$

and

$$\sup_{b\in\mathbb{R}_-} \left\|\left(D_{b-}^{3\alpha} f\right)(b)\right\| \leq \frac{4\sqrt[4]{2}\,\Gamma(3\alpha+1)\left(\Gamma(4\alpha+1)\right)^{-\frac{3}{4}}\left(2^{2\alpha}+1\right)}{\left(\sqrt[4]{3}\right)^3 \left(\sqrt{2}\right)^{\alpha} (2^{\alpha}-1)} \|\|f\|\|_{\infty,\mathbb{R}_-}^{\frac{1}{4}} K^{\frac{3}{4}}. \tag{11.9}$$

That is $\sup_{b\in\mathbb{R}_-} \left\|\left(D_{b-}^{2\alpha} f\right)(b)\right\|$, $\sup_{b\in\mathbb{R}_-} \left\|\left(D_{b-}^{3\alpha} f\right)(b)\right\| < \infty$.

Proof We notice again here that $\left(D_{b-}^{\alpha} f\right)(b) = 0$, $\forall\, b \in \mathbb{R}_-$.

We make use of Theorem 11.3 for $0 < \alpha < 1$ and $n = 3$, applied for any $b \in \mathbb{R}_-$ and $a = -\infty$.

Momentarily we fix $b \in \mathbb{R}_-$. Let $x_2 < x_1 < b$, then

$$f(x_1) - f(b) = \frac{(b - x_1)^{2\alpha}}{\Gamma(2\alpha + 1)} \left(D_{b-}^{2\alpha} f\right)(b) + \frac{(b - x_1)^{3\alpha}}{\Gamma(3\alpha + 1)} \left(D_{b-}^{3\alpha} f\right)(b) + \tag{11.10}$$

$$\frac{1}{\Gamma(4\alpha)} \int_{x_1}^{b} (t - x_1)^{4\alpha - 1} \left(D_{b-}^{4\alpha} f\right)(t)\, dt,$$

and

$$f(x_2) - f(b) = \frac{(b - x_2)^{2\alpha}}{\Gamma(2\alpha + 1)} \left(D_{b-}^{2\alpha} f\right)(b) + \frac{(b - x_2)^{3\alpha}}{\Gamma(3\alpha + 1)} \left(D_{b-}^{3\alpha} f\right)(b) + \tag{11.11}$$

$$\frac{1}{\Gamma(4\alpha)} \int_{x_2}^{b} (t - x_2)^{4\alpha - 1} \left(D_{b-}^{4\alpha} f\right)(t)\, dt.$$

That is

$$\frac{(b - x_1)^{2\alpha}}{\Gamma(2\alpha + 1)} \left(D_{b-}^{2\alpha} f\right)(b) + \frac{(b - x_1)^{3\alpha}}{\Gamma(3\alpha + 1)} \left(D_{b-}^{3\alpha} f\right)(b) = \tag{11.12}$$

$$f(x_1) - f(b) - \frac{1}{\Gamma(4\alpha)} \int_{x_1}^{b} (t - x_1)^{4\alpha - 1} \left(D_{b-}^{4\alpha} f\right)(t)\, dt =: A,$$

and

$$\frac{(b - x_2)^{2\alpha}}{\Gamma(2\alpha + 1)} \left(D_{b-}^{2\alpha} f\right)(b) + \frac{(b - x_2)^{3\alpha}}{\Gamma(3\alpha + 1)} \left(D_{b-}^{3\alpha} f\right)(b) = \tag{11.13}$$

$$f(x_2) - f(b) - \frac{1}{\Gamma(4\alpha)} \int_{x_2}^{b} (t - x_2)^{4\alpha - 1} \left(D_{b-}^{4\alpha} f\right)(t)\, dt =: B.$$

We are solving the above system of two equations with two unknowns $\left(D_{b-}^{2\alpha} f\right)(b)$, $\left(D_{b-}^{3\alpha} f\right)(b)$.

The main determinant of system is

$$D := \begin{vmatrix} \frac{(b-x_1)^{2\alpha}}{\Gamma(2\alpha+1)} & \frac{(b-x_1)^{3\alpha}}{\Gamma(3\alpha+1)} \\ \frac{(b-x_2)^{2\alpha}}{\Gamma(2\alpha+1)} & \frac{(b-x_2)^{3\alpha}}{\Gamma(3\alpha+1)} \end{vmatrix} =$$

$$\frac{1}{\Gamma(2\alpha + 1)\Gamma(3\alpha + 1)} \left[(b - x_1)^{2\alpha} (b - x_2)^{3\alpha} - (b - x_1)^{3\alpha} (b - x_2)^{2\alpha}\right] =$$

$$\frac{(b - x_1)^{2\alpha} (b - x_2)^{2\alpha}}{\Gamma(2\alpha + 1)\Gamma(3\alpha + 1)} \left[(b - x_2)^{\alpha} - (b - x_1)^{\alpha}\right] > 0.$$

I.e.

$$D = \frac{(b-x_1)^{2\alpha}(b-x_2)^{2\alpha}}{\Gamma(2\alpha+1)\Gamma(3\alpha+1)}\left[(b-x_2)^{\alpha}-(b-x_1)^{\alpha}\right] > 0. \tag{11.14}$$

We obtain the unique solution

$$\left(D_{b-}^{2\alpha}f\right)(b) = \frac{\begin{vmatrix} A & \frac{(b-x_1)^{3\alpha}}{\Gamma(3\alpha+1)} \\ B & \frac{(b-x_2)^{3\alpha}}{\Gamma(3\alpha+1)} \end{vmatrix}}{D},$$

$$\left(D_{b-}^{3\alpha}f\right)(b) = \frac{\begin{vmatrix} \frac{(b-x_1)^{2\alpha}}{\Gamma(2\alpha+1)} & A \\ \frac{(b-x_2)^{2\alpha}}{\Gamma(2\alpha+1)} & B \end{vmatrix}}{D}. \tag{11.15}$$

Therefore we have

$$\left(D_{b-}^{2\alpha}f\right)(b) = \frac{\frac{(b-x_2)^{3\alpha}}{\Gamma(3\alpha+1)}A - \frac{(b-x_1)^{3\alpha}}{\Gamma(3\alpha+1)}B}{D},$$

and

$$\left(D_{b-}^{3\alpha}f\right)(b) = \frac{\frac{(b-x_1)^{2\alpha}}{\Gamma(2\alpha+1)}B - \frac{(b-x_2)^{2\alpha}}{\Gamma(2\alpha+1)}A}{D}. \tag{11.16}$$

We have the following

$$\|A\| = \left\| f(x_1) - f(b) - \frac{1}{\Gamma(4\alpha)}\int_{x_1}^{b}(t-x_1)^{4\alpha-1}\left(D_{b-}^{4\alpha}f\right)(t)\,dt \right\| \leq$$

$$2\,\|\|f\|\|_{\infty,\mathbb{R}_-} + \frac{\left\|\left\|D_{b-}^{4\alpha}f(t)\right\|\right\|_{\infty,\mathbb{R}_-^2}}{\Gamma(4\alpha+1)}(b-x_1)^{4\alpha}, \tag{11.17}$$

under the assumption $\|\|f\|\|_{\infty,\mathbb{R}_-} < \infty$.

That is

$$\|A\| \leq 2\,\|\|f\|\|_{\infty,\mathbb{R}_-} + \frac{K}{\Gamma(4\alpha+1)}(b-x_1)^{4\alpha}, \tag{11.18}$$

and similarly,

$$\|B\| \leq 2\,\|\|f\|\|_{\infty,\mathbb{R}_-} + \frac{K}{\Gamma(4\alpha+1)}(b-x_2)^{4\alpha}, \tag{11.19}$$

where by assumption

$$K := \left\|\left\|D_{b-}^{4\alpha}f(t)\right\|\right\|_{\infty,\mathbb{R}_-^2} < \infty, \tag{11.20}$$

with $(b,t) \in \mathbb{R}_-^2$.

Consequently we have

$$\left\| \left(D_{b-}^{2\alpha} f \right)(b) \right\| \leq \frac{1}{\Gamma(3\alpha+1) D} \left[(b-x_2)^{3\alpha} \left(2 \left\| \| f \| \right\|_{\infty,\mathbb{R}_-} + \frac{K}{\Gamma(4\alpha+1)} (b-x_1)^{4\alpha} \right) \right.$$

$$\left. + (b-x_1)^{3\alpha} \left(2 \left\| \| f \| \right\|_{\infty,\mathbb{R}_-} + \frac{K}{\Gamma(4\alpha+1)} (b-x_2)^{4\alpha} \right) \right], \quad (11.21)$$

and

$$\left\| \left(D_{b-}^{3\alpha} f \right)(b) \right\| \leq \frac{1}{\Gamma(2\alpha+1) D} \left[(b-x_1)^{2\alpha} \left(2 \left\| \| f \| \right\|_{\infty,\mathbb{R}_-} + \frac{K}{\Gamma(4\alpha+1)} (b-x_2)^{4\alpha} \right) \right.$$

$$\left. + (b-x_2)^{2\alpha} \left(2 \left\| \| f \| \right\|_{\infty,\mathbb{R}_-} + \frac{K}{\Gamma(4\alpha+1)} (b-x_1)^{4\alpha} \right) \right]. \quad (11.22)$$

Set now $x_1 := b - h$, $x_2 := b - 2h$, where $h > 0$, so that $b - x_1 = h$, $b - x_2 = 2h$.
Hence we get

$$D = \frac{2^{2\alpha} h^{5\alpha} (2^{\alpha} - 1)}{\Gamma(2\alpha+1) \Gamma(3\alpha+1)} > 0. \quad (11.23)$$

Therefore we derive (from (11.16))

$$\left\| \left(D_{b-}^{2\alpha} f \right)(b) \right\| \leq \frac{\Gamma(2\alpha+1)}{2^{2\alpha} h^{5\alpha} (2^{\alpha} - 1)} \left[2^{3\alpha} h^{3\alpha} \left(2 \left\| \| f \| \right\|_{\infty,\mathbb{R}_-} + \frac{K}{\Gamma(4\alpha+1)} h^{4\alpha} \right) \right.$$

$$\left. + h^{3\alpha} \left(2 \left\| \| f \| \right\|_{\infty,\mathbb{R}_-} + \frac{K}{\Gamma(4\alpha+1)} 2^{4\alpha} h^{4\alpha} \right) \right] = \quad (11.24)$$

$$\frac{\Gamma(2\alpha+1)}{2^{2\alpha} h^{5\alpha} (2^{\alpha} - 1)} \left[2 \left\| \| f \| \right\|_{\infty,\mathbb{R}_-} \left(2^{3\alpha} + 1 \right) h^{3\alpha} + \frac{K}{\Gamma(4\alpha+1)} \left(2^{3\alpha} + 2^{4\alpha} \right) h^{7\alpha} \right]$$

$$= \left(\frac{\Gamma(2\alpha+1)}{2^{2\alpha} (2^{\alpha} - 1)} \right) \left[\frac{2 \left(2^{3\alpha} + 1 \right) \left\| \| f \| \right\|_{\infty,\mathbb{R}_-}}{h^{2\alpha}} + \frac{2^{3\alpha} (2^{\alpha} + 1)}{\Gamma(4\alpha+1)} K h^{2\alpha} \right]. \quad (11.25)$$

That is

$$\left\| \left(D_{b-}^{2\alpha} f \right)(b) \right\| \leq \left(\frac{\Gamma(2\alpha+1)}{2^{2\alpha} (2^{\alpha} - 1)} \right)$$

$$\left[\frac{2 \left(2^{3\alpha} + 1 \right) \left\| \| f \| \right\|_{\infty,\mathbb{R}_-}}{h^{2\alpha}} + \frac{2^{3\alpha} (2^{\alpha} + 1)}{\Gamma(4\alpha+1)} K h^{2\alpha} \right], \quad (11.26)$$

$\forall\, b \in \mathbb{R}_-$, $\forall\, h > 0$.
I.e. it holds

$$\sup_{b\in\mathbb{R}_-}\left\|\left(D_{b-}^{2\alpha}f\right)(b)\right\| \le \left(\frac{\Gamma(2\alpha+1)}{2^{2\alpha}(2^{\alpha}-1)}\right)$$

$$\left[\frac{2\left(2^{3\alpha}+1\right)\|\|f\|\|_{\infty,\mathbb{R}_-}}{h^{2\alpha}}+\frac{2^{3\alpha}(2^{\alpha}+1)}{\Gamma(4\alpha+1)}Kh^{2\alpha}\right]<\infty, \quad (11.27)$$

$\forall\, h>0,\ 0<\alpha<1.$

By (11.16) we derive

$$\left\|\left(D_{b-}^{3\alpha}f\right)(b)\right\| \le \frac{\Gamma(3\alpha+1)}{2^{2\alpha}h^{5\alpha}(2^{\alpha}-1)}\left[h^{2\alpha}\left(2\|\|f\|\|_{\infty,\mathbb{R}_-}+\frac{K}{\Gamma(4\alpha+1)}2^{4\alpha}h^{4\alpha}\right)\right.$$

$$\left.+2^{2\alpha}h^{2\alpha}\left(2\|\|f\|\|_{\infty,\mathbb{R}_-}+\frac{K}{\Gamma(4\alpha+1)}h^{4\alpha}\right)\right]=$$

$$\frac{\Gamma(3\alpha+1)}{2^{2\alpha}h^{5\alpha}(2^{\alpha}-1)}\left[2\|\|f\|\|_{\infty,\mathbb{R}_-}\left(2^{2\alpha}+1\right)h^{2\alpha}+\frac{K}{\Gamma(4\alpha+1)}\left(2^{4\alpha}+2^{2\alpha}\right)h^{6\alpha}\right] \quad (11.28)$$

$$=\left(\frac{\Gamma(3\alpha+1)}{2^{2\alpha}(2^{\alpha}-1)}\right)\left[\frac{2\left(2^{2\alpha}+1\right)\|\|f\|\|_{\infty,\mathbb{R}_-}}{h^{3\alpha}}+\frac{2^{2\alpha}\left(2^{2\alpha}+1\right)}{\Gamma(4\alpha+1)}Kh^{\alpha}\right]$$

$$=\frac{\Gamma(3\alpha+1)\left(2^{2\alpha}+1\right)}{2^{2\alpha}(2^{\alpha}-1)}\left[\frac{2\|\|f\|\|_{\infty,\mathbb{R}_-}}{h^{3\alpha}}+\frac{2^{2\alpha}K}{\Gamma(4\alpha+1)}h^{\alpha}\right].$$

That is

$$\left\|\left(D_{b-}^{3\alpha}f\right)(b)\right\| \le \frac{\Gamma(3\alpha+1)\left(2^{2\alpha}+1\right)}{2^{2\alpha}(2^{\alpha}-1)}$$

$$\left[\frac{2\|\|f\|\|_{\infty,\mathbb{R}_-}}{h^{3\alpha}}+\frac{2^{2\alpha}K}{\Gamma(4\alpha+1)}h^{\alpha}\right], \quad (11.29)$$

$\forall\, b\in\mathbb{R}_-,\ \forall\, h>0.$

I.e. it holds

$$\sup_{b\in\mathbb{R}_-}\left\|\left(D_{b-}^{3\alpha}f\right)(b)\right\| \le \frac{\Gamma(3\alpha+1)\left(2^{2\alpha}+1\right)}{2^{2\alpha}(2^{\alpha}-1)}$$

$$\left[\frac{2\|\|f\|\|_{\infty,\mathbb{R}_-}}{h^{3\alpha}}+\frac{2^{2\alpha}K}{\Gamma(4\alpha+1)}h^{\alpha}\right]<\infty, \quad (11.30)$$

$\forall\, h>0,\ 0<\alpha<1.$

Call

$$\mu := 2\left(2^{3\alpha}+1\right)\|\|f\|\|_{\infty,\mathbb{R}_-}, \quad (11.31)$$

$$\theta = \frac{2^{3\alpha}\left(2^{\alpha}+1\right)K}{\Gamma\left(4\alpha+1\right)}, \tag{11.32}$$

both are greater than zero.

Set also $\rho := 2\alpha$; $0 < \rho < 2$. We consider the function

$$y\left(h\right) := \mu h^{-\rho} + \theta h^{\rho}, \quad \forall\, h > 0. \tag{11.33}$$

We have

$$y'\left(h\right) = -\rho\mu h^{-\rho-1} + \rho\theta h^{\rho-1} = 0, \tag{11.34}$$

then

$$\theta h^{\rho-1} = \mu h^{-\rho-1}$$

and

$$\theta h^{2\rho} = \mu,$$

with a unique solution

$$h_0 := h_{crit.no.} = \left(\frac{\mu}{\theta}\right)^{\frac{1}{2\rho}}. \tag{11.35}$$

We have that

$$y''\left(h\right) = \rho\left(\rho+1\right)\mu h^{-\rho-2} + \rho\left(\rho-1\right)\theta h^{\rho-2}. \tag{11.36}$$

We see that

$$y''\left(h_0\right) = y''\left(\left(\frac{\mu}{\theta}\right)^{\frac{1}{2\rho}}\right) = \rho\left(\rho+1\right)\mu\left(\frac{\mu}{\theta}\right)^{\frac{-\rho-2}{2\rho}} + \rho\left(\rho-1\right)\theta\left(\frac{\mu}{\theta}\right)^{\frac{\rho-2}{2\rho}} =$$

$$\rho\left(\frac{\theta}{\mu}\right)^{\frac{1}{\rho}}\left(2\rho\sqrt{\mu\theta}\right) = 2\rho^2\sqrt{\mu\theta}\left(\frac{\theta}{\mu}\right)^{\frac{1}{\rho}} > 0.$$

Therefore y has a global minimum at $h_0 = \left(\frac{\mu}{\theta}\right)^{\frac{1}{2\rho}}$, which is

$$y\left(h_0\right) = \mu\left(\frac{\mu}{\theta}\right)^{-\frac{1}{2}} + \theta\left(\frac{\mu}{\theta}\right)^{\frac{1}{2}} = \mu\left(\frac{\theta}{\mu}\right)^{\frac{1}{2}} + \sqrt{\theta\mu} = 2\sqrt{\theta\mu}.$$

We have proved that (see (11.27))

$$\sup_{b\in\mathbb{R}_-}\left\|\left(D_{b-}^{2\alpha}f\right)(b)\right\| \leq \frac{\Gamma\left(2\alpha+1\right)}{2^{2\alpha-1}\left(2^{\alpha}-1\right)}$$

$$\sqrt{\frac{2^{3\alpha+1}\left(2^{3\alpha}+1\right)\left(2^{\alpha}+1\right)}{\Gamma\left(4\alpha+1\right)}\left\|\left\|f\right\|\right\|_{\infty,\mathbb{R}_-}K}. \tag{11.37}$$

Call

$$\begin{gathered}\xi := 2\left\|\left\|f\right\|\right\|_{\infty,\mathbb{R}_-},\\ \psi := \tfrac{2^{2\alpha}K}{\Gamma(4\alpha+1)},\end{gathered} \tag{11.38}$$

both are greater than zero.

We consider the function

$$\gamma\left(h\right) := \xi h^{-3\alpha} + \psi h^{\alpha}, \ \ \forall\, h > 0. \tag{11.39}$$

We have

$$\gamma'\left(h\right) = -3\alpha\xi h^{-3\alpha-1} + \alpha\psi h^{\alpha-1} = 0,$$

then

$$\psi h^{\alpha-1} = 3\xi h^{-3\alpha-1}$$

and

$$\psi h^{4\alpha} = 3\xi,$$

with unique solution

$$h_0 := h_{crit.no.} = \left(\frac{3\xi}{\psi}\right)^{\frac{1}{4\alpha}}. \tag{11.40}$$

We have that

$$\gamma''\left(h\right) = 3\alpha\left(3\alpha+1\right)\xi h^{-3\alpha-2} + \alpha\left(\alpha-1\right)\psi h^{\alpha-2}. \tag{11.41}$$

We see

$$\gamma''\left(h_0\right) = 3\alpha\left(3\alpha+1\right)\xi\left(\frac{3\xi}{\psi}\right)^{\frac{-3\alpha-2}{4\alpha}} + \alpha\left(\alpha-1\right)\psi\left(\frac{3\xi}{\psi}\right)^{\frac{\alpha-2}{4\alpha}} =$$

$$\alpha\left(\frac{3\xi}{\psi}\right)^{\frac{\alpha-2}{4\alpha}}\left(4\alpha\psi\right) = 4\alpha^2\psi\left(\frac{3\xi}{\psi}\right)^{\frac{\alpha-2}{4\alpha}} > 0. \tag{11.42}$$

Therefore γ has a global minimum at $h_0 = \left(\frac{3\xi}{\psi}\right)^{\frac{1}{4\alpha}}$, which is

$$\gamma\left(h_0\right) = \xi\left(\frac{3\xi}{\psi}\right)^{-\frac{3}{4}} + \psi\left(\frac{3\xi}{\psi}\right)^{\frac{1}{4}} =$$

$$\left(\frac{3\xi}{\psi}\right)^{\frac{1}{4}}\left(\xi\frac{\psi}{3\xi}+\psi\right)=\frac{4}{3}\psi\left(\frac{3\xi}{\psi}\right)^{\frac{1}{4}}. \tag{11.43}$$

Consequently,

$$\gamma\left(h_0\right)=\frac{4}{\left(\sqrt[4]{3}\right)^3}\psi^{\frac{3}{4}}\xi^{\frac{1}{4}}. \tag{11.44}$$

We have proved that (see (11.30))

$$\sup_{b\in\mathbb{R}_-}\left\|\left(D_{b-}^{3\alpha}f\right)(b)\right\|\leq\frac{4\Gamma\left(3\alpha+1\right)\left(2^{2\alpha}+1\right)}{\left(\sqrt[4]{3}\right)^3 2^{2\alpha}\left(2^{\alpha}-1\right)}$$

$$\left(2\left\|\left\|f\right\|\right\|_{\infty,\mathbb{R}_-}\right)^{\frac{1}{4}}\left(\frac{2^{2\alpha}K}{\Gamma\left(4\alpha+1\right)}\right)^{\frac{3}{4}}=$$

$$\frac{4\sqrt[4]{2}\Gamma\left(3\alpha+1\right)\Gamma\left(4\alpha+1\right)^{-\frac{3}{4}}\left(2^{2\alpha}+1\right)}{\left(\sqrt[4]{3}\right)^3 2^{\frac{\alpha}{2}}\left(2^{\alpha}-1\right)}\left\|\left\|f\right\|\right\|_{\infty,\mathbb{R}_-}^{\frac{1}{4}}K^{\frac{3}{4}}. \tag{11.45}$$

The theorem is proved. ■

We continue with abstract L_p right fractional Landau inequalities over $\mathbb{R}_-$.

Theorem 11.5 *Let* $p,q>1:\frac{1}{p}+\frac{1}{q}=1,\ 0<\alpha<1$. *Let* $f\in C^1\left(\mathbb{R}_-,X\right)$ *with* $\left\|\left\|f\right\|\right\|_{\infty,\mathbb{R}_-}$, $\left\|\left\|f'\right\|\right\|_{\infty,\mathbb{R}_-}<\infty$. *For* $k=1,2,3$, *we assume that* $D_{b-}^{k\alpha}f\in C^1\left(\left(-\infty,b\right],X\right)$ *and* $D_{b-}^{4\alpha}f\in C\left(\left(-\infty,b\right],X\right)$, $\forall\, b\in\mathbb{R}_-$. *We further assume that*

$$\left(\sup_{b\in\mathbb{R}_-}\left\|\left\|D_{b-}^{4\alpha}f\right\|\right\|_{p,\mathbb{R}_-}\right)<\infty. \tag{11.46}$$

Then

(1) under $\frac{1}{2p}<\alpha<1$, *we get*

$$\sup_{b\in\mathbb{R}_-}\left\|\left(D_{b-}^{2\alpha}f\right)(b)\right\|\leq\left[\left(\frac{2^{\alpha}\Gamma\left(2\alpha\right)\left(4\alpha-\frac{1}{p}\right)}{2^{\alpha}-1}\right)\left(\frac{4\alpha\left(1+2^{-3\alpha}\right)}{2\alpha-\frac{1}{p}}\right)^{\left(\frac{2\alpha-\frac{1}{p}}{4\alpha-\frac{1}{p}}\right)}\right.$$

$$\left.\left(\frac{1+2^{\alpha-\frac{1}{p}}}{\Gamma\left(4\alpha\right)\left(q\left(4\alpha-1\right)+1\right)^{\frac{1}{q}}}\right)^{\left(\frac{2\alpha}{4\alpha-\frac{1}{p}}\right)}\right]\left\|\left\|f\right\|\right\|_{\infty,\mathbb{R}_-}^{\left(\frac{2\alpha-\frac{1}{p}}{4\alpha-\frac{1}{p}}\right)}$$

$$\left(\sup_{b\in\mathbb{R}_-}\left\|\left\|D_{b-}^{4\alpha}f\right\|\right\|_{p,\mathbb{R}_-}\right)^{\left(\frac{2\alpha}{4\alpha-\frac{1}{p}}\right)}<\infty. \quad (11.47)$$

(2) under $\frac{1}{p}<\alpha<1$, *we get*

$$\sup_{b\in\mathbb{R}_-}\left\|\left(D_{b-}^{3\alpha}f\right)(b)\right\|\leq\left[\left(\frac{\Gamma(3\alpha)\left(4\alpha-\frac{1}{p}\right)}{2^{\alpha}-1}\right)\left(\frac{6\alpha\left(1+2^{-2\alpha}\right)}{\alpha-\frac{1}{p}}\right)^{\left(\frac{\alpha-\frac{1}{p}}{4\alpha-\frac{1}{p}}\right)}\right.$$

$$\left.\left(\frac{1+2^{2\alpha-\frac{1}{p}}}{\Gamma(4\alpha)\left(q(4\alpha-1)+1\right)^{\frac{1}{q}}}\right)^{\left(\frac{3\alpha}{4\alpha-\frac{1}{p}}\right)}\right]\||f|\|_{\infty,\mathbb{R}_-}^{\left(\frac{\alpha-\frac{1}{p}}{4\alpha-\frac{1}{p}}\right)}$$

$$\left(\sup_{b\in\mathbb{R}_-}\left\|\left\|D_{b-}^{4\alpha}f\right\|\right\|_{p,\mathbb{R}_-}\right)^{\left(\frac{3\alpha}{4\alpha-\frac{1}{p}}\right)}<\infty. \quad (11.48)$$

That is $\sup_{b\in\mathbb{R}_-}\left\|\left(D_{b-}^{2\alpha}f\right)(b)\right\|,\ \sup_{b\in\mathbb{R}_-}\left\|\left(D_{b-}^{3\alpha}f\right)(b)\right\|<\infty.$

Proof As in the proof of Theorem 11.4 we have that

$$\|A\|\overset{(11.12)}{=}\left\|f(x_1)-f(b)-\frac{1}{\Gamma(4\alpha)}\int_{x_1}^{b}(t-x_1)^{4\alpha-1}\left(D_{b-}^{4\alpha}f\right)(t)\,dt\right\|\leq$$

$$2\||f|\|_{\infty,\mathbb{R}_-}+\frac{1}{\Gamma(4\alpha)}\int_{x_1}^{b}(t-x_1)^{4\alpha-1}\left\|\left(D_{b-}^{4\alpha}f\right)(t)\right\|dt\leq$$

$$2\||f|\|_{\infty,\mathbb{R}_-}+\frac{1}{\Gamma(4\alpha)}\frac{(b-x_1)^{\frac{(q(4\alpha-1)+1)}{q}}}{\left(q(4\alpha-1)+1\right)^{\frac{1}{q}}}\left(\sup_{b\in\mathbb{R}_-}\left\|\left\|D_{b-}^{4\alpha}f\right\|\right\|_{p,\mathbb{R}_-}\right)\overset{(b-x_1=:h>0)}{=}$$

$$2\||f|\|_{\infty,\mathbb{R}_-}+\frac{1}{\Gamma(4\alpha)}\frac{h^{\left(4\alpha-\frac{1}{p}\right)}}{\left(q(4\alpha-1)+1\right)^{\frac{1}{q}}}\left(\sup_{b\in\mathbb{R}_-}\left\|\left\|D_{b-}^{4\alpha}f\right\|\right\|_{p,\mathbb{R}_-}\right), \quad (11.49)$$

with $\frac{1}{4p}<\alpha<1$.

That is

$$\|A\|\leq2\||f|\|_{\infty,\mathbb{R}_-}+\frac{h^{4\alpha-\frac{1}{p}}}{\Gamma(4\alpha)\left(q(4\alpha-1)+1\right)^{\frac{1}{q}}}\left(\sup_{b\in\mathbb{R}_-}\left\|\left\|D_{b-}^{4\alpha}f\right\|\right\|_{p,\mathbb{R}_-}\right), \quad (11.50)$$

where $\frac{1}{4p}<\alpha<1$.

We also have

$$\|B\| \overset{(11.13)}{=} \left\| f(x_2) - f(b) - \frac{1}{\Gamma(4\alpha)} \int_{x_2}^{b} (t - x_2)^{4\alpha-1} \left(D_{b-}^{4\alpha} f \right)(t)\, dt \right\| \leq$$

$$2 \left\| \|f\| \right\|_{\infty,\mathbb{R}_-} + \frac{(b - x_2)^{\frac{(q(4\alpha-1)+1)}{q}}}{\Gamma(4\alpha)\,(q(4\alpha - 1) + 1)^{\frac{1}{q}}} \left(\sup_{b \in \mathbb{R}_-} \left\| \left\| D_{b-}^{4\alpha} f \right\| \right\|_{p,\mathbb{R}_-} \right) \overset{(b-x_2=:2h)}{=}$$

$$2 \left\| \|f\| \right\|_{\infty,\mathbb{R}_-} + \frac{2^{4\alpha-\frac{1}{p}} h^{4\alpha-\frac{1}{p}}}{\Gamma(4\alpha)\,(q(4\alpha - 1) + 1)^{\frac{1}{q}}} \left(\sup_{b \in \mathbb{R}_-} \left\| \left\| D_{b-}^{4\alpha} f \right\| \right\|_{p,\mathbb{R}_-} \right). \tag{11.51}$$

That is

$$\|B\| \leq 2 \left\| \|f\| \right\|_{\infty,\mathbb{R}_-} + \frac{2^{4\alpha-\frac{1}{p}} h^{4\alpha-\frac{1}{p}}}{\Gamma(4\alpha)\,(q(4\alpha - 1) + 1)^{\frac{1}{q}}} \left(\sup_{b \in \mathbb{R}_-} \left\| \left\| D_{b-}^{4\alpha} f \right\| \right\|_{p,\mathbb{R}_-} \right), \tag{11.52}$$

where $\frac{1}{4p} < \alpha < 1$.

We have assumed that

$$M := \left(\sup_{b \in \mathbb{R}_-} \left\| \left\| D_{b-}^{4\alpha} f \right\| \right\|_{p,\mathbb{R}_-} \right) < \infty. \tag{11.53}$$

For convenience we call

$$c := \Gamma(4\alpha)\,(q(4\alpha - 1) + 1)^{\frac{1}{q}} > 0. \tag{11.54}$$

So we have

$$\begin{aligned} &\|A\| \leq 2 \left\| \|f\| \right\|_{\infty,\mathbb{R}_-} + \frac{h^{4\alpha-\frac{1}{p}}}{c} M, \\ &\text{and} \\ &\|B\| \leq 2 \left\| \|f\| \right\|_{\infty,\mathbb{R}_-} + \frac{2^{4\alpha-\frac{1}{p}} h^{4\alpha-\frac{1}{p}}}{c} M, \end{aligned} \tag{11.55}$$

where $\frac{1}{4p} < \alpha < 1$.

Next we estimate the (11.16)-quantities and we have:

$$\left\| \left(D_{b-}^{2\alpha} f \right)(b) \right\| \leq \frac{1}{D\Gamma(3\alpha + 1)} \left[2^{3\alpha} h^{3\alpha} \|A\| + h^{3\alpha} \|B\| \right] \overset{(11.23)}{=}$$

$$\frac{h^{3\alpha}\Gamma(2\alpha + 1)}{2^{2\alpha} h^{5\alpha} (2^{\alpha} - 1)} \left[2^{3\alpha} \|A\| + \|B\| \right] \overset{(11.55)}{\leq} \tag{11.56}$$

$$\frac{\Gamma(2\alpha + 1)}{2^{2\alpha} (2^{\alpha} - 1) h^{2\alpha}} \left[2^{3\alpha+1} \left\| \|f\| \right\|_{\infty,\mathbb{R}_-} + \frac{2^{3\alpha} h^{4\alpha-\frac{1}{p}}}{c} M + \right.$$

$$2 \left\|\left\| f \right\|\right\|_{\infty,\mathbb{R}_-} + \frac{2^{4\alpha-\frac{1}{p}} h^{4\alpha-\frac{1}{p}}}{c} M \Bigg] =$$

$$\frac{\Gamma(2\alpha+1)}{2^{2\alpha}(2^\alpha-1)h^{2\alpha}} \left[\left(2^{3\alpha+1}+2\right) \left\|\left\| f \right\|\right\|_{\infty,\mathbb{R}_-} + \frac{\left(2^{3\alpha}+2^{4\alpha-\frac{1}{p}}\right)}{c} M h^{4\alpha-\frac{1}{p}} \right] =$$

$$\frac{\Gamma(2\alpha+1)}{2^{2\alpha}(2^\alpha-1)} \left[\frac{\left(2^{3\alpha+1}+2\right) \left\|\left\| f \right\|\right\|_{\infty,\mathbb{R}_-}}{h^{2\alpha}} + \frac{\left(2^{3\alpha}+2^{4\alpha-\frac{1}{p}}\right)}{c} M h^{2\alpha-\frac{1}{p}} \right] =$$

$$\frac{2^\alpha \Gamma(2\alpha+1)}{(2^\alpha-1)} \left[\frac{2\left(1+2^{-3\alpha}\right) \left\|\left\| f \right\|\right\|_{\infty,\mathbb{R}_-}}{h^{2\alpha}} + \frac{\left(1+2^{\alpha-\frac{1}{p}}\right) M}{c} h^{2\alpha-\frac{1}{p}} \right]. \quad (11.57)$$

That is

$$\left\| \left(D_{b-}^{2\alpha} f\right)(b) \right\| \leq$$

$$\left(\frac{2^\alpha \Gamma(2\alpha+1)}{2^\alpha-1}\right) \left[\frac{2\left(1+2^{-3\alpha}\right) \left\|\left\| f \right\|\right\|_{\infty,\mathbb{R}_-}}{h^{2\alpha}} + \frac{\left(1+2^{\alpha-\frac{1}{p}}\right) M}{c} h^{2\alpha-\frac{1}{p}} \right], \quad (11.58)$$

$\forall\, b \in \mathbb{R}_-, \forall\, h > 0$.

I.e. it holds

$$\sup_{b\in\mathbb{R}_-} \left\| \left(D_{b-}^{2\alpha} f\right)(b) \right\| \leq \left(\frac{2^\alpha \Gamma(2\alpha+1)}{2^\alpha-1}\right)$$

$$\left[\frac{2\left(1+2^{-3\alpha}\right) \left\|\left\| f \right\|\right\|_{\infty,\mathbb{R}_-}}{h^{2\alpha}} + \frac{\left(1+2^{\alpha-\frac{1}{p}}\right) M}{c} h^{2\alpha-\frac{1}{p}} \right], \quad (11.59)$$

$\forall\, h > 0$, under $\frac{1}{4p} < \alpha < 1$.

Again from (11.16) we get

$$\left\| \left(D_{b-}^{3\alpha} f\right)(b) \right\| \leq \frac{1}{\Gamma(2\alpha+1) D} \left[h^{2\alpha} \|B\| + 2^{2\alpha} h^{2\alpha} \|A\| \right] =$$

$$\frac{h^{2\alpha} \Gamma(3\alpha+1)}{2^{2\alpha} h^{5\alpha} (2^\alpha-1)} \left[\|B\| + 2^{2\alpha} \|A\| \right] \overset{(11.55)}{\leq} \quad (11.60)$$

$$\left(\frac{h^{2\alpha} \Gamma(3\alpha+1)}{2^{2\alpha} h^{5\alpha} (2^\alpha-1)}\right) \Bigg[2 \left\|\left\| f \right\|\right\|_{\infty,\mathbb{R}_-} + \frac{2^{4\alpha-\frac{1}{p}} h^{4\alpha-\frac{1}{p}}}{c} M +$$

$$2^{2\alpha+1}\left\|\|f\|\right\|_{\infty,\mathbb{R}_-}+\frac{2^{2\alpha}h^{4\alpha-\frac{1}{p}}}{c}M\Bigg]=$$

$$\left(\frac{\Gamma(3\alpha+1)}{2^{2\alpha}(2^{\alpha}-1)h^{3\alpha}}\right)\left[\left(2+2^{2\alpha+1}\right)\left\|\|f\|\right\|_{\infty,\mathbb{R}_-}+\frac{\left(2^{2\alpha}+2^{4\alpha-\frac{1}{p}}\right)}{c}Mh^{4\alpha-\frac{1}{p}}\right]=$$

$$\frac{\Gamma(3\alpha+1)}{2^{2\alpha}(2^{\alpha}-1)}\left[\frac{\left(2+2^{2\alpha+1}\right)\left\|\|f\|\right\|_{\infty,\mathbb{R}_-}}{h^{3\alpha}}+\frac{\left(2^{2\alpha}+2^{4\alpha-\frac{1}{p}}\right)}{c}Mh^{\alpha-\frac{1}{p}}\right]= \tag{11.61}$$

$$\frac{\Gamma(3\alpha+1)}{(2^{\alpha}-1)}\left[\frac{2\left(1+2^{-2\alpha}\right)\left\|\|f\|\right\|_{\infty,\mathbb{R}_-}}{h^{3\alpha}}+\frac{\left(1+2^{2\alpha-\frac{1}{p}}\right)}{c}Mh^{\alpha-\frac{1}{p}}\right].$$

That is

$$\left\|\left(D_{b-}^{3\alpha}f\right)(b)\right\|\leq\left(\frac{\Gamma(3\alpha+1)}{2^{\alpha}-1}\right)\left[\frac{2\left(1+2^{-2\alpha}\right)\left\|\|f\|\right\|_{\infty,\mathbb{R}_-}}{h^{3\alpha}}+\frac{\left(1+2^{2\alpha-\frac{1}{p}}\right)}{c}Mh^{\alpha-\frac{1}{p}}\right], \tag{11.62}$$

$\forall\, b\in\mathbb{R}_-$, $\forall\, h>0$.

I.e. it holds

$$\sup_{b\in\mathbb{R}_-}\left\|\left(D_{b-}^{3\alpha}f\right)(b)\right\|\leq\left(\frac{\Gamma(3\alpha+1)}{2^{\alpha}-1}\right)$$

$$\left[\frac{2\left(1+2^{-2\alpha}\right)\left\|\|f\|\right\|_{\infty,\mathbb{R}_-}}{h^{3\alpha}}+\frac{\left(1+2^{2\alpha-\frac{1}{p}}\right)}{c}Mh^{\alpha-\frac{1}{p}}\right], \tag{11.63}$$

$\forall\, h>0$, $\frac{1}{4p}<\alpha<1$.

Call

$$\mu:=2\left(1+2^{-3\alpha}\right)\left\|\|f\|\right\|_{\infty,\mathbb{R}_-},$$

$$\theta:=\frac{\left(1+2^{\alpha-\frac{1}{p}}\right)M}{c}, \tag{11.64}$$

both are greater than zero.

We consider the function

$$y(h)=\mu h^{-2\alpha}+\theta h^{2\alpha-\frac{1}{p}},\quad \forall\, h>0. \tag{11.65}$$

We have

$$y'(h) = -2\alpha\mu h^{-2\alpha-1} + \left(2\alpha - \frac{1}{p}\right)\theta h^{2\alpha-\frac{1}{p}-1} = 0, \tag{11.66}$$

then

$$\left(2\alpha - \frac{1}{p}\right)\theta h^{2\alpha-\frac{1}{p}-1} = 2\alpha\mu h^{-2\alpha-1},$$

i.e.

$$\left(2\alpha - \frac{1}{p}\right)\theta h^{4\alpha-\frac{1}{p}} = 2\alpha\mu,$$

with a unique solution

$$h_0 := h_{crit.no.} = \left(\frac{2\alpha\mu}{\left(2\alpha - \frac{1}{p}\right)\theta}\right)^{\frac{1}{4\alpha-\frac{1}{p}}} \tag{11.67}$$

(assuming $\frac{1}{2p} < \alpha < 1$).

We have that

$$y''(h) = 2\alpha(2\alpha+1)\mu h^{-2\alpha-2} + \left(2\alpha - \frac{1}{p}\right)\left(2\alpha - \frac{1}{p} - 1\right)\theta h^{2\alpha-\frac{1}{p}-2}. \tag{11.68}$$

We see that

$$y''(h_0) = 2\alpha(2\alpha+1)\mu\left(\frac{2\alpha\mu}{\left(2\alpha - \frac{1}{p}\right)\theta}\right)^{\frac{-2\alpha-2}{4\alpha-\frac{1}{p}}} +$$

$$\left(2\alpha - \frac{1}{p}\right)\left(2\alpha - \frac{1}{p} - 1\right)\theta\left(\frac{2\alpha\mu}{\left(2\alpha - \frac{1}{p}\right)\theta}\right)^{\frac{2\alpha-\frac{1}{p}-2}{4\alpha-\frac{1}{p}}}$$

$$= 2\alpha\mu\left(\frac{\left(2\alpha - \frac{1}{p}\right)\theta}{2\alpha\mu}\right)^{\left(\frac{2(\alpha+1)}{4\alpha-\frac{1}{p}}\right)}\left(4\alpha - \frac{1}{p}\right) > 0. \tag{11.69}$$

Therefore y has a global minimum at

$$h_0 = \left(\frac{2\alpha\mu}{\left(2\alpha - \frac{1}{p}\right)\theta}\right)^{\frac{1}{4\alpha-\frac{1}{p}}},$$

which is

$$y(h_0) = \mu \left(\frac{2\alpha\mu}{\left(2\alpha - \frac{1}{p}\right)\theta} \right)^{\frac{-2\alpha}{4\alpha - \frac{1}{p}}} + \theta \left(\frac{2\alpha\mu}{\left(2\alpha - \frac{1}{p}\right)\theta} \right)^{\frac{2\alpha - \frac{1}{p}}{4\alpha - \frac{1}{p}}} = \quad (11.70)$$

$$\frac{\left(4\alpha - \frac{1}{p}\right)}{(2\alpha)^{\left(\frac{2\alpha}{4\alpha - \frac{1}{p}}\right)}} \left(2\alpha - \frac{1}{p}\right)^{\frac{-2\alpha + \frac{1}{p}}{4\alpha - \frac{1}{p}}} \mu^{\left(\frac{2\alpha - \frac{1}{p}}{4\alpha - \frac{1}{p}}\right)} \theta^{\left(\frac{2\alpha}{4\alpha - \frac{1}{p}}\right)}. \quad (11.71)$$

That is

$$y(h_0) = \frac{\left(4\alpha - \frac{1}{p}\right)\left(2\alpha - \frac{1}{p}\right)^{\left(\frac{-2\alpha + \frac{1}{p}}{4\alpha - \frac{1}{p}}\right)}}{(2\alpha)^{\left(\frac{2\alpha}{4\alpha - \frac{1}{p}}\right)}} \mu^{\left(\frac{2\alpha - \frac{1}{p}}{4\alpha - \frac{1}{p}}\right)} \theta^{\left(\frac{2\alpha}{4\alpha - \frac{1}{p}}\right)}. \quad (11.72)$$

Therefore we derive (see (11.59))

$$\sup_{b \in \mathbb{R}_-} \left\| \left(D_{b-}^{2\alpha} f\right)(b) \right\| \leq \left(\frac{2^{\alpha}\Gamma(2\alpha)}{2^{\alpha} - 1} \right) \left(\frac{2\alpha}{2\alpha - \frac{1}{p}} \right)^{\frac{\left(2\alpha - \frac{1}{p}\right)}{\left(4\alpha - \frac{1}{p}\right)}} \left(4\alpha - \frac{1}{p}\right)$$

$$\left(2\left(1 + 2^{-3\alpha}\right)\right)^{\left(\frac{2\alpha - \frac{1}{p}}{4\alpha - \frac{1}{p}}\right)} \left(\frac{\left(1 + 2^{\alpha - \frac{1}{p}}\right)}{\Gamma(4\alpha)\left(q(4\alpha - 1) + 1\right)^{\frac{1}{q}}} \right)^{\left(\frac{2\alpha}{4\alpha - \frac{1}{p}}\right)} \||| f |||_{\infty, \mathbb{R}_-}^{\left(\frac{2\alpha - \frac{1}{p}}{4\alpha - \frac{1}{p}}\right)}$$

$$\left(\sup_{b \in \mathbb{R}_-} \left\| \left\| D_{b-}^{4\alpha} f \right\| \right\|_{p, \mathbb{R}_-} \right)^{\left(\frac{2\alpha}{4\alpha - \frac{1}{p}}\right)} < \infty, \quad (11.73)$$

where $\frac{1}{2p} < \alpha < 1$.

Call

$$\xi := 2\left(1 + 2^{-2\alpha}\right) \||| f |||_{\infty, \mathbb{R}_-},$$

$$\psi := \frac{\left(1 + 2^{2\alpha - \frac{1}{p}}\right)}{c} M, \quad (11.74)$$

both are greater than zero.

We consider the function

$$\gamma(h) := \xi h^{-3\alpha} + \psi h^{\alpha - \frac{1}{p}}, \quad \forall\, h > 0. \quad (11.75)$$

We have

$$\gamma'(h) = -3\alpha\xi h^{-3\alpha-1} + \left(\alpha - \frac{1}{p}\right)\psi h^{\alpha-\frac{1}{p}-1} = 0,$$

then

$$\left(\alpha - \frac{1}{p}\right)\psi h^{\alpha-\frac{1}{p}-1} = 3\alpha\xi h^{-3\alpha-1}$$

and

$$\left(\alpha - \frac{1}{p}\right)\psi h^{4\alpha-\frac{1}{p}} = 3\alpha\xi,$$

with unique solution

$$h_0 := h_{crit.no.} = \left(\frac{3\alpha\xi}{\left(\alpha - \frac{1}{p}\right)\psi}\right)^{\frac{1}{4\alpha-\frac{1}{p}}} \tag{11.76}$$

(assuming $\frac{1}{p} < \alpha < 1$).

We have that

$$\gamma''(h) = 3\alpha(3\alpha+1)\xi h^{-3\alpha-2} + \left(\alpha - \frac{1}{p}\right)\left(\alpha - \frac{1}{p} - 1\right)\psi h^{\alpha-\frac{1}{p}-2}. \tag{11.77}$$

We observe that

$$\gamma''(h_0) = 3\alpha(3\alpha+1)\xi\left(\frac{3\alpha\xi}{\left(\alpha - \frac{1}{p}\right)\psi}\right)^{\left(\frac{-3\alpha-2}{4\alpha-\frac{1}{p}}\right)} +$$

$$\left(\alpha - \frac{1}{p}\right)\left(\alpha - \frac{1}{p} - 1\right)\psi\left(\frac{3\alpha\xi}{\left(\alpha - \frac{1}{p}\right)\psi}\right)^{\left(\frac{\alpha-\frac{1}{p}-2}{4\alpha-\frac{1}{p}}\right)} =$$

$$3\alpha\xi\left(\frac{3\alpha\xi}{\left(\alpha - \frac{1}{p}\right)\psi}\right)^{\left(\frac{-3\alpha-2}{4\alpha-\frac{1}{p}}\right)}\left(4\alpha - \frac{1}{p}\right) > 0. \tag{11.78}$$

Therefore y has a global minimum at

$$h_0 = \left(\frac{3\alpha\xi}{\left(\alpha - \frac{1}{p}\right)\psi}\right)^{\frac{1}{4\alpha-\frac{1}{p}}},$$

which is

$$\gamma(h_0) = \xi h_0^{-3\alpha} + \psi h_0^{\alpha-\frac{1}{p}} = h_0^{-3\alpha}\left(\xi + \psi h_0^{4\alpha-\frac{1}{p}}\right) = \tag{11.79}$$

$$\left(\frac{3\alpha\xi}{\left(\alpha-\frac{1}{p}\right)\psi}\right)^{\left(\frac{-3\alpha}{4\alpha-\frac{1}{p}}\right)}\left(\xi + \psi\left(\frac{3\alpha\xi}{\left(\alpha-\frac{1}{p}\right)\psi}\right)\right) =$$

$$\xi\left(\frac{3\alpha\xi}{\left(\alpha-\frac{1}{p}\right)\psi}\right)^{\left(\frac{-3\alpha}{4\alpha-\frac{1}{p}}\right)}\left(\frac{4\alpha-\frac{1}{p}}{\alpha-\frac{1}{p}}\right).$$

That is

$$\gamma(h_0) = \frac{\left(4\alpha-\frac{1}{p}\right)}{\left(\alpha-\frac{1}{p}\right)^{\left(\frac{\alpha-\frac{1}{p}}{4\alpha-\frac{1}{p}}\right)}(3\alpha)^{\left(\frac{3\alpha}{4\alpha-\frac{1}{p}}\right)}}\xi^{\left(\frac{\alpha-\frac{1}{p}}{4\alpha-\frac{1}{p}}\right)}\psi^{\left(\frac{3\alpha}{4\alpha-\frac{1}{p}}\right)}. \tag{11.80}$$

We have proved that (see (11.63))

$$\sup_{b\in\mathbb{R}_-}\left\|\left(D_{b-}^{3\alpha}f\right)(b)\right\| \leq \left(\frac{\left(4\alpha-\frac{1}{p}\right)\Gamma(3\alpha)}{2^{\alpha}-1}\right)\left(\frac{3\alpha}{\alpha-\frac{1}{p}}\right)^{\left(\frac{\alpha-\frac{1}{p}}{4\alpha-\frac{1}{p}}\right)} \tag{11.81}$$

$$\left(2\left(1+2^{-2\alpha}\right)\right)^{\left(\frac{\alpha-\frac{1}{p}}{4\alpha-\frac{1}{p}}\right)}\left(\frac{\left(1+2^{2\alpha-\frac{1}{p}}\right)}{\Gamma(4\alpha)\left(q(4\alpha-1)+1\right)^{\frac{1}{q}}}\right)^{\left(\frac{3\alpha}{4\alpha-\frac{1}{p}}\right)}\|\|f\|\|_{\infty,\mathbb{R}_-}^{\left(\frac{\alpha-\frac{1}{p}}{4\alpha-\frac{1}{p}}\right)}$$

$$\left(\sup_{b\in\mathbb{R}_-}\left\|\left\|D_{b-}^{4\alpha}f\right\|\right\|_{p,\mathbb{R}_-}\right)^{\left(\frac{3\alpha}{4\alpha-\frac{1}{p}}\right)} < \infty,$$

where $\frac{1}{p} < \alpha < 1$.

The theorem is proved. ■

We give an application when $\alpha = \frac{1}{2}$.

Corollary 11.6 *Let $f \in C^1(\mathbb{R}_-, X)$ with $\|\|f\|\|_{\infty,\mathbb{R}_-}$, $\left\|\left\|f'\right\|\right\|_{\infty,\mathbb{R}_-} < \infty$, where $(X, \|\cdot\|)$ is a Banach space. For k=1, 2, 3, we assume that $D_{b-}^{k\frac{1}{2}}f \in C^1((-\infty, b], X)$ and $D_{b-}^{4\frac{1}{2}}f \in C((-\infty, b], X)$, $\forall\, b \in \mathbb{R}_-$. We further assume that*

$$K := \left\| \left\| D_{b-}^{4\frac{1}{2}} f(t) \right\| \right\|_{\infty, \mathbb{R}_-^2} < \infty, \tag{11.82}$$

where $(b,t) \in \mathbb{R}_-^2$.

Then

$$\sup_{b \in \mathbb{R}_-} \left\| \left(D_{b-}^{2\frac{1}{2}} f \right)(b) \right\| \leq \left(\frac{\sqrt{12+6\sqrt{2}}}{\sqrt{2}-1} \right) \| \| f \| \|_{\infty, \mathbb{R}_-}^{\frac{1}{2}} \left(\left\| \left\| D_{b-}^{4\frac{1}{2}} f(t) \right\| \right\|_{\infty, \mathbb{R}_-^2} \right)^{\frac{1}{2}} < \infty, \tag{11.83}$$

and

$$\sup_{b \in \mathbb{R}_-} \left\| \left(D_{b-}^{3\frac{1}{2}} f \right)(b) \right\| \leq$$

$$\left(\frac{9\sqrt{\pi}}{\left(2-\sqrt{2}\right) \sqrt[4]{2} \left(\sqrt[4]{3}\right)^3} \right) \| \| f \| \|_{\infty, \mathbb{R}_-}^{\frac{1}{4}} \left(\left\| \left\| D_{b-}^{4\frac{1}{2}} f(t) \right\| \right\|_{\infty, \mathbb{R}_-^2} \right)^{\frac{3}{4}} < \infty. \tag{11.84}$$

That is $\sup_{b \in \mathbb{R}_-} \left\| \left(D_{b-}^{2\frac{1}{2}} f \right)(b) \right\|, \sup_{b \in \mathbb{R}_-} \left\| \left(D_{b-}^{3\frac{1}{2}} f \right)(b) \right\| < \infty.$

Proof By Theorem 11.4. ∎

Remark 11.7 All of the above results are also true for any half line $(-\infty, A]$, $A \in \mathbb{R}$, instead of only $\mathbb{R}_-$, just replace all $\mathbb{R}_-$ by $(-\infty, A]$.

References

1. Aglic Aljinovic, A., Marangunic, L., Pecaric, J.: On Landau type inequalities via Ostrowski inequalities. Nonlinear Funct. Anal. Appl. **10**(4), 565–579 (2005)
2. Anastassiou, G.: Fractional Differentiation Inequalities. Research Monograph. Springer, New York (2009)
3. Anastassiou, G.A.: Advances on Fractional Inequalities. Springer, New York (2011)
4. Anastassiou, G.A.: Intelligent Computations: Abstract Fractional Calculus, Inequalities, Approximations. Springer, Heidelberg (2018)
5. Anastassiou, G.A.: Iterated abstract right side fractional Landau inequalities. Submitted (2020)
6. Barnett, N.S., Dragomir, S.S.: Some Landau type inequalities for functions whose derivatives are of locally bounded variation. Tamkang J. Math. **37**(4), 301–308 (2006)
7. Ditzian, Z.: Remarks, questions and conjectures on Landau-Kolmogorov-type inequalities. Math. Inequal. Appl. **3**, 15–24 (2000)
8. Hardy, G.H., Littlewood, J.E.: Some integral inequalities connected with the calculus of variations. Q. J. Math. Oxford Ser. **3**, 241–252 (1932)
9. Hardy, G.H., Landau, E., Littlewood, J.E.: Some inequalities satisfied by the integrals or derivatives of real or analytic functions. Math. Z. **39**, 677–695 (1935)
10. Kallman, R.R., Rota, G.C.: On the inequality $\|f'\|^2 \leq 4\|f \cdot \|f''\|$. In: Shisha, O. (ed.) Inequalities, vol. II, pp. 187–192. Academic, New York (1970)

11. Landau, E.: Einige Ungleichungen für zweimal differentzierban funktionen. Proc. Lond. Math. Soc. **13**, 43–49 (1913)
12. Mikusinski, J.: The Bochner Integral. Academic, New York (1978)
13. Shilov, G.E.: Elementary Functional Analysis. Dover Publications Inc., New York (1996)

Chapter 12
Iterated Generalized Right Side Fractional Landau Inequalities

We give uniform and L_p Caputo–Bochner abstract iterated generalized right fractional Landau inequalities over $\mathbb{R}_-$. These estimate the size of second and third sequential abstract generalized right fractional derivatives of a Banach space valued function over $\mathbb{R}_-$. We give an application when the basic fractional order is $\frac{1}{2}$. It follows [5].

12.1 Introduction

Let $p \in [1, \infty]$, $I = \mathbb{R}_+$ or $I = \mathbb{R}$ and $f : I \to \mathbb{R}$ is twice differentiable with $f, f'' \in L_p(I)$, then $f' \in L_p(I)$. Moreover, there exists a constant $C_p(I) > 0$ independent of f, such that

$$\left\| f' \right\|_{p,I} \leq C_p(I) \left\| f \right\|_{p,I}^{\frac{1}{2}} \left\| f'' \right\|_{p,I}^{\frac{1}{2}}, \tag{12.1}$$

where $\|\cdot\|_{p,I}$ is the p-norm on the interval I, see [1, 6].

The research on these inequalities started by Landau [11] in 1913. For the case of $p = \infty$ he proved that

$$C_\infty(\mathbb{R}_+) = 2 \text{ and } C_\infty(\mathbb{R}) = \sqrt{2}, \tag{12.2}$$

are the best constants in (12.1).

In 1932, Hardy and Littlewood [8] proved (12.1) for $p = 2$, with the best constants

$$C_2(\mathbb{R}_+) = \sqrt{2} \text{ and } C_2(\mathbb{R}) = 1. \tag{12.3}$$

In 1935, Hardy, Landau and Littlewood [9] showed that the best constants $C_p(\mathbb{R}_+)$ in (12.1) satisfies the estimate

G. A. Anastassiou, *Constructive Fractional Analysis with Applications*, Studies in Systems, Decision and Control 362,
https://doi.org/10.1007/978-3-030-71481-9_12

$$C_p(\mathbb{R}_+) \leq 2, \text{ for } p \in [1, \infty), \tag{12.4}$$

which yields $C_p(\mathbb{R}) \leq 2$ for $p \in [1, \infty)$.

In fact, in [7, 10] was shown that $C_p(\mathbb{R}) \leq \sqrt{2}$.

We need the following concept from abstract fractional calculus.

Our integral next is of Bochner type [12].

We need

Definition 12.1 ([4], *p. 105*) Let $[a, b] \subset \mathbb{R}$, $(X, \|\cdot\|)$ a Banach space, $g \in C^1([a, b])$ and increasing, $f \in C([a, b], X)$, $\nu > 0$.

We define the right Riemann–Liouville generalized fractional Bochner integral operator

$$\left(J_{b-;g}^{\nu} f\right)(x) := \frac{1}{\Gamma(\nu)} \int_x^b (g(z) - g(x))^{\nu-1} g'(z) f(z) dz, \tag{12.5}$$

$\forall x \in [a, b]$, where Γ is the gamma function.

The last integral is of Bochner type. Since $f \in C([a, b], X)$, then $f \in L_\infty([a, b], X)$. By Theorem 4.11, p. 101, [4], we get that $\left(J_{b-;g}^{\nu} f\right) \in C([a, b], X)$. Above we set $J_{b-;g}^{0} f := f$ and see that $\left(J_{b-;g}^{\nu} f\right)(b) = 0$.

We also need

Definition 12.2 ([4], *p. 107*) Let $\alpha > 0$, $\lceil \alpha \rceil = n$, $\lceil \cdot \rceil$ the ceiling of the number. Let $f \in C^n([a, b], X)$, where $[a, b] \subset \mathbb{R}$, and $(X, \|\cdot\|)$ is a Banach space. Let $g \in C^1([a, b])$, strictly increasing, such that $g^{-1} \in C^n([g(a), g(b)])$.

We define the right generalized g-fractional derivative X-valued of f of order α as follows:

$$\left(D_{b-;g}^{\alpha} f\right)(x) := \frac{(-1)^n}{\Gamma(n-\alpha)} \int_x^b (g(t) - g(x))^{n-\alpha-1} g'(t) \left(f \circ g^{-1}\right)^{(n)} (g(t)) dt, \tag{12.6}$$

$\forall x \in [a, b]$. The last integral is of Bochner type.

Ordinary vector valued derivative is as in [13], similar to numerical one.

If $\alpha \notin \mathbb{N}$, by Theorem 4.11, p. 101, [4], we have that $\left(D_{b-;g}^{\alpha} f\right) \in C([a, b], X)$.

We see that

$$\left(J_{b;g}^{n-\alpha}\left((-1)^n \left(f \circ g^{-1}\right)^{(n)} \circ g\right)\right)(x) = \left(D_{b-;g}^{\alpha} f\right)(x), \quad \forall x \in [a, b]. \tag{12.7}$$

We set

$$D_{b-;g}^{n} f(x) := (-1)^n \left(\left(f \circ g^{-1}\right)^n \circ g\right)(x) \in C([a, b], X), \quad n \in \mathbb{N}, \tag{12.8}$$

$$D_{b-;g}^{0} f(x) = f(x), \quad \forall x \in [a, b].$$

When $g = id$, then

$$D_{b-;g}^{\alpha} f = D_{b-;id}^{\alpha} f = D_{b-}^{\alpha} f, \tag{12.9}$$

the usual left X-valued Caputo fractional derivative, see [4], Chap. 2.

By convention we suppose that

$$\left(D_{x_0-;g}^{\alpha} f\right)(x) = 0, \text{ for } x > x_0, \tag{12.10}$$

for any $x, x_0 \in [a, b]$.

Denote the sequential (also called iterated) generalized left fractional derivative by

$$D_{b-;g}^{n\alpha} := D_{b-;g}^{\alpha} D_{b-;g}^{\alpha} ... D_{b-;g}^{\alpha} \quad (n \text{ times}), n \in \mathbb{N}. \tag{12.11}$$

We need the following g-right generalized modified X-valued Taylor's formula.

Theorem 12.3 ([4], p. 120) *Let* $0 < \alpha \leq 1$, $n \in \mathbb{N}$, $f \in C^1([a, b], X)$, $g \in C^1([a, b])$ *strictly increasing, such that* $g^{-1} \in C^1([g(a), g(b)])$. *Let* $F_k := D_{b-;g}^{k\alpha} f$, $k = 1, ..., n$, *that fulfill* $F_k \in C^1([a, b], X)$, *and* $F_{n+1} \in C([a, b], X)$.

Then

$$f(x) = \sum_{i=0}^{n} \frac{(g(b) - g(x))^{i\alpha}}{\Gamma(i\alpha + 1)} \left(D_{b-;g}^{i\alpha} f\right)(b) + \tag{12.12}$$

$$\frac{1}{\Gamma((n+1)\alpha)} \int_x^b (g(t) - g(x))^{(n+1)\alpha - 1} g'(t) \left(D_{b-;g}^{(n+1)\alpha} f\right)(t)\, dt,$$

$\forall\, x \in [a, b]$.

We make

Remark 12.4 (*to Theorem* 12.3) When $0 < \alpha < 1$, by (12.6) we get

$$\left(D_{b-;g}^{\alpha} f\right)(x) = \frac{-1}{\Gamma(1-\alpha)} \int_x^b (g(t) - g(x))^{-\alpha} g'(t) \left(f \circ g^{-1}\right)'(g(t))\, dt, \tag{12.13}$$

$\forall\, x \in [a, b]$.

Hence

$$\left\|\left(D_{b-;g}^{\alpha} f\right)(x)\right\| \leq \frac{1}{\Gamma(1-\alpha)} \int_x^b (g(t) - g(x))^{-\alpha} g'(t) \left\|\left(f \circ g^{-1}\right)'(g(t))\right\| dt$$

$$\leq \frac{\left\|\left\|\left(f \circ g^{-1}\right)' \circ g\right\|\right\|_{\infty,[a,b]}}{\Gamma(1-\alpha)} \left(\int_x^b (g(t) - g(x))^{-\alpha} g'(t)\, dt\right) \tag{12.14}$$

$$= \frac{\left\| \left\| \left(f \circ g^{-1} \right)' \circ g \right\| \right\|_{\infty,[a,b]}}{\Gamma(2-\alpha)} (g(b) - g(x))^{1-\alpha}.$$

That is

$$\left\| \left(D_{b-;g}^{\alpha} f \right)(x) \right\| \leq \frac{\left\| \left\| \left(f \circ g^{-1} \right)' \circ g \right\| \right\|_{\infty,[a,b]}}{\Gamma(2-\alpha)} (g(b) - g(x))^{1-\alpha} < \infty, \quad (12.15)$$

$\forall\, x \in [a, b]$, $0 < \alpha < 1$.

Hence it holds

$$\left\| \left(D_{b-;g}^{\alpha} f \right)(b) \right\| = 0,$$

i.e.

$$\left(D_{b-;g}^{\alpha} f \right)(b) = 0, \quad (12.16)$$

when $0 < \alpha < 1$.

The author has already done an extensive amount of work on fractional Landau inequalities, see [3], and on abstract fractional Landau inequalities, see [4]. However there the proving methods came out of applications of fractional Ostrowski inequalities [2, 4]. Usually there the domains where $[A, +\infty)$ or $(-\infty, B]$, with $A, B \in \mathbb{R}$ and in one mixed case the domain was all of $\mathbb{R}$.

In this chapter with less assumptions we establish uniform and L_p type right Caputo–Bochner abstract sequential generalized fractional Landau inequalities over $\mathbb{R}_-$. The method of proving is based on right Caputo–Bochner sequential generalized fractional Taylor's formula with integral remainder, see Theorem 12.3.

We give also an application for $\alpha = \frac{1}{2}$. Clearly we are also inspired by [3, 4].

12.2 Main Results

We present the following abstract sequential generalized fractional Landau inequalities over $\mathbb{R}_-$.

Theorem 12.5 *Let $g \in C^1(\mathbb{R}_-)$ strictly increasing, with $g^{-1} \in C^1(g(\mathbb{R}_-))$. Let $0 < \alpha < 1$, $f \in C^1(\mathbb{R}_-, X)$ with $\| \|f\| \|_{\infty,\mathbb{R}_-}$, $\left\| \left\| \left(f \circ g^{-1} \right)' \circ g \right\| \right\|_{\infty,\mathbb{R}_-} < \infty$. For $k = 1, 2, 3$, we assume that $D_{b-;g}^{k\alpha} f \in C^1((-\infty, b], X)$ and $D_{b-;g}^{4\alpha} f \in C((-\infty, b], X)$, $\forall\, b \in \mathbb{R}_-$. We further assume that*

$$\overline{K_g} := \left\| \left\| D_{b-;g}^{4\alpha} f(t) \right\| \right\|_{\infty,\mathbb{R}_-^2} < \infty, \quad (12.17)$$

where $(b, t) \in \mathbb{R}_-^2$.

Then

$$\sup_{b\in\mathbb{R}_-}\left\|\left(D_{b-;g}^{2\alpha}f\right)(b)\right\|\leq\frac{\Gamma(2\alpha+1)}{2^{2\alpha-1}(2^{\alpha}-1)}\sqrt{\frac{2^{3\alpha+1}\left(2^{3\alpha}+1\right)(2^{\alpha}+1)}{\Gamma(4\alpha+1)}\left\|\|f\|\right\|_{\infty,\mathbb{R}_-}\overline{K_g}}, \tag{12.18}$$

and

$$\sup_{b\in\mathbb{R}_-}\left\|\left(D_{b-;g}^{3\alpha}f\right)(b)\right\|\leq\frac{4\sqrt[4]{2}\Gamma(3\alpha+1)(\Gamma(4\alpha+1))^{-\frac{3}{4}}\left(2^{2\alpha}+1\right)}{\left(\sqrt[4]{3}\right)^{3}\left(\sqrt{2}\right)^{\alpha}(2^{\alpha}-1)}\left\|\|f\|\right\|_{\infty,\mathbb{R}_-}^{\frac{1}{4}}\overline{K_g}^{\frac{3}{4}}. \tag{12.19}$$

That is $\sup_{b\in\mathbb{R}_-}\left\|\left(D_{b-;g}^{2\alpha}f\right)(b)\right\|,\ \sup_{b\in\mathbb{R}_-}\left\|\left(D_{b-;g}^{3\alpha}f\right)(b)\right\|<\infty.$

Proof We notice easily again here that $\left(D_{b-;g}^{\alpha}f\right)(b)=0,\ \forall\, b\in\mathbb{R}_-$.

We make use of Theorem 12.3 for $0<\alpha<1$ and $n=3$, applied for any $b\in\mathbb{R}_-$ and $a=-\infty$.

Momentarily we fix $b\in\mathbb{R}_-$. Let $x_2<x_1<b$, then $g(x_2)<g(x_1)<g(b)$, and

$$f(x_1)-f(b)=\frac{(g(b)-g(x_1))^{2\alpha}}{\Gamma(2\alpha+1)}\left(D_{b-;g}^{2\alpha}f\right)(b)+\frac{(g(b)-g(x_1))^{3\alpha}}{\Gamma(3\alpha+1)}\left(D_{b-;g}^{3\alpha}f\right)(b)$$

$$+\frac{1}{\Gamma(4\alpha)}\int_{x_1}^{b}(g(t)-g(x_1))^{4\alpha-1}g'(t)\left(D_{b-;g}^{4\alpha}f\right)(t)\,dt, \tag{12.20}$$

and

$$f(x_2)-f(b)=\frac{(g(b)-g(x_2))^{2\alpha}}{\Gamma(2\alpha+1)}\left(D_{b-;g}^{2\alpha}f\right)(b)+\frac{(g(b)-g(x_2))^{3\alpha}}{\Gamma(3\alpha+1)}\left(D_{b-;g}^{3\alpha}f\right)(b) \tag{12.21}$$

$$+\frac{1}{\Gamma(4\alpha)}\int_{x_2}^{b}(g(t)-g(x_2))^{4\alpha-1}g'(t)\left(D_{b-;g}^{4\alpha}f\right)(t)\,dt.$$

That is

$$\frac{(g(b)-g(x_1))^{2\alpha}}{\Gamma(2\alpha+1)}\left(D_{b-;g}^{2\alpha}f\right)(b)+\frac{(g(b)-g(x_1))^{3\alpha}}{\Gamma(3\alpha+1)}\left(D_{b-;g}^{3\alpha}f\right)(b)= \tag{12.22}$$

$$f(x_1)-f(b)-\frac{1}{\Gamma(4\alpha)}\int_{x_1}^{b}(g(t)-g(x_1))^{4\alpha-1}g'(t)\left(D_{b-;g}^{4\alpha}f\right)(t)\,dt=:A,$$

and

$$\frac{(g(b)-g(x_2))^{2\alpha}}{\Gamma(2\alpha+1)}\left(D_{b-;g}^{2\alpha}f\right)(b)+\frac{(g(b)-g(x_2))^{3\alpha}}{\Gamma(3\alpha+1)}\left(D_{b-;g}^{3\alpha}f\right)(b)= \tag{12.23}$$

$$f(x_2) - f(b) - \frac{1}{\Gamma(4\alpha)} \int_{x_2}^{b} (g(t) - g(x_2))^{4\alpha-1} g'(t) \left(D_{b-;g}^{4\alpha} f\right)(t)\, dt =: B.$$

We are solving the above system of two equations with two unknowns $\left(D_{b-;g}^{2\alpha} f\right)(b)$, $\left(D_{b-;g}^{3\alpha} f\right)(b)$.

The main determinant of system is

$$D := \begin{vmatrix} \frac{(g(b)-g(x_1))^{2\alpha}}{\Gamma(2\alpha+1)} & \frac{(g(b)-g(x_1))^{3\alpha}}{\Gamma(3\alpha+1)} \\ \frac{(g(b)-g(x_2))^{2\alpha}}{\Gamma(2\alpha+1)} & \frac{(g(b)-g(x_2))^{3\alpha}}{\Gamma(3\alpha+1)} \end{vmatrix} = \frac{1}{\Gamma(2\alpha+1)\Gamma(3\alpha+1)}$$

$$\left[(g(b) - g(x_1))^{2\alpha} (g(b) - g(x_2))^{3\alpha} - (g(b) - g(x_1))^{3\alpha} (g(b) - g(x_2))^{2\alpha}\right] =$$

$$\frac{(g(b) - g(x_1))^{2\alpha} (g(b) - g(x_2))^{2\alpha}}{\Gamma(2\alpha+1)\Gamma(3\alpha+1)} \left[(g(b) - g(x_2))^{\alpha} - (g(b) - g(x_1))^{\alpha}\right] > 0.$$

I.e.

$$D = \frac{(g(b) - g(x_1))^{2\alpha} (g(b) - g(x_2))^{2\alpha}}{\Gamma(2\alpha+1)\Gamma(3\alpha+1)} \left[(g(b) - g(x_2))^{\alpha} - (g(b) - g(x_1))^{\alpha}\right] > 0. \tag{12.24}$$

We obtain the unique solution

$$\left(D_{b-;g}^{2\alpha} f\right)(b) = \frac{\begin{vmatrix} A & \frac{(g(b)-g(x_1))^{3\alpha}}{\Gamma(3\alpha+1)} \\ B & \frac{(g(b)-g(x_2))^{3\alpha}}{\Gamma(3\alpha+1)} \end{vmatrix}}{D},$$

$$\left(D_{b-;g}^{3\alpha} f\right)(b) = \frac{\begin{vmatrix} \frac{(g(b)-g(x_1))^{2\alpha}}{\Gamma(2\alpha+1)} & A \\ \frac{(g(b)-g(x_2))^{2\alpha}}{\Gamma(2\alpha+1)} & B \end{vmatrix}}{D}. \tag{12.25}$$

Therefore we have

$$\left(D_{b-;g}^{2\alpha} f\right)(b) = \frac{\frac{(g(b)-g(x_2))^{3\alpha}}{\Gamma(3\alpha+1)} A - \frac{(g(b)-g(x_1))^{3\alpha}}{\Gamma(3\alpha+1)} B}{D},$$

and

$$\left(D_{b-;g}^{3\alpha} f\right)(b) = \frac{\frac{(g(b)-g(x_1))^{2\alpha}}{\Gamma(2\alpha+1)} B - \frac{(g(b)-g(x_2))^{2\alpha}}{\Gamma(2\alpha+1)} A}{D}. \tag{12.26}$$

We have the following

$$\|A\| = \left\| f(x_1) - f(b) - \frac{1}{\Gamma(4\alpha)} \int_{x_1}^{b} (g(t) - g(x_1))^{4\alpha-1} g'(t) \left(D_{b-;g}^{4\alpha} f\right)(t)\, dt \right\| \leq$$

$$2 \||| f ||\|_{\infty,\mathbb{R}_-} + \frac{\left\| \left\| D_{b-;g}^{4\alpha} f(t) \right\| \right\|_{\infty,\mathbb{R}_-^2}}{\Gamma(4\alpha+1)} (g(b) - g(x_1))^{4\alpha}, \qquad (12.27)$$

under the assumption $\||| f ||\|_{\infty,\mathbb{R}_-} < \infty$.

That is

$$\|A\| \leq 2 \||| f ||\|_{\infty,\mathbb{R}_-} + \frac{\overline{K_g}}{\Gamma(4\alpha+1)} (g(b) - g(x_1))^{4\alpha}, \qquad (12.28)$$

and similarly,

$$\|B\| \leq 2 \||| f ||\|_{\infty,\mathbb{R}_-} + \frac{\overline{K_g}}{\Gamma(4\alpha+1)} (g(b) - g(x_2))^{4\alpha}, \qquad (12.29)$$

where by assumption

$$\overline{K_g} := \left\| \left\| D_{b-;g}^{4\alpha} f(t) \right\| \right\|_{\infty,\mathbb{R}_-^2} < \infty, \qquad (12.30)$$

with $(b, t) \in \mathbb{R}_-^2$.

Consequently we have

$$\left\| \left(D_{b-;g}^{2\alpha} f\right)(b) \right\| \leq \frac{1}{\Gamma(3\alpha+1) D}$$

$$\left[(g(b) - g(x_2))^{3\alpha} \left(2 \||| f ||\|_{\infty,\mathbb{R}_-} + \frac{\overline{K_g}}{\Gamma(4\alpha+1)} (g(b) - g(x_1))^{4\alpha} \right) \right.$$

$$\left. + (g(b) - g(x_1))^{3\alpha} \left(2 \||| f ||\|_{\infty,\mathbb{R}_-} + \frac{\overline{K_g}}{\Gamma(4\alpha+1)} (g(b) - g(x_2))^{4\alpha} \right) \right], \qquad (12.31)$$

and

$$\left\| \left(D_{b-;g}^{3\alpha} f\right)(b) \right\| \leq \frac{1}{\Gamma(2\alpha+1) D}$$

$$\left[(g(b) - g(x_1))^{2\alpha} \left(2 \||| f ||\|_{\infty,\mathbb{R}_-} + \frac{\overline{K_g}}{\Gamma(4\alpha+1)} (g(b) - g(x_2))^{4\alpha} \right) \right.$$

$$+\left(g\left(b\right)-g\left(x_{2}\right)\right)^{2\alpha}\left(2\left\|\|f\|\right\|_{\infty,\mathbb{R}_{-}}+\frac{\overline{K_{g}}}{\Gamma\left(4\alpha+1\right)}\left(g\left(b\right)-g\left(x_{1}\right)\right)^{4\alpha}\right)\Bigg]. \tag{12.32}$$

Set now $g(x_1) := g(b) - h$, $g(x_2) := g(b) - 2h$, where $h > 0$, so that $g(b) - g(x_1) = h$, $g(b) - g(x_2) = 2h$.

Hence we get

$$D = \frac{2^{2\alpha}h^{5\alpha}\left(2^{\alpha}-1\right)}{\Gamma\left(2\alpha+1\right)\Gamma\left(3\alpha+1\right)} > 0. \tag{12.33}$$

Therefore we derive (from (12.26))

$$\left\|\left(D_{b-;g}^{2\alpha}f\right)(b)\right\| \leq \frac{\Gamma\left(2\alpha+1\right)}{2^{2\alpha}h^{5\alpha}\left(2^{\alpha}-1\right)}\Bigg[2^{3\alpha}h^{3\alpha}\left(2\left\|\|f\|\right\|_{\infty,\mathbb{R}_{-}}+\frac{\overline{K_{g}}}{\Gamma\left(4\alpha+1\right)}h^{4\alpha}\right)$$

$$+h^{3\alpha}\left(2\left\|\|f\|\right\|_{\infty,\mathbb{R}_{-}}+\frac{\overline{K_{g}}}{\Gamma\left(4\alpha+1\right)}2^{4\alpha}h^{4\alpha}\right)\Bigg] = \tag{12.34}$$

$$\frac{\Gamma\left(2\alpha+1\right)}{2^{2\alpha}h^{5\alpha}\left(2^{\alpha}-1\right)}\left[2\left\|\|f\|\right\|_{\infty,\mathbb{R}_{-}}\left(2^{3\alpha}+1\right)h^{3\alpha}+\frac{\overline{K_{g}}}{\Gamma\left(4\alpha+1\right)}\left(2^{3\alpha}+2^{4\alpha}\right)h^{7\alpha}\right]$$

$$=\left(\frac{\Gamma\left(2\alpha+1\right)}{2^{2\alpha}\left(2^{\alpha}-1\right)}\right)\left[\frac{2\left(2^{3\alpha}+1\right)\left\|\|f\|\right\|_{\infty,\mathbb{R}_{-}}}{h^{2\alpha}}+\frac{2^{3\alpha}\left(2^{\alpha}+1\right)}{\Gamma\left(4\alpha+1\right)}\overline{K_{g}}h^{2\alpha}\right]. \tag{12.35}$$

That is

$$\left\|\left(D_{b-;g}^{2\alpha}f\right)(b)\right\| \leq \left(\frac{\Gamma\left(2\alpha+1\right)}{2^{2\alpha}\left(2^{\alpha}-1\right)}\right)$$

$$\left[\frac{2\left(2^{3\alpha}+1\right)\left\|\|f\|\right\|_{\infty,\mathbb{R}_{-}}}{h^{2\alpha}}+\frac{2^{3\alpha}\left(2^{\alpha}+1\right)}{\Gamma\left(4\alpha+1\right)}\overline{K_{g}}h^{2\alpha}\right], \tag{12.36}$$

$\forall\, b \in \mathbb{R}_-, \forall\, h > 0$.

I.e. it holds

$$\sup_{b\in\mathbb{R}_{-}}\left\|\left(D_{b-;g}^{2\alpha}f\right)(b)\right\| \leq \left(\frac{\Gamma\left(2\alpha+1\right)}{2^{2\alpha}\left(2^{\alpha}-1\right)}\right)$$

$$\left[\frac{2\left(2^{3\alpha}+1\right)\left\|\|f\|\right\|_{\infty,\mathbb{R}_{-}}}{h^{2\alpha}}+\frac{2^{3\alpha}\left(2^{\alpha}+1\right)}{\Gamma\left(4\alpha+1\right)}\overline{K_{g}}h^{2\alpha}\right] < \infty, \tag{12.37}$$

$\forall\, h > 0$, $0 < \alpha < 1$.

By (12.26) we derive

$$\left\|\left(D_{b-;g}^{3\alpha}f\right)(b)\right\| \leq \frac{\Gamma(3\alpha+1)}{2^{2\alpha}h^{5\alpha}(2^{\alpha}-1)}\left[h^{2\alpha}\left(2\left\|\left\|f\right\|\right\|_{\infty,\mathbb{R}_-}+\frac{\overline{K_g}}{\Gamma(4\alpha+1)}2^{4\alpha}h^{4\alpha}\right)\right.$$

$$\left.+2^{2\alpha}h^{2\alpha}\left(2\left\|\left\|f\right\|\right\|_{\infty,\mathbb{R}_-}+\frac{\overline{K_g}}{\Gamma(4\alpha+1)}h^{4\alpha}\right)\right]=$$

$$\frac{\Gamma(3\alpha+1)}{2^{2\alpha}h^{5\alpha}(2^{\alpha}-1)}\left[2\left\|\left\|f\right\|\right\|_{\infty,\mathbb{R}_-}\left(2^{2\alpha}+1\right)h^{2\alpha}+\frac{\overline{K_g}}{\Gamma(4\alpha+1)}\left(2^{4\alpha}+2^{2\alpha}\right)h^{6\alpha}\right]$$

$$=\left(\frac{\Gamma(3\alpha+1)}{2^{2\alpha}(2^{\alpha}-1)}\right)\left[\frac{2\left(2^{2\alpha}+1\right)\left\|\left\|f\right\|\right\|_{\infty,\mathbb{R}_-}}{h^{3\alpha}}+\frac{2^{2\alpha}\left(2^{2\alpha}+1\right)}{\Gamma(4\alpha+1)}\overline{K_g}h^{\alpha}\right] \tag{12.38}$$

$$=\frac{\Gamma(3\alpha+1)\left(2^{2\alpha}+1\right)}{2^{2\alpha}(2^{\alpha}-1)}\left[\frac{2\left\|\left\|f\right\|\right\|_{\infty,\mathbb{R}_-}}{h^{3\alpha}}+\frac{2^{2\alpha}\overline{K_g}}{\Gamma(4\alpha+1)}h^{\alpha}\right].$$

That is

$$\left\|\left(D_{b-;g}^{3\alpha}f\right)(b)\right\| \leq \frac{\Gamma(3\alpha+1)\left(2^{2\alpha}+1\right)}{2^{2\alpha}(2^{\alpha}-1)}$$

$$\left[\frac{2\left\|\left\|f\right\|\right\|_{\infty,\mathbb{R}_-}}{h^{3\alpha}}+\frac{2^{2\alpha}\overline{K_g}}{\Gamma(4\alpha+1)}h^{\alpha}\right], \tag{12.39}$$

$\forall\, b \in \mathbb{R}_-$, $\forall\, h > 0$.

I.e. it holds

$$\sup_{b\in\mathbb{R}_-}\left\|\left(D_{b-;g}^{3\alpha}f\right)(b)\right\| \leq \frac{\Gamma(3\alpha+1)\left(2^{2\alpha}+1\right)}{2^{2\alpha}(2^{\alpha}-1)}$$

$$\left[\frac{2\left\|\left\|f\right\|\right\|_{\infty,\mathbb{R}_-}}{h^{3\alpha}}+\frac{2^{2\alpha}\overline{K_g}}{\Gamma(4\alpha+1)}h^{\alpha}\right]<\infty, \tag{12.40}$$

$\forall\, h > 0$, $0 < \alpha < 1$.

Call

$$\mu := 2\left(2^{3\alpha}+1\right)\left\|\left\|f\right\|\right\|_{\infty,\mathbb{R}_-}, \tag{12.41}$$

$$\theta = \frac{2^{3\alpha}\left(2^{\alpha}+1\right)\overline{K_g}}{\Gamma(4\alpha+1)}, \tag{12.42}$$

both are greater than zero.

Set also $\rho := 2\alpha$; $0 < \rho < 2$. We consider the function

$$y(h) := \mu h^{-\rho} + \theta h^{\rho}, \quad \forall\, h > 0. \tag{12.43}$$

We have

$$y'(h) = -\rho\mu h^{-\rho-1} + \rho\theta h^{\rho-1} = 0, \tag{12.44}$$

then

$$\theta h^{2\rho} = \mu,$$

with a unique solution

$$h_0 := h_{crit.no.} = \left(\frac{\mu}{\theta}\right)^{\frac{1}{2\rho}}. \tag{12.45}$$

We have that

$$y''(h) = \rho(\rho+1)\mu h^{-\rho-2} + \rho(\rho-1)\theta h^{\rho-2}. \tag{12.46}$$

We see that

$$y''(h_0) = y''\left(\left(\frac{\mu}{\theta}\right)^{\frac{1}{2\rho}}\right) = \rho(\rho+1)\mu\left(\frac{\mu}{\theta}\right)^{\frac{-\rho-2}{2\rho}} + \rho(\rho-1)\theta\left(\frac{\mu}{\theta}\right)^{\frac{\rho-2}{2\rho}} =$$

$$\rho\left(\frac{\theta}{\mu}\right)^{\frac{1}{\rho}}\left[(\rho+1)\sqrt{\mu\theta} + (\rho-1)\sqrt{\mu\theta}\right] =$$

$$\rho\left(\frac{\theta}{\mu}\right)^{\frac{1}{\rho}}\left(2\rho\sqrt{\mu\theta}\right) = 2\rho^2\sqrt{\mu\theta}\left(\frac{\theta}{\mu}\right)^{\frac{1}{\rho}} > 0.$$

Therefore y has a global minimum at $h_0 = \left(\frac{\mu}{\theta}\right)^{\frac{1}{2\rho}}$, which is

$$y(h_0) = \mu\left(\frac{\mu}{\theta}\right)^{-\frac{1}{2}} + \theta\left(\frac{\mu}{\theta}\right)^{\frac{1}{2}} = \mu\left(\frac{\theta}{\mu}\right)^{\frac{1}{2}} + \sqrt{\theta\mu} = 2\sqrt{\theta\mu}.$$

We have proved that (see (12.37))

$$\sup_{b\in\mathbb{R}_-}\left\|\left(D_{b-;g}^{2\alpha}f\right)(b)\right\| \leq \frac{\Gamma(2\alpha+1)}{2^{2\alpha-1}(2^{\alpha}-1)}$$

$$\sqrt{\frac{2^{3\alpha+1}\left(2^{3\alpha}+1\right)(2^{\alpha}+1)}{\Gamma(4\alpha+1)}\,\|\|f\|\|_{\infty,\mathbb{R}_-}\overline{K_g}}. \tag{12.47}$$

Call

$$\xi := 2\,\|\|f\|\|_{\infty,\mathbb{R}_-}, \qquad \psi := \frac{2^{2\alpha}\overline{K_g}}{\Gamma(4\alpha+1)}, \tag{12.48}$$

both are greater than zero.

We consider the function

$$\gamma(h) := \xi h^{-3\alpha} + \psi h^{\alpha}, \quad \forall\, h > 0. \tag{12.49}$$

We have

$$\gamma'(h) = -3\alpha\xi h^{-3\alpha-1} + \alpha\psi h^{\alpha-1} = 0,$$

then

$$\psi h^{4\alpha} = 3\xi,$$

with unique solution

$$h_0 := h_{crit.no.} = \left(\frac{3\xi}{\psi}\right)^{\frac{1}{4\alpha}}. \tag{12.50}$$

We have that

$$\gamma''(h) = 3\alpha(3\alpha+1)\xi h^{-3\alpha-2} + \alpha(\alpha-1)\psi h^{\alpha-2}. \tag{12.51}$$

We see

$$\gamma''(h_0) = 3\alpha(3\alpha+1)\xi\left(\frac{3\xi}{\psi}\right)^{\frac{-3\alpha-2}{4\alpha}} + \alpha(\alpha-1)\psi\left(\frac{3\xi}{\psi}\right)^{\frac{\alpha-2}{4\alpha}} =$$

$$\alpha\left(\frac{3\xi}{\psi}\right)^{\frac{\alpha-2}{4\alpha}}\left[3(3\alpha+1)\xi\frac{\psi}{3\xi} + (\alpha-1)\psi\right] =$$

$$\alpha\left(\frac{3\xi}{\psi}\right)^{\frac{\alpha-2}{4\alpha}}(4\alpha\psi) = 4\alpha^2\psi\left(\frac{3\xi}{\psi}\right)^{\frac{\alpha-2}{4\alpha}} > 0. \tag{12.52}$$

Therefore γ has a global minimum at $h_0 = \left(\frac{3\xi}{\psi}\right)^{\frac{1}{4\alpha}}$, which is

$$\gamma(h_0) = \xi\left(\frac{3\xi}{\psi}\right)^{-\frac{3}{4}} + \psi\left(\frac{3\xi}{\psi}\right)^{\frac{1}{4}} =$$

$$\left(\frac{3\xi}{\psi}\right)^{\frac{1}{4}}\left(\xi\frac{\psi}{3\xi} + \psi\right) = \frac{4}{3}\psi\left(\frac{3\xi}{\psi}\right)^{\frac{1}{4}}. \tag{12.53}$$

Consequently,

$$\gamma(h_0) = \frac{4}{3}\psi\left(\frac{3\xi}{\psi}\right)^{\frac{1}{4}} = \frac{4}{\left(\sqrt[4]{3}\right)^3}\psi^{\frac{3}{4}}\xi^{\frac{1}{4}}. \tag{12.54}$$

We have proved that (see (12.40))

$$\sup_{b\in\mathbb{R}_-}\left\|\left(D_{b-;g}^{3\alpha}f\right)(b)\right\|\leq\frac{4\Gamma(3\alpha+1)\left(2^{2\alpha}+1\right)}{\left(\sqrt[4]{3}\right)^3 2^{2\alpha}\left(2^{\alpha}-1\right)}$$

$$\left(2\left\|\|f\|\right\|_{\infty,\mathbb{R}_-}\right)^{\frac{1}{4}}\left(\frac{2^{2\alpha}\overline{K_g}}{\Gamma(4\alpha+1)}\right)^{\frac{3}{4}}=$$

$$\frac{4\sqrt[4]{2}\Gamma(3\alpha+1)\Gamma(4\alpha+1)^{-\frac{3}{4}}\left(2^{2\alpha}+1\right)}{\left(\sqrt[4]{3}\right)^3 2^{\frac{\alpha}{2}}\left(2^{\alpha}-1\right)}\left\|\|f\|\right\|_{\infty,\mathbb{R}_-}^{\frac{1}{4}}\overline{K_g}^{\frac{3}{4}}. \tag{12.55}$$

The theorem is proved. ■

We continue with abstract L_p right sequential generalized fractional Landau inequalities over $\mathbb{R}_-$.

Theorem 12.6 *Let $g\in C^1(\mathbb{R}_-)$ strictly increasing, with $g^{-1}\in C^1(g(\mathbb{R}_-))$. Let $p,q>1:\frac{1}{p}+\frac{1}{q}=1$, $0<\alpha<1$. Let $f\in C^1(\mathbb{R}_-,X)$ with $\left\|\|f\|\right\|_{\infty,\mathbb{R}_-}$, $\left\|\left\|\left(f\circ g^{-1}\right)'\circ g\right\|\right\|_{\infty,\mathbb{R}_-}<\infty$. For $k=1,2,3$, we assume that $D_{b-;g}^{k\alpha}f\in C^1((-\infty,b],X)$ and $D_{b-;g}^{4\alpha}f\in C((-\infty,b],X)$, $\forall\, b\in\mathbb{R}_-$. We further assume that*

$$\left(\sup_{b\in\mathbb{R}_-}\left\|\left\|D_{b-;g}^{4\alpha}f\right\|\right\|_{p,\mathbb{R}_-}\right)<\infty. \tag{12.56}$$

Then

(1) under $\frac{1}{2p}<\alpha<1$, we get

$$\sup_{b\in\mathbb{R}_-}\left\|\left(D_{b-;g}^{2\alpha}f\right)(b)\right\|\leq\left[\left(\frac{2^{\alpha}\Gamma(2\alpha)\left(4\alpha-\frac{1}{p}\right)}{2^{\alpha}-1}\right)\left(\frac{4\alpha\left(1+2^{-3\alpha}\right)}{2\alpha-\frac{1}{p}}\right)^{\left(\frac{2\alpha-\frac{1}{p}}{4\alpha-\frac{1}{p}}\right)}\right.$$

$$\left.\left(\frac{1+2^{\alpha-\frac{1}{p}}}{\Gamma(4\alpha)\left(q(4\alpha-1)+1\right)^{\frac{1}{q}}}\right)^{\left(\frac{2\alpha}{4\alpha-\frac{1}{p}}\right)}\right]\left\|\|f\|\right\|_{\infty,\mathbb{R}_-}^{\left(\frac{2\alpha-\frac{1}{p}}{4\alpha-\frac{1}{p}}\right)}$$

$$\left(\sup_{b\in\mathbb{R}_-}\left\|\left\|D_{b-;g}^{4\alpha}f\right\|\right\|_{p,\mathbb{R}_-}\right)^{\left(\frac{2\alpha}{4\alpha-\frac{1}{p}}\right)}<\infty. \tag{12.57}$$

(2) under $\frac{1}{p}<\alpha<1$, we get

$$\sup_{b\in\mathbb{R}_-}\left\|\left(D_{b-;g}^{3\alpha}f\right)(b)\right\|\leq\left[\left(\frac{\Gamma(3\alpha)\left(4\alpha-\frac{1}{p}\right)}{2^{\alpha}-1}\right)\left(\frac{6\alpha\left(1+2^{-2\alpha}\right)}{\alpha-\frac{1}{p}}\right)^{\left(\frac{\alpha-\frac{1}{p}}{4\alpha-\frac{1}{p}}\right)}\right.$$

$$\left.\left(\frac{1+2^{2\alpha-\frac{1}{p}}}{\Gamma(4\alpha)(q(4\alpha-1)+1)^{\frac{1}{q}}}\right)^{\left(\frac{3\alpha}{4\alpha-\frac{1}{p}}\right)}\right]\||f|\|_{\infty,\mathbb{R}_-}^{\left(\frac{\alpha-\frac{1}{p}}{4\alpha-\frac{1}{p}}\right)}$$

$$\left(\sup_{b\in\mathbb{R}_-}\left\|\left\|D_{b-;g}^{4\alpha}f\right\|\right\|_{p,\mathbb{R}_-}\right)^{\left(\frac{3\alpha}{4\alpha-\frac{1}{p}}\right)}<\infty. \tag{12.58}$$

That is $\sup_{b\in\mathbb{R}_-}\left\|\left(D_{b-;g}^{2\alpha}f\right)(b)\right\|,\ \sup_{b\in\mathbb{R}_-}\left\|\left(D_{b-;g}^{3\alpha}f\right)(b)\right\|<\infty.$

Proof As in the proof of Theorem 12.5 we have that

$$\|A\|\overset{(12.22.)}{=}\left\|f(x_1)-f(b)-\frac{1}{\Gamma(4\alpha)}\int_{x_1}^{b}(g(t)-g(x_1))^{4\alpha-1}g'(t)\left(D_{b-;g}^{4\alpha}f\right)(t)\,dt\right\|\leq$$

$$2\||f|\|_{\infty,\mathbb{R}_-}+\frac{1}{\Gamma(4\alpha)}\int_{x_1}^{b}(g(t)-g(x_1))^{4\alpha-1}g'(t)\left\|\left(D_{b-;g}^{4\alpha}f\right)(t)\right\|dt\leq$$

$$2\||f|\|_{\infty,\mathbb{R}_-}+\frac{1}{\Gamma(4\alpha)}\frac{(g(b)-g(x_1))^{\frac{(q(4\alpha-1)+1)}{q}}}{(q(4\alpha-1)+1)^{\frac{1}{q}}}\left(\sup_{b\in\mathbb{R}_-}\left\|\left\|D_{b-;g}^{4\alpha}f\right\|\right\|_{p,\mathbb{R}_-}\right)\overset{(g(b)-g(x_1)=:h>0)}{=}$$

$$2\||f|\|_{\infty,\mathbb{R}_-}+\frac{1}{\Gamma(4\alpha)}\frac{h^{\left(4\alpha-\frac{1}{p}\right)}}{(q(4\alpha-1)+1)^{\frac{1}{q}}}\left(\sup_{b\in\mathbb{R}_-}\left\|\left\|D_{b-;g}^{4\alpha}f\right\|\right\|_{p,\mathbb{R}_-}\right), \tag{12.59}$$

with $\frac{1}{4p}<\alpha<1$.

That is

$$\|A\|\leq 2\||f|\|_{\infty,\mathbb{R}_-}+\frac{h^{4\alpha-\frac{1}{p}}}{\Gamma(4\alpha)(q(4\alpha-1)+1)^{\frac{1}{q}}}\left(\sup_{b\in\mathbb{R}_-}\left\|\left\|D_{b-;g}^{4\alpha}f\right\|\right\|_{p,\mathbb{R}_-}\right), \tag{12.60}$$

where $\frac{1}{4p}<\alpha<1$.

We also have

$$\|B\|\overset{(12.23.)}{=}\left\|f(x_2)-f(b)-\frac{1}{\Gamma(4\alpha)}\int_{x_2}^{b}(g(t)-g(x_2))^{4\alpha-1}g'(t)\left(D_{b-;g}^{4\alpha}f\right)(t)\,dt\right\|\leq$$

$$2\,\| \| f \| \|_{\infty,\mathbb{R}_-} + \frac{(g(b) - g(x_2))^{\frac{(q(4\alpha-1)+1)}{q}}}{\Gamma(4\alpha)(q(4\alpha-1)+1)^{\frac{1}{q}}} \left(\sup_{b\in\mathbb{R}_-} \left\| \left\| D_{b-;g}^{4\alpha} f \right\| \right\|_{p,\mathbb{R}_-} \right) \overset{(g(b)-g(x_2)=:2h)}{=}$$

$$2\,\| \| f \| \|_{\infty,\mathbb{R}_-} + \frac{2^{4\alpha-\frac{1}{p}} h^{4\alpha-\frac{1}{p}}}{\Gamma(4\alpha)(q(4\alpha-1)+1)^{\frac{1}{q}}} \left(\sup_{b\in\mathbb{R}_-} \left\| \left\| D_{b-;g}^{4\alpha} f \right\| \right\|_{p,\mathbb{R}_-} \right). \tag{12.61}$$

That is

$$\|B\| \le 2\,\| \| f \| \|_{\infty,\mathbb{R}_-} + \frac{2^{4\alpha-\frac{1}{p}} h^{4\alpha-\frac{1}{p}}}{\Gamma(4\alpha)(q(4\alpha-1)+1)^{\frac{1}{q}}} \left(\sup_{b\in\mathbb{R}_-} \left\| \left\| D_{b-;g}^{4\alpha} f \right\| \right\|_{p,\mathbb{R}_-} \right), \tag{12.62}$$

where $\frac{1}{4p} < \alpha < 1$.

We have assumed that

$$\overline{M_g} := \left(\sup_{b\in\mathbb{R}_-} \left\| \left\| D_{b-;g}^{4\alpha} f \right\| \right\|_{p,\mathbb{R}_-} \right) < \infty. \tag{12.63}$$

For convenience we call

$$c := \Gamma(4\alpha)(q(4\alpha-1)+1)^{\frac{1}{q}} > 0. \tag{12.64}$$

So we have

$$\begin{aligned} &\|A\| \le 2\,\| \| f \| \|_{\infty,\mathbb{R}_-} + \frac{h^{4\alpha-\frac{1}{p}}}{c} \overline{M_g}, \\ &\text{and} \\ &\|B\| \le 2\,\| \| f \| \|_{\infty,\mathbb{R}_-} + \frac{2^{4\alpha-\frac{1}{p}} h^{4\alpha-\frac{1}{p}}}{c} \overline{M_g}, \end{aligned} \tag{12.65}$$

where $\frac{1}{4p} < \alpha < 1$.

Next we estimate the (12.26)-quantities and we have:

$$\left\| \left(D_{b-;g}^{2\alpha} f \right)(b) \right\| \le \frac{1}{D\Gamma(3\alpha+1)} \left[2^{3\alpha} h^{3\alpha} \|A\| + h^{3\alpha} \|B\| \right] \overset{(12.33.)}{=}$$

$$\frac{h^{3\alpha}\Gamma(2\alpha+1)}{2^{2\alpha}h^{5\alpha}(2^{\alpha}-1)} \left[2^{3\alpha}\|A\| + \|B\| \right] \overset{(12.65.)}{\le} \tag{12.66}$$

$$\frac{\Gamma(2\alpha+1)}{2^{2\alpha}(2^{\alpha}-1)h^{2\alpha}} \left[2^{3\alpha+1}\,\| \| f \| \|_{\infty,\mathbb{R}_-} + \frac{2^{3\alpha}h^{4\alpha-\frac{1}{p}}}{c}\overline{M_g} + \right.$$

$$\left. 2\,\| \| f \| \|_{\infty,\mathbb{R}_-} + \frac{2^{4\alpha-\frac{1}{p}} h^{4\alpha-\frac{1}{p}}}{c} \overline{M_g} \right] =$$

$$\frac{\Gamma(2\alpha+1)}{2^{2\alpha}(2^{\alpha}-1)}\left[\frac{\left(2^{3\alpha+1}+2\right)\||f|\|_{\infty,\mathbb{R}_-}}{h^{2\alpha}}+\frac{\left(2^{3\alpha}+2^{4\alpha-\frac{1}{p}}\right)}{c}\overline{M_g}h^{2\alpha-\frac{1}{p}}\right]=$$

$$\frac{2^{\alpha}\Gamma(2\alpha+1)}{(2^{\alpha}-1)}\left[\frac{2\left(1+2^{-3\alpha}\right)\||f|\|_{\infty,\mathbb{R}_-}}{h^{2\alpha}}+\frac{\left(1+2^{\alpha-\frac{1}{p}}\right)\overline{M_g}}{c}h^{2\alpha-\frac{1}{p}}\right]. \tag{12.67}$$

That is

$$\left\|\left(D_{b-;g}^{2\alpha}f\right)(b)\right\|\leq$$

$$\left(\frac{2^{\alpha}\Gamma(2\alpha+1)}{2^{\alpha}-1}\right)\left[\frac{2\left(1+2^{-3\alpha}\right)\||f|\|_{\infty,\mathbb{R}_-}}{h^{2\alpha}}+\frac{\left(1+2^{\alpha-\frac{1}{p}}\right)\overline{M_g}}{c}h^{2\alpha-\frac{1}{p}}\right], \tag{12.68}$$

$\forall\, b\in\mathbb{R}_-$, $\forall\, h>0$.

I.e. it holds

$$\sup_{b\in\mathbb{R}_-}\left\|\left(D_{b-;g}^{2\alpha}f\right)(b)\right\|\leq\left(\frac{2^{\alpha}\Gamma(2\alpha+1)}{2^{\alpha}-1}\right)$$

$$\left[\frac{2\left(1+2^{-3\alpha}\right)\||f|\|_{\infty,\mathbb{R}_-}}{h^{2\alpha}}+\frac{\left(1+2^{\alpha-\frac{1}{p}}\right)\overline{M_g}}{c}h^{2\alpha-\frac{1}{p}}\right], \tag{12.69}$$

$\forall\, h>0$, under $\frac{1}{4p}<\alpha<1$.

Again from (12.26) we get

$$\left\|\left(D_{b-;g}^{3\alpha}f\right)(b)\right\|\leq\frac{1}{\Gamma(2\alpha+1)\,D}\left[h^{2\alpha}\|B\|+2^{2\alpha}h^{2\alpha}\|A\|\right]=$$

$$\frac{h^{2\alpha}\Gamma(3\alpha+1)}{2^{2\alpha}h^{5\alpha}(2^{\alpha}-1)}\left[\|B\|+2^{2\alpha}\|A\|\right]\leq \tag{12.70}$$

$$\left(\frac{h^{2\alpha}\Gamma(3\alpha+1)}{2^{2\alpha}h^{5\alpha}(2^{\alpha}-1)}\right)\left[2\||f|\|_{\infty,\mathbb{R}_-}+\frac{2^{4\alpha-\frac{1}{p}}h^{4\alpha-\frac{1}{p}}}{c}\overline{M_g}+\right.$$

$$\left.2^{2\alpha+1}\||f|\|_{\infty,\mathbb{R}_-}+\frac{2^{2\alpha}h^{4\alpha-\frac{1}{p}}}{c}\overline{M_g}\right]=$$

$$\frac{\Gamma(3\alpha+1)}{2^{2\alpha}(2^{\alpha}-1)}\left[\frac{\left(2+2^{2\alpha+1}\right)\||f|\|_{\infty,\mathbb{R}_-}}{h^{3\alpha}}+\frac{\left(2^{2\alpha}+2^{4\alpha-\frac{1}{p}}\right)}{c}\overline{M_g}h^{\alpha-\frac{1}{p}}\right]= \tag{12.71}$$

$$\frac{\Gamma(3\alpha+1)}{(2^{\alpha}-1)}\left[\frac{2\left(1+2^{-2\alpha}\right)\||f|\|_{\infty,\mathbb{R}_-}}{h^{3\alpha}}+\frac{\left(1+2^{2\alpha-\frac{1}{p}}\right)}{c}\overline{M_g}h^{\alpha-\frac{1}{p}}\right].$$

That is

$$\left\|\left(D_{b-;g}^{3\alpha}f\right)(b)\right\|\leq$$

$$\left(\frac{\Gamma(3\alpha+1)}{2^{\alpha}-1}\right)\left[\frac{2\left(1+2^{-2\alpha}\right)\||f|\|_{\infty,\mathbb{R}_-}}{h^{3\alpha}}+\frac{\left(1+2^{2\alpha-\frac{1}{p}}\right)}{c}\overline{M_g}h^{\alpha-\frac{1}{p}}\right], \quad (12.72)$$

$\forall\, b\in\mathbb{R}_-, \forall\, h>0.$

I.e. it holds

$$\sup_{b\in\mathbb{R}_-}\left\|\left(D_{b-;g}^{3\alpha}f\right)(b)\right\|\leq\left(\frac{\Gamma(3\alpha+1)}{2^{\alpha}-1}\right)$$

$$\left[\frac{2\left(1+2^{-2\alpha}\right)\||f|\|_{\infty,\mathbb{R}_-}}{h^{3\alpha}}+\frac{\left(1+2^{2\alpha-\frac{1}{p}}\right)}{c}\overline{M_g}h^{\alpha-\frac{1}{p}}\right], \quad (12.73)$$

$\forall\, h>0,\ \frac{1}{4p}<\alpha<1.$

Call

$$\mu:=2\left(1+2^{-3\alpha}\right)\||f|\|_{\infty,\mathbb{R}_-},$$
$$\theta:=\frac{\left(1+2^{\alpha-\frac{1}{p}}\right)\overline{M_g}}{c}, \quad (12.74)$$

both are greater than zero.

We consider the function

$$y(h)=\mu h^{-2\alpha}+\theta h^{2\alpha-\frac{1}{p}},\quad \forall\, h>0. \quad (12.75)$$

We have

$$y'(h)=-2\alpha\mu h^{-2\alpha-1}+\left(2\alpha-\frac{1}{p}\right)\theta h^{2\alpha-\frac{1}{p}-1}=0, \quad (12.76)$$

then

$$\left(2\alpha-\frac{1}{p}\right)\theta h^{2\alpha-\frac{1}{p}-1}=2\alpha\mu h^{-2\alpha-1},$$

i.e.

$$\left(2\alpha-\frac{1}{p}\right)\theta h^{4\alpha-\frac{1}{p}}=2\alpha\mu,$$

with a unique solution

$$h_0 := h_{crit.no.} = \left(\frac{2\alpha\mu}{\left(2\alpha - \frac{1}{p}\right)\theta} \right)^{\frac{1}{4\alpha - \frac{1}{p}}} \tag{12.77}$$

(assuming $\frac{1}{2p} < \alpha < 1$).

We have that

$$y''(h) = 2\alpha(2\alpha + 1)\mu h^{-2\alpha-2} + \left(2\alpha - \frac{1}{p}\right)\left(2\alpha - \frac{1}{p} - 1\right)\theta h^{2\alpha - \frac{1}{p} - 2}. \tag{12.78}$$

We see that

$$y''(h_0) = 2\alpha(2\alpha + 1)\mu \left(\frac{2\alpha\mu}{\left(2\alpha - \frac{1}{p}\right)\theta} \right)^{\frac{-2\alpha-2}{4\alpha - \frac{1}{p}}} +$$

$$\left(2\alpha - \frac{1}{p}\right)\left(2\alpha - \frac{1}{p} - 1\right)\theta \left(\frac{2\alpha\mu}{\left(2\alpha - \frac{1}{p}\right)\theta} \right)^{\frac{2\alpha - \frac{1}{p} - 2}{4\alpha - \frac{1}{p}}} =$$

$$\left(\frac{2\alpha\mu}{\left(2\alpha - \frac{1}{p}\right)\theta} \right)^{\frac{-2\alpha-2}{4\alpha - \frac{1}{p}}} \left[2\alpha(2\alpha + 1)\mu + 2\alpha\mu\left(2\alpha - \frac{1}{p} - 1\right) \right] = \tag{12.79}$$

$$2\alpha\mu \left(\frac{\left(2\alpha - \frac{1}{p}\right)\theta}{2\alpha\mu} \right)^{\left(\frac{2(\alpha+1)}{4\alpha - \frac{1}{p}}\right)} \left(4\alpha - \frac{1}{p}\right) > 0.$$

Therefore y has a global minimum at

$$h_0 = \left(\frac{2\alpha\mu}{\left(2\alpha - \frac{1}{p}\right)\theta} \right)^{\frac{1}{4\alpha - \frac{1}{p}}},$$

which is

$$y(h_0) = \mu \left(\frac{2\alpha\mu}{\left(2\alpha - \frac{1}{p}\right)\theta} \right)^{\frac{-2\alpha}{4\alpha - \frac{1}{p}}} + \theta \left(\frac{2\alpha\mu}{\left(2\alpha - \frac{1}{p}\right)\theta} \right)^{\frac{2\alpha - \frac{1}{p}}{4\alpha - \frac{1}{p}}} = \tag{12.80}$$

$$\left(\frac{2\alpha\mu}{\left(2\alpha-\frac{1}{p}\right)\theta}\right)^{\frac{-2\alpha}{4\alpha-\frac{1}{p}}}\left[\mu+\theta\frac{2\alpha\mu}{\left(2\alpha-\frac{1}{p}\right)\theta}\right]=\mu\left(\frac{\left(2\alpha-\frac{1}{p}\right)\theta}{2\alpha\mu}\right)^{\left(\frac{2\alpha}{4\alpha-\frac{1}{p}}\right)}\left(1+\frac{2\alpha}{2\alpha-\frac{1}{p}}\right)$$

$$=\frac{\left(4\alpha-\frac{1}{p}\right)}{(2\alpha)^{\left(\frac{2\alpha}{4\alpha-\frac{1}{p}}\right)}}\left(2\alpha-\frac{1}{p}\right)^{\frac{-2\alpha+\frac{1}{p}}{4\alpha-\frac{1}{p}}}\mu^{\left(\frac{2\alpha-\frac{1}{p}}{4\alpha-\frac{1}{p}}\right)}\theta^{\left(\frac{2\alpha}{4\alpha-\frac{1}{p}}\right)}. \tag{12.81}$$

That is

$$y\left(h_0\right)=\frac{\left(4\alpha-\frac{1}{p}\right)\left(2\alpha-\frac{1}{p}\right)^{\left(\frac{-2\alpha+\frac{1}{p}}{4\alpha-\frac{1}{p}}\right)}}{(2\alpha)^{\left(\frac{2\alpha}{4\alpha-\frac{1}{p}}\right)}}\mu^{\left(\frac{2\alpha-\frac{1}{p}}{4\alpha-\frac{1}{p}}\right)}\theta^{\left(\frac{2\alpha}{4\alpha-\frac{1}{p}}\right)}. \tag{12.82}$$

Therefore we derive (see (12.69))

$$\sup_{b\in\mathbb{R}_-}\left\|\left(D_{b-;g}^{2\alpha}f\right)(b)\right\|\leq\left(\frac{2^{\alpha}\Gamma(2\alpha)}{2^{\alpha}-1}\right)\left(\frac{2\alpha}{2\alpha-\frac{1}{p}}\right)^{\frac{\left(2\alpha-\frac{1}{p}\right)}{\left(4\alpha-\frac{1}{p}\right)}}\left(4\alpha-\frac{1}{p}\right)$$

$$\left(2\left(1+2^{-3\alpha}\right)\right)^{\left(\frac{2\alpha-\frac{1}{p}}{4\alpha-\frac{1}{p}}\right)}\left(\frac{\left(1+2^{\alpha-\frac{1}{p}}\right)}{\Gamma(4\alpha)\left(q(4\alpha-1)+1\right)^{\frac{1}{q}}}\right)^{\left(\frac{2\alpha}{4\alpha-\frac{1}{p}}\right)}\||\|f\||\|_{\infty,\mathbb{R}_-}^{\left(\frac{2\alpha-\frac{1}{p}}{4\alpha-\frac{1}{p}}\right)}$$

$$\left(\sup_{b\in\mathbb{R}_-}\left\|\left\|D_{b-;g}^{4\alpha}f\right\|\right\|_{p,\mathbb{R}_-}\right)^{\left(\frac{2\alpha}{4\alpha-\frac{1}{p}}\right)}<\infty, \tag{12.83}$$

where $\frac{1}{2p}<\alpha<1$.

Call

$$\xi:=2\left(1+2^{-2\alpha}\right)\||\|f\||\|_{\infty,\mathbb{R}_-},$$
$$\psi:=\frac{\left(1+2^{2\alpha-\frac{1}{p}}\right)}{c}\overline{M_g}, \tag{12.84}$$

both are greater than zero.

We consider the function

$$\gamma(h):=\xi h^{-3\alpha}+\psi h^{\alpha-\frac{1}{p}},\quad\forall\, h>0. \tag{12.85}$$

We have

$$\gamma'(h) = -3\alpha\xi h^{-3\alpha-1} + \left(\alpha - \frac{1}{p}\right)\psi h^{\alpha-\frac{1}{p}-1} = 0,$$

then

$$\left(\alpha - \frac{1}{p}\right)\psi h^{\alpha-\frac{1}{p}-1} = 3\alpha\xi h^{-3\alpha-1},$$

and

$$\left(\alpha - \frac{1}{p}\right)\psi h^{4\alpha-\frac{1}{p}} = 3\alpha\xi,$$

with unique solution

$$h_0 := h_{crit.no.} = \left(\frac{3\alpha\xi}{\left(\alpha - \frac{1}{p}\right)\psi}\right)^{\frac{1}{4\alpha-\frac{1}{p}}} \tag{12.86}$$

(assuming $\frac{1}{p} < \alpha < 1$).

We have that

$$\gamma''(h) = 3\alpha(3\alpha+1)\xi h^{-3\alpha-2} + \left(\alpha - \frac{1}{p}\right)\left(\alpha - \frac{1}{p} - 1\right)\psi h^{\alpha-\frac{1}{p}-2}. \tag{12.87}$$

We observe

$$\gamma''(h_0) = 3\alpha(3\alpha+1)\xi\left(\frac{3\alpha\xi}{\left(\alpha - \frac{1}{p}\right)\psi}\right)^{\left(\frac{-3\alpha-2}{4\alpha-\frac{1}{p}}\right)} +$$

$$\left(\alpha - \frac{1}{p}\right)\left(\alpha - \frac{1}{p} - 1\right)\psi\left(\frac{3\alpha\xi}{\left(\alpha - \frac{1}{p}\right)\psi}\right)^{\left(\frac{\alpha-\frac{1}{p}-2}{4\alpha-\frac{1}{p}}\right)} =$$

$$\left(\frac{3\alpha\xi}{\left(\alpha - \frac{1}{p}\right)\psi}\right)^{\left(\frac{-3\alpha-2}{4\alpha-\frac{1}{p}}\right)}\left[3\alpha(3\alpha+1)\xi + \left(\alpha - \frac{1}{p} - 1\right)3\alpha\xi\right] = \tag{12.88}$$

$$3\alpha\xi\left(\frac{3\alpha\xi}{\left(\alpha - \frac{1}{p}\right)\psi}\right)^{\left(\frac{-3\alpha-2}{4\alpha-\frac{1}{p}}\right)}\left(4\alpha - \frac{1}{p}\right) > 0.$$

Therefore y has a global minimum at

$$h_0 = \left(\frac{3\alpha\xi}{\left(\alpha - \frac{1}{p}\right)\psi} \right)^{\frac{1}{4\alpha - \frac{1}{p}}},$$

which is

$$\gamma(h_0) = \xi h_0^{-3\alpha} + \psi h_0^{\alpha - \frac{1}{p}} = h_0^{-3\alpha} \left(\xi + \psi h_0^{4\alpha - \frac{1}{p}} \right) = \tag{12.89}$$

$$\left(\frac{3\alpha\xi}{\left(\alpha - \frac{1}{p}\right)\psi} \right)^{\left(\frac{-3\alpha}{4\alpha - \frac{1}{p}}\right)} \left(\xi + \psi \left(\frac{3\alpha\xi}{\left(\alpha - \frac{1}{p}\right)\psi} \right) \right) =$$

$$\xi \left(\frac{3\alpha\xi}{\left(\alpha - \frac{1}{p}\right)\psi} \right)^{\left(\frac{-3\alpha}{4\alpha - \frac{1}{p}}\right)} \left(\frac{4\alpha - \frac{1}{p}}{\alpha - \frac{1}{p}} \right).$$

That is

$$\gamma(h_0) = \xi \left(\frac{4\alpha - \frac{1}{p}}{\alpha - \frac{1}{p}} \right) \left(\frac{3\alpha\xi}{\left(\alpha - \frac{1}{p}\right)\psi} \right)^{\left(\frac{-3\alpha}{4\alpha - \frac{1}{p}}\right)} = \tag{12.90}$$

$$\left(\frac{4\alpha - \frac{1}{p}}{\alpha - \frac{1}{p}} \right) \xi^{\left(\frac{\alpha - \frac{1}{p}}{4\alpha - \frac{1}{p}}\right)} \left(\frac{\left(\alpha - \frac{1}{p}\right)}{3\alpha} \right)^{\left(\frac{3\alpha}{4\alpha - \frac{1}{p}}\right)} \psi^{\left(\frac{3\alpha}{4\alpha - \frac{1}{p}}\right)}.$$

I.e. we have found

$$\gamma(h_0) = \frac{\left(4\alpha - \frac{1}{p}\right)}{\left(\alpha - \frac{1}{p}\right)^{\left(\frac{\alpha - \frac{1}{p}}{4\alpha - \frac{1}{p}}\right)} (3\alpha)^{\left(\frac{3\alpha}{4\alpha - \frac{1}{p}}\right)}} \xi^{\left(\frac{\alpha - \frac{1}{p}}{4\alpha - \frac{1}{p}}\right)} \psi^{\left(\frac{3\alpha}{4\alpha - \frac{1}{p}}\right)}. \tag{12.91}$$

We have proved that (see (12.73))

$$\sup_{b \in \mathbb{R}_-} \left\| \left(D_{b-;g}^{3\alpha} f \right)(b) \right\| \leq \left(\frac{\left(4\alpha - \frac{1}{p}\right)\Gamma(3\alpha)}{2^\alpha - 1} \right) \left(\frac{3\alpha}{\alpha - \frac{1}{p}} \right)^{\left(\frac{\alpha - \frac{1}{p}}{4\alpha - \frac{1}{p}}\right)} \tag{12.92}$$

$$\left(2\left(1 + 2^{-2\alpha}\right)\right)^{\left(\frac{\alpha - \frac{1}{p}}{4\alpha - \frac{1}{p}}\right)} \left(\frac{\left(1 + 2^{2\alpha - \frac{1}{p}}\right)}{\Gamma(4\alpha)\left(q(4\alpha - 1) + 1\right)^{\frac{1}{q}}} \right)^{\left(\frac{3\alpha}{4\alpha - \frac{1}{p}}\right)} \| \| f \| \|_{\infty, \mathbb{R}_-}^{\left(\frac{\alpha - \frac{1}{p}}{4\alpha - \frac{1}{p}}\right)}$$

$$\left(\sup_{b\in\mathbb{R}_-}\left\|\left\|D_{b-;g}^{4\alpha}f\right\|\right\|_{p,\mathbb{R}_-}\right)^{\left(\frac{3\alpha}{4\alpha-\frac{1}{p}}\right)}<\infty,$$

where $\frac{1}{p}<\alpha<1$.

The theorem is proved. ■

We give an application when $\alpha=\frac{1}{2}$ and $g(t)=e^t|_{\mathbb{R}_-}$.

Corollary 12.7 *Let $f\in C^1(\mathbb{R}_-,X)$ with $\|\|f\|\|_{\infty,\mathbb{R}_-}$, $\left\|\left\|(f\circ\ln)'\circ e^t\right\|\right\|_{\infty,\mathbb{R}_-}<\infty$, where $(X,\|\cdot\|)$ is a Banach space. For $k=1,2,3$, we assume that $D_{b-;e^t}^{k\frac{1}{2}}f\in C^1((-\infty,b],X)$ and $D_{b-;e^t}^{4\frac{1}{2}}f\in C((-\infty,b],X)$, $\forall\, b\in\mathbb{R}_-$. We further assume that*

$$\left\|\left\|D_{b-;e^t}^{4\frac{1}{2}}f(t)\right\|\right\|_{\infty,\mathbb{R}_-^2}<\infty, \tag{12.93}$$

where $(b,t)\in\mathbb{R}_-^2$.

Then

$$\sup_{b\in\mathbb{R}_-}\left\|\left(D_{b-;e^t}^{2\frac{1}{2}}f\right)(b)\right\|\leq\left(\frac{\sqrt{12+6\sqrt{2}}}{\sqrt{2}-1}\right)\|\|f\|\|_{\infty,\mathbb{R}_-}^{\frac{1}{2}}\left(\left\|\left\|D_{b-;e^t}^{4\frac{1}{2}}f(t)\right\|\right\|_{\infty,\mathbb{R}_-^2}\right)^{\frac{1}{2}}<\infty, \tag{12.94}$$

and

$$\sup_{b\in\mathbb{R}_-}\left\|\left(D_{b-;e^t}^{3\frac{1}{2}}f\right)(b)\right\|\leq$$

$$\left(\frac{9\sqrt{\pi}}{\left(2-\sqrt{2}\right)\sqrt[4]{2}\left(\sqrt[4]{3}\right)^3}\right)\|\|f\|\|_{\infty,\mathbb{R}_-}^{\frac{1}{4}}\left(\left\|\left\|D_{b-;e^t}^{4\frac{1}{2}}f(t)\right\|\right\|_{\infty,\mathbb{R}_-^2}\right)^{\frac{3}{4}}<\infty. \tag{12.95}$$

That is $\sup_{b\in\mathbb{R}_-}\left\|\left(D_{b-;e^t}^{2\frac{1}{2}}f\right)(b)\right\|$, $\sup_{b\in\mathbb{R}_-}\left\|\left(D_{b-;e^t}^{3\frac{1}{2}}f\right)(b)\right\|<\infty$.

Proof By Theorem 12.5. ■

References

1. Aglic Aljinovic, A., Marangunic, L., Pecaric, J.: On Landau type inequalities via Ostrowski inequalities. Nonlinear Funct. Anal. Appl. **10**(4), 565–579 (2005)
2. Anastassiou, G.: Fractional Differentiation Inequalities. Research Monograph. Springer, New York (2009)
3. Anastassiou, G.A.: Advances on Fractional Inequalities. Springer, New York (2011)

4. Anastassiou, G.A.: Intelligent Computations: Abstract Fractional Calculus, Inequalities, Approximations. Springer, Heidelberg (2018)
5. Anastassiou, G.A.: Sequential abstract generalized right side fractional Landau inequalities. Submitted (2020)
6. Barnett, N.S., Dragomir, S.S.: Some Landau type inequalities for functions whose derivatives are of locally bounded variation. Tamkang J. Math. **37**(4), 301–308 (2006)
7. Ditzian, Z.: Remarks, questions and conjectures on Landau-Kolmogorov-type inequalities. Math. Inequal. Appl. **3**, 15–24 (2000)
8. Hardy, G.H., Littlewood, J.E.: Some integral inequalities connected with the calculus of variations. Q. J. Math. Oxford Ser. **3**, 241–252 (1932)
9. Hardy, G.H., Landau, E., Littlewood, J.E.: Some inequalities satisfied by the integrals or derivatives of real or analytic functions. Math. Z. **39**, 677–695 (1935)
10. Kallman, R.R., Rota, G.C.: On the inequality $\|'\|^2 \leq 4\| \| \cdot \|''\|$. In: Shisha, O. (ed.) Inequalities, vol. II, pp. 187–192. Academic, New York (1970)
11. Landau, E.: Einige Ungleichungen für zweimal differentzierban funktionen. Proc. Lond. Math. Soc. **13**, 43–49 (1913)
12. Mikusinski, J.: The Bochner Integral. Academic, New York (1978)
13. Shilov, G.E.: Elementary Functional Analysis. Dover Publications Inc., New York (1996)

Chapter 13
High Order Generalized Landau Inequalities

We present several high order abstract generalized Landau type inequalities involving the uniform and L_p norms. We mention also applications. Our estimates give best upper bounds. It follows [3].

13.1 Introduction

Let $p \in [1, \infty]$, $I = \mathbb{R}_+$ or $I = \mathbb{R}$ and $f : I \to \mathbb{R}$ is twice differentiable with $f, f'' \in L_p(I)$, then $f' \in L_p(I)$. Moreover, there exists a constant $C_p(I) > 0$ independent of f, such that

$$\left\| f' \right\|_{p,I} \leq C_p(I) \left\| f \right\|_{p,I}^{\frac{1}{2}} \left\| f'' \right\|_{p,I}^{\frac{1}{2}}, \tag{13.1}$$

where $\|\cdot\|_{p,I}$ is the p-norm on the interval I, see [1, 4].

The research on these inequalities started by Landau [9] in 1913. For the case of $p = \infty$ he proved that

$$C_\infty(\mathbb{R}_+) = 2 \text{ and } C_\infty(\mathbb{R}) = \sqrt{2}, \tag{13.2}$$

are the best constants in (13.1).

In 1932, Hardy and Littlewood [6] proved (13.1) for $p = 2$, with the best constants

$$C_2(\mathbb{R}_+) = \sqrt{2} \text{ and } C_2(\mathbb{R}) = 1. \tag{13.3}$$

In 1935, Hardy, Landau and Littlewood [7] showed that the best constants $C_p(\mathbb{R}_+)$ in (13.1) satisfies the estimate

$$C_p(\mathbb{R}_+) \leq 2, \text{ for } p \in [1, \infty), \tag{13.4}$$

which yields $C_p(\mathbb{R}) \leq 2$ for $p \in [1, \infty)$.

G. A. Anastassiou, *Constructive Fractional Analysis with Applications*, Studies in Systems, Decision and Control 362,
https://doi.org/10.1007/978-3-030-71481-9_13

In fact, in [5, 8] was shown that $C_p(\mathbb{R}) \leq \sqrt{2}$.

In this chapter we are greatly inspired by the interesting article [12].

We need

Theorem 13.1 ([2], p. 97) *Let $n \in \mathbb{N}$ and $f \in C^n([a,b], X)$, where $[a,b] \subset \mathbb{R}$ and $(X, \|\cdot\|)$ is a real or complex Banach space. Let $g \in C^1([a,b])$ strictly increasing, such that $g^{-1} \in C^n([g(a), g(b)])$. Let any $x, y \in [a,b]$. Then*

$$f(x) = f(y) + \sum_{i=1}^{n-1} \frac{(g(x) - g(y))^i}{i!} \left(f \circ g^{-1}\right)^{(i)} (g(y)) + R_n(y, x, g), \quad (13.5)$$

where

$$R_n(y, x, g) = \frac{1}{(n-1)!} \int_y^x (g(x) - g(t))^{n-1} \left(f \circ g^{-1}\right)^{(n)} (g(t)) g'(t) \, dt =$$

$$\frac{1}{(n-1)!} \int_{g(y)}^{g(x)} (g(x) - z)^{n-1} \left(f \circ g^{-1}\right)^{(n)} (z) \, dz. \quad (13.6)$$

Derivatives here are similar to numerical ones, see [11].

The above integrals are of Bochner type [10], see also the Riemann integral for Banach space valued functions [11].

In this chapter we establish high order abstract generalized Landau inequalities for $\|\cdot\|_\infty$ and $\|\cdot\|_p$, $p > 1$, norms.

We will use

Corollary 13.2 *Let $f \in C^5(\mathbb{R}, X)$ and $(X, \|\cdot\|)$ is a real or complex Banach space. Let $g \in C^1(\mathbb{R})$, strictly increasing, such that $g^{-1} \in C^5(g(\mathbb{R}))$. Let any $x, y \in \mathbb{R}$. Then*

$$f(x) = f(y) + \sum_{i=1}^{4} \frac{(g(x) - g(y))^i}{i!} \left(f \circ g^{-1}\right)^{(i)} (g(y))$$

$$+ \frac{1}{24} \int_{g(y)}^{g(x)} (g(x) - z)^4 \left(f \circ g^{-1}\right)^{(5)} (z) \, dz. \quad (13.7)$$

Proof Obvious, by Theorem 13.1. ■

13.2 Main Results

We present the following uniform high order abstract and generalized Landau type inequalities:

Theorem 13.3 *Let $f \in C^5(\mathbb{R}, X)$ and $(X, \|\cdot\|)$ is a real or complex Banach space. Let $g \in C^1(\mathbb{R})$, strictly increasing, such that $g^{-1} \in C^5(g(\mathbb{R}))$, and assume that $\|\|f\|\|_{\infty,\mathbb{R}}$, $\left\|\left\|\left(f \circ g^{-1}\right)^{(5)}\right\|\right\|_{\infty,g(\mathbb{R})} < +\infty$. Then we get the following best upper bounds*

(1)

$$\left\|\left\|\left(f \circ g^{-1}\right)^{(1)}\right\|\right\|_{\infty,g(\mathbb{R})} \leq \frac{5}{4}\sqrt[5]{\frac{9}{8}} \|\|f\|\|_{\infty,\mathbb{R}}^{\frac{4}{5}} \left\|\left\|\left(f \circ g^{-1}\right)^{(5)}\right\|\right\|_{\infty,g(\mathbb{R})}^{\frac{1}{5}} < +\infty, \tag{13.8}$$

(2)

$$\left\|\left\|\left(f \circ g^{-1}\right)^{(2)}\right\|\right\|_{\infty,g(\mathbb{R})} \leq \frac{20}{3\sqrt[5]{675}} \|\|f\|\|_{\infty,\mathbb{R}}^{\frac{3}{5}} \left\|\left\|\left(f \circ g^{-1}\right)^{(5)}\right\|\right\|_{\infty,g(\mathbb{R})}^{\frac{2}{5}} < +\infty, \tag{13.9}$$

(3)

$$\left\|\left\|\left(f \circ g^{-1}\right)^{(3)}\right\|\right\|_{\infty,g(\mathbb{R})} \leq \frac{5}{2}\sqrt[5]{\frac{4913}{81000}} \|\|f\|\|_{\infty,\mathbb{R}}^{\frac{2}{5}} \left\|\left\|\left(f \circ g^{-1}\right)^{(5)}\right\|\right\|_{\infty,g(\mathbb{R})}^{\frac{3}{5}} < +\infty, \tag{13.10}$$

and

(4)

$$\left\|\left\|\left(f \circ g^{-1}\right)^{(4)}\right\|\right\|_{\infty,g(\mathbb{R})} \leq 5\sqrt[5]{\frac{81}{10000}} \|\|f\|\|_{\infty,\mathbb{R}}^{\frac{1}{5}} \left\|\left\|\left(f \circ g^{-1}\right)^{(5)}\right\|\right\|_{\infty,g(\mathbb{R})}^{\frac{4}{5}} < +\infty. \tag{13.11}$$

Proof We apply (13.7) several times. Let $x_1 > y$ and $g(x_1) - g(y) = h > 0$, that is $g(x_1) = g(y) + h$. Thus

$$f(x_1) = f(y) + \sum_{i=1}^{4} \frac{h^i}{i!} \left(f \circ g^{-1}\right)^{(i)} (g(y)) +$$

$$\frac{1}{24} \int_{g(y)}^{g(y)+h} (g(y) + h - z)^4 \left(f \circ g^{-1}\right)^{(5)} (z)\, dz. \tag{13.12}$$

Let $x_2 > y$ and $g(x_2) - g(y) = 2h$, that is $g(x_2) = g(y) + 2h$. So that

$$f(x_2) = f(y) + \sum_{i=1}^{4} \frac{2^i h^i}{i!} \left(f \circ g^{-1}\right)^{(i)} (g(y)) +$$

$$\frac{1}{24} \int_{g(y)}^{g(y)+2h} (g(y) + 2h - z)^4 \left(f \circ g^{-1}\right)^{(5)} (z)\, dz. \tag{13.13}$$

Let $x_3 < y$ and $g(x_3) - g(y) = -h$, that is $g(x_3) = g(y) - h$. Hence

$$f(x_3) = f(y) + \sum_{i=1}^{4} \frac{(-1)^i h^i}{i!} \left(f \circ g^{-1}\right)^{(i)} (g(y)) -$$

$$\frac{1}{24} \int_{g(y)-h}^{g(y)} (g(y) - h - z)^4 \left(f \circ g^{-1}\right)^{(5)} (z)\, dz. \tag{13.14}$$

Finally, let $x_4 < y$ and $g(x_4) - g(y) = -2h$, that is $g(x_4) = g(y) - 2h$. Therefore it holds

$$f(x_4) = f(y) + \sum_{i=1}^{4} \frac{(-1)^i 2^i h^i}{i!} \left(f \circ g^{-1}\right)^{(i)} (g(y)) -$$

$$\frac{1}{24} \int_{g(y)-2h}^{g(y)} (g(y) - 2h - z)^4 \left(f \circ g^{-1}\right)^{(5)} (z)\, dz. \tag{13.15}$$

We get that

$$\sum_{i=1}^{4} \frac{h^i}{i!} \left(f \circ g^{-1}\right)^{(i)} (g(y)) \overset{(13.12)}{=} f(x_1) - f(y) -$$

$$\frac{1}{24} \int_{g(y)}^{g(y)+h} (g(y) + h - z)^4 \left(f \circ g^{-1}\right)^{(5)} (z)\, dz =: A, \tag{13.16}$$

$$\sum_{i=1}^{4} \frac{2^i h^i}{i!} \left(f \circ g^{-1}\right)^{(i)} (g(y)) \overset{(13.13)}{=} f(x_2) - f(y) -$$

$$\frac{1}{24} \int_{g(y)}^{g(y)+2h} (g(y) + 2h - z)^4 \left(f \circ g^{-1}\right)^{(5)} (z)\, dz =: B, \tag{13.17}$$

$$\sum_{i=1}^{4} \frac{(-1)^i h^i}{i!} \left(f \circ g^{-1}\right)^{(i)} (g(y)) \overset{(13.14)}{=} f(x_3) - f(y) +$$

$$\frac{1}{24} \int_{g(y)-h}^{g(y)} (g(y) - h - z)^4 \left(f \circ g^{-1}\right)^{(5)} (z)\, dz =: C, \tag{13.18}$$

and

$$\sum_{i=1}^{4} \frac{(-1)^i 2^i h^i}{i!} \left(f \circ g^{-1}\right)^{(i)} (g(y)) \overset{(13.15)}{=} f(x_4) - f(y) +$$

$$\frac{1}{24}\int_{g(y)-2h}^{g(y)} (g(y)-2h-z)^4 \left(f\circ g^{-1}\right)^{(5)}(z)\,dz =: D. \tag{13.19}$$

We solve the system of equation (13.16)–(13.19) for $\left(f\circ g^{-1}\right)^{(i)}(g(y))$, $i = 1, 2, 3, 4$.

Adding (13.16) and (13.18) we get

$$h^2\left(f\circ g^{-1}\right)^{(2)}(g(y)) + \frac{h^4}{12}\left(f\circ g^{-1}\right)^{(4)}(g(y)) = A + C. \tag{13.20}$$

Adding (13.17) and (13.19) we obtain

$$4h^2\left(f\circ g^{-1}\right)^{(2)}(g(y)) + \frac{4}{3}h^4\left(f\circ g^{-1}\right)^{(4)}(g(y)) = B + D. \tag{13.21}$$

Similarly we have

$$2h\left(f\circ g^{-1}\right)^{(1)}(g(y)) + \frac{h^3}{3}\left(f\circ g^{-1}\right)^{(3)}(g(y)) = A - C, \tag{13.22}$$

and

$$4h\left(f\circ g^{-1}\right)^{(1)}(g(y)) + \frac{8h^3}{3}\left(f\circ g^{-1}\right)^{(3)}(g(y)) = B - D. \tag{13.23}$$

We find

$$\begin{aligned}
\left(f\circ g^{-1}\right)^{(1)}(g(y)) &= \tfrac{8A-B+D-8C}{12h},\\
\left(f\circ g^{-1}\right)^{(2)}(g(y)) &= \tfrac{16A+16C-B-D}{12h^2},\\
\left(f\circ g^{-1}\right)^{(3)}(g(y)) &= \tfrac{B-D-2A+2C}{2h^3},\\
\left(f\circ g^{-1}\right)^{(4)}(g(y)) &= \tfrac{B+D-4A-4C}{h^4}.
\end{aligned} \tag{13.24}$$

We observe that

$$\|8A - B + D - 8C\| = \|8(A - C) + D - B\| =$$

$$\Bigg\| 8(f(x_1) - f(x_3)) - \frac{1}{3}\int_{g(y)}^{g(y)+h} (g(y)+h-z)^4\left(f\circ g^{-1}\right)^{(5)}(z)\,dz -$$

$$\frac{1}{3}\int_{g(y)-h}^{g(y)} (g(y)-h-z)^4\left(f\circ g^{-1}\right)^{(5)}(z)\,dz + f(x_4) - f(x_2) + \tag{13.25}$$

$$\frac{1}{24}\int_{g(y)-2h}^{g(y)} (g(y)-2h-z)^4\left(f\circ g^{-1}\right)^{(5)}(z)\,dz +$$

$$\frac{1}{24}\int_{g(y)}^{g(y)+2h}(g(y)+2h-z)^4\left(f\circ g^{-1}\right)^{(5)}(z)\,dz\Bigg\| \leq$$

$$18\,\|\|f\|\|_{\infty,\mathbb{R}}+\frac{1}{3}\int_{g(y)}^{g(y)+h}(g(y)+h-z)^4\left\|\left(f\circ g^{-1}\right)^{(5)}(z)\right\|dz+$$

$$\frac{1}{3}\int_{g(y)-h}^{g(y)}(g(y)-h-z)^4\left\|\left(f\circ g^{-1}\right)^{(5)}(z)\right\|dz+$$

$$\frac{1}{24}\int_{g(y)-2h}^{g(y)}(g(y)-2h-z)^4\left\|\left(f\circ g^{-1}\right)^{(5)}(z)\right\|dz+$$

$$\frac{1}{24}\int_{g(y)}^{g(y)+2h}(g(y)+2h-z)^4\left\|\left(f\circ g^{-1}\right)^{(5)}(z)\right\|dz\leq \tag{13.26}$$

$$18\,\|\|f\|\|_{\infty,\mathbb{R}}+\left\|\left\|\left(f\circ g^{-1}\right)^{(5)}(z)\right\|\right\|_{\infty,g(\mathbb{R})}\left[\frac{2h^5}{15}+\frac{8h^5}{15}\right]=$$

$$18\,\|\|f\|\|_{\infty,\mathbb{R}}+\frac{2}{3}\left\|\left\|\left(f\circ g^{-1}\right)^{(5)}\right\|\right\|_{\infty,g(\mathbb{R})}h^5.$$

That is

$$\|8A-B+D-8C\|\leq 18\,\|\|f\|\|_{\infty,\mathbb{R}}+\frac{2}{3}\left\|\left\|\left(f\circ g^{-1}\right)^{(5)}\right\|\right\|_{\infty,g(\mathbb{R})}h^5. \tag{13.27}$$

Therefore we find (by (13.24))

$$\left\|\left(f\circ g^{-1}\right)^{(1)}(g(y))\right\|\leq\frac{3\,\|\|f\|\|_{\infty,\mathbb{R}}}{2h}+\frac{\left\|\left\|\left(f\circ g^{-1}\right)^{(5)}\right\|\right\|_{\infty,g(\mathbb{R})}}{18}h^4, \tag{13.28}$$

$h>0$, for any $y\in\mathbb{R}$.

Hence it holds

$$\left\|\left\|\left(f\circ g^{-1}\right)^{(1)}\right\|\right\|_{\infty,g(\mathbb{R})}\leq\frac{3\,\|\|f\|\|_{\infty,\mathbb{R}}}{2h}+\frac{\left\|\left\|\left(f\circ g^{-1}\right)^{(5)}\right\|\right\|_{\infty,g(\mathbb{R})}}{18}h^4, \tag{13.29}$$

$\forall\, h>0$.

Next we estimate

$$\|16A+16C-B-D\|=\|16f(x_1)-16f(y)-$$

$$\frac{2}{3}\int_{g(y)}^{g(y)+h}(g(y)+h-z)^4\left(f\circ g^{-1}\right)^{(5)}(z)\,dz+16f(x_3)-16f(y)+$$

$$\frac{2}{3}\int_{g(y)-h}^{g(y)}(g(y)-h-z)^4\left(f\circ g^{-1}\right)^{(5)}(z)\,dz-f(x_2)+f(y)+$$

$$\frac{1}{24}\int_{g(y)}^{g(y)+2h}(g(y)+2h-z)^4\left(f\circ g^{-1}\right)^{(5)}(z)\,dz-f(x_4)+f(y)-$$

$$\left.\frac{1}{24}\int_{g(y)-2h}^{g(y)}(g(y)-2h-z)^4\left(f\circ g^{-1}\right)^{(5)}(z)\,dz\right\|\leq$$

$$64\,\|\|f\|\|_{\infty,\mathbb{R}}+\frac{4}{5}\left\|\left\|\left(f\circ g^{-1}\right)^{(5)}\right\|\right\|_{\infty,g(\mathbb{R})}h^5. \tag{13.30}$$

Hence (by (13.24)) we obtain

$$\left\|\left(f\circ g^{-1}\right)^{(2)}(g(y))\right\|\leq\frac{16\,\|\|f\|\|_{\infty,\mathbb{R}}}{3h^2}+\frac{\left\|\left\|\left(f\circ g^{-1}\right)^{(5)}\right\|\right\|_{\infty,g(\mathbb{R})}}{15}h^3, \tag{13.31}$$

$h>0$, for any $y\in\mathbb{R}$.

Therefore it holds

$$\left\|\left\|\left(f\circ g^{-1}\right)^{(2)}\right\|\right\|_{\infty,g(\mathbb{R})}\leq\frac{16\,\|\|f\|\|_{\infty,\mathbb{R}}}{3h^2}+\frac{\left\|\left\|\left(f\circ g^{-1}\right)^{(5)}\right\|\right\|_{\infty,g(\mathbb{R})}}{15}h^3, \tag{13.32}$$

$\forall\, h>0$.

Similarly, we find that

$$\|B-D-2A+2C\|\leq 6\,\|\|f\|\|_{\infty,\mathbb{R}}+\frac{17\left\|\left\|\left(f\circ g^{-1}\right)^{(5)}\right\|\right\|_{\infty,g(\mathbb{R})}}{30}h^5, \tag{13.33}$$

and by (13.24) we get that

$$\left\|\left(f\circ g^{-1}\right)^{(3)}(g(y))\right\|\leq\frac{3\,\|\|f\|\|_{\infty,\mathbb{R}}}{h^3}+\frac{17}{60}\left\|\left\|\left(f\circ g^{-1}\right)^{(5)}\right\|\right\|_{\infty,g(\mathbb{R})}h^2, \tag{13.34}$$

$h>0$, for any $y\in\mathbb{R}$.

Consequently we find

$$\left\|\left\|\left(f\circ g^{-1}\right)^{(3)}\right\|\right\|_{\infty,g(\mathbb{R})}\leq\frac{3\,\|\|f\|\|_{\infty,\mathbb{R}}}{h^3}+\frac{17}{60}\left\|\left\|\left(f\circ g^{-1}\right)^{(5)}\right\|\right\|_{\infty,g(\mathbb{R})}h^2, \tag{13.35}$$

$\forall\, h > 0$.

Finally we obtain

$$\|B + D - 4A - 4C\| \leq 16 \, \||f\||_{\infty,\mathbb{R}} + \frac{3}{5} \left\| \left(f \circ g^{-1} \right)^{(5)} \right\|_{\infty, g(\mathbb{R})} h^5. \quad (13.36)$$

By (13.24) we derive

$$\left\| \left(f \circ g^{-1} \right)^{(4)} (g(y)) \right\| \leq \frac{16 \, \||f\||_{\infty,\mathbb{R}}}{h^4} + \frac{3 \left\| \left(f \circ g^{-1} \right)^{(5)} \right\|_{\infty, g(\mathbb{R})}}{5} h, \quad (13.37)$$

$h > 0$, for any $y \in \mathbb{R}$.

The last implies

$$\left\| \left(f \circ g^{-1} \right)^{(4)} \right\|_{\infty, g(\mathbb{R})} \leq \frac{16 \, \||f\||_{\infty,\mathbb{R}}}{h^4} + \frac{3 \left\| \left(f \circ g^{-1} \right)^{(5)} \right\|_{\infty, g(\mathbb{R})}}{5} h, \quad (13.38)$$

$\forall\, h > 0$.

Next we work on (13.29).

By Theorem 13.6 we obtain the best upper bound:

$$\left\| \left(f \circ g^{-1} \right)^{(1)} \right\|_{\infty, g(\mathbb{R})} \leq \quad (13.39)$$

$$\frac{5}{4^{\frac{4}{5}}} \left(\frac{3}{2} \, \||f\||_{\infty,\mathbb{R}} \right)^{\frac{4}{5}} \left(\frac{\left\| \left(f \circ g^{-1} \right)^{(5)} \right\|_{\infty, g(\mathbb{R})}}{18} \right)^{\frac{1}{5}}$$

$$= \frac{5}{4} \sqrt[5]{\frac{9}{8}} \, \||f\||_{\infty,\mathbb{R}}^{\frac{4}{5}} \left\| \left(f \circ g^{-1} \right)^{(5)} \right\|_{\infty, g(\mathbb{R})}^{\frac{1}{5}}.$$

Next we work on (13.32).

Consequently by Theorem 13.6 we get the best upper bound:

$$\left\| \left(f \circ g^{-1} \right)^{(2)} \right\|_{\infty, g(\mathbb{R})} \leq \quad (13.40)$$

$$\frac{5}{3^{\frac{3}{5}} 2^{\frac{2}{5}}} \left(\frac{16 \, \||f\||_{\infty,\mathbb{R}}}{3} \right)^{\frac{3}{5}} \left(\frac{\left\| \left(f \circ g^{-1} \right)^{(5)} \right\|_{\infty, g(\mathbb{R})}}{15} \right)^{\frac{2}{5}}$$

$$= \frac{20}{3 \sqrt[5]{675}} \, \||f\||_{\infty,\mathbb{R}}^{\frac{3}{5}} \left\| \left(f \circ g^{-1} \right)^{(5)} \right\|_{\infty, g(\mathbb{R})}^{\frac{2}{5}}.$$

Next we work on (13.35).

Therefore by Theorem 13.6 we get the best upper bound:

$$\left\|\left\|\left(f\circ g^{-1}\right)^{(3)}\right\|\right\|_{\infty,g(\mathbb{R})}\leq \tag{13.41}$$

$$\frac{5\cdot 3^{\frac{2}{5}}}{2^{\frac{2}{5}}\cdot 3^{\frac{3}{5}}}\,\||f\||_{\infty,\mathbb{R}}^{\frac{2}{5}}\left(\frac{17}{60}\right)^{\frac{3}{5}}\left\|\left\|\left(f\circ g^{-1}\right)^{(5)}\right\|\right\|_{\infty,g(\mathbb{R})}^{\frac{3}{5}}$$

$$=\frac{5}{2}\sqrt[5]{\frac{4913}{81000}}\,\||f\||_{\infty,\mathbb{R}}^{\frac{2}{5}}\left\|\left\|\left(f\circ g^{-1}\right)^{(5)}\right\|\right\|_{\infty,g(\mathbb{R})}^{\frac{3}{5}}.$$

Finally we work on (13.38).

Therefore by Theorem 13.6 we find the best upper bound:

$$\left\|\left\|\left(f\circ g^{-1}\right)^{(4)}\right\|\right\|_{\infty,g(\mathbb{R})}\leq \tag{13.42}$$

$$\frac{5}{4^{\frac{4}{5}}}16^{\frac{1}{5}}\,\||f\||_{\infty,\mathbb{R}}^{\frac{1}{5}}\left(\frac{3}{5}\right)^{\frac{4}{5}}\left\|\left\|\left(f\circ g^{-1}\right)^{(5)}\right\|\right\|_{\infty,g(\mathbb{R})}^{\frac{4}{5}}$$

$$=5\sqrt[5]{\frac{81}{10000}}\,\||f\||_{\infty,\mathbb{R}}^{\frac{1}{5}}\left\|\left\|\left(f\circ g^{-1}\right)^{(5)}\right\|\right\|_{\infty,g(\mathbb{R})}^{\frac{4}{5}}.$$

The theorem is proved. ■

Next we give the following uniform- L_p high order abstract and generalized Landau type inequalities:

Theorem 13.4 *Let $f\in C^5(\mathbb{R},X)$ and $(X,\|\cdot\|)$ is a real or complex Banach space. Consider $g\in C^1(\mathbb{R})$, strictly increasing, such that $g^{-1}\in C^5(g(\mathbb{R}))$, and assume that $\||f\||_{\infty,\mathbb{R}}<+\infty$. Let $p,q>1:\frac{1}{p}+\frac{1}{q}=1$, and assume that $\left\|\left\|\left(f\circ g^{-1}\right)^{(5)}\right\|\right\|_{p,g(\mathbb{R})}<+\infty$. Then we get the following best upper bounds*

(1)

$$\left\|\left\|\left(f\circ g^{-1}\right)^{(1)}\right\|\right\|_{\infty,g(\mathbb{R})}\leq\frac{\left(4+\frac{1}{q}\right)}{\left(\frac{3}{2}\left(3+\frac{1}{q}\right)\right)^{\frac{\left(3+\frac{1}{q}\right)}{\left(4+\frac{1}{q}\right)}}} \tag{13.43}$$

$$\left(\frac{\left(1+2^{1+\frac{1}{q}}\right)}{18\,(4q+1)^{\frac{1}{q}}}\right)^{\left(\frac{1}{4+\frac{1}{q}}\right)}\||f\||_{\infty,\mathbb{R}}^{\left(\frac{3+\frac{1}{q}}{4+\frac{1}{q}}\right)}\left(\left\|\left\|\left(f\circ g^{-1}\right)^{(5)}\right\|\right\|_{p,g(\mathbb{R})}\right)^{\left(\frac{1}{4+\frac{1}{q}}\right)}<+\infty,$$

(2)

$$\left\|\left(f\circ g^{-1}\right)^{(2)}\right\|_{\infty,g(\mathbb{R})}\leq\frac{\left(4+\frac{1}{q}\right)}{2^{\left(\frac{2}{4+\frac{1}{q}}\right)}\left(\frac{16}{3}\left(2+\frac{1}{q}\right)\right)^{\left(\frac{2+\frac{1}{q}}{4+\frac{1}{q}}\right)}}$$

$$\left(\frac{\left(1+2^{\frac{1}{q}}\right)}{9\left(4q+1\right)^{\frac{1}{q}}}\right)^{\left(\frac{2}{4+\frac{1}{q}}\right)}\||f\||_{\infty,\mathbb{R}}^{\left(\frac{2+\frac{1}{q}}{4+\frac{1}{q}}\right)}\left\|\left(f\circ g^{-1}\right)^{(5)}\right\|_{p,g(\mathbb{R})}^{\left(\frac{2}{4+\frac{1}{q}}\right)}<+\infty, \tag{13.44}$$

(3)

$$\left\|\left(f\circ g^{-1}\right)^{(3)}\right\|_{\infty,g(\mathbb{R})}\leq\frac{\left(4+\frac{1}{q}\right)}{3^{\left(\frac{3}{4+\frac{1}{q}}\right)}\left(3\left(1+\frac{1}{q}\right)\right)^{\frac{\left(1+\frac{1}{q}\right)}{\left(4+\frac{1}{q}\right)}}}$$

$$\left(\frac{\left(1+8\cdot 2^{\frac{1}{q}}\right)}{12\left(4q+1\right)^{\frac{1}{q}}}\right)^{\frac{3}{\left(4+\frac{1}{q}\right)}}\||f\||_{\infty,\mathbb{R}}^{\left(\frac{1+\frac{1}{q}}{4+\frac{1}{q}}\right)}\left\|\left(f\circ g^{-1}\right)^{(5)}\right\|_{p,g(\mathbb{R})}^{\frac{3}{\left(4+\frac{1}{q}\right)}}<+\infty, \tag{13.45}$$

and

(4)

$$\left\|\left(f\circ g^{-1}\right)^{(4)}\right\|_{\infty,g(\mathbb{R})}\leq\frac{\left(4+\frac{1}{q}\right)}{4^{\left(\frac{4}{4+\frac{1}{q}}\right)}\left(\frac{16}{q}\right)^{\frac{1}{q\left(4+\frac{1}{q}\right)}}} \tag{13.46}$$

$$\left(\frac{\left(1+4\cdot 2^{\frac{1}{q}}\right)}{3\left(4q+1\right)^{\frac{1}{q}}}\right)^{\left(\frac{4}{4+\frac{1}{q}}\right)}\||f\||_{\infty,\mathbb{R}}^{\frac{1}{q\left(4+\frac{1}{q}\right)}}\left\|\left(f\circ g^{-1}\right)^{(5)}\right\|_{p,g(\mathbb{R})}^{\left(\frac{4}{4+\frac{1}{q}}\right)}<+\infty.$$

Proof We continue as in the proof of Theorem 13.3. As in (13.26) we have

$$\|8A-B+D-8C\|\leq 18\||f\||_{\infty,\mathbb{R}}+$$

$$\frac{1}{3}\int_{g(y)}^{g(y)+h}\left(g\left(y\right)+h-z\right)^{4}\left\|\left(f\circ g^{-1}\right)^{(5)}(z)\right\|dz+$$

$$\frac{1}{3}\int_{g(y)-h}^{g(y)}\left(z-\left(g\left(y\right)-h\right)\right)^{4}\left\|\left(f\circ g^{-1}\right)^{(5)}(z)\right\|dz+ \tag{13.47}$$

$$\frac{1}{24}\int_{g(y)-2h}^{g(y)}(z-(g(y)-2h))^{4}\left\|\left(f\circ g^{-1}\right)^{(5)}(z)\right\|dz+$$

$$\frac{1}{24}\int_{g(y)}^{g(y)+2h}(g(y)+2h-z)^{4}\left\|\left(f\circ g^{-1}\right)^{(5)}(z)\right\|dz\leq 18\left\|\|f\|\right\|_{\infty,\mathbb{R}}+$$

(we apply Hölder's inequality four times)

$$\frac{1}{3}\left(\int_{g(y)}^{g(y)+h}(g(y)+h-z)^{4q}dz\right)^{\frac{1}{q}}\left(\int_{g(y)}^{g(y)+h}\left\|\left(f\circ g^{-1}\right)^{(5)}(z)\right\|^{p}dz\right)^{\frac{1}{p}}+$$

$$\frac{1}{3}\left(\int_{g(y)-h}^{g(y)}(z-(g(y)-h))^{4q}dz\right)^{\frac{1}{q}}\left(\int_{g(y)-h}^{g(y)}\left\|\left(f\circ g^{-1}\right)^{(5)}(z)\right\|^{p}dz\right)^{\frac{1}{p}}+$$

$$\frac{1}{24}\left(\int_{g(y)-2h}^{g(y)}(z-(g(y)-2h))^{4q}dz\right)^{\frac{1}{q}}\left(\int_{g(y)-2h}^{g(y)}\left\|\left(f\circ g^{-1}\right)^{(5)}(z)\right\|^{p}dz\right)^{\frac{1}{p}}+$$

$$\frac{1}{24}\left(\int_{g(y)}^{g(y)+2h}(g(y)+2h-z)^{4q}dz\right)^{\frac{1}{q}}\left(\int_{g(y)}^{g(y)+2h}\left\|\left(f\circ g^{-1}\right)^{(5)}(z)\right\|^{p}dz\right)^{\frac{1}{p}}\leq$$

$$18\left\|\|f\|\right\|_{\infty,\mathbb{R}}+\frac{\left\|\left\|\left(f\circ g^{-1}\right)^{(5)}\right\|\right\|_{p,g(\mathbb{R})}}{(4q+1)^{\frac{1}{q}}}\left[\frac{2}{3}+\frac{2^{4+\frac{1}{q}}}{12}\right]h^{4+\frac{1}{q}}=$$

$$18\left\|\|f\|\right\|_{\infty,\mathbb{R}}+\frac{\left\|\left\|\left(f\circ g^{-1}\right)^{(5)}\right\|\right\|_{p,g(\mathbb{R})}2\left(1+2^{1+\frac{1}{q}}\right)}{3(4q+1)^{\frac{1}{q}}}h^{4+\frac{1}{q}}. \tag{13.48}$$

Therefore (by (13.24))

$$\left\|\left(f\circ g^{-1}\right)^{(1)}(g(y))\right\|\leq\frac{3\left\|\|f\|\right\|_{\infty,\mathbb{R}}}{2h}+\frac{\left(1+2^{1+\frac{1}{q}}\right)\left\|\left\|\left(f\circ g^{-1}\right)^{(5)}\right\|\right\|_{p,g(\mathbb{R})}}{18(4q+1)^{\frac{1}{q}}}h^{3+\frac{1}{q}}, \tag{13.49}$$

$h>0$, for any $y\in\mathbb{R}$.

Consequently it holds

$$\left\|\left\|\left(f\circ g^{-1}\right)^{(1)}\right\|\right\|_{\infty,g(\mathbb{R})}\leq\frac{3\,\|\|f\|\|_{\infty,\mathbb{R}}}{2h}+\frac{\left(1+2^{1+\frac{1}{q}}\right)}{18\,(4q+1)^{\frac{1}{q}}}\left\|\left\|\left(f\circ g^{-1}\right)^{(5)}\right\|\right\|_{p,g(\mathbb{R})}h^{3+\frac{1}{q}},\tag{13.50}$$

$\forall\, h>0$.

Working similarly over (13.30) we obtain

$$\|16A+16C-B-D\|\leq$$

$$64\,\|\|f\|\|_{\infty,\mathbb{R}}+\frac{4\left(1+2^{\frac{1}{q}}\right)}{3\,(4q+1)^{\frac{1}{q}}}\left\|\left\|\left(f\circ g^{-1}\right)^{(5)}\right\|\right\|_{p,g(\mathbb{R})}h^{4+\frac{1}{q}},\tag{13.51}$$

and

$$\left\|\left(f\circ g^{-1}\right)^{(2)}(g(y))\right\|\leq\frac{16\,\|\|f\|\|_{\infty,\mathbb{R}}}{3h^2}+\frac{\left(1+2^{\frac{1}{q}}\right)}{9\,(4q+1)^{\frac{1}{q}}}\left\|\left\|\left(f\circ g^{-1}\right)^{(5)}\right\|\right\|_{p,g(\mathbb{R})}h^{2+\frac{1}{q}},\tag{13.52}$$

$h>0$, for any $y\in\mathbb{R}$.

Consequently it holds

$$\left\|\left\|\left(f\circ g^{-1}\right)^{(2)}\right\|\right\|_{\infty,g(\mathbb{R})}\leq\frac{16\,\|\|f\|\|_{\infty,\mathbb{R}}}{3h^2}+\frac{\left(1+2^{\frac{1}{q}}\right)}{9\,(4q+1)^{\frac{1}{q}}}\left\|\left\|\left(f\circ g^{-1}\right)^{(5)}\right\|\right\|_{p,g(\mathbb{R})}h^{2+\frac{1}{q}},\tag{13.53}$$

$\forall\, h>0$.

Similarly, we find (case of (13.24), etc.)

$$\left\|\left\|\left(f\circ g^{-1}\right)^{(3)}\right\|\right\|_{\infty,g(\mathbb{R})}\leq\frac{3\,\|\|f\|\|_{\infty,\mathbb{R}}}{h^3}+\frac{\left(1+8\cdot 2^{\frac{1}{q}}\right)}{12\,(4q+1)^{\frac{1}{q}}}\left\|\left\|\left(f\circ g^{-1}\right)^{(5)}\right\|\right\|_{p,g(\mathbb{R})}h^{1+\frac{1}{q}},\tag{13.54}$$

$\forall\, h>0$.

Finally we obtain (case of (13.24), etc.)

$$\left\|\left\|\left(f\circ g^{-1}\right)^{(4)}\right\|\right\|_{\infty,g(\mathbb{R})}\leq\frac{16\,\|\|f\|\|_{\infty,\mathbb{R}}}{h^4}+\frac{\left(1+4\cdot 2^{\frac{1}{q}}\right)}{3\,(4q+1)^{\frac{1}{q}}}\left\|\left\|\left(f\circ g^{-1}\right)^{(5)}\right\|\right\|_{p,g(\mathbb{R})}h^{\frac{1}{q}},\tag{13.55}$$

$\forall\, h>0$.

Next we are working on (13.50) by using Theorem 13.6.

We have the best upper bound

$$\left\|\left|\left(f\circ g^{-1}\right)^{(1)}\right\|\right|_{\infty,g(\mathbb{R})}\leq\frac{\left(4+\frac{1}{q}\right)}{\left(3+\frac{1}{q}\right)^{\frac{\left(3+\frac{1}{q}\right)}{\left(4+\frac{1}{q}\right)}}}\tag{13.56}$$

$$\left(\frac{3\,\||f\||_{\infty,\mathbb{R}}}{2}\right)^{\left(\frac{3+\frac{1}{q}}{4+\frac{1}{q}}\right)}\left[\frac{\left(1+2^{1+\frac{1}{q}}\right)}{18\,(4q+1)^{\frac{1}{q}}}\left\|\left|\left(f\circ g^{-1}\right)^{(5)}\right\|\right|_{p,g(\mathbb{R})}\right]^{\left(\frac{1}{4+\frac{1}{q}}\right)}$$

$$=\frac{\left(4+\frac{1}{q}\right)}{\left(\frac{3}{2}\left(3+\frac{1}{q}\right)\right)^{\frac{\left(3+\frac{1}{q}\right)}{\left(4+\frac{1}{q}\right)}}}\left(\frac{\left(1+2^{1+\frac{1}{q}}\right)}{18\,(4q+1)^{\frac{1}{q}}}\right)^{\left(\frac{1}{4+\frac{1}{q}}\right)}$$

$$\||f\||_{\infty,\mathbb{R}}^{\left(\frac{3+\frac{1}{q}}{4+\frac{1}{q}}\right)}\left(\left\|\left|\left(f\circ g^{-1}\right)^{(5)}\right\|\right|_{p,g(\mathbb{R})}\right)^{\left(\frac{1}{4+\frac{1}{q}}\right)}.$$

Next we are working on (13.53) by using Theorem 13.6.

We have the best upper bound

$$\left\|\left|\left(f\circ g^{-1}\right)^{(2)}\right\|\right|_{\infty,g(\mathbb{R})}\leq\frac{\left(4+\frac{1}{q}\right)}{2^{\left(\frac{2}{4+\frac{1}{q}}\right)}\left(2+\frac{1}{q}\right)^{\left(\frac{2+\frac{1}{q}}{4+\frac{1}{q}}\right)}}\tag{13.57}$$

$$\left(\frac{16}{3}\,\||f\||_{\infty,\mathbb{R}}\right)^{\left(\frac{2+\frac{1}{q}}{4+\frac{1}{q}}\right)}\left(\frac{\left(1+2^{\frac{1}{q}}\right)}{9\,(4q+1)^{\frac{1}{q}}}\left\|\left|\left(f\circ g^{-1}\right)^{(5)}\right\|\right|_{p,g(\mathbb{R})}\right)^{\left(\frac{2}{4+\frac{1}{q}}\right)}$$

$$=\frac{\left(4+\frac{1}{q}\right)}{2^{\left(\frac{2}{4+\frac{1}{q}}\right)}\left(\frac{16}{3}\left(2+\frac{1}{q}\right)\right)^{\left(\frac{2+\frac{1}{q}}{4+\frac{1}{q}}\right)}}\left(\frac{\left(1+2^{\frac{1}{q}}\right)}{9\,(4q+1)^{\frac{1}{q}}}\right)^{\left(\frac{2}{4+\frac{1}{q}}\right)}$$

$$\||f\||_{\infty,\mathbb{R}}^{\left(\frac{2+\frac{1}{q}}{4+\frac{1}{q}}\right)}\left\|\left|\left(f\circ g^{-1}\right)^{(5)}\right\|\right|_{p,g(\mathbb{R})}^{\left(\frac{2}{4+\frac{1}{q}}\right)}.$$

Next we are working on (13.54) by using Theorem 13.6. We have the best upper bound

$$\left\|\left\|\left(f \circ g^{-1}\right)^{(3)}\right\|\right\|_{\infty, g(\mathbb{R})} \leq \frac{\left(4+\frac{1}{q}\right)}{3^{\left(\frac{3}{4+\frac{1}{q}}\right)}\left(3\left(1+\frac{1}{q}\right)\right)^{\frac{\left(1+\frac{1}{q}\right)}{\left(4+\frac{1}{q}\right)}}} \tag{13.58}$$

$$\left(\frac{\left(1+8 \cdot 2^{\frac{1}{q}}\right)}{12(4q+1)^{\frac{1}{q}}}\right)^{\frac{3}{\left(4+\frac{1}{q}\right)}} \||f|\|_{\infty, \mathbb{R}}^{\left(\frac{1+\frac{1}{q}}{4+\frac{1}{q}}\right)}\left\|\left\|\left(f \circ g^{-1}\right)^{(5)}\right\|\right\|_{p, g(\mathbb{R})}^{\frac{3}{\left(4+\frac{1}{q}\right)}}.$$

Finally we are working on (13.55) by using Theorem 13.6. We have the best upper bound

$$\left\|\left\|\left(f \circ g^{-1}\right)^{(4)}\right\|\right\|_{\infty, g(\mathbb{R})} \leq \frac{\left(4+\frac{1}{q}\right)}{4^{\left(\frac{4}{4+\frac{1}{q}}\right)}\left(\frac{16}{q}\right)^{\frac{1}{q\left(4+\frac{1}{q}\right)}}} \tag{13.59}$$

$$\left(\frac{\left(1+4 \cdot 2^{\frac{1}{q}}\right)}{3(4q+1)^{\frac{1}{q}}}\right)^{\left(\frac{4}{4+\frac{1}{q}}\right)} \||f|\|_{\infty, \mathbb{R}}^{\frac{1}{q\left(4+\frac{1}{q}\right)}}\left\|\left\|\left(f \circ g^{-1}\right)^{(5)}\right\|\right\|_{p, g(\mathbb{R})}^{\left(\frac{4}{4+\frac{1}{q}}\right)}.$$

The theorem is proved. ■

We make

Remark 13.5 Theorems 13.3 and 13.4 give nice applications when $g = id$, e^x, $\tanh(x)$, etc. Due to lack of space we omit writing down the exact inequalities.

13.3 Appendix

We mention

Theorem 13.6 *Let $\alpha, \beta, A, B > 0$ and the function $y(h) = \frac{A}{h^\alpha} + Bh^\beta$, $h > 0$. It has only one critical number $h_0 = \left(\frac{\alpha A}{\beta B}\right)^{\frac{1}{(\alpha+\beta)}}$, such that $y'(h_0) = 0$. Furthermore y has a global minimum which is*

$$y(h_0) = \frac{(\alpha+\beta)}{\alpha^{\frac{\alpha}{\alpha+\beta}} \beta^{\frac{\beta}{\alpha+\beta}}} A^{\frac{\beta}{\alpha+\beta}} B^{\frac{\alpha}{\alpha+\beta}}. \tag{13.60}$$

Proof We have $y(h) = Ah^{-\alpha} + Bh^\beta$, $\forall\, h > 0$, and setting $y'(h) = -\alpha A h^{-\alpha-1} + \beta B h^{\beta-1} = 0$, we find $h_0 := h_{crit.no.} = \left(\frac{\alpha A}{\beta B}\right)^{\frac{1}{\alpha+\beta}} > 0$. We notice that

$$y''(h_0) = \alpha(\alpha+1) A h_0^{-\alpha-2} + \beta(\beta-1) B h_0^{\beta-2} = \left(\frac{\alpha A}{\beta B}\right)^{\frac{\beta-2}{\alpha+\beta}} \beta B(\alpha+\beta) > 0. \tag{13.61}$$

That is $y''(h_0) > 0$, and y has a global minimum, which is

$$y(h_0) = A h_0^{-\alpha} + B h_0^{\beta} = h_0^{\beta}\left(A h_0^{-(\alpha+\beta)} + B\right) = \frac{(\alpha+\beta)}{\alpha^{\frac{\alpha}{\alpha+\beta}} \beta^{\frac{\beta}{\alpha+\beta}}} A^{\frac{\beta}{\alpha+\beta}} B^{\frac{\alpha}{\alpha+\beta}}. \tag{13.62}$$

■

References

1. Aglic Aljinovic, A., Marangunic, L., Pecaric, J.: On Landau type inequalities via Ostrowski inequalities. Nonlinear Funct. Anal. Appl. **10**(4), 565–579 (2005)
2. Anastassiou, G.A.: Intelligent Computations: Abstract Fractional Calculus, Inequalities, Approximations. Springer, Heidelberg (2018)
3. Anastassiou, G.A.: High order abstract generalized Landau inequalities. J. Appl. Pure Math. Accepted (2020)
4. Barnett, N.S., Dragomir, S.S.: Some Landau type inequalities for functions whose derivatives are of locally bounded variation. Tamkang J. Math. **37**(4), 301–308 (2006)
5. Ditzian, Z.: Remarks, questions and conjectures on Landau-Kolmogorov-type inequalities. Math. Inequal. Appl. **3**, 15–24 (2000)
6. Hardy, G.H., Littlewood, J.E.: Some integral inequalities connected with the calculus of variations. Q. J. Math. Oxford Ser. **3**, 241–252 (1932)
7. Hardy, G.H., Landau, E., Littlewood, J.E.: Some inequalities satisfied by the integrals or derivatives of real or analytic functions. Math. Z. **39**, 677–695 (1935)
8. Kallman, R.R., Rota, G.C.: On the inequality $\|f'\|^2 \leq 4\|f\| \cdot \|f''\|$. In: Shisha, O. (ed.) Inequalities, vol. II, pp. 187–192. Academic, New York (1970)
9. Landau, E.: Einige Ungleichungen für zweimal differentzierban funktionen. Proc. Lond. Math. Soc. **13**, 43–49 (1913)
10. Mikusinski, J.: The Bochner Integral. Academic, New York (1978)
11. Shilov, G.E.: Elementary Functional Analysis. Dover Publications Inc., New York (1996)
12. Xiao, Y.: Landau type inequalities for Banach space valued functions. J. Math. Inequal. **7**(1), 103–114 (2013)

Chapter 14
Multidimensional Caputo Left Side Fractional Landau Inequalities

Relied on author's first ever found multivariate Caputo fractional Taylor's formula ([1], Chap. 13), we develop and prove several multivariate left side Caputo fractional uniform Landau type inequalities. It follows [2].

14.1 Introduction

We are motivated by the following two results. First from [4], we have the following nice result:

If $\Delta f = \sum_{i=1}^{d} \left(\frac{\partial}{\partial x_i} \right)^2 f$ is the Laplacian of a function defined on $\mathbb{R}^d$, then for $0 < k < 2n$,

$$\left\| \frac{\partial}{\partial \xi_1} \cdots \frac{\partial}{\partial \xi_k} f \right\| \leq C(n,k) \|f\|^{1-\left(\frac{k}{2n}\right)} \left\| \Delta^n f \right\|^{\frac{k}{2n}}, \tag{14.1}$$

where $\|\cdot\|$ denotes the $L_\infty\left(\mathbb{R}^d\right)$ norm. $\Delta^n f$ needs to be defined only in the distributional sense and the result is valid in many other Banach spaces such as L^p spaces. $C(n,k)$ is an existential constant and $\Delta^k = \Delta\left(\Delta^{k-1}\right)$.

From [5], we have the following interesting result which is the sharp inequality

$$\|\Delta f\|_\infty \leq 2\sqrt{\frac{d}{(d+2)}} \sqrt{\|f\|_\infty \left\|\Delta^2 f\right\|_\infty}, \tag{14.2}$$

in $\mathbb{R}^d$, $d \geq 2$, where Δf is the Laplacian and $\Delta^2 = \Delta(\Delta)$.

This result is an analogue of the famous Landau inequality [6] in $\mathbb{R}$:

$$\left\|f'\right\|_\infty \leq \sqrt{2}\sqrt{\|f\|_\infty \|f''\|_\infty}, \tag{14.3}$$

G. A. Anastassiou, *Constructive Fractional Analysis with Applications*,
Studies in Systems, Decision and Control 362,
https://doi.org/10.1007/978-3-030-71481-9_14

where $\|\cdot\|_\infty$ means the L_∞ norm on $\mathbb{R}^d$ (respectively on $\mathbb{R}$).

In this chapter we present left Caputo fractional multivariate Landau type inequalities where all the constants are exactly calculated and we have best upper bounds. We need to mention the following background coming from [1], Chap. 13, pp. 269–278.

We need

Definition 14.1 Let $\nu \geq 0$; the operator J_a^ν, defined on $L_1(a,b)$ by

$$J_a^\nu f(x) := \frac{1}{\Gamma(\nu)} \int_a^x (x-t)^{\nu-1} f(t)\,dt, \tag{14.4}$$

for $a \leq x \leq b$, is called the Riemann–Liouville left fractional integral operator of order ν. For $\nu = 0$, we set $J_a^0 := I$, the identity operator. Here Γ stands for the gamma function.

By [3], p. 13, $J_a^\nu f(x), \nu > 0$, exists for almost all $x \in [a,b]$ and $J_a^\nu f \in L_1(a,b)$, where $f \in L_1(a,b)$.

Here $AC^n([a,b])$ is the space of functions with absolutely continuous $(n-1)$-st derivative.

We need to mention

Definition 14.2 ([3], *p. 49,* [1], *p. 394*) Let $\nu \geq 0$; $n := \lceil \nu \rceil$, $\lceil \cdot \rceil$ is the ceiling of the number, $f \in AC^n([a,b])$. We call the left Caputo fractional derivative

$$D_{*a}^\nu f(x) := \frac{1}{\Gamma(n-\nu)} \int_a^x (x-t)^{n-\nu-1} f^{(n)}(t)\,dt, \tag{14.5}$$

$\forall\, x \in [a,b]$.

The above function $D_{*a}^\nu f(x)$ exists almost everywhere for $x \in [a,b]$.

If $\nu \in \mathbb{N}$, then $D_{*a}^\nu f = f^{(\nu)}$ the ordinary derivative; it is also $D_{*a}^0 f = f$.

We also mention

Theorem 14.3 (Taylor expansion for left Caputo fractional derivatives, [3, p. 54]) *Assume* $\nu \geq 0$, $n = \lceil \nu \rceil$, *and* $f \in AC^n([a,b])$. *Then*

$$f(x) = \sum_{k=0}^{n-1} \frac{f^{(k)}(a)}{k!}(x-a)^k + \frac{1}{\Gamma(\nu)} \int_a^x (x-t)^{\nu-1} D_{*a}^\nu f(t)\,dt, \tag{14.6}$$

$\forall\, x \in [a,b]$.

Remark 14.4 Let Q be a compact and convex subset of $\mathbb{R}^k$, $k \geq 2$; $z := (z_1, ..., z_k)$, $x_0 := (x_{01}, ..., x_{0k}) \in Q$. Let $f \in C^n(Q)$, $n \in \mathbb{N}$.

Set

$$g_z(t) := f(x_0 + t(z - x_0)), \quad 0 \leq t \leq 1; \tag{14.7}$$

$$g_z(0) = f(x_0), \, g_z(1) = f(z).$$

Then

$$g_z^{(j)}(t) = \left[\left(\sum_{i=1}^{k}(z_i - x_{0i})\frac{\partial}{\partial x_i}\right)^j f\right](x_0 + t(z - x_0)), \tag{14.8}$$

$j = 0, 1, 2, ..., n,$
and

$$g_z^{(n)}(0) = \left[\left(\sum_{i=1}^{k}(z_i - x_{0i})\frac{\partial}{\partial x_i}\right)^n f\right](x_0). \tag{14.9}$$

rem
We mention the following basic multivariate Caputo left fractional fundamental theorem.

Theorem 14.5 ([1], p. 272) *Let Q be a compact and convex subset of $\mathbb{R}^k$, $k \geq 2$; $z := (z_1, ..., z_k)$, $x_0 := (x_{01}, ..., x_{0k}) \in Q$, $f \in C^1(Q)$, $0 < \nu \leq 1$. Then*

$$f(z) = f(x_0) +$$

$$\sum_{i=1}^{k}(z_i - x_{0i})\left[\frac{1}{\Gamma(\nu)}\int_0^1 (1-t)^{\nu-1} J_0^{1-\nu}\left(\frac{\partial}{\partial x_i}(x_0 + t(z - x_0))\right)dt\right]. \tag{14.10}$$

We also mention the following Caputo left fractional bivariate Taylor formula.

Theorem 14.6 ([1], p. 274) *Let $f \in C^2(Q)$, $Q \subseteq \mathbb{R}^2$ compact and convex, $z := (z_1, z_2)$, $x_0 := (x_{01}, x_{02}) \in Q$, and $1 < \nu \leq 2$. Then*
(1)

$$f(z_1, z_2) = f(x_{01}, x_{02}) + (z_1 - x_{01})\frac{\partial f}{\partial x_1}(x_{01}, x_{02}) + (z_2 - x_{02})\frac{\partial f}{\partial x_2}(x_{01}, x_{02}) +$$

$$(z_1 - x_{01})^2\left(\frac{1}{\Gamma(\nu)}\int_0^1 (1-t)^{\nu-1}\left(J_0^{2-\nu}\left(\frac{\partial^2 f}{\partial x_1^2}(x_0 + t(z - x_0))\right)\right)dt\right) +$$

$$2(z_1 - x_{01})(z_2 - x_{02})\left(\frac{1}{\Gamma(\nu)}\int_0^1 (1-t)^{\nu-1}\left(J_0^{2-\nu}\left(\frac{\partial^2 f}{\partial x_1 \partial x_2}(x_0 + t(z - x_0))\right)\right)dt\right)$$

$$+ (z_2 - x_{02})^2\left(\frac{1}{\Gamma(\nu)}\int_0^1 (1-t)^{\nu-1}\left(J_0^{2-\nu}\left(\frac{\partial^2 f}{\partial x_2^2}(x_0 + t(z - x_0))\right)\right)dt\right). \tag{14.11}$$

Additionally assume that

$$f(x_0) = \frac{\partial f}{\partial x_1}(x_0) = \frac{\partial f}{\partial x_2}(x_0) = 0,$$

then

(2)

$$f(z_1, z_2) =$$

$$(z_1 - x_{01})^2 \left(\frac{1}{\Gamma(\nu)} \int_0^1 (1-t)^{\nu-1} \left(J_0^{2-\nu} \left(\frac{\partial^2 f}{\partial x_1^2}(x_0 + t(z - x_0)) \right) \right) dt \right) +$$

$$2(z_1 - x_{01})(z_2 - x_{02}) \left(\frac{1}{\Gamma(\nu)} \int_0^1 (1-t)^{\nu-1} \left(J_0^{2-\nu} \left(\frac{\partial^2 f}{\partial x_1 \partial x_2}(x_0 + t(z - x_0)) \right) \right) dt \right)$$

$$+ (z_2 - x_{02})^2 \left(\frac{1}{\Gamma(\nu)} \int_0^1 (1-t)^{\nu-1} \left(J_0^{2-\nu} \left(\frac{\partial^2 f}{\partial x_2^2}(x_0 + t(z - x_0)) \right) \right) dt \right). \tag{14.12}$$

We finally mention the following general multivariate Caputo left fractional Taylor formula.

Theorem 14.7 ([1], p. 276) *Let $\nu > 0$, $n = \lceil \nu \rceil$, $f \in C^n(Q)$, where Q is a compact and convex subset of $\mathbb{R}^k$, $k \geq 2$; $z := (z_1, ..., z_k)$, $x_0 := (x_{01}, ..., x_{0k}) \in Q$. Then*

(1)

$$f(z) = f(x_0) + \sum_{i=1}^{k} (z_i - x_{0i}) \frac{\partial f(x_0)}{\partial x_i} +$$

$$\sum_{l=2}^{n-1} \frac{\left[\left(\sum_{i=1}^{k} (z_i - x_{0i}) \frac{\partial}{\partial x_i} \right)^l f \right](x_0)}{l!} +$$

$$\frac{1}{\Gamma(\nu)} \int_0^1 (1-t)^{\nu-1} \left[J_0^{n-\nu} \left\{ \left[\left(\sum_{i=1}^{k} (z_i - x_{0i}) \frac{\partial}{\partial x_i} \right)^n f \right] (x_0 + t(z - x_0)) \right\} \right] dt. \tag{14.13}$$

Additionally assume that $f_\alpha(x_0) = 0$, $\alpha := (\alpha_1, ..., \alpha_k)$, $\alpha_i \in \mathbb{Z}^+$, $i = 1, ..., k$, $|\alpha| := \sum_{i=1}^{k} \alpha_i =: r$, $r = 0, ..., n-1$; then

(2)

$$f(z) =$$

$$\frac{1}{\Gamma(\nu)} \int_0^1 (1-t)^{\nu-1} \left[J_0^{n-\nu} \left\{ \left[\left(\sum_{i=1}^{k} (z_i - x_{0i}) \frac{\partial}{\partial x_i} \right)^n f \right] (x_0 + t(z - x_0)) \right\} \right] dt. \tag{14.14}$$

14.2 Main Results

Here we present several left Caputo fractional multivariate Landau inequalities.

Theorem 14.8 *Let* $f \in C^2\left(\mathbb{R}_+^k\right)$, $k \geq 2$, *with* $\|f\|_{\infty,\mathbb{R}_+^k} < \infty$, *and* $1 < \nu < 2$. *Assume further that*

$$K_\nu := \max \left\{ \sup_{\substack{x_0, z \in \mathbb{R}_+^k, t \in [0,1], \\ i=1,\ldots,k}} \left| J_0^{2-\nu} \left(f_{x_i x_i} \left(x_0 + t \left(z - x_0 \right) \right) \right) \right|, \right.$$

$$\left. \sup_{\substack{x_0, z \in \mathbb{R}_+^k, t \in [0,1], \\ \left\{ l = l(i,j) = 1, \ldots, \frac{k(k-1)}{2};\ i < j \right.}} \left| J_0^{2-\nu} \left(f_{x_i x_j} \left(x_0 + t \left(z - x_0 \right) \right) \right) \right| \right\} < +\infty. \quad (14.15)$$

Then we derive the best upper bound

$$\left\| \sum_{i=1}^{k} \frac{\partial f}{\partial x_i} \right\|_{\infty,\mathbb{R}_+^k} \leq 2k \sqrt{\frac{2}{\Gamma(\nu+1)}} \|f\|_{\infty,\mathbb{R}_+^k}^{\frac{1}{2}} K_\nu^{\frac{1}{2}}. \quad (14.16)$$

That is

$$\left\| \sum_{i=1}^{k} \frac{\partial f}{\partial x_i} \right\|_{\infty,\mathbb{R}_+^k} < \infty.$$

Proof For any $x_0, z \in \mathbb{R}_+^k$ by Theorem 14.7 (see also Theorem 14.6) we have that

$$f(z) = f(x_0) + \sum_{i=1}^{k} (z_i - x_{0i}) \frac{\partial f(x_0)}{\partial x_i} +$$

$$\frac{1}{\Gamma(\nu)} \int_0^1 (1-t)^{\nu-1} \left[J_0^{2-\nu} \left\{ \left[\left(\sum_{i=1}^{k} (z_i - x_{0i}) \frac{\partial}{\partial x_i} \right)^2 f \right] (x_0 + t(z - x_0)) \right\} \right] dt$$

$$= f(x_0) + \sum_{i=1}^{k} (z_i - x_{0i}) \frac{\partial f(x_0)}{\partial x_i} + \quad (14.17)$$

$$\sum_{i=1}^{k}(z_i - x_{0i})^2 \frac{1}{\Gamma(\nu)} \int_0^1 (1-t)^{\nu-1} J_0^{2-\nu}\left(f_{x_i x_i}(x_0 + t(z - x_0))\right) dt +$$

$$2 \sum_{\begin{cases} l = l(i,j) = 1 \\ : i < j \end{cases}}^{\frac{k(k-1)}{2}} (z_i - x_{0i})(z_j - x_{0j}) \frac{1}{\Gamma(\nu)}$$

$$\int_0^1 (1-t)^{\nu-1} J_0^{2-\nu}\left(f_{x_i x_j}(x_0 + t(z - x_0))\right) dt.$$

We choose $z_i - x_{0i} = h > 0$, for all $i = 1, ..., k$.

Then

$$f(z) = f(x_0) + h \sum_{i=1}^{k} \frac{\partial f}{\partial x_i}(x_0) +$$

$$\frac{h^2}{\Gamma(\nu)} \left[\sum_{i=1}^{k} \int_0^1 (1-t)^{\nu-1} J_0^{2-\nu}\left(f_{x_i x_i}(x_0 + t(z - x_0))\right) dt + \right. \tag{14.18}$$

$$\left. 2 \sum_{\begin{cases} l = l(i,j) = 1 \\ : i < j \end{cases}}^{\frac{k(k-1)}{2}} \int_0^1 (1-t)^{\nu-1} J_0^{2-\nu}\left(f_{x_i x_j}(x_0 + t(z - x_0))\right) dt \right].$$

Hence

$$h \sum_{i=1}^{k} \frac{\partial f(x_0)}{\partial x_i} = f(z) - f(x_0) -$$

$$\frac{h^2}{\Gamma(\nu)} \left[\sum_{i=1}^{k} \int_0^1 (1-t)^{\nu-1} J_0^{2-\nu}\left(f_{x_i x_i}(x_0 + t(z - x_0))\right) dt + \right. \tag{14.19}$$

$$\left. 2 \sum_{\begin{cases} l = l(i,j) = 1 \\ : i < j \end{cases}}^{\frac{k(k-1)}{2}} \int_0^1 (1-t)^{\nu-1} J_0^{2-\nu}\left(f_{x_i x_j}(x_0 + t(z - x_0))\right) dt \right].$$

Therefore

$$h\left|\sum_{i=1}^{k}\frac{\partial f}{\partial x_i}(x_0)\right| \leq 2\|f\|_{\infty,\mathbb{R}_+^k}+$$

$$\frac{h^2}{\Gamma(\nu+1)}\left\{\sum_{i=1}^{k}\left\|J_0^{2-\nu}\left(f_{x_ix_i}(x_0+t(z-x_0))\right)\right\|_{\infty,[0,1]}+\right.$$

$$\left.2\sum_{\begin{cases} l=l(i,j)=1 \\ :i<j \end{cases}}^{\frac{k(k-1)}{2}}\left\|J_0^{2-\nu}\left(f_{x_ix_j}(x_0+t(z-x_0))\right)\right\|_{\infty,[0,1]}\right] \leq$$

$$2\|f\|_{\infty,\mathbb{R}_+^k}+\frac{K_\nu h^2}{\Gamma(\nu+1)}\left\{\sum_{i=1}^{k}1+2\sum_{l=1}^{\frac{k(k-1)}{2}}1\right\}=$$

$$2\|f\|_{\infty,\mathbb{R}_+^k}+\frac{k^2K_\nu h^2}{\Gamma(\nu+1)}. \tag{14.20}$$

Thus

$$\left|\sum_{i=1}^{k}\frac{\partial f}{\partial x_i}(x_0)\right| \leq \frac{2\|f\|_{\infty,\mathbb{R}_+^k}}{h}+\frac{k^2K_\nu}{\Gamma(\nu+1)}h, \tag{14.21}$$

$\forall\, h>0$, for any $x_0 \in \mathbb{R}_+^k$.

Consequently we find that

$$\left\|\sum_{i=1}^{k}\frac{\partial f}{\partial x_i}\right\|_{\infty,\mathbb{R}_+^k} \leq \frac{2\|f\|_{\infty,\mathbb{R}_+^k}}{h}+\frac{k^2K_\nu}{\Gamma(\nu+1)}h, \tag{14.22}$$

$\forall\, h>0$.

Set $A := 2\|f\|_{\infty,\mathbb{R}_+^k}$ and $B := \frac{k^2K_\nu}{\Gamma(\nu+1)}$, both are positive.

We consider the function

$$y(h)=\frac{A}{h}+Bh,\ \forall\, h>0. \tag{14.23}$$

We have that

$$y'(h)=-Ah^{-2}+B=0,$$

with only one critical number

$$h_0 = \sqrt{\frac{A}{B}}. \tag{14.24}$$

Notice that $y''(h) = 2Ah^{-3}$, and $y''(h_0) = 2A^{-\frac{1}{2}}B^{\frac{3}{2}} > 0$.

Therefore y has a global minimum which is

$$y(h_0) = 2\sqrt{AB}. \tag{14.25}$$

Hence

$$\left\| \sum_{i=1}^{k} \frac{\partial f}{\partial x_i} \right\|_{\infty,\mathbb{R}_+^k} \leq \frac{2\sqrt{2}k}{\sqrt{\Gamma(\nu+1)}} \sqrt{\|f\|_{\infty,\mathbb{R}_+^k} K_\nu}. \tag{14.26}$$

The theorem is proved. ■

We continue with

Theorem 14.9 *Let* $f \in C^3\left(\mathbb{R}_+^k\right)$, $k \geq 2$, *with* $\|f\|_{\infty,\mathbb{R}_+^k} < \infty$, *and* $2 < \nu < 3$. *Assume further that*

$$M_\nu :=$$

$$\sup_{x_0,z\in\mathbb{R}_+^k,t\in[0,1]} \left| \sum_{\substack{\lambda_1+\ldots+\lambda_k=3 \\ \lambda_i\in\mathbb{Z}^+,\, i=1,\ldots,k}} \frac{1}{\lambda_1!\ldots\lambda_k!} J_0^{3-\nu}\left(f_{x_1^{\lambda_1}\ldots x_k^{\lambda_k}}\left(x_0+t\left(z-x_0\right)\right)\right) \right| < +\infty. \tag{14.27}$$

Then we derive the best upper bounds:

(1)

$$\left\| \sum_{i=1}^{k} \frac{\partial f}{\partial x_i} \right\|_{\infty,\mathbb{R}_+^k} \leq 3\left(\frac{3}{2}\right)^{\frac{2}{3}} \left(\frac{12}{\Gamma(\nu+1)}\right)^{\frac{1}{3}} \|f\|_{\infty,\mathbb{R}_+^k}^{\frac{2}{3}} M_\nu^{\frac{1}{3}} < +\infty, \tag{14.28}$$

and

(2)

$$\left\| \frac{1}{2}\left(\sum_{i=1}^{k} \frac{\partial^2 f}{\partial x_i^2}\right) + \sum_{\substack{l=l(i,j)=1, \\ i<j}}^{\frac{k(k-1)}{2}} \frac{\partial^2 f}{\partial x_i \partial x_j} \right\| \leq 3\left(\frac{9}{\Gamma(\nu+1)}\right)^{\frac{2}{3}} \|f\|_{\infty,\mathbb{R}_+^k}^{\frac{1}{3}} M_\nu^{\frac{2}{3}} < +\infty. \tag{14.29}$$

Proof We set

$$\Delta_1(x_0) := \sum_{i=1}^{k} \frac{\partial f}{\partial x_i}(x_0),$$

and

$$\Delta_2(x_0) := \frac{1}{2}\left(\sum_{i=1}^{k} \frac{\partial^2 f}{\partial x_i^2}(x_0)\right) + \sum_{\substack{l = l(i,j) = 1, \\ i < j}}^{\frac{k(k-1)}{2}} \frac{\partial^2 f(x_0)}{\partial x_i \partial x_j} \tag{14.30}$$

By Theorem 14.7 we have, we choose also $z_i - x_{0i} = h > 0$, all $i = 1, ..., k$:

$$f(z) = f(x_0) + h\Delta_1(x_0) + h^2\Delta_2(x_0) +$$

$$\frac{1}{\Gamma(\nu)} \int_0^1 (1-t)^{\nu-1} \left\{ J_0^{3-\nu} \left[\left(\sum_{i=1}^{k} (z_i - x_{0i}) \frac{\partial}{\partial x_i} \right)^3 f \right] (x_0 + t(z - x_0)) \right\} dt$$

$$= f(x_0) + \Delta_1(x_0) h + \Delta_2(x_0) h^2 + h^3 \frac{6}{\Gamma(\nu)}$$

$$\left[\sum_{\substack{(\lambda_1 + ... + \lambda_k = 3 \\ \lambda_i \in \mathbb{Z}^+, i = 1, ..., k)}} \frac{1}{\lambda_1! ... \lambda_k!} \int_0^1 (1-t)^{\nu-1} J_0^{3-\nu} \left(f_{x_1^{\lambda_1} ... x_k^{\lambda_k}} (x_0 + t(z - x_0)) \right) dt \right]. \tag{14.31}$$

We set

$$R_3 := \frac{6}{\Gamma(\nu)}$$

$$\left[\sum_{\substack{(\lambda_1 + ... + \lambda_k = 3 \\ \lambda_i \in \mathbb{Z}^+, i = 1, ..., k)}} \frac{1}{\lambda_1! ... \lambda_k!} \int_0^1 (1-t)^{\nu-1} J_0^{3-\nu} \left(f_{x_1^{\lambda_1} ... x_k^{\lambda_k}} (x_0 + t(z - x_0)) \right) dt \right]. \tag{14.32}$$

That is ($h > 0$)

$$f(z) = f(x_0) + \Delta_1(x_0) h + \Delta_2(x_0) h^2 + R_3 h^3. \tag{14.33}$$

So that

$$\Delta_1(x_0) h + \Delta_2(x_0) h^2 = f(z) - f(x_0) - R_3 h^3 =: A_1, \tag{14.34}$$

and

$$2\Delta_1(x_0)h + 4\Delta_2(x_0)h^2 = f(z) - f(x_0) - 8R_3h^3 =: A_2. \tag{14.35}$$

Solving the system (14.34), (14.35) we find

$$\begin{array}{l} \Delta_1(x_0) = \frac{4A_1 - A_2}{2h}, \\ \text{and} \\ \Delta_2(x_0) = \frac{A_2 - 2A_1}{2h^2}. \end{array} \tag{14.36}$$

We notice that

$$|4A_1 - A_2| = \left|3(f(z) - f(x_0)) + 4R_3h^3\right| \le 6\|f\|_{\infty,\mathbb{R}_+^k} +$$

$$4h^3 \frac{6}{\Gamma(\nu+1)} \left\| \sum_{\substack{(\lambda_1 + ... + \lambda_k = 3 \\ \lambda_i \in \mathbb{Z}^+, i = 1, ..., k)}} \frac{1}{\lambda_1! ... \lambda_k!} J_0^{3-\nu} \left(f_{x_1^{\lambda_1} ... x_k^{\lambda_k}} (x_0 + t(z - x_0))\right) \right\|_{\infty,[0,1]}$$

$$\le 6\|f\|_{\infty,\mathbb{R}_+^k} + \frac{24}{\Gamma(\nu+1)} M_\nu h^3. \tag{14.37}$$

Hence

$$|\Delta_1(x_0)| \le \frac{3\|f\|_{\infty,\mathbb{R}_+^k}}{h} + \frac{12M_\nu}{\Gamma(\nu+1)} h^2, \tag{14.38}$$

$\forall\, h > 0$, for all $x_0 \in \mathbb{R}_+^k$.

Consequently it holds

$$\|\Delta_1\|_{\infty,\mathbb{R}_+^k} \le \frac{3\|f\|_{\infty,\mathbb{R}_+^k}}{h} + \frac{12M_\nu}{\Gamma(\nu+1)} h^2, \tag{14.39}$$

$\forall\, h > 0$.

Next we have

$$A_2 - 2A_1 = -\left((f(z) - f(x_0)) + 6R_3h^3\right). \tag{14.40}$$

So that

$$|A_2 - 2A_1| = \left|f(z) - f(x_0) + 6R_3h^3\right| \le 2\|f\|_{\infty,\mathbb{R}_+^k} + 6h^3 \frac{6}{\Gamma(\nu+1)} M_\nu. \tag{14.41}$$

Hence

$$|\Delta_2(x_0)| \le \frac{\|f\|_{\infty,\mathbb{R}_+^k}}{h^2} + \frac{18M_\nu}{\Gamma(\nu+1)} h, \tag{14.42}$$

$\forall\, h > 0, \forall\, x_0 \in \mathbb{R}_+^k$.

So that

$$\|\Delta_2\|_{\infty,\mathbb{R}_+^k} \leq \frac{\|f\|_{\infty,\mathbb{R}_+^k}}{h^2} + \left(\frac{18M_\nu}{\Gamma(\nu+1)}\right)h, \tag{14.43}$$

$\forall\, h > 0$.

Next we work on (14.39).

Call $\mu := 3\|f\|_{\infty,\mathbb{R}_+^k}$, $\theta := \frac{12M_\nu}{\Gamma(\nu+1)}$ that are both positive.

We study the function

$$y(h) = \frac{\mu}{h} + \theta h^2, \ \forall\, h > 0. \tag{14.44}$$

We have

$$y'(h) = -\mu h^{-2} + 2\theta h = 0.$$

Hence

$$h_0 = h_{crit.no.} = \left(\frac{\mu}{2\theta}\right)^{\frac{1}{3}} > 0. \tag{14.45}$$

We have $y''(h) = 2\mu h^{-3} + 2\theta$, and

$$y''(h_0) = 2\mu h_0^{-3} + 2\theta > 0. \tag{14.46}$$

Therefore y has a global minimum which is

$$y(h_0) = \mu h_0^{-1} + \theta h_0^2 = h_0^2\left(\mu h_0^{-3} + \theta\right) = \frac{3}{2^{\frac{2}{3}}}\theta^{\frac{1}{3}}\mu^{\frac{2}{3}}. \tag{14.47}$$

Hence

$$y(h_0) = \frac{3}{2^{\frac{2}{3}}}\theta^{\frac{1}{3}}\mu^{\frac{2}{3}} = 3\left(\frac{3}{2}\right)^{\frac{2}{3}}\left(\frac{12}{\Gamma(\nu+1)}\right)^{\frac{1}{3}}\|f\|_{\infty,\mathbb{R}_+^k}^{\frac{2}{3}} M_\nu^{\frac{1}{3}}. \tag{14.48}$$

We have proved that

$$\|\Delta_1\|_{\infty,\mathbb{R}_+^k} \leq 3\left(\frac{3}{2}\right)^{\frac{2}{3}}\left(\frac{12}{\Gamma(\nu+1)}\right)^{\frac{1}{3}}\|f\|_{\infty,\mathbb{R}_+^k}^{\frac{2}{3}} M_\nu^{\frac{1}{3}}. \tag{14.49}$$

Next we work on (14.43).

We call $\varphi := \|f\|_{\infty,\mathbb{R}_+^k}$, $\psi := \frac{18M_\nu}{\Gamma(\nu+1)}$, that are both positive.

We study

$$y(h) := \frac{\varphi}{h^2} + \psi h, \ \forall\, h > 0. \tag{14.50}$$

We have

$$y'(h) = -2\varphi h^{-3} + \psi = 0,$$

so that

$$h_0 := h_{crit.no.} = \left(\frac{2\varphi}{\psi}\right)^{\frac{1}{3}} > 0. \tag{14.51}$$

We have

$$y''(h) = 6\varphi h^{-4},$$

and

$$y''(h_0) = 6\varphi h_0^{-4} > 0. \tag{14.52}$$

Therefore y has a global minimum which is

$$y(h_0) = \varphi h_0^{-2} + \psi h_0 = h_0\left(\varphi h_0^{-3} + \psi\right) = h_0\left(\varphi\frac{\psi}{2\varphi} + \psi\right) = \frac{3}{2}2^{\frac{1}{3}}\psi^{\frac{2}{3}}\varphi^{\frac{1}{3}}. \tag{14.53}$$

That is

$$y(h_0) = \frac{3}{2}\sqrt[3]{2}\psi^{\frac{2}{3}}\varphi^{\frac{1}{3}}. \tag{14.54}$$

Consequently it holds

$$\|\Delta_2\|_{\infty,\mathbb{R}_+^k} \leq 3\left(\frac{9}{\Gamma(\nu+1)}\right)^{\frac{2}{3}} \|f\|_{\infty,\mathbb{R}_+^k}^{\frac{1}{3}} M_\nu^{\frac{2}{3}}. \tag{14.55}$$

The theorem is proved. ■

We also present

Theorem 14.10 *Let $f \in C^4\left(\mathbb{R}_+^k\right)$, $k \geq 2$, with $\|f\|_{\infty,\mathbb{R}_+^k} < \infty$, and $3 < \nu < 4$. Assume further that*

$$H_\nu :=$$

$$\sup_{x_0,z\in\mathbb{R}_+^k,t\in[0,1]} \left| \sum_{\left\{\begin{array}{c}\lambda_1+\ldots+\lambda_k=4\\ \lambda_i\in\mathbb{Z}^+,i=1,\ldots,k\end{array}\right\}} \frac{1}{\lambda_1!\ldots\lambda_k!} J_0^{4-\nu}\left(f_{x_1^{\lambda_1}\ldots x_k^{\lambda_k}}\left(x_0+t\left(z-x_0\right)\right)\right)\right| < +\infty. \tag{14.56}$$

Then we derive the best upper bounds:

(1)

$$\left\|\sum_{i=1}^k \frac{\partial f}{\partial x_i}\right\|_{\infty,\mathbb{R}_+^k} \leq 4\left(\frac{11}{9}\right)^{\frac{3}{4}}\left(\frac{144}{\Gamma(\nu+1)}\right)^{\frac{1}{4}} \|f\|_{\infty,\mathbb{R}_+^k}^{\frac{3}{4}} H_\nu^{\frac{1}{4}} < +\infty, \tag{14.57}$$

(2)

$$\left\| \frac{1}{2}\left(\sum_{i=1}^{k}\frac{\partial^2 f}{\partial x_i^2}\right) + \sum_{\left\{\begin{array}{c} l = l(i,j) = 1, \\ i < j \end{array}\right.}^{\frac{k(k-1)}{2}} \frac{\partial^2 f}{\partial x_i \partial x_j} \right\|_{\infty, \mathbb{R}_+^k} \leq$$

$$2\sqrt{\frac{528}{\Gamma(\nu+1)}}\sqrt{\|f\|_{\infty,\mathbb{R}_+^k} H_\nu} < +\infty, \tag{14.58}$$

and

(3)

$$\left\| \sum_{\left\{\begin{array}{c} \lambda_1 + ... + \lambda_k = 3 \\ \lambda_i \in \mathbb{Z}^+, i = 1, ..., k \end{array}\right\}} \frac{1}{\lambda_1!...\lambda_k!} f_{x_1^{\lambda_1}...x_k^{\lambda_k}} \right\|_{\infty, \mathbb{R}_+^k} \leq$$

$$\frac{4}{3}\left(\frac{144}{\Gamma(\nu+1)}\right)^{\frac{3}{4}} \|f\|_{\infty,\mathbb{R}_+^k}^{\frac{1}{4}} H_\nu^{\frac{3}{4}} < +\infty. \tag{14.59}$$

Proof We set

$$\Delta_1(x_0) := \sum_{i=1}^{k} \frac{\partial f}{\partial x_i}(x_0),$$

$$\Delta_2(x_0) := \frac{1}{2}\left(\sum_{i=1}^{k}\frac{\partial^2 f}{\partial x_i^2}(x_0)\right) + \sum_{\left\{\begin{array}{c} l = l(i,j) = 1, \\ i < j \end{array}\right.}^{\frac{k(k-1)}{2}} \frac{\partial^2 f(x_0)}{\partial x_i \partial x_j}, \tag{14.60}$$

and

$$\Delta_3(x_0) := \sum_{\left\{\begin{array}{c} \lambda_1 + ... + \lambda_k = 3 \\ \lambda_i \in \mathbb{Z}^+, i = 1, ..., k \end{array}\right\}} \frac{1}{\lambda_1!...\lambda_k!} f_{x_1^{\lambda_1}...x_k^{\lambda_k}}(x_0).$$

By Theorem 14.7 we have, we choose also $z_i - x_{0i} = h > 0$, all $i = 1, ..., k$:

$$f(z) = f(x_0) + \Delta_1(x_0)h + \Delta_2(x_0)h^2 + \Delta_3(x_0)h^3 + R_4 h^4, \tag{14.61}$$

where

$$R_4 := \frac{1}{\Gamma(\nu)} \left[\sum_{\begin{cases} \lambda_1 + ... + \lambda_k = 4, \\ \lambda_i \in \mathbb{Z}^+, i = 1, ..., k \end{cases}} \left(\frac{4!}{\lambda_1! ... \lambda_k!} \right) \right.$$

$$\left. \int_0^1 (1-t)^{\nu-1} J_0^{4-\nu} \left(f_{x_1^{\lambda_1} ... x_k^{\lambda_k}} (x_0 + t(z - x_0)) \right) dt \right]. \quad (14.62)$$

Therefore, $\forall\, h > 0$, we get

$$\begin{array}{l} \Delta_1(x_0) h + \Delta_2(x_0) h^2 + \Delta_3(x_0) h^3 = f(z) - f(x_0) - R_4 h^4 =: A_1, \\ \Delta_1(x_0) 2h + \Delta_2(x_0) 4h^2 + \Delta_3(x_0) 8h^3 = f(z) - f(x_0) - R_4 16 h^4 =: A_2, \\ \text{and} \\ \Delta_1(x_0) 3h + \Delta_2(x_0) 9h^2 + \Delta_3(x_0) 27h^3 = f(z) - f(x_0) - R_4 81 h^4 =: A_3. \end{array} \quad (14.63)$$

We solve the system of three equations (14.63) and we find:

$$\begin{array}{l} \Delta_1(x_0) = \frac{18A_1 - 9A_2 + 2A_3}{6h}, \\ \Delta_2(x_0) = \frac{4A_2 - 5A_1 - A_3}{2h^2}, \\ \text{and} \\ \Delta_3(x_0) = \frac{3A_1 - 3A_2 + A_3}{6h^3}. \end{array} \quad (14.64)$$

We have also that

$$|R_4| \leq \frac{24 H_\nu}{\Gamma(\nu + 1)}. \quad (14.65)$$

It is

$$18A_1 - 9A_2 + 2A_3 = 11(f(z) - f(x_0)) - 36 R_4 h^4. \quad (14.66)$$

Therefore it holds

$$|\Delta_1(x_0)| \leq \frac{11 \|f\|_{\infty, \mathbb{R}_+^k}}{3h} + \frac{144 H_\nu}{\Gamma(\nu + 1)} h^3, \quad (14.67)$$

$\forall\, h > 0, \forall\, x_0 \in \mathbb{R}_+^k$.

We get that

$$\|\Delta_1\|_{\infty, \mathbb{R}_+^k} \leq \frac{11 \|f\|_{\infty, \mathbb{R}_+^k}}{3h} + \left(\frac{144 H_\nu}{\Gamma(\nu + 1)} \right) h^3, \quad (14.68)$$

$\forall\, h > 0$.

It is

$$4A_2 - 5A_1 - A_3 = -2(f(z) - f(x_0)) + 22 R_4 h^4. \quad (14.69)$$

Hence

$$|\Delta_2(x_0)| \leq \frac{2\|f\|_{\infty,\mathbb{R}_+^k}}{h^2} + \left(\frac{264H_\nu}{\Gamma(\nu+1)}\right)h^2, \tag{14.70}$$

$\forall\, h > 0, \forall\, x_0 \in \mathbb{R}_+^k$.

The last implies

$$\|\Delta_2\|_{\infty,\mathbb{R}_+^k} \leq \frac{2\|f\|_{\infty,\mathbb{R}_+^k}}{h^2} + \left(\frac{264H_\nu}{\Gamma(\nu+1)}\right)h^2, \tag{14.71}$$

$\forall\, h > 0$.

It is

$$3A_1 - 3A_2 + A_3 = f(z) - f(x_0) - 36R_4h^4. \tag{14.72}$$

Therefore

$$|\Delta_3(x_0)| \leq \frac{\|f\|_{\infty,\mathbb{R}_+^k}}{3h^3} + \left(\frac{144H_\nu}{\Gamma(\nu+1)}\right)h, \tag{14.73}$$

$\forall\, h > 0, \forall\, x_0 \in \mathbb{R}_+^k$.

The last implies

$$\|\Delta_3\|_{\infty,\mathbb{R}_+^k} \leq \frac{\|f\|_{\infty,\mathbb{R}_+^k}}{3h^3} + \left(\frac{144H_\nu}{\Gamma(\nu+1)}\right)h, \tag{14.74}$$

$\forall\, h > 0$.

We will work on (14.68).

Call $\mu := \frac{11\|f\|_{\infty,\mathbb{R}_+^k}}{3}$, $\theta := \left(\frac{144H_\nu}{\Gamma(\nu+1)}\right)$, both are positive.

We study

$$y(h) = \frac{\mu}{h} + \theta h^3, \ \forall\, h > 0. \tag{14.75}$$

We have

$$y'(h) = -\mu h^{-2} + 3\theta h^2 = 0,$$

with one critical number

$$h_{crit.no.} = h_0 = \left(\frac{\mu}{3\theta}\right)^{\frac{1}{4}} > 0. \tag{14.76}$$

We have that

$$y''(h_0) = 2\mu h_0^{-3} + 6\theta h_0 > 0, \tag{14.77}$$

implying that y has a global minimum. We see that

$$y(h_0) = \mu h_0^{-1} + \theta h_0^3 = h_0^3 \left(\mu h_0^{-4} + \theta\right) = \frac{4}{3^{\frac{3}{4}}} \theta^{\frac{1}{4}} \mu^{\frac{3}{4}}. \tag{14.78}$$

I.e.

$$y(h_0) = \frac{4}{3^{\frac{3}{4}}} \theta^{\frac{1}{4}} \mu^{\frac{3}{4}}, \tag{14.79}$$

is the global minimum.

Consequently

$$\|\Delta_1\|_{\infty,\mathbb{R}_+^k} \leq 4 \left(\frac{11}{9}\right)^{\frac{3}{4}} \left(\frac{144}{\Gamma(\nu+1)}\right)^{\frac{1}{4}} \|f\|_{\infty,\mathbb{R}_+^k}^{\frac{3}{4}} H_\nu^{\frac{1}{4}}. \tag{14.80}$$

Next we work on (14.71).

Call $\varphi := 2\|f\|_{\infty,\mathbb{R}_+^k}$, $\psi := \frac{264 H_\nu}{\Gamma(\nu+1)} > 0$.

We study

$$y(h) := \frac{\varphi}{h^2} + \psi h^2, \ \forall\, h > 0. \tag{14.81}$$

We have

$$y'(h) = -2\varphi h^{-3} + 2\psi h = 0, \tag{14.82}$$

with

$$h_{crit.no.} = h_0 = \left(\frac{\varphi}{\psi}\right)^{\frac{1}{4}} > 0.$$

Notice that

$$y''(h_0) = 6\varphi h_0^{-4} + 2\psi > 0. \tag{14.83}$$

Thus we have that y has global minimum which is

$$y(h_0) = \varphi h_0^{-2} + \psi h_0^2 = h_0^2 \left(\varphi h_0^{-4} + \psi\right) = 2\psi^{\frac{1}{2}} \varphi^{\frac{1}{2}},$$

that is

$$y(h_0) = 2\psi^{\frac{1}{2}} \varphi^{\frac{1}{2}}. \tag{14.84}$$

Therefore we derive

$$\|\Delta_2\|_{\infty,\mathbb{R}_+^k} \leq 2\sqrt{\frac{528}{\Gamma(\nu+1)}} \sqrt{\|f\|_{\infty,\mathbb{R}_+^k} H_\nu}. \tag{14.85}$$

Finally we work on (14.74).

Let $A := \frac{\|f\|_{\infty,\mathbb{R}_+^k}}{3}$, $B := \frac{144 H_\nu}{\Gamma(\nu+1)}$, both are positive.

We study

$$y(h) := \frac{A}{h^3} + Bh, \ \forall h > 0. \tag{14.86}$$

Then

$$y'(h) = -3Ah^{-4} + B = 0,$$

so that

$$h_{crit.no.} = h_0 = \left(\frac{3A}{B}\right)^{\frac{1}{4}} > 0. \tag{14.87}$$

Here

$$y''(h_0) = 12Ah_0^{-5} > 0. \tag{14.88}$$

Hence y has global minimum which is

$$y(h_0) = Ah_0^{-3} + Bh_0 = h_0\left(Ah_0^{-4} + B\right) = \frac{4}{3^{\frac{3}{4}}} A^{\frac{1}{4}} B^{\frac{3}{4}}. \tag{14.89}$$

That is

$$y(h_0) = \frac{4}{3^{\frac{3}{4}}} A^{\frac{1}{4}} B^{\frac{3}{4}}, \tag{14.90}$$

the global minimum of y.

Consequently it holds

$$\|\Delta_3\|_{\infty,\mathbb{R}_+^k} \leq \frac{4}{3}\left(\frac{144}{\Gamma(\nu+1)}\right)^{\frac{3}{4}} \|f\|_{\infty,\mathbb{R}_+^k}^{\frac{1}{4}} H_\nu^{\frac{3}{4}}. \tag{14.91}$$

The theorem is proved. ∎

References

1. Anastassiou, G.A.: Fractional Differentiation Inequalities. Springer, Heidelberg (2009)
2. Anastassiou, G.A.: Multivariate Caputo left fractional Landau inequalities. Moroc. J. Pure Appl. Anal. **6**(2), 266–280 (2020)
3. Diethelm, K.: The Analysis of Fractional Differentiation Equations. Springer, Heidelberg (2010)
4. Ditzian, Z.: Multivariate Landau-Kolmogorov-type inequality. Math. Proc. Camb. Philos. Soc. **105**(2), 335–350 (1989)
5. Kounchev, O.: Extremizers for the multivariate Landau-Kolmogorov inequality. In: Multivariate Approximation (Witten-Bommerholz, 1996), pp. 123–132. Math. Res. 101. Akademie Verlag, Berlin (1997)
6. Landau, E.: Einige Ungleichungen für zweimal differentzierban funktionen. Proc. Lond. Math. Soc. **13**, 43–49 (1913)

Chapter 15
Multidimensional Left Canavati Fractional Landau Inequalities

Based on author's first ever found multivariate fractional Taylor's formula ([2], Chap. 12), we develop and prove several multivariate left side Canavati fractional uniform Landau type inequalities. It follows [4].

15.1 Introduction

We are motivated by the following two results. First from [6], we have the following nice result:

If $\Delta f = \sum_{i=1}^{d} \left(\frac{\partial}{\partial x_i}\right)^2 f$ is the Laplacian of a function defined on $\mathbb{R}^d$, then for $0 < k < 2n$,

$$\left\| \frac{\partial}{\partial \xi_1} \cdots \frac{\partial}{\partial \xi_k} f \right\| \leq C(n,k) \|f\|^{1-\left(\frac{k}{2n}\right)} \left\| \Delta^n f \right\|^{\frac{k}{2n}}, \tag{15.1}$$

where $\|\cdot\|$ denotes the $L_\infty\left(\mathbb{R}^d\right)$ norm. $\Delta^n f$ needs to be defined only in the distributional sense and the result is valid in many other Banach spaces such as L^p spaces. $C(n,k)$ is an existential constant and $\Delta^k = \Delta\left(\Delta^{k-1}\right)$.

From [7], we have the following interesting result which is the sharp inequality

$$\|\Delta f\|_\infty \leq 2\sqrt{\frac{d}{(d+2)}} \sqrt{\|f\|_\infty \left\|\Delta^2 f\right\|_\infty}, \tag{15.2}$$

in $\mathbb{R}^d$, $d \geq 2$, where Δf is the Laplacian and $\Delta^2 = \Delta(\Delta)$.

G. A. Anastassiou, *Constructive Fractional Analysis with Applications*,
Studies in Systems, Decision and Control 362,
https://doi.org/10.1007/978-3-030-71481-9_15

This result is an analogue of the famous Landau inequality [8] in $\mathbb{R}$:

$$\|f'\|_{\infty} \leq \sqrt{2}\sqrt{\|f\|_{\infty} \|f''\|_{\infty}}, \tag{15.3}$$

where $\|\cdot\|_{\infty}$ means the L_{∞} norm on $\mathbb{R}^d$ (respectively on $\mathbb{R}$).

In this chapter we present left Canavati [5] fractional multivariate Landau type inequalities where all the constants are exactly calculated and we have best upper bounds. We need to mention the following background coming from [2], Chap. 12, pp. 257–268.

We make

Remark 15.1 We follow Anastassiou [3, p. 540] (see also Canavati [5] and Anastassiou [1]).

Let $[a, b] \subseteq \mathbb{R}$. Let $x, x_0 \in [a, b]$ such that $x \geq x_0$, x_0 is fixed. Let $f \in C([a, b])$ and define

$$\left(J_{\nu}^{x_0} f\right)(x) = \frac{1}{\Gamma(\nu)} \int_{x_0}^{x} (x-t)^{\nu-1} f(t)\, dt, \ x_0 \leq x \leq b, \tag{15.4}$$

$\nu > 0$, the left generalized Riemann–Liouville integral, where Γ is the gamma function. We consider the subspace $C_{x_0}^{\nu}([a, b])$ of $C^n([a, b])$, $n := [\nu]$ (the integral part), $\alpha := \nu - n$ $(0 < \alpha < 1)$:

$$C_{x_0}^{\nu}([a, b]) := \left\{f \in C^n([a, b]) : J_{1-\alpha}^{x_0} f^{(n)} \in C^1([x_0, b])\right\}. \tag{15.5}$$

Hence, let $f \in C_{x_0}^{\nu}([a, b])$ we define the left generalized ν-fractional derivative of f over $[x_0, b]$ (see also Canavati [5] and Anastassiou [1]) as

$$D_{x_0}^{\nu} f := \left(J_{1-\alpha}^{x_0} f^{(n)}\right)'. \tag{15.6}$$

Notice that

$$\left(J_{1-\alpha}^{x_0} f^{(n)}\right)(x) = \frac{1}{\Gamma(1-\alpha)} \int_{x_0}^{x} (x-t)^{-\alpha} f^{(n)}(t)\, dt \tag{15.7}$$

exists for $f \in C_{x_0}^{\nu}([a, b])$.

Let $f_{x_0}(t) := f(x_0 + t)$, $0 \leq t \leq b - x_0$, $x \geq x_0$. By change of variable we obtain

$$\left(D_0^{\nu} f_{x_0}\right)(x - x_0) = \left(D_{x_0}^{\nu} f\right)(x). \tag{15.8}$$

When $\nu \in \mathbb{N}$ then the fractional derivative collapses to the usual one.

We mention the left fractional Taylor formula. See Anastassiou [3, p. 540], Canavati [5], and Anastassiou [2].

Theorem 15.2 *Let $f \in C_{x_0}^{\nu}([a, b])$, $x_0 \in [a, b]$ fixed.*

(i) If $\nu \geq 1$, then it holds

$$f(x) = f(x_0) + f'(x_0)(x - x_0) + f''(x_0)\frac{(x - x_0)^2}{2} + ... + f^{(n-1)}(x_0)\frac{(x - x_0)^{n-1}}{(n-1)!}$$

$$+ \left(J_\nu^{x_0} D_{x_0}^\nu f\right)(x), \text{ all } x \in [a, b] : x \geq x_0. \tag{15.9}$$

(ii) If $0 < \nu < 1$ we have

$$f(x) = \left(J_\nu^{x_0} D_{x_0}^\nu f\right)(x), \text{ all } x \in [a, b] : x \geq x_0. \tag{15.10}$$

Using Theorem 15.2 we established the basic multivariate Canavati left fractional Taylor formula.

Theorem 15.3 ([2]) *Let $f \in C^1(Q)$, Q compact and convex $\subseteq \mathbb{R}^k$, $k \geq 2$. For fixed $x_0, z \in Q$, assume that as a function of t : $f_{x_i}(x_0 + t(z - x_0)) \in C^{\nu-1}([0, 1])$, $1 \leq \nu < 2$, all $i = 1, ..., k$. Then*
(i)

$$f(z_1, ..., z_k) = f(x_{01}, ..., x_{0k}) +$$

$$\sum_{i=1}^{k} \frac{(z_i - x_{0i})}{\Gamma(\nu)} \int_0^1 (1 - t)^{\nu-1} \left(f_{x_i}(x_0 + t(z - x_0))\right)^{(\nu-1)} dt. \tag{15.11}$$

(ii) Given $f(x_0) = 0$, then

$$f(z) = \sum_{i=1}^{k} \frac{(z_i - x_{0i})}{\Gamma(\nu)} \int_0^1 (1 - t)^{\nu-1} \left(f_{x_i}(x_0 + t(z - x_0))\right)^{(\nu-1)} dt. \tag{15.12}$$

We mention the next left fractional Taylor result.

Theorem 15.4 ([2]) *Let $f \in C^2(Q)$, Q compact and convex $\subseteq \mathbb{R}^2$. For fixed $x_0, z \in Q$, assume that as functions of t : $f_{x_1 x_1}(x_0 + t(z - x_0))$, $f_{x_1 x_2}(x_0 + t(z - x_0))$, $f_{x_2 x_2}(x_0 + t(z - x_0)) \in C^{(\nu-2)}([0, 1])$, where $2 \leq \nu < 3$. Then*
(i)

$$f(z_1, z_2) = f(x_{01}, x_{02}) + (z_1 - x_{01})\frac{\partial f}{\partial x_1}(x_0) + (z_2 - x_{02})\frac{\partial f}{\partial x_2}(x_0) +$$

$$(z_1 - x_{01})^2 \frac{1}{\Gamma(\nu)} \int_0^1 (1 - t)^{\nu-1} \left(f_{x_1 x_1}(x_0 + t(z - x_0))\right)^{(\nu-2)} dt +$$

$$2(z_1 - x_{01})(z_2 - x_{02}) \frac{1}{\Gamma(\nu)} \int_0^1 (1 - t)^{\nu-1} \left(f_{x_1 x_2}(x_0 + t(z - x_0))\right)^{(\nu-2)} dt +$$

$$(z_2 - x_{02})^2 \frac{1}{\Gamma(\nu)} \int_0^1 (1-t)^{\nu-1} \left(f_{x_2 x_2}(x_0 + t(z - x_0))\right)^{(\nu-2)} dt. \tag{15.13}$$

(ii) When $f(x_0) = \frac{\partial f}{\partial x_1}(x_0) = \frac{\partial f}{\partial x_2}(x_0) = 0$, *then*

$$f(z_1, z_2) = (z_1 - x_{01})^2 \frac{1}{\Gamma(\nu)} \int_0^1 (1-t)^{\nu-1} \left(f_{x_1 x_1}(x_0 + t(z - x_0))\right)^{(\nu-2)} dt$$

$$+2(z_1 - x_{01})(z_2 - x_{02}) \frac{1}{\Gamma(\nu)} \int_0^1 (1-t)^{\nu-1} \left(f_{x_1 x_2}(x_0 + t(z - x_0))\right)^{(\nu-2)} dt$$

$$+ (z_2 - x_{02})^2 \frac{1}{\Gamma(\nu)} \int_0^1 (1-t)^{\nu-1} \left(f_{x_2 x_2}(x_0 + t(z - x_0))\right)^{(\nu-2)} dt. \tag{15.14}$$

We also mention the following general multivariate Canavati left fractional Taylor formula.

Theorem 15.5 ([2]) *Let* $f \in C^n(Q)$, Q *compact and convex* $\subseteq \mathbb{R}^k$, $k \geq 2$; *here* $\nu \geq 1$ *such that* $n = [\nu]$. *For fixed* $x_0, z \in Q$ *assume that as functions of* $t : f_\alpha(x_0 + t(z - x_0)) \in C^{(\nu-n)}([0,1])$, *for all* $\alpha := (\alpha_1, ..., \alpha_k)$, $\alpha_i \in \mathbb{Z}^+$, $i = 1, ..., k$; $|\alpha| := \sum_{i=1}^{k} \alpha_i = n$. *Then*

(i)

$$f(z_1, ..., z_k) = f(x_{01}, ..., x_{0k}) + \sum_{i=1}^{k} (z_i - x_{0i}) \frac{\partial f}{\partial x_i}(x_{01}, ..., x_{0k}) +$$

$$\frac{\left[\left(\sum_{i=1}^{k} (z_i - x_{0i}) \frac{\partial}{\partial x_i}\right)^2 f\right](x_{01}, ..., x_{0k})}{2} + ... +$$

$$\frac{\left[\left(\sum_{i=1}^{k} (z_i - x_{0i}) \frac{\partial}{\partial x_i}\right)^{n-1} f\right](x_{01}, ..., x_{0k})}{(n-1)!} +$$

$$\frac{1}{\Gamma(\nu)} \int_0^1 (1-t)^{\nu-1} \left\{\left[\left(\sum_{i=1}^{k} (z_i - x_{0i}) \frac{\partial}{\partial x_i}\right)^n f\right]^{(\nu-n)} (x_0 + t(z - x_0))\right\} dt. \tag{15.15}$$

(ii) If all $f_\alpha(x_0) = 0$, $\alpha := (\alpha_1, ..., \alpha_k)$, $\alpha_i \in \mathbb{Z}^+$, $i = 1, ..., k$, $|\alpha| := \sum_{i=1}^{k} \alpha_i = l$, $l = 0, ..., n-1$, *then*

$$f(z_1, ..., z_k) =$$

$$\frac{1}{\Gamma(\nu)}\int_0^1 (1-t)^{\nu-1}\left\{\left[\left(\sum_{i=1}^{k}(z_i-x_{0i})\frac{\partial}{\partial x_i}\right)^n f\right]^{(\nu-n)}(x_0+t(z-x_0))\right\}dt. \tag{15.16}$$

(Note that left fractional differentiation is a linear operation).

15.2 Main Results

Here we present several left Canavati fractional multivariate Landau inequalities.

Theorem 15.6 *Let $f\in C^2\left(\mathbb{R}_+^k\right)$, $k\geq 2$, with $\|f\|_{\infty,\mathbb{R}_+^k}<\infty$, and $2<\nu<3$. For any $x_0, z\in\mathbb{R}_+^k$, assume that as function of t: $f_\alpha(x_0+t(z-x_0))\in C^{(\nu-2)}([0,1])$, for all $\alpha:=(\alpha_1,...,\alpha_k)$, $\alpha_i\in\mathbb{Z}^+$, $i=1,...,k$; $|\alpha|:=\sum_{i=1}^{k}\alpha_i=2$. Assume further that*

$$K_\nu := \max\left\{\sup_{\substack{x_0,z\in\mathbb{R}_+^k, t\in[0,1],\\ i=1,\dots,k}} \left|\left(f_{x_ix_i}(x_0+t(z-x_0))\right)^{(\nu-2)}\right|,\right.$$

$$\left.\sup_{\substack{x_0,z\in\mathbb{R}_+^k, t\in[0,1],\\ \left\{\substack{l=l(i,j)=1,\dots,\frac{k(k-1)}{2};\\ i<j}\right.}} \left|\left(f_{x_ix_j}(x_0+t(z-x_0))\right)^{(\nu-2)}\right|\right\}<+\infty. \tag{15.17}$$

Then we derive the best upper bound

$$\left\|\sum_{i=1}^{k}\frac{\partial f}{\partial x_i}\right\|_{\infty,\mathbb{R}_+^k}\leq 2k\sqrt{\frac{2}{\Gamma(\nu+1)}}\,\|f\|_{\infty,\mathbb{R}_+^k}^{\frac{1}{2}}K_\nu^{\frac{1}{2}}. \tag{15.18}$$

That is

$$\left\|\sum_{i=1}^{k}\frac{\partial f}{\partial x_i}\right\|_{\infty,\mathbb{R}_+^k}<\infty.$$

Proof By Theorems 15.4 and 15.5 we have for any $x_0, z\in\mathbb{R}_+^k$ that

$$f(z)=f(x_0)+\sum_{i=1}^{k}(z_i-x_{0i})\frac{\partial f}{\partial x_i}(x_0)+$$

$$\frac{1}{\Gamma(\nu)}\int_0^1 (1-t)^{\nu-1}\left\{\left[\left(\sum_{i=1}^{k}(z_i-x_{0i})\frac{\partial}{\partial x_i}\right)^2 f\right]^{(\nu-2)}(x_0+t(z-x_0))\right\}dt=$$

$$f(x_0)+\sum_{i=1}^{k}(z_i-x_{0i})\frac{\partial f}{\partial x_i}(x_0)+ \tag{15.19}$$

$$\sum_{i=1}^{k}(z_i-x_{0i})^2\frac{1}{\Gamma(\nu)}\int_0^1(1-t)^{\nu-1}\left(f_{x_ix_i}(x_0+t(z-x_0))\right)^{(\nu-2)}dt+$$

$$2\sum_{\begin{cases} l=l(i,j)=1 \\ :i<j \end{cases}}^{\frac{k(k-1)}{2}}(z_i-x_{0i})(z_j-x_{0j})\frac{1}{\Gamma(\nu)}\int_0^1(1-t)^{\nu-1}\left(f_{x_ix_j}(x_0+t(z-x_0))\right)^{(\nu-2)}dt.$$

We choose $z_i - x_{0i} = h > 0$, for all $i = 1, ..., k$.

Then

$$f(z)=f(x_0)+h\sum_{i=1}^{k}\frac{\partial f}{\partial x_i}(x_0)+ \tag{15.20}$$

$$\frac{h^2}{\Gamma(\nu)}\left[\sum_{i=1}^{k}\int_0^1(1-t)^{\nu-1}\left(f_{x_ix_i}(x_0+t(z-x_0))\right)^{(\nu-2)}dt+\right.$$

$$\left.2\sum_{\begin{cases} l=l(i,j)=1 \\ :i<j \end{cases}}^{\frac{k(k-1)}{2}}\int_0^1(1-t)^{\nu-1}\left(f_{x_ix_j}(x_0+t(z-x_0))\right)^{(\nu-2)}dt\right].$$

Hence

$$h\sum_{i=1}^{k}\frac{\partial f}{\partial x_i}(x_0)=f(z)-f(x_0)-$$

$$\frac{h^2}{\Gamma(\nu)}\left[\sum_{i=1}^{k}\int_0^1(1-t)^{\nu-1}\left(f_{x_ix_i}(x_0+t(z-x_0))\right)^{(\nu-2)}dt+\right. \tag{15.21}$$

$$2 \sum_{\left\{\begin{array}{l} l=l(i,j)=1 \\ \quad : i<j \end{array}\right.}^{\frac{k(k-1)}{2}} \int_0^1 (1-t)^{\nu-1} \left(f_{x_i x_j}\left(x_0+t\left(z-x_0\right)\right)\right)^{(\nu-2)} dt \Bigg] .$$

Therefore

$$h \left| \sum_{i=1}^{k} \frac{\partial f}{\partial x_i}(x_0) \right| \leq 2 \|f\|_{\infty, \mathbb{R}_+^k} +$$

$$\frac{h^2}{\Gamma(\nu+1)} \left\{ \sum_{i=1}^{k} \left\| \left(f_{x_i x_i}\left(x_0+t\left(z-x_0\right)\right)\right)^{(\nu-2)} \right\|_{\infty,[0,1]} + \right.$$

$$2 \sum_{\left\{\begin{array}{l} l=l(i,j)=1 \\ \quad : i<j \end{array}\right.}^{\frac{k(k-1)}{2}} \left\| \left(f_{x_i x_j}\left(x_0+t\left(z-x_0\right)\right)\right)^{(\nu-2)} \right\|_{\infty,[0,1]} \Bigg] \leq$$

$$2 \|f\|_{\infty, \mathbb{R}_+^k} + \frac{K_\nu h^2}{\Gamma(\nu+1)} \left\{ \sum_{i=1}^{k} 1 + 2 \sum_{l=1}^{\frac{k(k-1)}{2}} 1 \right\} =$$

$$2 \|f\|_{\infty, \mathbb{R}_+^k} + \frac{k^2 K_\nu h^2}{\Gamma(\nu+1)}. \tag{15.22}$$

Thus

$$\left| \sum_{i=1}^{k} \frac{\partial f}{\partial x_i}(x_0) \right| \leq \frac{2 \|f\|_{\infty, \mathbb{R}_+^k}}{h} + \frac{k^2 K_\nu}{\Gamma(\nu+1)} h, \tag{15.23}$$

$\forall\, h>0$, for any $x_0 \in \mathbb{R}_+^k$.

Consequently we find that

$$\left\| \sum_{i=1}^{k} \frac{\partial f}{\partial x_i} \right\|_{\infty, \mathbb{R}_+^k} \leq \frac{2 \|f\|_{\infty, \mathbb{R}_+^k}}{h} + \frac{k^2 K_\nu}{\Gamma(\nu+1)} h, \tag{15.24}$$

$\forall\, h>0$.

Set $A := 2 \|f\|_{\infty, \mathbb{R}_+^k}$ and $B := \frac{k^2 K_\nu}{\Gamma(\nu+1)}$, both are positive.

We consider the function

$$y(h) = \frac{A}{h} + Bh = Ah^{-1} + Bh, \ \forall\, h > 0. \tag{15.25}$$

We have that

$$y'(h) = -Ah^{-2} + B = 0,$$

with only one critical number

$$h_0 = \sqrt{\frac{A}{B}}. \tag{15.26}$$

Notice that $y''(h) = 2Ah^{-3}$, and $y''(h_0) = 2A^{-\frac{1}{2}}B^{\frac{3}{2}} > 0$.
Therefore y has a global minimum which is

$$y(h_0) = y\left(\sqrt{\frac{A}{B}}\right) = 2\sqrt{AB}. \tag{15.27}$$

Hence

$$\left\| \sum_{i=1}^{k} \frac{\partial f}{\partial x_i} \right\|_{\infty,\mathbb{R}_+^k} \leq 2\sqrt{2\,\|f\|_{\infty,\mathbb{R}_+^k} \frac{k^2 K_\nu}{\Gamma(\nu+1)}} = \frac{2\sqrt{2}k}{\sqrt{\Gamma(\nu+1)}}\sqrt{\|f\|_{\infty,\mathbb{R}_+^k} K_\nu}. \tag{15.28}$$

The theorem is proved. ■

We continue with

Theorem 15.7 *Let* $f \in C^3\left(\mathbb{R}_+^k\right)$, $k \geq 2$, *with* $\|f\|_{\infty,\mathbb{R}_+^k} < \infty$, *and* $3 < \nu < 4$. *For any* $x_0, z \in \mathbb{R}_+^k$, *assume that as functions of* t : $f_\alpha(x_0 + t(z - x_0)) \in C^{(\nu-3)}([0,1])$, *for all* $\alpha := (\alpha_1, ..., \alpha_k)$, $\alpha_i \in \mathbb{Z}^+$, $i = 1, ..., k$; $|\alpha| := \sum_{i=1}^{k} \alpha_i = 3$. *Assume further that*

$$M_\nu :=$$

$$\sup_{x_0,z\in\mathbb{R}_+^k, t\in[0,1]} \left| \sum_{\left\{ \begin{array}{c} \lambda_1 + ... + \lambda_k = 3 \\ \lambda_i \in \mathbb{Z}^+, i = 1, ..., k \end{array} \right\}} \frac{1}{\lambda_1!...\lambda_k!} \left(f_{x_1^{\lambda_1}...x_k^{\lambda_k}}(x_0 + t(z - x_0)) \right)^{(\nu-3)} \right| < +\infty. \tag{15.29}$$

Then we derive the best upper bounds:

(1)

$$\left\| \sum_{i=1}^{k} \frac{\partial f}{\partial x_i} \right\|_{\infty,\mathbb{R}_+^k} \leq 3\left(\frac{3}{2}\right)^{\frac{2}{3}} \left(\frac{12}{\Gamma(\nu+1)}\right)^{\frac{1}{3}} \|f\|_{\infty,\mathbb{R}_+^k}^{\frac{2}{3}} M_\nu^{\frac{1}{3}} < +\infty, \tag{15.30}$$

and

(2)

$$\left\| \frac{1}{2}\left(\sum_{i=1}^{k}\frac{\partial^2 f}{\partial x_i^2}\right) + \sum_{\substack{l = l(i,j) = 1, \\ i<j}}^{\frac{k(k-1)}{2}} \frac{\partial^2 f}{\partial x_i \partial x_j} \right\| \leq 3\left(\frac{9}{\Gamma(\nu+1)}\right)^{\frac{2}{3}} \|f\|_{\infty,\mathbb{R}_+^k}^{\frac{1}{3}} M_\nu^{\frac{2}{3}} < +\infty. \tag{15.31}$$

Proof We set

$$\Delta_1(x_0) := \sum_{i=1}^{k}\frac{\partial f}{\partial x_i}(x_0),$$

and

$$\Delta_2(x_0) := \frac{1}{2}\left(\sum_{i=1}^{k}\frac{\partial^2 f}{\partial x_i^2}(x_0)\right) + \sum_{\substack{l = l(i,j) = 1, \\ i<j}}^{\frac{k(k-1)}{2}} \frac{\partial^2 f(x_0)}{\partial x_i \partial x_j} \tag{15.32}$$

By Theorem 15.5 we have, we choose also $z_i - x_{0i} = h > 0$, all $i = 1, ..., k$:

$$f(z) = f(x_0) + h\Delta_1(x_0) + h^2\Delta_2(x_0) +$$

$$\frac{1}{\Gamma(\nu)}\int_0^1 (1-t)^{\nu-1}\left\{\left[\left(\sum_{i=1}^{k}(z_i - x_{0i})\frac{\partial}{\partial x_i}\right)^3 f\right]^{(\nu-3)}(x_0 + t(z-x_0))\right\}dt$$

$$= f(x_0) + \Delta_1(x_0)h + \Delta_2(x_0)h^2 + h^3\frac{6}{\Gamma(\nu)}$$

$$\left[\sum_{\substack{(\lambda_1 + ... + \lambda_k = 3 \\ \lambda_i \in \mathbb{Z}^+, i = 1,...,k)}} \frac{1}{\lambda_1!...\lambda_k!}\int_0^1 (1-t)^{\nu-1}\left(f_{x_1^{\lambda_1}...x_k^{\lambda_k}}(x_0 + t(z-x_0))\right)^{(\nu-3)} dt\right]. \tag{15.33}$$

We set

$$R_3 := \frac{6}{\Gamma(\nu)}$$

$$\left[\sum_{\substack{(\lambda_1 + ... + \lambda_k = 3 \\ \lambda_i \in \mathbb{Z}^+, i = 1, ..., k)}} \frac{1}{\lambda_1! ... \lambda_k!} \int_0^1 (1-t)^{\nu-1} \left(f_{x_1^{\lambda_1} ... x_k^{\lambda_k}} (x_0 + t (z - x_0)) \right)^{(\nu-3)} dt \right]. \tag{15.34}$$

That is $(h > 0)$

$$f(z) = f(x_0) + \Delta_1 (x_0) h + \Delta_2 (x_0) h^2 + R_3 h^3. \tag{15.35}$$

So that

$$\Delta_1 (x_0) h + \Delta_2 (x_0) h^2 = f(z) - f(x_0) - R_3 h^3 =: A_1, \tag{15.36}$$

and

$$2\Delta_1 (x_0) h + 4\Delta_2 (x_0) h^2 = f(z) - f(x_0) - 8R_3 h^3 =: A_2. \tag{15.37}$$

Solving the system (15.36), (15.37) we find

$$\begin{gathered} \Delta_1 (x_0) = \frac{4A_1 - A_2}{2h}, \\ \text{and} \\ \Delta_2 (x_0) = \frac{A_2 - 2A_1}{2h^2}. \end{gathered} \tag{15.38}$$

We notice that

$$|4A_1 - A_2| = \left|4 (f(z) - f(x_0)) - 4R_3 h^3 - (f(z) - f(x_0)) + 8R_3 h^3\right| =$$

$$\left|3 (f(z) - f(x_0)) + 4R_3 h^3\right| \leq 6 \|f\|_{\infty, \mathbb{R}_+^k} +$$

$$4h^3 \frac{6}{\Gamma(\nu+1)} \left\| \sum_{\substack{(\lambda_1 + ... + \lambda_k = 3 \\ \lambda_i \in \mathbb{Z}^+, i = 1, ..., k)}} \frac{1}{\lambda_1! ... \lambda_k!} f_{x_1^{\lambda_1} ... x_k^{\lambda_k}} (x_0 + t (z - x_0))^{(\nu-3)} \right\|_{\infty, [0,1]}$$

$$\leq 6 \|f\|_{\infty, \mathbb{R}_+^k} + \frac{24}{\Gamma(\nu+1)} M_\nu h^3. \tag{15.39}$$

Hence

$$|\Delta_1 (x_0)| \leq \frac{3 \|f\|_{\infty, \mathbb{R}_+^k}}{h} + \frac{12 M_\nu}{\Gamma(\nu+1)} h^2, \tag{15.40}$$

$\forall\, h > 0$, for all $x_0 \in \mathbb{R}_+^k$.

Consequently it holds

$$\|\Delta_1\|_{\infty,\mathbb{R}_+^k} \leq \frac{3\|f\|_{\infty,\mathbb{R}_+^k}}{h} + \frac{12M_\nu}{\Gamma(\nu+1)}h^2, \tag{15.41}$$

$\forall\, h > 0$.

Next we have

$$\begin{aligned} A_2 - 2A_1 &= f(z) - f(x_0) - 8R_3h^3 - 2(f(z) - f(x_0)) + 2R_3h^3 \\ &= -\left((f(z) - f(x_0)) + 6R_3h^3\right). \end{aligned} \tag{15.42}$$

So that

$$|A_2 - 2A_1| = \left|f(z) - f(x_0) + 6R_3h^3\right| \leq 2\|f\|_{\infty,\mathbb{R}_+^k} + 6h^3\frac{6}{\Gamma(\nu+1)}M_\nu. \tag{15.43}$$

Hence

$$|\Delta_2(x_0)| \leq \frac{\|f\|_{\infty,\mathbb{R}_+^k}}{h^2} + \frac{18M_\nu}{\Gamma(\nu+1)}h, \tag{15.44}$$

$\forall\, h > 0, \forall\, x_0 \in \mathbb{R}_+^k$.

So that

$$\|\Delta_2\|_{\infty,\mathbb{R}_+^k} \leq \frac{\|f\|_{\infty,\mathbb{R}_+^k}}{h^2} + \left(\frac{18M_\nu}{\Gamma(\nu+1)}\right)h, \tag{15.45}$$

$\forall\, h > 0$.

Next we work on (15.41).

Call $\mu := 3\|f\|_{\infty,\mathbb{R}_+^k}$, $\theta := \frac{12M_\nu}{\Gamma(\nu+1)}$ that are both positive.

We study the function

$$y(h) = \frac{\mu}{h} + \theta h^2 = \mu h^{-1} + \theta h^2. \tag{15.46}$$

We have

$$y'(h) = -\mu h^{-2} + 2\theta h = 0.$$

Hence

$$2\theta h = \mu h^{-2} = \frac{\mu}{h^2},$$

$$2\theta h^3 = \mu,$$

$$h_0 = h_{crit.no.} = \left(\frac{\mu}{2\theta}\right)^{\frac{1}{3}} > 0. \tag{15.47}$$

We have $y''(h) = 2\mu h^{-3} + 2\theta$, and

$$y''(h_0) = 2\mu h_0^{-3} + 2\theta > 0. \tag{15.48}$$

Therefore y has a global minimum which is

$$y(h_0) = \mu h_0^{-1} + \theta h_0^2 = h_0^2\left(\mu h_0^{-3} + \theta\right) = h_0^2\left(\mu\frac{2\theta}{\mu} + \theta\right) \tag{15.49}$$

$$= 3\theta h_0^2 = 3\theta\left(\frac{\mu}{2\theta}\right)^{\frac{2}{3}} = \frac{3}{2^{\frac{2}{3}}}\theta^{\frac{1}{3}}\mu^{\frac{2}{3}}.$$

Hence

$$y(h_0) = \frac{3}{2^{\frac{2}{3}}}\theta^{\frac{1}{3}}\mu^{\frac{2}{3}} = 3\left(\frac{3}{2}\right)^{\frac{2}{3}}\left(\frac{12}{\Gamma(\nu+1)}\right)^{\frac{1}{3}}\|f\|_{\infty,\mathbb{R}_+^k}^{\frac{2}{3}} M_\nu^{\frac{1}{3}}. \tag{15.50}$$

We have proved that

$$\|\Delta_1\|_{\infty,\mathbb{R}_+^k} \leq 3\left(\frac{3}{2}\right)^{\frac{2}{3}}\left(\frac{12}{\Gamma(\nu+1)}\right)^{\frac{1}{3}}\|f\|_{\infty,\mathbb{R}_+^k}^{\frac{2}{3}} M_\nu^{\frac{1}{3}}. \tag{15.51}$$

Next we work on (15.45).

We call $\varphi := \|f\|_{\infty,\mathbb{R}_+^k}$, $\psi := \frac{18M_\nu}{\Gamma(\nu+1)}$, that are both positive.

We study

$$y(h) := \frac{\varphi}{h^2} + \psi h = \varphi h^{-2} + \psi h. \tag{15.52}$$

We have

$$y'(h) = -2\varphi h^{-3} + \psi = 0,$$

and

$$\psi = 2\varphi h^{-3} = \frac{2\varphi}{h^3},$$

$$\psi h^3 = 2\varphi,$$

so that

$$h_0 := h_{crit.no.} = \left(\frac{2\varphi}{\psi}\right)^{\frac{1}{3}} > 0. \tag{15.53}$$

We have

$$y''(h) = 6\varphi h^{-4},$$

and

$$y''(h_0) = 6\varphi h_0^{-4} > 0. \tag{15.54}$$

Therefore y has a global minimum which is

$$y(h_0) = \varphi h_0^{-2} + \psi h_0 = h_0\left(\varphi h_0^{-3} + \psi\right) = h_0\left(\varphi \frac{\psi}{2\varphi} + \psi\right)$$

$$= \frac{3}{2}\psi h_0 = \frac{3}{2}\psi\left(\frac{2\varphi}{\psi}\right)^{\frac{1}{3}} = \frac{3}{2} 2^{\frac{1}{3}} \psi^{\frac{2}{3}} \varphi^{\frac{1}{3}}. \tag{15.55}$$

That is

$$y(h_0) = \frac{3}{2}\sqrt[3]{2}\psi^{\frac{2}{3}}\varphi^{\frac{1}{3}}. \tag{15.56}$$

Consequently it holds

$$\|\Delta_2\|_{\infty,\mathbb{R}_+^k} \leq \frac{3\sqrt[3]{2}}{2}\|f\|_{\infty,\mathbb{R}_+^k}^{\frac{1}{3}}\left(\frac{18}{\Gamma(\nu+1)}\right)^{\frac{2}{3}} M_\nu^{\frac{2}{3}} =$$

$$3\left(\frac{9}{\Gamma(\nu+1)}\right)^{\frac{2}{3}}\|f\|_{\infty,\mathbb{R}_+^k}^{\frac{1}{3}} M_\nu^{\frac{2}{3}}. \tag{15.57}$$

The theorem is proved. ■

We also present

Theorem 15.8 *Let $f \in C^4\left(\mathbb{R}_+^k\right)$, $k \geq 2$, with $\|f\|_{\infty,\mathbb{R}_+^k} < \infty$, and $4 < \nu < 5$. For any $x_0, z \in \mathbb{R}_+^k$, assume that as functions of t : $f_\alpha(x_0 + t(z - x_0)) \in C^{(\nu-4)}([0,1])$, for all $\alpha := (\alpha_1, ..., \alpha_k)$, $\alpha_i \in \mathbb{Z}^+$, $i = 1, ..., k$; $|\alpha| := \sum_{i=1}^k \alpha_i = 4$. Assume further that*

$$H_\nu :=$$

$$\sup_{x_0,z\in\mathbb{R}_+^k,t\in[0,1]}\left|\sum_{\left\{\begin{array}{c}\lambda_1+...+\lambda_k=4\\ \lambda_i\in\mathbb{Z}^+, i=1,...,k\end{array}\right\}}\frac{1}{\lambda_1!...\lambda_k!}\left(f_{x_1^{\lambda_1}...x_k^{\lambda_k}}(x_0+t(z-x_0))\right)^{(\nu-4)}\right| < +\infty. \tag{15.58}$$

Then we derive the best upper bounds:

(1)

$$\left\|\sum_{i=1}^k \frac{\partial f}{\partial x_i}\right\|_{\infty,\mathbb{R}_+^k} \leq 4\left(\frac{11}{9}\right)^{\frac{3}{4}}\left(\frac{144}{\Gamma(\nu+1)}\right)^{\frac{1}{4}}\|f\|_{\infty,\mathbb{R}_+^k}^{\frac{3}{4}} H_\nu^{\frac{1}{4}} < +\infty, \tag{15.59}$$

(2)

$$\left\| \frac{1}{2}\left(\sum_{i=1}^{k}\frac{\partial^2 f}{\partial x_i^2}\right) + \sum_{\left\{\begin{array}{c} l=l(i,j)=1, \\ i<j \end{array}\right.}^{\frac{k(k-1)}{2}} \frac{\partial^2 f}{\partial x_i \partial x_j} \right\|_{\infty,\mathbb{R}_+^k} \leq$$

$$2\sqrt{\frac{528}{\Gamma(\nu+1)}}\sqrt{\|f\|_{\infty,\mathbb{R}_+^k} H_\nu} < +\infty, \tag{15.60}$$

and

(3)

$$\left\| \sum_{\left\{\begin{array}{c} \lambda_1+\ldots+\lambda_k=3 \\ \lambda_i\in\mathbb{Z}^+, i=1,\ldots,k \end{array}\right\}} \frac{1}{\lambda_1!\ldots\lambda_k!} f_{x_1^{\lambda_1}\ldots x_k^{\lambda_k}} \right\|_{\infty,\mathbb{R}_+^k} \leq$$

$$\frac{4}{3}\left(\frac{144}{\Gamma(\nu+1)}\right)^{\frac{3}{4}} \|f\|_{\infty,\mathbb{R}_+^k}^{\frac{1}{4}} H_\nu^{\frac{3}{4}} < +\infty. \tag{15.61}$$

Proof We set

$$\begin{aligned}
&\Delta_1(x_0) := \sum_{i=1}^{k}\frac{\partial f}{\partial x_i}(x_0), \\
&\Delta_2(x_0) := \frac{1}{2}\left(\sum_{i=1}^{k}\frac{\partial^2 f}{\partial x_i^2}(x_0)\right) + \sum_{\left\{\begin{array}{c} l=l(i,j)=1, \\ i<j \end{array}\right.}^{\frac{k(k-1)}{2}} \frac{\partial^2 f(x_0)}{\partial x_i \partial x_j}, \\
&\text{and} \\
&\Delta_3(x_0) := \sum_{\left\{\begin{array}{c} \lambda_1+\ldots+\lambda_k=3 \\ \lambda_i\in\mathbb{Z}^+, i=1,\ldots,k \end{array}\right\}} \frac{1}{\lambda_1!\ldots\lambda_k!} f_{x_1^{\lambda_1}\ldots x_k^{\lambda_k}}(x_0).
\end{aligned} \tag{15.62}$$

By Theorem 15.5 we have, we choose also $z_i - x_{0i} = h > 0$, all $i = 1, \ldots, k$:

$$f(z) = f(x_0) + \Delta_1(x_0)h + \Delta_2(x_0)h^2 + \Delta_3(x_0)h^3 + R_4 h^4, \tag{15.63}$$

where

$$R_4 := \frac{1}{\Gamma(\nu)} \left[\sum_{\begin{cases} \lambda_1 + ... + \lambda_k = 4, \\ \lambda_i \in \mathbb{Z}^+, i = 1, ..., k \end{cases}} \left(\frac{4!}{\lambda_1!...\lambda_k!} \right) \right.$$

$$\left. \int_0^1 (1-t)^{\nu-1} \left(f_{x_1^{\lambda_1}...x_k^{\lambda_k}} (x_0 + t(z - x_0)) \right)^{(\nu-4)} dt \right]. \tag{15.64}$$

Therefore, $\forall\, h > 0$, we get

$$\Delta_1(x_0)\, h + \Delta_2(x_0)\, h^2 + \Delta_3(x_0)\, h^3 = f(z) - f(x_0) - R_4 h^4 =: A_1,$$
$$\Delta_1(x_0)\, 2h + \Delta_2(x_0)\, 4h^2 + \Delta_3(x_0)\, 8h^3 = f(z) - f(x_0) - R_4 16 h^4 =: A_2,$$
and
$$\Delta_1(x_0)\, 3h + \Delta_2(x_0)\, 9h^2 + \Delta_3(x_0)\, 27h^3 = f(z) - f(x_0) - R_4 81 h^4 =: A_3. \tag{15.65}$$

We solve the system of three equations (15.65) and we find:

$$\Delta_1(x_0) = \tfrac{18A_1 - 9A_2 + 2A_3}{6h},$$
$$\Delta_2(x_0) = \tfrac{4A_2 - 5A_1 - A_3}{2h^2},$$
and
$$\Delta_3(x_0) = \tfrac{3A_1 - 3A_2 + A_3}{6h^3}. \tag{15.66}$$

We have also that

$$|R_4| \leq \frac{24 H_\nu}{\Gamma(\nu+1)}. \tag{15.67}$$

We calculate

$$18A_1 - 9A_2 + 2A_3 = 18(f(z) - f(x_0)) - 18R_4 h^4 - 9(f(z) - f(x_0)) + R_4 144 h^4$$

$$+\, 2(f(z) - f(x_0)) - R_4 162 h^4 = 11(f(z) - f(x_0)) - 36 R_4 h^4. \tag{15.68}$$

Therefore it holds

$$|\Delta_1(x_0)| \leq \frac{22 \|f\|_{\infty, \mathbb{R}_+^k}}{6h} + \frac{36 \left(\frac{24 H_\nu}{\Gamma(\nu+1)} \right) h^4}{6h} \tag{15.69}$$

$$= \frac{11 \|f\|_{\infty, \mathbb{R}_+^k}}{3h} + \frac{144 H_\nu}{\Gamma(\nu+1)} h^3,$$

$\forall\, h > 0, \forall\, x_0 \in \mathbb{R}_+^k$.

We get that

$$\|\Delta_1\|_{\infty,\mathbb{R}_+^k} \leq \frac{11\,\|f\|_{\infty,\mathbb{R}_+^k}}{3h} + \left(\frac{144H_\nu}{\Gamma(\nu+1)}\right)h^3, \tag{15.70}$$

$\forall\, h > 0$.

Next we find

$$4A_2 - 5A_1 - A_3 = 4\left(f(z) - f(z_0)\right) - R_4 64h^4 - 5\left(f(z) - f(x_0)\right) + R_4 5h^4$$
$$- \left(f(z) - f(x_0)\right) + R_4 81h^4 = -2\left(f(z) - f(x_0)\right) + 22R_4h^4. \tag{15.71}$$

Hence

$$|\Delta_2(x_0)| \leq \frac{4\,\|f\|_{\infty,\mathbb{R}_+^k}}{2h^2} + \frac{22\left(\frac{24H_\nu}{\Gamma(\nu+1)}\right)h^4}{2h^2}$$
$$= \frac{2\,\|f\|_{\infty,\mathbb{R}_+^k}}{h^2} + \left(\frac{264H_\nu}{\Gamma(\nu+1)}\right)h^2, \tag{15.72}$$

$\forall\, h > 0$, $\forall\, x_0 \in \mathbb{R}_+^k$.

The last implies

$$\|\Delta_2\|_{\infty,\mathbb{R}_+^k} \leq \frac{2\,\|f\|_{\infty,\mathbb{R}_+^k}}{h^2} + \left(\frac{264H_\nu}{\Gamma(\nu+1)}\right)h^2, \tag{15.73}$$

$\forall\, h > 0$.

We also calculate

$$3A_1 - 3A_2 + A_3 = 3\left(f(z) - f(x_0)\right) - 3R_4h^4 - 3\left(f(z) - f(x_0)\right) + 48R_4h^4$$
$$+ \left(f(z) - f(x_0)\right) - 81R_4h^4 = \left(f(z) - f(x_0)\right) - 36R_4h^4. \tag{15.74}$$

Therefore

$$|\Delta_3(x_0)| \leq \frac{2\,\|f\|_{\infty,\mathbb{R}_+^k}}{6h^3} + \frac{36\left(\frac{24H_\nu}{\Gamma(\nu+1)}\right)h^4}{6h^3} \tag{15.75}$$
$$= \frac{\|f\|_{\infty,\mathbb{R}_+^k}}{3h^3} + \left(\frac{144H_\nu}{\Gamma(\nu+1)}\right)h,$$

$\forall\, h > 0$, $\forall\, x_0 \in \mathbb{R}_+^k$.

The last implies

$$\|\Delta_3\|_{\infty,\mathbb{R}_+^k} \leq \frac{\|f\|_{\infty,\mathbb{R}_+^k}}{3h^3} + \left(\frac{144H_\nu}{\Gamma(\nu+1)}\right)h, \tag{15.76}$$

$\forall\, h > 0$.

We will work on (15.70).

Call $\mu := \frac{11\|f\|_{\infty,\mathbb{R}_+^k}}{3}$, $\theta := \left(\frac{144H_\nu}{\Gamma(\nu+1)}\right)$, both are positive.

We study

$$y(h) = \frac{\mu}{h} + \theta h^3 = \mu h^{-1} + \theta h^3. \tag{15.77}$$

We have

$$y'(h) = -\mu h^{-2} + 3\theta h^2 = 0,$$

implying

$$3\theta h^2 = \frac{\mu}{h^2},$$

and

$$3\theta h^4 = \mu,$$

with one critical number

$$h_{crit.no.} = h_0 = \left(\frac{\mu}{3\theta}\right)^{\frac{1}{4}} > 0. \tag{15.78}$$

We have that

$$y''(h_0) = 2\mu h_0^{-3} + 6\theta h_0 > 0, \tag{15.79}$$

implying that y has a global minimum. We see that

$$y(h_0) = \mu h_0^{-1} + \theta h_0^3 = h_0^3\left(\mu h_0^{-4} + \theta\right) = h_0^3\left(\mu\frac{3\theta}{\mu} + \theta\right)$$

$$= 4\theta h_0^3 = 4\theta\left(\frac{\mu}{3\theta}\right)^{\frac{3}{4}} = \frac{4}{3^{\frac{3}{4}}}\theta^{\frac{1}{4}}\mu^{\frac{3}{4}}. \tag{15.80}$$

I.e.

$$y(h_0) = \frac{4}{3^{\frac{3}{4}}}\theta^{\frac{1}{4}}\mu^{\frac{3}{4}}, \tag{15.81}$$

the global minimum only.

Consequently

$$\|\Delta_1\|_{\infty,\mathbb{R}_+^k} \leq \frac{4}{3^{\frac{3}{4}}}\frac{(144)^{\frac{1}{4}}H_\nu^{\frac{1}{4}}}{(\Gamma(\nu+1))^{\frac{1}{4}}}\left(\frac{11}{3}\right)^{\frac{3}{4}}\|f\|_{\infty,\mathbb{R}_+^k}^{\frac{3}{4}} =$$

$$4\left(\frac{11}{9}\right)^{\frac{3}{4}}\left(\frac{144}{\Gamma(\nu+1)}\right)^{\frac{1}{4}}\|f\|_{\infty,\mathbb{R}_+^k}^{\frac{3}{4}}H_\nu^{\frac{1}{4}}. \tag{15.82}$$

Next we work on (15.73).

Call $\varphi := 2\left\|f\right\|_{\infty,\mathbb{R}_+^k}$, $\psi := \frac{264H_\nu}{\Gamma(\nu+1)} > 0$.
We study

$$y(h) := \frac{\varphi}{h^2} + \psi h^2 = \varphi h^{-2} + \psi h^2,\ h > 0. \tag{15.83}$$

We have

$$y'(h) = -2\varphi h^{-3} + 2\psi h = 0, \tag{15.84}$$

then

$$\psi h = \varphi h^{-3}$$

and

$$\psi h = \frac{\varphi}{h^3},$$

so that

$$\psi h^4 = \varphi,$$

with

$$h_{crit.no.} = h_0 = \left(\frac{\varphi}{\psi}\right)^{\frac{1}{4}} > 0.$$

Notice that

$$y''(h_0) = 6\varphi h_0^{-4} + 2\psi > 0. \tag{15.85}$$

Thus we have that y has global minimum which is

$$y(h_0) = \varphi h_0^{-2} + \psi h_0^2 = h_0^2\left(\varphi h_0^{-4} + \psi\right) = h_0^2\left(\varphi\frac{\psi}{\varphi} + \psi\right)$$

$$= 2\psi h_0^2 = 2\psi\sqrt{\frac{\varphi}{\psi}} = 2\psi^{\frac{1}{2}}\varphi^{\frac{1}{2}},$$

that is

$$y(h_0) = 2\psi^{\frac{1}{2}}\varphi^{\frac{1}{2}}. \tag{15.86}$$

Therefore we derive

$$\left\|\Delta_2\right\|_{\infty,\mathbb{R}_+^k} \leq 2\sqrt{2}\left(\frac{264}{\Gamma(\nu+1)}\right)^{\frac{1}{2}}\sqrt{\left\|f\right\|_{\infty,\mathbb{R}_+^k} H_\nu} =$$

$$2\sqrt{\frac{528}{\Gamma(\nu+1)}}\sqrt{\left\|f\right\|_{\infty,\mathbb{R}_+^k} H_\nu}. \tag{15.87}$$

Finally we work on (15.76).

Let $A := \frac{\|f\|_{\infty,\mathbb{R}_+^k}}{3}$, $B := \frac{144H_\nu}{\Gamma(\nu+1)}$, both are positive.

We study

$$y(h) := \frac{A}{h^3} + Bh = Ah^{-3} + Bh. \tag{15.88}$$

Then

$$y'(h) = -3Ah^{-4} + B = 0,$$

and

$$B = \frac{3A}{h^4},$$

with

$$h^4 = \frac{3A}{B},$$

so that

$$h_{crit.no.} = h_0 = \left(\frac{3A}{B}\right)^{\frac{1}{4}} > 0. \tag{15.89}$$

Here

$$y''(h_0) = 12Ah_0^{-5} > 0. \tag{15.90}$$

Hence y has global minimum which is

$$y(h_0) = Ah_0^{-3} + Bh_0 = h_0\left(Ah_0^{-4} + B\right) = h_0\left(A\frac{B}{3A} + B\right)$$

$$= \frac{4}{3}Bh_0 = \frac{4}{3^{\frac{3}{4}}}A^{\frac{1}{4}}B^{\frac{3}{4}}. \tag{15.91}$$

That is

$$y(h_0) = \frac{4}{3^{\frac{3}{4}}}A^{\frac{1}{4}}B^{\frac{3}{4}}, \tag{15.92}$$

the global minimum of y.

Consequently it holds

$$\|\Delta_3\|_{\infty,\mathbb{R}_+^k} \leq \frac{4}{3^{\frac{3}{4}}}A^{\frac{1}{4}}B^{\frac{3}{4}} = \frac{4}{3}\left(\frac{144}{\Gamma(\nu+1)}\right)^{\frac{3}{4}} \|f\|_{\infty,\mathbb{R}_+^k}^{\frac{1}{4}} H_\nu^{\frac{3}{4}}. \tag{15.93}$$

The theorem is proved. ■

References

1. Anastassiou, G.A.: Opial type inequalities involving fractional derivatives of functions. Nonlinear Stud. **6**(2), 207–230 (1999)
2. Anastassiou, G.A.: Fractional Differentiation Inequalities. Springer, Heidelberg (2009)
3. Anastassiou, G.A.: Quantitative Approximations. Chapman & Hall/CRC, Boca Raron (2001)
4. Anastassiou, G.A.: Multivariate left side Canavati fractional Landau inequalities. Submitted (2020)
5. Canavati, J.A.: The Riemann-Liouville integral. Nieuw Archief Voor Wiskunde **5**(1), 53–75 (1987)
6. Ditzian, Z.: Multivariate Landau-Kolmogorov-type inequality. Math. Proc. Camb. Philos. Soc. **105**(2), 335–350 (1989)
7. Kounchev, O.: Extremizers for the multivariate Landau-Kolmogorov inequality. Multivariate Approximation (Witten-Bommerholz, 1996), pp. 123–132. Math. Res. 101. Akademie Verlag, Berlin (1997)
8. Landau, E.: Einige Ungleichungen für zweimal differentzierban funktionen. Proc. Lond. Math. Soc. **13**, 43–49 (1913)

Chapter 16
Multidimensional Right Caputo Fractional Taylor Formula and Landau Inequalities

Here we present a multivariate right side Caputo fractional Taylor's formula with fractional integral remainder. Based on this we give three multivariate right side Caputo fractional Landau's type inequalities. Their constants are precisely calculated and we give best upper bounds. It follows [2].

16.1 Introduction

We are inspired by the following two results. First from [3], we have the following nice result:

If $\Delta f = \sum_{i=1}^{d} \left(\frac{\partial}{\partial x_i} \right)^2 f$ is the Laplacian of a function defined on $\mathbb{R}^d$, then for $0 < k < 2n$,

$$\left\| \frac{\partial}{\partial \xi_1} \cdots \frac{\partial}{\partial \xi_k} f \right\| \leq C(n,k) \, \|f\|^{1-\left(\frac{k}{2n}\right)} \left\| \Delta^n f \right\|^{\frac{k}{2n}}, \tag{16.1}$$

where $\|\cdot\|$ denotes the $L_\infty \left(\mathbb{R}^d\right)$ norm. $\Delta^n f$ needs to be defined only in the distributional sense and the result is valid in many other Banach spaces such as L^p spaces. $C(n,k)$ is an existential constant and $\Delta^k = \Delta\left(\Delta^{k-1}\right)$.

From [4], we have the following interesting result which is the sharp inequality

$$\|\Delta f\|_\infty \leq 2\sqrt{\frac{d}{(d+2)}} \sqrt{\|f\|_\infty \left\|\Delta^2 f\right\|_\infty}, \tag{16.2}$$

in $\mathbb{R}^d$, $d \geq 2$, where Δf is the Laplacian and $\Delta^2 = \Delta(\Delta)$.

G. A. Anastassiou, *Constructive Fractional Analysis with Applications*,
Studies in Systems, Decision and Control 362,
https://doi.org/10.1007/978-3-030-71481-9_16

This result is an analogue of the famous Landau inequality [5] in $\mathbb{R}$:

$$\left\|f'\right\|_{\infty} \leq \sqrt{2}\sqrt{\|f\|_{\infty}\left\|f''\right\|_{\infty}}, \tag{16.3}$$

where $\|\cdot\|_{\infty}$ means the L_{∞} norm on $\mathbb{R}^d$ (respectively on $\mathbb{R}$).

In this chapter we first prove a multivariate right side Caputo fractional Taylor's formula and based on it we prove three multivariate right side Caputo fractional Landau's type inequalities where all the constants are precisely calculated and we obtain best upper bounds.

16.2 Background

Here we follow [1], Chap. 23.

We need

Definition 16.1 Let $f \in L_1([a, b])$, $\alpha > 0$. We define the right Riemann–Liouville fractional operator of order α by

$$I_{b-}^{\alpha} f(x) = \frac{1}{\Gamma(\alpha)} \int_x^b (s-x)^{\alpha-1} f(s)\, ds, \tag{16.4}$$

$\forall\, x \in [a, b]$, where Γ is the gamma function.

We set $I_{b-}^{0} := I$ (the identity operator).

By [1], p. 334 we get that $I_{b-}^{\alpha} f(x)$ exists almost everywhere on $[a, b]$ and $I_{b-}^{\alpha} f \in L_1([a, b])$. Also if $\alpha \geq 1$ then $I_{b-}^{\alpha} f \in C([a, b])$.

We need

Definition 16.2 Let $f \in AC^m([a, b])$ (space of functions from $[a, b]$ into $\mathbb{R}$ with $(m-1)$-derivative absolutely continuous function on $[a, b]$), $m \in \mathbb{N}$, where $m = \lceil \alpha \rceil$, $\alpha > 0$ ($\lceil \cdot \rceil$ the ceiling of the number).

We define the right Caputo fractional derivative of order $\alpha > 0$, by

$$D_{b-}^{\alpha} f(x) := (-1)^m I_{b-}^{m-\alpha} f^{(m)}(x), \tag{16.5}$$

that is

$$D_{b-}^{\alpha} f(x) = \frac{(-1)^m}{\Gamma(m-\alpha)} \int_x^b (s-x)^{m-\alpha-1} f^{(m)}(s)\, ds. \tag{16.6}$$

By [1], p. 337 we get that $D_{b-}^{\alpha} f(x)$ exists a.e. on $[a, b]$ and $D_{b-}^{\alpha} f \in L_1([a, b])$. When $\alpha = m \in \mathbb{N}$, then

$$D_{b-}^{m} f(x) = (-1)^m f^{(m)}(x), \tag{16.7}$$

$\forall\, x \in [a, b]$.

We also need the right Caputo fractional Taylor's formula with integral remainder.

Theorem 16.3 ([1], p. 341) *Let* $f \in AC^m([a,b])$, $x \in [a,b]$, $\alpha > 0$, $m = \lceil \alpha \rceil$. *Then*

$$f(x) = \sum_{\lambda=0}^{m-1} \frac{f^{(\lambda)}(b)}{\lambda!}(x-b)^{\lambda} + \frac{1}{\Gamma(\alpha)} \int_x^b (s-x)^{\alpha-1} D_{b-}^{\alpha} f(s)\, ds. \tag{16.8}$$

16.3 Main Results

We make

Remark 16.4 Let Q be a compact and convex subset of $\mathbb{R}^k$, $k \geq 2$; $z := (z_1, ..., z_k)$, $x_0 := (x_{01}, ..., x_{0k}) \in Q$. Let $f \in C^n(Q)$, $n \in \mathbb{N}$.

Set

$$g_z(t) := f(x_0 + t(z - x_0)), \ 0 \leq t \leq 1; \tag{16.9}$$

$$g_z(0) = f(x_0), \ \ g_z(1) = f(z).$$

Then

$$g_z^{(j)}(t) = \left[\left(\sum_{i=1}^{k}(z_i - x_{0i})\frac{\partial}{\partial x_i}\right)^j f\right](x_0 + t(z - x_0)), \tag{16.10}$$

and

$$g_z^{(j)}(1) = \left[\left(\sum_{i=1}^{k}(z_i - x_{0i})\frac{\partial}{\partial x_i}\right)^j f\right](z), \tag{16.11}$$

for $j = 0, 1, ..., n$.

If all $f_{\alpha}(z) := \frac{\partial^{\alpha} f}{\partial x^{\alpha}}(z) = 0$, $\alpha := (\alpha_1, ..., \alpha_k)$, $\alpha_i \in \mathbb{Z}^+$, $i = 1, ..., k$; $|\alpha| := \sum_{i=1}^{k} \alpha_i = l$, then $g_z^{(l)}(1) = 0$, where $l \in \{0, 1, ..., n\}$.

Clearly here $g_z \in C^n([0,1])$, in particular $g_z \in AC^n([0,1])$.

We also have that

$$g_z^{(j)}(t) = \sum_{\substack{\alpha:=(\alpha_1,...,\alpha_k),\alpha_i\in\mathbb{Z}^+ \\ i=1,...,k;\,|\alpha|:=\sum_{i=1}^{k}\alpha_i=j}} \left(\frac{j!}{\prod_{i=1}^{k}\alpha_i!}\right)\left(\prod_{i=1}^{k}(z_i - x_{0i})^{\alpha_i}\right) f_{\alpha}(x_0 + t(z - x_0)), \tag{16.12}$$

$0 \leq t \leq 1$, and

$$\frac{g_z^{(j)}(1)}{j!} = \sum_{\substack{\alpha:=(\alpha_1,\ldots,\alpha_k),\alpha_i\in\mathbb{Z}^+ \\ i=1,\ldots,k;|\alpha|:=\sum_{i=1}^{k}\alpha_i=j}} \left(\frac{\prod_{i=1}^{k}(z_i - x_{0i})^{\alpha_i}}{\prod_{i=1}^{k}\alpha_i!} \right) f_\alpha(z), \tag{16.13}$$

for all $j = 0, 1, \ldots, n$.

Let now $\nu > 0$ with $\lceil \nu \rceil = n$.

By applying (16.8) to g_z we find

$$g_z(0) = \sum_{j=0}^{n-1} (-1)^j \frac{g_z^{(j)}(1)}{j!} + \frac{1}{\Gamma(\nu)} \int_0^1 s^{\nu-1} D_{1-}^{\nu} g_z(s)\, ds. \tag{16.14}$$

Still we need to interpret $D_{1-}^{\nu} g_z$ as follows $(0 \leq s \leq 1)$:

$$D_{1-}^{\nu} g_z(s) \overset{(16.6)}{=} \frac{(-1)^n}{\Gamma(n-\nu)} \int_s^1 (y-s)^{n-\nu-1} g_z^{(n)}(y)\, dy = \tag{16.15}$$

$$\sum_{\substack{\alpha:=(\alpha_1,\ldots,\alpha_k),\alpha_i\in\mathbb{Z}^+ \\ i=1,\ldots,k;|\alpha|:=\sum_{i=1}^{k}\alpha_i=n}} \left(\frac{n!}{\prod_{i=1}^{k}\alpha_i!} \right) \left(\prod_{i=1}^{k}(z_i - x_{0i})^{\alpha_i} \right)$$

$$\frac{1}{\Gamma(n-\nu)} \int_s^1 (y-s)^{n-\nu-1} f_\alpha(x_0 + y(z-x_0))\, dy.$$

Thus it holds $(0 \leq s \leq 1)$

$$D_{1-}^{\nu} g_z(s) = (-1)^n \sum_{\substack{\alpha:=(\alpha_1,\ldots,\alpha_k),\alpha_i\in\mathbb{Z}^+ \\ i=1,\ldots,k;|\alpha|:=\sum_{i=1}^{k}\alpha_i=n}} \left(\frac{n!}{\prod_{i=1}^{k}\alpha_i!} \right) \left(\prod_{i=1}^{k}(z_i - x_{0i})^{\alpha_i} \right)$$

$$\left(I_{1-}^{n-\nu} f_\alpha(x_0 + t(z-x_0)) \right)(s). \tag{16.16}$$

Based on the above we have proved the following multivariate right side Caputo fractional Taylor's formula:

Theorem 16.5 *Let* $\nu > 0$, $n = \lceil \nu \rceil$, $f \in C^n(Q)$, *where* Q *is a compact and convex subset of* $\mathbb{R}^k$, $k \geq 2$; *with* $x_0 := (x_{01}, ..., x_{0k})$, $z := (z_1, ..., z_k) \in Q$. *Then*

(1)

$$f(x_0) = \sum_{j=0}^{n-1} (-1)^j \left(\sum_{\substack{\alpha:=(\alpha_1,...,\alpha_k), \alpha_i \in \mathbb{Z}^+ \\ i=1,...,k; |\alpha|:=\sum_{i=1}^k \alpha_i = j}} \left(\frac{\prod_{i=1}^k (z_i - x_{0i})^{\alpha_i}}{\prod_{i=1}^k \alpha_i!} \right) f_\alpha(z) \right) +$$

$$(-1)^n \sum_{\substack{\alpha:=(\alpha_1,...,\alpha_k), \alpha_i \in \mathbb{Z}^+ \\ i=1,...,k; |\alpha|:=\sum_{i=1}^k \alpha_i = n}} \left(\frac{n!}{\prod_{i=1}^k \alpha_i!} \right) \left(\prod_{i=1}^k (z_i - x_{0i})^{\alpha_i} \right) \tag{16.17}$$

$$\frac{1}{\Gamma(\nu)} \int_0^1 s^{\nu-1} \left(I_{1-}^{n-\nu} f_\alpha (x_0 + t(z - x_0)) \right)(s)\, ds.$$

(2) Additionally assume that $f_\alpha(z) = 0$, $\alpha := (\alpha_1, ..., \alpha_k)$, $\alpha_i \in \mathbb{Z}^+$, $i = 1, ..., k$, $|\alpha| := \sum_{i=1}^k \alpha_i =: r$, $r = 0, 1, ..., n-1$, *then*

$$f(x_0) = (-1)^n \sum_{\substack{\alpha:=(\alpha_1,...,\alpha_k), \alpha_i \in \mathbb{Z}^+ \\ i=1,...,k; |\alpha|:=\sum_{i=1}^k \alpha_i = n}} \left(\frac{n!}{\prod_{i=1}^k \alpha_i!} \right) \left(\prod_{i=1}^k (z_i - x_{0i})^{\alpha_i} \right)$$

$$\frac{1}{\Gamma(\nu)} \int_0^1 s^{\nu-1} \left(I_{1-}^{n-\nu} f_\alpha (x_0 + t(z - x_0)) \right)(s)\, ds. \tag{16.18}$$

Next we give three multivariate right side Caputo fractional Landaus' type inequalities.

Theorem 16.6 *Let* $f \in C^4(\mathbb{R}_-^k)$, $k \geq 2$, *with* $\|f\|_{\infty,\mathbb{R}_-^k} < \infty$, *and* $3 < \nu < 4$. *Assume further that*

$$\Phi_\nu := \sup_{x_0, z \in \mathbb{R}_-^k, s \in [0,1]} \left| \sum_{\substack{\alpha := (\alpha_1, \ldots, \alpha_k), \alpha_i \in \mathbb{Z}^+ \\ i=1,\ldots,k; |\alpha| := \sum_{i=1}^k \alpha_i = 4}} \frac{I_{1-}^{4-\nu} \left(f_\alpha \left(x_0 + t \left(z - x_0 \right) \right) \right) (s)}{\prod_{i=1}^k \alpha_i !} \right| < +\infty. \tag{16.19}$$

Then we derive the best upper bounds:

(1)

$$\left\| \sum_{i=1}^k \frac{\partial f}{\partial x_i} \right\|_{\infty, \mathbb{R}_-^k} \leq 4 \left(\frac{11}{9} \right)^{\frac{3}{4}} \left(\frac{144}{\Gamma(\nu+1)} \right)^{\frac{1}{4}} \| f \|_{\infty, \mathbb{R}_-^k}^{\frac{3}{4}} \Phi_\nu^{\frac{1}{4}} < +\infty, \tag{16.20}$$

(2)

$$\left\| \sum_{\substack{\alpha := (\alpha_1, \ldots, \alpha_k), \alpha_i \in \mathbb{Z}^+ \\ i=1,\ldots,k; |\alpha| := \sum_{i=1}^k \alpha_i = 2}} \frac{f_\alpha}{\prod_{i=1}^k \alpha_i !} \right\|_{\infty, \mathbb{R}_-^k} \leq 2 \sqrt{\frac{528}{\Gamma(\nu+1)}} \sqrt{\| f \|_{\infty, \mathbb{R}_-^k} \Phi_\nu} < +\infty, \tag{16.21}$$

and

(3)

$$\left\| \sum_{\substack{\alpha := (\alpha_1, \ldots, \alpha_k), \alpha_i \in \mathbb{Z}^+ \\ i=1,\ldots,k; |\alpha| := \sum_{i=1}^k \alpha_i = 3}} \frac{f_\alpha}{\prod_{i=1}^k \alpha_i !} \right\|_{\infty, \mathbb{R}_-^k} \leq \frac{4}{3} \left(\frac{144}{\Gamma(\nu+1)} \right)^{\frac{3}{4}} \| f \|_{\infty, \mathbb{R}_-^k}^{\frac{1}{4}} \Phi_\nu^{\frac{3}{4}} < +\infty. \tag{16.22}$$

Proof We set

$$\Delta_1(z) := \sum_{i=1}^k \frac{\partial f(z)}{\partial x_i},$$

$$\Delta_2(z) := \sum_{\substack{\alpha := (\alpha_1, \ldots, \alpha_k), \alpha_i \in \mathbb{Z}^+ \\ i=1,\ldots,k; |\alpha| := \sum_{i=1}^k \alpha_i = 2}} \frac{f_\alpha(z)}{\prod_{i=1}^k \alpha_i !}, \tag{16.23}$$

and

$$\Delta_3(z) := \sum_{\substack{\alpha := (\alpha_1, \ldots, \alpha_k), \alpha_i \in \mathbb{Z}^+ \\ i=1,\ldots,k; |\alpha| := \sum_{i=1}^k \alpha_i = 3}} \frac{f_\alpha(z)}{\prod_{i=1}^k \alpha_i !}.$$

We can choose $x_0, z \in \mathbb{R}^k_-$ so that $z_i - x_{0i} = -h$, all $i = 1, ..., k$, where $h > 0$. By Theorem 16.5 we get

$$f(x_0) = f(z) + \Delta_1(z)h + \Delta_2(z)h^2 + \Delta_3(z)h^3 + R_4h^4, \tag{16.24}$$

where

$$R_4 := \frac{1}{\Gamma(\nu)}\left[\sum_{\substack{\alpha:=(\alpha_1,...,\alpha_k),\alpha_i\in\mathbb{Z}^+\\ i=1,...,k;|\alpha|:=\sum_{i=1}^k\alpha_i=4}}\left(\frac{4!}{\prod_{i=1}^k\alpha_i!}\right)\right.$$

$$\left.\int_0^1 s^{\nu-1}I_{1-}^{4-\nu}\left(f_\alpha\left(x_0+t\left(z-x_0\right)\right)\right)(s)\,ds\right]. \tag{16.25}$$

Therefore, $\forall\, h > 0$, we get

$$\begin{aligned}
&\Delta_1(z)h + \Delta_2(z)h^2 + \Delta_3(z)h^3 = f(x_0) - f(z) - R_4h^4 =: A_1,\\
&\Delta_1(z)2h + \Delta_2(z)4h^2 + \Delta_3(z)8h^3 = f(x_0) - f(z) - R_416h^4 =: A_2,\\
&\text{and}\\
&\Delta_1(z)3h + \Delta_2(z)9h^2 + \Delta_3(z)27h^3 = f(x_0) - f(z) - R_481h^4 =: A_3.
\end{aligned} \tag{16.26}$$

We solve the system of three equations (16.26) and we find:

$$\begin{aligned}
&\Delta_1(z) = \tfrac{18A_1-9A_2+2A_3}{6h},\\
&\Delta_2(z) = \tfrac{4A_2-5A_1-A_3}{2h^2},\\
&\text{and}\\
&\Delta_3(z) = \tfrac{3A_1-3A_2+A_3}{6h^3}.
\end{aligned} \tag{16.27}$$

We have also that

$$|R_4| \leq \frac{24\Phi_\nu}{\Gamma(\nu+1)}. \tag{16.28}$$

We calculate

$$18A_1 - 9A_2 + 2A_3 = 11\left(f(x_0) - f(z)\right) - 36R_4h^4. \tag{16.29}$$

Therefore it holds

$$|\Delta_1(z)| \leq \frac{11\,\|f\|_{\infty,\mathbb{R}^k_-}}{3h} + \frac{144\Phi_\nu}{\Gamma(\nu+1)}h^3, \tag{16.30}$$

$\forall\, h > 0, \forall\, z \in \mathbb{R}^k_-$.

We get that

$$\|\Delta_1\|_{\infty,\mathbb{R}_-^k} \leq \frac{11\,\|f\|_{\infty,\mathbb{R}_-^k}}{3h} + \left(\frac{144\Phi_\nu}{\Gamma(\nu+1)}\right)h^3, \tag{16.31}$$

$\forall\, h > 0$.

Next we find

$$4A_2 - 5A_1 - A_3 = -2\,(f(x_0) - f(z)) + 22R_4h^4. \tag{16.32}$$

Hence

$$|\Delta_2(z)| \leq \frac{2\,\|f\|_{\infty,\mathbb{R}_-^k}}{h^2} + \left(\frac{264\Phi_\nu}{\Gamma(\nu+1)}\right)h^2, \tag{16.33}$$

$\forall\, h > 0, \forall\, z \in \mathbb{R}_-^k$.

The last implies

$$\|\Delta_2\|_{\infty,\mathbb{R}_-^k} \leq \frac{2\,\|f\|_{\infty,\mathbb{R}_-^k}}{h^2} + \left(\frac{264\Phi_\nu}{\Gamma(\nu+1)}\right)h^2, \tag{16.34}$$

$\forall\, h > 0$.

We also calculate

$$3A_1 - 3A_2 + A_3 = (f(x_0) - f(z)) - 36R_4h^4.$$

Therefore

$$|\Delta_3(z)| \leq \frac{\|f\|_{\infty,\mathbb{R}_-^k}}{3h^3} + \left(\frac{144\Phi_\nu}{\Gamma(\nu+1)}\right)h, \tag{16.35}$$

$\forall\, h > 0, \forall\, z \in \mathbb{R}_-^k$.

The last implies

$$\|\Delta_3\|_{\infty,\mathbb{R}_-^k} \leq \frac{\|f\|_{\infty,\mathbb{R}_-^k}}{3h^3} + \left(\frac{144\Phi_\nu}{\Gamma(\nu+1)}\right)h, \tag{16.36}$$

$\forall\, h > 0$.

We will work on (16.31).

Call $\mu := \frac{11\|f\|_{\infty,\mathbb{R}_-^k}}{3}$, $\theta := \left(\frac{144\Phi_\nu}{\Gamma(\nu+1)}\right)$, both are positive.

We study

$$y(h) = \frac{\mu}{h} + \theta h^3, \ \forall\, h > 0. \tag{16.37}$$

We have

$$y'(h) = -\mu h^{-2} + 3\theta h^2 = 0, \tag{16.38}$$

with one critical number

$$h_{crit.no.} = h_0 = \left(\frac{\mu}{3\theta}\right)^{\frac{1}{4}} > 0. \tag{16.39}$$

We have that

$$y''(h_0) = 2\mu h_0^{-3} + 6\theta h_0 > 0, \tag{16.40}$$

implying that y has a global minimum. We see that

$$y(h_0) = \mu h_0^{-1} + \theta h_0^3 = h_0^3\left(\mu h_0^{-4} + \theta\right) = \frac{4}{3^{\frac{3}{4}}}\theta^{\frac{1}{4}}\mu^{\frac{3}{4}}. \tag{16.41}$$

I.e.

$$y(h_0) = \frac{4}{3^{\frac{3}{4}}}\theta^{\frac{1}{4}}\mu^{\frac{3}{4}}, \tag{16.42}$$

is the global minimum.

Consequently

$$\|\Delta_1\|_{\infty,\mathbb{R}_-^k} \leq 4\left(\frac{11}{9}\right)^{\frac{3}{4}}\left(\frac{144}{\Gamma(\nu+1)}\right)^{\frac{1}{4}}\|f\|_{\infty,\mathbb{R}_-^k}^{\frac{3}{4}}\Phi_\nu^{\frac{1}{4}}. \tag{16.43}$$

Next we work on (16.34).

Call $\varphi := 2\|f\|_{\infty,\mathbb{R}_-^k}$, $\psi := \frac{264\Phi_\nu}{\Gamma(\nu+1)} > 0$.

We study

$$y(h) := \frac{\varphi}{h^2} + \psi h^2, \ \forall\, h > 0. \tag{16.44}$$

We have

$$y'(h) = -2\varphi h^{-3} + 2\psi h = 0, \tag{16.45}$$

with

$$h_{crit.no.} = h_0 = \left(\frac{\varphi}{\psi}\right)^{\frac{1}{4}} > 0. \tag{16.46}$$

Notice that

$$y''(h_0) = 6\varphi h_0^{-4} + 2\psi > 0. \tag{16.47}$$

Thus we have that y has global minimum which is

$$y(h_0) = \varphi h_0^{-2} + \psi h_0^2 = h_0^2\left(\varphi h_0^{-4} + \psi\right) = 2\psi^{\frac{1}{2}}\varphi^{\frac{1}{2}},$$

that is

$$y(h_0) = 2\psi^{\frac{1}{2}}\varphi^{\frac{1}{2}}. \tag{16.48}$$

Therefore we derive

$$\|\Delta_2\|_{\infty,\mathbb{R}_-^k} \leq 2\sqrt{\frac{528}{\Gamma(\nu+1)}}\sqrt{\|f\|_{\infty,\mathbb{R}_-^k}\Phi_\nu}. \tag{16.49}$$

Finally we work on (16.36).

Let $A := \frac{\|f\|_{\infty,\mathbb{R}_-^k}}{3}$, $B := \frac{144\Phi_\nu}{\Gamma(\nu+1)}$, both are positive.
We study

$$y(h) := \frac{A}{h^3} + Bh, \ \forall\, h > 0. \tag{16.50}$$

Then

$$y'(h) = -3Ah^{-4} + B = 0, \tag{16.51}$$

so that

$$h_{crit.no.} = h_0 = \left(\frac{3A}{B}\right)^{\frac{1}{4}} > 0. \tag{16.52}$$

Here

$$y''(h_0) = 12Ah_0^{-5} > 0. \tag{16.53}$$

Hence y has global minimum which is

$$y(h_0) = Ah_0^{-3} + Bh_0 = h_0\left(Ah_0^{-4} + B\right) = \frac{4}{3^{\frac{3}{4}}}A^{\frac{1}{4}}B^{\frac{3}{4}}.$$

That is

$$y(h_0) = \frac{4}{3^{\frac{3}{4}}}A^{\frac{1}{4}}B^{\frac{3}{4}}, \tag{16.54}$$

the global minimum of y.

Consequently it holds

$$\|\Delta_3\|_{\infty,\mathbb{R}_-^k} \leq \frac{4}{3}\left(\frac{144}{\Gamma(\nu+1)}\right)^{\frac{3}{4}}\|f\|_{\infty,\mathbb{R}_-^k}^{\frac{1}{4}}\Phi_\nu^{\frac{3}{4}}. \tag{16.55}$$

The theorem is proved. ■

References

1. Anastassiou, G.A.: Intelligent Mathematics: Computational Analysis. Springer, Heidelberg (2011)
2. Anastassiou, G.A.: Multivariate right side Caputo fractional Taylor formula and Landau inequalities. Ann. Commun. Math. Accepted (2020)
3. Ditzian, Z.: Multivariate Landau-Kolmogorov-type inequality. Math. Proc. Camb. Philos. Soc. **105**(2), 335–350 (1989)
4. Kounchev, O.: Extremizers for the multivariate Landau-Kolmogorov inequality. Multivariate Approximation (Witten-Bommerholz, 1996), pp. 123–132. Math. Res. 101. Akademie Verlag, Berlin (1997)
5. Landau, E.: Einige Ungleichungen für zweimal differentzierban funktionen. Proc. Lond. Math. Soc. **13**, 43–49 (1913)

Chapter 17
Multidimensional Generalized Right Fractional Taylor Formula and Landau Inequalities

Here we present a multivariate generalized right side fractional Taylor's formula with fractional integral remainder. Based on this we give three multivariate generalized right side fractional Landau's type inequalities. Their constants are exactly calculated and we give best upper bounds. It follows [2].

17.1 Introduction

We are motivated by the following two results. First from [3], we have the following nice result:

If $\Delta f = \sum_{i=1}^{d} \left(\frac{\partial}{\partial x_i} \right)^2 f$ is the Laplacian of a function defined on $\mathbb{R}^d$, then for $0 < k < 2n$,

$$\left\| \frac{\partial}{\partial \xi_1} \cdots \frac{\partial}{\partial \xi_k} f \right\| \leq C(n,k) \, \|f\|^{1-\left(\frac{k}{2n}\right)} \left\| \Delta^n f \right\|^{\frac{k}{2n}}, \tag{17.1}$$

where $\|\cdot\|$ denotes the $L_\infty\left(\mathbb{R}^d\right)$ norm. $\Delta^n f$ needs to be defined only in the distributional sense and the result is valid in many other Banach spaces such as L^p spaces. $C(n,k)$ is an existential constant and $\Delta^k = \Delta\left(\Delta^{k-1}\right)$.

From [4], we have the following interesting result which is the sharp inequality

$$\|\Delta f\|_\infty \leq 2\sqrt{\frac{d}{(d+2)}} \sqrt{\|f\|_\infty \left\|\Delta^2 f\right\|_\infty}, \tag{17.2}$$

in $\mathbb{R}^d$, $d \geq 2$, where Δf is the Laplacian and $\Delta^2 = \Delta(\Delta)$.

This result is an analogue of the famous Landau inequality [5] in $\mathbb{R}$:

G. A. Anastassiou, *Constructive Fractional Analysis with Applications*,
Studies in Systems, Decision and Control 362,
https://doi.org/10.1007/978-3-030-71481-9_17

$$\left\| f' \right\|_\infty \leq \sqrt{2} \sqrt{\| f \|_\infty \, \| f'' \|_\infty}, \tag{17.3}$$

where $\|\cdot\|_\infty$ means the L_∞ norm on $\mathbb{R}^d$ (respectively on $\mathbb{R}$).

In this chapter we first prove a multivariate generalized right side fractional Taylor's formula and based on it we prove three multivariate generalized right side fractional Landau's type inequalities where all the constants are exactly calculated and we get best upper bounds.

17.2 Background

Here we follow [1], pp. 345–348.

Let $\nu > 0, n := [\nu]$ ($[\cdot]$ the integral part), $\alpha := \nu - n, 0 < \alpha < 1, f \in C([a, b])$. We call the right Riemann–Liouville fractional integral operator by

$$\left(J_{b-}^{\nu} f\right)(x) := \frac{1}{\Gamma(\nu)} \int_x^b (z - x)^{\nu - 1} f(z)\, dz, \tag{17.4}$$

$\forall\, x \in [a, b]$, where Γ is the gamma function.

Define the subspace of functions

$$C_{b-}^{\nu}([a, b]) := \left\{ f \in C^n([a, b]) : J_{b-}^{1-\alpha} f^{(n)} \in C^1([a, b]) \right\}. \tag{17.5}$$

Define the right generalized ν-fractional derivative of f over $[a, b]$ as

$$D_{b-}^{\nu} f := (-1)^{n-1} \left(J_{b-}^{1-\alpha} f^{(n)} \right)'. \tag{17.6}$$

Notice that

$$J_{b-}^{1-\alpha} f^{(n)}(x) := \frac{1}{\Gamma(1-\alpha)} \int_x^b (z - x)^{-\alpha} f^{(n)}(z)\, dz, \tag{17.7}$$

exists for $f \in C_{b-}^{\nu}([a, b])$, and

$$D_{b-}^{\nu} f(x) = \frac{(-1)^{n-1}}{\Gamma(1-\alpha)} \frac{d}{dx} \int_x^b (z - x)^{-\alpha} f^{(n)}(z)\, dz. \tag{17.8}$$

That is

$$\left(D_{b-}^{\nu} f\right)(x) = \frac{(-1)^{n-1}}{\Gamma(n - \nu + 1)} \frac{d}{dx} \int_x^b (z - x)^{n-\nu} f^{(n)}(z)\, dz. \tag{17.9}$$

If $\nu \in \mathbb{N}$, then $\alpha = 0, n = \nu$, and

$$D_{b-}^{\nu} f(x) = (-1)^n f^{(n)}(x). \tag{17.10}$$

We mention the following right fractional Taylor formula:

Theorem 17.1 ([1], p. 348) *Let* $f \in C_{b-}^{\nu}([a, b])$, $\nu > 0$, $n := [\nu]$. *Then*
(1) If $\nu \geq 1$, *we get*

$$f(x) = \sum_{\lambda=0}^{n-1} \frac{f^{(\lambda)}(b)}{\lambda!} (x - b)^{\lambda} + \frac{1}{\Gamma(\nu)} \int_x^b (z - x)^{\nu-1} \left(D_{b-}^{\nu} f \right)(z) \, dz, \quad (17.11)$$

$\forall \, x \in [a, b]$.
(2) If $0 < \nu < 1$, *we obtain*

$$f(x) = \frac{1}{\Gamma(\nu)} \int_x^b (z - x)^{\nu-1} \left(D_{b-}^{\nu} f \right)(z) \, dz, \quad (17.12)$$

$\forall \, x \in [a, b]$.

17.3 Main Results

We make

Remark 17.2 Let $\nu \geq 1$ such that $n = [\nu]$, and let $f \in C^n(Q)$, where Q is compact and convex subset of $\mathbb{R}^k$, $k \geq 2$; $z := (z_1, ..., z_k)$, $x_0 := (x_{01}, ..., x_{0k}) \in Q$.
Call

$$g_z(t) := f(x_0 + t(z - x_0)), \ 0 \leq t \leq 1; \quad (17.13)$$

so that $g_z(0) = f(x_0)$, and $g_z(1) = f(z)$.
Then

$$g_z^{(j)}(t) = \left[\left(\sum_{i=1}^{k} (z_i - x_{0i}) \frac{\partial}{\partial x_i} \right)^j f \right] (x_0 + t(z - x_0)), \quad (17.14)$$

and

$$g_z^{(j)}(1) = \left[\left(\sum_{i=1}^{k} (z_i - x_{0i}) \frac{\partial}{\partial x_i} \right)^j f \right] (z), \quad (17.15)$$

for all $j = 0, 1, 2, ..., n$.
Here it is

$$\left(J_{1-}^{\nu} g_z \right)(x) = \frac{1}{\Gamma(\nu)} \int_x^1 (j - x)^{\nu-1} g_z(j) \, dj, \quad (17.16)$$

for all $0 \leq x \leq 1$.
Consider

$$C_{1-}^{\nu}([0,1]) := \left\{ g \in C^n([0,1]) : J_{1-}^{1-\alpha} g^{(n)} \in C^1([0,1]) \right\}. \tag{17.17}$$

The right generalized ν-fractional derivative of g over $[0, 1]$ is defined by

$$D_{1-}^{\nu} g := (-1)^{n-1} \left(J_{1-}^{1-\alpha} g^{(n)} \right)'. \tag{17.18}$$

For fixed $x_0, z \in Q$ we assume that as functions of t : $f_\alpha (x_0 + t (z - x_0)) \in C_{1-}^{(\nu-n)}([0, 1])$, for all $\alpha := (\alpha_1, ..., \alpha_k)$, $\alpha_i \in \mathbb{Z}^+$, $i = 1, ..., k$; $|\alpha| := \sum_{i=1}^{k} \alpha_i = n$.

We have that

$$g_z^{(n)}(t) = \sum_{\substack{\alpha:=(\alpha_1,\ldots,\alpha_k), \alpha_i \in \mathbb{Z}^+ \\ i=1,\ldots,k; |\alpha|:=\sum_{i=1}^{k} \alpha_i = n}} \left(\frac{n!}{\prod_{i=1}^{k} \alpha_i!} \right) \left(\prod_{i=1}^{k} (z_i - x_{0i})^{\alpha_i} \right) f_\alpha (x_0 + t (z - x_0)), \tag{17.19}$$

$0 \leq t \leq 1$.

Here it is $1 - \alpha = n + 1 - \nu$ and $[\nu - n] = 0$. We also have that $(0 \leq x \leq 1)$

$$\left(J_{1-}^{n+1-\nu} g_z^{(n)} \right)(x) = \sum_{\substack{\alpha:=(\alpha_1,\ldots,\alpha_k), \alpha_i \in \mathbb{Z}^+ \\ i=1,\ldots,k; |\alpha|:=\sum_{i=1}^{k} \alpha_i = n}} \left(\frac{n!}{\prod_{i=1}^{k} \alpha_i!} \right)$$

$$\left(\prod_{i=1}^{k} (z_i - x_{0i})^{\alpha_i} \right) \left(J_{1-}^{n+1-\nu} (f_\alpha (x_0 + t (z - x_0))) \right)(x), \tag{17.20}$$

and

$$\left(D_{1-}^{\nu} g_z \right)(x) = (-1)^{n-1} \left(J_{1-}^{1-\alpha} g_z^{(n)} \right)'(x) = (-1)^{n-1} \left(J_{1-}^{n+1-\nu} g_z^{(n)} \right)'(x) =$$

$$\sum_{\substack{\alpha:=(\alpha_1,\ldots,\alpha_k), \alpha_i \in \mathbb{Z}^+ \\ i=1,\ldots,k; |\alpha|:=\sum_{i=1}^{k} \alpha_i = n}} \left(\frac{n!}{\prod_{i=1}^{k} \alpha_i!} \right) \left(\prod_{i=1}^{k} (z_i - x_{0i})^{\alpha_i} \right)$$

$$(-1)^{n-1} \left(J_{1-}^{n+1-\nu} (f_\alpha (x_0 + t (z - x_0))) \right)'(x), \tag{17.21}$$

for all $0 \leq x \leq 1$.

That is

$$\left(D_{1-}^{\nu} g_z\right)(x) = (-1)^n \sum_{\substack{\alpha:=(\alpha_1,\ldots,\alpha_k),\alpha_i\in\mathbb{Z}^+ \\ i=1,\ldots,k;|\alpha|:=\sum_{i=1}^{k}\alpha_i=n}} \left(\frac{n!}{\prod_{i=1}^{k}\alpha_i!}\right)\left(\prod_{i=1}^{k}(z_i - x_{0i})^{\alpha_i}\right)$$

$$D_{1-}^{\nu-n}\left(f_\alpha\left(x_0 + t\left(z - x_0\right)\right)\right)(x), \tag{17.22}$$

for all $0 \leq x \leq 1$.

Clearly then there exists $\left(D_{1-}^{\nu} g_z\right)(t)$, $\forall\, t \in [0, 1]$, i.e. $g_z \in C_{1-}^{\nu}([0, 1])$.

By Theorem 17.1, (17.11), we obtain

$$g_z(0) = \sum_{\lambda=0}^{n-1} \frac{(-1)^\lambda g_z^{(\lambda)}(1)}{\lambda!} + \frac{1}{\Gamma(\nu)} \int_0^1 j^{\nu-1}\left(D_{1-}^{\nu} g_z\right)(j)\, dj. \tag{17.23}$$

Based on the above we have proved the following multivariate right generalized fractional Taylor's formula:

Theorem 17.3 *Let $f \in C^n(Q)$, Q is a compact and convex subset of $\mathbb{R}^k$, $k \geq 2$; here $\nu \geq 1$ such that $n = [\nu]$. For fixed $x_0, z \in Q$ assume that as functions of $t: f_\alpha(x_0 + t(z - x_0)) \in C_{1-}^{(\nu-n)}([0, 1])$, for all $\alpha := (\alpha_1, \ldots, \alpha_k)$, $\alpha_i \in \mathbb{Z}^+$, $i = 1, \ldots, k$; $|\alpha| := \sum_{i=1}^{k} \alpha_i = n$. Then*

(i)

$$f(x_0) = f(z) - \sum_{i=1}^{k}(z_i - x_{0i})\frac{\partial f}{\partial x_i}(z) +$$

$$\frac{\left[\left(\sum_{i=1}^{k}(z_i - x_{0i})\frac{\partial}{\partial x_i}\right)^2 f\right](z)}{2} + \ldots + \frac{(-1)^{n-1}\left[\left(\sum_{i=1}^{k}(z_i - x_{0i})\frac{\partial}{\partial x_i}\right)^{n-1} f\right](z)}{(n-1)!}$$

$$+(-1)^n \sum_{\substack{\alpha:=(\alpha_1,\ldots,\alpha_k),\alpha_i\in\mathbb{Z}^+ \\ i=1,\ldots,k;|\alpha|:=\sum_{i=1}^{k}\alpha_i=n}} \left(\frac{n!}{\prod_{i=1}^{k}\alpha_i!}\right)\left(\prod_{i=1}^{k}(z_i - x_{0i})^{\alpha_i}\right)$$

$$\frac{1}{\Gamma(\nu)} \int_0^1 j^{\nu-1} D_{1-}^{\nu-n}\left(f_\alpha\left(x_0 + t\left(z - x_0\right)\right)\right)(j)\, dj. \tag{17.24}$$

(ii) *If all* $f_\alpha(z) = 0$, $\alpha := (\alpha_1, ..., \alpha_k)$, $\alpha_i \in \mathbb{Z}^+$, $i = 1, ..., k$, $|\alpha| := \sum_{i=1}^{k} \alpha_i =: l$, $l = 0, ..., n-1$, *then*

$$f(x_0) = (-1)^n \sum_{\substack{\alpha:=(\alpha_1,...,\alpha_k), \alpha_i \in \mathbb{Z}^+ \\ i=1,...,k; |\alpha|:=\sum_{i=1}^{k} \alpha_i = n}} \left(\frac{n!}{\prod_{i=1}^{k} \alpha_i!} \right) \left(\prod_{i=1}^{k} (z_i - x_{0i})^{\alpha_i} \right)$$

$$\frac{1}{\Gamma(\nu)} \int_0^1 j^{\nu-1} D_{1-}^{\nu-n} \left(f_\alpha (x_0 + t(z - x_0)) \right)(j) \, dj. \quad (17.25)$$

Additionally we make

Remark 17.4 Continuing from Remark 17.2. Let here $0 < \nu < 1$. Assume now that $f(x_0 + t(z - x_0)) \in C_{1-}^{\nu}([0, 1])$ as a function of t. Then by $g_z(t) := f(x_0 + t(z - x_0))$ we have that

$$D_{1-}^{\nu} g_z(t) = D_{1-}^{\nu} f(x_0 + t(z - x_0)),$$

and

$$\left(J_{1-}^{\nu} D_{1-}^{\nu} g_z \right)(t) = \left(J_{1-}^{\nu} D_{1-}^{\nu} f(x_0 + t(z - x_0)) \right)(t), \quad (17.26)$$

$\forall \, t \in [0, 1]$. Hence by (17.12)

$$f(x_0) = g_z(0) = \left(J_{1-}^{\nu} D_{1-}^{\nu} g_z \right)(0) = \frac{1}{\Gamma(\nu)} \int_0^1 j^{\nu-1} D_{1-}^{\nu} f(x_0 + j(z - x_0)) \, dj. \quad (17.27)$$

We have established the next multivariate generalized right side fractional Taylor's formula when $0 < \nu < 1$.

Theorem 17.5 *Let continuous* $f : Q \to \mathbb{R}$, *where* Q *is compact and convex* $\subseteq \mathbb{R}^k$, $k \geq 2$, *such that as a function of* $t : f(x_0 + t(z - x_0)) \in C_{1-}^{\nu}([0, 1])$, *where* $x_0, z \in Q$ *are being fixed, and* $0 < \nu < 1$. *Then*

$$f(x_0) = \frac{1}{\Gamma(\nu)} \int_0^1 j^{\nu-1} D_{1-}^{\nu} f(x_0 + j(z - x_0)) \, dj. \quad (17.28)$$

We need

Corollary 17.6 (to Theorem *17.3*, $n = 4$ case) *Let* $f \in C^4(Q)$, Q *is a compact and convex subset of* $\mathbb{R}^k$, $k \geq 2$; *here* $4 < \nu < 5$. *For fixed* $x_0, z \in Q$ *assume that as functions of* $t : f_\alpha(x_0 + t(z - x_0)) \in C_{1-}^{(\nu-4)}([0, 1])$, *for all* $\alpha := (\alpha_1, ..., \alpha_k)$, $\alpha_i \in \mathbb{Z}^+$, $i = 1, ..., k$; $|\alpha| := \sum_{i=1}^{k} \alpha_i = 4$. *Then*

$$f(x_0) = f(z) - \sum_{i=1}^{k} (z_i - x_{0i}) \frac{\partial f}{\partial x_i}(z) +$$

$$\left(\sum_{\substack{\alpha:=(\alpha_1,\ldots,\alpha_k),\alpha_i\in\mathbb{Z}^+ \\ i=1,\ldots,k;|\alpha|:=\sum_{i=1}^{k}\alpha_i=2}} \left(\frac{\prod_{i=1}^{k} (z_i - x_{0i})^{\alpha_i}}{\prod_{i=1}^{k} \alpha_i!} \right) f_\alpha(z) \right) -$$

$$\left(\sum_{\substack{\alpha:=(\alpha_1,\ldots,\alpha_k),\alpha_i\in\mathbb{Z}^+ \\ i=1,\ldots,k;|\alpha|:=\sum_{i=1}^{k}\alpha_i=3}} \left(\frac{\prod_{i=1}^{k} (z_i - x_{0i})^{\alpha_i}}{\prod_{i=1}^{k} \alpha_i!} \right) f_\alpha(z) \right) +$$

$$\sum_{\substack{\alpha:=(\alpha_1,\ldots,\alpha_k),\alpha_i\in\mathbb{Z}^+ \\ i=1,\ldots,k;|\alpha|:=\sum_{i=1}^{k}\alpha_i=4}} \left(\frac{4!}{\prod_{i=1}^{k} \alpha_i!} \right) \left(\prod_{i=1}^{k} (z_i - x_{0i})^{\alpha_i} \right)$$

$$\frac{1}{\Gamma(\nu)} \int_0^1 j^{\nu-1} D_{1-}^{\nu-4} (f_\alpha (x_0 + t(z - x_0)))(j)\, dj. \tag{17.29}$$

Proof By (17.24) for $n = 4$. ■

Next follow some multivariate right side generalized fractional Landau type inequalities.

Theorem 17.7 *Let $f \in C^4(\mathbb{R}_-^k)$, $k \geq 2$, with $\|f\|_{\infty,\mathbb{R}_-^k} < +\infty$, and $4 < \nu < 5$. For any $x_0, z \in \mathbb{R}_-^k$ assume that as functions of t : $f_\alpha(x_0 + t(z - x_0)) \in C_{1-}^{(\nu-4)}([0, 1])$, for all $\alpha := (\alpha_1, \ldots, \alpha_k)$, $\alpha_i \in \mathbb{Z}^+$, $i = 1, \ldots, k$; $|\alpha| := \sum_{i=1}^{k} \alpha_i = 4$. Assume further that*

$$H_\nu := \sup_{x_0, z \in \mathbb{R}_-^k, j \in [0,1]} \left| \sum_{\substack{\alpha:=(\alpha_1,\ldots,\alpha_k),\alpha_i\in\mathbb{Z}^+ \\ i=1,\ldots,k;|\alpha|:=\sum_{i=1}^{k}\alpha_i=4}} \frac{D_{1-}^{\nu-4} (f_\alpha (x_0 + t(z - x_0)))(j)}{\prod_{i=1}^{k} \alpha_i!} \right| < +\infty. \tag{17.30}$$

Then we derive the best upper bounds:

(1)

$$\left\| \sum_{i=1}^{k} \frac{\partial f}{\partial x_i} \right\|_{\infty,\mathbb{R}_-^k} \leq 4 \left(\frac{11}{9} \right)^{\frac{3}{4}} \left(\frac{144}{\Gamma(\nu+1)} \right)^{\frac{1}{4}} \|f\|_{\infty,\mathbb{R}_-^k}^{\frac{3}{4}} H_\nu^{\frac{1}{4}} < +\infty, \tag{17.31}$$

(2)

$$\left\| \sum_{\substack{\alpha:=(\alpha_1,\dots,\alpha_k),\alpha_i\in\mathbb{Z}^+ \\ i=1,\dots,k;\,|\alpha|:=\sum_{i=1}^{k}\alpha_i=2}} \frac{f_\alpha}{\prod_{i=1}^{k}\alpha_i!} \right\|_{\infty,\mathbb{R}_-^k} \leq 2\sqrt{\frac{528}{\Gamma(\nu+1)}} \sqrt{\|f\|_{\infty,\mathbb{R}_-^k} H_\nu} < +\infty, \tag{17.32}$$

and

(3)

$$\left\| \sum_{\substack{\alpha:=(\alpha_1,\dots,\alpha_k),\alpha_i\in\mathbb{Z}^+ \\ i=1,\dots,k;\,|\alpha|:=\sum_{i=1}^{k}\alpha_i=3}} \frac{f_\alpha}{\prod_{i=1}^{k}\alpha_i!} \right\|_{\infty,\mathbb{R}_-^k} \leq \frac{4}{3} \left(\frac{144}{\Gamma(\nu+1)} \right)^{\frac{3}{4}} \|f\|_{\infty,\mathbb{R}_-^k}^{\frac{1}{4}} H_\nu^{\frac{3}{4}} < +\infty. \tag{17.33}$$

Proof We set

$$\begin{gathered} \Delta_1(z) := \sum_{i=1}^{k} \frac{\partial f(z)}{\partial x_i}, \\ \Delta_2(z) := \sum_{\substack{\alpha:=(\alpha_1,\dots,\alpha_k),\alpha_i\in\mathbb{Z}^+ \\ i=1,\dots,k;\,|\alpha|:=\sum_{i=1}^{k}\alpha_i=2}} \frac{f_\alpha(z)}{\prod_{i=1}^{k}\alpha_i!}, \\ \text{and} \\ \Delta_3(z) := \sum_{\substack{\alpha:=(\alpha_1,\dots,\alpha_k),\alpha_i\in\mathbb{Z}^+ \\ i=1,\dots,k;\,|\alpha|:=\sum_{i=1}^{k}\alpha_i=3}} \frac{f_\alpha(z)}{\prod_{i=1}^{k}\alpha_i!}. \end{gathered} \tag{17.34}$$

We can choose $x_0, z \in \mathbb{R}_-^k$ so that $z_i - x_{0i} = -h$, all $i = 1, \dots, k$, where $h > 0$. By Corollary 17.6 we get

$$f(x_0) = f(z) + \Delta_1(z)h + \Delta_2(z)h^2 + \Delta_3(z)h^3 + R_4 h^4, \tag{17.35}$$

where

$$R_4 := \frac{1}{\Gamma(\nu)} \left[\sum_{\substack{\alpha:=(\alpha_1,\ldots,\alpha_k),\alpha_i\in\mathbb{Z}^+ \\ i=1,\ldots,k;|\alpha|:=\sum_{i=1}^{k}\alpha_i=4}} \left(\frac{4!}{\prod_{i=1}^{k}\alpha_i!} \right) \right.$$

$$\left. \int_0^1 j^{\nu-1} D_{1-}^{\nu-4} \left(f_\alpha\left(x_0 + t\left(z - x_0\right)\right)\right)(j)\, dj \right]. \quad (17.36)$$

Therefore, $\forall\, h > 0$, we get

$$\begin{aligned}
&\Delta_1(z)\, h + \Delta_2(z)\, h^2 + \Delta_3(z)\, h^3 = f(x_0) - f(z) - R_4 h^4 =: A_1,\\
&\Delta_1(z)\, 2h + \Delta_2(z)\, 4h^2 + \Delta_3(z)\, 8h^3 = f(x_0) - f(z) - R_4 16 h^4 =: A_2,\\
&\text{and}\\
&\Delta_1(z)\, 3h + \Delta_2(z)\, 9h^2 + \Delta_3(z)\, 27h^3 = f(x_0) - f(z) - R_4 81 h^4 =: A_3.
\end{aligned} \quad (17.37)$$

We solve the system of three equations (17.37) and we find:

$$\begin{aligned}
\Delta_1(z) &= \tfrac{18A_1 - 9A_2 + 2A_3}{6h},\\
\Delta_2(z) &= \tfrac{4A_2 - 5A_1 - A_3}{2h^2},\\
\text{and}&\\
\Delta_3(z) &= \tfrac{3A_1 - 3A_2 + A_3}{6h^3}.
\end{aligned} \quad (17.38)$$

We have also that

$$|R_4| \leq \frac{24 H_\nu}{\Gamma(\nu + 1)}. \quad (17.39)$$

We calculate

$$18A_1 - 9A_2 + 2A_3 = 11\left(f(x_0) - f(z)\right) - 36 R_4 h^4. \quad (17.40)$$

Therefore it holds

$$|\Delta_1(z)| \leq \frac{11\, \|f\|_{\infty,\mathbb{R}_-^k}}{3h} + \frac{144 H_\nu}{\Gamma(\nu + 1)} h^3, \quad (17.41)$$

$\forall\, h > 0, \forall\, z \in \mathbb{R}_-^k$.

We get that

$$\|\Delta_1\|_{\infty,\mathbb{R}_-^k} \leq \frac{11\, \|f\|_{\infty,\mathbb{R}_-^k}}{3h} + \left(\frac{144 H_\nu}{\Gamma(\nu + 1)}\right) h^3, \quad (17.42)$$

$\forall\, h > 0$.

Next we find

$$4A_2 - 5A_1 - A_3 = -2\left(f\left(x_0\right) - f\left(z\right)\right) + 22R_4h^4. \tag{17.43}$$

Hence

$$\left|\Delta_2\left(z\right)\right| \leq \frac{2\left\|f\right\|_{\infty,\mathbb{R}_-^k}}{h^2} + \left(\frac{264H_\nu}{\Gamma\left(\nu+1\right)}\right)h^2, \tag{17.44}$$

$\forall\, h > 0, \forall\, z \in \mathbb{R}_-^k$.

The last implies

$$\left\|\Delta_2\right\|_{\infty,\mathbb{R}_-^k} \leq \frac{2\left\|f\right\|_{\infty,\mathbb{R}_-^k}}{h^2} + \left(\frac{264H_\nu}{\Gamma\left(\nu+1\right)}\right)h^2, \tag{17.45}$$

$\forall\, h > 0$.

We also calculate

$$3A_1 - 3A_2 + A_3 = \left(f\left(x_0\right) - f\left(z\right)\right) - 36R_4h^4.$$

Therefore

$$\left|\Delta_3\left(z\right)\right| \leq \frac{\left\|f\right\|_{\infty,\mathbb{R}_-^k}}{3h^3} + \left(\frac{144H_\nu}{\Gamma\left(\nu+1\right)}\right)h, \tag{17.46}$$

$\forall\, h > 0, \forall\, z \in \mathbb{R}_-^k$.

The last implies

$$\left\|\Delta_3\right\|_{\infty,\mathbb{R}_-^k} \leq \frac{\left\|f\right\|_{\infty,\mathbb{R}_-^k}}{3h^3} + \left(\frac{144H_\nu}{\Gamma\left(\nu+1\right)}\right)h, \tag{17.47}$$

$\forall\, h > 0$.

We will work on (17.42).

Call $\mu := \frac{11\left\|f\right\|_{\infty,\mathbb{R}_-^k}}{3}$, $\theta := \left(\frac{144H_\nu}{\Gamma(\nu+1)}\right)$, both are positive.

We study

$$y\left(h\right) = \frac{\mu}{h} + \theta h^3, \ \forall\, h > 0. \tag{17.48}$$

We have

$$y'\left(h\right) = -\mu h^{-2} + 3\theta h^2 = 0, \tag{17.49}$$

with one critical number

$$h_{crit.no.} = h_0 = \left(\frac{\mu}{3\theta}\right)^{\frac{1}{4}} > 0. \tag{17.50}$$

We have that

$$y''(h_0) = 2\mu h_0^{-3} + 6\theta h_0 > 0, \tag{17.51}$$

implying that y has a global minimum. We see that

$$y(h_0) = \mu h_0^{-1} + \theta h_0^3 = h_0^3 \left(\mu h_0^{-4} + \theta\right) = \frac{4}{3^{\frac{3}{4}}} \theta^{\frac{1}{4}} \mu^{\frac{3}{4}}. \tag{17.52}$$

I.e.

$$y(h_0) = \frac{4}{3^{\frac{3}{4}}} \theta^{\frac{1}{4}} \mu^{\frac{3}{4}}, \tag{17.53}$$

is the global minimum.

Consequently

$$\|\Delta_1\|_{\infty,\mathbb{R}_-^k} \leq 4 \left(\frac{11}{9}\right)^{\frac{3}{4}} \left(\frac{144}{\Gamma(\nu+1)}\right)^{\frac{1}{4}} \|f\|_{\infty,\mathbb{R}_-^k}^{\frac{3}{4}} H_\nu^{\frac{1}{4}}. \tag{17.54}$$

Next we work on (17.45).

Call $\varphi := 2\|f\|_{\infty,\mathbb{R}_-^k}$, $\psi := \frac{264 H_\nu}{\Gamma(\nu+1)} > 0$.

We study

$$y(h) := \frac{\varphi}{h^2} + \psi h^2, \ \forall\, h > 0. \tag{17.55}$$

We have

$$y'(h) = -2\varphi h^{-3} + 2\psi h = 0, \tag{17.56}$$

with

$$h_{crit.no.} = h_0 = \left(\frac{\varphi}{\psi}\right)^{\frac{1}{4}} > 0. \tag{17.57}$$

Notice that

$$y''(h_0) = 6\varphi h_0^{-4} + 2\psi > 0. \tag{17.58}$$

Thus we have that y has global minimum which is

$$y(h_0) = \varphi h_0^{-2} + \psi h_0^2 = h_0^2 \left(\varphi h_0^{-4} + \psi\right) = 2\psi^{\frac{1}{2}} \varphi^{\frac{1}{2}},$$

that is

$$y(h_0) = 2\psi^{\frac{1}{2}} \varphi^{\frac{1}{2}}. \tag{17.59}$$

Therefore we derive

$$\|\Delta_2\|_{\infty,\mathbb{R}_-^k} \leq 2\sqrt{\frac{528}{\Gamma(\nu+1)}}\sqrt{\|f\|_{\infty,\mathbb{R}_-^k} H_\nu}. \tag{17.60}$$

Finally we work on (17.47).

Let $A := \frac{\|f\|_{\infty,\mathbb{R}_-^k}}{3}$, $B := \frac{144H_\nu}{\Gamma(\nu+1)}$, both are positive.

We study

$$y(h) := \frac{A}{h^3} + Bh, \ \forall\, h > 0. \tag{17.61}$$

Then

$$y'(h) = -3Ah^{-4} + B = 0, \tag{17.62}$$

so that

$$h_{crit.no.} = h_0 = \left(\frac{3A}{B}\right)^{\frac{1}{4}} > 0. \tag{17.63}$$

Here

$$y''(h_0) = 12Ah_0^{-5} > 0. \tag{17.64}$$

Hence y has global minimum which is

$$y(h_0) = Ah_0^{-3} + Bh_0 = h_0\left(Ah_0^{-4} + B\right) = \frac{4}{3^{\frac{3}{4}}}A^{\frac{1}{4}}B^{\frac{3}{4}}.$$

That is

$$y(h_0) = \frac{4}{3^{\frac{3}{4}}}A^{\frac{1}{4}}B^{\frac{3}{4}}, \tag{17.65}$$

the global minimum of y.

Consequently it holds

$$\|\Delta_3\|_{\infty,\mathbb{R}_-^k} \leq \frac{4}{3}\left(\frac{144}{\Gamma(\nu+1)}\right)^{\frac{3}{4}} \|f\|_{\infty,\mathbb{R}_-^k}^{\frac{1}{4}} H_\nu^{\frac{3}{4}}. \tag{17.66}$$

The theorem is proved. ■

References

1. Anastassiou, G.A.: Intelligent Mathematics: Computational Analysis. Springer, Heidelberg (2011)
2. Anastassiou, G.A.: Multivariate generalized right side fractional Taylor formula and Landau inequalities. Submitted (2020)
3. Ditzian, Z.: Multivariate Landau-Kolmogorov-type inequality. Math. Proc. Camb. Philos. Soc. **105**(2), 335–350 (1989)
4. Kounchev, O.: Extremizers for the multivariate Landau-Kolmogorov inequality. Multivariate Approximation (Witten-Bommerholz, 1996), pp. 123–132. Math. Res. 101. Akademie Verlag, Berlin (1997)
5. Landau, E.: Einige Ungleichungen für zweimal differentzierban funktionen. Proc. Lond. Math. Soc. **13**, 43–49 (1913)

Chapter 18
Landau's Inequality for Semigroups

Kraljevič and Kurepa in 1970 [14] proved a nice Landau type inequality for semigroups by making use of Taylor's formula with integral remainder for semigroups, which has been a traditional method for proving such inequalities. The author offers another method in the semigroups and reproves the above result by using his earlier result about Ostrowski inequalities for semigroups, see [1], Chap. 16, pp. 259–289. He finds his method easier and convenient. An application is given at the end. It follows [2].

18.1 Introduction

We state Kraljevič and Kurepa, 1970, interesting theorem [14]:

Theorem 18.1 *Let A be an infinitesimal generator of a strongly continuous semigroup $\{T(t) : t > 0\}$ of bounded linear operators on a Banach space X. Let M, ω be real numbers with the property:*

$$\|T(t)\| \leq Me^{\omega t}, \forall\, t > 0. \tag{18.1}$$

Then for any $f \in D\left(A^2\right)$ such that $(A - \omega I)^2 f \neq 0$, the inequality

$$\|(A - \omega I) f\|^2 \leq 2M(M+1)\|f\| \left\|(A - \omega I)^2 f\right\| \tag{18.2}$$

holds.

Other inspirations follow:

G. A. Anastassiou, *Constructive Fractional Analysis with Applications*,
Studies in Systems, Decision and Control 362,
https://doi.org/10.1007/978-3-030-71481-9_18

Theorem 18.2 (Kallman and Rota [11]) *Let X be a complex Banach space and $t \to T(t)$ $(t \geq 0)$ be a strongly continuous semigroup of linear contractions on X. Let A be its infinitesimal generator. Then, for every $f \in D\left(A^2\right)$, it holds*

$$\|Af\|^2 \leq 4 \|f\| \left\|A^2 f\right\|. \tag{18.3}$$

We also mention

Theorem 18.3 (Hille 1992, [8]) *If A is the infinitesimal generator of a strongly continuous contraction semigroup, if $f \in D(A^n)$ and $1 \leq k < n$, then there exist constants $C_{n,k}$ independent of A, so that*

$$\left\|A^k(f)\right\|^n \leq C_{n,k}^n \|f\|^{n-k} \left\|A^n(f)\right\|^k. \tag{18.4}$$

This type of research started by Landau [16]. He proved first that if $f \in C^2([0,1])$, $\|f\|_\infty = 1$, $\left\|f''\right\|_\infty = 4$, then $\left\|f'\right\|_\infty \leq 4$, with 4 the best constant and the result is not necessarily true for an interval of length < 1. It is customary now to write Landau's inequality as

$$\left\|f'\right\|_{p,I} \leq C_p(I) \|f\|_{p,I}^{\frac{1}{2}} \left\|f''\right\|_{p,I}^{\frac{1}{2}}, \tag{18.5}$$

where $\|\cdot\|_{p,I}$ is the p-norm on the interval I; $p \in [1,\infty]$, $I = \mathbb{R}_+$ or $I = \mathbb{R}$, and $f : I \to \mathbb{R}$ is twice differentiable with $f, f'' \in L_p(I)$. $C_p(I) > 0$ is independent of f.

Landau proved the best constants

$$C_\infty(\mathbb{R}_+) = 2 \text{ and } C_\infty(\mathbb{R}) = \sqrt{2},$$

see also [3].

Using Theorem 18.2 we can derive a Landau's inequality, see (18.5), by talking $X = C([0,\infty))$ or $L_p([0,\infty))$ and defining the shift operator

$$T(s) f(t) = f(t+s), \; s \geq 0, \; Af(t) = f'(t),$$

see [7, 8].

Of inspiration are also the articles [9, 13, 15, 17].

18.2 Background

We will be using the following notions. Here all this background comes from [4] (in general see also [6, 18]).

Let X a real or complex Banach space with elements $f, g, \ldots$ having norm $\|f\|, \|g\|, \ldots$ and let $\varepsilon(X)$ be the Banach algebra of endomorphisms of X.

If $T \in \varepsilon(X)$, $\|T\|$ denotes the norm of T.

Definition 18.4 If $T(t)$ is an operator function on the non-negative real axis $0 \leq t < \infty$ to the Banach algebra $\varepsilon(X)$ satisfying the following conditions:

$$\begin{cases} \text{(i) } T(t_1 + t_2) = T(t_1)\,T(t_2), \quad (t_1, t_2 \geq 0) \\ \text{(ii) } T(0) = I \ (I = \text{identity operator}), \end{cases} \tag{18.6}$$

then $\{T(t) : 0 \leq t < \infty\}$ is called a one-parameter semi-group of operators in $\varepsilon(X)$.

The semi-group $\{T(t) : 0 \leq t < \infty\}$ is said to be of class C_0 if it satisfies the further property

$$\text{(iii) } s - \lim_{t \to 0+} T(t)\, f = f \ \ (f \in X) \tag{18.7}$$

referred to as the strong continuity of $T(t)$ at the origin.

In this chapter we shall assume that the family of bounded linear operators $\{T(t) : 0 \leq t < \infty\}$ mapping X to itself is a semi-group of class C_0, thus all three conditions of the above definition are satisfied.

Proposition 18.5 *(a)* $\|T(t)\|$ *is bounded on every finite subinterval of* $[0, \infty)$.

(b) For each $f \in X$*, the vector-valued function* $T(t)\, f$ *on* $[0, \infty)$ *is strongly continuous, thus vector-Riemann integrable on* $[0, a]$, $a > 0$.

Definition 18.6 The infinitesimal generator A of the semi-group $\{T(t) : 0 \leq t < \infty\}$ is defined by

$$Af = s - \lim_{\tau \to 0+} A_\tau f, \ \ A_\tau f = \frac{1}{\tau}\,[T(\tau) - I]\, f \tag{18.8}$$

whenever the limit exists; the domain of A, in symbols $D(A)$, is the set of elements f for which the limits exists.

Proposition 18.7 *(a)* $D(A)$ *is a linear manifold in* X *and* A *is a linear operator.*

(b) If $f \in D(A)$*, then* $T(t)\, f \in D(A)$ *for each* $t \geq 0$ *and*

$$\frac{d}{dt} T(t)\, f = AT(t)\, f = T(t)\, Af \ \ (t \geq 0); \tag{18.9}$$

furthermore,

$$T(t)\, f - f = \int_0^t T(u)\, Af du \ \ (t > 0). \tag{18.10}$$

(c) $D(A)$ *is dense in* X*, i.e.* $\overline{D(A)} = X$*, and* A *is a closed operator.*

Definition 18.8 For $r = 0, 1, 2, \ldots$ the operator A^r is defined inductively by the relations $A^0 = I$, $A^1 = A$, and

$$D(A^r) = \{f : f \in D(A^{r-1}) \, and \, A^{r-1} f \in D(A)\}$$

$$A^r f = A\left(A^{r-1} f\right) = s - \lim_{\tau \to 0+} A_\tau \left(A^{r-1} f\right) \ \ (f \in D(A^r)). \tag{18.11}$$

For the operator A^r and its domain $D(A^r)$ we have the following

Proposition 18.9 *(a) $D(A^r)$ is a linear subspace in X and A^r is a linear operator.*
(b) If $f \in D(A^r)$, so does $T(t) f$ for each $t \geq 0$ and

$$\frac{d^r}{dt^r} T(t) f = A^r T(t) f = T(t) A^r f. \tag{18.12}$$

Furthermore

$$T(t) f - \sum_{k=0}^{r-1} \frac{t^k}{k!} A^k f = \frac{1}{(r-1)!} \int_0^t (t-u)^{r-1} T(u) A^r f du, \tag{18.13}$$

the Taylor's formula for semigroups.
Additionally it holds

$$(T(t) - I)^r f = \int_0^t \int_0^t \dots \int_0^t T(u_1 + u_2 + \dots + u_r) A^r f du_1 du_2 \dots du_r. \tag{18.14}$$

(c) $D(A^r)$ is dense in X for $r = 1, 2, \dots,$; moreover, $\cap_{r=1}^{\infty} D(A^r)$ is dense in X, A^r is a closed operator.

Integrals in (18.13) and (18.14) are vector valued Riemann integrals, see [4, 12].

Here we will assume that $f \in D(A^2)$. Clearly here $\int_0^a T(t) f dt \in X$, where $a > 0$, see [18].

We mention an Ostrowski type inequality for semigroups

Theorem 18.10 ([1], Chap. 13, p. 219) *Let $f \in D(A)$, $a > 0$, and denote*

$$\| \|T(\cdot) Af\| \|_{\infty,[0,a]} := \sup_{u \in [0,a]} \|T(u) Af\|. \tag{18.15}$$

Then

$$\left\| \frac{1}{a} \int_0^a T(t) f dt - T(t_0) f \right\| \leq \left(\frac{t_0^2 + (a - t_0)^2}{2a} \right) \| \|T(\cdot) Af\| \|_{\infty,[0,a]}, \tag{18.16}$$

for a fixed $t_0 \in [0, a]$.

We will be using Theorem 18.10 to reprove Theorem 18.1 by another method.

18.3 Main Results

We give the following Landau inequality for semigroups.

Theorem 18.11 *Let the family of bounded linear operator* $\{T(t) : 0 \leq t < \infty\}$ *mapping* X *into itself be a semigroup of class* C_0. *We assume here that* $\|T(t)\| \leq M$, $\forall\, t \in \mathbb{R}_+$, *where* $M > 0$. *Let* $f \in D\left(A^2\right)$. *Then*

$$\|Af\|^2 \leq 2M(M+1)\|f\|\left\|A^2 f\right\|, \tag{18.17}$$

where A *is the infinitesimal generator.*

Proof Let $f \in D(A)$, $t > 0$, and denote $\|\|T(\cdot)Af\|\|_{\infty,[0,t]} := \sup_{u\in[0,t]} \|T(u)Af\|$. By Theorem 18.10 we have

$$\left\|\frac{1}{t}\int_0^t T(z) f dz - T(z_0) f\right\| \leq \frac{t}{2}\|\|T(\cdot)Af\|\|_{\infty,[0,t]}, \tag{18.18}$$

for any fixed $z_0 \in [0,t]$.

Here let $f \in D\left(A^2\right)$, and we denote

$$\left\|\left\|T(\cdot)A^2 f\right\|\right\|_{\infty,[0,t]} := \sup_{u\in[0,t]}\left\|T(u)A^2 f\right\|. \tag{18.19}$$

By (18.18) we obtain

$$\left\|\frac{1}{t}\int_0^t T(z) Af dz - T(z_0) Af\right\| \leq \frac{t}{2}\left\|\left\|T(\cdot)A^2 f\right\|\right\|_{\infty,[0,t]}, \tag{18.20}$$

for any fixed $z_0 \in [0,t]$.

Notice here that $\left\|\left\|T(\cdot)A^2 f\right\|\right\|_{\infty,[0,t]} < \infty$. By (18.10) we get

$$\left\|\frac{1}{t}(T(t) f - f) - T(z_0) Af\right\| \leq \frac{t}{2}\left\|\left\|T(\cdot)A^2 f\right\|\right\|_{\infty,[0,t]}, \tag{18.21}$$

for any fixed $z_0 \in [0,t]$.

Hence

$$\|T(z_0) Af\| - \frac{\|T(t) f - f\|}{t} \leq \frac{t}{2}\left\|\left\|T(\cdot)A^2 f\right\|\right\|_{\infty,[0,t]}. \tag{18.22}$$

Furthermore we have

$$\|T(z_0)Af\| \leq \frac{\|T(t)f - f\|}{t} + \frac{t}{2}\left\|\|T(\cdot)A^2 f\|\right\|_{\infty,[0,t]} \leq$$

$$\frac{(\|T(t)f\| + \|f\|)}{t} + \frac{t}{2}\left\|\|T(\cdot)A^2 f\|\right\|_{\infty,[0,t]}. \tag{18.23}$$

By assumption $\|T(t)\| \leq M$, $\forall\, t \geq 0$, so we have that $\|T(t)f\| \leq \|T(t)\|\,\|f\| \leq M\,\|f\|$.

Similarly it holds

$$\left\|T(z)A^2 f\right\| \leq M\left\|A^2 f\right\|, \forall\, z \geq 0.$$

That is

$$\left\|\left\|T(\cdot)A^2 f\right\|\right\|_{\infty,[0,t]} \leq M\left\|A^2 f\right\|, \ \forall\, t \geq 0.$$

So we have that

$$\|T(z_0)Af\| \leq \frac{(M+1)}{t}\|f\| + \frac{tM}{2}\left\|A^2 f\right\|, \tag{18.24}$$

$\forall\, t \in \mathbb{R}_+$, and $\forall\, z_0 \in [0, t]$.

Choosing $z_0 = 0$, we get

$$\|Af\| \leq \frac{(M+1)}{t}\|f\| + \frac{tM}{2}\left\|A^2 f\right\|, \ \forall\, t \in \mathbb{R}_+. \tag{18.25}$$

If $A^2 f = 0$, then $\left\|A^2 f\right\| = 0$, and clearly it holds by (18.25) that $\|Af\| = 0$, that is $Af = 0$.

So next we will assume that $f \neq 0$ and $A^2 f \neq 0$.

We consider the function

$$y(t) := \frac{(M+1)\|f\|}{t} + t\frac{M\left\|A^2 f\right\|}{2}, \ \forall\, t \in \mathbb{R}_+. \tag{18.26}$$

Set

$$A := (M+1)\|f\|, \ B := \frac{M\left\|A^2 f\right\|}{2}.$$

That is

$$y(t) = \frac{A}{t} + Bt, \forall\, t \in \mathbb{R}_+. \tag{18.27}$$

Hence

$$y'(t) = -At^{-2} + B = 0,$$

and the critical number is

$$t_0 = \sqrt{\frac{A}{B}}.$$

Notice that $y''(t) = 2At^{-3}$, and

$$y''(t_0) = y''\left(\sqrt{\frac{A}{B}}\right) = 2A\left(\sqrt{\frac{A}{B}}\right)^{-3} = 2A\left(\frac{A}{B}\right)^{-\frac{3}{2}} = 2A\left(\frac{B}{A}\right)^{\frac{3}{2}} = 2A^{-\frac{1}{2}}B^{\frac{3}{2}} > 0. \quad (18.28)$$

Hence y has a global minimum which is

$$y(t_0) = y\left(\sqrt{\frac{A}{B}}\right) = 2\sqrt{AB} = \sqrt{2M(M+1)\|f\|\left\|A^2 f\right\|}. \quad (18.29)$$

We have proved that

$$\|Af\| \leq \sqrt{2M(M+1)\|f\|\left\|A^2 f\right\|}, \quad (18.30)$$

when $A^2 f \neq 0$.

The last inequality is true even when $A^2 f = 0$.

If $f = 0$ then $Af = 0$, so that (18.30) is trivially again true.

The theorem is proved. ■

Next comes a more general Landau inequality for semigroups.

Corollary 18.12 *Let the family of bounded linear operators* $\{T(t) : 0 \leq t < \infty\}$ *mapping* X *into itself be a semigroup of class* C_0. *Let* $M > 0$ *and* ω *be a real number with the property* $\|T(t)\| \leq Me^{\omega t}$, $\forall\, t \geq 0$. *Then, for any* $f \in D\left(A^2\right)$, *it holds*

$$\|(A - \omega I) f\|^2 \leq 2M(M+1)\|f\|\left\|(A - \omega I)^2 f\right\|. \quad (18.31)$$

Proof We are acting as in the proof of Theorem 18.1. We have that

$$\left\|e^{-\omega t}T(t)\right\| \leq M, \forall\, t \geq 0,$$

and $t \to e^{-\omega t}T(t)$ is a semigroup with $A - \omega I$ as the infinitesimal generator. Hence by applying (18.17) we derive (18.31). ■

Notice that our results are valid without the restriction $(A - \omega I)^2 f \neq 0$, $\omega \in \mathbb{R}$, which is assumed in Theorem 18.1.

18.4 Application

Here we follow [5]. It is known that the classical diffusion equation

$$\frac{\partial W}{\partial t} = \frac{\partial^2 W}{\partial x^2}, \quad -\infty < x < \infty, \, t > 0, \tag{18.32}$$

with initial condition

$$\lim_{t \to 0+} W(x,t) = f(x), \tag{18.33}$$

has under general conditions its solution given by

$$W(x,t,f) = (T(t) f)(x) = \frac{1}{2\sqrt{\pi t}} \int_{-\infty}^{\infty} f(x+u) e^{-\frac{u^2}{4t}} du, \tag{18.34}$$

the so called Gauss–Weierstrass singular integral.

The infinitesimal generator of the semigroup $\{T(t) : 0 \le t < \infty\}$ is $A = \frac{\partial^2}{\partial x^2}$ ([10], p. 578).

Here we suppose that f belongs to the Banach space $X = UCB(\mathbb{R})$, the space of bounded and uniformly continuous functions from $\mathbb{R}$ into itself, with norm $\|f\|_C := \sup_{x \in \mathbb{R}} |f(x)|$.

We notice that

$$|W(x,t,f)| = \frac{1}{2\sqrt{\pi t}} \left| \int_{-\infty}^{\infty} f(x+u) e^{-\frac{u^2}{4t}} du \right| \le$$

$$\frac{1}{2\sqrt{\pi t}} \int_{-\infty}^{\infty} |f(x+u)| e^{-\frac{u^2}{4t}} du \le \|f\|_C \left(\frac{1}{2\sqrt{\pi t}} \int_{-\infty}^{\infty} e^{-\frac{u^2}{4t}} du \right) \tag{18.35}$$

$$= \|f\|_C \cdot 1 = \|f\|_C < \infty, \ \forall \, t \in \mathbb{R}_+.$$

That is

$$\|W(\cdot,t,f)\|_C \le 1 \cdot \|f\|_C, \ \forall \, t \in \mathbb{R}_+. \tag{18.36}$$

So in Theorem 18.11, $M = 1$.

Let now $f \in D\left(\left(\frac{\partial^2}{\partial x^2}\right)^2\right) = D\left(\frac{\partial^4}{\partial x^4}\right)$. Then, by (18.17), we derive

$$\left\| \frac{\partial^2 f}{\partial x^2} \right\|_C^2 \le 4 \|f\|_C \left\| \frac{\partial^4 f}{\partial x^4} \right\|_C. \tag{18.37}$$

References

1. Anastassiou, G.A.: Intelligent Comparisons: Analytic Inequalities. Springer, Heidelberg (2016)
2. Anastassiou, G.A.: A Landau's inequality for semigroups revisited. Transylv. J. Math. Mech. Accepted (2020)
3. Barnett, N.S., Dragomir, S.S.: Some Landau type inequalities for functions whose derivatives are of locally bounded variation. Tamkang J. Math. **37**(4), 301–308 (2006)
4. Butzer, P.L., Berens, H.: Semi-Groups of Operators and Approximation. Springer, New York (1967)
5. Butzer, P.L., Tillmann, H.G.: Approximation theorems for semi-groups of bounded linear transformations. Math. Ann. **140**, 256–262 (1960)
6. Goldstein, J.A.: Semigroups of Linear Operators and Applications. Oxford University Press, Oxford (1985)
7. Hille, E.: Remark on the Landau-Kallman-Rota inequality. In: Reports of Meetings, 7th International Symposium on Functional Equations, September 1–13 (1969). Aequ. Math. **4**, 239–240 (1970)
8. Hille, E.: Generalizations of Landau's inequality to linear operators. In: Butzer, P.L., Kahane, J.-P., Nagy, B.S. (eds.) Linear Operators and Approximation. Proceedings of the Conference at Oberwolfach Mathematical Research Institute, August 14–22, 1971, pp. 20–32. Birkhauser, Basel (1972)
9. Hille, E.: On the Landau-Kallman-Rota inequality. J. Approx. Theory **6**, 117–122 (1972)
10. Hille, E., Phillips, R.S.: Functional Analysis and Semigroups, revised edn., pp. XII a. 808. American Mathematical Society Colloquium Publications, vol. 31. American Mathematical Society, Providence (1957)
11. Kallman, R.R., Rota, G.-C.: On the inequality $\left\|f'\right\|^2 \leq 4\left\|f\right\|\left\|f''\right\|$. In: Shisha, O. (ed.) Inequalities II, pp. 187–192. Academic, New York (1970)
12. Katznelson, Y.: An Introduction to Harmonic Analysis. Dover, New York (1976)
13. Kolmogorov, A.N.: On inequalities between the upper bounds of the successive derivatives of an arbitrary function on an infinite interval. (Russian.) Užen. Zap. Moskov. Gos. Univ. Matematika **30**, 3–13 (1939) (Am. Math. Soc. Transl. (1) 2, 233–243 (1962))
14. Kraljevič, H., Kurepa, S.: Semi-groups on Banach spaces. Glasnik Mat. **5**(25), 109–117 (1970)
15. Kurepa, S.: Remark on the Landau inequality. In: Reports of Meetings, 7th International Symposium on Functional Equations, September 1–13 (1969). Aequ. Math. **4**, 240–241 (1970)
16. Landau, E.: Einige Ungleichungen für zweimal differentzierban funktionen. Proc. Lond. Math. Soc. **13**, 43–49 (1913)
17. Schoenberg, I.J., Cavaretta, A.: Solution of Landau's problem concerning higher derivatives on the halfline. MRC Technical Summary Report, No. 1050, Madison, Wisconsin (1970)
18. Shilov, G.: Elementary Functional Analysis. Dover Publications Inc, New York (1996)

Chapter 19
Fractional Variable Order Gronwall Inequality

This chapter presents the first generalized fractional variable order Gronwall inequality. It follows [1].

19.1 Introduction

The following generalized Gronwall inequality for fractional differential equations (of constant order) was established in [2]:

Theorem 19.1 *Suppose that $a(t)$ is a nonnegative function locally integrable on $[0, T)$ (for some $T \leq \infty$), $g(t)$ is a nonnegative, nondecreasing, and bounded continuous function defined on $[0, T)$ and $\beta_0 > 0$. If $u(t)$ is nonnegative and locally integrable on $[0, T)$ satisfying*

$$u(t) \leq a(t) + g(t) \int_0^t (t-s)^{\beta_0 - 1} u(s) ds, \qquad 0 \leq t < T, \tag{19.1}$$

then

$$u(t) \leq a(t) + \int_0^t \left\{ \sum_{n=1}^{\infty} \frac{\left[\Gamma(\beta_0) g(t)\right]^n}{\Gamma(n\beta_0)} (t-s)^{n\beta_0 - 1} a(s) \right\} ds, \qquad 0 \leq t < T, \tag{19.2}$$

where $\Gamma(\cdot)$ is the Gamma function.

G. A. Anastassiou, *Constructive Fractional Analysis with Applications*,
Studies in Systems, Decision and Control 362,
https://doi.org/10.1007/978-3-030-71481-9_19

The idea of the proof is to introduce the Volterra-type (linear) operator

$$B\phi(t) := g(t) \int_0^t (t-s)^{\beta_0 - 1} \phi(s) ds, \qquad 0 \le t < T, \tag{19.3}$$

so that (19.1) can be written as

$$u(t) \le a(t) + Bu(t), \qquad 0 \le t < T, \tag{19.4}$$

and hence, by repeated iteration of (19.4),

$$u(t) \le \sum_{k=0}^{n-1} B^k a(t) + B^n u(t), \qquad 0 \le t < T. \tag{19.5}$$

The remaining part of the proof of Theorem 19.1 [3] is the inductive justification of the inequality

$$B^n \phi(t) \le \frac{\left[\Gamma(\beta_0) g(t)\right]^n}{\Gamma(n\beta_0)} \int_0^t (t-s)^{n\beta_0 - 1} \phi(s) ds, \qquad 0 \le t < T, \quad n = 1, 2, \ldots, \tag{19.6}$$

for any function $\phi(t) \ge 0$ which is locally integrable on $[0, T)$. An immediate implication of (19.6) is that

$$B^n \phi(t) \to 0 \quad \text{as } n \to +\infty \qquad \text{for any } t \in [0, T), \tag{19.7}$$

and the validity of (19.2) follows, by using (19.6) and (19.7) in (19.5).

Remark 19.2 In the statement of Theorem 19.1, the assumption that $g(t)$ is bounded is not necessary, since for any fixed $t \in [0, T)$ the monotonicity of g implies that $g(s) \le g(t) < \infty$ for $0 \le s \le t$. Also, the assumption that $g(t)$ is continuous is not used in the proof of the theorem, hence it, too, is not necessary. Therefore, the theorem remains true for any nonnegative increasing function $g(t)$ defined on $[0, T)$. Actually, there is an immediate extension of Theorem 19.1 to the case where $g(t)$ is any function which is locally bounded in $[0, T)$. In this case we can just set

$$G(t) := \sup_{s \in [0,t]} \{g(s), 0\}, \qquad 0 \le t < T, \tag{19.8}$$

so that $G(t)$ is nonnegative, increasing, and satisfies $G(t) \ge g(t)$ for $t \in [0, T)$, and then apply Theorem 19.1 with $G(t)$ in place of $g(t)$.

19.2 Main Results

In this short note we propose an extension of Theorem 19.1 to the case where the constant order β_0 is replaced by a strictly positive variable order $\beta(t)$. Our motivation came from the recent monograph [2], which contains an extensive discussion on fractional integrals and derivatives of variable order and their applications.

Theorem 19.3 *Suppose that $a(t)$ is a nonnegative function locally integrable on $[0, T)$ for some $T < \infty$, $g(t)$ is a nonnegative and nondecreasing function defined on $[0, T)$, and $\beta(t)$ is a (Lebesgue) measurable function satisfying*

$$0 < \beta_0 \leq \beta(t) \leq A < \infty, \qquad 0 \leq t < T. \tag{19.9}$$

If $u(t)$ is nonnegative and locally integrable on $[0, T)$ satisfying the inequality

$$u(t) \leq a(t) + g(t) \int_0^t (t-s)^{\beta(s)-1} u(s)ds, \qquad 0 \leq t < T, \tag{19.10}$$

then

$$\begin{aligned} u(t) &\leq a(t) + \sum_{k=1}^{\infty} L^k a(t) \\ &\leq a(t) + \int_0^t \left\{ \sum_{n=1}^{\infty} \frac{\left[\Gamma(\beta_0) K g(t)\right]^n}{\Gamma(n\beta_0)} (t-s)^{n\beta_0 - 1} a(s) \right\} ds, \qquad 0 \leq t < T, \end{aligned} \tag{19.11}$$

where L is the Volterra-type (linear) operator

$$L\phi(t) := g(t) \int_0^t (t-s)^{\beta(s)-1} \phi(s)ds, \qquad 0 \leq t < T, \tag{19.12}$$

and

$$K := \max\{1, T\}^{A-\beta_0}. \tag{19.13}$$

Proof Observe that from (19.9) we get (since $\beta(s) \geq \beta_0$)

$$\frac{(t-s)^{\beta(s)-1}}{(t-s)^{\beta_0 - 1}} = (t-s)^{\beta(s)-\beta_0} \leq \max\{1, T\}^{A-\beta_0}, \qquad 0 \leq s < t < T,$$

that is

$$(t-s)^{\beta(s)-1} \leq \max\{1, T\}^{A-\beta_0} (t-s)^{\beta_0 - 1}, \qquad 0 \leq s < t < T. \tag{19.14}$$

Therefore, (19.10) implies

$$u(t) \leq a(t) + Kg(t) \int_0^t (t-s)^{\beta_0 - 1} u(s) ds, \qquad 0 \leq t < T, \tag{19.15}$$

where K is given by (19.13). Since $\beta_0 > 0$ is a constant, we can apply Theorem 19.1–(19.15), where the operator B now has the slightly different form:

$$B\phi(t) = Kg(t) \int_0^t (t-s)^{\beta_0 - 1} \phi(s) ds, \qquad 0 \leq t < T. \tag{19.16}$$

As in the proof of Theorem 19.1,

$$\lim_{n \to +\infty} B^n u(t) = 0, \qquad 0 \leq t < T. \tag{19.17}$$

Now, it is clear from (19.12), (19.14), and (19.16) that for $u(t) \geq 0$ we have

$$0 \leq Lu(t) \leq Bu(t), \qquad 0 \leq t < T. \tag{19.18}$$

Hence, by using (19.17) in (19.18) we get

$$\lim_{n \to +\infty} L^n u(t) = 0, \qquad 0 \leq t < T, \tag{19.19}$$

and since (19.10) can be written as

$$u(t) \leq a(t) + Lu(t), \qquad 0 \leq t < T, \tag{19.20}$$

formula (19.11) follows easily from (19.19), (19.18) and Theorem 19.1. ■

Notice that, as in the standard Gronwall inequality, the value of (19.11) lies in the fact that it gives a bound for $u(t)$ in terms of $a(t)$, $g(t)$, and $\beta(t)$.

As explained in Remark 19.2, in the case where $g(t)$ is any locally bounded function in $[0, T)$, Theorem 19.3 holds by replacing $g(t)$ with $G(t)$ of (19.8).

Corollary 19.4 *All as in Theorem 19.3 with $g(t) = b \geq 0$ constant. If*

$$u(t) \leq a(t) + b \int_0^t (t-s)^{\beta(s) - 1} u(s) ds, \qquad 0 \leq t < T, \tag{19.21}$$

then

$$\begin{aligned} u(t) &\leq a(t) + \sum_{k=1}^{\infty} L_1^k a(t) \\ &\leq a(t) + \int_0^t \left\{ \sum_{n=1}^{\infty} \frac{[\Gamma(\beta_0) K b]^n}{\Gamma(n\beta_0)} (t-s)^{n\beta_0 - 1} a(s) \right\} ds, \qquad 0 \leq t < T, \end{aligned} \tag{19.22}$$

where L_1 *is the Volterra operator*

$$L_1\phi(t) := b\int_0^t (t-s)^{\beta(s)-1}\phi(s)ds, \qquad 0 \le t < T. \tag{19.23}$$

Corollary 19.5 *All as in Theorem 19.3 with* $a(t)$ *be a nondecreasing function on* $[0, T)$. *Then*

$$u(t) \le a(t)E_{\beta_0}\left(Kg(t)\Gamma(\beta_0)t^{\beta_0}\right), \qquad 0 \le t < T, \tag{19.24}$$

where $E_{\beta_0}(\cdot)$ *is the Mittag-Leffler function defined by*

$$E_{\beta_0}(z) := \sum_{k=0}^{\infty} \frac{z^k}{\Gamma(k\beta_0 + 1)}.$$

Proof The assumptions of Theorem 19.3 and (19.11) imply

$$\begin{aligned} u(t) &\le a(t)\left(1 + \int_0^t \left\{\sum_{n=1}^{\infty} \frac{[\Gamma(\beta_0)Kg(t)]^n}{\Gamma(n\beta_0)}(t-s)^{n\beta_0-1}\right\}ds\right) \\ &= a(t)\sum_{n=0}^{\infty} \frac{\left[\Gamma(\beta_0)Kg(t)t^{\beta_0}\right]^n}{\Gamma(n\beta_0+1)} = a(t)E_{\beta_0}\left(Kg(t)\Gamma(\beta_0)t^{\beta_0}\right). \end{aligned} \tag{19.25}$$

■

The Gronwall inequality of fractional variable order is expected to find wide applications in the forthcoming studies of fractional differential equations of variable order.

References

1. Anastassiou, G.A., Papanicolaou, V.G.: A Gronwall inequality of fractional variable order. Prog. Fract. Differ. Appl. Accepted (2020)
2. Yang, X.-J.: General Fractional Derivatives. Theory, Methods and Applications. CRC Press, Taylor & Francis Group, Boca Raton (2019)
3. Ye, H., Gao, J., Ding, Y.: A generalized Gronwall inequality and its application to a fractional differential equation. J. Math. Anal. Appl. **328**(4), 1075–1081 (2007)

Chapter 20
A Study of Gronwall Inequalities of Fractional Variable Order

A wide variety of Gronwall inequalities of fractional variable order is presented. These are of left and right sides, applications follow. This new research strongly supports the new trend of Fractional Calculus of variable order, see [7]. It follows [4].

20.1 Introduction

We are motivated by the following results. The next generalized Gronwall inequality for fractional differential equations (of constant) order was established in [8]:

Theorem 20.1 *Suppose that $a(t)$ is a nonnegative function locally integrable on $[0, T)$ (for some $T \leq \infty$), $g(t)$ is a nonnegative, nondecreasing and bounded continuous function defined on $[0, T)$ and $\beta_0 > 0$. If $u(t)$ is nonnegative and locally integrable on $[0, T)$ satisfying*

$$u(t) \leq a(t) + g(t) \int_0^t (t-s)^{\beta_0 - 1} u(s)\, ds, \quad 0 \leq t < T, \tag{20.1}$$

then

$$u(t) \leq a(t) + \int_0^t \left\{ \sum_{n=1}^{\infty} \frac{[\Gamma(\beta_0)\, g(t)]^n}{\Gamma(n\beta_0)} (t-s)^{n\beta_0 - 1} a(s) \right\} ds, \quad 0 \leq t < T, \tag{20.2}$$

where $\Gamma(\cdot)$ is the Gamma function.

G. A. Anastassiou, *Constructive Fractional Analysis with Applications*,
Studies in Systems, Decision and Control 362,
https://doi.org/10.1007/978-3-030-71481-9_20

In [3] the authors generalized Theorem 20.1 for a fractional variable order, the first work in the literature of this kind:

Theorem 20.2 ([3]) *Suppose that $a(t)$ is a nonnegative function locally integrable on $[0, T)$ for some $T < \infty$, $g(t)$ is a nonnegative and nondecreasing function defined on $[0, T)$, and $\beta(t)$ is a (Lebesgue) measurable function satisfying*

$$0 < \beta_0 \leq \beta(t) \leq A < \infty, \quad 0 \leq t < T. \tag{20.3}$$

If $u(t)$ is nonnegative and locally integrable on $[0, T)$ satisfying the inequality

$$u(t) \leq a(t) + g(t) \int_0^t (t-s)^{\beta(s)-1} u(s)\, ds, \quad 0 \leq t < T, \tag{20.4}$$

then

$$u(t) \leq a(t) + \sum_{k=1}^{\infty} L_1^k a(t) \leq \tag{20.5}$$

$$a(t) + \int_0^t \left\{ \sum_{n=1}^{\infty} \frac{[\Gamma(\beta_0) K g(t)]^n}{\Gamma(n\beta_0)} (t-s)^{n\beta_0 - 1} a(s) \right\} ds, \quad 0 \leq t < T,$$

where L_1 is the Volterra-type (positive linear) operator

$$L_1 \phi(t) := g(t) \int_0^t (t-s)^{\beta(s)-1} \phi(s)\, ds, \quad 0 \leq t < T, \tag{20.6}$$

and

$$K := \max\{1, T\}^{A-\beta_0}. \tag{20.7}$$

In this chapter the author applies the proving method of Theorem 20.2, and produces several interesting Gronwall type inequalities of fractional variable order covering other various important cases. Gronwall inequalities of fractional variable order are meant to find wide applications in the forthcoming new research of fractional differential equations of variable order, see also [7].

20.2 Main Results

It is all about the Gronwall fractional variable order inequalities. We need to mention:

Theorem 20.3 (Lemma 7.1.2, p. 189 in [5]) *Suppose $\beta, \gamma > 0$, $\beta + \gamma > 1$ and $a, b \geq 0$, u is nonnegative, $t^{\gamma-1} u(t)$ is locally integrable on $[0, T)$, and*

$$u(t) \leq a + b \int_0^t (t-s)^{\beta-1} s^{\gamma-1} u(s)\, ds, \tag{20.8}$$

a.e. in $(0, T)$*; then*

$$u(t) \leq a E_{\beta,\gamma}\left((b\Gamma(\beta))^{\frac{1}{\nu}} t \right), \tag{20.9}$$

where

$$\nu = \beta + \gamma - 1 > 0, \ E_{\beta,\gamma}(s) = \sum_{m=0}^{\infty} c_m s^{m\nu}, \tag{20.10}$$

with $c_0 = 1$, $\frac{c_{m+1}}{c_m} = \frac{\Gamma(m\nu+\gamma)}{\Gamma(m\nu+\gamma+\beta)}$, *for* $m \geq 0$. *As* $s \to \infty$, *we get that*

$$E_{\beta,\gamma}(s) = O\left(s^{\frac{1}{2}\left(\frac{\nu}{\beta}-\gamma\right)} \exp\left(\frac{\beta}{\nu} s^{\frac{\nu}{\beta}} \right) \right). \tag{20.11}$$

We present the following left inequality:

Theorem 20.4 *Suppose* $\beta, \gamma > 0$, $\beta + \gamma > 1$ *and* $a, b \geq 0$; u *is nonnegative and (Lebesgue) measurable and* $t^{\gamma-1} u(t)$ *is locally integrable: both on* $[0, T)$. *Let also* $0 < \beta \leq \beta(t) \leq A < \infty$ *and* $0 < \gamma \leq \gamma(t) \leq B < \infty$, *for* $0 \leq t < T$. *Denote by* $K_1 := \max\{1, T\}^{A-\beta}$, $K_2 := \max\{1, T\}^{B-\gamma}$. *The above functions* $\beta(t)$ *and* $\gamma(t)$ *are assumed to be (Lebesgue) measurable functions over* $[0, T)$. *Let also the positive linear operator*

$$L_2 \phi(t) := b \int_0^t (t-s)^{\beta(s)-1} s^{\gamma(s)-1} \phi(s)\, ds, \quad 0 \leq t < T. \tag{20.12}$$

If

$$u(t) \leq a + b \int_0^t (t-s)^{\beta(s)-1} s^{\gamma(s)-1} u(s)\, ds, \tag{20.13}$$

then

$$u(t) \leq a + \sum_{k=1}^{\infty} \left(L_2^k(a) \right)(t) \leq a E_{\beta,\gamma}\left((b K_1 K_2 \Gamma(\beta))^{\frac{1}{\nu}} t \right), \tag{20.14}$$

for every $t \in [0, T)$, *where* $\nu = \beta + \gamma - 1$ *and* $E_{\beta,\gamma}(s)$ *as in (20.10).*

Proof We observe that

$$(t-s)^{\beta(s)-1} \leq \max\{1, T\}^{A-\beta} (t-s)^{\beta-1} = K_1 (t-s)^{\beta-1}, \tag{20.15}$$

and

$$s^{\gamma(s)-1} \leq \max\{1, T\}^{B-\gamma} s^{\gamma-1} = K_2 s^{\gamma-1}, \text{ for all } 0 \leq s < t < T. \tag{20.16}$$

Therefore we get that

$$u(t) \leq a + b \int_0^t (t-s)^{\beta(s)-1} s^{\gamma(s)-1} u(s)\, ds \leq \tag{20.17}$$

$$a + bK_1K_2 \int_0^t (t-s)^{\beta-1} s^{\gamma-1} u(s)\, ds.$$

By Theorem 20.3 we obtain that

$$u(t) \leq aE_{\beta,\gamma}\left(\left(bK_1K_2\Gamma(\beta)\right)^{\frac{1}{\nu}} t \right). \tag{20.18}$$

We call

$$B^*\phi(t) := bK_1K_2 \int_0^t (t-s)^{\beta-1} s^{\gamma-1} \phi(s)\, ds. \tag{20.19}$$

As in the proof of Theorem 20.3 we have that

$$B^{*n}u(t) \to 0, \text{ as } n \to \infty.$$

Also we can write

$$u(t) \leq a + B^*u(t), \tag{20.20}$$

and by induction as in the proof of Theorem 20.3 we derive that

$$u(t) \leq a \sum_{m=0}^{n} c_m \left(bK_1K_2\Gamma(\beta)\right)^m t^{m\nu} + B^{*n+1}u(t). \tag{20.21}$$

But here it is

$$L_2u(t) = b \int_0^t (t-s)^{\beta(s)-1} s^{\gamma(s)-1} u(s)\, ds \leq B^*u(t). \tag{20.22}$$

Consequently it holds

$$\lim_{n\to\infty} L_2^n u(t) = 0, \quad 0 \leq t < T. \tag{20.23}$$

Also we have

$$u(t) \leq a + L_2u(t), \quad 0 \leq t < T. \tag{20.24}$$

It is clear now that

$$u(t) \leq a + \sum_{k=1}^{\infty} L_2^k a \leq a + \sum_{k=1}^{\infty} B^{*k} a \leq aE_{\beta,\gamma}\left(\left(bK_1K_2\Gamma(\beta)\right)^{\frac{1}{\nu}} t \right), \tag{20.25}$$

proving the claim. ■

Next we mention

Theorem 20.5 ([9]) *Suppose that $\alpha, \beta > 0$, $a(t)$ is a nonnegative function locally integrable on $\mathbb{R}^+$, $g(t)$ and $h(t)$ are nonnegative, nondecreasing continuous functions defined on $\mathbb{R}^+$, and $u(t)$ is a nonnegative and locally integrable function on $\mathbb{R}^+$ such that*

$$u(t) \leq a(t) + g(t) \int_0^t (t-s)^{\alpha-1} u(s)\, ds + h(t) \int_0^t (t-s)^{\beta-1} u(s)\, ds, \quad t \in \mathbb{R}^+. \tag{20.26}$$

Then, for each constant $b \geq 0$, $t \in [0, b]$, we have

$$u(t) \leq a(t) +$$

$$\int_0^t \sum_{n=1}^{\infty} \sum_{k=0}^{n} \binom{n}{k} \frac{(g(t)\Gamma(\alpha))^k (h(t)\Gamma(\beta))^{n-k}}{\Gamma(k\alpha + (n-k)\beta)} (t-s)^{k\alpha+(n-k)\beta-1} a(s)\, ds. \tag{20.27}$$

We present the following left inequality:

Theorem 20.6 *Suppose that $\alpha, \beta > 0$, $a(t)$ is a nonnegative function locally integrable on $\mathbb{R}^+$, $g(t)$ and $h(t)$ are nonnegative, nondecreasing continuous functions defined on $\mathbb{R}^+$, and $u(t)$ is a nonnegative and locally integrable functions on $\mathbb{R}^+$. Here $\alpha(t)$, $\beta(t)$ are (Lebesgue) measurable function on $[0, T)$, $0 < T < \infty$, such that*

$$0 < \alpha \leq \alpha(t) \leq A < \infty, \quad 0 < \beta \leq \beta(t) \leq B < \infty, \ t \in [0, T). \tag{20.28}$$

We call

$$K_1 := \max\{1, T\}^{A-\alpha}, \quad K_2 := \max\{1, T\}^{B-\beta}. \tag{20.29}$$

Let also the positive linear operator

$$L_3\phi(t) := g(t) \int_0^t (t-s)^{\alpha(s)-1} \phi(s)\, ds + h(t) \int_0^t (t-s)^{\beta(s)-1} \phi(s)\, ds, \tag{20.30}$$

$t \in [0, T)$.

If

$$u(t) \leq a(t) + g(t) \int_0^t (t-s)^{\alpha(s)-1} u(s)\, ds + h(t) \int_0^t (t-s)^{\beta(s)-1} u(s)\, ds, \tag{20.31}$$

$t \in [0, T)$,

then

$$u(t) \leq a(t) + \sum_{k=1}^{\infty} L_3^k a(t) \leq a(t) +$$

$$\int_0^t \sum_{n=1}^{\infty} \sum_{k=0}^{n} \binom{n}{k} \frac{(K_1 g(t) \Gamma(\alpha))^k (K_2 h(t) \Gamma(\beta))^{n-k}}{\Gamma(k\alpha + (n-k)\beta)} (t-s)^{k\alpha + (n-k)\beta - 1} a(s) \, ds, \tag{20.32}$$

for any $t \in [0, T)$.

Proof We observe that

$$(t-s)^{\alpha(s)-1} \leq K_1 (t-s)^{\alpha-1}, \text{ and } (t-s)^{\beta(s)-1} \leq K_2 (t-s)^{\beta-1}, \tag{20.33}$$

$0 \leq s < t < T$.

Therefore it holds ($t \in [0, T)$)

$$u(t) \leq a(t) + g(t) \int_0^t (t-s)^{\alpha(s)-1} u(s) \, ds + h(t) \int_0^t (t-s)^{\beta(s)-1} u(s) \, ds \leq$$

$$a(t) + K_1 g(t) \int_0^t (t-s)^{\alpha-1} u(s) \, ds + K_2 h(t) \int_0^t (t-s)^{\beta-1} u(s) \, ds. \tag{20.34}$$

By Theorem 20.5 we get that

$$u(t) \leq a(t) + \int_0^t \sum_{n=1}^{\infty} \sum_{k=0}^{n} \binom{n}{k} \frac{(K_1 g(t) \Gamma(\alpha))^k (K_2 h(t) \Gamma(\beta))^{n-k}}{\Gamma(k\alpha + (n-k)\beta)}$$

$$(t-s)^{k\alpha + (n-k)\beta - 1} a(s) \, ds =: \psi(t), \quad t \in [0, T). \tag{20.35}$$

We call

$$B^* \phi(t) := K_1 g(t) \int_0^t (t-s)^{\alpha-1} \phi(s) \, ds + K_2 h(t) \int_0^t (t-s)^{\beta-1} \phi(s) \, ds, \quad t \in \mathbb{R}^+. \tag{20.36}$$

As in the proof of Theorem 20.5 we get for each $t \in [0, T)$, that $\lim_{n \to \infty} B^{*n} u(t) = 0$.

Furthermore from there we have

$$B^{*n} u(t) \leq \int_0^t \sum_{k=0}^{n} \binom{n}{k} \frac{(K_1 g(t) \Gamma(\alpha))^k (K_2 h(t) \Gamma(\beta))^{n-k}}{\Gamma(k\alpha + (n-k)\beta)}$$

$$(t-s)^{k\alpha + (n-k)\beta - 1} u(s) \, ds, \quad t \in [0, T).$$

Also we can write

$$u(t) \overset{(20.34)}{\leq} a(t) + B^* u(t), \tag{20.37}$$

and by iteration we have that

$$u(t) \leq \sum_{k=0}^{n-1} B^{*k} a(t) + B^{*n} u(t), \quad t \in [0, T). \tag{20.38}$$

But here it is

$$L_3 u(t) \leq B^{*} u(t), \ t \in [0, T). \tag{20.39}$$

Consequently it holds

$$\lim_{n \to \infty} L_3^n u(t) = 0, \quad 0 \leq t < T. \tag{20.40}$$

Also we have

$$u(t) \overset{(20.30),(20.31)}{\leq} a(t) + L_3 u(t), \quad 0 \leq t < T. \tag{20.41}$$

It is clear now that

$$u(t) \leq a(t) + \sum_{k=1}^{\infty} L_3^k a(t) \leq a(t) + \sum_{k=1}^{\infty} B^{*k} a(t) \leq \psi(t), \quad t \in [0, T). \tag{20.42}$$

The claim is proved. ■

We mention

Theorem 20.7 ([6]) *Let u, v be two integrable functions and g continuous, with domain $[a, b]$. Let $\psi \in C^1([a, b])$ an increasing function such that $\psi'(t) \neq 0$, $\forall$ $t \in [a, b]$. Assume that*

1. *u and v are nonnegative;*
2. *g is nonnegative and nondecreasing; $\alpha > 0$.*

If

$$u(t) \leq v(t) + g(t) + \int_a^t \psi'(\tau)(\psi(t) - \psi(\tau))^{\alpha-1} u(\tau) d\tau, \tag{20.43}$$

then

$$u(t) \leq v(t) + \int_a^t \sum_{k=1}^{\infty} \frac{[g(t)\Gamma(\alpha)]^k}{\Gamma(\alpha k)} \psi'(\tau)[\psi(t) - \psi(\tau)]^{\alpha k-1} v(\tau) d\tau, \tag{20.44}$$

$\forall t \in [a, b]$.

We give the following left inequality:

Theorem 20.8 *Let u, v be two integrable functions and g continuous, with domain $[a, b]$. Let $\psi \in C^1([a, b])$ a strictly increasing function such that $\psi'(t) \neq 0$, $\forall t \in [a, b]$. Assume that*

1. u and v are nonnegative; $\alpha > 0$,

2. g is nonnegative and nondecreasing; $\alpha(t)$ *is a (Lebesgue) measurable function over* $[a, b]$ *such that* $0 < \alpha \leq \alpha(t) \leq A < \infty$, $t \in [a, b]$.

We call

$$K_{\psi} := \max\{1, \psi(b) - \psi(a)\}^{A-\alpha}. \tag{20.45}$$

Let also the positive linear operator

$$L_4\phi(t) = g(t) \int_a^t \psi'(\tau)(\psi(t) - \psi(\tau))^{\alpha(\tau)-1} \phi(\tau) d\tau, \quad t \in [a, b]. \tag{20.46}$$

If

$$u(t) \leq v(t) + g(t) \int_a^t \psi'(\tau)(\psi(t) - \psi(\tau))^{\alpha(\tau)-1} u(\tau) d\tau, \tag{20.47}$$

then

$$u(t) \leq v(t) + \sum_{k=1}^{\infty} L_4^k v(t) \leq v(t) +$$

$$\int_a^t \sum_{k=1}^{\infty} \frac{\left[K_{\psi} g(t) \Gamma(\alpha)\right]^k}{\Gamma(\alpha k)} \psi'(\tau) [\psi(t) - \psi(\tau)]^{\alpha k-1} v(\tau) d\tau, \tag{20.48}$$

$\forall t \in [a, b]$.

Proof Let $a \leq \tau < t \leq b$. Since ψ is strictly increasing over $[a, b]$, when $\tau < t$ we get that $\psi(t) - \psi(\tau) > 0$. In this case we have

$$\frac{(\psi(t) - \psi(\tau))^{\alpha(\tau)-1}}{(\psi(t) - \psi(\tau))^{\alpha-1}} = (\psi(t) - \psi(\tau))^{\alpha(\tau)-\alpha}. \tag{20.49}$$

If $\alpha(\tau) = \alpha$, then $(\psi(t) - \psi(\tau))^{\alpha(\tau)-\alpha} = 1 = 1^{A-\alpha} \leq \max\{1, \psi(b) - \psi(a)\}^{A-\alpha}$.

Let now $a(\tau) > \alpha$, i.e. $\alpha(\tau) - \alpha > 0$.

If $\psi(t) - \psi(\tau) < 1$, then $(\psi(t) - \psi(\tau))^{\alpha(\tau)-\alpha} \leq 1 = 1^{A-\alpha} \leq \max\{1, \psi(b) - \psi(a)\}^{A-\alpha}$.

If $\psi(t) - \psi(\tau) > 1$, then $(\psi(t) - \psi(\tau))^{\alpha(\tau)-\alpha} \leq (\psi(t) - \psi(\tau))^{A-\alpha} \leq (\psi(b) - \psi(a))^{A-\alpha} \leq \max\{1, \psi(b) - \psi(a)\}^{A-\alpha}$.

If $\psi(t) - \psi(\tau) = 1$, then $(\psi(t) - \psi(\tau))^{\alpha(\tau)-\alpha} = 1 = 1^{A-\alpha} \leq \max\{1, \psi(b) - \psi(a)\}^{A-\alpha}$.

So we have proved that

$$\frac{(\psi(t) - \psi(\tau))^{\alpha(\tau)-1}}{(\psi(t) - \psi(\tau))^{\alpha-1}} \leq \max\{1, \psi(b) - \psi(a)\}^{A-\alpha} = K_{\psi},$$

that is

$$\left(\psi\left(t\right)-\psi\left(\tau\right)\right)^{\alpha(\tau)-1}\leq K_{\psi}\left(\psi\left(t\right)-\psi\left(\tau\right)\right)^{\alpha-1}, \tag{20.50}$$

when $a\leq\tau<t\leq b$.

Therefore it holds ($t\in[a,b]$)

$$u\left(t\right)\leq v\left(t\right)+g\left(t\right)\int_{a}^{t}\psi'\left(\tau\right)\left(\psi\left(t\right)-\psi\left(\tau\right)\right)^{\alpha(\tau)-1}u\left(\tau\right)d\tau\overset{(20.50)}{\leq}$$

$$v\left(t\right)+K_{\psi}g\left(t\right)\int_{a}^{t}\psi'\left(\tau\right)\left(\psi\left(t\right)-\psi\left(\tau\right)\right)^{\alpha-1}u\left(\tau\right)d\tau. \tag{20.51}$$

By Theorem 20.7 we obtain

$$u\left(t\right)\leq v\left(t\right)+\int_{a}^{t}\sum_{k=1}^{\infty}\frac{\left[K_{\psi}g\left(t\right)\Gamma\left(\alpha\right)\right]^{k}}{\Gamma\left(\alpha k\right)}\psi'\left(\tau\right)\left[\psi\left(t\right)-\psi\left(\tau\right)\right]^{\alpha k-1}v\left(\tau\right)d\tau=:\varphi\left(t\right), \tag{20.52}$$

$t\in[a,b]$.

We call

$$B\phi\left(t\right):=K_{\psi}g\left(t\right)\int_{a}^{t}\psi'\left(\tau\right)\left[\psi\left(t\right)-\psi\left(\tau\right)\right]^{\alpha-1}\phi\left(\tau\right)d\tau, \tag{20.53}$$

$\forall\, t\in[a,b]$.

Then

$$u\left(t\right)\leq v\left(t\right)+Bu\left(t\right), \tag{20.54}$$

and by iterating ($n\in\mathbb{N}$) we get

$$u\left(t\right)\leq\sum_{k=0}^{n-1}B^{k}v\left(t\right)+B^{n}u\left(t\right),\quad\forall\, t\in[a,b]. \tag{20.55}$$

As in the proof of Theorem 20.7 [6] by mathematical induction we obtain

$$B^{*}u\left(t\right)\leq\int_{a}^{t}\frac{\left[K_{\psi}g\left(t\right)\Gamma\left(\alpha\right)\right]^{k}}{\Gamma\left(\alpha k\right)}\psi'\left(\tau\right)\left[\psi\left(t\right)-\psi\left(\tau\right)\right]^{\alpha k-1}u\left(\tau\right)d\tau, \tag{20.56}$$

$\forall\, k\in\mathbb{N}, t\in[a,b]$.

Also from that proof we get $B^{n}u\left(t\right)\to0$, as $n\to\infty$.

Here we have that

$$L_{4}u\left(t\right)\leq Bu\left(t\right),\quad t\in[a,b]. \tag{20.57}$$

Consequently it holds

$$\lim_{n\to\infty}L_{4}^{n}u\left(t\right)=0,\quad t\in[a,b]. \tag{20.58}$$

Also we have

$$u(t) \leq v(t) + L_4 u(t), \quad t \in [a,b]. \tag{20.59}$$

It is clear now that

$$u(t) \leq v(t) + \sum_{k=1}^{\infty} L_4^k v(t) \leq v(t) + \sum_{k=1}^{\infty} B^k v(t) \overset{(20.52)}{\leq} \varphi(t), \; t \in [a,b]. \tag{20.60}$$

The claim is proved. ■

We need

Theorem 20.9 ([2]) *Let $u, v : [a,b] \to \mathbb{R}$ be two integrable functions, $g, h : [a,b] \to \mathbb{R}$ be two continuous functions, $\psi \in C^1([a,b])$ be increasing function such that $\psi'(t) \neq 0$, for all $t \in [a,b]$, and α, β two positive real numbers. Suppose that*

1. *u and v are nonnegative;*
2. *g and h are nonnegative and nondecreasing;*
3. *the following inequality is true:*

$$u(t) \leq v(t) + g(t) \int_a^t \psi'(\tau)(\psi(t) - \psi(\tau))^{\alpha-1} u(\tau) d\tau +$$

$$h(t) \int_a^t \psi'(\tau)(\psi(t) - \psi(\tau))^{\beta-1} u(\tau) d\tau, \tag{20.61}$$

for all $t \in [a,b]$.

Then for all $t \in [a,b]$

$$u(t) \leq v(t) +$$

$$\int_a^t \sum_{n=1}^{\infty} \sum_{k=0}^{n} \binom{n}{k} \frac{(g(t)\Gamma(\alpha))^k (h(t)\Gamma(\beta))^{n-k}}{\Gamma(k\alpha + (n-k)\beta)} \psi'(\tau)[\psi(t) - \psi(\tau)]^{k\alpha+(n-k)\beta-1} v(\tau) d\tau. \tag{20.62}$$

We give the following left inequality:

Theorem 20.10 *Let $u, v : [a,b] \to \mathbb{R}$ be two integrable functions, $g, h : [a,b] \to \mathbb{R}$ be two continuous functions, $\psi \in C^1([a,b])$ be a strictly increasing function such that $\psi'(t) \neq 0$, for all $t \in [a,b]$, and $\alpha, \beta > 0$. Suppose that*

1. *u and v are nonnegative;*
2. *g and h are nonnegative and nondecreasing;*
3. *$\alpha(t)$ and $\beta(t)$ are (Lebesgue) measurable functions on $[a,b]$, such that $0 < \alpha \leq \alpha(t) \leq A < \infty$, and $0 < \beta \leq \beta(t) \leq B < \infty$, $t \in [a,b]$.*

We call

$$K_{1\psi} := \max\{1, \psi(b) - \psi(a)\}^{A-\alpha}, \quad K_{2\psi} := \max\{1, \psi(b) - \psi(a)\}^{B-\beta}. \tag{20.63}$$

Let also the positive linear operator

$$L_5\phi(t) = g(t) \int_a^t \psi'(\tau)(\psi(t) - \psi(\tau))^{\alpha(\tau)-1} \phi(\tau) d\tau + \tag{20.64}$$

$$h(t) \int_a^t \psi'(\tau)(\psi(t) - \psi(\tau))^{\beta(\tau)-1} \phi(\tau) d\tau, \quad t \in [a,b].$$

If

$$u(t) \leq v(t) + g(t) \int_a^t \psi'(\tau)(\psi(t) - \psi(\tau))^{\alpha(\tau)-1} u(\tau) d\tau + \tag{20.65}$$

$$h(t) \int_a^t \psi'(\tau)(\psi(t) - \psi(\tau))^{\beta(\tau)-1} u(\tau) d\tau,$$

then

$$u(t) \leq v(t) + \sum_{k=1}^{\infty} L_5^k v(t) \leq v(t) +$$

$$\int_a^t \sum_{n=1}^{\infty} \sum_{k=0}^{n} \binom{n}{k} \frac{\left(K_{1\psi} g(t) \Gamma(\alpha)\right)^k \left(K_{2\psi} h(t) \Gamma(\beta)\right)^{n-k}}{\Gamma(k\alpha + (n-k)\beta)}$$

$$\psi'(\tau)[\psi(t) - \psi(\tau)]^{k\alpha + (n-k)\beta - 1} v(\tau) d\tau, \tag{20.66}$$

$\forall\, t \in [a,b]$.

Proof Clearly here we have that (see proof of Theorem 20.8)

$$(\psi(t) - \psi(\tau))^{\alpha(\tau)-1} \leq K_{1\psi}(\psi(t) - \psi(\tau))^{\alpha-1}, \text{ and } (\psi(t) - \psi(\tau))^{\beta(\tau)-1} \leq K_{2\psi}(\psi(t) - \psi(\tau))^{\beta-1}, \tag{20.67}$$

for all $a \leq \tau < t \leq b$.

Therefore it holds

$$u(t) \leq v(t) + g(t) \int_a^t \psi'(\tau)(\psi(t) - \psi(\tau))^{\alpha(\tau)-1} u(\tau) d\tau +$$

$$h(t) \int_a^t \psi'(\tau)(\psi(t) - \psi(\tau))^{\beta(\tau)-1} u(\tau) d\tau \overset{(20.67)}{\leq}$$

$$v(t) + K_{1\psi} g(t) \int_a^t \psi'(\tau)(\psi(t) - \psi(\tau))^{\alpha-1} u(\tau) d\tau +$$

$$K_{2\psi} h(t) \int_a^t \psi'(\tau)(\psi(t) - \psi(\tau))^{\beta-1} u(\tau) d\tau, \tag{20.68}$$

$\forall\, t \in [a, b]$.

By Theorem 20.9 we derive

$$u(t) \overset{(20.68)}{\leq} v(t) + \int_a^t \sum_{n=1}^{\infty} \sum_{k=0}^{n} \binom{n}{k} \frac{(K_{1\psi} g(t) \Gamma(\alpha))^k (K_{2\psi} h(t) \Gamma(\beta))^{n-k}}{\Gamma(k\alpha + (n-k)\beta)}$$

$$\psi'(\tau)[\psi(t) - \psi(\tau)]^{k\alpha+(n-k)\beta-1} v(\tau) d\tau =: \lambda(t), \tag{20.69}$$

$\forall\, t \in [a, b]$.

We call

$$\overline{B}\phi(t) := K_{1\psi} g(t) \int_a^t \psi'(\tau)(\psi(t) - \psi(\tau))^{\alpha-1} \phi(\tau) d\tau +$$

$$K_{2\psi} h(t) \int_a^t \psi'(\tau)(\psi(t) - \psi(\tau))^{\beta-1} \phi(\tau) d\tau, \tag{20.70}$$

$t \in [a, b]$.

Then

$$u(t) \overset{(20.68)}{\leq} v(t) + \overline{B}u(t), \tag{20.71}$$

and by iterating ($n \in \mathbb{N}$) we get

$$u(t) \leq \sum_{k=0}^{n-1} \overline{B}^k v(t) + \overline{B}^n u(t), \quad \forall\, t \in [a, b]. \tag{20.72}$$

As in the proof of Theorem 20.9 [2] we get that

$$\overline{B}^n u(t) \leq \int_a^t \sum_{k=0}^{n} \binom{n}{k} \frac{[K_{1\psi} g(t) \Gamma(\alpha)]^k [K_{2\psi} h(t) \Gamma(\beta)]^{n-k}}{\Gamma(k\alpha + (n-k)\beta)}$$

$$\psi'(\tau)[\psi(t) - \psi(\tau)]^{k\alpha+(n-k)\beta-1} u(\tau) d\tau, \tag{20.73}$$

$\forall\, n \in \mathbb{N}$, $t \in [a, b]$.

Again as in the proof of Theorem 20.9 [2], we also derive that $\overline{B}^n u(t) \to 0$, as $n \to \infty$.

Also here we have

$$L_5 u(t) \leq \overline{B} u(t), \ t \in [a,b]. \tag{20.74}$$

Consequently it holds

$$\lim_{n \to \infty} L_5^n u(t) = 0, \ t \in [a,b]. \tag{20.75}$$

Also we have

$$u(t) \overset{((20.64),\,(20.65))}{\leq} v(t) + L_5 u(t), \ t \in [a,b]. \tag{20.76}$$

It is clear now that

$$u(t) \leq v(t) + \sum_{k=1}^{\infty} L_5^k v(t) \leq v(t) + \sum_{k=1}^{\infty} \overline{B}^k v(t) \overset{(20.69)}{\leq} \lambda(t), \ t \in [a,b]. \tag{20.77}$$

The theorem is proved. ■

We mention a right fractional Gronwall inequality.

Theorem 20.11 ([1], case of $\rho = 1$ in Theorem 9 there) *Let u, v be two integrable functions and g a continuous function, with domain $[a,b]$. Assume that*

1. *u and v are nonnegative;*
2. *g is nonnegative and nondecreasing, $\alpha > 0$.*

If

$$u(t) \leq v(t) + g(t) \int_t^b (\tau - t)^{\alpha - 1} u(\tau) d\tau, \tag{20.78}$$

then

$$u(t) \leq v(t) + \int_t^b \sum_{k=1}^{\infty} \frac{[g(t)\Gamma(\alpha)]^k}{\Gamma(k\alpha)} (\tau - t)^{k\alpha - 1} v(\tau) d\tau, \tag{20.79}$$

$\forall \, t \in [a,b]$.

In addition, if v is nondecreasing, then

$$u(t) \leq v(t) E_\alpha \left[g(t) \Gamma(\alpha) (b - t)^\alpha \right], \ \ \forall \, t \in [a,b]. \tag{20.80}$$

Above E_α is the Mittag–Leffler function

$$E_\alpha(t) := \sum_{k=0}^{\infty} \frac{t^k}{\Gamma(\alpha k + 1)}, \ \ t \in \mathbb{R}. \tag{20.81}$$

We give the following right Gronwall inequality of fractional variable order.

Theorem 20.12 *All as Theorem 20.11. Let also $\alpha(t)$ a (Lebesgue) measurable function on $[a,b]$, such that $0 < \alpha \leq \alpha(t) \leq A < \infty$, $t \in [a,b]$.*

Set

$$K := \max\{1, b-a\}^{A-\alpha}. \tag{20.82}$$

Let also the positive linear operator

$$L_6\phi(t) = g(t)\int_t^b (\tau - t)^{\alpha(\tau)-1}\phi(\tau)\,d\tau, \quad t \in [a,b]. \tag{20.83}$$

If

$$u(t) \le v(t) + g(t)\int_t^b (\tau - t)^{\alpha(\tau)-1} u(\tau)\,d\tau, \tag{20.84}$$

then

$$u(t) \le v(t) + \sum_{k=1}^{\infty} L_6^k v(t) \le v(t) +$$

$$\int_t^b \sum_{k=1}^{\infty} \frac{(Kg(t)\Gamma(\alpha))^k}{\Gamma(k\alpha)} (\tau - t)^{k\alpha-1} v(\tau)\,d\tau, \tag{20.85}$$

$\forall\, t \in [a,b]$.

Proof Let $a \le t < \tau \le b$. Then

$$\frac{(\tau - t)^{\alpha(\tau)-1}}{(\tau - t)^{\alpha-1}} = (\tau - t)^{\alpha(\tau)-\alpha}. \tag{20.86}$$

If $\alpha(\tau) = \alpha$, then $(\tau - t)^{\alpha(\tau)-\alpha} = 1 = 1^{A-\alpha} \le \max\{1, b-a\}^{A-\alpha}$.

Let now $a(\tau) > \alpha$, i.e. $\alpha(\tau) - \alpha > 0$. If $\tau - t < 1$, then $(\tau - t)^{\alpha(\tau)-\alpha} \le 1 = 1^{A-\alpha} \le \max\{1, b-a\}^{A-\alpha}$.

If $\tau - t > 1$, then $(\tau - t)^{\alpha(\tau)-\alpha} \le (\tau - t)^{A-\alpha} \le (b-a)^{A-\alpha} \le \max\{1, b-a\}^{A-\alpha}$.

If $\tau - t = 1$, then $(\tau - t)^{\alpha(\tau)-\alpha} = 1 = 1^{A-\alpha} \le \max\{1, b-a\}^{A-\alpha}$.

So we have proved that

$$\frac{(\tau - t)^{\alpha(\tau)-1}}{(\tau - t)^{\alpha-1}} \le \max\{1, b-a\}^{A-\alpha} = K,$$

that is

$$(\tau - t)^{\alpha(\tau)-1} \le K(\tau - t)^{\alpha-1}, \tag{20.87}$$

when $a \le t < \tau \le b$.

By (20.87), (20.84) we get

$$u(t) \le v(t) + g(t)\int_t^b (\tau - t)^{\alpha(\tau)-1} u(\tau)\,d\tau \le$$

$$v(t) + g(t) K \int_t^b (\tau - t)^{\alpha-1} u(\tau) d\tau, \tag{20.88}$$

$\forall\, t \in [a, b]$.

By Theorem 20.11 we obtain

$$u(t) \leq v(t) + \int_t^b \sum_{k=1}^{\infty} \frac{[K g(t) \Gamma(\alpha)]^k}{\Gamma(k\alpha)} (\tau - t)^{k\alpha-1} v(\tau) d\tau =: Q(t), \tag{20.89}$$

$\forall\, t \in [a, b]$.

Call

$$\Delta\phi(t) := K g(t) \int_t^b (\tau - t)^{\alpha-1} \phi(\tau) d\tau, \tag{20.90}$$

$\forall\, t \in [a, b]$.

Then

$$u(t) \leq v(t) + \Delta u(t), \tag{20.91}$$

and by iterating ($n \in \mathbb{N}$) we find

$$u(t) \leq \sum_{k=0}^{n-1} \Delta^k v(t) + \Delta^n u(t), \quad \forall\, t \in [a, b]. \tag{20.92}$$

As in the proof of Theorem 20.1 [8] we have (inductively) that

$$\Delta^k u(t) \leq \int_t^b \frac{(K g(t) \Gamma(\alpha))^k}{\Gamma(k\alpha)} (\tau - t)^{k\alpha-1} u(\tau) d\tau, \tag{20.93}$$

$\forall\, k \in \mathbb{N}$, and $\Delta^n u(t) \to 0$, as $n \to \infty$, $\forall\, t \in [a, b]$.

Also we have that

$$L_6 u(t) \leq \Delta u(t), \quad t \in [a, b]. \tag{20.94}$$

Therefore it holds

$$\lim_{n\to\infty} L_6^n u(t) = 0, \; t \in [a, b]. \tag{20.95}$$

Notice also that

$$u(t) \leq v(t) + L_6 u(t), \quad t \in [a, b]. \tag{20.96}$$

It is clear now that

$$u(t) \leq v(t) + \sum_{k=1}^{\infty} L_6^k v(t) \leq v(t) + \sum_{k=1}^{\infty} \Delta^k v(t) \overset{(20.89)}{\leq} Q(t), \; t \in [a, b]. \tag{20.97}$$

The theorem is proved. ■

Two applications follow:

Corollary 20.13 (to Theorem 20.10) *All as in Theorem 20.10, with* $\psi(t) = e^t$. *Here*

$$K_{1e^x} := \max\left\{1, e^b - e^a\right\}^{A-\alpha}, \quad \text{and} \quad K_{2e^x} := \max\left\{1, e^b - e^a\right\}^{B-\beta}. \tag{20.98}$$

Let also the positive linear operator

$$L_5^{e^x}\phi(t) = g(t)\int_a^t e^\tau\left(e^t - e^\tau\right)^{\alpha(\tau)-1}\phi(\tau)\,d\tau + \tag{20.99}$$

$$h(t)\int_a^t e^\tau\left(e^t - e^\tau\right)^{\beta(\tau)-1}\phi(\tau)\,d\tau, \quad t \in [a,b].$$

If

$$u(t) \leq v(t) + g(t)\int_a^t e^\tau\left(e^t - e^\tau\right)^{\alpha(\tau)-1}u(\tau)\,d\tau + \tag{20.100}$$

$$h(t)\int_a^t e^\tau\left(e^t - e^\tau\right)^{\beta(\tau)-1}u(\tau)\,d\tau,$$

then

$$u(t) \leq v(t) + \sum_{k=1}^{\infty}\left(L_5^{e^x}\right)^k v(t) \leq v(t) +$$

$$\int_a^t \sum_{n=1}^{\infty}\sum_{k=0}^{n}\binom{n}{k}\frac{\left(K_{1e^x}g(t)\Gamma(\alpha)\right)^k\left(K_{2e^x}h(t)\Gamma(\beta)\right)^{n-k}}{\Gamma(k\alpha + (n-k)\beta)}$$

$$e^\tau\left(e^t - e^\tau\right)^{k\alpha+(n-k)\beta-1}v(\tau)\,d\tau, \tag{20.101}$$

$\forall\, t \in [a,b]$.

Corollary 20.14 (to Theorem 20.10) *All as in Theorem 20.10, with* $\psi(t) = \ln t$, $[a,b] \subseteq \mathbb{R}_+ - \{0\}$. *Here*

$$K_{1\ln x} := \max\left\{1, \ln\frac{b}{a}\right\}^{A-\alpha}, \quad \text{and} \quad K_{2\ln x} := \max\left\{1, \ln\frac{b}{a}\right\}^{B-\beta}. \tag{20.102}$$

Let also the positive linear operator

$$L_5^{\ln x}\phi(t) = g(t)\int_a^t \left(\ln\frac{t}{\tau}\right)^{\alpha(\tau)-1}\frac{\phi(\tau)}{\tau}d\tau + \tag{20.103}$$

$$h(t)\int_a^t \left(\ln\frac{t}{\tau}\right)^{\beta(\tau)-1}\frac{\phi(\tau)}{\tau}d\tau,\quad t\in[a,b].$$

If

$$u(t) \leq v(t) + g(t)\int_a^t \left(\ln\frac{t}{\tau}\right)^{\alpha(\tau)-1}\frac{u(\tau)}{\tau}d\tau +$$

$$h(t)\int_a^t \left(\ln\frac{t}{\tau}\right)^{\beta(\tau)-1}\frac{u(\tau)}{\tau}d\tau,$$

then

$$u(t) \leq v(t) + \sum_{k=1}^{\infty}\left(L_5^{\ln x}\right)^k v(t) \leq v(t) +$$

$$\int_a^t \sum_{n=1}^{\infty}\sum_{k=0}^{n}\binom{n}{k}\frac{\left(K_{1\ln x}g(t)\Gamma(\alpha)\right)^k\left(K_{2\ln x}h(t)\Gamma(\beta)\right)^{n-k}}{\Gamma(k\alpha+(n-k)\beta)}$$

$$\left(\ln\frac{t}{\tau}\right)^{k\alpha+(n-k)\beta-1}\frac{v(\tau)}{\tau}d\tau, \tag{20.104}$$

$\forall\, t\in[a,b]$.

References

1. Almeida, R.: A Gronwall inequality for a general Caputo fractional operator. Math. Inequal. Appl. **20**(4), 1089–1105 (2017)
2. Almeida, R., Malinowska, A., Odzijewicz, T.: An extension of the fractional Gronwall inequality. In: Malinowska, A., Mozyrska, D., Sajewski, L. (eds.) Advances in non-integer order calculus and its Applications, Proceedings of the 10th International Conference of Non-integer Order Calculus and Its Applications, pp. 20–28. Springer (2020)
3. Anastassiou, G.A., Papanicolaou, V.G.: A Gronwall inequality of fractional variable order. Progr. Fract. Diff. Appl. accepted for publication (2020)
4. Anastassiou, G.A.: A variety of Gronwall Inequalities of Fractional Variable Order, submitted (2020)
5. Henry, D.: Geometric Theory of Semilinear Parabolic Equations. Springer, Heidelberg (1981)
6. Sousa, J., Oliveira, E.: A Gronwall inequality and the Cauchy-type problem by means of ψ-Hilfer operator. Diff. Equ. & Appl. **11**(1), 87–106 (2019)
7. Yang, X.-I.: General Fractional Derivatives Theory, Methods and Applications. CRC Press, Taylor & Francis Group, Boca Raton (2019)

8. Ye, H., Gao, J., Ding, Y.: A generalized Gronwall inequality and its application to a fractional differential equation. J. Math. Anal. Appl. **328**(4), 1075–1081 (2007)
9. Zhang, Z., Wei, Z.: A generalized Gronwall inequality and its application to fractional neutral evolution inclusions. J. Inequal. Appl. **2016**, 45 (2016)

Chapter 21
General Ordinary and Fractional Approximation with Positive Sublinear Operators

Here we consider the ordinary and fractional approximation of functions by sublinear positive operators with applications to generalized convolution type operators expressed by sublinear integrals such as of Choquet and Shilkret ones. The fractional approximation is under fractional differentiability of Caputo, Canavati and Iterated-Caputo types. We produce Jackson type inequalities under basic initial conditions. So our way is quantitative by producing inequalities with their right hand sides involving the modulus of continuity of ordinary and fractional derivatives of the function under approximation. We give also an application related to Picard singular integral operators. It follows [5].

21.1 Background—I

Here we follow [3], pp. 1–17.

Let $I \subset \mathbb{R}$ be a bounded or unbounded interval, $n \in \mathbb{N}$, and

$$CB_{+}^{n}(I) = \left\{ f : I \to \mathbb{R}_{+} : f^{(i)} \text{ is continuous and bounded on } I, \text{ for both } i = 0, n \right\}. \tag{21.1}$$

We define for

$$f \in CB_{+}(I) = \{ f : I \to \mathbb{R}_{+} : f \text{ is continuous and bounded on } I \},$$

the first modulus of continuity

$$\omega_1(f, \delta) = \sup_{\substack{x, y \in I: \\ |x-y| \leq \delta}} |f(x) - f(y)|, \tag{21.2}$$

G. A. Anastassiou, *Constructive Fractional Analysis with Applications*, Studies in Systems, Decision and Control 362,
https://doi.org/10.1007/978-3-030-71481-9_21

where $0 < \delta \leq diameter\,(I)$, also defined the same for just uniformly continuous functions $f : I \to \mathbb{R}_+$.

Call $C_+ (I) = \{f : I \to \mathbb{R}_+ : f \text{ is continuous on } I\}$.

Let $L_N : C_+ (I) \to CB_+ (I), n, N \in \mathbb{N}$ be a sequence of operators satisfying the following properties (see also [6], p. 17):

(i) (positive homogeneous)

$$L_N (\alpha f) = \alpha L_N (f), \ \forall \alpha \geq 0, f \in C_+ (I), \tag{21.3}$$

(ii) (Monotonicity)

$$\text{if } f, g \in C_+ (I) \text{ satisfy } f \leq g, \text{ then } L_N (f) \leq L_N (g), \ \ \forall N \in \mathbb{N}, \tag{21.4}$$

and

(iii) (Subadditivity)

$$L_N (f + g) \leq L_N (f) + L_N (g), \ \ \forall f, g \in C_+ (I). \tag{21.5}$$

We call L_N positive sublinear operators.

In particular we consider the restrictions $L_N|_{CB_+^n(I)} : CB_+^n (I) \to CB_+ (I)$.

As in [6], p. 17, we get that for $f, g \in CB_+ (I)$,

$$|L_N (f) (x) - L_N (g) (x)| \leq L_N (|f - g|) (x), \ \ \forall x \in I. \tag{21.6}$$

Furthermore, also from [6], p. 17, we have

$$|L_N (f) (x) - f (x)| \leq L_N (|f (\cdot) - f (x)|) (x) + |f (x)| \, |L_N (1) (x) - 1|, \ \ \forall x \in I. \tag{21.7}$$

Given that $L_N (1) = 1, \forall N \in \mathbb{N}$, we get

$$|L_N (f) (x) - f (x)| \leq L_N (|f (\cdot) - f (x)|) (x), \ \ \forall x \in I, \ \forall N \in \mathbb{N}. \tag{21.8}$$

We mention Hölder's inequality for positive sublinear operators

Theorem 21.1 ([3], p. 6) *Let $L : C_+ (I) \to CB_+ (I)$, be a positive sublinear operator and $f, g \in C_+ (I)$, furthermore let $p, q > 1 : \frac{1}{p} + \frac{1}{q} = 1$. Assume that $L\left((f (\cdot))^p\right)(s_*), L\left((g (\cdot))^q\right)(s_*) > 0$ for some $s_* \in I$. Then*

$$L (f (\cdot) g (\cdot)) (s_*) \leq \left(L\left((f (\cdot))^p\right)(s_*)\right)^{\frac{1}{p}} \left(L\left((g (\cdot))^q\right)(s_*)\right)^{\frac{1}{q}}. \tag{21.9}$$

By assuming $L_N\left(|\cdot - x|^{n+1}\right)(x) > 0$, (21.9) and $L_N (1) = 1$, we obtain

$$L_N\left(|\cdot - x|^n\right)(x) \leq \left(L_N\left(|\cdot - x|^{n+1}\right)(x)\right)^{\frac{n}{n+1}}, \tag{21.10}$$

in case of $n = 1$ we derive

$$L_N(|\cdot - x|)(x) \leq \sqrt{\left(L_N\left((\cdot - x)^2\right)(x)\right)}. \tag{21.11}$$

We mention also the following result.

Theorem 21.2 ([3], p. 7) *Let $(L_N)_{N \in \mathbb{N}}$ be a sequence of positive sublinear operators from $C_+(I)$ into $CB_+(I)$, and $f \in CB_+^n(I)$, $f^{(n)}$ could be only uniformly continuous, where $n \in \mathbb{N}$ and $I \subset \mathbb{R}$ a bounded or unbounded interval. Assume $L_N(1) = 1$, $\forall N \in \mathbb{N}$, and $f^{(i)}(x) = 0$, $i = 1, ..., n$, for a fixed $x \in I$. Furthermore assume that $L_N\left(|\cdot - x|^{n+1}\right)(x) > 0$, $\forall N \in \mathbb{N}$.*

Then

$$|L_N(f)(x) - f(x)| \leq \frac{\omega_1\left(f^{(n)}, \left(L_N\left(|\cdot - x|^{n+1}\right)(x)\right)^{\frac{1}{n+1}}\right)}{n!} \cdot$$

$$\left[L_N\left(|\cdot - x|^n\right)(x) + \frac{\left(L_N\left(|\cdot - x|^{n+1}\right)(x)\right)^{\frac{n}{n+1}}}{(n+1)}\right], \quad \forall N \in \mathbb{N}. \tag{21.12}$$

We mention ($n = 1$ case)

Corollary 21.3 ([3], p. 7) *Let $(L_N)_{N \in \mathbb{N}}$ be a sequence of positive sublinear operators from $C_+(I)$ into $CB_+(I)$, and $f \in CB_+^1(I)$, f' could be only uniformly continuous, and $I \subset \mathbb{R}$ a bounded or unbounded interval. Assume $L_N(1) = 1$, $\forall N \in \mathbb{N}$, and $f'(x) = 0$, for a fixed $x \in I$. Furthermore assume that $L_N\left((\cdot - x)^2\right)(x) > 0$, $\forall N \in \mathbb{N}$.*

Then

$$|L_N(f)(x) - f(x)| \leq \omega_1\left(f', \sqrt{\left(L_N\left((\cdot - x)^2\right)(x)\right)}\right) \cdot$$

$$\left[L_N(|\cdot - x|)(x) + \frac{\sqrt{\left(L_N\left((\cdot - x)^2\right)(x)\right)}}{2}\right], \quad \forall N \in \mathbb{N}. \tag{21.13}$$

Remark 21.4 ([3], p. 7) (i) to Theorem 21.2: Assuming $L_N\left(|\cdot - x|^{n+1}\right)(x) \to 0$, as $N \to \infty$, using (21.10), we get that $(L_N(f))(x) \to f(x)$, as $N \to \infty$.

(ii) to Corollary 21.3: Assuming $L_N\left((\cdot - x)^2\right)(x) \to 0$, as $N \to \infty$, using (21.11), we get that $(L_N(f))(x) \to f(x)$, as $N \to \infty$.

(iii) The right hand sides of (21.12), (21.13) are finite.

We also mention the basic result ($n = 0$ case).

Theorem 21.5 ([3], p. 8) *Let $(L_N)_{N \in \mathbb{N}}$ be a sequence of positive sublinear operators from $C_+(I)$ into $CB_+(I)$, and $f \in CB_+(I)$, f could be only uniformly continuous,* *where $I \subset \mathbb{R}$ a bounded or unbounded interval. Assume that $L_N(|\cdot - x|)(x) > 0$, for some fixed $x \in I$, $\forall\, N \in \mathbb{N}$. Then*

(1)

$$|L_N(f)(x) - f(x)| \leq f(x)|L_N(1)(x) - 1| +$$

$$[L_N(1)(x) + 1]\,\omega_1(f, L_N(|\cdot - x|)(x)), \quad \forall\, N \in \mathbb{N}, \tag{21.14}$$

(2) when $L_N(1) = 1$, we get

$$|L_N(f)(x) - f(x)| \leq 2\omega_1(f, L_N(|\cdot - x|)(x)), \quad \forall\, N \in \mathbb{N}. \tag{21.15}$$

Remark 21.6 ([3], p. 8) (to Theorem 21.5) Here $x \in I$ is fixed.

(i) Assume $L_N(1)(x) \to 1$, as $N \to \infty$, and $L_N(|\cdot - x|)(x) \to 0$, as $N \to \infty$, given that f is uniformly continuous we get that $L_n(f)(x) \to f(x)$, as $N \to \infty$ (use of (21.14)). Notice here that $L_N(1)(x)$ is bounded.

(ii) Assume that $L_N(1) = 1$, and $L_N(|\cdot - x|)(x) \to 0$, as $N \to \infty$, and f is uniformly continuous on I, then $L_n(f)(x) \to f(x)$, as $N \to \infty$ (use of (21.15)).

(iii) The right hand sides of (21.14) and (21.15) are finite.

21.2 Background—II [4]

Consider $\Omega \neq \varnothing$ and let $\mathcal{F}$ be a σ-algebra in Ω. Here μ is a set function $\mu : \mathcal{F} \to [0, +\infty)$ which is monotone, i.e. for $A, B \in \Omega : A \subset B$ we have $\mu(A) \leq \mu(B)$, furthermore it holds $\mu(\varnothing) = 0$.

Here $f, g : \Omega \to \mathbb{R}_+ = [0, +\infty)$ are $\mathcal{F}$-measurable, we write it as $f, g \in M(\Omega, \mathbb{R}_+)$.

We consider a functional denoted by the integral symbol $(SL)\int_A f d\mu$, $\forall\, A \in \mathcal{F}$, which is positive, i.e. $\int_A f d\mu \geq 0$.

We assume the following properties:

(i) (positive homogeneous)

$$(SL)\int_A \alpha f d\mu = \alpha (SL)\int_A f d\mu, \ \forall\, \alpha \geq 0, \forall\, f \in M(\Omega, \mathbb{R}_+).$$

(ii) (Monotonicity) if $f, g \in M(\Omega, \mathbb{R}_+)$ satisfy $f \leq g$, then $(SL)\int_A f d\mu \leq (SL)\int_A g d\mu$, $\forall\, A \in \mathcal{F}$.

And

(iii) (Subadditivity)

$$(SL)\int_A (f + g)\, d\mu \leq (SL)\int_A f d\mu + (SL)\int_A g d\mu, \ \forall\, A \in \mathcal{F}.$$

(iv)

$$(SL)\int_A 1 d\mu = \mu(A), \ \forall\, A \in \mathcal{F}.$$

(v) If $\Omega = \mathbb{R}^d$, $d \in \mathbb{N}$, we assume that μ is strictly positive, i.e. $\mu(A) > 0$, for any A compact subset of $\mathbb{R}^d$. Here $\mathcal{F} = \mathcal{B}$ the Borel σ -algebra.

We call $(SL)\int_A f d\mu$ a sublinear integral.

We notice the following:

$$f(x) = f(x) - g(x) + g(x) \leq |f(x) - g(x)| + g(x),$$

hence

$$(SL)\int_A f(x)\,d\mu(x) \leq (SL)\int_A (|f(x) - g(x)| + g(x))\,d\mu(x) \leq$$

$$(SL)\int_A |f(x) - g(x)|\,d\mu(x) + (SL)\int_A g(x)\,d\mu(x),$$

i.e.

$$(SL)\int_A f(x)\,d\mu(x) - (SL)\int_A g(x)\,d\mu(x) \leq (SL)\int_A |f(x) - g(x)|\,d\mu(x).$$

Similarly, we get that

$$(SL)\int_A g(x)\,d\mu(x) - (SL)\int_A f(x)\,d\mu(x) \leq (SL)\int_A |f(x) - g(x)|\,d\mu(x).$$

In conclusion, it holds

$$\left|(SL)\int_A f(x)\,d\mu(x) - (SL)\int_A g(x)\,d\mu(x)\right| \leq (SL)\int_A |f(x) - g(x)|\,d\mu(x), \tag{21.16}$$

$\forall\, A \in \mathcal{F}$ and $\forall\, f, g \in M(\Omega, \mathbb{R}_+)$.

21.3 Background—III

About the Choquet integral:

We make

Definition 21.7 Consider $\Omega \neq \varnothing$ and let $\mathcal{C}$ be a σ-algebra of subsets in Ω.

(i) (see, e.g., [14], p. 63) The set function $\mu : \mathcal{C} \to [0, +\infty]$ is called a monotone set function (or capacity) if $\mu(\varnothing) = 0$ and $\mu(A) \leq \mu(B)$ for all $A, B \in \mathcal{C}$, with $A \subset B$. Also, μ is called submodular if

$$\mu(A \cup B) + \mu(A \cap B) \leq \mu(A) + \mu(B), \text{ for all } A, B \in \mathcal{C}. \tag{21.17}$$

μ is called bounded if $\mu(\Omega) < +\infty$ and normalized if $\mu(\Omega) = 1$.

(ii) (see, e.g., [14], p. 233, or [8]) If μ is a monotone set function on $\mathcal{C}$ and if $f : \Omega \to \mathbb{R}$ is $\mathcal{C}$-measurable (that is, for any Borel subset $B \subset \mathbb{R}$ it follows $f^{-1}(B) \in \mathcal{C}$), then for any $A \in \mathcal{C}$, the Choquet integral is defined by

$$(C)\int_A f d\mu = \int_0^{+\infty} \mu\left(F_\beta(f) \cap A\right) d\beta + \int_{-\infty}^0 \left[\mu\left(F_\beta(f) \cap A\right) - \mu(A)\right] d\beta, \tag{21.18}$$

where we used the notation $F_\beta(f) = \{\omega \in \Omega : f(\omega) \geq \beta\}$. Notice that if $f \geq 0$ on A, then in the above formula we get $\int_{-\infty}^0 = 0$.

The integrals on the right-hand side are the usual Riemann integral.

The function f will be called Choquet integrable on A if $(C)\int_A f d\mu \in \mathbb{R}$.

Next we list some well known properties of the Choquet integral.

Remark 21.8 If $\mu : \mathcal{C} \to [0, +\infty]$ is a monotone set function, then the following properties hold:

(i) For all $a \geq 0$ we have $(C)\int_A af d\mu = a \cdot (C)\int_A f d\mu$ (if $f \geq 0$ then see, e.g., [14], Theorem 11.2, (5), p. 228 and if f is arbitrary sign, then see, e.g., [9], p. 64, Proposition 5.1, (ii)).

(ii) For all $c \in \mathbb{R}$ and f of arbitrary sign, we have (see, e.g., [14], pp. 232–233, or [9], p. 65) $(C)\int_A (f + c) d\mu = (C)\int_A f d\mu + c \cdot \mu(A)$.

If μ is submodular too, then for all f, g of arbitrary sign and lower bounded, we have (see, e.g., [9], p. 75, Theorem 6.3)

$$(C)\int_A (f + g) d\mu \leq (C)\int_A f d\mu + (C)\int_A g d\mu. \tag{21.19}$$

(iii) If $f \leq g$ on A then $(C)\int_A f d\mu \leq (C)\int_A g d\mu$ (see, e.g., [14], p. 228, Theorem 11.2, (3) if $f, g \geq 0$ and p. 232 if f, g are of arbitrary sign).

(iv) Let $f \geq 0$. If $A \subset B$ then $(C)\int_A f d\mu \leq (C)\int_B f d\mu$. In addition, if μ is finitely subadditive, then

$$(C)\int_{A \cup B} f d\mu \leq (C)\int_A f d\mu + (C)\int_B f d\mu. \tag{21.20}$$

(v) It is immediate that $(C)\int_A 1 \cdot d\mu(t) = \mu(A)$.

(vi) If μ is a countably additive bounded measure, then the Choquet integral $(C)\int_A f d\mu$ reduces to the usual Lebesgue type integral (see, e.g., [9], p. 62, or [14], p. 226).

(vii) If $\Omega = \mathbb{R}^d$, $d \in \mathbb{N}$, we assume μ is strictly positive, i.e. $\mu(A) > 0$, for every A compact subset of $\mathbb{R}^d$. Here $\mathcal{C} = \mathcal{B}$ the Borel σ-algebra.

Clearly here, for μ being submodular, we get

$$\left|(C)\int_A f(x)\,d\mu(x) - (C)\int_A g(x)\,d\mu(x)\right| \le (C)\int_A |f(x) - g(x)|\,d\mu(x), \tag{21.21}$$

$\forall A \in \mathcal{C}$ and $\forall f, g \in M(\Omega, \mathbb{R}_+)$ (f, g are measurable with respect to $\mathcal{C}$ σ-algebra).
(viii) If $f \ge 0$, then $(C)\int_A f d\mu \ge 0$.

From now on in this chapter we assume that $\mu : \mathcal{C} \to [0, +\infty)$ and is submodular.

21.4 Background—IV

Here we follow [13].

Let $\mathcal{F}$ be a σ-field of subsets of an arbitrary set Ω. An extended non-negative real valued function μ on $\mathcal{F}$ is called maxitive if $\mu(\varnothing) = 0$ and

$$\mu(\cup_{i\in I} E_i) = \sup_{i\in I} \mu(E_i), \tag{21.22}$$

where the set I is of cardinality at most countable, where $\{E_i\}_{i\in I}$ is a disjoint collection of sets from $\mathcal{F}$. We notice that μ is monotone and (21.22) is true even $\{E_i\}_{i\in I}$ are not disjoint. For more properties of μ see [13]. We also call μ a maxitive measure. Here f stands for a non-negative measurable function on Ω. In [13], Niel Shilkret developed his non-additive integral defined as follows:

$$\left(N^*\right)\int_D f d\mu := \sup_{y\in Y}\{y \cdot \mu(D \cap \{f \ge y\})\}, \tag{21.23}$$

where $Y = [0, m]$ or $Y = [0, m)$ with $0 < m \le \infty$, and $D \in \mathcal{F}$. Here we take $Y = [0, \infty)$.

It is easily proved that

$$\left(N^*\right)\int_D f d\mu = \sup_{y>0}\{y \cdot \mu(D \cap \{f > y\})\}. \tag{21.24}$$

The Shilkret integral takes values in $[0, \infty]$.

The Shilkret integral [13]) has the following properties:

$$\left(N^*\right)\int_\Omega \chi_E d\mu = \mu(E), \tag{21.25}$$

where χ_E is the indicator function on $E \in \mathcal{F}$,

$$\left(N^*\right)\int_D c f d\mu = c\left(N^*\right)\int_D f d\mu, \quad c \ge 0, \tag{21.26}$$

$$\left(N^*\right)\int_D \sup_{n\in\mathbb{N}} f_n d\mu = \sup_{n\in\mathbb{N}} \left(N^*\right)\int_D f_n d\mu, \tag{21.27}$$

where f_n, $n \in \mathbb{N}$, is an increasing sequence of elementary (countably valued) functions converging uniformly to f. Furthermore we have

$$\left(N^*\right)\int_D f d\mu \geq 0, \tag{21.28}$$

$$f \geq g \text{ implies } \left(N^*\right)\int_D f d\mu \geq \left(N^*\right)\int_D g d\mu, \tag{21.29}$$

where $f, g : \Omega \to [0, \infty]$ are measurable.

Let $a \leq f(\omega) \leq b$ for almost every $\omega \in E$, then

$$a\mu(E) \leq \left(N^*\right)\int_E f d\mu \leq b\mu(E); \tag{21.30}$$

$$\left(N^*\right)\int_E 1 d\mu = \mu(E); \tag{21.31}$$

$f > 0$ almost everywhere and $(N^*)\int_E f d\mu = 0$ imply $\mu(E) = 0$;

$(N^*)\int_\Omega f d\mu = 0$ if and only $f = 0$ almost everywhere;

$(N^*)\int_\Omega f d\mu < \infty$ implies that

$$\overline{N}(f) := \{\omega \in \Omega | f(\omega) \neq 0\} \text{ has } \sigma\text{-finite measure};$$

$$\left(N^*\right)\int_D (f+g) d\mu \leq \left(N^*\right)\int_D f d\mu + \left(N^*\right)\int_D g d\mu; \tag{21.32}$$

and

$$\left|\left(N^*\right)\int_D f d\mu - \left(N^*\right)\int_D g d\mu\right| \leq \left(N^*\right)\int_D |f-g| d\mu. \tag{21.33}$$

From now on in this chapter we assume that $\mu : \mathcal{F} \to [0, +\infty)$.

If $\Omega = \mathbb{R}^d$, $d \in \mathbb{N}$, we assume μ is strictly positive, i.e. $\mu(A) > 0$, for every A compact subset of $\mathbb{R}^d$. Here $\mathcal{F} = \mathcal{B}$ the Borel σ-algebra.

Conclusion 21.9 We observe that the Choquet integral $(C)\int_A f d\mu$ and Shilkret integral $(N^*)\int_A f d\mu$ are perfect examples of the sublinear integral $(SL)\int_A f d\mu$ of Sect. 21.2, fulfilling all properties and they have great applications in many areas of pure and applied mathematics and mathematical economics.

Therefore, all the results presented in this chapter which are for the general integral $(SL)\int_A f d\mu$ are of course valid for the Choquet and Shilkret integrals.

21.5 Main Results

21.5.1 Ordinary Approximation

All terms, notations and assumptions here will be as in Backgrounds I–IV.

So here it is $f \in C^n(\mathbb{R}, \mathbb{R}_+)$ with f and $f^{(n)}$ are being bounded, but $f^{(n)}$ could be uniformly continuous regardless if it is bounded or not, $n \in \mathbb{Z}_+$.

Let $\mathcal{B}$ be the Borel σ-algebra on $\mathbb{R}$ and μ_N ($N \in \mathbb{N}$) be a sequence of monotone set functions from $\mathcal{B}$ into $\mathbb{R}_+$, i.e. for $A, B \in \mathcal{B} : A \subset B$ we have $\mu_N(A) \leq \mu_N(B)$, and $\mu_N(\varnothing) = 0$, with $\mu_N(\mathbb{R}) = 1$, $\forall\, N \in \mathbb{N}$.

We will study here the approximation properties of the following sequence of positive sublinear convolution type operators

$$P_N(f)(x) = (SL)\int_{\mathbb{R}} f(x+t)\, d\mu_N(t), \tag{21.34}$$

$\forall\, N \in \mathbb{N}$, to $f(x)$, where $x \in \mathbb{R}$ is fixed, pointwise and uniform in a quantitative way.

We would assume that $P_N(f) \in CB_+(\mathbb{R})$, $\forall\, N \in \mathbb{N}$. Clearly it holds $P_N(1) = 1$. Notice here that

$$P_N\left(|\cdot - x|^{n+1}\right)(x) = (SL)\int_{\mathbb{R}} |t|^{n+1}\, d\mu_N(t), \tag{21.35}$$

$\forall\, N \in \mathbb{N}$, where $n \in \mathbb{Z}_+$.

Based on the above we present

Theorem 21.10 *Assume further that $f^{(i)}(x) = 0$, $i = 1, ..., n$, for a fixed $x \in \mathbb{R}$ and that*

$$(SL)\int_{\mathbb{R}} |t|^{n+1}\, d\mu_N(t) > 0, \ \ \forall\, N \in \mathbb{N}.$$

Then

$$|P_N(f)(x) - f(x)| = \left|(SL)\int_{\mathbb{R}} f(x+t)\, d\mu_N(t) - f(x)\right| \leq$$

$$\frac{(n+2)}{(n+1)!}\omega_1\left(f^{(n)}, \left((SL)\int_{\mathbb{R}} |t|^{n+1}\, d\mu_N(t)\right)^{\frac{1}{n+1}}\right)$$

$$\left((SL)\int_{\mathbb{R}} |t|^{n+1}\, d\mu_N(t)\right)^{\frac{n}{n+1}}, \ \ \forall\, N \in \mathbb{N}. \tag{21.36}$$

Proof By Theorem 21.2 and (21.10). ■

Remark 21.11 If $(SL)\int_{\mathbb{R}}|t|^{n+1}d\mu_N(t)\to 0$, then $P_N(f)(x)\to f(x)$, as $N\to +\infty$.

The $n=1$ case follows:

Corollary 21.12 *Assume further that $f'(x)=0$, for a fixed $x\in\mathbb{R}$ and that*

$$(SL)\int_{\mathbb{R}}t^2d\mu_N(t)>0,\ \ \forall\, N\in\mathbb{N}.$$

Then

$$|P_N(f)(x)-f(x)|\leq\frac{3}{2}\omega_1\left(f',\left((SL)\int_{\mathbb{R}}t^2d\mu_N(t)\right)^{\frac{1}{2}}\right)\left((SL)\int_{\mathbb{R}}t^2d\mu_N(t)\right)^{\frac{1}{2}},\tag{21.37}$$

$\forall\, N\in\mathbb{N}$.

If $(SL)\int_{\mathbb{R}}t^2d\mu_N(t)\to 0$, then $P_N(f)(x)\to f(x)$, as $N\to+\infty$.

Proof By Corollary 21.3 and (21.11). ■

The case $n=0$ comes next.

Theorem 21.13 *Assume that $(SL)\int_{\mathbb{R}}|t|\,d\mu_N(t)>0$, $\forall\, N\in\mathbb{N}$. Then*

$$\|P_N(f)-f\|_{\infty}\leq 2\omega_1\left(f,(SL)\int_{\mathbb{R}}|t|\,d\mu_N(t)\right),\tag{21.38}$$

$\forall\, N\in\mathbb{N}$.

Given that f is uniformly continuous and not necessarily bounded, from $\mathbb{R}$ into $\mathbb{R}_+$, then (21.38) is again valid and if $(SL)\int_{\mathbb{R}}t^2d\mu_N(t)\to 0$, then $P_N(f)\to f$, uniformly, as $N\to+\infty$.

Proof By Theorem 21.5 (21.15). ■

Application 21.14 Consider the well-known Picard singular integral operators:

$$P_{\xi}^{*}(f)(x):=\frac{1}{2\xi}\int_{-\infty}^{\infty}f(x+t)\,e^{-\frac{|t|}{\xi}}dt,\tag{21.39}$$

where $\xi>0$. Here f is chosen so that $P_{\xi}^{*}(f)(x)\in\mathbb{R}$, $\forall\, x\in\mathbb{R}$, e.g. f is bounded. Also $P_{\xi}^{*}(f)$ is continuous when f is uniformly continuous.

We notice that

$$\frac{1}{2\xi}\int_{-\infty}^{\infty}e^{-\frac{|t|}{\xi}}dt=1,\ \ \xi>0.\tag{21.40}$$

In [12] they obtained the degree of convergence of the operators P_{ξ}^{*} to the unit operator I with rates over the class of Hölder-continuous functions as $\xi\to 0$. In [11] they derived some more refined convergence to I (as $\xi \to 0$), however only over the set of $(C_{2\pi})$ 2π-periodic continuous functions on $\mathbb{R}$. See also [2], pp. 127–129 for bounded and uniformly continuous functions over $\mathbb{R}$,we get uniform convergence of $P_\xi^* \to I$.

We consider here only $\xi > 0$ such that $\frac{1}{\xi} = N \in \mathbb{N}$, as $\xi \to 0$.

Thus, it holds

$$\frac{N}{2}\int_{-\infty}^{\infty} e^{-|t|N}dt = 1, \ \forall\, N \in \mathbb{N}. \tag{21.41}$$

Clearly here according to our theory

$$d\mu_N(t) = \frac{N}{2}e^{-|t|N}dt, \quad \forall\, N \in \mathbb{N},$$

and $P_{N^{-1}}^*(f)$ is a special case of $P_N(f)$.

We observe here that ($n \in \mathbb{Z}_+$)

$$(SL)\int_{\mathbb{R}} |t|^{n+1}\, d\mu_N(t) = \frac{N}{2}\int_{-\infty}^{\infty} |t|^{n+1}\, e^{-|t|N}dt = N\int_0^{\infty} t^{n+1}e^{-tN}dt \tag{21.42}$$

$$= \frac{N(n+1)!}{N^{n+2}} = \frac{(n+1)!}{N^{n+1}} \to 0, \text{ as } N \to +\infty.$$

Therefore our special case of $(SL)\int_{\mathbb{R}} |t|^{n+1}\, d\mu_N(t)$ generated from the Picard operators converges to zero as $N \to +\infty$, where $n \in \mathbb{Z}_+$. Furthermore our results apply to Picard operators convergence to the unit I.

This application realizes our general convergence theory making possible our assumptions.

21.5.2 Fractional Approximation

We need

Definition 21.15 Let $\nu \geq 0$, $n = \lceil \nu \rceil$ ($\lceil \cdot \rceil$ is the ceiling of the number), $f \in AC^n([a,b])$ (space of functions f with $f^{(n-1)} \in AC([a,b])$, absolutely continuous functions). We call left Caputo fractional derivative of order $\nu > 0$ (see [10], p. 49, [1], p. 394) the function

$$D_{*a}^{\nu} f(x) = \frac{1}{\Gamma(n-\nu)}\int_a^x (x-t)^{n-\nu-1} f^{(n)}(t)\,dt, \quad \forall\, x \in [a,b], \tag{21.43}$$

where Γ is the gamma function $\Gamma(v) = \int_0^{\infty} e^{-t}t^{v-1}dt$, $v > 0$.

We set $D_{*a}^0 f(x) = f(x)$, $\forall\, x \in [a,b]$.

Exactly the same way one can define $D_{*x_0}^{\nu} f$ over $[x_0, +\infty)$, $x_0 \in \mathbb{R}$, for $f \in AC^n([x_0, b])$, $\forall\, b \in \mathbb{R}$, $b \geq x_0$.

We also need

Definition 21.16 (*see also* [2], *p. 336*) Let $f \in AC^r([a, b])$, $r = \lceil \alpha \rceil$, $\alpha \geq 0$. We right Caputo fractional derivative of order $\alpha > 0$ is given by

$$D_{b-}^{\alpha} f(x) = \frac{(-1)^r}{\Gamma(r - \alpha)} \int_x^b (\zeta - x)^{r-\alpha-1} f^{(r)}(\zeta)\, d\zeta, \quad \forall\, x \in [a, b]. \tag{21.44}$$

We set $D_{b-}^{0} f(x) = f(x)$.

Exactly the same way one can define $D_{x_0-}^{\alpha} f$ over $(-\infty, x_0]$, $x_0 \in \mathbb{R}$, for $f \in AC^r([a, x_0])$, $\forall\, a \in \mathbb{R} : a \leq x_0$.

We make

Convention 21.17 We assume that

$$D_{*x_0}^{a} f(x) = 0, \text{ for } x < x_0, \tag{21.45}$$

and

$$D_{x_0-}^{\alpha} f(x) = 0, \text{ for } x > x_0, \tag{21.46}$$

for all $x, x_0 \in \mathbb{R}$.

We also make

Convention 21.18 Let a real number $m > 0$, from now on we assume that $D_{x_0-}^{m} f$ is either bounded or uniformly continuous function on $(-\infty, x_0]$, similarly from now on we assume that $D_{*x_0}^{m} f$ is either bounded or uniformly continuous function on $[x_0, +\infty)$.

We need

Definition 21.19 Let $D_{x_0}^{m} f$ (real number $m > 0$) denote any of $D_{x_0-}^{m} f$, $D_{*x_0}^{m} f$ and $\delta > 0$. We set

$$\omega_1 \left(D_{x_0}^{m} f, \delta \right)_{\mathbb{R}} := \max \left\{ \omega_1 \left(D_{x_0-}^{m} f, \delta \right)_{(-\infty, x_0]}, \omega_1 \left(D_{*x_0}^{m} f, \delta \right)_{[x_0, +\infty)} \right\}, \tag{21.47}$$

where $x_0 \in \mathbb{R}$. Notice that $\omega_1 \left(D_{x_0}^{m} f, \delta \right)_{\mathbb{R}} < +\infty$.

We will use

Theorem 21.20 ([3], p. 89) *Let the real number $m > 0$, $m \notin \mathbb{N}$, $\lambda = \lceil m \rceil$, $x_0 \in \mathbb{R}$, $f \in AC^{\lambda}([a, b], \mathbb{R}_+)$ (i.e. $f^{(\lambda-1)} \in AC[a, b]$, absolutely continuous functions on $[a, b]$), $\forall\, [a, b] \subset \mathbb{R}$, and $f^{(\lambda)} \in L_{\infty}(\mathbb{R})$. Furthermore we assume that $f^{(k)}(x_0) = 0$, $k = 1, ..., \lambda - 1$. The Convention 21.18 is imposed here. Then*

$$|f(x) - f(x_0)| \leq \frac{\omega_1\left(D_{x_0}^m f, \delta\right)_{\mathbb{R}}}{\Gamma(m+1)}\left[|x - x_0|^m + \frac{|x - x_0|^{m+1}}{(m+1)\delta}\right], \quad \delta > 0, \quad (21.48)$$

for all $x \in \mathbb{R}$, $\delta > 0$.

If $0 < m < 1$, *then we do not need initial conditions.*

We make

Remark 21.21 Let L_N, $N \in \mathbb{N}$, be a sequence of positive sublinear operators from $C_+(\mathbb{R})$ into $CB_+(\mathbb{R})$. Here all are as in Theorem 21.20 for $x = x_0$ and we can rewrite (21.48) as follows:

$$|f(\cdot) - f(x)| \leq \frac{\omega_1\left(D_x^m f, \delta\right)_{\mathbb{R}}}{\Gamma(m+1)}\left[|\cdot - x|^m + \frac{|\cdot - x|^{m+1}}{(m+1)\delta}\right], \quad (21.49)$$

valid over $\mathbb{R}$. Assume that $L_N(1) = 1$, $\forall\, N \in \mathbb{N}$.

By (21.8) we obtain

$$|L_N(f)(x) - f(x)| \leq$$

$$\frac{\omega_1\left(D_x^m f, \delta\right)_{\mathbb{R}}}{\Gamma(m+1)}\left[L_N\left(|\cdot - x|^m\right)(x) + \frac{L_N\left(|\cdot - x|^{m+1}\right)(x)}{(m+1)\delta}\right] =: (\xi). \quad (21.50)$$

We also assume that $L_N\left(|\cdot - x|^{m+1}\right)(x) > 0$, $\forall\, N \in \mathbb{N}$.

By (21.9) we get that

$$L_N\left(|\cdot - x|^m\right)(x) \leq \left(L_N\left(|\cdot - x|^{m+1}\right)(x)\right)^{\frac{m}{m+1}}. \quad (21.51)$$

Choose

$$\delta := \left(L_N\left(|\cdot - x|^{m+1}\right)(x)\right)^{\frac{1}{m+1}} > 0, \quad (21.52)$$

i.e.

$$\delta^{m+1} = L_N\left(|\cdot - x|^{m+1}\right)(x).$$

Therefore we have

$$(\xi) \leq \frac{\omega_1\left(D_x^m f, \left(L_N\left(|\cdot - x|^{m+1}\right)(x)\right)^{\frac{1}{m+1}}\right)_{\mathbb{R}}}{\Gamma(m+1)}\left[\delta^m + \frac{\delta^{m+1}}{(m+1)\delta}\right] = \quad (21.53)$$

$$\frac{(m+2)}{\Gamma(m+2)}\omega_1\left(D_x^m f, \left(L_N\left(|\cdot - x|^{m+1}\right)(x)\right)^{\frac{1}{m+1}}\right)_{\mathbb{R}}\left(L_N\left(|\cdot - x|^{m+1}\right)(x)\right)^{\frac{m}{m+1}},$$

$\forall\, N \in \mathbb{N}$.

We have proved

Theorem 21.22 *Let $m > 0$, $m \notin \mathbb{N}$, $\lambda = \lceil m \rceil$, $x \in \mathbb{R}$, $f \in AC^{\lambda}([a,b], \mathbb{R}_+)$, $\forall$ $[a,b] \subset \mathbb{R}$, and $f^{(\lambda)} \in L_{\infty}(\mathbb{R})$. Furthermore we assume that $f^{(k)}(x) = 0$, $k = 1, ..., \lambda - 1$. We assume that $D^m_{x-} f$, $D^m_{*x} f$ are either bounded or uniformly continuous over $(-\infty, x]$, $[x, +\infty)$, respectively. Let L_N ($N \in \mathbb{N}$) be a sequence of positive sublinear operators from $C_+(\mathbb{R})$ into $CB_+(\mathbb{R})$. Assume that $L_N(1) = 1$, and $L_N\left(|\cdot - x|^{m+1}\right)(x) > 0$, $\forall\, N \in \mathbb{N}$. Then*

$$|L_N(f)(x) - f(x)| \leq \frac{(m+2)}{\Gamma(m+2)} \omega_1\left(D^m_x f, \left(L_N\left(|\cdot - x|^{m+1}\right)(x)\right)^{\frac{1}{m+1}}\right)_{\mathbb{R}}$$

$$\left(L_N\left(|\cdot - x|^{m+1}\right)(x)\right)^{\frac{m}{m+1}}, \quad \forall\, N \in \mathbb{N}. \tag{21.54}$$

If $L_N\left(|\cdot - x|^{m+1}\right)(x) \to 0$, then $L_N(f)(x) \to f(x)$, as $N \to +\infty$.

Next we specialize for $L_N = P_N$, $\forall\ N \in \mathbb{N}$, where P_N is as in (21.34), see Sect. 21.5.1 for the full description.

We give

Corollary 21.23 *All are as in Theorem 21.22 regarding f.*

Assume that $(SL) \int_{\mathbb{R}} |t|^{m+1} d\mu_N(t) > 0$, and $P_N(f) \in CB_+(\mathbb{R})$, $\forall\, N \in \mathbb{N}$. Then

$$|P_N(f)(x) - f(x)| = \left|(SL) \int_{\mathbb{R}} f(x+t)\, d\mu_N(t) - f(x)\right| \leq$$

$$\frac{(m+2)}{\Gamma(m+2)} \omega_1\left(D^m_x f, \left((SL) \int_{\mathbb{R}} |t|^{m+1} d\mu_N(t)\right)^{\frac{1}{m+1}}\right)_{\mathbb{R}}$$

$$\left((SL) \int_{\mathbb{R}} |t|^{m+1} d\mu_N(t)\right)^{\frac{m}{m+1}}, \quad \forall\, N \in \mathbb{N}. \tag{21.55}$$

If $(SL) \int_{\mathbb{R}} |t|^{m+1} d\mu_N(t) \to 0$, then $P_N(f)(x) \to f(x)$, as $N \to +\infty$.

Proof By Theorem 21.22. Notice also that

$P_N\left(|\cdot - x|^{m+1}\right)(x) = (SL) \int_{\mathbb{R}} |t|^{m+1} d\mu_N(t)$, $\forall\, N \in \mathbb{N}$. ■

We need

Definition 21.24 (*see* [1], *p. 24, and* [2], *p. 334*) Let $x, x_0 \in \mathbb{R}$ be such that $x \geq x_0$, $\nu > 0$, $\nu \notin \mathbb{N}$, such that $p = [\nu]$, $[\cdot]$ the integral part, $\alpha = \nu - p$ $(0 < \alpha < 1)$.

Let $f \in C^p(\mathbb{R})$ and define

$$\left(J^{x_0}_{\nu} f\right)(x) := \frac{1}{\Gamma(\nu)} \int_{x_0}^{x} (x-t)^{\nu-1} f(t)\, dt, \quad x_0 \leq x < +\infty. \tag{21.56}$$

the left generalized Riemann-Liouville fractional integral.

Let $x, x_0 \in \mathbb{R}$ be such that $x \leq x_0$, $\nu > 0$, $\nu \notin \mathbb{N}$, such that $p = [\nu]$, $\alpha = \nu - p$ $(0 < \alpha < 1)$.

Let $f \in C^p(\mathbb{R})$ and define

$$\left(J_{x_0-}^{\nu} f\right)(x) := \frac{1}{\Gamma(\nu)} \int_x^{x_0} (z - x)^{\nu-1} f(z)\, dz, \quad -\infty < x \leq x_0. \tag{21.57}$$

the right generalized Riemann-Liouville fractional integral.

We need

Definition 21.25 (*see also* [7]) Let $x, x_0 \in \mathbb{R}$, $x \geq x_0$, $\nu > 0$, $\nu \notin \mathbb{N}$, $p = [\nu]$, $\alpha = \nu - p$.

Let $f \in C_b^p(\mathbb{R})$, i.e. $f \in C^p(\mathbb{R})$ with $\left\| f^{(p)} \right\|_{\infty} < +\infty$, where $\|\cdot\|_{\infty}$ is the supremum norm.

Here $\left(J_{\nu}^{x_0} f\right)(x)$ is defined via (21.56) over $[x_0, +\infty)$.

We define the subspace $C_{x_0+}^{\nu}(\mathbb{R})$ of $C_b^p(\mathbb{R})$:

$$C_{x_0+}^{\nu}(\mathbb{R}) := \left\{ f \in C_b^p(\mathbb{R}) : J_{1-\alpha}^{x_0} f^{(p)} \in C^1([x_0, +\infty)) \right\}.$$

For $f \in C_{x_0+}^{\nu}(\mathbb{R})$, we define the left generalized ν-fractional derivative of f over $[x_0, +\infty)$ as

$$D_{x_0+}^{\nu} f = \left(J_{1-\alpha}^{x_0} f^{(p)}\right)'. \tag{21.58}$$

We need

Definition 21.26 (*see also* [2], *p. 345*) Let $x, x_0 \in \mathbb{R}$, $x \leq x_0$, $\nu > 0$, $\nu \notin \mathbb{N}$, $p = [\nu]$, $\alpha = \nu - p$. Let $f \in C_b^p(\mathbb{R})$. Here $\left(J_{x_0-}^{\nu} f\right)(x)$ is defined via (21.57) over $(-\infty, x_0]$.

We define the subspace of $C_{x_0-}^{\nu}(\mathbb{R})$ of $C_b^p(\mathbb{R})$:

$$C_{x_0-}^{\nu}(\mathbb{R}) := \left\{ f \in C_b^p(\mathbb{R}) : \left(J_{x_0-}^{1-\alpha} f^{(p)}\right) \in C^1((-\infty, x_0]) \right\}.$$

For $f \in C_{x_0-}^{\nu}(\mathbb{R})$, we define the right generalized ν-fractional derivative of f over $(-\infty, x_0]$ as

$$\overline{D}_{x_0-}^{\nu} f = (-1)^{p-1} \left(J_{x_0-}^{1-\alpha} f^{(p)}\right)'. \tag{21.59}$$

We make

Convention 21.27 Let a real number $m > 1$, from now on we assume that $\overline{D}_{x_0-}^{m} f$ is either bounded or uniformly continuous function on $(-\infty, x_0]$, similarly from now on we assume that $D_{x_0+}^{m} f$ is either bounded or uniformly continuous function on $[x_0, +\infty)$.

We use

Definition 21.28 Let $\overline{D}_{x_0}^m f$ (real number $m > 1$) denote any of $\overline{D}_{x_0-}^m f$, $D_{x_0+}^m f$ and $\delta > 0$. We set

$$\omega_1\left(\overline{D}_{x_0}^m f, \delta\right)_{\mathbb{R}} := \max\left\{\omega_1\left(\overline{D}_{x_0-}^m f, \delta\right)_{(-\infty, x_0]}, \omega_1\left(D_{x_0+}^m f, \delta\right)_{[x_0, +\infty)}\right\}, \quad (21.60)$$

where $x_0 \in \mathbb{R}$. Notice that $\omega_1\left(\overline{D}_{x_0}^m f, \delta\right)_{\mathbb{R}} < +\infty$.

We give

Theorem 21.29 ([3], p. 113) *Let $m > 1$, $m \notin \mathbb{N}$, $p = [m]$, $x_0 \in \mathbb{R}$, and $f \in C_{x_0+}^m(\mathbb{R}) \cap C_{x_0-}^m(\mathbb{R})$. Assume that $f^{(k)}(x_0) = 0$, $k = 1, \ldots, p-1$, and $\left(D_{x_0+}^m f\right)(x_0) = \left(\overline{D}_{x_0-}^m f\right)(x_0) = 0$. The Convention 21.27 is imposed. Then*

$$|f(x) - f(x_0)| \leq \frac{\omega_1\left(\overline{D}_{x_0}^m f, \delta\right)_{\mathbb{R}}}{\Gamma(m+1)}\left[|x - x_0|^m + \frac{|x - x_0|^{m+1}}{(m+1)\delta}\right], \quad \delta > 0, \quad (21.61)$$

for all $x \in \mathbb{R}$.

We give

Theorem 21.30 *Let $m>1$, $m \notin \mathbb{N}$, $p=\lceil m \rceil$, $x \in \mathbb{R}$, $f : \mathbb{R} \to \mathbb{R}_+$ with $f \in C_{x+}^m(\mathbb{R}) \cap C_{x-}^m(\mathbb{R})$. Assume that $f^{(k)}(x)=0$, $k=1, \ldots, p-1$, and $\left(D_{x+}^m f\right)(x) = \left(\overline{D}_{x-}^m f\right)(x)=0$. We assume that $\overline{D}_{x-}^m f$, $D_{x+}^m f$ are either bounded or uniformly continuous over $(-\infty, x]$, $[x, +\infty)$, respectively.*

Let L_N ($N \in \mathbb{N}$) be a sequence of positive sublinear operators from $C_+(\mathbb{R})$ into $CB_+(\mathbb{R})$. Assume that $L_N(1) = 1$, and $L_N\left(|\cdot - x|^{m+1}\right)(x) > 0$, $\forall\, N \in \mathbb{N}$. Then

$$|L_N(f)(x) - f(x)| \leq \frac{(m+2)}{\Gamma(m+2)}\omega_1\left(\overline{D}_x^m f, \left(L_N\left(|\cdot - x|^{m+1}\right)(x)\right)^{\frac{1}{m+1}}\right)_{\mathbb{R}}$$

$$\left(L_N\left(|\cdot - x|^{m+1}\right)(x)\right)^{\frac{m}{m+1}}, \quad \forall\, N \in \mathbb{N}. \quad (21.62)$$

If $L_N\left(|\cdot - x|^{m+1}\right)(x) \to 0$, then $L_N(f)(x) \to f(x)$, as $N \to +\infty$.

Proof Similar to the proof of Theorem 21.22, by using (21.61). ∎

We have

Corollary 21.31 *All are as in Theorem 21.30 regarding f, with $L_N = P_N$, $\forall\, N \in \mathbb{N}$. Assume that $(SL)\int_{\mathbb{R}} |t|^{m+1} d\mu_N(t) > 0$, and $P_N(f) \in CB_+(\mathbb{R})$, $\forall\, N \in \mathbb{N}$. Then*

$$|P_N(f)(x) - f(x)| = \left|(SL)\int_{\mathbb{R}} f(x+t)\,d\mu_N(t) - f(x)\right| \leq$$

$$\frac{(m+2)}{\Gamma(m+2)}\omega_1\left(\overline{D}_x^m f, \left((SL)\int_{\mathbb{R}} |t|^{m+1}\,d\mu_N(t)\right)^{\frac{1}{m+1}}\right)_{\mathbb{R}}$$

$$\left((SL)\int_{\mathbb{R}} |t|^{m+1}\,d\mu_N(t)\right)^{\frac{m}{m+1}}, \quad \forall\, N \in \mathbb{N}. \tag{21.63}$$

If $(SL)\int_{\mathbb{R}} |t|^{m+1}\,d\mu_N(t) \to 0$, *then* $P_N(f)(x) \to f(x)$, *as* $N \to +\infty$.

Proof By Theorem 21.30. ■

21.5.3 Iterated Fractional Approximation

Here $D^{\alpha}_{*x_0}$, $D^{\alpha}_{x_0-}$ stand for the Caputo left and right fractional derivatives, see (21.43), (21.44). For $n \in \mathbb{N}$, denote the iterated left and right fractional derivatives as $D^{n\alpha}_{*x_0} = D^{\alpha}_{*x_0} D^{\alpha}_{*x_0} \dots D^{\alpha}_{*x_0}$ and $D^{n\alpha}_{x_0-} = D^{\alpha}_{x_0-} D^{\alpha}_{x_0-} \dots D^{\alpha}_{x_0-}$ (n-times).

We make

Remark 21.32 Let $f : \mathbb{R} \to \mathbb{R}$ such that $f' \in L_{\infty}(\mathbb{R})$, $x_0 \in \mathbb{R}$, $0 < \alpha < 1$. The left Caputo fractional derivative $\left(D^{\alpha}_{*x_0} f\right)(x)$ is given for $x \geq x_0$. Clearly it holds $\left(D^{\alpha}_{*x_0} f\right)(x_0) = 0$, and we define $\left(D^{\alpha}_{*x_0} f\right)(x) = 0$, for $x < x_0$.

Let us assume that $D^{k\alpha}_{*x_0} f \in C([x_0, +\infty))$, $k = 0, 1, \dots, n+1$; $n \in \mathbb{N}$.

The right Caputo fractional derivative $\left(D^{\alpha}_{x_0-} f\right)(x)$ is given for $x \leq x_0$. Clearly it holds $\left(D^{\alpha}_{x_0-} f\right)(x_0) = 0$, and define $\left(D^{\alpha}_{x_0-} f\right)(x) = 0$, for $x > x_0$.

Let us assume that $D^{k\alpha}_{x_0-} f \in C((-\infty, x_0])$, $k = 0, 1, \dots, n+1$.

Here we restrict ourselves to $\frac{1}{n+1} < \alpha < 1$, that is $\overline{\lambda} := (n+1)\alpha > 1$. We denote $D^{\overline{\lambda}}_{*x_0} f := D^{(n+1)\alpha}_{*x_0} f$, and $D^{\overline{\lambda}}_{x_0-} f := D^{(n+1)\alpha}_{x_0-} f$.

We make

Convention 21.33 We assume that $D^{\overline{\lambda}}_{x_0-} f$ is either bounded or uniformly continuous function on $(-\infty, x_0]$, similarly we assume that $D^{\overline{\lambda}}_{*x_0} f$ is either bounded or uniformly continuous function on $[x_0, +\infty)$.

We need

Definition 21.34 Let $D^{\overline{\lambda}}_{x_0} f$ denote any of $D^{\overline{\lambda}}_{x_0-} f$, $D^{\overline{\lambda}}_{*x_0} f$ and $\delta > 0$. We set

$$\omega_1\left(D^{\overline{\lambda}}_{x_0} f, \delta\right)_{\mathbb{R}} := \max\left\{\omega_1\left(D^{\overline{\lambda}}_{x_0-} f, \delta\right)_{(-\infty, x_0]}, \omega_1\left(D^{\overline{\lambda}}_{*x_0} f, \delta\right)_{[x_0, +\infty)}\right\}, \tag{21.64}$$

where $x_0 \in \mathbb{R}$. Notice that $\omega_1 \left(D_{x_0}^{\overline{\lambda}} f, \delta \right)_{\mathbb{R}} < +\infty$.

We mention

Theorem 21.35 ([3], p. 137) *Let* $\frac{1}{n+1} < \alpha < 1$, $n \in \mathbb{N}$, $\overline{\lambda} := (n+1)\alpha > 1$, $f : \mathbb{R} \to \mathbb{R}$, $f' \in L_\infty(\mathbb{R})$, $x_0 \in \mathbb{R}$. *Assume that* $D_{*x_0}^{k\alpha} f \in C([x_0, +\infty))$, $k = 0, 1, ..., n+1$, *and* $\left(D_{*x_0}^{i\alpha} f \right)(x_0) = 0$, $i = 2, 3, ..., n+1$. *Suppose that* $D_{x_0-}^{k\alpha} f \in C((-\infty, x_0])$, *for* $k = 0, 1, ..., n+1$, *and* $\left(D_{x_0-}^{i\alpha} f \right)(x_0) = 0$, *for* $i = 2, 3, ..., n+1$. *Convention 21.33 is imposed. Then*

$$|f(x) - f(x_0)| \leq \frac{\omega_1 \left(D_{x_0}^{\overline{\lambda}} f, \delta \right)_{\mathbb{R}}}{\Gamma\left(\overline{\lambda}+1\right)} \left[|x - x_0|^{\overline{\lambda}} + \frac{|x - x_0|^{\overline{\lambda}+1}}{\left(\overline{\lambda}+1\right)\delta} \right], \tag{21.65}$$

$\forall\, x \in \mathbb{R}$, $\delta > 0$.

We present

Theorem 21.36 *Let* $\frac{1}{n+1} < \alpha < 1$, $n \in \mathbb{N}$, $\overline{\lambda} := (n+1)\alpha > 1$, $f : \mathbb{R} \to \mathbb{R}_+$, $f' \in L_\infty(\mathbb{R})$, $x_0 \in \mathbb{R}$. *Assume that* $D_{*x}^{k\alpha} f \in C([x, +\infty))$, $k = 0, 1, ..., n+1$, *and* $\left(D_{*x}^{i\alpha} f \right)(x) = 0$, $i = 2, 3, ..., n+1$. *Suppose that* $D_{x-}^{k\alpha} f \in C((-\infty, x])$, *for* $k = 0, 1, ..., n+1$, *and* $\left(D_{x-}^{i\alpha} f \right)(x) = 0$, *for* $i = 2, 3, ..., n+1$. *We assume that* $D_{x-}^{\overline{\lambda}} f$, $D_{*x}^{\overline{\lambda}} f$ *are either bounded or uniformly continuous over* $(-\infty, x]$, $[x, +\infty)$, *respectively. Let* L_N *(*$N \in \mathbb{N}$*) be a sequence of positive sublinear operators from* $C_+(\mathbb{R})$ *into* $CB_+(\mathbb{R})$. *Assume that* $L_N(1) = 1$, *and* $L_N\left(|\cdot - x|^{\overline{\lambda}+1} \right)(x) > 0$, $\forall\, N \in \mathbb{N}$. *Then*

$$|L_N(f)(x) - f(x)| \leq \frac{\left(\overline{\lambda}+2\right)}{\Gamma\left(\overline{\lambda}+2\right)} \omega_1 \left(D_x^{\overline{\lambda}} f, \left(L_N\left(|\cdot - x|^{\overline{\lambda}+1} \right)(x) \right)^{\frac{1}{\overline{\lambda}+1}} \right)_{\mathbb{R}}$$

$$\left(L_N\left(|\cdot - x|^{\overline{\lambda}+1} \right)(x) \right)^{\frac{\overline{\lambda}}{\overline{\lambda}+1}}, \quad \forall\, N \in \mathbb{N}. \tag{21.66}$$

If $L_N\left(|\cdot - x|^{\overline{\lambda}+1} \right)(x) \to 0$, *then* $L_N(f)(x) \to f(x)$, *as* $N \to +\infty$.

Proof Similar to the proof of Theorem 21.22, by using (21.65). ■

We finish with

Corollary 21.37 *All are as in Theorem 21.36 regarding* f *and* $L_N = P_N$, $\forall\, N \in \mathbb{N}$. *Assume that* $(SL) \int_{\mathbb{R}} |t|^{\overline{\lambda}+1} d\mu_N(t) > 0$, *and* $P_N(f) \in CB_+(\mathbb{R})$, $\forall\, N \in \mathbb{N}$. *Then*

$$|P_N(f)(x) - f(x)| \leq \frac{\left(\overline{\lambda}+2\right)}{\Gamma\left(\overline{\lambda}+2\right)} \omega_1 \left(D_x^{\overline{\lambda}} f, \left((SL) \int_{\mathbb{R}} |t|^{\overline{\lambda}+1} d\mu_N(t) \right)^{\frac{1}{\overline{\lambda}+1}} \right)_{\mathbb{R}}$$

$$\left((SL) \int_{\mathbb{R}} |t|^{\overline{\lambda}+1} d\mu_N(t) \right)^{\frac{\overline{\lambda}}{\overline{\lambda}+1}}, \quad \forall\, N \in \mathbb{N}. \tag{21.67}$$

If $(SL) \int_{\mathbb{R}} |t|^{\overline{\lambda}+1} d\mu_N(t) \to 0$, *then* $P_N(f)(x) \to f(x)$, *as* $N \to +\infty$.

Proof By Theorem 21.36. ∎

References

1. Anastassiou, G.A.: Fractional Differentiation Inequalities. Springer, New York (2009)
2. Anastassiou, G.A.: Intelligent Mathematics: Computational Analysis. Springer, Heidelberg (2011)
3. Anastassiou, G.A.: Nonlinearity: Ordinary and Fractional Approximation by Sublinear and Max-Product Operators. Springer, Heidelberg (2018)
4. Anastassiou, G.A.: Ostrowski type inequalities involving sublinear integrals. In: Anastassiou, G.A., Rassias, J.M. (eds.) Frontiers in Functional Equations and Analytic Inequalities, pp. 325–355. Springer, New York (2019)
5. Anastassiou, G.A.: Advanced Ordinary and Fractional Approximation by Positive Sublinear Operators, Filomat, accepted (2020)
6. Bede, B., Coroianu, L., Gal, S.: Approximation by Max-Product type Operators. Springer, Heidelberg (2016)
7. Canavati, J.A.: The Riemann-Liouville integral. Nieuw Archief Voor Wiskunde **5**(1), 53–75 (1987)
8. Choquet, G.: Theory of capacities. Ann. Inst. Fourier (Grenoble) **5**, 131–295 (1954)
9. Denneberg, D.: Non-additive Measure and Integral. Kluwer, Dordrecht (1994)
10. Diethelm, K.: The Analysis of Fractional Differential Equations. Springer, Heidelberg (2010)
11. Gal, S.G.: Remark on the degree of approximation of functions by singular integrals. Math. Nachr. **164**, 197–199 (1993)
12. Mohapatra, R.N., Rodriguez, R.S.: On the rate of convergence of singular integrals for Hölder continuous functions. Math. Nachr. **149**, 117–124 (1990)
13. Shilkret, Niel: Maxitive measure and integration. Indag. Math. **33**, 109–116 (1971)
14. Wang, Z., Klir, G.J.: Fuzzy Measure Theory. Plenum, New York (1992)

Chapter 22
Complete Approximations with Multivariate Generalized Picard Singular Integrals

This research and survey chapter deals exclusively with the study of the approximation of generalized multivariate Picard singular integrals to the identity-unit operator. Here we study quantitatively most of their approximation properties. These operators are not in general positive linear operators. In particular we study the rate of convergence of these operators to the unit operator, as well as the related simultaneous approximation. These are given via Jackson type inequalities and by the use of multivariate high order modulus of smoothness of the high order partial derivatives of the involved function. Also we study the global smoothness preservation properties of these operators. These multivariate inequalities are nearly sharp and in one case the inequality is attained, that is sharp. Furthermore we give asymptotic expansions of Voronovskaya type for the error of approximation. The above properties are studied with respect to L_p norm, $1 \leq p \leq \infty$. It follows [3].

22.1 Introduction

We start with our motivation for this chapter. The following come from [1], Chap. 10.

In the next we mention the following univariate smooth Picard singular integral operators $P_{r,\xi}(f,x)$ defined as follows.

For $r \in \mathbb{N}$ and $n \in \mathbb{Z}_+$ we set

$$\alpha_j = \begin{cases} (-1)^{r-j} \binom{r}{j} j^{-n}, & j = 1, ..., r, \\ 1 - \sum_{j=1}^{r} (-1)^{r-j} \binom{r}{j} j^{-n}, & j = 0, \end{cases} \tag{22.1}$$

G. A. Anastassiou, *Constructive Fractional Analysis with Applications*,
Studies in Systems, Decision and Control 362,
https://doi.org/10.1007/978-3-030-71481-9_22

that is $\sum_{j=0}^{r} \alpha_j = 1$. Let $f : \mathbb{R} \to \mathbb{R}$ be Lebesgue measurable, we define for $x \in \mathbb{R}$, $\xi > 0$ the Lebesgue integral

$$P_{r,\xi}(f;x) := \frac{1}{2\xi} \int_{-\infty}^{\infty} \left(\sum_{j=0}^{r} \alpha_j f(x+jt) \right) e^{-\frac{|t|}{\xi}} dt. \tag{22.2}$$

We assume that $P_{r,\xi}(f;x) \in \mathbb{R}$ for all $x \in \mathbb{R}$.

We notice by $\frac{1}{2\xi}\int_{-\infty}^{\infty} e^{-\frac{|t|}{\xi}} dt = 1$ that $P_{r,\xi}(c;x) = c$, c constant and

$$P_{r,\xi}(f;x) - f(x) = \frac{1}{2\xi} \left(\sum_{j=0}^{r} \alpha_j \int_{-\infty}^{\infty} (f(x+jt) - f(x)) \right) e^{-\frac{|t|}{\xi}} dt. \tag{22.3}$$

Let $f \in C^n(\mathbb{R})$, $n \in \mathbb{Z}^+$ with the r th modulus of smoothness finite, i.e.

$$\omega_r\left(f^{(n)}, h\right) := \sup_{|t| \le h} \left\| \Delta_t^r f^{(n)}(x) \right\|_{\infty,x} < \infty, \quad h > 0, \tag{22.4}$$

where

$$\Delta_t^r f^{(n)}(x) := \sum_{j=0}^{r} (-1)^{r-j} \binom{r}{j} f^{(n)}(x+jt), \tag{22.5}$$

see [6], p. 44.

We need to introduce

$$\delta_k := \sum_{j=1}^{r} \alpha_j j^k, \quad k = 1, \ldots, n \in \mathbb{N}, \tag{22.6}$$

and the even function

$$G_n(t) := \int_0^{|t|} \frac{(|t|-w)^{n-1}}{(n-1)!} \omega_r\left(f^{(n)}, w\right) dw, \quad n \in \mathbb{N}, \tag{22.7}$$

with

$$G_0(t) := \omega_r(f, |t|), \quad t \in \mathbb{R}. \tag{22.8}$$

Denote by $\lfloor \cdot \rfloor$ the integral part.

We mention

Theorem 22.1 *It holds that*

$$\left| P_{r,\xi}(f;x) - f(x) - \sum_{m=1}^{\lfloor \frac{n}{2} \rfloor} f^{(2m)}(x)\, \delta_{2m} \xi^{2m} \right| \le \frac{1}{\xi} \int_0^\infty G_n(t)\, e^{-\frac{t}{\xi}} dt, \quad n \in \mathbb{N}. \tag{22.9}$$

In L.H.S. (22.9) the sum collapse when $n = 1$.

Corollary 22.2 *Assume $\omega_r(f, \xi) < \infty$, $\xi > 0$. Then it holds for $n = 0$ that*

$$\left| P_{r,\xi}(f;x) - f(x) \right| \le \frac{1}{\xi} \int_0^\infty \omega_r(f,t)\, e^{-\frac{t}{\xi}} dt. \tag{22.10}$$

Theorem 22.3 *Inequality (22.9) at $x = 0$ is attained by $f(x) = x^{r+n}$, $r, n \in \mathbb{N}$ with $r + n$ even.*

Corollary 22.4 *Inequality (22.10) is attained at $x = 0$ by $f(x) = x^r$, r even.*

Note 22.5 The operators $P_{r,\xi}$ are not in general positive and they are of convolution type.

Let $r = 2$, $n = 3$. Then $\alpha_0 = \frac{23}{8}$, $\alpha_1 = -2$, $\alpha_2 = \frac{1}{8}$. Consider $f(t) = t^2 \ge 0$ and $x = 0$. Then

$$P_{r,\xi}\left(t^2; 0\right) = -3\xi^2 < 0.$$

A related convergence theorem follows.

Theorem 22.6 *Let $f \in C(\mathbb{R})$ uniformly continuous. Then*

$$\left\| P_{r,\xi}(f) - f \right\|_\infty \le 2^r \omega_1(f, \xi). \tag{22.11}$$

I.e. as $\xi \to 0$ we get $P_{r,\xi} \xrightarrow{u} I$ uniformly, where I is the unit operator.

In this chapter we study the approximation properties of multivariate smooth Picard singular integral operators

$$P_{r,n}^{[m]}(f; x_1, ..., x_N) :=$$

$$\frac{1}{(2\xi_n)^N} \sum_{j=0}^{r} \alpha_{j,r}^{[m]} \int_{\mathbb{R}^N} f(x_1 + s_1 j, x_2 + s_2 j, ..., x_N + s_N j)\, e^{-\frac{\left(\sum_{i=1}^{N} |s_i|\right)}{\xi_n}} ds_1 ... ds_N, \tag{22.12}$$

$r, n \in \mathbb{N}$, $m \in \mathbb{Z}_+$.

Observe that

$$\frac{1}{(2\xi_n)^N}\int_{\mathbb{R}^N} e^{-\frac{\left(\sum_{i=1}^{N}|s_i|\right)}{\xi_n}} ds_1...ds_N = 1, \tag{22.13}$$

see [1], Chap. 10.

Here $r \in \mathbb{N}$, $m \in Z_+$, and

$$\alpha_{j,r}^{[m]} := \begin{cases} (-1)^{r-j}\binom{r}{j} j^{-m}, & \text{if } j = 1, 2, ..., r, \\ 1 - \sum_{i=1}^{r}(-1)^{r-i}\binom{r}{i} i^{-m}, & \text{if } j = 0, \end{cases} \tag{22.14}$$

and

$$\delta_{k,r}^{[m]} := \sum_{j=1}^{r} \alpha_{j,r}^{[m]} j^k, \quad k = 1, 2, ..., m \in \mathbb{N}. \tag{22.15}$$

See that $\sum_{j=0}^{r} \alpha_{j,r}^{[m]} = 1$.

Also here $\xi_n \in (0, 1]$, $n \in \mathbb{N}$, and $f : \mathbb{R}^N \to \mathbb{R}$ is a Borel measurable function. The above operator $P_{r,n}^{[m]}$ is a special case of a more general operator $\theta_{r,n}^{[m]}$ studied in general in [2] by the author.

We mention next about $\theta_{r,n}^{[m]}$.

Let μ_{ξ_n} be a probability Borel measure on $\mathbb{R}^N$, $N \geq 1$.

We define the multiple smooth singular integral operators

$$\theta_{r,n}^{[m]}(f; x_1, ..., x_N) := \sum_{j=0}^{r} \alpha_{j.r}^{[m]} \int_{\mathbb{R}^N} f(x_1 + s_1 j, x_2 + s_2 j, ..., x_N + s_N j)\, d\mu_{\xi_n}(s), \tag{22.16}$$

where $s := (s_1, ..., s_N)$, $x := (x_1, ..., x_N) \in \mathbb{R}^N$.

The operators $\theta_{r,n}^{[m]}$ are not in general positive. For example, consider the function $\varphi(u_1, ..., u_N) = \sum_{i=1}^{N} u_i^2$ and also take $r = 2$, $m = 3$; $x_i = 0$, $i = 1, ..., N$. See that $\varphi \geq 0$, however

$$\theta_{2,n}^{[3]}(\varphi; 0, 0, ..., 0) = \left(\sum_{j=1}^{2} j^2 \alpha_{j,2}^{[3]}\right) \int_{\mathbb{R}^N} \left(\sum_{i=1}^{N} s_i^2\right) d\mu_{\xi_n}(s) =$$

$$\left(\alpha_{1,2}^{[3]} + 4\alpha_{2,2}^{[3]}\right) \int_{\mathbb{R}^N} \left(\sum_{i=1}^{N} s_i^2\right) d\mu_{\xi_n}(s) = \left(-2 + \frac{1}{2}\right) \int_{\mathbb{R}^N} \left(\sum_{i=1}^{N} s_i^2\right) d\mu_{\xi_n}(s) < 0. \tag{22.17}$$

assuming that $\int_{\mathbb{R}^N} \left(\sum_{i=1}^{N} s_i^2\right) d\mu_{\xi_n}(s) < \infty$.

Clearly in the case of $P_{r,n}^{[m]}$ we have that

$$d\mu_{\xi_n}(s) = \frac{1}{(2\xi_n)^N} e^{-\frac{\left(\sum_{i=1}^{N} |s_i|\right)}{\xi_n}} ds_1 ... ds_N, \quad s \in \mathbb{R}^N.$$

Lemma 22.7 *The operator* $\theta_{r,n}^{[m]}$ *preserve the constant functions in N variables.*

We need

Definition 22.8 Let $f \in C_B(\mathbb{R}^N)$, the space of all bounded and continuous functions or uniformly continuous on $\mathbb{R}^N$. Then, the rth multivariate modulus of smoothness of f is given by (see, e.g. [4])

$$\omega_r(f; h) := \sup_{\sqrt{u_1^2 + ... + u_N^2} \leq h} \left\| \Delta_{u_1, u_2, ..., u_N}^r (f) \right\|_\infty < \infty, \quad h > 0, \tag{22.18}$$

where $\|\cdot\|_\infty$ is the sup-norm and

$$\Delta_u^r f(x) := \Delta_{u_1, u_2, ..., u_N}^r f(x_1, ..., x_N) =$$

$$\sum_{j=0}^{r} (-1)^{r-j} \binom{r}{j} f(x_1 + ju_1, x_2 + ju_2, ..., x_N + ju_N). \tag{22.19}$$

Let $m \in \mathbb{N}$ and let $f \in C^m(\mathbb{R}^N)$.

Suppose that all partial derivatives of f of order m are bounded, i.e.

$$\left\| \frac{\partial^m f(\cdot, \cdot, ..., \cdot)}{\partial x_1^{\alpha_1} ... \partial x_N^{\alpha_N}} \right\|_\infty < \infty, \tag{22.20}$$

for all $\alpha_j \in \mathbb{Z}^+$, $j = 1, ..., N$; $\sum_{j=1}^{N} \alpha_j = m$.

In this chapter we apply the general theory developed in [2] about $\theta_{r,n}^{[m]}$ to the operators $P_{r,n}^{[m]}$, so we can obtain computationally specific results and show that the general theory has applications and it is a valid theory.

So for the very important in various branches of mathematics operators $P_{r,n}^{[m]}$ we prove the very essential properties of uniform approximation, L_p approximation, global smoothness preservation and simultaneously approximation, Voronovskaya asymptotic expansions and complex simultaneous approximation.

22.2 Auxiliary Essential Results

We need the following:

Theorem 22.9 *Let* $r, N \in \mathbb{N}$, $\alpha_j \in \mathbb{Z}^+$, $j = 1, ..., N : |\alpha| := \sum_{j=1}^{N} \alpha_j = m \in \mathbb{N}$, $\xi_n \in (0, 1]$, $n \in \mathbb{N}$. *Then*

$$u_{\xi_n}^* (\alpha) := \frac{1}{(2\xi_n)^N} \int_{\mathbb{R}^N} \left(\prod_{i=1}^{N} |s_i|^{\alpha_i} \right) \left(1 + \frac{\|s\|_2}{\xi_n} \right)^r e^{-\frac{\left(\sum_{i=1}^{N} |s_i| \right)}{\xi_n}} ds_1 ... ds_N \leq$$

$$\xi_n^m \left[(1 + N)^r + 2^r \left(\frac{\lfloor e (m + r)! \rfloor}{e} \right)^N \right] \leq (1 + N)^r + 2^r \left(\frac{\lfloor e (m + r)! \rfloor}{e} \right)^N, \tag{22.21}$$

are uniformly bounded.

Proof We have here that

$$u_{\xi_n}^* (\alpha) := \frac{1}{(2\xi_n)^N} \int_{\mathbb{R}^N} \left(\prod_{i=1}^{N} |s_i|^{\alpha_i} \right) \left(1 + \frac{\|s\|_2}{\xi_n} \right)^r e^{-\frac{\left(\sum_{i=1}^{N} |s_i| \right)}{\xi_n}} ds_1 ... ds_N =$$

$$\frac{1}{\xi_n^N} \int_{\mathbb{R}_+^N} \left(\prod_{i=1}^{N} s_i^{\alpha_i} \right) \left(1 + \frac{\|s\|_2}{\xi_n} \right)^r e^{-\frac{\left(\sum_{i=1}^{N} s_i \right)}{\xi_n}} ds_1 ... ds_N \leq \tag{22.22}$$

$$\frac{1}{\xi_n^N} \int_{\mathbb{R}_+^N} \left(\prod_{i=1}^{N} s_i^{\alpha_i} \right) \left(1 + \frac{\sum_{i=1}^{N} s_i}{\xi_n} \right)^r e^{-\frac{\left(\sum_{i=1}^{N} s_i \right)}{\xi_n}} ds_1 ... ds_N =$$

$$\xi_n^m \int_{\mathbb{R}_+^N} \prod_{i=1}^{N} \left(\frac{s_i}{\xi_n} \right)^{\alpha_i} \left(1 + \sum_{i=1}^{N} \left(\frac{s_i}{\xi_n} \right) \right)^r e^{-\sum_{i=1}^{N} \left(\frac{s_i}{\xi_n} \right)} d \left(\frac{s_1}{\xi_n} \right) ... d \left(\frac{s_N}{\xi_n} \right) =$$

$$\xi_n^m \int_{\mathbb{R}_+^N} \prod_{i=1}^{N} z_i^{\alpha_i} \left(1 + \sum_{i=1}^{N} z_i \right)^r e^{-\sum_{i=1}^{N} z_i} dz_1 ... dz_N =$$

$$\xi_n^m \left[\int_{[0,1]^N} \prod_{i=1}^{N} z_i^{\alpha_i} \left(1 + \sum_{i=1}^{N} z_i \right)^r e^{-\sum_{i=1}^{N} z_i} dz_1 ... dz_N + \right.$$

$$\int_{(\mathbb{R}_+-[0,1])^N} \prod_{i=1}^{N} z_i^{\alpha_i} \left(1+\sum_{i=1}^{N} z_i\right)^r e^{-\sum_{i=1}^{N} z_i} dz_1...dz_N \Bigg] \leq$$

$$\xi_n^m \left[(1+N)^r + \int_{(\mathbb{R}_+-[0,1])^N} \prod_{i=1}^{N} z_i^{\alpha_i} \left(1+\sum_{i=1}^{N} z_i\right)^r e^{-\sum_{i=1}^{N} z_i} dz_1...dz_N\right] \leq \quad (22.23)$$

$$\xi_n^m \left[(1+N)^r + 2^r \int_{(\mathbb{R}_+-[0,1])^N} \prod_{i=1}^{N} z_i^{\alpha_i} \left(\sum_{i=1}^{N} z_i\right)^r e^{-\sum_{i=1}^{N} z_i} dz_1...dz_N\right] \leq$$

$$\xi_n^m \left[(1+N)^r + 2^r \int_{(\mathbb{R}_+-[0,1])^N} \prod_{i=1}^{N} z_i^{\alpha_i} \left(\prod_{i=1}^{N} z_i^r\right) \prod_{i=1}^{N} e^{-z_i} \prod_{i=1}^{N} dz_i\right] =$$

$$\xi_n^m \left[(1+N)^r + 2^r \prod_{i=1}^{N} \int_1^{\infty} z_i^{\alpha_i+r} e^{-z_i} dz_i\right] =$$

$$\xi_n^m \left[(1+N)^r + 2^r \prod_{i=1}^{N} \int_1^{\infty} z_i^{((\alpha_i+r)+1)-1} e^{-z_i} dz_i\right] \quad (22.24)$$

(by [5], p. 348)

$$\xi_n^m \left[(1+N)^r + 2^r \prod_{i=1}^{N} \Gamma\left((\alpha_i+r)+1, 1\right)\right],$$

where $\Gamma(\cdot, \cdot)$ is the upper incomplete gamma function.

We have proved that

$$u_{\xi_n}^*(\alpha) \leq \xi_n^m \left[(1+N)^r + 2^r \prod_{i=1}^{N} \Gamma\left((\alpha_i+r)+1, 1\right)\right] \leq$$

$$\xi_n^m \left[(1+N)^r + 2^r \left(\frac{\lfloor e(m+r)! \rfloor}{e}\right)^N\right] \leq (1+N)^r + 2^r \left(\frac{\lfloor e(m+r)! \rfloor}{e}\right)^N < +\infty, \quad (22.25)$$

therefore $u_{\xi_n}^*(\alpha)$ are uniformly bounded.

Above we used the formula

$$\Gamma(s+1, 1) = \frac{\lfloor es! \rfloor}{e}, \quad s \in \mathbb{N}. \quad (22.26)$$

Here $\alpha_i + r \in \mathbb{N}$, hence

$$\Gamma\left((\alpha_i + r) + 1, 1\right) = \frac{\lfloor e\,(\alpha_i + r)! \rfloor}{e} \leq \frac{\lfloor e\,(m + r)! \rfloor}{e}, \quad i = 1, ..., N. \tag{22.27}$$

The claim is proved. ■

We continue with

Theorem 22.10 *Let $r, n, N \in \mathbb{N}$, $\xi_n \in (0, 1]$. Then*

$$\Phi^*_{\xi_n} := \frac{1}{(2\xi_n)^N} \int_{\mathbb{R}^N} \left(1 + \frac{\|s\|_2}{\xi_n}\right)^r e^{-\frac{\left(\sum_{i=1}^N |s_i|\right)}{\xi_n}} ds_1 ... ds_N \leq$$

$$(1 + N)^r + 2^r \left(\frac{\lfloor er! \rfloor}{e}\right)^N < \infty, \tag{22.28}$$

are uniformly bounded.

Proof We have that

$$\Phi^*_{\xi_n} = \frac{1}{\xi_n^N} \int_{\mathbb{R}^N_+} \left(1 + \frac{\|s\|_2}{\xi_n}\right)^r e^{-\frac{\left(\sum_{i=1}^N s_i\right)}{\xi_n}} ds_1 ... ds_N \leq$$

$$\frac{1}{\xi_n^N} \int_{\mathbb{R}^N_+} \left(1 + \frac{\sum_{i=1}^N s_i}{\xi_n}\right)^r e^{-\frac{\left(\sum_{i=1}^N s_i\right)}{\xi_n}} ds_1 ... ds_N =$$

$$\int_{\mathbb{R}^N_+} \left(1 + \sum_{i=1}^N \left(\frac{s_i}{\xi_n}\right)\right)^r e^{-\sum_{i=1}^N \left(\frac{s_i}{\xi_n}\right)} d\left(\frac{s_1}{\xi_n}\right) ... d\left(\frac{s_N}{\xi_n}\right) =$$

$$\int_{\mathbb{R}^N_+} \left(1 + \sum_{i=1}^N z_i\right)^r e^{-\sum_{i=1}^N z_i} dz_1 ... dz_N =$$

$$\left[\int_{[0,1]^N} \left(1 + \sum_{i=1}^N z_i\right)^r e^{-\sum_{i=1}^N z_i} dz_1 ... dz_N + \right.$$

$$\left. \int_{(\mathbb{R}_+ - [0,1])^N} \left(1 + \sum_{i=1}^N z_i\right)^r e^{-\sum_{i=1}^N z_i} dz_1 ... dz_N \right] \leq \tag{22.29}$$

$$\left[(1+N)^r + \int_{(\mathbb{R}_+-[0,1])^N}\left(1+\sum_{i=1}^N z_i\right)^r e^{-\sum_{i=1}^N z_i} dz_1...dz_N\right] \leq$$

$$(1+N)^r + 2^r \int_{(\mathbb{R}_+-[0,1])^N}\left(\sum_{i=1}^N z_i\right)^r e^{-\sum_{i=1}^N z_i} dz_1...dz_N \leq$$

$$(1+N)^r + 2^r \int_{(\mathbb{R}_+-[0,1])^N}\left(\prod_{i=1}^N z_i^r\right)\left(\prod_{i=1}^N e^{-z_i}\right)\prod_{i=1}^N dz_i =$$

$$(1+N)^r + 2^r \prod_{i=1}^N \int_1^\infty z_i^{(r+1)-1} e^{-z_i} dz_i =$$

(by [5], p. 348)

$$(1+N)^r + 2^r \Gamma^N(r+1,1) = (1+N)^r + 2^r\left(\frac{\lfloor er! \rfloor}{e}\right)^N < +\infty. \tag{22.30}$$

That is, $\Phi^*_{\xi_n}$ are uniformly bounded. ∎

We give

Theorem 22.11 *All as in Theorem 22.9 and $p > 1$. Then*

$$A_{\xi_n}(\alpha) := \frac{1}{(2\xi_n)^N}\int_{\mathbb{R}^N}\left(\left(\prod_{i=1}^N |s_i|^{\alpha_i}\right)\left(1+\frac{\|s\|_2}{\xi_n}\right)^r\right)^p e^{-\frac{\left(\sum_{i=1}^N |s_i|\right)}{\xi_n}} ds_1...ds_N \leq$$
$$\xi_n^{mp}\left[(1+N)^{rp} + 2^{rp}\Gamma^N((m+r)p+1,1)\right] \leq \tag{22.31}$$
$$(1+N)^{rp} + 2^{rp}\Gamma^N((m+r)p+1,1) < +\infty,$$

are uniformly bounded.

Proof We have here that

$$A_{\xi_n}(\alpha) := \frac{1}{(2\xi_n)^N}\int_{\mathbb{R}^N}\left(\left(\prod_{i=1}^N |s_i|^{\alpha_i}\right)\left(1+\frac{\|s\|_2}{\xi_n}\right)^r\right)^p e^{-\frac{\left(\sum_{i=1}^N |s_i|\right)}{\xi_n}} ds_1...ds_N =$$

$$\frac{1}{\xi_n^N}\int_{\mathbb{R}_+^N}\prod_{i=1}^N s_i^{\alpha_i p}\left(1+\frac{\|s\|_2}{\xi_n}\right)^{rp} e^{-\frac{\left(\sum_{i=1}^N s_i\right)}{\xi_n}} ds_1...ds_N \leq \tag{22.32}$$

$$\frac{1}{\xi_n^N}\int_{\mathbb{R}_+^N}\prod_{i=1}^{N}s_i^{\alpha_i p}\left(1+\frac{\sum_{i=1}^{N}s_i}{\xi_n}\right)^{rp}e^{-\frac{\left(\sum_{i=1}^{N}s_i\right)}{\xi_n}}ds_1...ds_N=$$

$$\xi_n^{mp}\int_{\mathbb{R}_+^N}\prod_{i=1}^{N}\left(\frac{s_i}{\xi_n}\right)^{\alpha_i p}\left(1+\sum_{i=1}^{N}\frac{s_i}{\xi_n}\right)^{rp}e^{-\sum_{i=1}^{N}\left(\frac{s_i}{\xi_n}\right)}d\frac{s_1}{\xi_n}...d\frac{s_N}{\xi_n}=$$

$$\xi_n^{mp}\int_{\mathbb{R}_+^N}\prod_{i=1}^{N}z_i^{\alpha_i p}\left(1+\sum_{i=1}^{N}z_i\right)^{rp}e^{-\sum_{i=1}^{N}z_i}dz_1...dz_N= \tag{22.33}$$

$$\xi_n^{mp}\left[\int_{[0,1]^N}\prod_{i=1}^{N}z_i^{\alpha_i p}\left(1+\sum_{i=1}^{N}z_i\right)^{rp}e^{-\sum_{i=1}^{N}z_i}dz_1...dz_N+\right.$$

$$\left.\int_{(\mathbb{R}_+-[0,1])^N}\prod_{i=1}^{N}z_i^{\alpha_i p}\left(1+\sum_{i=1}^{N}z_i\right)^{rp}e^{-\sum_{i=1}^{N}z_i}dz_1...dz_N\right]\leq$$

$$\xi_n^{mp}\left[(1+N)^{rp}+2^{rp}\int_{(\mathbb{R}_+-[0,1])^N}\prod_{i=1}^{N}z_i^{\alpha_i p}\left(\sum_{i=1}^{N}z_i\right)^{rp}\prod_{i=1}^{N}e^{-z_i}\prod_{i=1}^{N}dz_i\right]\leq$$

$$\xi_n^{mp}\left[(1+N)^{rp}+2^{rp}\int_{(\mathbb{R}_+-[0,1])^N}\prod_{i=1}^{N}z_i^{\alpha_i p}\left(\prod_{i=1}^{N}z_i^{rp}\right)\prod_{i=1}^{N}e^{-z_i}\prod_{i=1}^{N}dz_i\right]=$$

$$\xi_n^{mp}\left[(1+N)^{rp}+2^{rp}\prod_{i=1}^{N}\int_1^{\infty}z_i^{(\alpha_i+r)p}e^{-z_i}dz_i\right]= \tag{22.34}$$

(by [5], p. 348)

$$\xi_n^{mp}\left[(1+N)^{rp}+2^{rp}\prod_{i=1}^{N}\Gamma\left((\alpha_i+r)p+1,1\right)\right]\leq$$

(by [5], p. 909)

$$\xi_n^{mp}\left[(1+N)^{rp}+2^{rp}\Gamma^N\left((m+r)p+1,1\right)\right].$$

We have proved that

$$A_{\xi_n}(\alpha) \leq \xi_n^{mp}\left[(1+N)^{rp} + 2^{rp}\Gamma^N((m+r)p+1,1)\right] \leq \tag{22.35}$$

$$(1+N)^{rp} + 2^{rp}\Gamma^N((m+r)p+1,1) < +\infty,$$

are uniformly bounded.

Above we used ([5], p. 909) the formula

$$\Gamma(\alpha, xy) = y^{\alpha}e^{-xy}\int_0^{\infty} e^{-ty}(t+x)^{\alpha-1}\,dt, \tag{22.36}$$

where $Rey > 0, x > 0, Re\alpha > 1$.

We notice that $(x = y = 1, \alpha_i \leq m)$

$$\Gamma((\alpha_i+r)p+1,1) = e^{-1}\int_0^{\infty} e^{-t}(t+1)^{(\alpha_i+r)p}\,dt \leq \tag{22.37}$$

$$e^{-1}\int_0^{\infty} e^{-t}(t+1)^{(m+r)p}\,dt = \Gamma((m+r)p+1,1).$$

That is

$$\Gamma((\alpha_i+r)p+1,1) \leq \Gamma((m+r)p+1,1), \tag{22.38}$$

for all $i = 1, ..., N$.

The theorem is proved. ∎

We give

Theorem 22.12 *Let $r, n, N \in \mathbb{N}$, $\xi_n \in (0,1]$, $p > 1$. Then*

$$A_{\xi_n}^{*} := \frac{1}{(2\xi_n)^N}\int_{\mathbb{R}^N}\left(1+\frac{\|s\|_2}{\xi_n}\right)^{rp} e^{-\frac{\left(\sum_{i=1}^{N}|s_i|\right)}{\xi_n}}\,ds_1...ds_N \leq \tag{22.39}$$

$$(1+N)^{rp} + 2^{rp}\Gamma^N(rp+1,1) < +\infty,$$

are uniformly bounded.

Proof We have

$$A_{\xi_n}^{*} = \frac{1}{\xi_n^N}\int_{\mathbb{R}_+^N}\left(1+\frac{\|s\|_2}{\xi_n}\right)^{rp} e^{-\frac{\left(\sum_{i=1}^{N}s_i\right)}{\xi_n}}\,ds_1...ds_N \leq \tag{22.40}$$

$$\frac{1}{\xi_n^N}\int_{\mathbb{R}_+^N}\left(1+\frac{\sum_{i=1}^N s_i}{\xi_n}\right)^{rp} e^{-\frac{\left(\sum_{i=1}^N s_i\right)}{\xi_n}} ds_1...ds_N =$$

$$\int_{\mathbb{R}_+^N}\left(1+\sum_{i=1}^N z_i\right)^{rp} e^{-\sum_{i=1}^N z_i} dz_1...dz_N =$$

$$\int_{[0,1]^N}\left(1+\sum_{i=1}^N z_i\right)^{rp} e^{-\sum_{i=1}^N z_i} dz_1...dz_N +$$

$$\int_{(\mathbb{R}_+-[0,1])^N}\left(1+\sum_{i=1}^N z_i\right)^{rp} e^{-\sum_{i=1}^N z_i} dz_1...dz_N \leq \tag{22.41}$$

$$(1+N)^{rp}+2^{rp}\int_{(\mathbb{R}_+-[0,1])^N}\left(\sum_{i=1}^N z_i\right)^{rp}\prod_{i=1}^N e^{-z_i}\prod_{i=1}^N dz_i \leq$$

$$(1+N)^{rp}+2^{rp}\int_{(\mathbb{R}_+-[0,1])^N}\left(\prod_{i=1}^N z_i^{rp}\right)\left(\prod_{i=1}^N e^{-z_i}\right)\prod_{i=1}^N dz_i =$$

$$(1+N)^{rp}+2^{rp}\prod_{i=1}^N\int_1^\infty z_i^{rp}e^{-z_i}dz_i =$$

(by [5], p. 348)

$$(1+N)^{rp}+2^{rp}\Gamma^N(rp+1,1).$$

We have proved that

$$A_{\xi_n}^* \leq (1+N)^{rp}+2^{rp}\Gamma^N(rp+1,1) < +\infty, \tag{22.42}$$

are uniformly bounded. ■

We finish this section with

Theorem 22.13 *Let* $n, N \in \mathbb{N}$, $\xi_n \in (0,1]$, $\alpha_j \in \mathbb{Z}^+$, $j=1,...,N : |\alpha| := \sum_{j=1}^N \alpha_j = m \in \mathbb{N}$. *Then*

$$B_{\xi_n}^*(\alpha) := \xi_n^{-m}\left(\frac{1}{(2\xi_n)^N}\int_{\mathbb{R}^N}\left(\prod_{i=1}^N |s_i|^{\alpha_i}\right)e^{-\frac{\left(\sum_{i=1}^N |s_i|\right)}{\xi_n}} ds_1...ds_N\right) \leq (m!)^N. \tag{22.43}$$

Proof We have that

$$B^*_{\xi_n}(\alpha) \le \frac{1}{\xi_n^m \xi_n^N} \int_{\mathbb{R}_+^N} \left(\prod_{i=1}^{N} s_i^{\alpha_i} \right) e^{-\left(\sum_{i=1}^{N} \frac{s_i}{\xi_n} \right)} ds_1 ... ds_N = \tag{22.44}$$

$$\int_{\mathbb{R}_+^N} \prod_{i=1}^{N} \left(\frac{s_i}{\xi_n} \right)^{\alpha_i} \prod_{i=1}^{N} e^{-\frac{s_i}{\xi_n}} d\frac{s_1}{\xi_n} ... d\frac{s_N}{\xi_n} = \int_{\mathbb{R}_+^N} \prod_{i=1}^{N} z_i^{\alpha_i} \prod_{i=1}^{N} e^{-z_i} \prod_{i=1}^{N} dz_i =$$

$$\prod_{i=1}^{N} \int_0^\infty z_i^{\alpha_i} e^{-z_i} dz_i = \prod_{i=1}^{N} \int_0^\infty z_i^{(\alpha_i+1)-1} e^{-z_i} dz_i \tag{22.45}$$

(by [5], p. 348)

$$= \prod_{i=1}^{N} \Gamma(\alpha_i + 1) \le \Gamma^N(m+1) = (m!)^N,$$

where Γ is the gamma function.

The last is valid because: if $\alpha_i = 0$, then $\Gamma(0+1) = 1$; if $\alpha_i \in \mathbb{N}$, then $2 \le \alpha_i + 1 \le m + 1$, and $\Gamma(\alpha_i + 1) \le \Gamma(m+1) = m!$.

We took into account that Γ has minimum at $x_{\min} = 1.46...$, with $\Gamma(x_{\min}) = 0.885...$, and $\Gamma(x)$ is increasing for $x \ge x_{\min}$.

The theorem is proved. ■

22.3 Main Results for $P_{r,n}^{[m]}$

22.3.1 Uniform Approximation

We start with an application to $P_{r,n}^{[m]}$ of the following theorem.

Theorem 22.14 ([2], p. 11) *Let $m \in \mathbb{N}$, $f \in C^m(\mathbb{R}^N)$, $N \ge 1$, $x \in \mathbb{R}^N$. Assume $\left\| \frac{\partial^m f(\cdot,\cdot,...,\cdot)}{\partial x_1^{\alpha_1} ... \partial x_N^{\alpha_N}} \right\|_\infty < \infty$, for all $\alpha_j \in \mathbb{Z}^+$, $j = 1, ..., N : |\alpha| := \sum_{i=1}^{N} \alpha_i = m$. Let μ_{ξ_n} be a Borel probability measure on $\mathbb{R}^N$, for $\xi_n > 0$, $(\xi_n)_{n\in\mathbb{N}}$ bounded sequence.*

Suppose that for all $\alpha := (\alpha_1, ..., \alpha_N)$, $\alpha_i \in \mathbb{Z}^+$, $i = 1, ..., N$, $|\alpha| := \sum_{i=1}^{N} \alpha_i = m$ we have that

$$u_{\xi_n}(\alpha) := \int_{\mathbb{R}^N} \left(\prod_{i=1}^{N} |s_i|^{\alpha_i} \right) \left(1 + \frac{\|s\|_2}{\xi_n} \right)^r d\mu_{\xi_n}(s) < \infty. \tag{22.46}$$

For $\widetilde{j} = 1, ..., m$, *and* $\alpha := (\alpha_1, ..., \alpha_N)$, $\alpha_i \in \mathbb{Z}^+$, $i = 1, ..., N$, $|\alpha| := \sum_{j=1}^{N} \alpha_j = \widetilde{j}$, *call*

$$c_{\alpha,n} := c_{\alpha,n,\widetilde{j}} := \int_{\mathbb{R}^N} \prod_{i=1}^{N} s_i^{\alpha_i} d\mu_{\xi_n}(s_1, ..., s_N). \tag{22.47}$$

Then

(i)

$$E_{r,n}^{[m]}(x) := \left| \theta_{r,n}^{[m]}(f; x) - f(x) - \sum_{\widetilde{j}=1}^{m} \delta_{\widetilde{j},r}^{[m]} \left(\sum_{\substack{\alpha_1,...,\alpha_N \geq 0; \\ |\alpha|=\widetilde{j}}} \frac{c_{\alpha,n,\widetilde{j}} f_\alpha(x)}{\prod_{i=1}^{N} \alpha_i!} \right) \right|$$

$$\leq \sum_{\substack{\alpha_1,...,\alpha_N \geq 0; \\ |\alpha|=m}} \frac{(\omega_r(f_\alpha, \xi_n))}{\left(\prod_{i=1}^{N} \alpha_i \right)} \left(\int_{\mathbb{R}^N} \left(\prod_{i=1}^{N} |s_i|^{\alpha_i} \right) \left(1 + \frac{\|s\|_2}{\xi_n} \right)^r d\mu_{\xi_n}(s) \right). \tag{22.48}$$

$\forall\, x \in \mathbb{R}^N$.

(ii)

$$\left\| E_{r,n}^{[m]} \right\|_\infty \leq R.H.S.(22.48). \tag{22.49}$$

Given that $\xi_n \to 0$, *as* $n \to \infty$, *and* u_{ξ_n} *is uniformly bounded, then we derive that* $\left\| E_{r,n}^{[m]} \right\| \to 0$ *with rates.*

(iii) It holds also that

$$\left\| \theta_{r,n}^{[m]}(f) - f \right\|_\infty \leq \sum_{\widetilde{j}=1}^{m} \delta_{\widetilde{j},r}^{[m]} \left(\sum_{\substack{\alpha_1,...,\alpha_N \geq 0; \\ |\alpha|=\widetilde{j}}} \frac{\left| c_{\alpha,n,\widetilde{j}} \right| \|f_\alpha\|_\infty}{\prod_{i=1}^{N} \alpha_i!} \right) + R.H.S.(22.48), \tag{22.50}$$

given that $\|f_\alpha\|_\infty < \infty$, *for all* $\alpha : |\alpha| = \widetilde{j}$, $\widetilde{j} = 1, ..., m$. *Furthermore, as* $\xi_n \to 0$ *when* $n \to \infty$, *assuming that* $c_{\alpha,n,\widetilde{j}} \to 0$, *while* u_{ξ_n} *is uniformly bounded, we conclude that*

$$\left\| \theta_{r,n}^{[m]}(f) - f \right\|_\infty \to 0 \tag{22.51}$$

with rates.

A uniform approximation result for $P_{r,n}^{[m]}$ follows:

Theorem 22.15 *Let* $m \in \mathbb{N}$, $f \in C^m\left(\mathbb{R}^N\right)$, $N \geq 1$, $x \in \mathbb{R}^N$. *Assume* $\left\| \frac{\partial^m f(\cdot,\cdot,...,\cdot)}{\partial x_1^{\alpha_1} ... \partial x_N^{\alpha_N}} \right\|_\infty < \infty$, *for all* $\alpha_j \in \mathbb{Z}^+$, $j = 1, ..., N : |\alpha| := \sum_{j=1}^{N} \alpha_j = m$. *Here* $\xi_n \in (0, 1]$, $n \in \mathbb{N}$.

For all $\alpha := (\alpha_1, ..., \alpha_N)$, $\alpha_i \in \mathbb{Z}^+$, $i = 1, ..., N$, $|\alpha| := \sum_{i=1}^{N} \alpha_i = m$, *we denote*

$$u_{\xi_n}^{*}(\alpha) := \frac{1}{(2\xi_n)^N} \int_{\mathbb{R}^N} \left(\prod_{i=1}^{N} |s_i|^{\alpha_i} \right) \left(1 + \frac{\|s\|_2}{\xi_n} \right)^r e^{-\frac{\left(\sum_{i=1}^{N} |s_i|\right)}{\xi_n}} ds_1 ... ds_N. \quad (22.52)$$

For $\widetilde{j} = 1, ..., m$, *and* $\alpha := (\alpha_1, ..., \alpha_N)$, $\alpha_i \in \mathbb{Z}^+$, $i = 1, ..., N$, $|\alpha| := \sum_{i=1}^{N} \alpha_i = \widetilde{j}$, *call*

$$c_{\alpha,n}^{*} := c_{\alpha,n,\widetilde{j}}^{*} := \frac{1}{(2\xi_n)^N} \int_{\mathbb{R}^N} \prod_{i=1}^{N} s_i^{\alpha_i} e^{-\frac{\left(\sum_{i=1}^{N} |s_i|\right)}{\xi_n}} ds_1 ... ds_N. \quad (22.53)$$

Then

(i)

$$E_{r,n}^{*[m]}(x) := \left| P_{r,n}^{[m]}(f; x) - f(x) - \sum_{\widetilde{j}=1}^{m} \delta_{\widetilde{j},r}^{[m]} \left(\sum_{\substack{\alpha_1,...,\alpha_N \geq 0; \\ |\alpha| = \widetilde{j}}} \frac{c_{\alpha,n,\widetilde{j}}^{*} f_\alpha(x)}{\prod_{i=1}^{N} \alpha_i!} \right) \right|$$

$$\leq \sum_{\substack{\alpha_1,...,\alpha_N \geq 0; \\ |\alpha| = m}} \frac{(\omega_r(f_\alpha, \xi_n))}{\left(\prod_{i=1}^{N} \alpha_i! \right)} u_{\xi_n}^{*}(\alpha), \quad (22.54)$$

$\forall\, x \in \mathbb{R}^N$.

(ii)

$$\left\| E_{r,n}^{*[m]} \right\|_\infty \leq R.H.S.(22.54). \quad (22.55)$$

Given that $\xi_n \to 0$, *as* $n \to +\infty$, *we have that* $u_{\xi_n}^{*}(\alpha) \to 0$ *and are uniformly bounded, and then we derive that* $\left\| E_{r,n}^{*[m]} \right\|_\infty \to 0$ *with rates.*

(iii) It holds also that

$$\left\| P_{r,n}^{[m]}(f) - f \right\|_\infty \leq \sum_{\widetilde{j}=1}^{m} \left| \delta_{\widetilde{j},r}^{[m]} \right| \left(\sum_{\substack{\alpha_1,...,\alpha_N \geq 0; \\ |\alpha| = \widetilde{j}}} \frac{\left| c_{\alpha,n,\widetilde{j}}^{*} \right| \|f_\alpha\|_\infty}{\prod_{i=1}^{N} \alpha_i!} \right) + R.H.S.(22.54), \quad (22.56)$$

given that $\|f_\alpha\|_\infty < +\infty$, *for all* $\alpha : |\alpha| = \widetilde{j}$, $\widetilde{j} = 1, ..., m$. *Furthermore, as* $\xi_n \to 0$ *when* $n \to +\infty$, *we have that* $c_{\alpha,n,\widetilde{j}}^{*} \to 0$ *and* $u_{\xi_n}^{*}(\alpha) \to 0$, *and both are uniformly bounded, and we conclude that*

$$\left\| P_{r,n}^{[m]}(f) - f \right\|_{\infty} \to 0 \tag{22.57}$$

with rates.

Proof Mainly by applying Theorem 22.14. By Theorem 22.9 we get that $u_{\xi_n}^{*}(\alpha) \to 0$ and $u_{\xi_n}^{*}(\alpha)$ are uniformly bounded. By Theorem 22.13 we get that

$$\left|c_{\alpha,n}^{*}\right| = \left|c_{\alpha,n,\tilde{j}}^{*}\right| \leq \frac{1}{(2\xi_n)^N} \int_{\mathbb{R}^N} \left(\prod_{i=1}^{N} |s_i|^{\alpha_i}\right) e^{-\frac{\left(\sum_{i=1}^{N} |s_i|\right)}{\xi_n}} ds_1 ... ds_N$$

$$\leq (m!)^N \xi_n^m \leq (m!)^N .$$

That is $c_{\alpha,n,\tilde{j}}^{*} \to 0$ and are uniformly bounded. The proof is complete. ■

We mention

Theorem 22.16 ([2], p. 14) *Let $f \in C_B(\mathbb{R}^N)$, uniformly continuous, $N \geq 1$, $\xi_n \in (0, 1]$. Then*

$$\left\|\theta_{r,n}^{[0]} f - f\right\|_{\infty} \leq \left(\int_{\mathbb{R}^N} \left(1 + \frac{\|s\|_2}{\xi_n}\right)^r d\mu_{\xi_n}(s)\right) \omega_r(f, \xi_n), \tag{22.58}$$

under the assumption

$$\Phi_{\xi_n} := \int_{\mathbb{R}^N} \left(1 + \frac{\|s\|_2}{\xi_n}\right)^r d\mu_{\xi_n}(s) < \infty. \tag{22.59}$$

As $n \to \infty$ and $\xi_n \to 0$, given that Φ_{ξ_n} are uniformly bounded, we derive

$$\left\|\theta_{r,n}^{[0]} f - f\right\|_{\infty} \to 0 \tag{22.60}$$

with rates.

We give

Theorem 22.17 *Let $f \in C_B(\mathbb{R}^N)$, uniformly continuous, $N \geq 1$, $\xi_n \in (0, 1]$. Then*

$$\left\| P_{r,n}^{[0]} f - f \right\|_{\infty} \leq \left[(1+N)^r + 2^r \left(\frac{\lfloor er! \rfloor}{e}\right)^N\right] \omega_r(f, \xi_n), \tag{22.61}$$

where $r, n \in \mathbb{N}$.

As $n \to \infty$ and $\xi_n \to 0$, we derive

$$\left\| P_{r,n}^{[0]} f - f \right\|_{\infty} \to 0 \tag{22.62}$$

with rates.

Proof Use of Theorems 22.10 and 22.16. ∎

22.3.2 L_p Approximation for $P_{r,n}^{[m]}$

We need

Definition 22.18 ([4, 6]) We call

$$\Delta_u^r f(x) := \Delta_{u_1,u_2,\ldots,u_N}^r f(x_1, \ldots, x_N) := \tag{22.63}$$

$$\sum_{j=0}^{r} (-1)^{r-j} \binom{r}{j} f(x_1 + ju_1, x_2 + ju_2, \ldots, x_N + ju_N).$$

Let $p \geq 1$, the modulus of smoothness of order r is given by

$$\omega_r(f; h)_p := \sup_{\|u\|_2 \leq h} \left\| \Delta_u^r(f) \right\|_p, \tag{22.64}$$

$h > 0$.

We will apply

Theorem 22.19 ([2], p. 24) *Let $f \in C^m(\mathbb{R}^N)$, $m \in \mathbb{N}$, $N \geq 1$, with $f_\alpha \in L_p(\mathbb{R}^N)$, $|\alpha| = m$, $x \in \mathbb{R}^N$. Let $p, q > 1 : \frac{1}{p} + \frac{1}{q} = 1$. Here, μ_{ξ_n} is a Borel probability measure on $\mathbb{R}^N$ for $\xi_n > 0$, $(\xi_n)_{n\in\mathbb{N}}$ bounded sequence. Assume for all $\alpha := (\alpha_1, \ldots, \alpha_N)$, $\alpha_i \in \mathbb{Z}^+$, $i = 1, \ldots, N$, $|\alpha| := \sum_{i=1}^{N} \alpha_i = m$ that we have*

$$\int_{\mathbb{R}^N} \left(\left(\prod_{i=1}^{N} |s_i|^{\alpha_i} \right) \left(1 + \frac{\|s\|_2}{\xi_n} \right)^r \right)^p d\mu_{\xi_n}(s) < \infty. \tag{22.65}$$

For $\widetilde{j} = 1, \ldots, m$, and $\alpha := (\alpha_1, \ldots, \alpha_N)$, $\alpha_i \in \mathbb{Z}^+$, $i = 1, \ldots, N$, $|\alpha| := \sum_{j=1}^{N} \alpha_j = \widetilde{j}$, call

$$c_{\alpha,n,\widetilde{j}} := \int_{\mathbb{R}^N} \prod_{i=1}^{N} s_i^{\alpha_i} d\mu_{\xi_n}(s). \tag{22.66}$$

Then

$$\left\| E_{r,n}^{[m]} \right\|_p = \left\| \theta_{r,n}^{[m]}(f;x) - f(x) - \sum_{\widetilde{j}=1}^{m} \delta_{\widetilde{j},r}^{[m]} \left(\sum_{|\alpha|=\widetilde{j}} \frac{c_{\alpha,n,\widetilde{j}} f_\alpha(x)}{\left(\prod_{i=1}^{N} \alpha_i!\right)} \right) \right\|_{p,x} \tag{22.67}$$

$$\leq \left(\frac{m}{(q(m-1)+1)^{\frac{1}{q}}} \right) \left(\sum_{|\alpha|=m} \frac{1}{\prod_{i=1}^{N} \alpha_i!} \right)$$

$$\left[\int_{\mathbb{R}^N} \left[\left(\prod_{i=1}^{N} |s_i|^{\alpha_i} \right) \left(1 + \frac{\|s\|_2}{\xi_n} \right)^r \right]^p d\mu_{\xi_n}(s) \right]^{\frac{1}{p}} \omega_r(f_\alpha, \xi_n)_p .$$

As $n \to \infty$ and $\xi_n \to 0$, by (22.67) we obtain that $\left\| E_{r,n}^{[m]} \right\| \to 0$ with rates.

One also finds by (22.67) that

$$\left\| \theta_{r,n}^{[m]}(f;x) - f(x) \right\|_{p,x} \leq \sum_{\widetilde{j}=1}^{m} \left| \delta_{\widetilde{j},r}^{[m]} \right| \left(\sum_{|\alpha|=\widetilde{j}} \frac{\left| c_{\alpha,n,\widetilde{j}} \right|}{\prod_{i=1}^{N} \alpha_i!} \|f_\alpha\|_p \right) + R.H.S.(22.67),$$

given that $\|f_\alpha\|_p < \infty$, $|\alpha| = \widetilde{j}$, $\widetilde{j} = 1, ..., m$.

Assuming that $c_{\alpha,n,\widetilde{j}} \to 0$, $\xi_n \to 0$, as $n \to \infty$, we get $\left\| \theta_{r,n}^{[m]}(f) - f \right\|_p \to 0$, that is $\theta_{r,n}^{[m]} \to I$ the unit operator, in L_p norm, with rates.

We present

Theorem 22.20 *Let $f \in C^m(\mathbb{R}^N)$, $m \in \mathbb{N}$, $N \geq 1$, with $f_\alpha \in L_p(\mathbb{R}^N)$, $|\alpha| = m$, $x \in \mathbb{R}^N$. Let $p, q > 1 : \frac{1}{p} + \frac{1}{q} = 1$, $\xi_n \in (0,1]$. For $\widetilde{j} = 1, ..., m$, and $\alpha := (\alpha_1, ..., \alpha_N)$, $\alpha_i \in \mathbb{Z}^+$, $i = 1, ..., N$, $|\alpha| := \sum_{i=1}^{N} \alpha_i = \widetilde{j}$, call*

$$c_{\alpha,n,\widetilde{j}}^{*} = \frac{1}{(2\xi_n)^N} \int_{\mathbb{R}^N} \prod_{i=1}^{N} s_i^{\alpha_i} e^{-\frac{\left(\sum_{i=1}^{N}|s_i|\right)}{\xi_n}} ds_1 ... ds_N . \tag{22.68}$$

Then

$$\left\| E_{r,n}^{*[m]} \right\|_p = \left\| P_{r,n}^{[m]}(f;x) - f(x) - \sum_{\widetilde{j}=1}^{m} \delta_{\widetilde{j},r}^{[m]} \left(\sum_{|\alpha|=\widetilde{j}} \frac{c_{\alpha,n,\widetilde{j}}^{*} f_\alpha(x)}{\left(\prod_{i=1}^{N} \alpha_i!\right)} \right) \right\|_{p,x}$$

$$\leq \left(\frac{m}{(q(m-1)+1)^{\frac{1}{q}}} \right) \left[(1+N)^{rp} + 2^{rp} \Gamma^N ((m+r)p+1,1) \right]^{\frac{1}{p}} \tag{22.69}$$

$$\left(\sum_{|\alpha|=m} \frac{1}{\prod_{i=1}^{N} \alpha_i!} \right) \xi_n^m \omega_r (f_\alpha, \xi_n)_p .$$

As $n \to +\infty$ and $\xi_n \to 0$, by (22.69) we obtain that $\left\| E_{r,n}^{*[m]} \right\|_p \to 0$ with rates.

One also finds by (22.69) that

$$\left\| P_{r,n}^{[m]} (f; x) - f(x) \right\|_{p,x} \leq \sum_{\widetilde{j}=1}^{m} \left| \delta_{\widetilde{j},r}^{[m]} \right| \left(\sum_{|\alpha|=\widetilde{j}} \frac{\left| c_{\alpha,n,\widetilde{j}}^{*} \right|}{\prod_{i=1}^{N} \alpha_i!} \| f_\alpha \|_p \right) + R.H.S.(22.69), \tag{22.70}$$

given that $\| f_\alpha \|_p < \infty$, $|\alpha| = \widetilde{j}$, $\widetilde{j} = 1, ..., m$.

Assuming that $\xi_n \to 0$, as $n \to \infty$, we get $\left\| P_{r,n}^{[m]} (f) - f \right\|_p \to 0$, that is $P_{r,n}^{[m]} \to I$ the unit operator, in L_p norm, with rates.

Proof By Theorem 22.13 we get that $\left| c_{\alpha,n,\widetilde{j}}^{*} \right| \leq (m!)^N \xi_n$ and $c_{\alpha,n,\widetilde{j}}^{*} \to 0$. We use also Theorem 22.11 to uniformly bound $\frac{1}{(2\xi_n)^N} \int_{\mathbb{R}^N} \left(\left(\prod_{i=1}^{N} |s_i|^{\alpha_i} \right) \left(1 + \frac{\|s\|_2}{\xi_n} \right)^r \right)^p$ $e^{-\frac{\left(\sum_{i=1}^{N} |s_i| \right)}{\xi_n}} ds_1 ... ds_N$. Then we apply directly Theorem 22.19. ■

We continue with an application of

Theorem 22.21 ([2], p. 26) *Let $f \in \left(C\left(\mathbb{R}^N\right) \cap L_p\left(\mathbb{R}^N\right) \right)$; $N \geq 1$; $p, q > 1 : \frac{1}{p} + \frac{1}{q} = 1$. Assume μ_{ξ_n} probability Borel measure on $\mathbb{R}^N$, $(\xi_n)_{n \in \mathbb{N}} > 0$ and bounded. Also suppose*

$$\int_{\mathbb{R}^N} \left(1 + \frac{\|s\|_2}{\xi_n} \right)^{rp} d\mu_{\xi_n}(s) < \infty. \tag{22.71}$$

Then

$$\left\| \theta_{r,n}^{[0]} (f) - f \right\|_p \leq \tag{22.72}$$

$$\left(\int_{\mathbb{R}^N} \left(1 + \frac{\|s\|_2}{\xi_n} \right)^{rp} d\mu_{\xi_n}(s) \right)^{\frac{1}{p}} \omega_r (f, \xi_n)_p .$$

As $\xi_n \to 0$, when $n \to \infty$, we derive $\left\| \theta_{r,n}^{[0]} (f) - f \right\|_p \to 0$, i.e. $\theta_{r,n}^{[0]} \to I$, the unit operator, in L_p norm.

We give

Theorem 22.22 *Let* $f \in \left(C\left(\mathbb{R}^N\right) \cap L_p\left(\mathbb{R}^N\right)\right)$*;* $N \geq 1$*;* $p, q > 1 : \frac{1}{p} + \frac{1}{q} = 1$*,* $\xi_n \in (0, 1]$*,* $n \in \mathbb{N}$*. Then*

$$\left\| P_{r,n}^{[0]}(f) - f \right\|_p \leq \tag{22.73}$$

$$\left[(1+N)^{rp} + 2^{rp}\Gamma^N(rp+1, 1)\right]^{\frac{1}{p}} \omega_r(f, \xi_n)_p .$$

As $\xi_n \to 0$*, when* $n \to \infty$*, we derive* $\left\| P_{r,n}^{[0]}(f) - f \right\|_p \to 0$*, i.e.* $P_{r,n}^{[0]} \to I$*, the unit operator, in* L_p *norm.*

Proof By Theorems 22.21 and 22.12. ■

We mention

Theorem 22.23 ([2], p. 27) *Let* $f \in \left(C\left(\mathbb{R}^N\right) \cap L_1\left(\mathbb{R}^N\right)\right)$*;* $N \geq 1$*. Assume* μ_{ξ_n} *probability Borel measure on* $\mathbb{R}^N$*,* $(\xi_n)_{n\in\mathbb{N}} > 0$ *and bounded. Also suppose*

$$\int_{\mathbb{R}^N} \left(1 + \frac{\|s\|_2}{\xi_n}\right)^r d\mu_{\xi_n}(s) < \infty. \tag{22.74}$$

Then

$$\left\| \theta_{r,n}^{[0]}(f) - f \right\|_1 \leq \tag{22.75}$$

$$\left(\int_{\mathbb{R}^N} \left(1 + \frac{\|s\|_2}{\xi_n}\right)^r d\mu_{\xi_n}(s)\right) \omega_r(f, \xi_n)_1 .$$

As $\xi_n \to 0$*, we get* $\theta_{r,n}^{[0]} \to I$*, in* L_1 *norm.*

We give

Theorem 22.24 *Let* $f \in \left(C\left(\mathbb{R}^N\right) \cap L_1\left(\mathbb{R}^N\right)\right)$*;* $N \geq 1$*,* $\xi_n \in (0, 1]$*,* $n \in \mathbb{N}$*. Then*

$$\left\| P_{r,n}^{[0]}(f) - f \right\|_1 \leq \left[(1+N)^r + 2^r \left(\frac{\lfloor er! \rfloor}{e}\right)^N\right] \omega_r(f, \xi_n)_1 . \tag{22.76}$$

As $\xi_n \to 0$*, we get* $P_{r,n}^{[0]} \to I$*, in* L_1 *norm.*

Proof By Theorems 22.10 and 22.23. ■

We mention

Theorem 22.25 ([2], p. 29) *Let* $f \in C^m\left(\mathbb{R}^N\right)$*,* $m, N \in \mathbb{N}$*, with* $f_\alpha \in L_1\left(\mathbb{R}^N\right)$*,* $|\alpha| = m$*,* $x \in \mathbb{R}^N$ *. Here,* μ_{ξ_n} *is a Borel probability measure on* $\mathbb{R}^N$ *for* $\xi_n > 0$*,* $(\xi_n)_{n\in\mathbb{N}}$ *is a bounded sequence. Suppose for all* $\alpha := (\alpha_1, ..., \alpha_N)$*,* $\alpha_i \in \mathbb{Z}^+$*,* $i = 1, ..., N$*,* $|\alpha| := \sum_{i=1}^{N} \alpha_i = m$ *that we have*

$$\int_{\mathbb{R}^N} \left(\left(\prod_{i=1}^{N} |s_i|^{\alpha_i} \right) \left(1 + \frac{\|s\|_2}{\xi_n} \right)^r \right) d\mu_{\xi_n}(s) < \infty. \tag{22.77}$$

For $\widetilde{j} = 1, ..., m$, *and* $\alpha := (\alpha_1, ..., \alpha_N)$, $\alpha_i \in \mathbb{Z}^+$, $i = 1, ..., N$, $|\alpha| := \sum_{i=1}^{N} \alpha_i = \widetilde{j}$, *call*

$$c_{\alpha,n,\widetilde{j}} := \int_{\mathbb{R}^N} \prod_{i=1}^{N} s_i^{\alpha_i} d\mu_{\xi_n}(s). \tag{22.78}$$

Then

$$\left\| E_{r,n}^{[m]} \right\|_1 = \left\| \theta_{r,n}^{[m]}(f; x) - f(x) - \sum_{\widetilde{j}=1}^{m} \delta_{\widetilde{j},r}^{[m]} \left(\sum_{|\alpha|=\widetilde{j}} \frac{c_{\alpha,n,\widetilde{j}} f_\alpha(x)}{\prod_{i=1}^{N} \alpha_i!} \right) \right\|_{1,x} \tag{22.79}$$

$$\leq \sum_{|\alpha|=m} \left(\frac{1}{\prod_{i=1}^{N} \alpha_i!} \right) \omega_r(f_\alpha, \xi_n)_1 \int_{\mathbb{R}^N} \left(\prod_{i=1}^{N} |s_i|^{\alpha_i} \right) \left(1 + \frac{\|s\|_2}{\xi_n} \right)^r d\mu_{\xi_n}(s).$$

As $\xi_n \to 0$, *we get* $\left\| E_{r,n}^{[m]} \right\|_1 \to 0$ *with rates.*

From (22.79) we get

$$\left\| \theta_{r,n}^{[m]} f - f \right\|_1 \leq \sum_{\widetilde{j}=1}^{m} \left| \delta_{\widetilde{j},r}^{[m]} \right| \left(\sum_{|\alpha|=\widetilde{j}} \frac{\left| c_{\alpha,n,\widetilde{j}} \right|}{\prod_{i=1}^{N} \alpha_i!} \|f_\alpha\|_1 \right) + R.H.S.(22.79), \tag{22.80}$$

given that $\|f_\alpha\|_1 < \infty$, $|\alpha| = \widetilde{j}$, $\widetilde{j} = 1, ..., m$.

As $n \to \infty$, *assuming* $\xi_n \to 0$ *and* $c_{\alpha,n,\widetilde{j}} \to 0$, *we obtain* $\left\| \theta_{r,n}^{[m]}(f) - f \right\|_1 \to 0$, *that is* $\theta_{r,n}^{[m]} \to I$ *in* L_1 *norm, with rates.*

We present

Theorem 22.26 *Let* $f \in C^m(\mathbb{R}^N)$, $m, N \in \mathbb{N}$, *with* $f_\alpha \in L_1(\mathbb{R}^N)$, $|\alpha| = m$, $x \in \mathbb{R}^N$, $\xi_n \in (0, 1]$, $n \in \mathbb{N}$. *For* $\widetilde{j} = 1, ..., m$, *and* $\alpha := (\alpha_1, ..., \alpha_N)$, $\alpha_i \in \mathbb{Z}^+$, $i = 1, ..., N$, $|\alpha| := \sum_{i=1}^{N} \alpha_i = \widetilde{j}$, *call*

$$c_{\alpha,n,\widetilde{j}}^{*} := \frac{1}{(2\xi_n)^N} \int_{\mathbb{R}^N} \prod_{i=1}^{N} s_i^{\alpha_i} e^{-\frac{\left(\sum_{i=1}^{N} |s_i|\right)}{\xi_n}} ds_1 ... ds. \tag{22.81}$$

Then

$$\left\| E_{r,n}^{*[m]} \right\|_1 = \left\| P_{r,n}^{[m]}(f;x) - f(x) - \sum_{\widetilde{j}=1}^{m} \delta_{\widetilde{j},r}^{[m]} \left(\sum_{|\alpha|=\widetilde{j}} \frac{c_{\alpha,n,\widetilde{j}}^{*} f_\alpha(x)}{\prod_{i=1}^{N} \alpha_i!} \right) \right\|_{1,x}$$

$$\leq \left[(1+N)^r + 2^r \left(\frac{\lfloor e(m+r)! \rfloor}{e} \right)^N \right] \sum_{|\alpha|=m} \left(\frac{1}{\prod_{i=1}^{N} \alpha_i!} \right) \xi_n^m \omega_r(f_\alpha, \xi_n)_1 . \quad (22.82)$$

As $\xi_n \to 0$, *we get* $\left\| E_{r,n}^{*[m]} \right\|_1 \to 0$ *with rates.*

From (22.82) we get

$$\left\| P_{r,n}^{[m]} f - f \right\|_1 \leq \sum_{\widetilde{j}=1}^{m} \left| \delta_{\widetilde{j},r}^{[m]} \right| \left(\sum_{|\alpha|=\widetilde{j}} \frac{\left| c_{\alpha,n,\widetilde{j}}^{*} \right|}{\prod_{i=1}^{N} \alpha_i!} \| f_\alpha \|_1 \right) + R.H.S.(22.82), \quad (22.83)$$

given that $\| f_\alpha \|_1 < \infty$, $|\alpha| = \widetilde{j}$, $\widetilde{j} = 1, \ldots, m$.

As $n \to \infty$, *assuming* $\xi_n \to 0$, *we obtain* $\left\| P_{r,n}^{[m]}(f) - f \right\|_1 \to 0$, *that is* $P_{r,n}^{[m]} \to I$ *in* L_1 *norm, with rates.*

Proof We apply Theorems 22.9, 22.13 and 22.25. ■

22.3.3 Global Smoothness Preservation and Simultaneous Approximation of $P_{r,n}^{[m]}$

We need

Definition 22.27 ([2], *p. 34*) Let $f \in C(\mathbb{R}^N)$, $N \geq 1$, $m \in \mathbb{N}$, the mth modulus of smoothness for $1 \leq p \leq \infty$, is given by

$$\omega_m(f;h)_p := \sup_{\|t\|_2 \leq h} \left\| \Delta_t^m(f) \right\|_{p,x}, \quad (22.84)$$

$h > 0$, where

$$\Delta_t^m f(x) := \sum_{j=0}^{m} (-1)^{m-j} \binom{m}{j} f(x+jt). \quad (22.85)$$

Denote

$$\omega_m(f;h)_\infty = \omega_m(f,h). \tag{22.86}$$

Above, $x, t \in \mathbb{R}^N$.

We present the related global smoothness preservation result

Theorem 22.28 *We assume $P_{r,n}^{[\widetilde{m}]}(f;x) \in \mathbb{R}$, $\widetilde{m} \in \mathbb{Z}_+$, $\forall\, x \in \mathbb{R}$. Let $h > 0$, $f \in C(\mathbb{R}^N)$, $N \geq 1$.*

(i) Assume $\omega_m(f,h) < \infty$. Then

$$\omega_m\left(P_{r,n}^{[\widetilde{m}]} f, h\right) \leq \left(\sum_{\widetilde{j}=0}^{r} \left|\alpha_{\widetilde{j},r}^{[\widetilde{m}]}\right|\right) \omega_m(f,h). \tag{22.87}$$

(ii) Assume $f \in \left(C(\mathbb{R}^N) \cap L_1(\mathbb{R}^N)\right)$. Then

$$\omega_m\left(P_{r,n}^{[\widetilde{m}]} f, h\right)_1 \leq \left(\sum_{\widetilde{j}=0}^{r} \left|\alpha_{\widetilde{j},r}^{[\widetilde{m}]}\right|\right) \omega_m(f,h)_1. \tag{22.88}$$

(iii) Assume $f \in \left(C(\mathbb{R}^N) \cap L_p(\mathbb{R}^N)\right)$, $p > 1$. Then

$$\omega_m\left(P_{r,n}^{[\widetilde{m}]} f, h\right)_p \leq \left(\sum_{\widetilde{j}=0}^{r} \left|\alpha_{\widetilde{j},r}^{[\widetilde{m}]}\right|\right) \omega_m(f,h)_p. \tag{22.89}$$

Proof Direct application of ([2]) Theorem 3.2, p. 35. ∎

We make

Remark 22.29 Let $r = 1$, $\widetilde{m} \in \mathbb{Z}_+$, then $\alpha_{0,1}^{[\widetilde{m}]} = 0$, $\alpha_{1,1}^{[\widetilde{m}]} = 1$. Hence

$$P_{1,n}^{[\widetilde{m}]}(f;x) = \frac{1}{(2\xi_n)^N} \int_{\mathbb{R}^N} f(x+s)\, e^{-\left(\frac{\sum_{i=1}^{N}|s_i|}{\xi_n}\right)} ds_1 \ldots ds_N =: P_n(f;x). \tag{22.90}$$

By Theorem 22.28, we get

Theorem 22.30 *We suppose $P_n(f;x) \in \mathbb{R}$, $\forall\, x \in \mathbb{R}$. Let $h > 0$, $f \in C(\mathbb{R}^N)$, $N \geq 1$.*

(i) Assume $\omega_m(f,h) < \infty$. Then

$$\omega_m(P_n f, h) \leq \omega_m(f,h). \tag{22.91}$$

(ii) Assume $f \in \left(C\left(\mathbb{R}^N\right) \cap L_1\left(\mathbb{R}^N\right)\right)$. *Then*

$$\omega_m\left(P_n f, h\right)_1 \leq \omega_m\left(f, h\right)_1 . \tag{22.92}$$

(iii) Assume $f \in \left(C\left(\mathbb{R}^N\right) \cap L_p\left(\mathbb{R}^N\right)\right)$, $p > 1$. *Then*

$$\omega_m\left(P_n f, h\right)_p \leq \omega_m\left(f, h\right)_p . \tag{22.93}$$

Next, we get an optimality result

Proposition 22.31 *Above inequality (22.91):*

$$\omega_m\left(P_n f, h\right) \leq \omega_m\left(f, h\right)$$

is sharp, namely it is attained by any

$$f_j^*(x) = x_j^m, \ \ j = 1, ..., N, \ \ x = \left(x_1, ..., x_j, ..., x_N\right) \in \mathbb{R}^N. \tag{22.94}$$

Proof Apply Proposition 3.5, p. 38, of [2]. ■

We need

Theorem 22.32 ([2], p. 39) *Let* $f \in C^l\left(\mathbb{R}^N\right)$, $l, N \in \mathbb{N}$. *Here,* μ_{ξ_n} *is a Borel probability measure on* $\mathbb{R}^N$, $\xi_n > 0$, $(\xi_n)_{n\in\mathbb{N}}$ *a bounded sequence. Let* $\beta := (\beta_1, ..., \beta_N)$, $\beta_i \in \mathbb{Z}^+$, $i=1, ..., N$; $|\beta| := \sum_{i=1}^{N} \beta_i = l$. *Here* $f(x + sj)$, $x, s \in \mathbb{R}^N$, *is* μ_{ξ_n}*-integrable wrt* s, *for* $j = 1, ..., r$. *There exist* μ_{ξ_n}*-integrable functions* $h_{i_1,j}$, $h_{\beta_1,i_2,j}$, $h_{\beta_1,\beta_2,i_3,j}$, ..., $h_{\beta_1,\beta_2,...,\beta_{N-1},i_N,j} \geq 0$ *(*$j = 1, ..., r$*) on* $\mathbb{R}^N$ *such that*

$$\left|\frac{\partial^{i_1} f(x + sj)}{\partial x_1^{i_1}}\right| \leq h_{i_1,j}(s), \quad i_1 = 1, ..., \beta_1, \tag{22.95}$$

$$\left|\frac{\partial^{\beta_1+i_2} f(x + sj)}{\partial x_2^{i_2} \partial x_1^{\beta_1}}\right| \leq h_{\beta_1,i_2,j}(s), \quad i_2 = 1, ..., \beta_2,$$

$$\vdots$$

$$\left|\frac{\partial^{\beta_1+\beta_2+...+\beta_{N-1}+i_N} f(x + sj)}{\partial x_N^{i_N} \partial x_{N-1}^{\beta_{N-1}} ... \partial x_2^{\beta_2} \partial x_1^{\beta_1}}\right| \leq h_{\beta_1,\beta_2,...,\beta_{N-1},i_N,j}(s), \quad i_N = 1, ..., \beta_N,$$

$\forall \, x, s \in \mathbb{R}^N$.

Then, both of the next exist and

$$\left(\theta_{r,n}^{[\widetilde{m}]}(f; x)\right)_\beta = \theta_{r,n}^{[\widetilde{m}]}\left(f_\beta; x\right), \quad \widetilde{m} \in \mathbb{Z}_+. \tag{22.96}$$

In particular it holds

$$\left(P_{r,n}^{[\widetilde{m}]}(f;x)\right)_{\beta} = P_{r,n}^{[\widetilde{m}]}\left(f_{\beta};x\right), \tag{22.97}$$

when

$$d\mu_{\xi_n} = \frac{1}{(2\xi_n)^N} e^{-\left(\frac{\sum_{i=1}^{N}|s_i|}{\xi_n}\right)} ds_1...ds_N.$$

Corollary 22.33 (by Theorem *22.32, r* = 1) *It holds*

$$\left(P_n(f;x)\right)_{\beta} = P_n\left(f_{\beta};x\right). \tag{22.98}$$

We present simultaneous global smoothness results.

Theorem 22.34 *Let $h > 0$ and the assumptions of Theorem 22.32 are valid for $d\mu_{\xi_n} = \frac{1}{(2\xi_n)^N} e^{-\left(\frac{\sum_{i=1}^{N}|s_i|}{\xi_n}\right)} ds_1...ds_N$. Here $\gamma = 0, \beta$ $(0 = (0, ..., 0))$, $\widetilde{m} \in \mathbb{Z}_+$.*

(i) Assume $\omega_m\left(f_{\gamma}, h\right) < \infty$. Then

$$\omega_m\left(\left(P_{r,n}^{[\widetilde{m}]}(f)\right)_{\gamma}, h\right) \leq \left(\sum_{\widetilde{j}=0}^{r}\left|\alpha_{\widetilde{j},r}^{[\widetilde{m}]}\right|\right)\omega_m\left(f_{\gamma}, h\right). \tag{22.99}$$

(ii) Additionally suppose $f_{\gamma} \in L_1\left(\mathbb{R}^N\right)$. Then

$$\omega_m\left(\left(P_{r,n}^{[\widetilde{m}]}(f)\right)_{\gamma}, h\right)_1 \leq \left(\sum_{\widetilde{j}=0}^{r}\left|\alpha_{\widetilde{j},r}^{[\widetilde{m}]}\right|\right)\omega_m\left(f_{\gamma}, h\right)_1. \tag{22.100}$$

(iii) Additionally suppose $f_{\gamma} \in L_p\left(\mathbb{R}^N\right)$, $p > 1$. Then

$$\omega_m\left(\left(P_{r,n}^{[\widetilde{m}]}(f)\right)_{\gamma}, h\right)_p \leq \left(\sum_{\widetilde{j}=0}^{r}\left|\alpha_{\widetilde{j},r}^{[\widetilde{m}]}\right|\right)\omega_m\left(f_{\gamma}, h\right)_p. \tag{22.101}$$

It follows

Corollary 22.35 (to Theorem *22.34) Let $h > 0$, $r = 1$ and $\gamma = 0, \beta$.*

(i) Assume $\omega_m\left(f_{\gamma}, h\right) < \infty$. Then

$$\omega_m\left(\left(P_n(f)\right)_{\gamma}, h\right) \leq \omega_m\left(f_{\gamma}, h\right). \tag{22.102}$$

(ii) Additionally suppose $f_\gamma \in L_1(\mathbb{R}^N)$. *Then*

$$\omega_m\left(\left(P_n(f)\right)_\gamma, h\right)_1 \leq \omega_m\left(f_\gamma, h\right)_1. \tag{22.103}$$

(iii) Additionally suppose $f_\gamma \in L_p(\mathbb{R}^N)$, $p > 1$. *Then*

$$\omega_m\left(\left(P_n(f)\right)_\gamma, h\right)_p \leq \omega_m\left(f_\gamma, h\right)_p. \tag{22.104}$$

Next comes multi-simultaneous approximation.

Theorem 22.36 *Let* $f \in C^{m+l}(\mathbb{R}^N)$, $m, l, N \in \mathbb{N}$. *The assumptions of Theorem 22.32 are valid for* $d\mu_{\xi_n} = \frac{1}{(2\xi_n)^N} e^{-\left(\frac{\sum_{i=1}^N |s_i|}{\xi_n}\right)} ds_1...ds_N$. *Call* $\gamma = 0, \beta$. *Assume* $\left\| f_{\gamma+\alpha} \right\|_\infty < \infty$, *and let*

$$u^*_{\xi_n}(\alpha) := \frac{1}{(2\xi_n)^N} \int_{\mathbb{R}^N} \left(\prod_{i=1}^N |s_i|^{\alpha_i}\right) \left(1 + \frac{\|s\|_2}{\xi_n}\right)^r e^{-\frac{\left(\sum_{i=1}^N |s_i|\right)}{\xi_n}} ds_1...ds_N,$$

for all $\alpha_j \in \mathbb{Z}^+$, $j = 1, ..., N$, $|\alpha| := \sum_{j=1}^N \alpha_j = m$, $\xi_n \in (0, 1]$.

For $\widetilde{j} = 1, ..., m$, *and* $\alpha := (\alpha_1, ..., \alpha_N)$, $\alpha_i \in \mathbb{Z}^+$, $i = 1, ..., N$, $|\alpha| := \sum_{i=1}^N \alpha_i = \widetilde{j}$, *call*

$$c^*_{\alpha,n,\widetilde{j}} := \frac{1}{(2\xi_n)^N} \int_{\mathbb{R}^N} \prod_{i=1}^N s_i^{\alpha_i} e^{-\frac{\left(\sum_{i=1}^N |s_i|\right)}{\xi_n}} ds_1...ds_N. \tag{22.105}$$

Then

$$\left\| \left(P^{[m]}_{r,n}(f;\cdot)\right)_\gamma - f_\gamma(\cdot) - \sum_{\widetilde{j}=1}^m \delta^{[m]}_{\widetilde{j},r} \left(\sum_{\substack{\alpha_1,...,\alpha_N \geq 0; \\ |\alpha| = \widetilde{j}}} \frac{c^*_{\alpha,n,\widetilde{j}} f_{\gamma+\alpha}(\cdot)}{\prod_{i=1}^N \alpha_i!} \right) \right\|_\infty \tag{22.106}$$

$$\leq \sum_{\substack{\alpha_1,...,\alpha_N \geq 0; \\ |\alpha| = m}} \frac{\left(\omega_r\left(f_{\gamma+\alpha}, \xi_n\right)\right)}{\left(\prod_{i=1}^N \alpha_i!\right)} u^*_{\xi_n}(\alpha).$$

Proof Based on Theorems 22.15 and 22.32. ■

We continue with

Theorem 22.37 *Let $f \in C_B^l(\mathbb{R}^N)$, $l, N \in \mathbb{N}$ (functions l-times continuously differentiable and bounded). The assumptions of Theorem 22.32 are valid for $d\mu_{\xi_n} = \frac{1}{(2\xi_n)^N} e^{-\left(\frac{\sum_{i=1}^N |s_i|}{\xi_n}\right)} ds_1 ... ds_N$. Call $\gamma = 0, \beta$, $\xi_n \in (0, 1]$. Then*

$$\left\| \left(P_{r,n}^{[0]} f\right)_\gamma - f_\gamma \right\|_\infty \leq \left[(1+N)^r + 2^r \left(\frac{\lfloor er! \rfloor}{e} \right)^N \right] \omega_r \left(f_\gamma, \xi_n \right). \tag{22.107}$$

If additionally f_γ is uniformly continuous and $\xi_n \to 0$, the $\left(P_{r,n}^{[0]} f\right)_\gamma \to f_\gamma$ uniformly, as $n \to \infty$.

Proof By Theorems 22.17 and 22.32. ■

We continue with

Theorem 22.38 *Let $f \in C^{m+l}(\mathbb{R}^N)$, $m, l, N \in \mathbb{N}$. The assumptions of Theorem 22.32 are valid for $d\mu_{\xi_n} = \frac{1}{(2\xi_n)^N} e^{-\left(\frac{\sum_{i=1}^N |s_i|}{\xi_n}\right)} ds_1 ... ds_N$. Call $\gamma = 0, \beta$. Let $f_{(\gamma+\alpha)} \in L_p(\mathbb{R}^N)$, $|\alpha| = m$, $x \in \mathbb{R}^N$, and $p, q > 1 : \frac{1}{p} + \frac{1}{q} = 1$, $\xi_n \in (0, 1]$, $n \in \mathbb{N}$. For $\widetilde{j} = 1, ..., m$, and $\alpha := (\alpha_1, ..., \alpha_N)$, $\alpha_i \in \mathbb{Z}^+$, $i = 1, ..., N$, $|\alpha| := \sum_{i=1}^N \alpha_i = \widetilde{j}$, call*

$$c_{\alpha,n,\widetilde{j}}^* := \frac{1}{(2\xi_n)^N} \int_{\mathbb{R}^N} \prod_{i=1}^N s_i^{\alpha_i} e^{-\frac{\left(\sum_{i=1}^N |s_i|\right)}{\xi_n}} ds_1 ... ds_N. \tag{22.108}$$

Then

$$\left\| \left(P_{r,n}^{[m]}(f;x)\right)_\gamma - f_\gamma(x) - \sum_{\widetilde{j}=1}^m \delta_{\widetilde{j},r}^{[m]} \left(\sum_{|\alpha|=\widetilde{j}} \frac{c_{\alpha,n,\widetilde{j}}^* f_{\gamma+\alpha}(x)}{\prod_{i=1}^N \alpha_i!} \right) \right\|_{p,x}$$

$$\leq \left(\frac{m}{(q(m-1)+1)^{\frac{1}{q}}} \right) \left[(1+N)^{rp} + 2^{rp} \Gamma^N((m+r)p+1, 1) \right]^{\frac{1}{p}}$$

$$\left(\sum_{|\alpha|=m} \frac{1}{\prod_{i=1}^N \alpha_i!} \right) \xi_n^m \omega_r \left(f_{\gamma+\alpha}, \xi_n \right)_p. \tag{22.109}$$

Proof By Theorems 22.20 and 22.32. ■

We give also

Theorem 22.39 *Let $f \in C^l(\mathbb{R}^N)$, $l, N \in \mathbb{N}$. The assumptions of Theorem 22.32 are valid for* $d\mu_{\xi_n} = \frac{1}{(2\xi_n)^N} e^{-\left(\frac{\sum_{i=1}^N |s_i|}{\xi_n}\right)} ds_1 ... ds_N$. *Call* $\gamma = 0, \beta$. *Let* $f_\gamma \in L_p(\mathbb{R}^N)$, $x \in \mathbb{R}^N$; $p, q > 1 : \frac{1}{p} + \frac{1}{q} = 1$, $\xi_n \in (0, 1]$, $n \in \mathbb{N}$. *Then*

$$\left\| \left(P_{r,n}^{[0]} f\right)_\gamma - f_\gamma \right\|_p \leq$$

$$\left[(1+N)^{rp} + 2^{rp} \Gamma^N (rp+1, 1)\right]^{\frac{1}{p}} \omega_r \left(f_\gamma, \xi_n\right)_p . \tag{22.110}$$

As $n \to +\infty$ *and* $\xi_n \to 0$, *then* $\left(P_{r,n}^{[0]}(f)\right)_\gamma \overset{\|\cdot\|_p}{\to} f_\gamma$.

Proof By Theorems 22.22 and 22.32. ■

We continue with

Theorem 22.40 *Let $f \in C^l(\mathbb{R}^N)$, $l, N \in \mathbb{N}$. The assumptions of Theorem 22.32 are valid for* $d\mu_{\xi_n} = \frac{1}{(2\xi_n)^N} e^{-\left(\frac{\sum_{i=1}^N |s_i|}{\xi_n}\right)} ds_1 ... ds_N$. *Call* $\gamma = 0, \beta$. *Let* $f_\gamma \in L_1(\mathbb{R}^N)$, $x \in \mathbb{R}^N$, $\xi_n \in (0, 1]$, $n \in \mathbb{N}$. *Then*

$$\left\| \left(P_{r,n}^{[0]}(f)\right)_\gamma - f_\gamma \right\|_1 \leq$$

$$\left[(1+N)^r + 2^r \left(\frac{\lfloor er! \rfloor}{e}\right)^N\right] \omega_r \left(f_\gamma, \xi_n\right)_1 . \tag{22.111}$$

As $n \to +\infty$ *and* $\xi_n \to 0$, *then* $\left(P_{r,n}^{[0]}(f)\right)_\gamma \overset{\|\cdot\|_1}{\to} f_\gamma$.

Proof By Theorems 22.24 and 22.32. ■

We continue with

Theorem 22.41 *Let $f \in C^{m+l}(\mathbb{R}^N)$, $m, l, N \in \mathbb{N}$. The assumptions of Theorem 22.32 for* $d\mu_{\xi_n} = \frac{1}{(2\xi_n)^N} e^{-\left(\frac{\sum_{i=1}^N |s_i|}{\xi_n}\right)} ds_1 ... ds_N$ *are valid. Call* $\gamma = 0, \beta$. *Let* $f_{(\gamma+\alpha)} \in L_1(\mathbb{R}^N)$, $|\alpha| = m$, $x \in \mathbb{R}^N$, $\xi_n \in (0, 1]$, $n \in \mathbb{N}$. *For* $\widetilde{j} = 1, ..., m$, *and* $\alpha := (\alpha_1, ..., \alpha_N)$, $\alpha_i \in \mathbb{Z}^+$, $i = 1, ..., N$, $|\alpha| := \sum_{i=1}^N \alpha_i = \widetilde{j}$, *call*

$$c^{*}_{\alpha,n,\widetilde{j}} := \frac{1}{(2\xi_n)^N} \int_{\mathbb{R}^N} \prod_{i=1}^{N} s_i^{\alpha_i} e^{-\frac{\left(\sum_{i=1}^{N}|s_i|\right)}{\xi_n}} ds_1 ... ds_N. \tag{22.112}$$

Then

$$\left\| \left(P_{r,n}^{[m]}(f;x)\right)_{\gamma} - f_{\gamma}(x) - \sum_{\widetilde{j}=1}^{m} \delta_{\widetilde{j},r}^{[m]} \left(\sum_{|\alpha|=\widetilde{j}} \frac{c^{*}_{\alpha,n,\widetilde{j}} f_{(\gamma+\alpha)}(x)}{\prod_{i=1}^{N} \alpha_i!} \right) \right\|_{1,x}$$

$$\leq \left(\sum_{|\alpha|=m} \left(\frac{1}{\prod_{i=1}^{N} \alpha_i!} \right) \omega_r \left(f_{\gamma+\alpha}, \xi_n\right)_1 \right) \xi_n^m \left[(1+N)^r + 2^r \left(\frac{\lfloor e(m+r)! \rfloor}{e} \right)^N \right]. \tag{22.113}$$

Proof By Theorems 22.26 and 22.32. ∎

22.3.4 Voronovskaya Asymptotic Expansions for $P_{r,n}^{[m]}$

We will apply

Theorem 22.42 ([2], p. 53) *Let $f \in C^m\left(\mathbb{R}^N\right)$, $m, N \in \mathbb{N}$, with all $\|f_{\alpha}\|_{\infty} \leq M$, $M > 0$, all $\alpha : |\alpha| = m$. Let $\xi_n > 0$, $(\xi_n)_{n\in\mathbb{N}}$ bounded sequence, μ_{ξ_n} probability Borel measures on $\mathbb{R}^N$.*

Call $c_{\alpha,n,\widetilde{j}} := \int_{\mathbb{R}^N} \left(\prod_{i=1}^{N} s_i^{\alpha_i}\right) d\mu_{\xi_n}(s)$, all $|\alpha| = \widetilde{j} = 1, ..., m-1$. Suppose $\xi_n^{-m} \int_{\mathbb{R}^N} \left(\prod_{i=1}^{N} |s_i|^{\alpha_i}\right) d\mu_{\xi_n}(s) \leq \rho$, all $\alpha : |\alpha| = m$, $\rho > 0$, for any such $(\xi_n)_{n\in\mathbb{N}}$. Also $0 < \gamma \leq 1$, $x \in \mathbb{R}^N$. Then

$$\theta_{r,n}^{[m]}(f;x) - f(x) = \sum_{\widetilde{j}=1}^{m-1} \delta_{\widetilde{j},r}^{[m]} \left(\sum_{|\alpha|=\widetilde{j}} \frac{c_{\alpha,n,\widetilde{j}} f_{\alpha}(x)}{\left(\prod_{i=1}^{N} \alpha_i!\right)} \right) + 0\left(\xi_n^{m-\gamma}\right). \tag{22.114}$$

When $m = 1$, the sum collapses.

Above we assume $\theta_{r,n}^{[m]}(f;x) \in \mathbb{R}$, $\forall\, x \in \mathbb{R}^N$.

We give

Theorem 22.43 *Let $f \in C^m\left(\mathbb{R}^N\right)$, $m, N \in \mathbb{N}$, with all $\|f_\alpha\|_\infty \leq M$, $M > 0$, all $\alpha : |\alpha| = m$, $\xi_n \in (0,1]$, and $d\mu_{\xi_n} = \frac{1}{(2\xi_n)^N} e^{-\left(\frac{\sum_{i=1}^N |s_i|}{\xi_n}\right)} ds_1 \ldots ds_N$ on $\mathbb{R}^N$. Assume $P_{r,n}^{[m]}(f;x) \in \mathbb{R}$, $\forall\, x \in \mathbb{R}^N$.*

Call $c^*_{\alpha,n,\widetilde{j}} = \frac{1}{(2\xi_n)^N} \int_{\mathbb{R}^N} \left(\prod_{i=1}^N s_i^{\alpha_i}\right) e^{-\left(\frac{\sum_{i=1}^N |s_i|}{\xi_n}\right)} ds_1 \ldots ds_N$, all $|\alpha| = \widetilde{j} = 1, \ldots, m-1$. Let $0 < \gamma \leq 1$, $x \in \mathbb{R}^N$. Then

$$P_{r,n}^{[m]}(f;x) - f(x) = \sum_{\widetilde{j}=1}^{m-1} \delta_{\widetilde{j},r}^{[m]} \left(\sum_{|\alpha|=\widetilde{j}} \frac{c^*_{\alpha,n,\widetilde{j}} f_\alpha(x)}{\left(\prod_{i=1}^N \alpha_i!\right)} \right) + 0\left(\xi_n^{m-\gamma}\right). \quad (22.115)$$

When $m = 1$, the sum collapses.

Proof By Theorems 22.13 and 22.42. Here $\rho = (m!)^N$. ■

Corollary 22.44 (to Theorem 22.43) *Let $f \in C^1\left(\mathbb{R}^N\right)$, $N \geq 1$, with all $\left\|\frac{\partial f}{\partial x_i}\right\|_\infty \leq M$, $M > 0$, $i = 1, \ldots, N$. Let $0 < \gamma \leq 1$, $x \in \mathbb{R}^N$. Then*

$$P_{r,n}^{[1]}(f;x) - f(x) = 0\left(\xi_n^{1-\gamma}\right). \quad (22.116)$$

Above we assume $P_{r,n}^{[1]}(f;x) \in \mathbb{R}$, $\forall\, x \in \mathbb{R}^N$.

Proof By Theorems 22.13 and 22.42, here it is $\rho = 1$. ■

Corollary 22.45 (to Theorem 22.43) *Let $f \in C^2\left(\mathbb{R}^2\right)$, with all $\left\|\frac{\partial^2 f}{\partial x_1^2}\right\|_\infty$, $\left\|\frac{\partial^2 f}{\partial x_2^2}\right\|_\infty$, $\left\|\frac{\partial^2 f}{\partial x_1 \partial x_2}\right\|_\infty \leq M$, $M > 0$, $\xi_n \in (0,1]$, $n \in \mathbb{N}$. Call*

$$c_1 = \int_{\mathbb{R}^2} s_1 d\mu_{\xi_n}(s), \quad c_2 = \int_{\mathbb{R}^2} s_2 d\mu_{\xi_n}(s), \quad (22.117)$$

where

$$d\mu_{\xi_n} = \frac{1}{4\xi_n^2} e^{-\frac{(|s_1|+|s_2|)}{\xi_n}} ds_1 ds_2.$$

Let $0 < \gamma \leq 1$, $x \in \mathbb{R}^2$. Then

$$P_{r,n}^{[2]}(f;x) - f(x) = \left(\sum_{j=1}^r \alpha_{j,r}^{[2]} j\right) \left(c_1 \frac{\partial f}{\partial x_1}(x) + c_2 \frac{\partial f}{\partial x_2}(x)\right) + 0\left(\xi_n^{2-\gamma}\right). \quad (22.118)$$

Proof By Theorems 22.13 and 22.42, here it is $\rho = 4$. ■

We also give

Theorem 22.46 *Let $f \in C^{m+l}\left(\mathbb{R}^N\right)$, $m, l, N \in \mathbb{N}$. Assumptions of Theorem 22.32 are valid for $d\mu_{\xi_n}(s) = \frac{1}{(2\xi_n)^N} e^{-\left(\frac{\sum_{i=1}^N |s_i|}{\xi_n}\right)} ds_1 \dots ds_N$. Call $\gamma = 0, \beta$. Suppose $\left\| f_{\gamma+\alpha} \right\|_\infty \leq M$, $M > 0$, for all $\alpha : |\alpha| = m$, $\xi_n \in (0, 1]$, $n \in \mathbb{N}$.*

Call $c^*_{\alpha,n,\widetilde{j}} := \frac{1}{(2\xi_n)^N} \int_{\mathbb{R}^N} \left(\prod_{i=1}^N s_i^{\alpha_i}\right) e^{-\left(\frac{\sum_{i=1}^N |s_i|}{\xi_n}\right)} ds_1 \dots ds_N$, all $|\alpha| = \widetilde{j} = 1, \dots, m - 1$; $0 < \gamma \leq 1$, $x \in \mathbb{R}^N$. Then

$$\left(P_{r,n}^{[m]}(f; x)\right)_\gamma - f_\gamma(x) = \sum_{\widetilde{j}=1}^{m-1} \delta_{\widetilde{j},r}^{[m]} \left(\sum_{|\alpha|=\widetilde{j}} \frac{c^*_{\alpha,n,\widetilde{j}} f_{\gamma+\alpha}(x)}{\left(\prod_{i=1}^N \alpha_i!\right)} \right) + 0\left(\xi_n^{m-\gamma}\right). \tag{22.119}$$

When $m = 1$, the sum collapses.

Proof Use of Theorem 22.13 and Theorem 4.6, p. 54 of [2], here it is $\rho = (m!)^N$. ■

22.3.5 Simultaneous Approximation by Multivariate Complex $P_{r,n}^{[m]}$

We make

Remark 22.47 We consider here complex valued Borel measurable functions $f : \mathbb{R}^N \to \mathbb{C}$ such that $f = f_1 + if_2$, $i = \sqrt{-1}$, where $f_1, f_2 : \mathbb{R}^N \to \mathbb{R}$ are implied to be real valued Borel measurable functions.

We define the multivariate complex Picard singular operators

$$P_{r,n}^{[m]}(f; x) := P_{r,n}^{[m]}(f_1; x) + i P_{r,n}^{[m]}(f_2; x), \quad x \in \mathbb{R}^N. \tag{22.120}$$

We assume that $P_{r,n}^{[m]}\left(f_j; x\right) \in \mathbb{R}$, $\forall\, x \in \mathbb{R}^N$, $j = 1, 2$.

One notices easily that

$$\left|P_{r,n}^{[m]}(f; x) - f(x)\right| \leq \left|P_{r,n}^{[m]}(f_1; x) - f_1(x)\right| + \left|P_{r,n}^{[m]}(f_2; x) - f_2(x)\right| \tag{22.121}$$

also

$$\left\|P_{r,n}^{[m]}(f; x) - f(x)\right\|_{\infty,x} \leq \left\|P_{r,n}^{[m]}(f_1; x) - f_1(x)\right\|_{\infty,x} + \left\|P_{r,n}^{[m]}(f_2; x) - f_2(x)\right\|_{\infty,x} \tag{22.122}$$

and

$$\left\| P_{r,n}^{[m]}(f) - f \right\|_p \leq \left\| P_{r,n}^{[m]}(f_1) - f_1 \right\|_p + \left\| P_{r,n}^{[m]}(f_2) - f_2 \right\|_p, \quad p \geq 1. \quad (22.123)$$

Furthermore, it holds

$$f_\alpha(x) = f_{1,\alpha}(x) + i f_{2,\alpha}(x), \quad (22.124)$$

where α denotes a partial derivative of any order and arrangement.

We give

Theorem 22.48 *Let $f : \mathbb{R}^N \to \mathbb{C}$, $N \geq 1$, such that $f = f_1 + i f_2$, $j = 1, 2$. Here $m \in \mathbb{N}$, $f_j \in C^m(\mathbb{R}^N)$, $x \in \mathbb{R}^N$. Assume $\left\| \frac{\partial^m f_j(\cdot,\cdot,\ldots,\cdot)}{\partial x_1^{\alpha_1} \ldots \partial x_N^{\alpha_N}} \right\|_\infty < \infty$, for all $\alpha_i \in \mathbb{Z}^+$, $i = 1, \ldots, N : |\alpha| := \sum_{i=1}^N \alpha_i = m$. Here $\xi_n \in (0, 1]$, $n \in \mathbb{N}$. For all $\alpha := (\alpha_1, \ldots, \alpha_N)$, $\alpha_i \in \mathbb{Z}^+$, $i = 1, \ldots, N$, $|\alpha| := \sum_{i=1}^N \alpha_i = m$ we denote*

$$u_{\xi_n}^*(\alpha) := \frac{1}{(2\xi_n)^N} \int_{\mathbb{R}^N} \left(\prod_{i=1}^N |s_i|^{\alpha_i} \right) \left(1 + \frac{\|s\|_2}{\xi_n} \right)^r e^{-\left(\frac{\sum_{i=1}^N |s_i|}{\xi_n} \right)} ds_1 \ldots ds_N. \quad (22.125)$$

For $\widetilde{j} = 1, \ldots, m$, and $\alpha := (\alpha_1, \ldots, \alpha_N)$, $\alpha_i \in \mathbb{Z}^+$, $i = 1, \ldots, N$, $|\alpha| := \sum_{i=1}^N \alpha_i = \widetilde{j}$, call

$$c_{\alpha,n}^* := c_{\alpha,n,\widetilde{j}}^* := \frac{1}{(2\xi_n)^N} \int_{\mathbb{R}^N} \prod_{i=1}^N s_i^{\alpha_i} e^{-\left(\frac{\sum_{i=1}^N |s_i|}{\xi_n} \right)} ds_1 \ldots ds_N. \quad (22.126)$$

Then

$$\left\| P_{r,n}^{[m]}(f; x) - f(x) - \sum_{\widetilde{j}=1}^m \delta_{\widetilde{j},r}^{[m]} \left(\sum_{\substack{\alpha_1,\ldots,\alpha_N \geq 0; \\ |\alpha| = \widetilde{j}}} \frac{c_{\alpha,n,\widetilde{j}}^* f_\alpha(x)}{\prod_{i=1}^N \alpha_i!} \right) \right\|_{\infty,x}$$

$$\leq \sum_{\substack{\alpha_1,\ldots,\alpha_N \geq 0; \\ |\alpha| = m}} \frac{\left(\omega_r(f_{1,\alpha}, \xi_n) + \omega_r(f_{2,\alpha}, \xi_n) \right)}{\left(\prod_{i=1}^N \alpha_i! \right)} u_{\xi_n}^*(\alpha). \quad (22.127)$$

Proof By Theorem 22.15. ■

We proceed with

Theorem 22.49 *Let $f : \mathbb{R}^N \to \mathbb{C} : f = f_1 + i f_2$, $N \geq 1$, $j = 1, 2$. Here $f_j \in C_B(\mathbb{R}^N)$ uniformly continuous, $\xi_n \in (0, 1]$. Then*

$$\left\| P_{r,n}^{[0]} f - f \right\|_\infty \leq \left[(1+N)^r + 2^r \left(\frac{\lfloor er! \rfloor}{e} \right)^N \right] (\omega_r(f_1, \xi_n) + \omega_r(f_2, \xi_n)), \tag{22.128}$$

where $r, n \in \mathbb{N}$.

As $n \to \infty$ and $\xi_n \to 0$, we derive

$$\left\| P_{r,n}^{[0]} f - f \right\|_\infty \to 0 \tag{22.129}$$

with rates.

Proof Use of Theorem 22.17. ■

Next comes multi-simultaneous approximation

Theorem 22.50 *Let $f : \mathbb{R}^N \to \mathbb{C} : f = f_1 + i f_2$, $j = 1, 2$. Here $f_j \in C^{m+l}(\mathbb{R}^N)$, $m, l, N \in \mathbb{N}$. The assumptions of Theorem 22.32 are valid for $d\mu_{\xi_n} = \frac{1}{(2\xi_n)^N} e^{-\left(\frac{\sum_{i=1}^N |s_i|}{\xi_n}\right)} ds_1 ... ds_N$ and f_j. Call $\gamma = 0, \beta$. Assume $\left\| f_{j,\gamma+\alpha} \right\|_\infty < \infty$, and let*

$$u_{\xi_n}^*(\alpha) := \frac{1}{(2\xi_n)^N} \int_{\mathbb{R}^N} \left(\prod_{i=1}^N |s_i|^{\alpha_i} \right) \left(1 + \frac{\|s\|_2}{\xi_n} \right)^r e^{-\left(\frac{\sum_{i=1}^N |s_i|}{\xi_n}\right)} ds_1 ... ds_N. \tag{22.130}$$

for all $\alpha_i \in \mathbb{Z}^+$, $i = 1, ..., N : |\alpha| := \sum_{i=1}^N \alpha_i = m$, $\xi_n \in (0, 1]$.

For $\widetilde{j} = 1, ..., m$, and $\alpha := (\alpha_1, ..., \alpha_N)$, $\alpha_i \in \mathbb{Z}^+$, $i = 1, ..., N$, $|\alpha| := \sum_{i=1}^N \alpha_i = \widetilde{j}$, call

$$c_{\alpha,n,\widetilde{j}}^* := \frac{1}{(2\xi_n)^N} \int_{\mathbb{R}^N} \prod_{i=1}^N s_i^{\alpha_i} e^{-\left(\frac{\sum_{i=1}^N |s_i|}{\xi_n}\right)} ds_1 ... ds_N. \tag{22.131}$$

Then

$$\left\| \left(P_{r,n}^{[m]}(f;\cdot)\right)_\gamma - f_\gamma(\cdot) - \sum_{\widetilde{j}=1}^{m} \delta_{\widetilde{j},r}^{[m]} \left(\sum_{\substack{\alpha_1,\dots,\alpha_N \geq 0; \\ |\alpha| = \widetilde{j}}} \frac{c_{\alpha,n,\widetilde{j}}^{*} f_{\gamma+\alpha}(\cdot)}{\prod_{i=1}^{N} \alpha_i!} \right) \right\|_\infty$$

$$\leq \sum_{\substack{\alpha_1,\dots,\alpha_N \geq 0; \\ |\alpha| = m}} \frac{\left(\omega_r\left(f_{1,\gamma+\alpha}, \xi_n\right) + \omega_r\left(f_{2,\gamma+\alpha}, \xi_n\right)\right)}{\left(\prod_{i=1}^{N} \alpha_i!\right)} u_{\xi_n}^{*}(\alpha). \quad (22.132)$$

Proof Based on Theorems 22.32 and 22.36. ■

We continue with

Theorem 22.51 *Let $f : \mathbb{R}^N \to \mathbb{C} : f = f_1 + i f_2$, $j = 1, 2$. Here $f_j \in C_B^l\left(\mathbb{R}^N\right)$, $l, N \in \mathbb{N}$ (functions l-times continuously differentiable and bounded). The assumptions of Theorem 22.32 are valid for $d\mu_{\xi_n} = \frac{1}{(2\xi_n)^N} e^{-\left(\frac{\sum_{i=1}^{N}|s_i|}{\xi_n}\right)} ds_1 \dots ds_N$ and f_j, $j = 1, 2$. Call $\gamma = 0, \beta$, $\xi_n \in (0, 1]$. Then*

$$\left\| \left(P_{r,n}^{[0]} f\right)_\gamma - f_\gamma \right\|_\infty \leq \left[(1+N)^r + 2^r \left(\frac{\lfloor er! \rfloor}{e}\right)^N \right] \left(\omega_r\left(f_{1,\gamma}, \xi_n\right) + \omega_r\left(f_{2,\gamma}, \xi_n\right)\right). \quad (22.133)$$

If additionally $f_{j,\gamma}$, $j = 1, 2$, is uniformly continuous and $\xi_n \to 0$, then $\left(P_{r,n}^{[0]} f\right)_\gamma \to f_\gamma$ uniformly, as $n \to \infty$.

Proof By Theorems 22.32 and 22.37. ■

We proceed with L_p approximations

Theorem 22.52 *Let $f : \mathbb{R}^N \to \mathbb{C} : f = f_1 + i f_2$, $j = 1, 2$. Here $f_j \in C^m\left(\mathbb{R}^N\right)$, $m \in \mathbb{N}$, $N \geq 1$, with $f_{j,\alpha} \in L_p\left(\mathbb{R}^N\right)$, $|\alpha| = m$, $x \in \mathbb{R}^N$. Let $p, q > 1 : \frac{1}{p} + \frac{1}{q} = 1$, $\xi_n \in (0, 1]$. For $\widetilde{j} = 1, \dots, m$, and $\alpha := (\alpha_1, \dots, \alpha_N)$, $\alpha_i \in \mathbb{Z}^+$, $i = 1, \dots, N$, $|\alpha| := \sum_{i=1}^{N} \alpha_i = \widetilde{j}$, call*

$$c_{\alpha,n,\widetilde{j}}^{*} = \frac{1}{(2\xi_n)^N} \int_{\mathbb{R}^N} \prod_{i=1}^{N} s_i^{\alpha_i} e^{-\frac{\left(\sum_{i=1}^{N}|s_i|\right)}{\xi_n}} ds_1 \dots ds. \quad (22.134)$$

Then

$$\left\| P_{r,n}^{[m]}(f;x) - f(x) - \sum_{\widetilde{j}=1}^{m} \delta_{\widetilde{j},r}^{[m]} \left(\sum_{|\alpha|=\widetilde{j}} \frac{c_{\alpha,n,\widetilde{j}}^{*} f_{\alpha}(x)}{\left(\prod_{i=1}^{N} \alpha_i!\right)} \right) \right\|_{p,x}$$

$$\leq \left(\frac{m}{(q(m-1)+1)^{\frac{1}{q}}} \right) \left[(1+N)^{rp} + 2^{rp} \Gamma^{N}((m+r)p+1,1) \right]^{\frac{1}{p}}$$

$$\left(\sum_{|\alpha|=m} \frac{1}{\prod_{i=1}^{N} \alpha_i!} \right) \xi_n^m \left[\omega_r(f_{1,\alpha},\xi_n)_p + \omega_r(f_{2,\alpha},\xi_n)_p \right]. \quad (22.135)$$

Proof By Theorem 22.20. ∎

We continue with

Theorem 22.53 *Let* $f : \mathbb{R}^N \to \mathbb{C} : f = f_1 + i f_2$, $j = 1, 2$. *Here* $f_j \in \left(C\left(\mathbb{R}^N\right) \cap L_p\left(\mathbb{R}^N\right)\right)$*;* $N \geq 1$*;* $p, q > 1 : \frac{1}{p} + \frac{1}{q} = 1$, $\xi_n \in (0,1]$, $n \in \mathbb{N}$. *Then*

$$\left\| P_{r,n}^{[0]}(f) - f \right\|_p \leq \quad (22.136)$$

$$\left[(1+N)^{rp} + 2^{rp} \Gamma^{N}(rp+1,1) \right]^{\frac{1}{p}} \left(\omega_r(f_1,\xi_n)_p + \omega_r(f_2,\xi_n)_p \right).$$

As $\xi_n \to 0$*, when* $n \to \infty$*, we derive* $\left\| P_{r,n}^{[0]}(f) - f \right\|_p \to 0$*, i.e.* $P_{r,n}^{[0]} \to I$*, the unit operator, in* L_p *norm.*

Proof By Theorem 22.22. ∎

We also give

Theorem 22.54 *Let* $f : \mathbb{R}^N \to \mathbb{C} : f = f_1 + i f_2$, $j = 1, 2$. *Here* $f_j \in \left(C\left(\mathbb{R}^N\right) \cap L_1\left(\mathbb{R}^N\right)\right)$*;* $N \geq 1$, $\xi_n \in (0,1]$, $n \in \mathbb{N}$. *Then*

$$\left\| P_{r,n}^{[0]}(f) - f \right\|_1 \leq \left[(1+N)^r + 2^r \left(\frac{\lfloor er! \rfloor}{e} \right)^N \right] \left[\omega_r(f_1,\xi_n)_1 + \omega_r(f_2,\xi_n)_1 \right]. \quad (22.137)$$

As $\xi_n \to 0$*, we get* $P_{r,n}^{[0]} \to I$*, in* L_1 *norm.*

Proof By Theorem 22.24. ∎

We further present

Theorem 22.55 *Let* $f : \mathbb{R}^N \to \mathbb{C} : f = f_1 + i f_2$, $j = 1, 2$. *Here* $f_j \in C^m(\mathbb{R}^N)$, $m, N \in \mathbb{N}$, *with* $f_{j,\alpha} \in L_1(\mathbb{R}^N)$, $|\alpha| = m$, $x \in \mathbb{R}^N$, $\xi_n \in (0, 1]$, $n \in \mathbb{N}$. *For* $\widetilde{j} = 1, \ldots, m$, *and* $\alpha := (\alpha_1, \ldots, \alpha_N)$, $\alpha_i \in \mathbb{Z}^+$, $i = 1, \ldots, N$, $|\alpha| := \sum_{i=1}^{N} \alpha_i = \widetilde{j}$, *call*

$$c^*_{\alpha,n,\widetilde{j}} := \frac{1}{(2\xi_n)^N} \int_{\mathbb{R}^N} \prod_{i=1}^{N} s_i^{\alpha_i} e^{-\frac{\left(\sum_{i=1}^{N} |s_i|\right)}{\xi_n}} ds_1 \ldots ds. \tag{22.138}$$

Then

$$\left\| P_{r,n}^{[m]}(f; x) - f(x) - \sum_{\widetilde{j}=1}^{m} \delta_{\widetilde{j},r}^{[m]} \left(\sum_{|\alpha|=\widetilde{j}} \frac{c^*_{\alpha,n,\widetilde{j}} f_\alpha(x)}{\prod_{i=1}^{N} \alpha_i!} \right) \right\|_{1,x} \tag{22.139}$$

$$\leq \left[(1+N)^r + 2^r \left(\frac{\lfloor e(m+r)! \rfloor}{e} \right)^N \right] \sum_{|\alpha|=m} \left(\frac{1}{\prod_{i=1}^{N} \alpha_i!} \right)$$

$$\xi_n^m \left[\omega_r(f_{1,\alpha}, \xi_n)_1 + \omega_r(f_{2,\alpha}, \xi_n)_1 \right].$$

Proof By Theorem 22.26. ∎

We continue with simultaneous L_p approximations.

Theorem 22.56 *Let* $f : \mathbb{R}^N \to \mathbb{C} : f = f_1 + i f_2$, $j = 1, 2$. *Here* $f_j \in C^{m+l}(\mathbb{R}^N)$, $m, l, N \in \mathbb{N}$. *The assumptions of Theorem 22.32 are valid for* $d\mu_{\xi_n} = \frac{1}{(2\xi_n)^N} e^{-\left(\frac{\sum_{i=1}^{N} |s_i|}{\xi_n}\right)} ds_1 \ldots ds_N$ *and* f_j. *Call* $\gamma = 0, \beta$. *Let* $f_{j,(\gamma+\alpha)} \in L_p(\mathbb{R}^N)$, $|\alpha| = m$, $x \in \mathbb{R}^N$, *and* $p, q > 1 : \frac{1}{p} + \frac{1}{q} = 1$, $\xi_n \in (0, 1]$, $n \in \mathbb{N}$. *For* $\widetilde{j} = 1, \ldots, m$, *and* $\alpha := (\alpha_1, \ldots, \alpha_N)$, $\alpha_i \in \mathbb{Z}^+$, $i = 1, \ldots, N$, $|\alpha| := \sum_{i=1}^{N} \alpha_i = \widetilde{j}$, *call*

$$c^*_{\alpha,n,\widetilde{j}} := \frac{1}{(2\xi_n)^N} \int_{\mathbb{R}^N} \prod_{i=1}^{N} s_i^{\alpha_i} e^{-\frac{\left(\sum_{i=1}^{N} |s_i|\right)}{\xi_n}} ds_1 \ldots ds_N. \tag{22.140}$$

Then

$$\left\| \left(P_{r,n}^{[m]}(f;x)\right)_{\gamma} - f_{\gamma}(x) - \sum_{\widetilde{j}=1}^{m} \delta_{\widetilde{j},r}^{[m]} \left(\sum_{|\alpha|=\widetilde{j}} \frac{c_{\alpha,n,\widetilde{j}}^{*} f_{\gamma+\alpha}(x)}{\prod_{i=1}^{N} \alpha_i!} \right) \right\|_{p,x}$$

$$\leq \left(\frac{m}{(q(m-1)+1)^{\frac{1}{q}}} \right) \left[(1+N)^{rp} + 2^{rp}\Gamma^{N}((m+r)p+1,1)\right]^{\frac{1}{p}} \quad (22.141)$$

$$\left(\sum_{|\alpha|=m} \frac{1}{\prod_{i=1}^{N} \alpha_i!} \right) \xi_n^{m} \left[\omega_r\left(f_{1,\gamma+\alpha},\xi_n\right)_p + \omega_r\left(f_{2,\gamma+\alpha},\xi_n\right)_p\right].$$

Proof By Theorems 22.32 and 22.38. ■

We give also

Theorem 22.57 *Let $f:\mathbb{R}^N \to \mathbb{C}: f = f_1 + if_2,\ j=1,2$. Here $f_j \in C^l\left(\mathbb{R}^N\right)$, $l, N \in \mathbb{N}$. The assumptions of Theorem 22.32 are valid for $d\mu_{\xi_n} = \frac{1}{(2\xi_n)^N} e^{-\left(\frac{\sum_{i=1}^{N}|s_i|}{\xi_n}\right)} ds_1...ds_N$ and f_j. Call $\gamma = 0, \beta$. Let $f_{j,\gamma} \in L_p\left(\mathbb{R}^N\right)$, $x \in \mathbb{R}^N$; $p, q > 1: \frac{1}{p} + \frac{1}{q} = 1$, $\xi_n \in (0,1]$, $n \in \mathbb{N}$. Then*

$$\left\| \left(P_{r,n}^{[0]}(f)\right)_{\gamma} - f_{\gamma} \right\|_{p} \leq \quad (22.142)$$

$$\left[(1+N)^{rp} + 2^{rp}\Gamma^{N}(rp+1,1)\right]^{\frac{1}{p}} \left[\omega_r\left(f_{1,\gamma},\xi_n\right)_p + \omega_r\left(f_{2,\gamma},\xi_n\right)_p\right].$$

As $n \to +\infty$ and $\xi_n \to 0$, then $\left(P_{r,n}^{[0]} f\right)_{\gamma} \overset{\|\cdot\|_p}{\to} f_{\gamma}$.

Proof By Theorems 22.32 and 22.39. ■

We continue with

Theorem 22.58 *Let $f:\mathbb{R}^N \to \mathbb{C}: f = f_1 + if_2,\ j=1,2$. Here $f_j \in C^l\left(\mathbb{R}^N\right)$, $l, N \in \mathbb{N}$. The assumptions of Theorem 22.32 are valid for $d\mu_{\xi_n} = \frac{1}{(2\xi_n)^N} e^{-\left(\frac{\sum_{i=1}^{N}|s_i|}{\xi_n}\right)} ds_1...ds_N$ and f_j. Call $\gamma = 0, \beta$. Let $f_{j,\gamma} \in L_1\left(\mathbb{R}^N\right)$, $x \in \mathbb{R}^N$, $\xi_n \in (0,1]$, $n \in \mathbb{N}$. Then*

$$\left\| \left(P_{r,n}^{[0]} (f) \right)_\gamma - f_\gamma \right\|_1 \leq \tag{22.143}$$

$$\left[(1+N)^r + 2^r \left(\frac{\lfloor er! \rfloor}{e} \right)^N \right] \left[\omega_r \left(f_{1,\gamma}, \xi_n \right)_1 + \omega_r \left(f_{2,\gamma}, \xi_n \right)_1 \right].$$

As $n \to +\infty$ *and* $\xi_n \to 0$, *then* $\left(P_{r,n}^{[0]} (f) \right)_\gamma \overset{\|\cdot\|_1}{\to} f_\gamma$.

Proof By Theorems 22.32 and 22.40. ■

We finish with

Theorem 22.59 *Let* $f : \mathbb{R}^N \to \mathbb{C} : f = f_1 + i f_2$, $j = 1, 2$. *Here* $f_j \in C^{m+l} \left(\mathbb{R}^N \right)$, $m, l, N \in \mathbb{N}$. *The assumptions of Theorem 22.32 for* $d\mu_{\xi_n} = \frac{1}{(2\xi_n)^N} e^{-\left(\frac{\sum_{i=1}^N |s_i|}{\xi_n} \right)} ds_1 \ldots ds_N$ *and* f_j *are valid. Call* $\gamma = 0, \beta$. *Let* $f_{j,(\gamma+\alpha)} \in L_1 \left(\mathbb{R}^N \right)$, $|\alpha| = m$, $x \in \mathbb{R}^N$, $\xi_n \in (0, 1]$, $n \in \mathbb{N}$. *For* $\widetilde{j} = 1, \ldots, m$, *and* $\alpha := (\alpha_1, \ldots, \alpha_N)$, $\alpha_i \in \mathbb{Z}^+$, $i = 1, \ldots, N$, $|\alpha| := \sum_{i=1}^N \alpha_i = \widetilde{j}$, *call*

$$c_{\alpha,n,\widetilde{j}}^* := \frac{1}{(2\xi_n)^N} \int_{\mathbb{R}^N} \prod_{i=1}^N s_i^{\alpha_i} e^{-\frac{\left(\sum_{i=1}^N |s_i| \right)}{\xi_n}} ds_1 \ldots ds_N. \tag{22.144}$$

Then

$$\left\| \left(P_{r,n}^{[m]} (f; x) \right)_\gamma - f_\gamma (x) - \sum_{\widetilde{j}=1}^m \delta_{\widetilde{j},r}^{[m]} \left(\sum_{|\alpha| = \widetilde{j}} \frac{c_{\alpha,n,\widetilde{j}}^* f_{\gamma+\alpha} (x)}{\prod_{i=1}^N \alpha_i!} \right) \right\|_{1,x}$$

$$\leq \left(\sum_{|\alpha|=m} \left(\frac{1}{\prod_{i=1}^N \alpha_i!} \right) \left[\omega_r \left(f_{1,\gamma+\alpha}, \xi_n \right)_1 + \omega_r \left(f_{2,\gamma+\alpha}, \xi_n \right)_1 \right] \right)$$

$$\xi_n^m \left[(1+N)^r + 2^r \left(\frac{\lfloor e (m+r)! \rfloor}{e} \right)^N \right]. \tag{22.145}$$

Proof By Theorems 22.32 and 22.41. ■

References

1. Anastassiou, G.A.: Intelligent Mathematics: Computational Analysis. Springer, Heidelberg (2011)
2. Anastassiou, G.A.: Approximation by Multivariate Singular Integrals. Springer, New York (2011)
3. Anastassiou, G.A.: High order approximation by multivariate generalized Picard singular integrals. Submitted (2020)
4. Anastassiou, G., Gal, S.: Approximation Theory. Birkhaüser, Boston (2000)
5. Gradshteyn, I.S., Ryzhik, I.M.: Table of Integrals, Series and Products, 8th edn. Elsevier, Amsterdam (2015)
6. DeVore, R.A., Lorentz, G.G.: Constructive Approximation, vol. 303. Springer, Berlin (1993)

Chapter 23
High Order Approximation with Multivariate Generalized Gauss–Weierstrass Singular Integrals

This research and survey chapter deals exclusively with the study of the approximation of generalized multivariate Gauss–Weierstrass singular integrals to the identity-unit operator. Here we study quantitatively most of their approximation properties. The multivariate generalized Gauss–Weierstrass operators are not in general positive linear operators. In particular we study the rate of convergence of these operators to the unit operator, as well as the related simultaneous approximation. These are given via Jackson type inequalities and by the use of multivariate high order modulus of smoothness of the high order partial derivatives of the involved function. Also we study the global smoothness preservation properties of these operators. These multivariate inequalities are nearly sharp and in one case the inequality is attained, that is sharp. Furthermore we give asymptotic expansions of Voronovskaya type for the error of multivariate approximation. The above properties are studied with respect to L_p norm, $1 \leq p \leq \infty$. It follows [4].

23.1 Introduction

We start with our motivation for this chapter which comes from [7].

In the next we introduce and deal with the smooth Gauss–Weierstrass singular integral operators $W_{r,\xi}(f, x)$ defined as follows.

For $r \in \mathbb{N}$ and $n \in \mathbb{Z}_+$ we set

$$\alpha_j = \begin{cases} (-1)^{r-j} \binom{r}{j} j^{-n}, \ j = 1, ..., r, \\ 1 - \sum\limits_{j=1}^{r} (-1)^{r-j} \binom{r}{j} j^{-n}, \ \ j = 0, \end{cases} \tag{23.1}$$

G. A. Anastassiou, *Constructive Fractional Analysis with Applications*,
Studies in Systems, Decision and Control 362,
https://doi.org/10.1007/978-3-030-71481-9_23

that is $\sum_{j=0}^{r} \alpha_j = 1$.

Let $f \in C^n(\mathbb{R})$ and $f^{(n)} \in L_p(\mathbb{R})$, $1 \leq p < \infty$, we define for $x \in \mathbb{R}$, $\xi > 0$ the Lebesgue integral

$$W_{r,\xi}(f; x) = \frac{1}{\sqrt{\pi \xi}} \int_{-\infty}^{\infty} \left(\sum_{j=0}^{r} \alpha_j f(x + jt) \right) e^{-\frac{t^2}{\xi}} dt. \tag{23.2}$$

Note 23.1 *The operators $W_{r,\xi}$ are not, in general, positive.*

Let $r = 2, n = 3$. Then $\alpha_0 = \frac{23}{8}, \alpha_1 = -2, \alpha_2 = \frac{1}{8}$. Consider $f(t) = t^2 \geq 0$ and $x = 0$. Then

$$W_{r,\xi}\left(t^2; 0\right) = -\frac{3\xi}{4} < 0.$$

We notice by $\frac{1}{\sqrt{\pi\xi}} \int_{-\infty}^{\infty} e^{-\frac{t^2}{\xi}} dt = 1$, that $W_{r,\xi}(c, x) = c$, c constant, and

$$W_{r,\xi}(f; x) - f(x) = \frac{1}{\sqrt{\pi \xi}} \left(\sum_{j=0}^{r} \alpha_j \int_{-\infty}^{\infty} (f(x + jt) - f(x)) e^{-\frac{t^2}{\xi}} dt \right). \tag{23.3}$$

We need the rth L_p- modulus of smoothness

$$\omega_r\left(f^{(n)}, h\right)_p := \sup_{|t| \leq h} \left\| \Delta_t^r f^{(n)}(x) \right\|_{p,x}, \quad h > 0, \tag{23.4}$$

where

$$\Delta_t^r f^{(n)}(x) := \sum_{j=0}^{r} (-1)^{r-j} \binom{r}{j} f^{(n)}(x + jt), \tag{23.5}$$

see [9], p. 44. Here we have that $\omega_r\left(f^{(n)}, h\right)_p < \infty, \ h > 0$.

We need to introduce

$$\delta_k := \sum_{j=1}^{r} \alpha_j j^k, \ k = 1, ..., n \in \mathbb{N}, \tag{23.6}$$

and denote by $\lfloor \cdot \rfloor$ the integral part. Call

$$r(w, x) := \sum_{j=0}^{r} \alpha_j j^n f^{(n)}(x + jw) - \delta_n f^{(n)}(x). \tag{23.7}$$

According to [2], p. 306, [1], we get

$$\tau(w,x) = \Delta_w^r f^{(n)}(x). \tag{23.8}$$

Thus

$$\|\tau(w,x)\|_{p,x} \le w_r\left(f^{(n)}, |w|\right)_p, \quad w \in \mathbb{R}. \tag{23.9}$$

Define

$$R_n(0,t,x) := \int_0^t \frac{(t-w)^{n-1}}{(n-1)!}\tau(w,x)\,dw, \quad n \in \mathbb{N}. \tag{23.10}$$

Using the above terminology [7] we obtain

$$\Delta(x) := W_{r,\xi}(f;x) - f(x) -$$

$$\sum_{m=1}^{\lfloor \frac{n}{2} \rfloor} f^{(2m)}(x)\,\delta_{2m}\frac{(2m-1)(2m-3)\cdot\ldots\cdot 3\cdot 1}{(2m)!}\left(\frac{\xi}{2}\right)^m = R_n^*(x), \tag{23.11}$$

where

$$R_n^*(x) := \frac{1}{\sqrt{\pi\xi}}\int_{-\infty}^{\infty} R_n(0,t,x)\,e^{-\frac{t^2}{\xi}}dt, \quad n \in \mathbb{N}. \tag{23.12}$$

In $\Delta(x)$, see (23.11), the sum collapses when $n = 1$.

We mention

Theorem 23.2 ([7]) *Let $p, q > 1$ such that $\frac{1}{p} + \frac{1}{q} = 1$, $n \in \mathbb{N}$ and the rest as above. Then*

$$\|\Delta(x)\|_p \le \frac{2^{\frac{p+1}{2p}}\tau^{\frac{1}{p}}\xi^{\frac{n}{2}}}{q^{\frac{1}{2q}}\pi^{\frac{1}{2p}}(rp+1)^{\frac{1}{p}}((n-1)!)(q(n-1)+1)^{\frac{1}{q}}}\omega_r\left(f^{(n)}, \sqrt{\xi}\right)_p, \tag{23.13}$$

where

$$\tau := \left[\int_0^{\infty}(1+u)^{rp+1}u^{np-1}e^{-\frac{pu^2}{2}}du - \int_0^{\infty}u^{np-1}e^{-\frac{pu^2}{2}}du\right] < \infty. \tag{23.14}$$

Hence as $\xi \to 0$ we obtain $\|\Delta(x)\|_p \to 0$.

If additionally $f^{(2m)} \in L_p(\mathbb{R})$, $m = 1, 2, \ldots, \lfloor \frac{n}{2} \rfloor$ then $\left\|W_{r,\xi}(f) - f\right\|_p \to 0$, as $\xi \to 0$.

The counterpart of Theorem 23.2 follows, case of $p = 1$.

Theorem 23.3 ([7]) *Let $f \in C^n(\mathbb{R})$ and $f^{(n)} \in L_1(\mathbb{R})$, $n \in \mathbb{N}$. Then*

$$\|\Delta(x)\|_1 \le \frac{2}{(r+1)(n-1)!\sqrt{\pi}}$$

$$\left(\sum_{k=1}^{r+1}\binom{r+1}{k}\right)\left(\prod_{r=1}^{\lfloor\frac{n+k}{2}\rfloor-1}(n+k-2r)\right)\cdot M\cdot\left(\frac{1}{2}\right)^{\lfloor\frac{n+k}{2}\rfloor}\right)\omega_r\left(f^{(n)},\sqrt{\xi}\right)_1\xi^{\frac{n}{2}}, \tag{23.15}$$

where

$$M=\begin{cases}\frac{\sqrt{\pi}}{2}, & n+k \text{ is even}\\ 1, & n+k \text{ is odd}\end{cases}.$$

Hence as $\xi\to 0$ *we obtain* $\|\Delta(x)\|_1\to 0$.

If additionally $f^{(2m)}\in L_1(\mathbb{R})$, $m=1,2,\ldots,\lfloor\frac{n}{2}\rfloor$ *then* $\|W_{r,\xi}(f)-f\|_1\to 0$, *as* $\xi\to 0$.

The case $n=0$ is mentioned next.

Proposition 23.4 ([7]) *Let* $p,q>1$ *such that* $\frac{1}{p}+\frac{1}{q}=1$ *and the rest as above. Then*

$$\|W_{r,\xi}(f)-f\|_p\le\frac{2^{\frac{1}{2q}+\frac{1}{p}}\pi^{\frac{1}{2q}-\frac{1}{2}}}{q^{\frac{1}{2q}}}\theta^{\frac{1}{p}}\omega_r\left(f,\sqrt{\xi}\right)_p, \tag{23.16}$$

where

$$\theta:=\int_0^\infty(1+t)^{rp}e^{-\frac{pt^2}{2}}dt<\infty.$$

Hence as $\xi\to 0$ *we obtain* $W_{r,\xi}(f)\to$ *unit operator* I *in the* L_p *norm,* $p>1$.

We also mention

Proposition 23.5 ([7]) *It holds*

$$\|W_{r,\xi}f-f\|_1\le\frac{2}{\sqrt{\pi}}\left(\int_0^\infty(1+t)^r e^{-t^2}dt\right)\omega_r\left(f,\sqrt{\xi}\right)_1. \tag{23.17}$$

Hence as $\xi\to 0$ *we get* $W_{r,\xi}\to I$ *in the* L_1 *norm.*

In this chapter we study the approximation properties of multivariate smooth general Gauss–Weierstrass singular integral operators:

$$W_{r,n}^{[m]}(f;x_1,\ldots,x_N):=$$

$$\frac{1}{\left(\sqrt{\pi\xi_n}\right)^N}\sum_{j=0}^r\alpha_{j,r}^{[m]}\int_{\mathbb{R}^N}f(x_1+s_1j,x_2+s_2j,\ldots,x_N+s_Nj)\,e^{-\frac{\left(\sum_{i=1}^N s_i^2\right)}{\xi_n}}ds_1\ldots ds_N. \tag{23.18}$$

Notice that

$$\frac{1}{\left(\sqrt{\pi\xi_n}\right)^N}\int_{\mathbb{R}^N}e^{-\frac{\left(\sum_{i=1}^N s_i^2\right)}{\xi_n}}ds_1\ldots ds_N=1, \tag{23.19}$$

see [6].

Here $r \in \mathbb{N}$, $m \in Z_+$, and

$$\alpha_{j,r}^{[m]} := \begin{cases} (-1)^{r-j} \binom{r}{j} j^{-m}, & \text{if } j = 1, 2, ..., r, \\ 1 - \sum_{i=1}^{r} (-1)^{r-i} \binom{r}{i} i^{-m}, & \text{if } j = 0, \end{cases} \tag{23.20}$$

and

$$\delta_{k,r}^{[m]} := \sum_{j=1}^{r} \alpha_{j,r}^{[m]} j^k, \quad k = 1, 2, ..., m \in \mathbb{N}. \tag{23.21}$$

See that $\sum_{j=0}^{r} \alpha_{j,r}^{[m]} = 1$.

Also here $\xi_n \in (0, 1]$, $n \in \mathbb{N}$, and $f : \mathbb{R}^N \to \mathbb{R}$ is a Borel measurable function. The above operator $W_{r,n}^{[m]}$ is a special case of a more general operator $\theta_{r,n}^{[m]}$ studied in general in [3] by the author.

We mention next about $\theta_{r,n}^{[m]}$.

Let μ_{ξ_n} be a probability Borel measure on $\mathbb{R}^N$, $N \geq 1$.

We define the multiple smooth singular integral operators

$$\theta_{r,n}^{[m]} (f; x_1, ..., x_N) := \sum_{j=0}^{r} \alpha_{j,r}^{[m]} \int_{\mathbb{R}^N} f (x_1 + s_1 j, x_2 + s_2 j, ..., x_N + s_N j) \, d\mu_{\xi_n} (s), \tag{23.22}$$

where $s := (s_1, ..., s_N)$, $x := (x_1, ..., x_N) \in \mathbb{R}^N$.

The operators $\theta_{r,n}^{[m]}$ are not in general positive. For example, consider the function $\varphi (u_1, ..., u_N) = \sum_{i=1}^{N} u_i^2$ and also take $r = 2$, $m = 3$; $x_i = 0$, $i = 1, ..., N$. See that $\varphi \geq 0$, however

$$\theta_{2,n}^{[3]} (\varphi; 0, 0, ..., 0) = \left(\sum_{j=1}^{2} j^2 \alpha_{j,2}^{[3]} \right) \int_{\mathbb{R}^N} \left(\sum_{i=1}^{N} s_i^2 \right) d\mu_{\xi_n} (s) =$$

$$\left(\alpha_{1,2}^{[3]} + 4\alpha_{2,2}^{[3]} \right) \int_{\mathbb{R}^N} \left(\sum_{i=1}^{N} s_i^2 \right) d\mu_{\xi_n} (s) = \left(-2 + \frac{1}{2} \right) \int_{\mathbb{R}^N} \left(\sum_{i=1}^{N} s_i^2 \right) d\mu_{\xi_n} (s) < 0. \tag{23.23}$$

assuming that $\int_{\mathbb{R}^N} \left(\sum_{i=1}^{N} s_i^2 \right) d\mu_{\xi_n} (s) < \infty$.

Clearly in the case of $W_{r,n}^{[m]}$ we have that

$$d\mu_{\xi_n}(s) = \frac{1}{\left(\sqrt{\pi\xi_n}\right)^N} e^{-\frac{\left(\sum_{i=1}^N s_i^2\right)}{\xi_n}} ds_1...ds_N, \quad s \in \mathbb{R}^N.$$

Lemma 23.6 *The operator $\theta_{r,n}^{[m]}$ preserve the constant functions in N variables.*

We need

Definition 23.7 Let $f \in C_B\left(\mathbb{R}^N\right)$, the space of all bounded and continuous functions or uniformly continuous on $\mathbb{R}^N$. Then, the rth multivariate modulus of smoothness of f is given by (see, e.g. [5])

$$\omega_r(f;h) := \sup_{\sqrt{u_1^2+...+u_N^2}\leq h} \left\|\Delta_{u_1,u_2,...,u_N}^r(f)\right\|_\infty < \infty, \quad h>0, \tag{23.24}$$

where $\|\cdot\|_\infty$ is the sup-norm and

$$\Delta_u^r f(x) := \Delta_{u_1,u_2,...,u_N}^r f(x_1,...,x_N) =$$

$$\sum_{j=0}^r (-1)^{r-j}\binom{r}{j} f(x_1+ju_1, x_2+ju_2, ..., x_N+ju_N). \tag{23.25}$$

Let $m \in \mathbb{N}$ and let $f \in C^m\left(\mathbb{R}^N\right)$.

Suppose that all partial derivatives of f of order m are bounded, i.e.

$$\left\|\frac{\partial^m f(\cdot,\cdot,...,\cdot)}{\partial x_1^{\alpha_1}...\partial x_N^{\alpha_N}}\right\|_\infty < \infty, \tag{23.26}$$

for all $\alpha_j \in \mathbb{Z}^+$, $j=1,...,N$; $\sum_{j=1}^N \alpha_j = m$.

In this chapter we apply the general theory developed in [3] about $\theta_{r,n}^{[m]}$ to the operators $W_{r,n}^{[m]}$, so we can obtain computationally specific results and show that the general theory has applications and it is a valid theory.

So for the very important in various branches of mathematics operators $W_{r,n}^{[m]}$ we prove the very essential properties of uniform approximation, L_p approximation, global smoothness preservation and simultaneously approximation, Voronovskaya asymptotic expansions and complex simultaneous approximation.

23.2 Auxiliary Essential Results

We need the following:

Theorem 23.8 *Let* $r, N, m \in \mathbb{N}$, *with* $m > r$; $\alpha_j \in \mathbb{Z}^+$, $j = 1, ..., N : |\alpha| := \sum_{j=1}^{N} \alpha_j = m$, $\xi_n \in (0, 1]$, $n \in \mathbb{N}$. *Then*

$$\widetilde{u}_{\xi_n}(\alpha) := \frac{1}{\left(\sqrt{\pi\xi_n}\right)^N} \int_{\mathbb{R}^N} \left(\prod_{i=1}^{N} |s_i|^{\alpha_i}\right) \left(1 + \frac{\|s\|_2}{\xi_n}\right)^r e^{-\frac{\left(\sum_{i=1}^{N} s_i^2\right)}{\xi_n}} ds_1...ds_N$$

$$\leq \left(\sqrt{\xi_n}\right)^{m-r} \left(\frac{2}{\sqrt{\pi}}\right)^N \left[(1+N)^r + 2^r \left(\frac{\lfloor e(m+r)! \rfloor}{e}\right)^N\right]$$

$$\leq \left(\frac{2}{\sqrt{\pi}}\right)^N \left[(1+N)^r + 2^r \left(\frac{\lfloor e(m+r)! \rfloor}{e}\right)^N\right] < +\infty, \tag{23.27}$$

are uniformly bounded.

Proof We estimate

$$\widetilde{u}_{\xi_n}(\alpha) := \frac{1}{\left(\sqrt{\pi\xi_n}\right)^N} \int_{\mathbb{R}^N} \left(\prod_{i=1}^{N} |s_i|^{\alpha_i}\right) \left(1 + \frac{\|s\|_2}{\xi_n}\right)^r e^{-\frac{\left(\sum_{i=1}^{N} s_i^2\right)}{\xi_n}} ds_1...ds_N =$$

$$\frac{2^N}{\left(\sqrt{\pi}\right)^N \left(\sqrt{\xi_n}\right)^N} \int_{\mathbb{R}_+^N} \left(\prod_{i=1}^{N} s_i^{\alpha_i}\right) \left(1 + \frac{\|s\|_2}{\xi_n}\right)^r e^{-\frac{\left(\sum_{i=1}^{N} s_i^2\right)}{\xi_n}} ds_1...ds_N \leq \tag{23.28}$$

$$\frac{2^N}{\left(\sqrt{\pi}\right)^N \left(\sqrt{\xi_n}\right)^N} \int_{\mathbb{R}_+^N} \left(\prod_{i=1}^{N} s_i^{\alpha_i}\right) \left(1 + \frac{\sum_{i=1}^{N} s_i}{\xi_n}\right)^r e^{-\frac{\left(\sum_{i=1}^{N} s_i^2\right)}{\xi_n}} ds_1...ds_N \leq$$

$$\frac{2^N \left(\sqrt{\xi_n}\right)^m}{\left(\sqrt{\pi}\right)^N} \int_{\mathbb{R}_+^N} \left(\prod_{i=1}^{N} \left(\frac{s_i}{\sqrt{\xi_n}}\right)^{\alpha_i}\right) \left(\frac{1}{\sqrt{\xi_n}} + \frac{1}{\sqrt{\xi_n}} \sum_{i=1}^{N} \left(\frac{s_i}{\sqrt{\xi_n}}\right)\right)^r$$

$$e^{-\sum_{i=1}^{N} \left(\frac{s_i}{\sqrt{\xi_n}}\right)^2} \frac{ds_1}{\sqrt{\xi_n}} \cdots \frac{ds_N}{\sqrt{\xi_n}} =$$

$$\left(\frac{2}{\sqrt{\pi}}\right)^N \left(\sqrt{\xi_n}\right)^{m-r} \int_{\mathbb{R}_+^N} \left(\prod_{i=1}^N z_i^{\alpha_i}\right) \left(1+\sum_{i=1}^N z_i\right)^r e^{-\sum_{i=1}^N z_i^2} dz_1...dz_N = \tag{23.29}$$

$$\left(\sqrt{\xi_n}\right)^{m-r} \left(\frac{2}{\sqrt{\pi}}\right)^N \left[\int_{[0,1]^N} \left(\prod_{i=1}^N z_i^{\alpha_i}\right) \left(1+\sum_{i=1}^N z_i\right)^r e^{-\sum_{i=1}^N z_i^2} dz_1...dz_N + \right.$$

$$\left. \int_{(\mathbb{R}_+-[0,1])^N} \left(\prod_{i=1}^N z_i^{\alpha_i}\right) \left(1+\sum_{i=1}^N z_i\right)^r e^{-\sum_{i=1}^N z_i^2} dz_1...dz_N \right] \leq$$

$$\left(\sqrt{\xi_n}\right)^{m-r} \left(\frac{2}{\sqrt{\pi}}\right)^N \left[\int_{[0,1]^N} \left(\prod_{i=1}^N z_i^{\alpha_i}\right) \left(1+\sum_{i=1}^N z_i\right)^r dz_1...dz_N + \right.$$

$$\left. 2^r \int_{(\mathbb{R}_+-[0,1])^N} \left(\prod_{i=1}^N z_i^{\alpha_i}\right) \left(\sum_{i=1}^N z_i\right)^r e^{-\sum_{i=1}^N z_i} dz_1...dz_N \right] \leq$$

$$\left(\sqrt{\xi_n}\right)^{m-r} \left(\frac{2}{\sqrt{\pi}}\right)^N \left[(1+N)^r + \right.$$

$$\left. 2^r \int_{(\mathbb{R}_+-[0,1])^N} \left(\prod_{i=1}^N z_i^{\alpha_i}\right) \left(\prod_{i=1}^N z_i^r\right) \left(\prod_{i=1}^N e^{-z_i}\right) \prod_{i=1}^N dz_i \right] = \tag{23.30}$$

$$\left(\sqrt{\xi_n}\right)^{m-r} \left(\frac{2}{\sqrt{\pi}}\right)^N \left[(1+N)^r + 2^r \prod_{i=1}^N \int_1^\infty z_i^{((\alpha_i+r)+1)-1} e^{-z_i} dz_i \right] =$$

(by [8], p. 348)

$$\left(\sqrt{\xi_n}\right)^{m-r} \left(\frac{2}{\sqrt{\pi}}\right)^N \left[(1+N)^r + 2^r \prod_{i=1}^N \Gamma\left((\alpha_i+r)+1, 1\right)\right],$$

where $\Gamma(\cdot,\cdot)$ is the upper incomplete gamma function.

We have proved that

$$\widetilde{u}_{\xi_n}(\alpha) \leq \left(\sqrt{\xi_n}\right)^{m-r} \left(\frac{2}{\sqrt{\pi}}\right)^N \left[(1+N)^r + 2^r \prod_{i=1}^N \Gamma\left((\alpha_i+r)+1, 1\right)\right] \leq$$

$$\left(\sqrt{\xi_n}\right)^{m-r} \left(\frac{2}{\sqrt{\pi}}\right)^N \left[(1+N)^r + 2^r \left(\frac{\lfloor e(m+r)! \rfloor}{e}\right)^N\right] \leq \tag{23.31}$$

$$\left(\frac{2}{\sqrt{\pi}}\right)^N \left[(1+N)^r + 2^r \left(\frac{\lfloor e\,(m+r)!\rfloor}{e}\right)^N\right] < +\infty,$$

therefore $\widetilde{u}_{\xi_n}(\alpha)$ are uniformly bounded.

Above we used the formula

$$\Gamma(s+1,1) = \frac{\lfloor es!\rfloor}{e}, \quad s \in \mathbb{N}.$$

Here $\alpha_i + r \in \mathbb{N}$, hence

$$\Gamma((\alpha_i + r) + 1, 1) = \frac{\lfloor e\,(\alpha_i + r)!\rfloor}{e} \leq \frac{\lfloor e\,(m+r)!\rfloor}{e}, \quad i = 1, ..., N. \tag{23.32}$$

The claim is proved. ■

We continue with

Theorem 23.9 *Let* $r, n, N \in \mathbb{N}$, $\xi_n \in (0, 1]$. *Then*

$$\widetilde{\Phi}_{\xi_n} := \frac{1}{\left(\sqrt{\pi\xi_n}\right)^N} \int_{\mathbb{R}^N} \left(1 + \frac{\|s\|_2}{\xi_n}\right)^r e^{-\frac{\left(\sum_{i=1}^N s_i^2\right)}{\xi_n}} ds_1 ... ds_N \leq$$

$$\frac{1}{\left(\sqrt{\xi_n}\right)^r} \left(\frac{2}{\sqrt{\pi}}\right)^N \left[(1+N)^r + 2^r \left(\frac{\lfloor er!\rfloor}{e}\right)^N\right]. \tag{23.33}$$

Proof We estimate

$$\widetilde{\Phi}_{\xi_n} = \frac{1}{\left(\sqrt{\pi\xi_n}\right)^N} \int_{\mathbb{R}^N} \left(1 + \frac{\|s\|_2}{\xi_n}\right)^r e^{-\frac{\left(\sum_{i=1}^N s_i^2\right)}{\xi_n}} ds_1 ... ds_N =$$

$$\frac{2^N}{\left(\sqrt{\pi}\right)^N \left(\sqrt{\xi_n}\right)^N} \int_{\mathbb{R}_+^N} \left(1 + \frac{\|s\|_2}{\xi_n}\right)^r e^{-\frac{\left(\sum_{i=1}^N s_i^2\right)}{\xi_n}} ds_1 ... ds_N \leq$$

$$\left(\frac{2}{\sqrt{\pi}}\right)^N \int_{\mathbb{R}_+^N} \left(\frac{1}{\sqrt{\xi_n}} + \frac{\sum_{i=1}^N s_i}{\sqrt{\xi_n}\sqrt{\xi_n}}\right)^r e^{-\frac{\left(\sum_{i=1}^N s_i^2\right)}{\xi_n}} \frac{ds_1}{\sqrt{\xi_n}} ... \frac{ds_N}{\sqrt{\xi_n}} =$$

$$\left(\frac{2}{\sqrt{\pi}}\right)^N \frac{1}{\left(\sqrt{\xi_n}\right)^r} \int_{\mathbb{R}_+^N} \left(1 + \sum_{i=1}^N \left(\frac{s_i}{\sqrt{\xi_n}}\right)\right)^r e^{-\sum_{i=1}^N \left(\frac{s_i}{\sqrt{\xi_n}}\right)^2} \frac{ds_1}{\sqrt{\xi_n}} ... \frac{ds_N}{\sqrt{\xi_n}} = \tag{23.34}$$

$$\left(\frac{2}{\sqrt{\pi}}\right)^N \frac{1}{\left(\sqrt{\xi_n}\right)^r} \int_{\mathbb{R}_+^N} \left(1+\sum_{i=1}^N z_i\right)^r e^{-\sum_{i=1}^N z_i^2} dz_1...dz_N =$$

$$\frac{1}{\left(\sqrt{\xi_n}\right)^r}\left(\frac{2}{\sqrt{\pi}}\right)^N \left[\int_{[0,1]^N} \left(1+\sum_{i=1}^N z_i\right)^r e^{-\sum_{i=1}^N z_i^2} dz_1...dz_N + \right.$$

$$\left. \int_{(\mathbb{R}_+-[0,1])^N} \left(1+\sum_{i=1}^N z_i\right)^r e^{-\sum_{i=1}^N z_i^2} dz_1...dz_N \right] \leq$$

$$\frac{1}{\left(\sqrt{\xi_n}\right)^r}\left(\frac{2}{\sqrt{\pi}}\right)^N \left[\int_{[0,1]^N} \left(1+\sum_{i=1}^N z_i\right)^r dz_1...dz_N + \right.$$

$$\left. \int_{(\mathbb{R}_+-[0,1])^N} \left(1+\sum_{i=1}^N z_i\right)^r e^{-\sum_{i=1}^N z_i} dz_1...dz_N \right] \leq \tag{23.35}$$

$$\frac{1}{\left(\sqrt{\xi_n}\right)^r}\left(\frac{2}{\sqrt{\pi}}\right)^N \left[(1+N)^r + 2^r \int_{(\mathbb{R}_+-[0,1])^N} \left(\sum_{i=1}^N z_i\right)^r e^{-\sum_{i=1}^N z_i} dz_1...dz_N \right] \leq$$

$$\frac{1}{\left(\sqrt{\xi_n}\right)^r}\left(\frac{2}{\sqrt{\pi}}\right)^N \left[(1+N)^r + 2^r \int_{(\mathbb{R}_+-[0,1])^N} \left(\prod_{i=1}^N z_i^r\right) \prod_{i=1}^N e^{-z_i} \prod_{i=1}^N dz_i \right] =$$

$$\frac{1}{\left(\sqrt{\xi_n}\right)^r}\left(\frac{2}{\sqrt{\pi}}\right)^N \left[(1+N)^r + 2^r \prod_{i=1}^N \int_1^\infty z_i^r e^{-z_i} dz_i \right] =$$

$$\frac{1}{\left(\sqrt{\xi_n}\right)^r}\left(\frac{2}{\sqrt{\pi}}\right)^N \left[(1+N)^r + 2^r \left(\int_1^\infty z^{(r+1)-1} e^{-z} dz\right)^N \right] = \tag{23.36}$$

(by [8], p. 348)

$$\frac{1}{\left(\sqrt{\xi_n}\right)^r}\left(\frac{2}{\sqrt{\pi}}\right)^N \left[(1+N)^r + 2^r \Gamma^N(r+1,1)\right] =$$

$$\frac{1}{\left(\sqrt{\xi_n}\right)^r}\left(\frac{2}{\sqrt{\pi}}\right)^N \left[(1+N)^r + 2^r \left(\frac{\lfloor er! \rfloor}{e}\right)^N \right],$$

proving the claim. ■

We give

Theorem 23.10 *All as in Theorem 23.8 and $p > 1$. Then*

$$\widetilde{A}_{\xi_n}(\alpha) := \frac{1}{\left(\sqrt{\pi}\xi_n\right)^N} \int_{\mathbb{R}^N} \left(\left(\prod_{i=1}^N |s_i|^{\alpha_i} \right) \left(1 + \frac{\|s\|_2}{\xi_n} \right)^r \right)^p e^{-\frac{\left(\sum_{i=1}^N s_i^2\right)}{\xi_n}} ds_1 ... ds_N \leq$$

$$\left(\sqrt{\xi_n}\right)^{p(m-r)} \left(\frac{2}{\sqrt{\pi}} \right)^N \left[(1+N)^{rp} + 2^{rp} \Gamma^N \left((m+r) p + 1, 1 \right) \right] \leq \qquad (23.37)$$

$$\left(\frac{2}{\sqrt{\pi}} \right)^N \left[(1+N)^{rp} + 2^{rp} \Gamma^N \left((m+r) p + 1, 1 \right) \right] < +\infty,$$

are uniformly bounded, where $m > r$.

Proof We estimate

$$\widetilde{A}_{\xi_n}(\alpha) = \frac{1}{\left(\sqrt{\pi}\xi_n\right)^N} \int_{\mathbb{R}^N} \left(\left(\prod_{i=1}^N |s_i|^{\alpha_i} \right) \left(1 + \frac{\|s\|_2}{\xi_n} \right)^r \right)^p e^{-\frac{\left(\sum_{i=1}^N s_i^2\right)}{\xi_n}} ds_1 ... ds_N =$$

$$\frac{2^N}{\left(\sqrt{\pi}\right)^N \left(\sqrt{\xi_n}\right)^N} \int_{\mathbb{R}_+^N} \left(\prod_{i=1}^N s_i^{\alpha_i p} \right) \left(1 + \frac{\|s\|_2}{\xi_n} \right)^{rp} e^{-\frac{\left(\sum_{i=1}^N s_i^2\right)}{\xi_n}} ds_1 ... ds_N \leq \qquad (23.38)$$

$$\left(\frac{2}{\sqrt{\pi}} \right)^N \left(\sqrt{\xi_n}\right)^{mp} \int_{\mathbb{R}_+^N} \left(\prod_{i=1}^N \left(\frac{s_i}{\sqrt{\xi_n}} \right)^{\alpha_i p} \right)$$

$$\left(\frac{1}{\sqrt{\xi_n}} + \frac{\sum_{i=1}^N s_i}{\sqrt{\xi_n}\sqrt{\xi_n}} \right)^{rp} e^{-\sum_{i=1}^N \left(\frac{s_i}{\sqrt{\xi_n}} \right)^2} \frac{ds_1}{\sqrt{\xi_n}} ... \frac{ds_N}{\sqrt{\xi_n}} =$$

(set $z_i = \frac{s_i}{\sqrt{\xi_n}}$, $i = 1, ..., N$)

$$\left(\frac{2}{\sqrt{\pi}} \right)^N \left(\sqrt{\xi_n}\right)^{p(m-r)} \int_{\mathbb{R}_+^N} \left(\prod_{i=1}^N z_i^{\alpha_i p} \right) \left(1 + \sum_{i=1}^N z_i \right)^{rp} e^{-\sum_{i=1}^N z_i^2} dz_1 ... dz_N =$$

$$\left(\frac{2}{\sqrt{\pi}} \right)^N \left(\sqrt{\xi_n}\right)^{p(m-r)} \left[\int_{[0,1]^N} \left(\prod_{i=1}^N z_i^{\alpha_i p} \right) \left(1 + \sum_{i=1}^N z_i \right)^{rp} e^{-\sum_{i=1}^N z_i^2} dz_1 ... dz_N + \right.$$

$$\int_{(\mathbb{R}_+-[0,1])^N}\left(\prod_{i=1}^{N} z_i^{\alpha_i p}\right)\left(1+\sum_{i=1}^{N} z_i\right)^{rp} e^{-\sum_{i=1}^{N} z_i^2} dz_1...dz_N\Bigg] \leq$$

$$\left(\frac{2}{\sqrt{\pi}}\right)^N \left(\sqrt{\xi_n}\right)^{p(m-r)}\left[\int_{[0,1]^N}\left(\prod_{i=1}^{N} z_i^{\alpha_i p}\right)\left(1+\sum_{i=1}^{N} z_i\right)^{rp} dz_1...dz_N+ \right. \quad (23.39)$$

$$\left. 2^{rp}\int_{(\mathbb{R}_+-[0,1])^N}\left(\prod_{i=1}^{N} z_i^{\alpha_i p}\right)\left(\sum_{i=1}^{N} z_i\right)^{rp} e^{-\sum_{i=1}^{N} z_i} dz_1...dz_N\right] \leq$$

$$\left(\frac{2}{\sqrt{\pi}}\right)^N \left(\sqrt{\xi_n}\right)^{p(m-r)}$$

$$\left[\frac{(1+N)^{rp}}{\prod_{i=1}^{N}(p\alpha_i+1)} + 2^{rp}\int_{(\mathbb{R}_+-[0,1])^N}\left(\prod_{i=1}^{N} z_i^{\alpha_i p}\right)\left(\prod_{i=1}^{N} z_i^{rp}\right)\prod_{i=1}^{N} e^{-z_i}\prod_{i=1}^{N} dz_i\right]$$

$$= \left(\sqrt{\xi_n}\right)^{p(m-r)}\left(\frac{2}{\sqrt{\pi}}\right)^N\left[\frac{(1+N)^{rp}}{\prod_{i=1}^{N}(p\alpha_i+1)} + 2^{rp}\prod_{i=1}^{N}\int_1^{\infty} z_i^{(\alpha_i+r)p} e^{-z_i} dz_i\right] =$$

(by [8], p. 348)

$$\left(\sqrt{\xi_n}\right)^{p(m-r)}\left(\frac{2}{\sqrt{\pi}}\right)^N\left[\frac{(1+N)^{rp}}{\prod_{i=1}^{N}(p\alpha_i+1)} + 2^{rp}\prod_{i=1}^{N}\Gamma\left(\left(\alpha_i+r\right)p+1,1\right)\right] \leq$$

(by [8], p. 909)

$$\left(\sqrt{\xi_n}\right)^{p(m-r)}\left(\frac{2}{\sqrt{\pi}}\right)^N\left[(1+N)^{rp} + 2^{rp}\Gamma^N\left((m+r)p+1,1\right)\right]. \quad (23.40)$$

We have proved that

$$\widetilde{A}_{\xi_n}(\alpha) \leq \left(\sqrt{\xi_n}\right)^{p(m-r)}\left(\frac{2}{\sqrt{\pi}}\right)^N\left[(1+N)^{rp} + 2^{rp}\Gamma^N\left((m+r)p+1,1\right)\right] \leq \quad (23.41)$$

$$\left(\frac{2}{\sqrt{\pi}}\right)^N \left[(1+N)^{rp} + 2^{rp}\Gamma^N\left((m+r)\,p+1,1\right)\right] < +\infty,$$

are uniformly bounded, where $m > r$.

Above we used ([8], p. 909) the formula

$$\Gamma(\alpha, xy) = y^{\alpha} e^{-xy} \int_0^{\infty} e^{-ty} (t+x)^{\alpha-1}\, dt, \tag{23.42}$$

where $Rey > 0$, $x > 0$, $Re\alpha > 1$.

We notice that $(x = y = 1, \alpha_i \le m)$

$$\Gamma\left((\alpha_i + r)\,p+1,1\right) = e^{-1} \int_0^{\infty} e^{-t} (t+1)^{(\alpha_i+r)p}\, dt \le$$

$$e^{-1} \int_0^{\infty} e^{-t} (t+1)^{(m+r)p}\, dt = \Gamma\left((m+r)\,p+1,1\right). \tag{23.43}$$

That is

$$\Gamma\left((\alpha_i + r)\,p+1,1\right) \le \Gamma\left((m+r)\,p+1,1\right), \tag{23.44}$$

for all $i = 1, \ldots, N$.

The theorem is proved. ■

We continue with

Theorem 23.11 *Let $r, n, N \in \mathbb{N}$, $\xi_n \in (0,1]$, $p > 1$. Then*

$$B_{\xi_n} := \frac{1}{\left(\sqrt{\pi \xi_n}\right)^N} \int_{\mathbb{R}^N} \left(1 + \frac{\|s\|_2}{\xi_n}\right)^{rp} e^{-\frac{\left(\sum_{i=1}^N s_i^2\right)}{\xi_n}} ds_1 \ldots ds_N \le$$

$$\frac{1}{\left(\sqrt{\xi_n}\right)^{pr}} \left(\frac{2}{\sqrt{\pi}}\right)^N \left[(1+N)^{pr} + 2^{pr}\Gamma^N(rp+1,1)\right]. \tag{23.45}$$

Proof We estimate

$$B_{\xi_n} = \frac{1}{\left(\sqrt{\pi \xi_n}\right)^N} \int_{\mathbb{R}^N} \left(1 + \frac{\|s\|_2}{\xi_n}\right)^{pr} e^{-\frac{\left(\sum_{i=1}^N s_i^2\right)}{\xi_n}} ds_1 \ldots ds_N =$$

$$\frac{2^N}{\left(\sqrt{\pi}\right)^N \left(\sqrt{\xi_n}\right)^N} \int_{\mathbb{R}_+^N} \left(1 + \frac{\|s\|_2}{\xi_n}\right)^{pr} e^{-\frac{\left(\sum_{i=1}^N s_i^2\right)}{\xi_n}} ds_1 \ldots ds_N \le \tag{23.46}$$

$$\left(\frac{2}{\sqrt{\pi}}\right)^N \int_{\mathbb{R}_+^N} \left(\frac{1}{\sqrt{\xi_n}} + \frac{\sum_{i=1}^N s_i}{\sqrt{\xi_n}\sqrt{\xi_n}}\right)^{pr} e^{-\frac{\left(\sum_{i=1}^N s_i^2\right)}{\xi_n}} \frac{ds_1}{\sqrt{\xi_n}} \ldots \frac{ds_N}{\sqrt{\xi_n}} =$$

$$\left(\frac{2}{\sqrt{\pi}}\right)^N \frac{1}{\left(\sqrt{\xi_n}\right)^{pr}} \int_{\mathbb{R}_+^N} \left(1 + \sum_{i=1}^N \frac{s_i}{\sqrt{\xi_n}}\right)^{pr} e^{-\left(\sum_{i=1}^N \frac{s_i}{\sqrt{\xi_n}}\right)^2} \frac{ds_1}{\sqrt{\xi_n}} \ldots \frac{ds_N}{\sqrt{\xi_n}} =$$

$$\left(\frac{2}{\sqrt{\pi}}\right)^N \frac{1}{\left(\sqrt{\xi_n}\right)^{pr}} \int_{\mathbb{R}_+^N} \left(1 + \sum_{i=1}^N z_i\right)^{pr} e^{-\sum_{i=1}^N z_i^2} dz_1 \ldots dz_N =$$

$$\frac{1}{\left(\sqrt{\xi_n}\right)^{pr}} \left(\frac{2}{\sqrt{\pi}}\right)^N \left[\int_{[0,1]^N} \left(1 + \sum_{i=1}^N z_i\right)^{pr} e^{-\sum_{i=1}^N z_i^2} dz_1 \ldots dz_N + \right.$$

$$\left. \int_{(\mathbb{R}_+ - [0,1])^N} \left(1 + \sum_{i=1}^N z_i\right)^{pr} e^{-\sum_{i=1}^N z_i^2} dz_1 \ldots dz_N \right] \leq \tag{23.47}$$

$$\frac{1}{\left(\sqrt{\xi_n}\right)^{pr}} \left(\frac{2}{\sqrt{\pi}}\right)^N \left[(1+N)^{pr} + 2^{pr} \int_{(\mathbb{R}_+ - [0,1])^N} \left(\sum_{i=1}^N z_i\right)^{pr} e^{-\sum_{i=1}^N z_i} dz_1 \ldots dz_N \right] \leq$$

$$\frac{1}{\left(\sqrt{\xi_n}\right)^{pr}} \left(\frac{2}{\sqrt{\pi}}\right)^N \left[(1+N)^{pr} + 2^{pr} \int_{(\mathbb{R}_+ - [0,1])^N} \left(\prod_{i=1}^N z_i^{pr}\right) \prod_{i=1}^N e^{-z_i} \prod_{i=1}^N dz_i \right] =$$

$$\frac{1}{\left(\sqrt{\xi_n}\right)^{pr}} \left(\frac{2}{\sqrt{\pi}}\right)^N \left[(1+N)^{pr} + 2^{pr} \prod_{i=1}^N \int_1^\infty z_i^{pr} e^{-z_i} dz_i \right] =$$

(by [8], p. 348)

$$\frac{1}{\left(\sqrt{\xi_n}\right)^{pr}} \left(\frac{2}{\sqrt{\pi}}\right)^N \left[(1+N)^{pr} + 2^{pr} \Gamma^N (rp+1, 1)\right],$$

proving the claim. ■

We finish this section with

Theorem 23.12 *Let* $n, N \in \mathbb{N}, \xi_n \in (0,1], \alpha_j \in \mathbb{Z}^+, j = 1, \ldots, N : |\alpha| := \sum_{j=1}^N \alpha_j = m \in \mathbb{N}$. *Then*

$$D_{\xi_n}(\alpha) :=$$

$$\left(\sqrt{\xi_n}\right)^{-m}\left(\frac{1}{\left(\sqrt{\pi\xi_n}\right)^N}\int_{\mathbb{R}^N}\left(\prod_{i=1}^N |s_i|^{\alpha_i}\right)e^{-\frac{\left(\sum_{i=1}^N s_i^2\right)}{\xi_n}}ds_1...ds_N\right)\leq\left(\frac{\mu}{\sqrt{\pi}}\right)^N, \tag{23.48}$$

where

$$\mu := \max\left\{\sqrt{\pi}, \Gamma\left(\frac{m+1}{2}\right)\right\}, \tag{23.49}$$

where Γ *is the gamma function.*

Proof We have that

$$D_{\xi_n}(\alpha)\leq\left(\sqrt{\xi_n}\right)^{-m}\left(\frac{1}{\left(\sqrt{\pi\xi_n}\right)^N}\int_{\mathbb{R}^N}\left(\prod_{i=1}^N |s_i|^{\alpha_i}\right)e^{-\frac{\left(\sum_{i=1}^N s_i^2\right)}{\xi_n}}ds_1...ds_N\right)=$$

$$\left(\sqrt{\xi_n}\right)^{-m}\left(\frac{2^N}{\left(\sqrt{\pi\xi_n}\right)^N}\int_{\mathbb{R}_+^N}\left(\prod_{i=1}^N s_i^{\alpha_i}\right)e^{-\frac{\left(\sum_{i=1}^N s_i^2\right)}{\xi_n}}ds_1...ds_N\right)= \tag{23.50}$$

$$\left(\sqrt{\xi_n}\right)^{-m}\left(\left(\frac{2}{\sqrt{\pi}}\right)^N\left(\sqrt{\xi_n}\right)^m\int_{\mathbb{R}_+^N}\left(\prod_{i=1}^N\left(\frac{s_i}{\sqrt{\xi_n}}\right)^{\alpha_i}\right)e^{-\sum_{i=1}^N\left(\frac{s_i}{\sqrt{\xi_n}}\right)^2}\frac{ds_1}{\sqrt{\xi_n}}...\frac{ds_N}{\sqrt{\xi_n}}\right)=$$

$$\left(\frac{2}{\sqrt{\pi}}\right)^N\int_{\mathbb{R}_+^N}\left(\prod_{i=1}^N z_i^{\alpha_i}\right)e^{-\sum_{i=1}^N z_i^2}dz_1...dz_n=$$

$$\left(\frac{2}{\sqrt{\pi}}\right)^N\prod_{i=1}^N\int_0^\infty z_i^{\alpha_i}e^{-z_i^2}dz_i=$$

(by [8], p. 349)

$$\left(\left(\frac{2}{\sqrt{\pi}}\right)^N\prod_{i=1}^N\left(\frac{\Gamma\left(\frac{\alpha_i+1}{2}\right)}{2}\right)\right)=\frac{1}{\left(\sqrt{\pi}\right)^N}\left(\prod_{i=1}^N\Gamma\left(\frac{\alpha_i+1}{2}\right)\right)\leq(*)\,. \tag{23.51}$$

Some important values of $\Gamma\left(\frac{\alpha_i+1}{2}\right)$ follow:
if $\alpha_i = 0$, then $\Gamma\left(\frac{1}{2}\right) = \sqrt{\pi} \approx 1.77245...$,
if $\alpha_i = 1$, then $\Gamma(1) = 1$,
if $\alpha_i = 2$, then $\Gamma\left(\frac{3}{2}\right) = \frac{\sqrt{\pi}}{2} \approx 0.886226...$,
if $\alpha_i = 3$, then $\Gamma(2) = 1$,
if $\alpha_i = 4$, then $\Gamma\left(\frac{5}{2}\right) = \frac{3\sqrt{\pi}}{4} \approx 1.32934$,
if $\alpha_i = 5$, then $\Gamma(3) = 2$.

The function Γ has minimum at $x_{\min} = 1.46...$, with $\Gamma(x_{\min}) = 0.885...$, and $\Gamma(x)$ is increasing for $x \geq x_{\min}$. To have $\frac{\alpha_i+1}{2} > 1.46...$, we need $\alpha_i > 1.92....$ Thus when $\alpha_i \geq 2$, then $\Gamma\left(\frac{\alpha_i+1}{2}\right)$ increases, and $\Gamma\left(\frac{\alpha_i+1}{2}\right) \leq \Gamma\left(\frac{m+1}{2}\right)$, when $m \geq 2$.

Hence it holds $\Gamma\left(\frac{\alpha_i+1}{2}\right) \leq \mu$, for $i = 1, ..., N$, and

$$(*) \leq \left(\frac{\mu}{\sqrt{\pi}}\right)^N, \tag{23.52}$$

proving the claim. ∎

23.3 Main Results for $W_{r,n}^{[m]}$

23.3.1 Uniform Approximation

We start with an application to $W_{r,n}^{[m]}$ of the following theorem.

Theorem 23.13 ([3], p. 11) *Let $m \in \mathbb{N}$, $f \in C^m\left(\mathbb{R}^N\right)$, $N \geq 1$, $x \in \mathbb{R}^N$. Assume $\left\| \frac{\partial^m f(\cdot,\cdot,...,\cdot)}{\partial x_1^{\alpha_1} ... \partial x_N^{\alpha_N}} \right\|_\infty < \infty$, for all $\alpha_j \in \mathbb{Z}^+$, $j = 1, ..., N : |\alpha| := \sum_{i=1}^N \alpha_i = m$. Let μ_{ξ_n} be a Borel probability measure on $\mathbb{R}^N$, for $\xi_n > 0$, $(\xi_n)_{n \in \mathbb{N}}$ bounded sequence.*

Suppose that for all $\alpha := (\alpha_1, ..., \alpha_N)$, $\alpha_i \in \mathbb{Z}^+$, $i = 1, ..., N$, $|\alpha| := \sum_{i=1}^N \alpha_i = m$ we have that

$$u_{\xi_n}(\alpha) := \int_{\mathbb{R}^N} \left(\prod_{i=1}^N |s_i|^{\alpha_i}\right) \left(1 + \frac{\|s\|_2}{\xi_n}\right)^r d\mu_{\xi_n}(s) < \infty. \tag{23.53}$$

For $\widetilde{j} = 1, ..., m$, and $\alpha := (\alpha_1, ..., \alpha_N)$, $\alpha_i \in \mathbb{Z}^+$, $i = 1, ..., N$, $|\alpha| := \sum_{j=1}^N \alpha_j = \widetilde{j}$, call

$$c_{\alpha,n} := c_{\alpha,n,\widetilde{j}} := \int_{\mathbb{R}^N} \prod_{i=1}^N s_i^{\alpha_i} d\mu_{\xi_n}(s_1, ..., s_N). \tag{23.54}$$

Then

(i)

$$E_{r,n}^{[m]}(x) := \left| \theta_{r,n}^{[m]}(f; x) - f(x) - \sum_{\widetilde{j}=1}^m \delta_{\widetilde{j},r}^{[m]} \left(\sum_{\substack{\alpha_1,...,\alpha_N \geq 0; \\ |\alpha| = \widetilde{j}}} \frac{c_{\alpha,n,\widetilde{j}} f_\alpha(x)}{\prod_{i=1}^N \alpha_i!} \right) \right|$$

$$\leq \sum_{\substack{\alpha_1,\ldots,\alpha_N \geq 0; \\ |\alpha|=m}} \frac{(\omega_r(f_\alpha, \xi_n))}{\left(\prod_{i=1}^{N} \alpha_i\right)} \left(\int_{\mathbb{R}^N} \left(\prod_{i=1}^{N} |s_i|^{\alpha_i}\right) \left(1 + \frac{\|s\|_2}{\xi_n}\right)^r d\mu_{\xi_n}(s)\right). \tag{23.55}$$

$\forall\, x \in \mathbb{R}^N$.

(ii)

$$\left\| E_{r,n}^{[m]} \right\|_\infty \leq R.H.S.(23.55). \tag{23.56}$$

Given that $\xi_n \to 0$, as $n \to \infty$, and u_{ξ_n} is uniformly bounded, then we derive that $\left\| E_{r,n}^{[m]} \right\| \to 0$ with rates.

(iii) It holds also that

$$\left\| \theta_{r,n}^{[m]}(f) - f \right\|_\infty \leq \sum_{\widetilde{j}=1}^{m} \delta_{\widetilde{j},r}^{[m]} \left(\sum_{\substack{\alpha_1,\ldots,\alpha_N \geq 0; \\ |\alpha|=\widetilde{j}}} \frac{\left|c_{\alpha,n,\widetilde{j}}\right| \|f_\alpha\|_\infty}{\prod_{i=1}^{N} \alpha_i!} \right) + R.H.S.(23.55), \tag{23.57}$$

given that $\|f_\alpha\|_\infty < \infty$, for all $\alpha : |\alpha| = \widetilde{j}$, $\widetilde{j} = 1, \ldots, m$. Furthermore, as $\xi_n \to 0$ when $n \to \infty$, assuming that $c_{\alpha,n,\widetilde{j}} \to 0$, while u_{ξ_n} is uniformly bounded, we conclude that

$$\left\| \theta_{r,n}^{[m]}(f) - f \right\|_\infty \to 0 \tag{23.58}$$

with rates.

A uniform approximation result for $W_{r,n}^{[m]}$ follows:

Theorem 23.14 *Let $m \in \mathbb{N}$, $f \in C^m(\mathbb{R}^N)$, $N \geq 1$, $x \in \mathbb{R}^N$. Assume $\left\| \frac{\partial^m f(\cdot,\cdot,\ldots,\cdot)}{\partial x_1^{\alpha_1} \ldots \partial x_N^{\alpha_N}} \right\|_\infty < \infty$, for all $\alpha_j \in \mathbb{Z}^+$, $j = 1, \ldots, N : |\alpha| := \sum_{j=1}^{N} \alpha_j = m$. Here $\xi_n \in (0, 1]$, $n \in \mathbb{N}$, $m > r \in \mathbb{N}$. For all $\alpha := (\alpha_1, \ldots, \alpha_N)$, $\alpha_i \in \mathbb{Z}^+$, $i = 1, \ldots, N$, $|\alpha| := \sum_{i=1}^{N} \alpha_i = m$, we denote*

$$\widetilde{u}_{\xi_n}(\alpha) := \frac{1}{\left(\sqrt{\pi \xi_n}\right)^N} \int_{\mathbb{R}^N} \left(\prod_{i=1}^{N} |s_i|^{\alpha_i}\right) \left(1 + \frac{\|s\|_2}{\xi_n}\right)^r e^{-\frac{\left(\sum_{i=1}^{N} s_i^2\right)}{\xi_n}} ds_1 \ldots ds_N. \tag{23.59}$$

For $\widetilde{j} = 1, \ldots, m$, and $\alpha := (\alpha_1, \ldots, \alpha_N)$, $\alpha_i \in \mathbb{Z}^+$, $i = 1, \ldots, N$, $|\alpha| := \sum_{i=1}^{N} \alpha_i = \widetilde{j}$, call

$$\widetilde{c}_{\alpha,n} := \widetilde{c}_{\alpha,n,\widetilde{j}} := \frac{1}{\left(\sqrt{\pi \xi_{nn}}\right)^N} \int_{\mathbb{R}^N} \prod_{i=1}^{N} s_i^{\alpha_i} e^{-\frac{\left(\sum_{i=1}^{N} s_i^2\right)}{\xi_n}} ds_1 \ldots ds_N. \tag{23.60}$$

Then

(i)

$$\widetilde{E}_{r,n}^{[m]}(x) := \left| W_{r,n}^{[m]}(f;x) - f(x) - \sum_{\widetilde{j}=1}^{m} \delta_{\widetilde{j},r}^{[m]} \left(\sum_{\substack{\alpha_1,\ldots,\alpha_N \geq 0; \\ |\alpha| = \widetilde{j}}} \frac{\widetilde{c}_{\alpha,n,\widetilde{j}} f_\alpha(x)}{\prod_{i=1}^{N} \alpha_i!} \right) \right|$$

$$\leq \sum_{\substack{\alpha_1,\ldots,\alpha_N \geq 0; \\ |\alpha| = m}} \frac{(\omega_r(f_\alpha, \xi_n))}{\left(\prod_{i=1}^{N} \alpha_i!\right)} \widetilde{u}_{\xi_n}(\alpha), \tag{23.61}$$

$\forall\, x \in \mathbb{R}^N$.

(ii)

$$\left\| \widetilde{E}_{r,n}^{[m]} \right\|_\infty \leq R.H.S.(23.61). \tag{23.62}$$

Given that $\xi_n \to 0$, *as* $n \to +\infty$, *we have that* $\widetilde{u}_{\xi_n}(\alpha) \to 0$ *and are uniformly bounded, and then we derive that* $\left\| \widetilde{E}_{r,n}^{[m]} \right\|_\infty \to 0$ *with rates.*

(iii) It holds also that

$$\left\| W_{r,n}^{[m]}(f) - f \right\|_\infty \leq \sum_{\widetilde{j}=1}^{m} \left| \delta_{\widetilde{j},r}^{[m]} \right| \left(\sum_{\substack{\alpha_1,\ldots,\alpha_N \geq 0; \\ |\alpha| = \widetilde{j}}} \frac{\left| \widetilde{c}_{\alpha,n,\widetilde{j}} \right| \|f_\alpha\|_\infty}{\prod_{i=1}^{N} \alpha_i!} \right) + R.H.S.(23.61), \tag{23.63}$$

given that $\|f_\alpha\|_\infty < +\infty$, *for all* $\alpha : |\alpha| = \widetilde{j}$, $\widetilde{j} = 1, \ldots, m$. *Furthermore, as* $\xi_n \to 0$ *when* $n \to +\infty$, *we have that* $\widetilde{c}_{\alpha,n,\widetilde{j}} \to 0$ *and* $\widetilde{u}_{\xi_n}(\alpha) \to 0$, *and both are uniformly bounded, and we conclude that*

$$\left\| W_{r,n}^{[m]}(f) - f \right\|_\infty \to 0 \tag{23.64}$$

with rates.

Proof Mainly by applying Theorem 23.13. By Theorem 23.8 we get that $\widetilde{u}_{\xi_n}(\alpha) \to 0$ and $\widetilde{u}_{\xi_n}(\alpha)$ are uniformly bounded. By Theorem 23.12 we get that

$$\left| \widetilde{c}_{\alpha,n} \right| = \left| \widetilde{c}_{\alpha,n,\widetilde{j}} \right| \leq \frac{1}{\left(\sqrt{\pi \xi_n}\right)^N} \int_{\mathbb{R}^N} \left(\prod_{i=1}^{N} |s_i|^{\alpha_i} \right) e^{-\frac{\left(\sum_{i=1}^{N} s_i^2\right)}{\xi_n}} ds_1 \ldots ds_N$$

$$\leq \left(\frac{\mu}{\sqrt{\pi}} \right)^N \left(\sqrt{\xi_n} \right)^m \leq \left(\frac{\mu}{\sqrt{\pi}} \right)^N.$$

That is $\widetilde{c}_{\alpha,n,\widetilde{j}} \to 0$ and are uniformly bounded. The proof is complete. ■

We mention

Theorem 23.15 ([3], p. 14) *Let* $f \in C_B(\mathbb{R}^N)$, $N \geq 1$, $\xi_n \in (0, 1]$. *Then*

$$\left\| \theta_{r,n}^{[0]} f - f \right\|_\infty \leq \left(\int_{\mathbb{R}^N} \left(1 + \frac{\|s\|_2}{\xi_n} \right)^r d\mu_{\xi_n}(s) \right) \omega_r(f, \xi_n), \tag{23.65}$$

under the assumption

$$\Phi_{\xi_n} := \int_{\mathbb{R}^N} \left(1 + \frac{\|s\|_2}{\xi_n} \right)^r d\mu_{\xi_n}(s) < \infty. \tag{23.66}$$

In case of f *being uniformly, continuous and as* $n \to \infty$ *and* $\xi_n \to 0$, *given that* Φ_{ξ_n} *are uniformly bounded, we derive*

$$\left\| \theta_{r,n}^{[0]} f - f \right\|_\infty \to 0 \tag{23.67}$$

with rates.

We give

Theorem 23.16 *Let* $f \in C_B(\mathbb{R}^N)$, $N \geq 1$, $\xi_n \in (0, 1]$. *Then*

(1)

$$\left\| W_{r,n}^{[0]} f - f \right\|_\infty \leq \left(\frac{2}{\sqrt{\pi}} \right)^N \left[(1+N)^r + 2^r \left(\frac{\lfloor er! \rfloor}{e} \right)^N \right] \frac{\omega_r(f, \xi_n)}{\left(\sqrt{\xi_n}\right)^r}, \tag{23.68}$$

where $r, n \in \mathbb{N}$.

(2) Let $r \in \mathbb{N} - \{1\}$, $0 < \alpha \leq 1$, *and consider* $f \in C_B(\mathbb{R}^N)$ *such that* $\omega_r(f, t) \leq K t^{r-1+\alpha}$, $\forall\, t > 0$, *where* $K > 0$ *is independent of* t.
In this case

$$\frac{\omega_r(f, \xi_n)}{\left(\sqrt{\xi_n}\right)^r} \leq K \xi_n^{\left(\frac{r}{2}+\alpha-1\right)}, \tag{23.69}$$

and as $n \to \infty$ *and* $\xi_n \to 0$, *we derive*

$$\left\| W_{r,n}^{[0]} f - f \right\|_\infty \to 0. \tag{23.70}$$

Proof By Theorems 23.9 and 23.15. For (23.69), since $r \geq 2$, we get that $\frac{r}{2} + \alpha - 1 > 0$. ■

23.3.2 L_p Approximation for $W_{r,n}^{[m]}$

We need

Definition 23.17 ([5, 9]) We call

$$\Delta_u^r f(x) := \Delta_{u_1,u_2,\ldots,u_N}^r f(x_1, \ldots, x_N) := \tag{23.71}$$

$$\sum_{j=0}^{r} (-1)^{r-j} \binom{r}{j} f(x_1 + ju_1, x_2 + ju_2, \ldots, x_N + ju_N).$$

Let $p \geq 1$, the modulus of smoothness of order r is given by

$$\omega_r(f; h)_p := \sup_{\|u\|_2 \leq h} \left\| \Delta_u^r(f) \right\|_p, \tag{23.72}$$

$h > 0$.

We will apply.

Theorem 23.18 ([3], p. 24) *Let $f \in C^m(\mathbb{R}^N)$, $m \in \mathbb{N}$, $N \geq 1$, with $f_\alpha \in L_p(\mathbb{R}^N)$, $|\alpha| = m$, $x \in \mathbb{R}^N$. Let $p, q > 1 : \frac{1}{p} + \frac{1}{q} = 1$. Here, μ_{ξ_n} is a Borel probability measure on $\mathbb{R}^N$ for $\xi_n > 0$, $(\xi_n)_{n\in\mathbb{N}}$ bounded sequence. Assume for all $\alpha := (\alpha_1, \ldots, \alpha_N)$, $\alpha_i \in \mathbb{Z}^+$, $i = 1, \ldots, N$, $|\alpha| := \sum_{i=1}^{N} \alpha_i = m$ that we have*

$$\int_{\mathbb{R}^N} \left(\left(\prod_{i=1}^{N} |s_i|^{\alpha_i} \right) \left(1 + \frac{\|s\|_2}{\xi_n} \right)^r \right)^p d\mu_{\xi_n}(s) < \infty. \tag{23.73}$$

For $\widetilde{j} = 1, \ldots, m$, and $\alpha := (\alpha_1, \ldots, \alpha_N)$, $\alpha_i \in \mathbb{Z}^+$, $i = 1, \ldots, N$, $|\alpha| := \sum_{j=1}^{N} \alpha_j = \widetilde{j}$, call

$$c_{\alpha,n,\widetilde{j}} := \int_{\mathbb{R}^N} \prod_{i=1}^{N} s_i^{\alpha_i} d\mu_{\xi_n}(s). \tag{23.74}$$

Then

$$\left\| E_{r,n}^{[m]} \right\|_p = \left\| \theta_{r,n}^{[m]}(f; x) - f(x) - \sum_{\widetilde{j}=1}^{m} \delta_{\widetilde{j},r}^{[m]} \left(\sum_{|\alpha|=\widetilde{j}} \frac{c_{\alpha,n,\widetilde{j}} f_\alpha(x)}{\left(\prod_{i=1}^{N} \alpha_i! \right)} \right) \right\|_{p,x} \tag{23.75}$$

$$\leq \left(\frac{m}{(q(m-1)+1)^{\frac{1}{q}}} \right) \left(\sum_{|\alpha|=m} \frac{1}{\prod_{i=1}^{N} \alpha_i!} \right)$$

$$\left[\int_{\mathbb{R}^N}\left[\left(\prod_{i=1}^{N}|s_i|^{\alpha_i}\right)\left(1+\frac{\|s\|_2}{\xi_n}\right)^r\right]^p d\mu_{\xi_n}(s)\right]^{\frac{1}{p}} \omega_r(f_\alpha,\xi_n)_p .$$

As $n \to \infty$ and $\xi_n \to 0$, by (23.75) we obtain that $\left\|E_{r,n}^{[m]}\right\| \to 0$ with rates.

One also finds by (23.75) that

$$\left\|\theta_{r,n}^{[m]}(f;x)-f(x)\right\|_{p,x} \le \sum_{\widetilde{j}=1}^{m}\left|\delta_{\widetilde{j},r}^{[m]}\right|\left(\sum_{|\alpha|=\widetilde{j}}\frac{\left|c_{\alpha,n,\widetilde{j}}\right|}{\prod_{i=1}^{N}\alpha_i!}\|f_\alpha\|_p\right)+R.H.S.(23.75), \tag{23.76}$$

given that $\|f_\alpha\|_p < \infty$, $|\alpha|=\widetilde{j}$, $\widetilde{j}=1,...,m$.

Assuming that $c_{\alpha,n,\widetilde{j}} \to 0$, $\xi_n \to 0$, as $n \to \infty$, we get $\left\|\theta_{r,n}^{[m]}(f)-f\right\|_p \to 0$, that is $\theta_{r,n}^{[m]} \to I$ the unit operator, in L_p norm, with rates.

We present.

Theorem 23.19 *Let $f \in C^m\left(\mathbb{R}^N\right)$, $m \in \mathbb{N}$, $m > r$, $N \ge 1$, with $f_\alpha \in L_p\left(\mathbb{R}^N\right)$, $|\alpha|=m$, $x \in \mathbb{R}^N$. Let $p,q>1 : \frac{1}{p}+\frac{1}{q}=1$, $\xi_n \in (0,1]$. For $\widetilde{j}=1,...,m$, and $\alpha := (\alpha_1,...,\alpha_N)$, $\alpha_i \in \mathbb{Z}^+$, $i=1,...,N$, $|\alpha| := \sum_{i=1}^{N}\alpha_i = \widetilde{j}$, call*

$$\widetilde{c}_{\alpha,n,\widetilde{j}} = \frac{1}{\left(\sqrt{\pi\xi_n}\right)^N}\int_{\mathbb{R}^N}\prod_{i=1}^{N}s_i^{\alpha_i}e^{-\frac{\left(\sum_{i=1}^{N}s_i^2\right)}{\xi_n}}ds_1...ds_N. \tag{23.77}$$

Then

$$\left\|\widetilde{E}_{r,n}^{[m]}\right\|_p = \left\|W_{r,n}^{[m]}(f;x)-f(x)-\sum_{\widetilde{j}=1}^{m}\delta_{\widetilde{j},r}^{[m]}\left(\sum_{|\alpha|=\widetilde{j}}\frac{\widetilde{c}_{\alpha,n,\widetilde{j}}f_\alpha(x)}{\left(\prod_{i=1}^{N}\alpha_i!\right)}\right)\right\|_{p,x}$$

$$\le \left(\frac{2}{\sqrt{\pi}}\right)^{\frac{N}{p}}\left(\frac{m}{(q(m-1)+1)^{\frac{1}{q}}}\right)\left[(1+N)^{rp}+2^{rp}\Gamma^N((m+r)p+1,1)\right]^{\frac{1}{p}} \tag{23.78}$$

$$\left(\sum_{|\alpha|=m}\frac{1}{\prod_{i=1}^{N}\alpha_i!}\right)\left(\sqrt{\xi_n}\right)^{(m-r)}\omega_r(f_\alpha,\xi_n)_p .$$

As $n \to +\infty$ and $\xi_n \to 0$, by (23.78) we obtain that $\left\|\widetilde{E}_{r,n}^{[m]}\right\|_p \to 0$ with rates.

One also finds by (23.78) that

$$\left\| W_{r,n}^{[m]}(f;x) - f(x) \right\|_{p,x} \leq \sum_{\widetilde{j}=1}^{m} \left| \delta_{\widetilde{j},r}^{[m]} \right| \left(\sum_{|\alpha|=\widetilde{j}} \frac{\left| \widetilde{c}_{\alpha,n,\widetilde{j}} \right|}{\prod_{i=1}^{N} \alpha_i!} \| f_\alpha \|_p \right) + R.H.S.(23.78), \tag{23.79}$$

given that $\| f_\alpha \|_p < \infty$, $|\alpha| = \widetilde{j}$, $\widetilde{j} = 1, ..., m$.

Assuming that $\xi_n \to 0$, *as* $n \to \infty$, *we get* $\left\| W_{r,n}^{[m]}(f) - f \right\|_p \to 0$, *that is* $W_{r,n}^{[m]} \to I$ *the unit operator, in* L_p *norm, with rates.*

Proof By Theorem 23.12 we get that $\left| \widetilde{c}_{\alpha,n,\widetilde{j}} \right| \leq \left(\frac{\mu}{\sqrt{\pi}} \right)^N \left(\sqrt{\xi_n} \right)^m \leq \left(\frac{\mu}{\sqrt{\pi}} \right)^N$ and $\widetilde{c}_{\alpha,n,\widetilde{j}} \to 0$. We use also Theorem 23.10 to uniformly bound $\frac{1}{\left(\sqrt{\pi \xi_n}\right)^N} \int_{\mathbb{R}^N} \left(\left(\prod_{i=1}^{N} |s_i|^{\alpha_i} \right) \left(1 + \frac{\|s\|_2}{\xi_n} \right)^r \right)^p e^{-\frac{\left(\sum_{i=1}^{N} s_i^2 \right)}{\xi_n}} ds_1 ... ds_N$. Then we apply directly Theorem 23.18. ■

We continue with an application of

Theorem 23.20 ([3], p. 26) *Let* $f \in \left(C\left(\mathbb{R}^N\right) \cap L_p\left(\mathbb{R}^N\right) \right)$; $N \geq 1$; $p, q > 1 : \frac{1}{p} + \frac{1}{q} = 1$. *Assume* μ_{ξ_n} *probability Borel measure on* $\mathbb{R}^N$, $(\xi_n)_{n \in \mathbb{N}} > 0$ *and bounded. Also suppose*

$$\int_{\mathbb{R}^N} \left(1 + \frac{\|s\|_2}{\xi_n} \right)^{rp} d\mu_{\xi_n}(s) < \infty. \tag{23.80}$$

Then

$$\left\| \theta_{r,n}^{[0]}(f) - f \right\|_p \leq \tag{23.81}$$

$$\left(\int_{\mathbb{R}^N} \left(1 + \frac{\|s\|_2}{\xi_n} \right)^{rp} d\mu_{\xi_n}(s) \right)^{\frac{1}{p}} \omega_r(f, \xi_n)_p .$$

As $\xi_n \to 0$, *when* $n \to \infty$, *we derive* $\left\| \theta_{r,n}^{[0]}(f) - f \right\|_p \to 0$, *i.e.* $\theta_{r,n}^{[0]} \to I$, *the unit operator, in* L_p *norm.*

We give

Theorem 23.21 *Let* $f \in \left(C\left(\mathbb{R}^N\right) \cap L_p\left(\mathbb{R}^N\right) \right)$; $N \geq 1$; $p, q > 1 : \frac{1}{p} + \frac{1}{q} = 1$, $\xi_n \in (0, 1]$, $n \in \mathbb{N}$. *Then*

(1)

$$\left\| W_{r,n}^{[0]}(f) - f \right\|_p \leq \tag{23.82}$$

$$\left(\frac{2}{\sqrt{\pi}} \right)^{\frac{N}{p}} \left[(1+N)^{rp} + 2^{rp} \Gamma^N (rp+1, 1) \right]^{\frac{1}{p}} \frac{\omega_r(f, \xi_n)_p}{\left(\sqrt{\xi_n}\right)^r}.$$

(2) *Let* $r \in \mathbb{N} - \{1\}$, $0 < \alpha \leq 1$, *and consider* $f \in \left(C\left(\mathbb{R}^N\right) \cap L_p\left(\mathbb{R}^N\right)\right) : \omega_r(f,t)_p \leq K t^{r-1+\alpha}$, $\forall\, t > 0$, *where* $K > 0$ *is independent of* t. *In this case*

$$\frac{\omega_r(f,\xi_n)_p}{\left(\sqrt{\xi_n}\right)^r} \leq K \xi_n^{\left(\frac{r}{2}+\alpha-1\right)}, \tag{23.83}$$

and as $n \to \infty$ *and* $\xi_n \to 0$, *we obtain*

$$\left\| W_{r,n}^{[0]} f - f \right\|_p \to 0, \tag{23.84}$$

i.e. $W_{r,n}^{[0]} \to I$, *the unit operator in* L_p *norm.*

Proof By Theorems 23.11 and 23.20. For $r \geq 2$ we get $\frac{r}{2} + \alpha - 1 > 0$. ■

We mention

Theorem 23.22 ([3], p. 27) *Let* $f \in \left(C\left(\mathbb{R}^N\right) \cap L_1\left(\mathbb{R}^N\right)\right)$; $N \geq 1$. *Assume* μ_{ξ_n} *probability Borel measure on* $\mathbb{R}^N$, $(\xi_n)_{n\in\mathbb{N}} > 0$ *and bounded. Also suppose*

$$\int_{\mathbb{R}^N} \left(1 + \frac{\|s\|_2}{\xi_n}\right)^r d\mu_{\xi_n}(s) < \infty. \tag{23.85}$$

Then

$$\left\| \theta_{r,n}^{[0]}(f) - f \right\|_1 \leq \tag{23.86}$$

$$\left(\int_{\mathbb{R}^N} \left(1 + \frac{\|s\|_2}{\xi_n}\right)^r d\mu_{\xi_n}(s) \right) \omega_r(f,\xi_n)_1 .$$

As $\xi_n \to 0$, *we get* $\theta_{r,n}^{[0]} \to I$, *in* L_1 *norm.*

We give

Theorem 23.23 *Let* $f \in \left(C\left(\mathbb{R}^N\right) \cap L_1\left(\mathbb{R}^N\right)\right)$; $N \geq 1$, $\xi_n \in (0,1]$, $n \in \mathbb{N}$. *Then*

(1)

$$\left\| W_{r,n}^{[0]}(f) - f \right\|_1 \leq \left(\frac{2}{\sqrt{\pi}}\right)^N \left[(1+N)^r + 2^r \left(\frac{\lfloor er! \rfloor}{e}\right)^N \right] \frac{\omega_r(f,\xi_n)_1}{\left(\sqrt{\xi_n}\right)^r}. \tag{23.87}$$

(2) *Let* $r \in \mathbb{N} - \{1\}$, $0 < \alpha \leq 1$, *and consider* $f \in \left(C\left(\mathbb{R}^N\right) \cap L_1\left(\mathbb{R}^N\right)\right) : \omega_r(f,t)_1 \leq K t^{r-1+\alpha}$, $\forall\, t > 0$, *where* $K > 0$ *is independent of* t. *In this case*

$$\frac{\omega_r(f,\xi_n)_1}{\left(\sqrt{\xi_n}\right)^r} \leq K \xi_n^{\left(\frac{r}{2}+\alpha-1\right)}, \tag{23.88}$$

and as $n \to \infty$ *and* $\xi_n \to 0$, *we derive* $\left\| W_{r,n}^{[0]}(f) - f \right\|_1 \to 0$, *i.e.* $W_{r,n}^{[0]} \to I$ *in* L_1 *norm.*

Proof By Theorems 23.9 and 23.22. Notice $\frac{r}{2}+\alpha-1>0$. ■

We need

Theorem 23.24 ([3], p. 29) *Let $f \in C^m(\mathbb{R}^N)$, $m, N \in \mathbb{N}$, with $f_\alpha \in L_1(\mathbb{R}^N)$, $|\alpha| = m$, $x \in \mathbb{R}^N$. Here, μ_{ξ_n} is a Borel probability measure on $\mathbb{R}^N$ for $\xi_n > 0$, $(\xi_n)_{n\in\mathbb{N}}$ is a bounded sequence. Suppose for all $\alpha := (\alpha_1, ..., \alpha_N)$, $\alpha_i \in \mathbb{Z}^+$, $i = 1, ..., N$, $|\alpha| := \sum_{i=1}^{N} \alpha_i = m$ that we have*

$$\int_{\mathbb{R}^N} \left(\left(\prod_{i=1}^{N} |s_i|^{\alpha_i} \right) \left(1 + \frac{\|s\|_2}{\xi_n} \right)^r \right) d\mu_{\xi_n}(s) < \infty. \tag{23.89}$$

For $\widetilde{j} = 1, ..., m$, and $\alpha := (\alpha_1, ..., \alpha_N)$, $\alpha_i \in \mathbb{Z}^+$, $i = 1, ..., N$, $|\alpha| := \sum_{i=1}^{N} \alpha_i = \widetilde{j}$, call

$$c_{\alpha,n,\widetilde{j}} := \int_{\mathbb{R}^N} \prod_{i=1}^{N} s_i^{\alpha_i} d\mu_{\xi_n}(s). \tag{23.90}$$

Then

$$\left\| E_{r,n}^{[m]} \right\|_1 = \left\| \theta_{r,n}^{[m]}(f; x) - f(x) - \sum_{\widetilde{j}=1}^{m} \delta_{\widetilde{j},r}^{[m]} \left(\sum_{|\alpha|=\widetilde{j}} \frac{c_{\alpha,n,\widetilde{j}} f_\alpha(x)}{\prod_{i=1}^{N} \alpha_i!} \right) \right\|_{1,x} \tag{23.91}$$

$$\leq \sum_{|\alpha|=m} \left(\frac{1}{\prod_{i=1}^{N} \alpha_i!} \right) \omega_r(f_\alpha, \xi_n)_1 \int_{\mathbb{R}^N} \left(\prod_{i=1}^{N} |s_i|^{\alpha_i} \right) \left(1 + \frac{\|s\|_2}{\xi_n} \right)^r d\mu_{\xi_n}(s).$$

As $\xi_n \to 0$, we get $\left\| E_{r,n}^{[m]} \right\|_1 \to 0$ with rates.
From (23.91) we get

$$\left\| \theta_{r,n}^{[m]} f - f \right\|_1 \leq \sum_{\widetilde{j}=1}^{m} \left| \delta_{\widetilde{j},r}^{[m]} \right| \left(\sum_{|\alpha|=\widetilde{j}} \frac{\left| c_{\alpha,n,\widetilde{j}} \right|}{\prod_{i=1}^{N} \alpha_i!} \|f_\alpha\|_1 \right) + R.H.S.(23.91), \tag{23.92}$$

given that $\|f_\alpha\|_1 < \infty$, $|\alpha| = \widetilde{j}$, $\widetilde{j} = 1, ..., m$.
As $n \to \infty$, assuming $\xi_n \to 0$ and $c_{\alpha,n,\widetilde{j}} \to 0$, we obtain $\left\| \theta_{r,n}^{[m]}(f) - f \right\|_1 \to 0$, that is $\theta_{r,n}^{[m]} \to I$ in L_1 norm, with rates.

We present

Theorem 23.25 *Let $f \in C^m(\mathbb{R}^N)$, $m, N \in \mathbb{N}$, $m > r$, with $f_\alpha \in L_1(\mathbb{R}^N)$, $|\alpha| = m$, $x \in \mathbb{R}^N$, $\xi_n \in (0, 1]$, $n \in \mathbb{N}$. For $\widetilde{j} = 1, ..., m$, and $\alpha := (\alpha_1, ..., \alpha_N)$, $\alpha_i \in \mathbb{Z}^+$, $i = 1, ..., N$, $|\alpha| := \sum_{i=1}^{N} \alpha_i = \widetilde{j}$, call*

$$\widetilde{c}_{\alpha,n,\widetilde{j}} := \frac{1}{\left(\sqrt{\pi\xi_n}\right)^N} \int_{\mathbb{R}^N} \prod_{i=1}^{N} s_i^{\alpha_i} e^{-\frac{\left(\sum_{i=1}^{N} s_i^2\right)}{\xi_n}} ds_1...ds. \tag{23.93}$$

Then

$$\left\| \widetilde{E}_{r,n}^{[m]} \right\|_1 = \left\| W_{r,n}^{[m]}(f; x) - f(x) - \sum_{\widetilde{j}=1}^{m} \delta_{\widetilde{j},r}^{[m]} \left(\sum_{|\alpha|=\widetilde{j}} \frac{\widetilde{c}_{\alpha,n,\widetilde{j}} f_\alpha(x)}{\prod_{i=1}^{N} \alpha_i!} \right) \right\|_{1,x} \leq$$

$$\left(\frac{2}{\sqrt{\pi}}\right)^N \left[(1+N)^r + 2^r \left(\frac{\lfloor e(m+r)! \rfloor}{e} \right)^N \right] \sum_{|\alpha|=m} \left(\frac{1}{\prod_{i=1}^{N} \alpha_i!} \right) \left(\sqrt{\xi_n}\right)^{m-r} \omega_r(f_\alpha, \xi_n)_1 . \tag{23.94}$$

As $\xi_n \to 0$, we get $\left\| \widetilde{E}_{r,n}^{[m]} \right\|_1 \to 0$ with rates.

From (23.94) we get

$$\left\| W_{r,n}^{[m]} f - f \right\|_1 \leq \sum_{\widetilde{j}=1}^{m} \left| \delta_{\widetilde{j},r}^{[m]} \right| \left(\sum_{|\alpha|=\widetilde{j}} \frac{\left| \widetilde{c}_{\alpha,n,\widetilde{j}} \right|}{\prod_{i=1}^{N} \alpha_i!} \| f_\alpha \|_1 \right) + R.H.S.(23.94), \tag{23.95}$$

given that $\| f_\alpha \|_1 < \infty$, $|\alpha| = \widetilde{j}$, $\widetilde{j} = 1, ..., m$.

As $n \to \infty$, assuming $\xi_n \to 0$, we obtain $\left\| W_{r,n}^{[m]}(f) - f \right\|_1 \to 0$, that is $W_{r,n}^{[m]} \to I$ in L_1 norm, with rates.

Proof We apply Theorems 23.8, 23.12 and 23.24. ∎

23.3.3 Global Smoothness Preservation and Simultaneous Approximation of $W_{r,n}^{[m]}$

We need

Definition 23.26 ([3], p. 34) Let $f \in C(\mathbb{R}^N)$, $N \geq 1$, $m \in \mathbb{N}$, the mth modulus of smoothness for $1 \leq p \leq \infty$, is given by

$$\omega_m(f;h)_p := \sup_{\|t\|_2 \le h} \left\| \Delta_t^m(f) \right\|_{p,x}, \tag{23.96}$$

$h > 0$, where

$$\Delta_t^m f(x) := \sum_{j=0}^{m} (-1)^{m-j} \binom{m}{j} f(x + jt). \tag{23.97}$$

Denote

$$\omega_m(f;h)_\infty = \omega_m(f,h). \tag{23.98}$$

Above, $x, t \in \mathbb{R}^N$.

We present the related global smoothness preservation result

Theorem 23.27 *We assume $W_{r,n}^{[\widetilde{m}]}(f;x) \in \mathbb{R}$, $\widetilde{m} \in \mathbb{Z}_+$, $\forall\, x \in \mathbb{R}$. Let $h > 0$, $f \in C(\mathbb{R}^N)$, $N \ge 1$.*

(i) Assume $\omega_m(f,h) < \infty$. Then

$$\omega_m\left(W_{r,n}^{[\widetilde{m}]} f, h\right) \le \left(\sum_{\widetilde{j}=0}^{r} \left|\alpha_{\widetilde{j},r}^{[\widetilde{m}]}\right|\right) \omega_m(f,h). \tag{23.99}$$

(ii) Assume $f \in \left(C\left(\mathbb{R}^N\right) \cap L_1\left(\mathbb{R}^N\right)\right)$. Then

$$\omega_m\left(W_{r,n}^{[\widetilde{m}]} f, h\right)_1 \le \left(\sum_{\widetilde{j}=0}^{r} \left|\alpha_{\widetilde{j},r}^{[\widetilde{m}]}\right|\right) \omega_m(f,h)_1. \tag{23.100}$$

(iii) Assume $f \in \left(C\left(\mathbb{R}^N\right) \cap L_p\left(\mathbb{R}^N\right)\right)$, $p > 1$. Then

$$\omega_m\left(W_{r,n}^{[\widetilde{m}]} f, h\right)_p \le \left(\sum_{\widetilde{j}=0}^{r} \left|\alpha_{\widetilde{j},r}^{[\widetilde{m}]}\right|\right) \omega_m(f,h)_p. \tag{23.101}$$

Proof Direct application of [3] Theorem 3.2, p. 35. ■

We make

Remark 23.28 Let $r = 1$, $\widetilde{m} \in \mathbb{Z}_+$, then $\alpha_{0,1}^{[\widetilde{m}]} = 0$, $\alpha_{1,1}^{[\widetilde{m}]} = 1$. Hence

$$W_{1,n}^{[\widetilde{m}]}(f;x) = \frac{1}{\left(\sqrt{\pi \xi_n}\right)^N} \int_{\mathbb{R}^N} f(x+s)\, e^{-\left(\frac{\sum_{i=1}^{N} s_i^2}{\xi_n}\right)} ds_1 ... ds_N =: W_n(f;x). \tag{23.102}$$

By Theorem 23.27, we get

Theorem 23.29 *We suppose $W_n(f;x) \in \mathbb{R}$, $\forall\, x \in \mathbb{R}$. Let $h > 0$, $f \in C\left(\mathbb{R}^N\right)$, $N \geq 1$.*

(i) Assume $\omega_m(f,h) < \infty$. Then

$$\omega_m(W_n f, h) \leq \omega_m(f,h). \tag{23.103}$$

(ii) Assume $f \in \left(C\left(\mathbb{R}^N\right) \cap L_1\left(\mathbb{R}^N\right)\right)$. Then

$$\omega_m(W_n f, h)_1 \leq \omega_m(f,h)_1. \tag{23.104}$$

(iii) Assume $f \in \left(C\left(\mathbb{R}^N\right) \cap L_p\left(\mathbb{R}^N\right)\right)$, $p > 1$. Then

$$\omega_m(W_n f, h)_p \leq \omega_m(f,h)_p. \tag{23.105}$$

Next, we get an optimality result

Proposition 23.30 *Above inequality (23.103):*

$$\omega_m(W_n f, h) \leq \omega_m(f,h)$$

is sharp, namely it is attained by any

$$f_j^*(x) = x_j^m,\ j = 1, ..., N,\ x = \left(x_1, ..., x_j, ..., x_N\right) \in \mathbb{R}^N. \tag{23.106}$$

Proof Apply Proposition 3.5, p. 38, of [3]. ■

We need

Theorem 23.31 ([3], p. 39) *Let $f \in C^l\left(\mathbb{R}^N\right)$, $l, N \in \mathbb{N}$. Here, μ_{ξ_n} is a Borel probability measure on $\mathbb{R}^N$, $\xi_n > 0$, $(\xi_n)_{n\in\mathbb{N}}$ a bounded sequence. Let $\beta := (\beta_1, ..., \beta_N)$, $\beta_i \in \mathbb{Z}^+, i = 1, ..., N$; $|\beta| := \sum_{i=1}^{N} \beta_i = l$. Here $f(x+sj)$, $x, s \in \mathbb{R}^N$, is μ_{ξ_n}-integrable w.r.t. s, for $j = 1, ..., r$. There exist μ_{ξ_n}-integrable functions $h_{i_1,j}$, $h_{\beta_1,i_2,j}$, $h_{\beta_1,\beta_2,i_3,j}$, ..., $h_{\beta_1,\beta_2,...,\beta_{N-1},i_N,j} \geq 0$ $(j = 1, ..., r)$ on $\mathbb{R}^N$ such that*

$$\left|\frac{\partial^{i_1} f(x+sj)}{\partial x_1^{i_1}}\right| \leq h_{i_1,j}(s),\quad i_1 = 1, ..., \beta_1, \tag{23.107}$$

$$\left|\frac{\partial^{\beta_1+i_2} f(x+sj)}{\partial x_2^{i_2}\partial x_1^{\beta_1}}\right| \leq h_{\beta_1,i_2,j}(s),\quad i_2 = 1, ..., \beta_2,$$

$$\vdots$$

$$\left|\frac{\partial^{\beta_1+\beta_2+...+\beta_{N-1}+i_N} f(x+sj)}{\partial x_N^{i_N} \partial x_{N-1}^{\beta_{N-1}} ... \partial x_2^{\beta_2} \partial x_1^{\beta_1}}\right| \leq h_{\beta_1,\beta_2,...,\beta_{N-1},i_N,j}(s), \quad i_N = 1, ..., \beta_N,$$

$\forall\, x, s \in \mathbb{R}^N$.

Then, both of the next exist and

$$\left(\theta_{r,n}^{[\widetilde{m}]}(f;x)\right)_\beta = \theta_{r,n}^{[\widetilde{m}]}\left(f_\beta; x\right), \quad \widetilde{m} \in \mathbb{Z}_+. \tag{23.108}$$

In particular it holds

$$\left(W_{r,n}^{[\widetilde{m}]}(f;x)\right)_\beta = W_{r,n}^{[\widetilde{m}]}\left(f_\beta; x\right), \tag{23.109}$$

when

$$d\mu_{\xi_n} = \frac{1}{\left(\sqrt{\pi\xi_n}\right)^N} e^{-\left(\frac{\sum_{i=1}^N s_i^2}{\xi_n}\right)} ds_1...ds_N.$$

Corollary 23.32 (by Theorem 23.31, $r = 1$) *It holds*

$$\left(W_n(f;x)\right)_\beta = W_n\left(f_\beta; x\right). \tag{23.110}$$

We present simultaneous global smoothness results.

Theorem 23.33 *Let $h > 0$ and the assumptions of Theorem 23.31 are valid for $d\mu_{\xi_n} = \frac{1}{\left(\sqrt{\pi\xi_n}\right)^N} e^{-\left(\frac{\sum_{i=1}^N s_i^2}{\xi_n}\right)} ds_1...ds_N$. Here $\gamma = 0, \beta$ $(0 = (0, ..., 0))$, $\widetilde{m} \in \mathbb{Z}_+$.*

(i) Assume $\omega_m\left(f_\gamma, h\right) < \infty$. Then

$$\omega_m\left(\left(W_{r,n}^{[\widetilde{m}]}(f)\right)_\gamma, h\right) \leq \left(\sum_{\widetilde{j}=0}^r \left|\alpha_{\widetilde{j},r}^{[\widetilde{m}]}\right|\right) \omega_m\left(f_\gamma, h\right). \tag{23.111}$$

(ii) Additionally suppose $f_\gamma \in L_1\left(\mathbb{R}^N\right)$. Then

$$\omega_m\left(\left(W_{r,n}^{[\widetilde{m}]}(f)\right)_\gamma, h\right)_1 \leq \left(\sum_{\widetilde{j}=0}^r \left|\alpha_{\widetilde{j},r}^{[\widetilde{m}]}\right|\right) \omega_m\left(f_\gamma, h\right)_1. \tag{23.112}$$

(iii) Additionally suppose $f_\gamma \in L_p\left(\mathbb{R}^N\right)$, $p > 1$. Then

$$\omega_m\left(\left(W_{r,n}^{[\widetilde{m}]}(f)\right)_\gamma, h\right)_p \leq \left(\sum_{\widetilde{j}=0}^r \left|\alpha_{\widetilde{j},r}^{[\widetilde{m}]}\right|\right) \omega_m\left(f_\gamma, h\right)_p. \tag{23.113}$$

It follows

Corollary 23.34 (to Theorem 23.33) *Let $h > 0$, $r = 1$ and $\gamma = 0, \beta$.*

(i) Assume $\omega_m\left(f_\gamma, h\right) < \infty$. Then

$$\omega_m\left(\left(W_n\left(f\right)\right)_\gamma, h\right) \leq \omega_m\left(f_\gamma, h\right). \tag{23.114}$$

(ii) Additionally suppose $f_\gamma \in L_1\left(\mathbb{R}^N\right)$. Then

$$\omega_m\left(\left(W_n\left(f\right)\right)_\gamma, h\right)_1 \leq \omega_m\left(f_\gamma, h\right)_1. \tag{23.115}$$

(iii) Additionally suppose $f_\gamma \in L_p\left(\mathbb{R}^N\right)$, $p > 1$. Then

$$\omega_m\left(\left(W_n\left(f\right)\right)_\gamma, h\right)_p \leq \omega_m\left(f_\gamma, h\right)_p. \tag{23.116}$$

Next comes multi-simultaneous approximation.

Theorem 23.35 *Let $f \in C^{m+l}\left(\mathbb{R}^N\right)$, $m, l, N \in \mathbb{N}$, $m > r$. The assumptions of Theorem 23.31 are valid for $d\mu_{\xi_n} = \frac{1}{\left(\sqrt{\pi\xi_n}\right)^N} e^{-\left(\frac{\sum_{i=1}^N s_i^2}{\xi_n}\right)} ds_1 \ldots ds_N$. Call $\gamma = 0, \beta$. Assume $\left\| f_{\gamma+\alpha}\right\|_\infty < \infty$, and let*

$$\widetilde{u}_{\xi_n}\left(\alpha\right) := \frac{1}{\left(\sqrt{\pi\xi_n}\right)^N} \int_{\mathbb{R}^N} \left(\prod_{i=1}^N \left|s_i\right|^{\alpha_i}\right)\left(1 + \frac{\|s\|_2}{\xi_n}\right)^r e^{-\frac{\left(\sum_{i=1}^N s_i^2\right)}{\xi_n}} ds_1 \ldots ds_N,$$

for all $\alpha_j \in \mathbb{Z}^+$, $j = 1, \ldots, N$, $|\alpha| := \sum_{j=1}^N \alpha_j = m$, $\xi_n \in (0, 1]$.

For $\widetilde{j} = 1, \ldots, m$, and $\alpha := (\alpha_1, \ldots, \alpha_N)$, $\alpha_i \in \mathbb{Z}^+$, $i = 1, \ldots, N$, $|\alpha| := \sum_{i=1}^N \alpha_i = \widetilde{j}$, call

$$\widetilde{c}_{\alpha,n,\widetilde{j}} := \frac{1}{\left(\sqrt{\pi\xi_n}\right)^N} \int_{\mathbb{R}^N} \prod_{i=1}^N s_i^{\alpha_i} e^{-\frac{\left(\sum_{i=1}^N s_i^2\right)}{\xi_n}} ds_1 \ldots ds_N. \tag{23.117}$$

Then

$$\left\| \left(W_{r,n}^{[m]}\left(f;\cdot\right)\right)_\gamma - f_\gamma\left(\cdot\right) - \sum_{\widetilde{j}=1}^m \delta_{\widetilde{j},r}^{[m]} \left(\sum_{\substack{\alpha_1,\ldots,\alpha_N \geq 0;\\ |\alpha| = \widetilde{j}}} \frac{\widetilde{c}_{\alpha,n,\widetilde{j}} f_{\gamma+\alpha}\left(\cdot\right)}{\prod_{i=1}^N \alpha_i!}\right)\right\|_\infty \tag{23.118}$$

$$\leq \sum_{\substack{\alpha_1,\ldots,\alpha_N \geq 0; \\ |\alpha|=m}} \frac{\left(\omega_r\left(f_{\gamma+\alpha}, \xi_n\right)\right)}{\left(\prod_{i=1}^{N} \alpha_i!\right)} \widetilde{u}_{\xi_n}(\alpha).$$

Proof Based on Theorems 23.14, 23.31. ■

We continue with

Theorem 23.36 *Let $f \in C_B^l\left(\mathbb{R}^N\right)$, $l, N \in \mathbb{N}$ (functions l-times continuously differentiable and bounded). The assumptions of Theorem 23.31 are valid for $d\mu_{\xi_n} = \frac{1}{\left(\sqrt{\pi \xi_n}\right)^N} e^{-\left(\frac{\sum_{i=1}^{N} s_i^2}{\xi_n}\right)} ds_1 \ldots ds_N$. Call $\gamma = 0, \beta$, $\xi_n \in (0,1]$. Then*

(1)

$$\left\|\left(W_{r,n}^{[0]} f\right)_\gamma - f_\gamma\right\|_\infty \leq \left(\frac{2}{\sqrt{\pi}}\right)^N \left[(1+N)^r + 2^r \left(\frac{\lfloor er! \rfloor}{e}\right)^N\right] \frac{\omega_r\left(f_\gamma, \xi_n\right)}{\left(\sqrt{\xi_n}\right)^r}. \tag{23.119}$$

(2) Let $r \in \mathbb{N} - \{1\}$, $0 < \alpha \leq 1$, and consider $f_\gamma : \omega_r\left(f_\gamma, t\right) \leq K t^{r-1+\alpha}$, $\forall\, t > 0$, where $K > 0$ is independent of t. We get

$$\frac{\omega_r\left(f_\gamma, \xi_n\right)}{\left(\sqrt{\xi_n}\right)^r} \leq K \xi_n^{\left(\frac{r}{2}+\alpha-1\right)}, \tag{23.120}$$

and as $n \to \infty$ and $\xi_n \to 0$, we derive $\left(W_{r,n}^{[0]} f\right)_\gamma \to f_\gamma$ uniformly.

Proof By Theorems 23.16 and 23.31. ■

We continue with

Theorem 23.37 *Let $f \in C^{m+l}\left(\mathbb{R}^N\right)$, $m, l, N \in \mathbb{N}$, $m > r$. The assumptions of Theorem 23.31 are valid for $d\mu_{\xi_n} = \frac{1}{\left(\sqrt{\pi \xi_n}\right)^N} e^{-\left(\frac{\sum_{i=1}^{N} s_i^2}{\xi_n}\right)} ds_1 \ldots ds_N$. Call $\gamma = 0, \beta$. Let $f_{(\gamma+\alpha)} \in L_p\left(\mathbb{R}^N\right)$, $|\alpha| = m$, $x \in \mathbb{R}^N$, and $p, q > 1 : \frac{1}{p} + \frac{1}{q} = 1$, $\xi_n \in (0,1]$, $n \in \mathbb{N}$. For $\widetilde{j} = 1, \ldots, m$, and $\alpha := (\alpha_1, \ldots, \alpha_N)$, $\alpha_i \in \mathbb{Z}^+$, $i = 1, \ldots, N$, $|\alpha| := \sum_{i=1}^{N} \alpha_i = \widetilde{j}$, call*

$$\widetilde{c}_{\alpha,n,\widetilde{j}} := \frac{1}{\left(\sqrt{\pi \xi_n}\right)^N} \int_{\mathbb{R}^N} \prod_{i=1}^{N} s_i^{\alpha_i} e^{-\frac{\left(\sum_{i=1}^{N} s_i^2\right)}{\xi_n}} ds_1 \ldots ds_N. \tag{23.121}$$

Then

$$\left\| \left(W_{r,n}^{[m]}(f;x)\right)_{\gamma} - f_{\gamma}(x) - \sum_{\widetilde{j}=1}^{m} \delta_{\widetilde{j},r}^{[m]} \left(\sum_{|\alpha|=\widetilde{j}} \frac{\widetilde{c}_{\alpha,n,\widetilde{j}} f_{\gamma+\alpha}(x)}{\prod_{i=1}^{N} \alpha_i!} \right) \right\|_{p,x}$$

$$\leq \left(\frac{2}{\sqrt{\pi}}\right)^{\frac{N}{p}} \left(\frac{m}{(q(m-1)+1)^{\frac{1}{q}}}\right) \left[(1+N)^{rp} + 2^{rp}\Gamma^N((m+r)p+1,1)\right]^{\frac{1}{p}}$$

$$\left(\sum_{|\alpha|=m} \frac{1}{\prod_{i=1}^{N} \alpha_i!}\right) \left(\sqrt{\xi_n}\right)^{(m-r)} \omega_r\left(f_{\gamma+\alpha}, \xi_n\right)_p. \tag{23.122}$$

Proof By Theorems 23.19, 23.31. ■

We give also

Theorem 23.38 *Let $f \in C^l\left(\mathbb{R}^N\right)$, $l, N \in \mathbb{N}$. The assumptions of Theorem 23.31 are valid for $d\mu_{\xi_n} = \frac{1}{\left(\sqrt{\pi\xi_n}\right)^N} e^{-\left(\frac{\sum_{i=1}^{N} s_i^2}{\xi_n}\right)} ds_1...ds_N$. Call $\gamma = 0, \beta$. Let $f_{\gamma} \in L_p\left(\mathbb{R}^N\right)$, $x \in \mathbb{R}^N$; $p, q > 1 : \frac{1}{p} + \frac{1}{q} = 1$, $\xi_n \in (0,1]$, $n \in \mathbb{N}$. Then*

(1)

$$\left\| \left(W_{r,n}^{[0]} f\right)_{\gamma} - f_{\gamma} \right\|_p \leq$$

$$\left(\frac{2}{\sqrt{\pi}}\right)^{\frac{N}{p}} \left[(1+N)^{rp} + 2^{rp}\Gamma^N(rp+1,1)\right]^{\frac{1}{p}} \frac{\omega_r\left(f_{\gamma}, \xi_n\right)_p}{\left(\sqrt{\xi_n}\right)^r}. \tag{23.123}$$

(2) Let $r \in \mathbb{N} - \{1\}$, $0 < \alpha \leq 1$, and $f_{\gamma} : \omega_r\left(f_{\gamma}, t\right)_p \leq K t^{r-1+\alpha}$, $\forall\, t > 0$, $K > 0$ independent of t. In this case

$$\frac{\omega_r\left(f_{\gamma}, \xi_n\right)_p}{\left(\sqrt{\xi_n}\right)^r} \leq K \xi_n^{\left(\frac{r}{2}+\alpha-1\right)}. \tag{23.124}$$

As $n \to +\infty$ and $\xi_n \to 0$, then $\left(W_{r,n}^{[0]}(f)\right)_{\gamma} \overset{\|\cdot\|_p}{\to} f_{\gamma}$.

Proof By Theorems 23.21 and 23.31. ■

We continue with

Theorem 23.39 *Let $f \in C^{l}\left(\mathbb{R}^{N}\right)$, $l, N \in \mathbb{N}$. The assumptions of Theorem 23.31 are valid for* $d\mu_{\xi_n} = \frac{1}{\left(\sqrt{\pi \xi_n}\right)^N} e^{-\left(\frac{\sum_{i=1}^{N} s_i^2}{\xi_n}\right)} ds_1...ds_N$. *Call $\gamma = 0, \beta$. Let $f_\gamma \in L_1\left(\mathbb{R}^{N}\right)$, $x \in \mathbb{R}^{N}$, $\xi_n \in (0,1]$, $n \in \mathbb{N}$. Then*

(1)

$$\left\| \left(W_{r,n}^{[0]}(f)\right)_\gamma - f_\gamma \right\|_1 \leq$$

$$\left(\frac{2}{\sqrt{\pi}}\right)^N \left[(1+N)^r + 2^r \left(\frac{\lfloor er! \rfloor}{e}\right)^N\right] \frac{\omega_r\left(f_\gamma, \xi_n\right)_1}{\left(\sqrt{\xi_n}\right)^r}. \tag{23.125}$$

(2) Let $r \in \mathbb{N} - \{1\}$, $0 < \alpha \leq 1$, and $f_\gamma : \omega_r\left(f_\gamma, t\right)_1 \leq K t^{r-1+\alpha}$, $\forall\, t > 0$, $K > 0$ independent of t. In this case

$$\frac{\omega_r\left(f_\gamma, \xi_n\right)_1}{\left(\sqrt{\xi_n}\right)^r} \leq K \xi_n^{\left(\frac{r}{2}+\alpha-1\right)}, \tag{23.126}$$

and as $n \to \infty$ and $\xi_n \to 0$, we get $\left(W_{r,n}^{[0]}(f)\right)_\gamma \overset{\|\cdot\|_1}{\to} f_\gamma$.

Proof By Theorems 23.23 and 23.31. ∎

We continue with

Theorem 23.40 *Let $f \in C^{m+l}\left(\mathbb{R}^{N}\right)$, $m, l, N \in \mathbb{N}$, $m > r$. The assumptions of Theorem 23.31 for* $d\mu_{\xi_n} = \frac{1}{\left(\sqrt{\pi \xi_n}\right)^N} e^{-\left(\frac{\sum_{i=1}^{N} s_i^2}{\xi_n}\right)} ds_1...ds_N$ *are valid. Call $\gamma = 0, \beta$. Let $f_{(\gamma+\alpha)} \in L_1\left(\mathbb{R}^{N}\right)$, $|\alpha| = m$, $x \in \mathbb{R}^{N}$, $\xi_n \in (0,1]$, $n \in \mathbb{N}$. For $\widetilde{j} = 1, ..., m$, and $\alpha := (\alpha_1, ..., \alpha_N)$, $\alpha_i \in \mathbb{Z}^{+}$, $i = 1, ..., N$, $|\alpha| := \sum_{i=1}^{N} \alpha_i = \widetilde{j}$, call*

$$\widetilde{c}_{\alpha,n,\widetilde{j}} := \frac{1}{\left(\sqrt{\pi \xi_n}\right)^N} \int_{\mathbb{R}^N} \prod_{i=1}^{N} s_i^{\alpha_i} e^{-\frac{\left(\sum_{i=1}^{N} s_i^2\right)}{\xi_n}} ds_1...ds_N. \tag{23.127}$$

Then

$$\left\| \left(W_{r,n}^{[m]}(f;x)\right)_\gamma - f_\gamma(x) - \sum_{\widetilde{j}=1}^{m} \delta_{\widetilde{j},r}^{[m]} \left(\sum_{|\alpha|=\widetilde{j}} \frac{\widetilde{c}_{\alpha,n,\widetilde{j}} f_{(\gamma+\alpha)}(x)}{\prod_{i=1}^{N} \alpha_i!} \right) \right\|_{1,x}$$

$$\leq \left(\frac{2}{\sqrt{\pi}}\right)^{N}\left(\sum_{|\alpha|=m}\left(\frac{1}{\prod_{i=1}^{N}\alpha_i!}\right)\omega_r\left(f_{\gamma+\alpha},\xi_n\right)_1\right)$$

$$\left(\sqrt{\xi_n}\right)^{m-r}\left[(1+N)^r+2^r\left(\frac{\lfloor e\,(m+r)!\rfloor}{e}\right)^{N}\right]. \quad (23.128)$$

Proof By Theorems 23.25, 23.31. ■

23.3.4 Voronovskaya Asymptotic Expansions for $W_{r,n}^{[m]}$

Here we will apply the following general theorem.

Theorem 23.41 *Let $f \in C^m\left(\mathbb{R}^N\right)$, $m, N \in \mathbb{N}$, with all $\|f_\alpha\|_\infty \leq M$, $M > 0$, all $\alpha : |\alpha| = m$. Let $\xi_n > 0$, $(\xi_n)_{n\in\mathbb{N}}$ bounded sequence, μ_{ξ_n} probability Borel measures on $\mathbb{R}^N$.*

Call $c_{\alpha,n,\widetilde{j}} := \int_{\mathbb{R}^N}\left(\prod_{i=1}^{N} s_i^{\alpha_i}\right)d\mu_{\xi_n}(s)$, *all* $|\alpha| = \widetilde{j} = 1, ..., m-1$. *Suppose* $\left(\sqrt{\xi_n}\right)^{-m}\int_{\mathbb{R}^N}\left(\prod_{i=1}^{N}|s_i|^{\alpha_i}\right)d\mu_{\xi_n}(s) \leq \rho$, *all* $\alpha : |\alpha| = m$, $\rho > 0$, *for any such* $(\xi_n)_{n\in\mathbb{N}}$. *Also* $0 < \gamma \leq 1$, $x \in \mathbb{R}^N$. *Then*

$$\theta_{r,n}^{[m]}(f;x) - f(x) = \sum_{\widetilde{j}=1}^{m-1}\delta_{\widetilde{j},r}^{[m]}\left(\sum_{|\alpha|=\widetilde{j}}\frac{c_{\alpha,n,\widetilde{j}}\,f_\alpha(x)}{\left(\prod_{i=1}^{N}\alpha_i!\right)}\right) + 0\left(\left(\sqrt{\xi_n}\right)^{m-\gamma}\right). \quad (23.129)$$

When $m = 1$, the sum collapses.

Above we assume $\theta_{r,n}^{[m]}(f;x) \in \mathbb{R}$, $\forall\, x \in \mathbb{R}^N$.

Proof The proof of Theorem 23.41 is the same as the proof of Theorem 4.2, p. 53 of [3], just replace there ξ_n by $\sqrt{\xi_n}$. ■

We give

Theorem 23.42 *Let $f \in C^m\left(\mathbb{R}^N\right)$, $m, N \in \mathbb{N}$, with all $\|f_\alpha\|_\infty \leq M$, $M > 0$, all $\alpha : |\alpha| = m$, $\xi_n \in (0, 1]$, and $d\mu_{\xi_n} = \frac{1}{\left(\sqrt{\pi\xi_n}\right)^N}e^{-\left(\frac{\sum_{i=1}^{N}s_i^2}{\xi_n}\right)}ds_1...ds_N$ on $\mathbb{R}^N$. Assume $W_{r,n}^{[m]}(f;x) \in \mathbb{R}$, $\forall\, x \in \mathbb{R}^N$.*

Call $\widetilde{c}_{\alpha,n,\widetilde{j}} = \frac{1}{\left(\sqrt{\pi\xi_n}\right)^N} \int_{\mathbb{R}^N} \left(\prod_{i=1}^{N} s_i^{\alpha_i}\right) e^{-\left(\frac{\sum_{i=1}^{N} s_i^2}{\xi_n}\right)} ds_1 ... ds_N$, *all* $|\alpha| = \widetilde{j} = 1, ..., m - 1$. *Let* $0 < \gamma \leq 1$, $x \in \mathbb{R}^N$. *Then*

$$W_{r,n}^{[m]}(f;x) - f(x) = \sum_{\widetilde{j}=1}^{m-1} \delta_{\widetilde{j},r}^{[m]} \left(\sum_{|\alpha|=\widetilde{j}} \frac{\widetilde{c}_{\alpha,n,\widetilde{j}} f_\alpha(x)}{\left(\prod_{i=1}^{N} \alpha_i!\right)} \right) + 0\left(\left(\sqrt{\xi_n}\right)^{m-\gamma}\right). \tag{23.130}$$

When $m = 1$, *the sum collapses.*

Proof By Theorems 23.12 and 23.41. Here $\rho = \left(\frac{\mu}{\sqrt{\pi}}\right)^N$. ■

Corollary 23.43 (to Theorem 23.42) *Let* $f \in C^1(\mathbb{R}^N)$, $N \geq 1$, *with all* $\left\|\frac{\partial f}{\partial x_i}\right\|_\infty \leq M$, $M > 0$, $i = 1, ..., N$. *Let* $0 < \gamma \leq 1$, $x \in \mathbb{R}^N$. *Then*

$$W_{r,n}^{[1]}(f;x) - f(x) = 0\left(\left(\sqrt{\xi_n}\right)^{1-\gamma}\right). \tag{23.131}$$

Above we assume $W_{r,n}^{[1]}(f;x) \in \mathbb{R}$, $\forall\, x \in \mathbb{R}^N$.

Proof By Theorems 23.12 and 23.41, here it is $\rho = 1$. ■

Corollary 23.44 (to Theorem 23.42) *Let* $f \in C^2(\mathbb{R}^2)$, *with all* $\left\|\frac{\partial^2 f}{\partial x_1^2}\right\|_\infty$, $\left\|\frac{\partial^2 f}{\partial x_2^2}\right\|_\infty$, $\left\|\frac{\partial^2 f}{\partial x_1 \partial x_2}\right\|_\infty \leq M$, $M > 0$, $\xi_n \in (0, 1]$, $n \in \mathbb{N}$. *Call*

$$c_1 = \int_{\mathbb{R}^2} s_1 d\mu_{\xi_n}(s), \quad c_2 = \int_{\mathbb{R}^2} s_2 d\mu_{\xi_n}(s), \tag{23.132}$$

where

$$d\mu_{\xi_n} = \frac{1}{\sqrt{\pi\xi_n}} e^{-\frac{(s_1^2+s_2^2)}{\xi_n}} ds_1 ds_2.$$

Let $0 < \gamma \leq 1$, $x \in \mathbb{R}^2$. *Then*

$$W_{r,n}^{[2]}(f;x) - f(x) = \left(\sum_{j=1}^{r} \alpha_{j,r}^{[2]} j\right)\left(c_1 \frac{\partial f}{\partial x_1}(x) + c_2 \frac{\partial f}{\partial x_2}(x)\right) + 0\left(\left(\sqrt{\xi_n}\right)^{2-\gamma}\right). \tag{23.133}$$

Proof By Theorems 23.12 and 23.41, here it is $\rho = 1$. ■

We also give

Theorem 23.45 *Let $f \in C^{m+l}\left(\mathbb{R}^N\right)$, $m, l, N \in \mathbb{N}$. Assumptions of Theorem 23.31 are valid for $d\mu_{\xi_n}(s) = \frac{1}{\left(\sqrt{\pi\xi_n}\right)^N} e^{-\left(\frac{\sum_{i=1}^N s_i^2}{\xi_n}\right)} ds_1 ... ds_N$. Call $\gamma = 0, \beta$. Suppose $\left\| f_{\gamma+\alpha} \right\|_\infty \leq M$, $M > 0$, for all $\alpha : |\alpha| = m$, $\xi_n \in (0, 1]$, $n \in \mathbb{N}$.*

Call $\widetilde{c}_{\alpha,n,\widetilde{j}} := \frac{1}{\left(\sqrt{\pi\xi_n}\right)^N} \int_{\mathbb{R}^N} \left(\prod_{i=1}^N s_i^{\alpha_i} \right) e^{-\left(\frac{\sum_{i=1}^N s_i^2}{\xi_n}\right)} ds_1 ... ds_N$, all $|\alpha| = \widetilde{j} = 1, ..., m-1$; $0 < \gamma \leq 1$, $x \in \mathbb{R}^N$. Then

$$\left(W_{r,n}^{[m]}(f; x)\right)_\gamma - f_\gamma(x) = \sum_{\widetilde{j}=1}^{m-1} \delta_{\widetilde{j},r}^{[m]} \left(\sum_{|\alpha|=\widetilde{j}} \frac{\widetilde{c}_{\alpha,n,\widetilde{j}} f_{\gamma+\alpha}(x)}{\left(\prod_{i=1}^N \alpha_i!\right)} \right) + 0\left(\left(\sqrt{\xi_n}\right)^{m-\gamma}\right). \tag{23.134}$$

When $m = 1$, the sum collapses.

Proof Use of Theorem 23.12 and Theorem 4.6, p. 54 of [3], which is true also for $\sqrt{\xi_n}$ instead of ξ_n in the main assumption and conclusion. Here it is $\rho = \left(\frac{\mu}{\sqrt{\pi}}\right)^N$. ■

23.3.5 Simultaneous Approximation by Multivariate Complex $W_{r,n}^{[m]}$

We make

Remark 23.46 We consider here complex valued Borel measurable functions $f : \mathbb{R}^N \to \mathbb{C}$ such that $f = f_1 + i f_2$, $i = \sqrt{-1}$, where $f_1, f_2 : \mathbb{R}^N \to \mathbb{R}$ are implied to be real valued Borel measurable functions.

We define the multivariate complex Gauss–Weierstrass singular operators

$$W_{r,n}^{[m]}(f; x) := W_{r,n}^{[m]}(f_1; x) + i W_{r,n}^{[m]}(f_2; x), \quad x \in \mathbb{R}^N. \tag{23.135}$$

We assume that $W_{r,n}^{[m]}\left(f_j; x\right) \in \mathbb{R}$, $\forall\, x \in \mathbb{R}^N$, $j = 1, 2$.

One notices easily that

$$\left| W_{r,n}^{[m]}(f; x) - f(x) \right| \leq \left| W_{r,n}^{[m]}(f_1; x) - f_1(x) \right| + \left| W_{r,n}^{[m]}(f_2; x) - f_2(x) \right| \tag{23.136}$$

also

$$\left\| W_{r,n}^{[m]}(f; x) - f(x) \right\|_{\infty,x} \leq \left\| W_{r,n}^{[m]}(f_1; x) - f_1(x) \right\|_{\infty,x} + \left\| W_{r,n}^{[m]}(f_2; x) - f_2(x) \right\|_{\infty,x} \tag{23.137}$$

and

$$\left\| W_{r,n}^{[m]}(f) - f \right\|_p \leq \left\| W_{r,n}^{[m]}(f_1) - f_1 \right\|_p + \left\| W_{r,n}^{[m]}(f_2) - f_2 \right\|_p, \ p \geq 1. \quad (23.138)$$

Furthermore, it holds

$$f_\alpha(x) = f_{1,\alpha}(x) + i f_{2,\alpha}(x), \quad (23.139)$$

where α denotes a partial derivative of any order and arrangement.

We give

Theorem 23.47 *Let $f : \mathbb{R}^N \to \mathbb{C}$, $N \geq 1$, such that $f = f_1 + i f_2$, $j = 1, 2$. Here $m \in \mathbb{N}$, $f_j \in C^m(\mathbb{R}^N)$, $x \in \mathbb{R}^N$. Assume $\left\| \frac{\partial^m f_j(\cdot,\cdot,\ldots,\cdot)}{\partial x_1^{\alpha_1} \ldots \partial x_N^{\alpha_N}} \right\|_\infty < \infty$, for all $\alpha_i \in \mathbb{Z}^+$, $i = 1, ..., N : |\alpha| := \sum_{i=1}^N \alpha_i = m$, $m > r$. Here $\xi_n \in (0, 1]$, $n \in \mathbb{N}$. For all $\alpha := (\alpha_1, ..., \alpha_N)$, $\alpha_i \in \mathbb{Z}^+$, $i = 1, ..., N$, $|\alpha| := \sum_{i=1}^N \alpha_i = m$ we denote*

$$\widetilde{u}_{\xi_n}(\alpha) := \frac{1}{\left(\sqrt{\pi \xi_n}\right)^N} \int_{\mathbb{R}^N} \left(\prod_{i=1}^N |s_i|^{\alpha_i} \right) \left(1 + \frac{\|s\|_2}{\xi_n} \right)^r e^{-\left(\frac{\sum_{i=1}^N s_i^2}{\xi_n} \right)} ds_1 ... ds_N. \quad (23.140)$$

For $\widetilde{j} = 1, ..., m$, and $\alpha := (\alpha_1, ..., \alpha_N)$, $\alpha_i \in \mathbb{Z}^+$, $i = 1, ..., N$, $|\alpha| := \sum_{i=1}^N \alpha_i = \widetilde{j}$, call

$$\widetilde{c}_{\alpha,n} := \widetilde{c}_{\alpha,n,\widetilde{j}} := \frac{1}{\left(\sqrt{\pi \xi_n}\right)^N} \int_{\mathbb{R}^N} \prod_{i=1}^N s_i^{\alpha_i} e^{-\left(\frac{\sum_{i=1}^N s_i^2}{\xi_n} \right)} ds_1 ... ds_N. \quad (23.141)$$

Then

$$\left\| W_{r,n}^{[m]}(f; x) - f(x) - \sum_{\widetilde{j}=1}^m \delta_{\widetilde{j},r}^{[m]} \left(\sum_{\substack{\alpha_1,...,\alpha_N \geq 0; \\ |\alpha| = \widetilde{j}}} \frac{\widetilde{c}_{\alpha,n,\widetilde{j}} f_\alpha(x)}{\prod_{i=1}^N \alpha_i!} \right) \right\|_{\infty,x}$$

$$\leq \sum_{\substack{\alpha_1,...,\alpha_N \geq 0; \\ |\alpha| = m}} \frac{\left(\omega_r\left(f_{1,\alpha}, \xi_n\right) + \omega_r\left(f_{2,\alpha}, \xi_n\right) \right)}{\left(\prod_{i=1}^N \alpha_i! \right)} \widetilde{u}_{\xi_n}(\alpha). \quad (23.142)$$

Proof By Theorem 23.14. ■

We proceed with

Theorem 23.48 *Let $f:\mathbb{R}^N\to\mathbb{C}: f=f_1+if_2$, $N\geq 1$, $j=1,2$. Here $f_j\in C_B\left(\mathbb{R}^N\right)$ uniformly continuous, $\xi_n\in(0,1]$. Then*

(1)

$$\left\| W_{r,n}^{[0]}f-f\right\|_\infty \leq \left(\frac{2}{\sqrt{\pi}}\right)^N\left[(1+N)^r+2^r\left(\frac{\lfloor er!\rfloor}{e}\right)^N\right]\frac{(\omega_r(f_1,\xi_n)+\omega_r(f_2,\xi_n))}{\left(\sqrt{\xi_n}\right)^r}, \tag{23.143}$$

where $r,n\in\mathbb{N}$.

(2) Let $r\in\mathbb{N}-\{1\}$, $0<\alpha\leq 1$, and $\omega_r\left(f_j,t\right)\leq Kt^{r-1+\alpha}$, $\forall\, t>0$, $K>0$ is independent of t, $j=1,2$. We get

$$\frac{\omega_r\left(f_j,\xi_n\right)}{\left(\sqrt{\xi_n}\right)^r}\leq K\xi_n^{\left(\frac{r}{2}+\alpha-1\right)},\quad j=1,2 \tag{23.144}$$

and as $n\to\infty$ and $\xi_n\to 0$, we get $\left\| W_{r,n}^{[0]}(f)-f\right\|_\infty\to 0$.

Proof Use of Theorem 23.16. ■

Next comes multi-simultaneous approximation

Theorem 23.49 *Let $f:\mathbb{R}^N\to\mathbb{C}: f=f_1+if_2$, $j=1,2$. Here $f_j\in C^{m+l}\left(\mathbb{R}^N\right)$, $m,l,N\in\mathbb{N}$, $m>r$. The assumptions of Theorem 23.31 are valid for $d\mu_{\xi_n}=\frac{1}{\left(\sqrt{\pi\xi_n}\right)^N}e^{-\left(\frac{\sum_{i=1}^N s_i^2}{\xi_n}\right)}ds_1...ds_N$ and f_j. Call $\gamma=0,\beta$. Assume $\left\| f_{j,\gamma+\alpha}\right\|_\infty<\infty$, and let*

$$\widetilde{u}_{\xi_n}(\alpha):=\frac{1}{\left(\sqrt{\pi\xi_n}\right)^N}\int_{\mathbb{R}^N}\left(\prod_{i=1}^N|s_i|^{\alpha_i}\right)\left(1+\frac{\|s\|_2}{\xi_n}\right)^r e^{-\left(\frac{\sum_{i=1}^N s_i^2}{\xi_n}\right)}ds_1...ds_N. \tag{23.145}$$

for all $\alpha_i\in\mathbb{Z}^+$, $i=1,...,N:|\alpha|:=\sum_{i=1}^N\alpha_i=m$, $\xi_n\in(0,1]$.

For $\widetilde{j}=1,...,m$, and $\alpha:=(\alpha_1,...,\alpha_N)$, $\alpha_i\in\mathbb{Z}^+$, $i=1,...,N$, $|\alpha|:=\sum_{i=1}^N\alpha_i=\widetilde{j}$, call

$$\widetilde{c}_{\alpha,n,\widetilde{j}}:=\frac{1}{\left(\sqrt{\pi\xi_n}\right)^N}\int_{\mathbb{R}^N}\prod_{i=1}^N s_i^{\alpha_i}e^{-\left(\frac{\sum_{i=1}^N s_i^2}{\xi_n}\right)}ds_1...ds_N. \tag{23.146}$$

Then

$$\left\| \left(W_{r,n}^{[m]}(f;\cdot)\right)_{\gamma} - f_{\gamma}(\cdot) - \sum_{\widetilde{j}=1}^{m} \delta_{\widetilde{j},r}^{[m]} \left(\sum_{\substack{\alpha_1,\dots,\alpha_N \geq 0; \\ |\alpha| = \widetilde{j}}} \frac{\widetilde{c}_{\alpha,n,\widetilde{j}} f_{\gamma+\alpha}(\cdot)}{\prod_{i=1}^{N} \alpha_i!} \right) \right\|_{\infty}$$

$$\leq \sum_{\substack{\alpha_1,\dots,\alpha_N \geq 0; \\ |\alpha| = m}} \frac{\left(\omega_r\left(f_{1,\gamma+\alpha}, \xi_n\right) + \omega_r\left(f_{2,\gamma+\alpha}, \xi_n\right)\right)}{\left(\prod_{i=1}^{N} \alpha_i!\right)} \widetilde{u}_{\xi_n}(\alpha). \tag{23.147}$$

Proof Based on Theorems 23.31, 23.35. ∎

We continue with

Theorem 23.50 *Let $f : \mathbb{R}^N \to \mathbb{C} : f = f_1 + i f_2$, $j = 1, 2$. Here $f_j \in C_B^l\left(\mathbb{R}^N\right)$, $l, N \in \mathbb{N}$ (functions l-times continuously differentiable and bounded). The assumptions of Theorem 23.31 are valid for $d\mu_{\xi_n} = \frac{1}{\left(\sqrt{\pi \xi_n}\right)^N} e^{-\left(\frac{\sum_{i=1}^{N} s_i^2}{\xi_n}\right)} ds_1 \dots ds_N$ and f_j, $j = 1, 2$. Call $\gamma = 0, \beta$, $\xi_n \in (0, 1]$. Then*

(1)

$$\left\| \left(W_{r,n}^{[0]} f\right)_{\gamma} - f_{\gamma} \right\|_{\infty} \leq$$

$$\left(\frac{2}{\sqrt{\pi}}\right)^N \left[(1+N)^r + 2^r \left(\frac{\lfloor er! \rfloor}{e}\right)^N \right] \frac{\left(\omega_r\left(f_{1,\gamma}, \xi_n\right) + \omega_r\left(f_{2,\gamma}, \xi_n\right)\right)}{\left(\sqrt{\xi_n}\right)^r}. \tag{23.148}$$

(2) Let $r \in \mathbb{N} - \{1\}$, $0 < \alpha \leq 1$, and consider $f_{j,\gamma} : \omega_r\left(f_{j,\gamma}, t\right) \leq K t^{r-1+\alpha}$, $\forall\, t > 0$, $j = 1, 2$, where $K > 0$ is independent of t. We get

$$\frac{\omega_r\left(f_{j,\gamma}, \xi_n\right)}{\left(\sqrt{\xi_n}\right)^r} \leq K \xi_n^{\left(\frac{r}{2} + \alpha - 1\right)}, \tag{23.149}$$

$j = 1, 2$, and as $n \to \infty$ and $\xi_n \to 0$, we find $\left(W_{r,n}^{[0]} f\right)_{\gamma} \to f_{\gamma}$ uniformly.

Proof By Theorems 23.31 and 23.36. ∎

We proceed with L_p approximations

Theorem 23.51 *Let $f : \mathbb{R}^N \to \mathbb{C} : f = f_1 + i f_2$, $j = 1, 2$. Here $f_j \in C^m\left(\mathbb{R}^N\right)$, $m \in \mathbb{N}$, $m > r$, $N \geq 1$, with $f_{j,\alpha} \in L_p\left(\mathbb{R}^N\right)$, $|\alpha| = m$, $x \in \mathbb{R}^N$. Let $p, q > 1 : \frac{1}{p} + \frac{1}{q} = 1$, $\xi_n \in (0, 1]$. For $\widetilde{j} = 1, \dots, m$, and $\alpha := (\alpha_1, \dots, \alpha_N)$, $\alpha_i \in \mathbb{Z}^+$, $i = 1, \dots, N$, $|\alpha| := \sum_{i=1}^{N} \alpha_i = \widetilde{j}$, call*

$$\widetilde{c}_{\alpha,n,\widetilde{j}} = \frac{1}{\left(\sqrt{\pi \xi_n}\right)^N} \int_{\mathbb{R}^N} \prod_{i=1}^{N} s_i^{\alpha_i} e^{-\frac{\left(\sum_{i=1}^{N} s_i^2\right)}{\xi_n}} ds_1 ... ds. \tag{23.150}$$

Then

$$\left\| W_{r,n}^{[m]}(f; x) - f(x) - \sum_{\widetilde{j}=1}^{m} \delta_{\widetilde{j},r}^{[m]} \left(\sum_{|\alpha|=\widetilde{j}} \frac{\widetilde{c}_{\alpha,n,\widetilde{j}} f_\alpha(x)}{\left(\prod_{i=1}^{N} \alpha_i!\right)} \right) \right\|_{p,x}$$

$$\leq \left(\frac{2}{\sqrt{\pi}}\right)^{\frac{N}{p}} \left(\frac{m}{(q(m-1)+1)^{\frac{1}{q}}}\right) \left[(1+N)^{rp} + 2^{rp} \Gamma^N ((m+r)p+1, 1)\right]^{\frac{1}{p}}$$

$$\left(\sum_{|\alpha|=m} \frac{1}{\prod_{i=1}^{N} \alpha_i!} \right) \left(\sqrt{\xi_n}\right)^{(m-r)} \left[\omega_r\left(f_{1,\alpha}, \xi_n\right)_p + \omega_r\left(f_{2,\alpha}, \xi_n\right)_p\right]. \tag{23.151}$$

Proof By Theorem 23.19. ■

We continue with

Theorem 23.52 *Let* $f : \mathbb{R}^N \to \mathbb{C} : f = f_1 + i f_2$, $j = 1, 2$. *Here* $f_j \in \left(C\left(\mathbb{R}^N\right) \cap L_p\left(\mathbb{R}^N\right)\right)$; $N \geq 1$; $p, q > 1 : \frac{1}{p} + \frac{1}{q} = 1$, $\xi_n \in (0, 1]$, $n \in \mathbb{N}$. *Then*

(1)

$$\left\| W_{r,n}^{[0]}(f) - f \right\|_p \leq \tag{23.152}$$

$$\left(\frac{2}{\sqrt{\pi}}\right)^{\frac{N}{p}} \left[(1+N)^{rp} + 2^{rp} \Gamma^N (rp+1, 1)\right]^{\frac{1}{p}} \frac{\left(\omega_r(f_1, \xi_n)_p + \omega_r(f_2, \xi_n)_p\right)}{\left(\sqrt{\xi_n}\right)^r}.$$

(2) Let $r \in \mathbb{N} - \{1\}$, $0 < \alpha \leq 1$, *and assume that* $\omega_r\left(f_j, t\right)_p \leq K t^{r-1+\alpha}$, $\forall\, t > 0$, $j = 1, 2$, *where* $K > 0$ *is independent of* t. *In this case*

$$\frac{\omega_r\left(f_j, \xi_n\right)_p}{\left(\sqrt{\xi_n}\right)^r} \leq K \xi_n^{\left(\frac{r}{2}+\alpha-1\right)}, \quad j = 1, 2, \tag{23.153}$$

and as $n \to \infty$ *and* $\xi_n \to 0$, *we get* $\left\| W_{r,n}^{[0]} f - f \right\|_p \to 0$.

Proof By Theorem 23.21. ■

We also give

Theorem 23.53 *Let* $f : \mathbb{R}^N \to \mathbb{C} : f = f_1 + i f_2$, $j = 1, 2$. *Here* $f_j \in \left(C\left(\mathbb{R}^N\right) \cap L_1\left(\mathbb{R}^N\right)\right)$; $N \geq 1$, $\xi_n \in (0, 1]$, $n \in \mathbb{N}$. *Then*

(1)

$$\left\| W_{r,n}^{[0]}(f) - f \right\|_1 \leq$$

$$\left(\frac{2}{\sqrt{\pi}}\right)^N \left[(1+N)^r + 2^r \left(\frac{\lfloor er! \rfloor}{e}\right)^N\right] \frac{\left[\omega_r(f_1,\xi_n)_1 + \omega_r(f_2,\xi_n)_1\right]}{\left(\sqrt{\xi_n}\right)^r}. \tag{23.154}$$

(2) Let $r \in \mathbb{N} - \{1\}$, $0 < \alpha \leq 1$, and assume for $j = 1, 2$ that $\omega_r\left(f_j, t\right)_1 \leq K t^{r-1+\alpha}$, $\forall\, t > 0$, $K > 0$ independent of t. We find that

$$\frac{\omega_r\left(f_j,\xi_n\right)_1}{\left(\sqrt{\xi_n}\right)^r} \leq K \xi_n^{\left(\frac{r}{2}+\alpha-1\right)}, \quad j = 1, 2, \tag{23.155}$$

and as $n \to \infty$ and $\xi_n \to 0$, we derive $\left\| W_{r,n}^{[0]}(f) - f \right\|_1 \to 0$.

Proof By Theorem 23.23. ■

We further present

Theorem 23.54 *Let $f : \mathbb{R}^N \to \mathbb{C} : f = f_1 + i f_2$, $j = 1, 2$. Here $f_j \in C^m\left(\mathbb{R}^N\right)$, $m, N \in \mathbb{N}$, with $f_{j,\alpha} \in L_1\left(\mathbb{R}^N\right)$, $|\alpha| = m$, $x \in \mathbb{R}^N$, $\xi_n \in (0, 1]$, $n \in \mathbb{N}$, $m > r$. For $\widetilde{j} = 1, \ldots, m$, and $\alpha := (\alpha_1, \ldots, \alpha_N)$, $\alpha_i \in \mathbb{Z}^+$, $i = 1, \ldots, N$, $|\alpha| := \sum_{i=1}^{N} \alpha_i = \widetilde{j}$, call*

$$\widetilde{c}_{\alpha,n,\widetilde{j}} := \frac{1}{\left(\sqrt{\pi \xi_n}\right)^N} \int_{\mathbb{R}^N} \prod_{i=1}^{N} s_i^{\alpha_i} e^{-\frac{\left(\sum_{i=1}^{N} s_i^2\right)}{\xi_n}} ds_1 \ldots ds. \tag{23.156}$$

Then

$$\left\| W_{r,n}^{[m]}(f;x) - f(x) - \sum_{\widetilde{j}=1}^{m} \delta_{\widetilde{j},r}^{[m]} \left(\sum_{|\alpha|=\widetilde{j}} \frac{\widetilde{c}_{\alpha,n,\widetilde{j}} f_\alpha(x)}{\prod_{i=1}^{N} \alpha_i!} \right) \right\|_{1,x} \tag{23.157}$$

$$\leq \left(\frac{2}{\sqrt{\pi}}\right)^N \left[(1+N)^r + 2^r \left(\frac{\lfloor e(m+r)! \rfloor}{e}\right)^N\right] \sum_{|\alpha|=m} \left(\frac{1}{\prod_{i=1}^{N} \alpha_i!} \right)$$

$$\left(\sqrt{\xi_n}\right)^{m-r} \left[\omega_r\left(f_{1,\alpha},\xi_n\right)_1 + \omega_r\left(f_{2,\alpha},\xi_n\right)_1\right].$$

Proof By Theorem 23.25. ■

We continue with simultaneous L_p approximations.

Theorem 23.55 *Let $f : \mathbb{R}^N \to \mathbb{C} : f = f_1 + if_2$, $j = 1, 2$. Here $f_j \in C^{m+l}\left(\mathbb{R}^N\right)$, $m, l, N \in \mathbb{N}$, $m > r$. The assumptions of Theorem 23.31 are valid for $d\mu_{\xi_n} = \frac{1}{\left(\sqrt{\pi\xi_n}\right)^N} e^{-\left(\frac{\sum_{i=1}^N s_i^2}{\xi_n}\right)} ds_1...ds_N$ and f_j. Call $\gamma = 0, \beta$. Let $f_{j,(\gamma+\alpha)} \in L_p\left(\mathbb{R}^N\right)$, $|\alpha| = m$, $x \in \mathbb{R}^N$, and $p, q > 1 : \frac{1}{p} + \frac{1}{q} = 1$, $\xi_n \in (0, 1]$, $n \in \mathbb{N}$. For $\widetilde{j} = 1, ..., m$, and $\alpha := (\alpha_1, ..., \alpha_N)$, $\alpha_i \in \mathbb{Z}^+$, $i = 1, ..., N$, $|\alpha| := \sum_{i=1}^N \alpha_i = \widetilde{j}$, call*

$$\widetilde{c}_{\alpha,n,\widetilde{j}} := \frac{1}{\left(\sqrt{\pi\xi_n}\right)^N} \int_{\mathbb{R}^N} \prod_{i=1}^N s_i^{\alpha_i} e^{-\frac{\left(\sum_{i=1}^N s_i^2\right)}{\xi_n}} ds_1...ds_N. \tag{23.158}$$

Then

$$\left\| \left(W_{r,n}^{[m]}(f; x)\right)_\gamma - f_\gamma(x) - \sum_{\widetilde{j}=1}^m \delta_{\widetilde{j},r}^{[m]} \left(\sum_{|\alpha|=\widetilde{j}} \frac{\widetilde{c}_{\alpha,n,\widetilde{j}} f_{\gamma+\alpha}(x)}{\prod_{i=1}^N \alpha_i!} \right) \right\|_{p,x}$$

$$\leq \left(\frac{2}{\sqrt{\pi}}\right)^{\frac{N}{p}} \left(\frac{m}{(q(m-1)+1)^{\frac{1}{q}}}\right) \left[(1+N)^{rp} + 2^{rp}\Gamma^N((m+r)p+1, 1)\right]^{\frac{1}{p}} \tag{23.159}$$

$$\left(\sum_{|\alpha|=m} \frac{1}{\prod_{i=1}^N \alpha_i!}\right) \left(\sqrt{\xi_n}\right)^{(m-r)} \left[\omega_r\left(f_{1,\gamma+\alpha}, \xi_n\right)_p + \omega_r\left(f_{2,\gamma+\alpha}, \xi_n\right)_p\right].$$

Proof By Theorems 23.31, 23.37. ∎

We give also

Theorem 23.56 *Let $f : \mathbb{R}^N \to \mathbb{C} : f = f_1 + if_2$, $j = 1, 2$. Here $f_j \in C^l\left(\mathbb{R}^N\right)$, $l, N \in \mathbb{N}$. The assumptions of Theorem 23.31 are valid for $d\mu_{\xi_n} = \frac{1}{\left(\sqrt{\pi\xi_n}\right)^N} e^{-\left(\frac{\sum_{i=1}^N s_i^2}{\xi_n}\right)} ds_1...ds_N$ and f_j. Call $\gamma = 0, \beta$. Let $f_{j,\gamma} \in L_p\left(\mathbb{R}^N\right)$, $x \in \mathbb{R}^N$; $p, q > 1 : \frac{1}{p} + \frac{1}{q} = 1$, $\xi_n \in (0, 1]$, $n \in \mathbb{N}$. Then*

(1)

$$\left\| \left(W_{r,n}^{[0]}(f)\right)_\gamma - f_\gamma \right\|_p \leq \tag{23.160}$$

$$\left(\frac{2}{\sqrt{\pi}}\right)^{\frac{N}{p}}\left[(1+N)^{rp}+2^{rp}\Gamma^N(rp+1,1)\right]^{\frac{1}{p}}\frac{\left[\omega_r\left(f_{1,\gamma},\xi_n\right)_p+\omega_r\left(f_{2,\gamma},\xi_n\right)_p\right]}{\left(\sqrt{\xi_n}\right)^r}.$$

(2) Let $r\in\mathbb{N}-\{1\}$, $0<\alpha\leq 1$, with $\omega_r\left(f_{j,\gamma},t\right)_p\leq Kt^{r-1+\alpha}$, $\forall t>0$, $K>0$ independent of t, $j=1,2$. In this case

$$\frac{\omega_r\left(f_{j,\gamma},\xi_n\right)_p}{\left(\sqrt{\xi_n}\right)^r}\leq K\xi_n^{\left(\frac{r}{2}+\alpha-1\right)},\quad j=1,2. \tag{23.161}$$

As $n\to+\infty$ and $\xi_n\to 0$, then $\left(W_{r,n}^{[0]}(f)\right)_\gamma\overset{\|\cdot\|_p}{\to}f_\gamma$.

Proof By Theorems 23.31 and 23.38. ■

We continue with

Theorem 23.57 *Let $f:\mathbb{R}^N\to\mathbb{C}: f=f_1+if_2$, $j=1,2$. Here $f_j\in C^l\left(\mathbb{R}^N\right)$, $l,N\in\mathbb{N}$. The assumptions of Theorem 23.31 are valid for $d\mu_{\xi_n}=\frac{1}{\left(\sqrt{\pi\xi_n}\right)^N}e^{-\left(\frac{\sum_{i=1}^N s_i^2}{\xi_n}\right)}ds_1...ds_N$ and f_j. Call $\gamma=0,\beta$. Let $f_{j,\gamma}\in L_1\left(\mathbb{R}^N\right)$, $x\in\mathbb{R}^N$, $\xi_n\in(0,1]$, $n\in\mathbb{N}$. Then*

(1)

$$\left\|\left(W_{r,n}^{[0]}(f)\right)_\gamma-f_\gamma\right\|_1\leq \tag{23.162}$$

$$\left(\frac{2}{\sqrt{\pi}}\right)^N\left[(1+N)^r+2^r\left(\frac{\lfloor er!\rfloor}{e}\right)^N\right]\frac{\left[\omega_r\left(f_{1,\gamma},\xi_n\right)_1+\omega_r\left(f_{2,\gamma},\xi_n\right)_1\right]}{\left(\sqrt{\xi_n}\right)^r}.$$

(2) Let $r\in\mathbb{N}-\{1\}$, $0<\alpha\leq 1$, with $\omega_r\left(f_{j,\gamma},t\right)_1\leq Kt^{r-1+\alpha}$, $\forall t>0$, $K>0$ independent of t, $j=1,2$. In this case

$$\frac{\omega_r\left(f_{j,\gamma},\xi_n\right)_1}{\left(\sqrt{\xi_n}\right)^r}\leq K\xi_n^{\left(\frac{r}{2}+\alpha-1\right)},\quad j=1,2 \tag{23.163}$$

and as $n\to\infty$ and $\xi_n\to 0$, we get $\left(W_{r,n}^{[0]}(f)\right)_\gamma\overset{\|\cdot\|_1}{\to}f_\gamma$.

Proof By Theorems 23.31 and 23.39. ■

We finish with

Theorem 23.58 *Let $f:\mathbb{R}^N\to\mathbb{C}: f=f_1+if_2$, $j=1,2$. Here $f_j\in C^{m+l}\left(\mathbb{R}^N\right)$, $m,l,N\in\mathbb{N}$. The assumptions of Theorem 23.31 for $d\mu_{\xi_n}=\frac{1}{\left(\sqrt{\pi\xi_n}\right)^N}e^{-\left(\frac{\sum_{i=1}^N s_i^2}{\xi_n}\right)}ds_1...ds_N$*

and f_j *are valid. Call* $\gamma = 0, \beta$. *Let* $f_{j,(\gamma+\alpha)} \in L_1\left(\mathbb{R}^N\right)$, $|\alpha| = m$, $x \in \mathbb{R}^N$, $\xi_n \in (0, 1]$, $n \in \mathbb{N}$. *For* $\widetilde{j} = 1, ..., m$, *and* $\alpha := (\alpha_1, ..., \alpha_N)$, $\alpha_i \in \mathbb{Z}^+$, $i = 1, ..., N$, $|\alpha| := \sum_{i=1}^{N} \alpha_i = \widetilde{j}$, *call*

$$\widetilde{c}_{\alpha,n,\widetilde{j}} := \frac{1}{\left(\sqrt{\pi \xi_n}\right)^N} \int_{\mathbb{R}^N} \prod_{i=1}^{N} s_i^{\alpha_i} e^{-\frac{\left(\sum_{i=1}^{N} s_i^2\right)}{\xi_n}} ds_1 ... ds_N. \tag{23.164}$$

Then

$$\left\| \left(W_{r,n}^{[m]}(f; x)\right)_\gamma - f_\gamma(x) - \sum_{\widetilde{j}=1}^{m} \delta_{\widetilde{j},r}^{[m]} \left(\sum_{|\alpha|=\widetilde{j}} \frac{\widetilde{c}_{\alpha,n,\widetilde{j}} f_{\gamma+\alpha}(x)}{\prod_{i=1}^{N} \alpha_i!} \right) \right\|_{1,x}$$

$$\leq \left(\frac{2}{\sqrt{\pi}}\right)^N \left(\sum_{|\alpha|=m} \left(\frac{1}{\prod_{i=1}^{N} \alpha_i!} \right) \left[\omega_r\left(f_{1,\gamma+\alpha}, \xi_n\right)_1 + \omega_r\left(f_{2,\gamma+\alpha}, \xi_n\right)_1 \right] \right.$$

$$\left. \left(\sqrt{\xi_n}\right)^{m-r} \left[(1+N)^r + 2^r \left(\frac{\lfloor e(m+r)! \rfloor}{e} \right)^N \right] \right). \tag{23.165}$$

Proof By Theorems 23.31 and 23.40. ∎

References

1. Anastassiou, G.A.: Rate of convergence of non-positive linear convolution type operators. A sharp inequality. J. Math. Anal. Appl. **142**, 441–451 (1989)
2. Anastassiou, G.A.: Moments in Probability and Approximation Theory, Pitman Research Notes in Math, vol. 287. Longman Scientific & Technical, Harlow (1993)
3. Anastassiou, G.A.: Approximation by Multivariate Singular Integrals. Springer, New York (2011)
4. Anastassiou, G.A.: Complete approximations by multivariate generalized Gauss-Weierstrass singular integrals. Maroccan J. Pure & Appl. Anal. (Accepted) (2020)
5. Anastassiou, G., Gal, S.: Approximation Theory. Birkhaüser, Boston (2000)
6. Anastassiou, G., Mezei, R.: Uniform convergence with rates of smooth Gauss-Weierstrass singular integral operators. Appl. Anal. **88**(7), 1015–1037 (2009)
7. Anastassiou, G.A., Mezei, R.A.: L_p convergence with rates of smooth Gauss-Weierstrass singular operators. Nonlinear Stud. **17**(4), 373–386 (2010)
8. Gradshteyn, I.S., Ryzhik, I.M.: Table of Integrals, Series and Products, 8th edn. Elsevier, Amsterdam (2015)
9. DeVore, R.A., Lorentz, G.G.: Constructive Approximation, vol. 303. Springer, Berlin (1993)

Chapter 24
Complete Approximations with Multivariate Generalized Poisson–Cauchy Type Singular Integral Operators

This research and survey chapter deals exclusively with the study of the approximation of generalized multivariate Poisson–Cauchy type singular integrals to the identity-unit operator. Here we study quantitatively most of their approximation properties. These operators are not in general positive linear operators. In particular we study the rate of convergence of these integral operators to the unit operator, as well as the related simultaneous approximation. These are given via Jackson type inequalities and by the use of multivariate high order modulus of smoothness of the high order partial derivatives of the involved function. Also we study the global smoothness preservation properties of these integral operators. These multivariate inequalities are nearly sharp and in one case the inequality is attained, that is sharp. Furthermore we give asymptotic expansions of Voronovskaya type for the error of approximation. The above properties are studied with respect to L_p norm, $1 \leq p \leq \infty$. It follows [4].

24.1 Introduction

We start with our motivation for this chapter. The following come from [9].

In the next we mention and deal with the smooth univariate Poisson–Cauchy Type singular integral operators $M_{r,\xi}(f, x)$ defined as follows.

For $r \in \mathbb{N}$ and $n \in \mathbb{Z}_+$ we set

$$\alpha_j = \begin{cases} (-1)^{r-j} \binom{r}{j} j^{-n}, & j = 1, ..., r, \\ 1 - \sum\limits_{j=1}^{r} (-1)^{r-j} \binom{r}{j} j^{-n}, & j = 0, \end{cases} \tag{24.1}$$

G. A. Anastassiou, *Constructive Fractional Analysis with Applications*,
Studies in Systems, Decision and Control 362,
https://doi.org/10.1007/978-3-030-71481-9_24

that is $\sum_{j=0}^{r} \alpha_j = 1$.

Let $f \in C^n(\mathbb{R})$ and $f^{(n)} \in L_p(\mathbb{R})$, $1 \leq p < \infty$, $\alpha \in \mathbb{N}$, $\beta > \frac{1}{2\alpha}$, we define for $x \in \mathbb{R}$, $\xi > 0$ the Lebesgue integral

$$M_{r,\xi}(f;x) = W \int_{-\infty}^{\infty} \frac{\sum_{j=0}^{r} \alpha_j f(x+jt)}{\left(t^{2\alpha}+\xi^{2\alpha}\right)^{\beta}} dt, \tag{24.2}$$

where the constant is defined as

$$W = \frac{\Gamma(\beta)\,\alpha \xi^{2\alpha\beta-1}}{\Gamma\left(\frac{1}{2\alpha}\right)\Gamma\left(\beta-\frac{1}{2\alpha}\right)},$$

where Γ is the gamma function.

Note 24.1 *The operators $M_{r,\xi}$ are not, in general, positive, see [6].*

We notice by $W \int_{-\infty}^{\infty} \frac{1}{\left(t^{2\alpha}+\xi^{2\alpha}\right)^{\beta}} dt = 1$, that $M_{r,\xi}(c;x) = c$, c constant, see also [6, 8], and

$$M_{r,\xi}(f;x) - f(x) = W \left(\sum_{j=0}^{r} \alpha_j \int_{-\infty}^{\infty} |f(x+jt) - f(x)| \right) \frac{1}{\left(t^{2\alpha}+\xi^{2\alpha}\right)^{\beta}} dt. \tag{24.3}$$

We need the rth L_p-modulus of smoothness

$$\omega_r\left(f^{(n)}, h\right)_p := \sup_{|t| \leq h} \left\| \Delta_t^r f^{(n)}(x) \right\|_{p,x}, \quad h > 0, \tag{24.4}$$

where

$$\Delta_t^r f^{(n)}(x) := \sum_{j=0}^{r} (-1)^{r-j} \binom{r}{j} f^{(n)}(x+jt), \tag{24.5}$$

see [10], p. 44. Here we have that $\omega_r\left(f^{(n)}, h\right)_p < \infty$, $h > 0$.

We need to introduce

$$\delta_k := \sum_{j=1}^{r} \alpha_j j^k, \quad k = 1, ..., n \in \mathbb{N}, \tag{24.6}$$

and denote by $\lfloor \cdot \rfloor$ the integral part. Call

$$\tau(w,x) := \sum_{j=0}^{r} \alpha_j j^n f^{(n)}(x+jw) - \delta_n f^{(n)}(x). \tag{24.7}$$

According to [2], p. 306, [1], we get

$$\tau(w,x)=\Delta_w^r f^{(n)}(x). \tag{24.8}$$

Thus

$$\|\tau(w,x)\|_{p,x}\le\omega_r\left(f^{(n)},|w|\right)_p,\ \ w\in\mathbb{R}. \tag{24.9}$$

Denote by

$$R_n(0,t,x):=\int_0^t\frac{(t-w)^{n-1}}{(n-1)!}\tau(w,x)\,dw,\ n\in\mathbb{N}. \tag{24.10}$$

Using the above terminology we obtain for $\beta>\frac{2\lfloor\frac{n}{2}\rfloor+1}{2\alpha}$ that

$$\Delta(x):=M_{r,\xi}(f;x)-f(x)-$$

$$\sum_{m=1}^{\lfloor\frac{n}{2}\rfloor}\frac{f^{(2m)}(x)\,\delta_{2m}}{(2m)!}\frac{\Gamma\left(\frac{2m+1}{2\alpha}\right)\Gamma\left(\beta-\frac{2m+1}{2\alpha}\right)}{\Gamma\left(\frac{1}{2\alpha}\right)\Gamma\left(\beta-\frac{1}{2\alpha}\right)}\xi^{2m}=R_n^*(x), \tag{24.11}$$

where

$$R_n^*(x):=W\int_{-\infty}^{\infty}R_n(0,t,x)\frac{1}{\left(t^{2\alpha}+\xi^{2\alpha}\right)^{\beta}}dt,\ \ n\in\mathbb{N}. \tag{24.12}$$

In $\Delta(x)$, see (24.11), the sum collapses when $n=1$.

We mention the following result.

Theorem 24.2 ([9]) *Let $p,q>1$ such that $\frac{1}{p}+\frac{1}{q}=1$, $n\in\mathbb{N}$, $\alpha\in\mathbb{N}$, $\beta>\frac{1}{\alpha}\left(\frac{1}{p}+n+r\right)$ and the rest as above. Then*

$$\|\Delta(x)\|_p\le$$

$$\frac{(2\alpha)^{\frac{1}{p}}\Gamma(\beta)\Gamma\left(\frac{q\beta}{2}-\frac{1}{2\alpha}\right)^{1/q}\xi^n\tau^{\frac{1}{p}}}{\Gamma\left(\frac{q\beta}{2}\right)^{\frac{1}{q}}\Gamma\left(\frac{1}{2\alpha}\right)^{\frac{1}{p}}\Gamma\left(\beta-\frac{1}{2\alpha}\right)(\tau p+1)^{\frac{1}{p}}[(n-1)!](q(n-1)+1)^{\frac{1}{q}}}\omega_r\left(f^{(n)},\xi\right)_p, \tag{24.13}$$

where

$$0<\tau:=\left[\int_0^{\infty}(1+u)^{rp+1}\frac{u^{np-1}}{\left(u^{2\alpha}+1\right)^{\frac{p\beta}{2}}}du-\int_0^{\infty}\frac{u^{np-1}}{\left(u^{2\alpha}+1\right)^{\frac{p\beta}{2}}}du\right]<\infty. \tag{24.14}$$

Hence as $\xi\to0$ we obtain $\|\Delta(x)\|_p\to0$.

If additionally $f^{(2m)}\in L_p(\mathbb{R})$, $m=1,2,\ldots,\lfloor\frac{n}{2}\rfloor$ then $\left\|M_{r,\xi}(f)-f\right\|_p\to0$, as $\xi\to0$.

The counterpart of Theorem 24.2 follows, case of $p = 1$.

Theorem 24.3 ([9]) *Let $f \in C^n(\mathbb{R})$ and $f^{(n)} \in L_1(\mathbb{R})$, $n \in \mathbb{N}$, $\alpha \in \mathbb{N}$, $\beta > \frac{n+r+1}{2\alpha}$. Then*

$$\|\Delta(x)\|_1 \leq \frac{1}{(r+1)(n-1)!\Gamma\left(\frac{1}{2\alpha}\right)\Gamma\left(\beta-\frac{1}{2\alpha}\right)} \tag{24.15}$$

$$\left[\sum_{k=1}^{r+1}\binom{r+1}{k}\Gamma\left(\frac{n+k}{2\alpha}\right)\Gamma\left(\beta-\frac{n+k}{2\alpha}\right)\right]\omega_r\left(f^{(n)},\xi\right)_1\xi^n.$$

Hence as $\xi \to 0$ we obtain $\|\Delta(x)\|_1 \to 0$.

If additionally $f^{(2m)} \in L_1(\mathbb{R})$, $m = 1, 2, ..., \left\lfloor \frac{n}{2} \right\rfloor$ then $\left\|M_{r,\xi}(f) - f\right\|_1 \to 0$, as $\xi \to 0$.

The case $n = 0$ is mentioned next.

Proposition 24.4 ([9]) *Let $p, q > 1$ such that $\frac{1}{p} + \frac{1}{q} = 1$, $\alpha \in \mathbb{N}$, $\beta > \frac{1}{\alpha}\left(r + \frac{1}{p}\right)$ and the rest as above. Then*

$$\left\|M_{r,\xi}(f) - f\right\|_p \leq \frac{(2\alpha)^{\frac{1}{p}}\left[\Gamma(\beta)\right]\Gamma\left(\frac{q\beta}{2}-\frac{1}{2\alpha}\right)^{\frac{1}{q}}\theta^{\frac{1}{p}}}{\Gamma\left(\frac{1}{2\alpha}\right)^{\frac{1}{p}}\Gamma\left(\beta-\frac{1}{2\alpha}\right)\Gamma\left(\frac{q\beta}{2}\right)^{\frac{1}{q}}}\omega_r(f,\xi)_p, \tag{24.16}$$

where

$$0 < \theta := \int_0^\infty (1+t)^{rp}\frac{1}{\left(t^{2\alpha}+1\right)^{\frac{p\beta}{2}}}dt < \infty. \tag{24.17}$$

Hence as $\xi \to 0$ we obtain $M_{r,\xi} \to$ unit operator I in the L_p norm, $p > 1$.

We also mention the following.

Proposition 24.5 ([9]) *Assume $\beta > \frac{r+1}{2\alpha}$. Then*

$$\left\|M_{r,\xi}f - f\right\|_1 \leq \frac{2\alpha\Gamma(\beta)}{\Gamma\left(\frac{1}{2\alpha}\right)\Gamma\left(\beta-\frac{1}{2\alpha}\right)}\left(\int_0^\infty (1+t)^r\frac{1}{\left(t^{2\alpha}+1\right)^{\beta}}dt\right)\omega_r(f,\xi)_1. \tag{24.18}$$

Hence as $\xi \to 0$ we get $M_{r,\xi} \to I$ in the L_1 norm.

In this chapter we study the approximation properties of general multivariate smooth Poisson–Cauchy singular integral operators:

$$U_{r,n}^{[m]}(f; x_1, ..., x_N) :=$$

$$W_n^N \sum_{j=0}^{r} \alpha_{j,r}^{[m]} \int_{\mathbb{R}^N} f\left(x_1 + s_1 j, ..., x_N + s_N j\right) \prod_{i=1}^{N} \frac{1}{\left(s_i^{2\alpha} + \xi_n^{2\alpha}\right)^{\beta}} ds_1 ... ds_N, \quad (24.19)$$

with $\alpha \in N$, $\beta > \frac{1}{2\alpha}$ and

$$W_n := \frac{\Gamma(\beta)\,\alpha \xi_n^{2\alpha\beta-1}}{\Gamma\left(\frac{1}{2\alpha}\right)\Gamma\left(\beta - \frac{1}{2\alpha}\right)}, \quad (24.20)$$

see [7].

We observe that

$$W_n^N \int_{\mathbb{R}^N} \prod_{i=1}^{N} \frac{1}{\left(s_i^{2\alpha} + \xi_n^{2\alpha}\right)^{\beta}} ds_1 ... ds_N = 1, \quad (24.21)$$

see [7], [11, p. 397, formula 595].

Here $r \in \mathbb{N}$, $m \in Z_+$, and

$$\alpha_{j,r}^{[m]} := \begin{cases} (-1)^{r-j} \binom{r}{j} j^{-m}, & if\, j = 1, 2, ..., r, \\ 1 - \sum_{i=1}^{r} (-1)^{r-i} \binom{r}{i} i^{-m}, & if\, j = 0, \end{cases} \quad (24.22)$$

and

$$\delta_{k,r}^{[m]} := \sum_{j=1}^{r} \alpha_{j,r}^{[m]} j^k, \quad k = 1, 2, ..., m \in \mathbb{N}. \quad (24.23)$$

See that $\sum_{j=0}^{r} \alpha_{j,r}^{[m]} = 1$.

Also here $\xi_n \in (0, 1]$, $n \in \mathbb{N}$, and $f : \mathbb{R}^N \to \mathbb{R}$ is a Borel measurable function. The above operator $U_{r,n}^{[m]}$ is a special case of a more general operator $\theta_{r,n}^{[m]}$ studied in general in [3] by the author.

We mention next about $\theta_{r,n}^{[m]}$.

Let μ_{ξ_n} be a probability Borel measure on $\mathbb{R}^N$, $N \geq 1$.

We define the multiple smooth singular integral operators

$$\theta_{r,n}^{[m]}\left(f; x_1, ..., x_N\right) := \sum_{j=0}^{r} \alpha_{j,r}^{[m]} \int_{\mathbb{R}^N} f\left(x_1 + s_1 j, x_2 + s_2 j, ..., x_N + s_N j\right) d\mu_{\xi_n}(s), \quad (24.24)$$

where $s := (s_1, ..., s_N)$, $x := (x_1, ..., x_N) \in \mathbb{R}^N$.

The operators $\theta_{r,n}^{[m]}$ are not in general positive. For example, consider the function $\varphi(u_1, ..., u_N) = \sum_{i=1}^{N} u_i^2$ and also take $r = 2$, $m = 3$; $x_i = 0$, $i = 1, ..., N$. See that

$\varphi \geq 0$, however

$$\theta_{2,n}^{[3]}(\varphi; 0, 0, ..., 0) = \left(\sum_{j=1}^{2} j^2 \alpha_{j,2}^{[3]} \right) \int_{\mathbb{R}^N} \left(\sum_{i=1}^{N} s_i^2 \right) d\mu_{\xi_n}(s) =$$

$$\left(\alpha_{1,2}^{[3]} + 4\alpha_{2,2}^{[3]} \right) \int_{\mathbb{R}^N} \left(\sum_{i=1}^{N} s_i^2 \right) d\mu_{\xi_n}(s) = \left(-2 + \frac{1}{2} \right) \int_{\mathbb{R}^N} \left(\sum_{i=1}^{N} s_i^2 \right) d\mu_{\xi_n}(s) < 0. \tag{24.25}$$

assuming that $\int_{\mathbb{R}^N} \left(\sum_{i=1}^{N} s_i^2 \right) d\mu_{\xi_n}(s) < \infty$.

Clearly in the case of $U_{r,n}^{[m]}$ we have that

$$d\mu_{\xi_n}(s) = W_n^N \prod_{i=1}^{N} \frac{1}{\left(s_i^{2\alpha} + \xi_n^{2\alpha} \right)^{\beta}} \prod_{i=1}^{N} ds_i =: d\lambda_{\xi_n}(s), \quad s \in \mathbb{R}^N. \tag{24.26}$$

Lemma 24.6 *The operator $\theta_{r,n}^{[m]}$ preserve the constant functions in N variables.*

We need the following definition.

Definition 24.7 Let $f \in C_B\left(\mathbb{R}^N\right)$, the space of all bounded and continuous functions or uniformly continuous on $\mathbb{R}^N$. Then, the rth multivariate modulus of smoothness of f is given by (see, e.g. [5])

$$\omega_r(f; h) := \sup_{\sqrt{u_1^2 + ... + u_N^2} \leq h} \left\| \Delta_{u_1, u_2, ..., u_N}^{r}(f) \right\|_\infty < \infty, \; h > 0, \tag{24.27}$$

where $\|\cdot\|_\infty$ is the sup-norm and

$$\Delta_u^r f(x) := \Delta_{u_1, u_2, ..., u_N}^{r} f(x_1, ..., x_N) =$$

$$\sum_{j=0}^{r} (-1)^{r-j} \binom{r}{j} f(x_1 + ju_1, x_2 + ju_2, ..., x_N + ju_N). \tag{24.28}$$

Let $m \in \mathbb{N}$ and let $f \in C^m\left(\mathbb{R}^N\right)$.

Suppose that all partial derivatives of f of order m are bounded, i.e.

$$\left\| \frac{\partial^m f(\cdot, \cdot, ..., \cdot)}{\partial x_1^{\alpha_1} ... \partial x_N^{\alpha_N}} \right\|_\infty < \infty, \tag{24.29}$$

for all $\alpha_j \in \mathbb{Z}^+$, $j = 1, ..., N$; $\sum_{j=1}^{N} \alpha_j = m$.

In this chapter we apply the general theory developed in [3] about $\theta_{r,n}^{[m]}$ to the operators $U_{r,n}^{[m]}$, so we can obtain computationally specific results and show that the general theory has applications and it is a valid theory.

So for the very important in various branches of mathematics operators $U_{r,n}^{[m]}$ we prove the very essential properties of uniform approximation, L_p approximation, global smoothness preservation and simultaneously approximation, Voronovskaya asymptotic expansions and complex simultaneous approximation.

24.2 Auxiliary Essential Results

We need the following:

Theorem 24.8 *Let* $r, m \in \mathbb{N}$, $N \in \mathbb{N} - \{1\}$, $\alpha \in \mathbb{N}$, $\beta > \frac{m+r+1}{2\alpha}$, $\alpha_j \in \mathbb{Z}^+$, $j = 1, ..., N$: $|\overline{\alpha}| := \sum_{j=1}^{N} \alpha_j = m$, $\xi_n \in (0, 1]$, $n \in \mathbb{N}$. *Denote* $W_n := \frac{\alpha\Gamma(\beta)\xi_n^{2\alpha\beta-1}}{\Gamma\left(\frac{1}{2\alpha}\right)\Gamma\left(\beta-\frac{1}{2\alpha}\right)}$. *Then*

$$w_{\xi_n}(\overline{\alpha}) := W_n^N \int_{\mathbb{R}^N} \left(\prod_{i=1}^{N} |s_i|^{\alpha_i}\right) \left(1 + \frac{\|s\|_2}{\xi_n}\right)^r \prod_{i=1}^{N} \frac{1}{\left(s_i^{2\alpha} + \xi_n^{2\alpha}\right)^{\beta}} \prod_{i=1}^{N} ds_i \le \tag{24.30}$$

$$\xi_n^{2\alpha\beta(N-1)+m} \left(\frac{2\alpha\Gamma(\beta)}{\Gamma\left(\frac{1}{2\alpha}\right)\Gamma\left(\beta-\frac{1}{2\alpha}\right)}\right)^N$$

$$\left[(1+N)^r + 2^r \left(\frac{\Gamma\left(\frac{m+r+1}{2\alpha}\right)\Gamma\left(\beta - \left(\frac{m+r+1}{2\alpha}\right)\right)}{2\alpha\Gamma(\beta)}\right)^N\right] \le$$

$$\left(\frac{2\alpha\Gamma(\beta)}{\Gamma\left(\frac{1}{2\alpha}\right)\Gamma\left(\beta-\frac{1}{2\alpha}\right)}\right)^N \left[(1+N)^r + 2^r \left(\frac{\Gamma\left(\frac{m+r+1}{2\alpha}\right)\Gamma\left(\beta - \left(\frac{m+r+1}{2\alpha}\right)\right)}{2\alpha\Gamma(\beta)}\right)^N\right] < +\infty,$$

where Γ *is the gamma function.*

Proof We estimate

$$w_{\xi_n}(\overline{\alpha}) = W_n^N \int_{\mathbb{R}^N} \left(\prod_{i=1}^{N} |s_i|^{\alpha_i}\right) \left(1 + \frac{\|s\|_2}{\xi_n}\right)^r \prod_{i=1}^{N} \frac{1}{\left(s_i^{2\alpha} + \xi_n^{2\alpha}\right)^{\beta}} \prod_{i=1}^{N} ds_i =$$

$$2^N W_n^N \int_{\mathbb{R}_+^N} \left(\prod_{i=1}^{N} s_i^{\alpha_i}\right) \left(1 + \frac{\|s\|_2}{\xi_n}\right)^r \prod_{i=1}^{N} \frac{1}{\left(s_i^{2\alpha} + \xi_n^{2\alpha}\right)^{\beta}} \prod_{i=1}^{N} ds_i \le \tag{24.31}$$

[call

$$c := \frac{\Gamma(\beta)\,\alpha}{\Gamma\left(\frac{1}{2\alpha}\right)\Gamma\left(\beta-\frac{1}{2\alpha}\right)},\ i.e. W_n = c\xi^{2\alpha\beta-1}\,] \tag{24.32}$$

$$(2c)^N \xi_n^{2\alpha\beta N-N} \int_{\mathbb{R}_+^N} \left(\prod_{i=1}^N s_i^{\alpha_i}\right) \left(1+\frac{\sum_{i=1}^N s_i}{\xi_n}\right)^r \prod_{i=1}^N \frac{1}{\left(s_i^{2\alpha}+\xi_n^{2\alpha}\right)^\beta} \prod_{i=1}^N ds_i =$$

$$(2c)^N \xi_n^{2\alpha\beta N-N} \xi_n^N \xi_n^m \xi_n^{-2\alpha\beta}$$

$$\int_{\mathbb{R}_+^N} \left(\prod_{i=1}^N \left(\frac{s_i}{\xi_n}\right)^{\alpha_i}\right) \left(1+\sum_{i=1}^N \left(\frac{s_i}{\xi_n}\right)\right)^r \prod_{i=1}^N \frac{1}{\left(\left(\frac{s_i}{\xi_n}\right)^{2\alpha}+1\right)^\beta} \prod_{i=1}^N d\left(\frac{s_i}{\xi_n}\right) =$$

$$(2c)^N \xi_n^{2\alpha\beta(N-1)+m} \int_{\mathbb{R}_+^N} \left(\prod_{i=1}^N z_i^{\alpha_i}\right) \left(1+\sum_{i=1}^N z_i\right)^r \prod_{i=1}^N \frac{1}{\left(z_i^{2\alpha}+1\right)^\beta} \prod_{i=1}^N dz_i = \tag{24.33}$$

$$(2c)^N \xi_n^{2\alpha\beta(N-1)+m} \left[\int_{[0,1]^N} \left(\prod_{i=1}^N z_i^{\alpha_i}\right) \left(1+\sum_{i=1}^N z_i\right)^r \prod_{i=1}^N \frac{1}{\left(z_i^{2\alpha}+1\right)^\beta} \prod_{i=1}^N dz_i + \right.$$

$$\left. \int_{(\mathbb{R}_+-[0,1])^N} \left(\prod_{i=1}^N z_i^{\alpha_i}\right) \left(1+\sum_{i=1}^N z_i\right)^r \prod_{i=1}^N \frac{1}{\left(z_i^{2\alpha}+1\right)^\beta} \prod_{i=1}^N dz_i \right] \leq$$

$$(2c)^N \xi_n^{2\alpha\beta(N-1)+m}$$

$$\left[(1+N)^r + 2^r \int_{(\mathbb{R}_+-[0,1])^N} \left(\prod_{i=1}^N z_i^{\alpha_i}\right) \left(\sum_{i=1}^N z_i\right)^r \prod_{i=1}^N \frac{1}{\left(z_i^{2\alpha}+1\right)^\beta} \prod_{i=1}^N dz_i \right] \leq$$

$$(2c)^N \xi_n^{2\alpha\beta(N-1)+m}$$

$$\left[(1+N)^r + 2^r \int_{(\mathbb{R}_+-[0,1])^N} \left(\prod_{i=1}^N z_i^{\alpha_i}\right) \left(\prod_{i=1}^N z_i^r\right) \prod_{i=1}^N \frac{1}{\left(z_i^{2\alpha}+1\right)^\beta} \prod_{i=1}^N dz_i \right] = \tag{24.34}$$

$$(2c)^N \xi_n^{2\alpha\beta(N-1)+m} \left[(1+N)^r + 2^r \prod_{i=1}^N \left(\int_1^\infty z_i^{\alpha_i+r} \frac{1}{\left(z_i^{2\alpha}+1\right)^\beta} dz_i\right)\right] \leq$$

$$(2c)^N \xi_n^{2\alpha\beta(N-1)+m} \left[(1+N)^r + 2^r \left(\int_1^\infty \frac{z^{m+r}}{\left(z^{2\alpha}+1\right)^\beta} dz\right)^N\right] \leq$$

$$(2c)^N \xi_n^{2\alpha\beta(N-1)+m} \left[(1+N)^r + 2^r \left(\int_0^{\infty} \frac{z^{m+r}}{\left(1+z^{2\alpha}\right)^{\beta}} dz \right)^N \right]$$

(by [11], p. 397, formula 595)

$$= (2c)^N \xi_n^{2\alpha\beta(N-1)+m} \left[(1+N)^r + 2^r \left(\frac{\Gamma\left(\frac{m+r+1}{2\alpha}\right)\Gamma\left(\beta - \left(\frac{m+r+1}{2\alpha}\right)\right)}{2\alpha\Gamma(\beta)} \right)^N \right]$$

$$= \left(\frac{2\alpha\Gamma(\beta)}{\Gamma\left(\frac{1}{2\alpha}\right)\Gamma\left(\beta - \frac{1}{2\alpha}\right)} \right)^N \xi_n^{2\alpha\beta(N-1)+m}$$

$$\left[(1+N)^r + 2^r \left(\frac{\Gamma\left(\frac{m+r+1}{2\alpha}\right)\Gamma\left(\beta - \left(\frac{m+r+1}{2\alpha}\right)\right)}{2\alpha\Gamma(\beta)} \right)^N \right],$$

where $N \in \mathbb{N} - \{1\}$ and $\beta > \frac{m+r+1}{2\alpha}$.

The claim is proved. ■

We continue with

Theorem 24.9 *Let* $r, N, \alpha \in \mathbb{N}$, $\beta > \frac{r+1}{2\alpha}$, $\xi_n \in (0, 1]$, $n \in \mathbb{N}$. *Denote* $W_n := \frac{\alpha\Gamma(\beta)\xi_n^{2\alpha\beta-1}}{\Gamma\left(\frac{1}{2\alpha}\right)\Gamma\left(\beta-\frac{1}{2\alpha}\right)}$. *Then*

$$\psi_{\xi_n} := W_n^N \int_{\mathbb{R}^N} \left(1 + \frac{\|s\|_2}{\xi_n}\right)^r \prod_{i=1}^N \frac{1}{\left(s_i^{2\alpha} + \xi_n^{2\alpha}\right)^{\beta}} \prod_{i=1}^N ds_i \leq$$

$$\xi_n^{2\alpha\beta(N-1)} \left(\frac{2\alpha\Gamma(\beta)}{\Gamma\left(\frac{1}{2\alpha}\right)\Gamma\left(\beta - \frac{1}{2\alpha}\right)} \right)^N$$

$$\left[(1+N)^r + 2^r \left(\frac{\Gamma\left(\frac{r+1}{2\alpha}\right)\Gamma\left(\beta - \left(\frac{r+1}{2\alpha}\right)\right)}{2\alpha\Gamma(\beta)} \right)^N \right] \leq$$

$$\left(\frac{2\alpha\Gamma(\beta)}{\Gamma\left(\frac{1}{2\alpha}\right)\Gamma\left(\beta - \frac{1}{2\alpha}\right)} \right)^N \left[(1+N)^r + 2^r \left(\frac{\Gamma\left(\frac{r+1}{2\alpha}\right)\Gamma\left(\beta - \left(\frac{r+1}{2\alpha}\right)\right)}{2\alpha\Gamma(\beta)} \right)^N \right] < +\infty. \quad (24.36)$$

Proof We estimate

$$\psi_{\xi_n} = W_n^N \int_{\mathbb{R}^N} \left(1 + \frac{\|s\|_2}{\xi_n}\right)^r \prod_{i=1}^N \frac{1}{\left(s_i^{2\alpha} + \xi_n^{2\alpha}\right)^\beta} \prod_{i=1}^N ds_i =$$

$$2^N W_n^N \int_{\mathbb{R}_+^N} \left(1 + \frac{\|s\|_2}{\xi_n}\right)^r \prod_{i=1}^N \frac{1}{\left(s_i^{2\alpha} + \xi_n^{2\alpha}\right)^\beta} \prod_{i=1}^N ds_i \leq$$

$$2^N W_n^N \int_{\mathbb{R}_+^N} \left(1 + \sum_{i=1}^N \frac{s_i}{\xi_n}\right)^r \prod_{i=1}^N \frac{1}{\left(s_i^{2\alpha} + \xi_n^{2\alpha}\right)^\beta} \prod_{i=1}^N ds_i =$$

[call $c := \frac{\Gamma(\beta)\alpha}{\Gamma\left(\frac{1}{2\alpha}\right)\Gamma\left(\beta - \frac{1}{2\alpha}\right)}$, i.e. $W_n = c\xi^{2\alpha\beta-1}$]

$$(2c)^N \xi_n^{2\alpha\beta(N-1)} \int_{\mathbb{R}_+^N} \left(1 + \sum_{i=1}^N z_i\right)^r \prod_{i=1}^N \frac{1}{\left(z_i^{2\alpha} + 1\right)^\beta} \prod_{i=1}^N dz_i = \tag{24.37}$$

$$(2c)^N \xi_n^{2\alpha\beta(N-1)} \left[\int_{[0,1]^N} \left(1 + \sum_{i=1}^N z_i\right)^r \prod_{i=1}^N \frac{1}{\left(1 + z_i^{2\alpha}\right)^\beta} \prod_{i=1}^N dz_i + \right.$$

$$\left. \int_{(\mathbb{R}_+ - [0,1])^N} \left(1 + \sum_{i=1}^N z_i\right)^r \prod_{i=1}^N \frac{1}{\left(1 + z_i^{2\alpha}\right)^\beta} \prod_{i=1}^N dz_i \right] \leq$$

$$(2c)^N \xi_n^{2\alpha\beta(N-1)} \left[(1+N)^r + 2^r \int_{(\mathbb{R}_+ - [0,1])^N} \left(\sum_{i=1}^N z_i\right)^r \prod_{i=1}^N \frac{1}{\left(1 + z_i^{2\alpha}\right)^\beta} \prod_{i=1}^N dz_i \right] \leq$$

$$(2c)^N \xi_n^{2\alpha\beta(N-1)} \left[(1+N)^r + 2^r \int_{(\mathbb{R}_+ - [0,1])^N} \prod_{i=1}^N z_i^r \prod_{i=1}^N \frac{1}{\left(1 + z_i^{2\alpha}\right)^\beta} \prod_{i=1}^N dz_i \right] =$$

$$(2c)^N \xi_n^{2\alpha\beta(N-1)} \left[(1+N)^r + 2^r \left(\int_1^\infty \frac{z^r}{\left(1 + z^{2\alpha}\right)^\beta} dz\right)^N\right] \leq$$

$$(2c)^N \xi_n^{2\alpha\beta(N-1)} \left[(1+N)^r + 2^r \left(\int_0^\infty \frac{z^r}{\left(1 + z^{2\alpha}\right)^\beta} dz\right)^N\right] = \tag{24.38}$$

(by [11], p. 397, formula 595)

$$(2c)^N \xi_n^{2\alpha\beta(N-1)} \left[(1+N)^r + 2^r \left(\frac{\Gamma\left(\frac{r+1}{2\alpha}\right)\Gamma\left(\beta - \left(\frac{r+1}{2\alpha}\right)\right)}{2\alpha\Gamma(\beta)} \right)^N \right] =$$

$$\left(\frac{2\alpha\Gamma(\beta)}{\Gamma\left(\frac{1}{2\alpha}\right)\Gamma\left(\beta - \frac{1}{2\alpha}\right)} \right)^N \xi_n^{2\alpha\beta(N-1)}$$

$$\left[(1+N)^r + 2^r \left(\frac{\Gamma\left(\frac{r+1}{2\alpha}\right)\Gamma\left(\beta - \left(\frac{r+1}{2\alpha}\right)\right)}{2\alpha\Gamma(\beta)} \right)^N \right],$$

where $N \geq 1$, $\beta > \frac{r+1}{2\alpha}$. The claim is proved. ∎

We give

Theorem 24.10 *Let* $r, m \in \mathbb{N}$, $N \in \mathbb{N} - \{1\}$, $\alpha \in \mathbb{N}$, $\beta > \frac{(m+r)p+1}{2\alpha}$, $\alpha_j \in \mathbb{Z}^+$, $j = 1, \ldots, N : |\overline{\alpha}| := \sum_{j=1}^{N} \alpha_j = m$, $\xi_n \in (0, 1]$, $n \in \mathbb{N}$. *Denote* $W_n := \frac{\alpha\Gamma(\beta)\xi_n^{2\alpha\beta-1}}{\Gamma\left(\frac{1}{2\alpha}\right)\Gamma\left(\beta-\frac{1}{2\alpha}\right)}$. *Then*

$$\overline{\theta}_{\xi_n}(\overline{\alpha}) := W_n^N \int_{\mathbb{R}^N} \left(\left(\prod_{i=1}^{N} |s_i|^{\alpha_i} \right) \left(1 + \frac{\|s\|_2}{\xi_n} \right)^r \right)^p \prod_{i=1}^{N} \frac{1}{\left(s_i^{2\alpha} + \xi_n^{2\alpha}\right)^{\beta}} \prod_{i=1}^{N} ds_i \leq$$

$$\xi_n^{2\alpha\beta(N-1)+mp} \left(\frac{2\alpha\Gamma(\beta)}{\Gamma\left(\frac{1}{2\alpha}\right)\Gamma\left(\beta - \frac{1}{2\alpha}\right)} \right)^N$$

$$\left[(1+N)^{rp} + 2^{rp} \left(\frac{\Gamma\left(\frac{(m+r)p+1}{2\alpha}\right)\Gamma\left(\beta - \left(\frac{(m+r)p+1}{2\alpha}\right)\right)}{2\alpha\Gamma(\beta)} \right)^N \right] \leq$$

$$\left(\frac{2\alpha\Gamma(\beta)}{\Gamma\left(\frac{1}{2\alpha}\right)\Gamma\left(\beta - \frac{1}{2\alpha}\right)} \right)^N$$

$$\left[(1+N)^{rp} + 2^{rp} \left(\frac{\Gamma\left(\frac{(m+r)p+1}{2\alpha}\right)\Gamma\left(\beta - \left(\frac{(m+r)p+1}{2\alpha}\right)\right)}{2\alpha\Gamma(\beta)} \right)^N \right] < +\infty. \quad (24.39)$$

Proof We estimate

$$\overline{\theta}_{\xi_n}(\overline{\alpha}) = W_n^N \int_{\mathbb{R}^N} \left(\left(\prod_{i=1}^{N} |s_i|^{\alpha_i} \right) \left(1 + \frac{\|s\|_2}{\xi_n} \right)^r \right)^p \prod_{i=1}^{N} \frac{1}{\left(s_i^{2\alpha} + \xi_n^{2\alpha}\right)^{\beta}} \prod_{i=1}^{N} ds_i =$$

$$2^N W_n^N \int_{\mathbb{R}_+^N} \left(\left(\prod_{i=1}^N s_i^{\alpha_i} \right) \left(1 + \frac{\|s\|_2}{\xi_n} \right)^r \right)^p \prod_{i=1}^N \frac{1}{\left(s_i^{2\alpha} + \xi_n^{2\alpha} \right)^\beta} \prod_{i=1}^N ds_i \le$$

$(c := \frac{\Gamma(\beta)\alpha}{\Gamma(\frac{1}{2\alpha})\Gamma(\beta - \frac{1}{2\alpha})})$

$$(2c)^N \xi_n^{2\alpha\beta N - N} \int_{\mathbb{R}_+^N} \left(\prod_{i=1}^N s_i^{\alpha_i p} \right) \left(1 + \sum_{i=1}^N \left(\frac{s_i}{\xi_n} \right) \right)^{rp} \prod_{i=1}^N \frac{1}{\left(s_i^{2\alpha} + \xi_n^{2\alpha} \right)^\beta} \prod_{i=1}^N ds_i =$$

$$(2c)^N \xi_n^{2\alpha\beta(N-1)+mp} \int_{\mathbb{R}_+^N} \left(\prod_{i=1}^N z_i^{\alpha_i p} \right) \left(1 + \sum_{i=1}^N z_i \right)^{rp} \prod_{i=1}^N \frac{1}{\left(z_i^{2\alpha} + 1 \right)^\beta} \prod_{i=1}^N dz_i = \tag{24.40}$$

$$\xi_n^{2\alpha\beta(N-1)+mp} (2c)^N \left[\int_{[0,1]^N} \left(\prod_{i=1}^N z_i^{\alpha_i p} \right) \left(1 + \sum_{i=1}^N z_i \right)^{rp} \prod_{i=1}^N \frac{1}{\left(1 + z_i^{2\alpha} \right)^\beta} \prod_{i=1}^N dz_i + \right.$$

$$\left. \int_{(\mathbb{R}_+ - [0,1])^N} \left(\prod_{i=1}^N z_i^{\alpha_i p} \right) \left(1 + \sum_{i=1}^N z_i \right)^{rp} \prod_{i=1}^N \frac{1}{\left(1 + z_i^{2\alpha} \right)^\beta} \prod_{i=1}^N dz_i \right] \le$$

$$\xi_n^{2\alpha\beta(N-1)+mp} (2c)^N \left[(1+N)^{rp} + \right.$$

$$\left. 2^{rp} \int_{(\mathbb{R}_+ - [0,1])^N} \left(\prod_{i=1}^N z_i^{\alpha_i p} \right) \left(\sum_{i=1}^N z_i \right)^{rp} \prod_{i=1}^N \frac{1}{\left(1 + z_i^{2\alpha} \right)^\beta} \prod_{i=1}^N dz_i \right] \le$$

$$\xi_n^{2\alpha\beta(N-1)+mp} (2c)^N \left[(1+N)^{rp} + \right.$$

$$\left. 2^{rp} \int_{(\mathbb{R}_+ - [0,1])^N} \left(\prod_{i=1}^N z_i^{\alpha_i p} \right) \left(\prod_{i=1}^N z_i^{rp} \right) \prod_{i=1}^N \frac{1}{\left(1 + z_i^{2\alpha} \right)^\beta} \prod_{i=1}^N dz_i \right] =$$

$$\xi_n^{2\alpha\beta(N-1)+mp} (2c)^N \left[(1+N)^{rp} + 2^{rp} \prod_{i=1}^N \int_1^\infty z_i^{(\alpha_i + r)p} \frac{1}{\left(1 + z_i^{2\alpha} \right)^\beta} dz_i \right] \le$$

$$\xi_n^{2\alpha\beta(N-1)+mp} (2c)^N \left[(1+N)^{rp} + 2^{rp} \left(\int_1^\infty z^{(m+r)p} \frac{1}{\left(1 + z^{2\alpha} \right)^\beta} dz \right)^N \right] \le \tag{24.41}$$

$$\xi_n^{2\alpha\beta(N-1)+mp} (2c)^N \left[(1+N)^{rp} + 2^{rp} \left(\int_0^\infty \frac{z^{(m+r)p}}{\left(1 + z^{2\alpha} \right)^\beta} dz \right)^N \right] =$$

(by [11], p. 397, formula 595)

$$\xi_n^{2\alpha\beta(N-1)+mp} (2c)^N \left[(1+N)^{rp} + 2^{rp} \left(\frac{\Gamma\left(\frac{(m+r)p+1}{2\alpha}\right) \Gamma\left(\beta - \left(\frac{(m+r)p+1}{2\alpha}\right)\right)}{2\alpha\Gamma(\beta)} \right)^N \right]$$

$$= \xi_n^{2\alpha\beta(N-1)+mp} \left(\frac{2\alpha\Gamma(\beta)}{\Gamma\left(\frac{1}{2\alpha}\right)\Gamma\left(\beta - \frac{1}{2\alpha}\right)} \right)^N$$

$$\left[(1+N)^{rp} + 2^{rp} \left(\frac{\Gamma\left(\frac{(m+r)p+1}{2\alpha}\right) \Gamma\left(\beta - \left(\frac{(m+r)p+1}{2\alpha}\right)\right)}{2\alpha\Gamma(\beta)} \right)^N \right] \leq$$

$$\left(\frac{2\alpha\Gamma(\beta)}{\Gamma\left(\frac{1}{2\alpha}\right)\Gamma\left(\beta - \frac{1}{2\alpha}\right)} \right)^N$$

$$\left[(1+N)^{rp} + 2^{rp} \left(\frac{\Gamma\left(\frac{(m+r)p+1}{2\alpha}\right) \Gamma\left(\beta - \left(\frac{(m+r)p+1}{2\alpha}\right)\right)}{2\alpha\Gamma(\beta)} \right)^N \right] < +\infty,$$

where $N \geq 2$ and $\beta > \frac{(m+r)p+1}{2\alpha}$.

The claim is proved. ■

We give

Theorem 24.11 *Let* $r, \alpha, N \in \mathbb{N}$, $p > 1$, $\beta > \frac{rp+1}{2\alpha}$, $\xi_n \in (0, 1]$, $n \in \mathbb{N}$. *Denote* $W_n := \frac{\alpha\Gamma(\beta)\xi_n^{2\alpha\beta-1}}{\Gamma\left(\frac{1}{2\alpha}\right)\Gamma\left(\beta-\frac{1}{2\alpha}\right)}$. *Then*

$$T_{\xi_n} := W_n^N \int_{\mathbb{R}^N} \left(1 + \frac{\|s\|_2}{\xi_n}\right)^{rp} \prod_{i=1}^N \frac{1}{\left(s_i^{2\alpha} + \xi_n^{2\alpha}\right)^\beta} \prod_{i=1}^N ds_i \leq$$

$$\xi_n^{2\alpha\beta(N-1)} \left(\frac{2\alpha\Gamma(\beta)}{\Gamma\left(\frac{1}{2\alpha}\right)\Gamma\left(\beta - \frac{1}{2\alpha}\right)} \right)^N$$

$$\left[(1+N)^{rp} + 2^{rp} \left(\frac{\Gamma\left(\frac{rp+1}{2\alpha}\right) \Gamma\left(\beta - \left(\frac{rp+1}{2\alpha}\right)\right)}{2\alpha\Gamma(\beta)} \right)^N \right] \leq$$

$$\left(\frac{2\alpha\Gamma(\beta)}{\Gamma\left(\frac{1}{2\alpha}\right)\Gamma\left(\beta-\frac{1}{2\alpha}\right)}\right)^{N}\left[(1+N)^{rp}+2^{rp}\left(\frac{\Gamma\left(\frac{rp+1}{2\alpha}\right)\Gamma\left(\beta-\left(\frac{rp+1}{2\alpha}\right)\right)}{2\alpha\Gamma(\beta)}\right)^{N}\right]<+\infty. \tag{24.42}$$

Proof We estimate

$$T_{\xi_n}=W_n^N\int_{\mathbb{R}^N}\left(1+\frac{\|s\|_2}{\xi_n}\right)^{rp}\prod_{i=1}^{N}\frac{1}{\left(s_i^{2\alpha}+\xi_n^{2\alpha}\right)^{\beta}}\prod_{i=1}^{N}ds_i=$$

$$2^NW_n^N\int_{\mathbb{R}_+^N}\left(1+\frac{\|s\|_2}{\xi_n}\right)^{rp}\prod_{i=1}^{N}\frac{1}{\left(s_i^{2\alpha}+\xi_n^{2\alpha}\right)^{\beta}}\prod_{i=1}^{N}ds_i\leq$$

$(c:=\frac{\Gamma(\beta)\alpha}{\Gamma(\frac{1}{2\alpha})\Gamma(\beta-\frac{1}{2\alpha})})$

$$(2c)^N\xi_n^{2\alpha\beta N-N}\int_{\mathbb{R}_+^N}\left(1+\sum_{i=1}^{N}\left(\frac{s_i}{\xi_n}\right)\right)^{rp}\prod_{i=1}^{N}\frac{1}{\left(s_i^{2\alpha}+\xi_n^{2\alpha}\right)^{\beta}}\prod_{i=1}^{N}ds_i=$$

$$(2c)^N\xi_n^{2\alpha\beta(N-1)}\int_{\mathbb{R}_+^N}\left(1+\sum_{i=1}^{N}z_i\right)^{rp}\prod_{i=1}^{N}\frac{1}{\left(z_i^{2\alpha}+1\right)^{\beta}}\prod_{i=1}^{N}dz_i= \tag{24.43}$$

$$\xi_n^{2\alpha\beta(N-1)}(2c)^N\left[\int_{[0,1]^N}\left(1+\sum_{i=1}^{N}z_i\right)^{rp}\prod_{i=1}^{N}\frac{1}{\left(z_i^{2\alpha}+1\right)^{\beta}}\prod_{i=1}^{N}dz_i+\right.$$

$$\left.\int_{(\mathbb{R}_+-[0,1])^N}\left(1+\sum_{i=1}^{N}z_i\right)^{rp}\prod_{i=1}^{N}\frac{1}{\left(z_i^{2\alpha}+1\right)^{\beta}}\prod_{i=1}^{N}dz_i\right]\leq$$

$$\xi_n^{2\alpha\beta(N-1)}(2c)^N\left[(1+N)^{rp}+\right.$$

$$\left.2^{rp}\int_{(\mathbb{R}_+-[0,1])^N}\left(\sum_{i=1}^{N}z_i\right)^{rp}\prod_{i=1}^{N}\frac{1}{\left(z_i^{2\alpha}+1\right)^{\beta}}\prod_{i=1}^{N}dz_i\right]\leq$$

$$\xi_n^{2\alpha\beta(N-1)}(2c)^N\left[(1+N)^{rp}+\right.$$

$$\left.2^{rp}\int_{(\mathbb{R}_+-[0,1])^N}\prod_{i=1}^{N}z_i^{rp}\prod_{i=1}^{N}\frac{1}{\left(z_i^{2\alpha}+1\right)^{\beta}}\prod_{i=1}^{N}dz_i\right]=$$

$$\xi_n^{2\alpha\beta(N-1)} (2c)^N \left[(1+N)^{rp} + 2^{rp} \left(\int_1^{\infty} \frac{z^{rp}}{\left(z^{2\alpha}+1\right)^{\beta}} dz \right)^N \right] \leq$$

$$\xi_n^{2\alpha\beta(N-1)} (2c)^N \left[(1+N)^{rp} + 2^{rp} \left(\int_0^{\infty} \frac{z^{rp}}{\left(1+z^{2\alpha}\right)^{\beta}} dz \right)^N \right] =$$

(by [11], p. 397, formula 595)

$$\xi_n^{2\alpha\beta(N-1)} (2c)^N \left[(1+N)^{rp} + 2^{rp} \left(\frac{\Gamma\left(\frac{rp+1}{2\alpha}\right)\Gamma\left(\beta - \left(\frac{rp+1}{2\alpha}\right)\right)}{2\alpha\Gamma(\beta)} \right)^N \right] = \quad (24.44)$$

$$\xi_n^{2\alpha\beta(N-1)} \left(\frac{2\alpha\Gamma(\beta)}{\Gamma\left(\frac{1}{2\alpha}\right)\Gamma\left(\beta - \frac{1}{2\alpha}\right)} \right)^N$$

$$\left[(1+N)^{rp} + 2^{rp} \left(\frac{\Gamma\left(\frac{rp+1}{2\alpha}\right)\Gamma\left(\beta - \left(\frac{rp+1}{2\alpha}\right)\right)}{2\alpha\Gamma(\beta)} \right)^N \right] \leq$$

$$\left(\frac{2\alpha\Gamma(\beta)}{\Gamma\left(\frac{1}{2\alpha}\right)\Gamma\left(\beta - \frac{1}{2\alpha}\right)} \right)^N \left[(1+N)^{rp} + 2^{rp} \left(\frac{\Gamma\left(\frac{rp+1}{2\alpha}\right)\Gamma\left(\beta - \left(\frac{rp+1}{2\alpha}\right)\right)}{2\alpha\Gamma(\beta)} \right)^N \right] < +\infty,$$

where $N \geq 1$, $\beta > \frac{rp+1}{2\alpha}$.

The claim is proved. ■

It follows

Theorem 24.12 *Let $r, m, \alpha, N \in \mathbb{N}$, $\beta > \frac{m+1}{2\alpha}$, $\xi_n \in (0,1]$, $n \in \mathbb{N}$. Denote $W_n := \frac{\alpha\Gamma(\beta)\xi_n^{2\alpha\beta-1}}{\Gamma\left(\frac{1}{2\alpha}\right)\Gamma\left(\beta-\frac{1}{2\alpha}\right)}$. Also $\alpha_j \in \mathbb{Z}^+$, $j = 1, ..., N : |\overline{\alpha}| := \sum_{j=1}^{N} \alpha_j = m$. Then*

$$K_{\xi_n}(\overline{\alpha}) := \xi_n^{-m} \left(W_n^N \int_{\mathbb{R}^N} \left(\prod_{i=1}^{N} |s_i|^{\alpha_i} \right) \prod_{i=1}^{N} \frac{1}{\left(s_i^{2\alpha} + \xi_n^{2\alpha}\right)^{\beta}} \prod_{i=1}^{N} ds_i \right) \leq$$

$$\left[\frac{2\alpha\Gamma(\beta) + \Gamma\left(\frac{m+1}{2\alpha}\right)\Gamma\left(\beta - \left(\frac{m+1}{2\alpha}\right)\right)}{\Gamma\left(\frac{1}{2\alpha}\right)\Gamma\left(\beta - \frac{1}{2\alpha}\right)} \right]^N =: \varphi < +\infty. \quad (24.45)$$

Proof We estimate

$$K_{\xi_n}(\overline{\alpha}) = \xi_n^{-m}\left(W_n^N \int_{\mathbb{R}^N}\left(\prod_{i=1}^N |s_i|^{\alpha_i}\right)\prod_{i=1}^N \frac{1}{\left(s_i^{2\alpha}+\xi_n^{2\alpha}\right)^{\beta}}\prod_{i=1}^N ds_i\right) =$$

$$2^N \xi_n^{-m}\left(W_n^N \int_{\mathbb{R}_+^N}\left(\prod_{i=1}^N s_i^{\alpha_i}\right)\prod_{i=1}^N \frac{1}{\left(s_i^{2\alpha}+\xi_n^{2\alpha}\right)^{\beta}}\prod_{i=1}^N ds_i\right) =$$

$(c := \frac{\Gamma(\beta)\alpha}{\Gamma\left(\frac{1}{2\alpha}\right)\Gamma\left(\beta-\frac{1}{2\alpha}\right)})$

$$(2c)^N \xi_n^{2\alpha\beta N-N-m}\xi_n^{m-2\alpha\beta+N}\left(\int_{\mathbb{R}_+^N}\left(\prod_{i=1}^N\left(\frac{s_i}{\xi_n}\right)^{\alpha_i}\right)\prod_{i=1}^N \frac{1}{\left(\left(\frac{s_i}{\xi_n}\right)^{2\alpha}+1\right)^{\beta}}\prod_{i=1}^N d\frac{s_i}{\xi_n}\right) =$$

$$(2c)^N \xi_n^{2\alpha\beta(N-1)}\left(\int_{\mathbb{R}_+^N}\left(\prod_{i=1}^N z_i^{\alpha_i}\right)\prod_{i=1}^N \frac{1}{\left(1+z_i^{2\alpha}\right)^{\beta}}\prod_{i=1}^N dz_i\right) = \tag{24.46}$$

$$(2c)^N \xi_n^{2\alpha\beta(N-1)}\left(\prod_{i=1}^N \int_0^\infty z^{\alpha_i}\frac{1}{\left(1+z^{2\alpha}\right)^{\beta}}dz\right) =$$

$$(2c)^N \xi_n^{2\alpha\beta(N-1)}\left\{\prod_{i=1}^N\left[\int_0^1 \frac{z^{\alpha_i}}{\left(1+z^{2\alpha}\right)^{\beta}}dz+\int_1^\infty \frac{z^{\alpha_i}}{\left(1+z^{2\alpha}\right)^{\beta}}dz\right]\right\} \leq$$

$$(2c)^N \xi_n^{2\alpha\beta(N-1)}\left[1+\int_1^\infty \frac{z^m}{\left(1+z^{2\alpha}\right)^{\beta}}dz\right]^N \leq$$

$$(2c)^N \xi_n^{2\alpha\beta(N-1)}\left[1+\int_0^\infty \frac{z^m}{\left(1+z^{2\alpha}\right)^{\beta}}dz\right]^N =$$

(by [11], p. 397, formula 595)

$$(2c)^N \xi_n^{2\alpha\beta(N-1)}\left[1+\frac{\Gamma\left(\frac{m+1}{2\alpha}\right)\Gamma\left(\beta-\left(\frac{m+1}{2\alpha}\right)\right)}{2\alpha\Gamma(\beta)}\right]^N \leq$$

$$(2c)^N\left[1+\frac{\Gamma\left(\frac{m+1}{2\alpha}\right)\Gamma\left(\beta-\left(\frac{m+1}{2\alpha}\right)\right)}{2\alpha\Gamma(\beta)}\right]^N =$$

$$\left\{\frac{2\alpha\Gamma(\beta)}{\Gamma\left(\frac{1}{2\alpha}\right)\Gamma\left(\beta-\frac{1}{2\alpha}\right)}\left[1+\frac{\Gamma\left(\frac{m+1}{2\alpha}\right)\Gamma\left(\beta-\left(\frac{m+1}{2\alpha}\right)\right)}{2\alpha\Gamma(\beta)}\right]\right\}^N=$$

$$\left[\frac{2\alpha\Gamma(\beta)+\Gamma\left(\frac{m+1}{2\alpha}\right)\Gamma\left(\beta-\left(\frac{m+1}{2\alpha}\right)\right)}{\Gamma\left(\frac{1}{2\alpha}\right)\Gamma\left(\beta-\frac{1}{2\alpha}\right)}\right]^N=\varphi<+\infty, \tag{24.47}$$

where $\beta > \frac{m+1}{2\alpha}$.

The claim is proved. ■

We make

Remark 24.13 All as in Theorem 24.12. We denote

$$\overline{c}_{\overline{\alpha},n}:=\overline{c}_{\overline{\alpha},n,\widetilde{j}}:=W_n^N\int_{\mathbb{R}^N}\left(\prod_{i=1}^N s_i^{\alpha_i}\right)\prod_{i=1}^N\frac{1}{\left(s_i^{2\alpha}+\xi_n^{2\alpha}\right)^\beta}\prod_{i=1}^N ds_i, \tag{24.48}$$

where $\widetilde{j}=1,...,m$, and $\overline{\alpha}:=(\alpha_1,...,\alpha_N)$, $\alpha_i\in\mathbb{Z}^+$, $i=1,...,N$, $|\overline{\alpha}|:=\sum_{j=1}^N\alpha_j=\widetilde{j}$.

By (24.45) we obtain

$$\left|\overline{c}_{\overline{\alpha},n}\right|=\left|\overline{c}_{\overline{\alpha},n,\widetilde{j}}\right|\le\varphi\xi_n^m\le\varphi. \tag{24.49}$$

24.3 Main Results for $U_{r,n}^{[m]}$

24.3.1 Uniform Approximation

We start with an application to $U_{r,n}^{[m]}$ of the following theorem.

Theorem 24.14 ([3], p. 11) *Let $m\in\mathbb{N}$, $f\in C^m\left(\mathbb{R}^N\right)$, $N\ge1$, $x\in\mathbb{R}^N$. Assume $\left\|\frac{\partial^m f(\cdot,\cdot,...,\cdot)}{\partial x_1^{\alpha_1}...\partial x_N^{\alpha_N}}\right\|_\infty<\infty$, for all $\alpha_j\in\mathbb{Z}^+$, $j=1,...,N:|\overline{\alpha}|:=\sum_{i=1}^N\alpha_i=m$. Let μ_{ξ_n} be a Borel probability measure on $\mathbb{R}^N$, for $\xi_n>0$, $(\xi_n)_{n\in\mathbb{N}}$ bounded sequence.*

Suppose that for all $\overline{\alpha}:=(\alpha_1,...,\alpha_N)$, $\alpha_i\in\mathbb{Z}^+$, $i=1,...,N$, $|\overline{\alpha}|:=\sum_{i=1}^N\alpha_i=m$ we have that

$$u_{\xi_n}(\overline{\alpha}):=\int_{\mathbb{R}^N}\left(\prod_{i=1}^N|s_i|^{\alpha_i}\right)\left(1+\frac{\|s\|_2}{\xi_n}\right)^r d\mu_{\xi_n}(s)<\infty. \tag{24.50}$$

For $\widetilde{j} = 1, ..., m$, *and* $\overline{\alpha} := (\alpha_1, ..., \alpha_N)$, $\alpha_i \in \mathbb{Z}^+$, $i = 1, ..., N$, $|\overline{\alpha}| := \sum_{j=1}^{N} \alpha_j = \widetilde{j}$, *call*

$$c_{\overline{\alpha},n} := c_{\overline{\alpha},n,\widetilde{j}} := \int_{\mathbb{R}^N} \prod_{i=1}^{N} s_i^{\alpha_i} d\mu_{\xi_n}(s_1, ..., s_N). \tag{24.51}$$

Then

(i)

$$E_{r,n}^{[m]}(x) := \left| \theta_{r,n}^{[m]}(f; x) - f(x) - \sum_{\widetilde{j}=1}^{m} \delta_{\widetilde{j},r}^{[m]} \left(\sum_{\substack{\alpha_1,...,\alpha_N \geq 0; \\ |\overline{\alpha}| = \widetilde{j}}} \frac{c_{\overline{\alpha},n,\widetilde{j}} f_{\overline{\alpha}}(x)}{\prod_{i=1}^{N} \alpha_i!} \right) \right|$$

$$\leq \sum_{\substack{\alpha_1,...,\alpha_N \geq 0; \\ |\overline{\alpha}| = m}} \frac{(\omega_r(f_{\overline{\alpha}}, \xi_n))}{\left(\prod_{i=1}^{N} \alpha_i \right)} \left(\int_{\mathbb{R}^N} \left(\prod_{i=1}^{N} |s_i|^{\alpha_i} \right) \left(1 + \frac{\|s\|_2}{\xi_n} \right)^r d\mu_{\xi_n}(s) \right). \tag{24.52}$$

$\forall\, x \in \mathbb{R}^N$.

(ii)

$$\left\| E_{r,n}^{[m]} \right\|_\infty \leq R.H.S.(24.52). \tag{24.53}$$

Given that $\xi_n \to 0$, *as* $n \to \infty$, *and* u_{ξ_n} *is uniformly bounded, then we derive that* $\left\| E_{r,n}^{[m]} \right\| \to 0$ *with rates.*

(iii) It holds also that

$$\left\| \theta_{r,n}^{[m]}(f) - f \right\|_\infty \leq \sum_{\widetilde{j}=1}^{m} \delta_{\widetilde{j},r}^{[m]} \left(\sum_{\substack{\alpha_1,...,\alpha_N \geq 0; \\ |\overline{\alpha}| = \widetilde{j}}} \frac{\left| c_{\overline{\alpha},n,\widetilde{j}} \right| \|f_{\overline{\alpha}}\|_\infty}{\prod_{i=1}^{N} \alpha_i!} \right) + R.H.S.(24.52), \tag{24.54}$$

given that $\|f_{\overline{\alpha}}\|_\infty < \infty$, *for all* $\overline{\alpha} : |\overline{\alpha}| = \widetilde{j}$, $\widetilde{j} = 1, ..., m$. *Furthermore, as* $\xi_n \to 0$ *when* $n \to \infty$, *assuming that* $c_{\overline{\alpha},n,\widetilde{j}} \to 0$, *while* u_{ξ_n} *is uniformly bounded, we conclude that*

$$\left\| \theta_{r,n}^{[m]}(f) - f \right\|_\infty \to 0 \tag{24.55}$$

with rates.

A uniform approximation result for $U_{r,n}^{[m]}$ follows:

Theorem 24.15 *Let* $m \in \mathbb{N}$, $f \in C^m(\mathbb{R}^N)$, $N \geq 2$, $x \in \mathbb{R}^N$. *Assume* $\left\| \frac{\partial^m f(\cdot,\cdot,...,\cdot)}{\partial x_1^{\alpha_1} ... \partial x_N^{\alpha_N}} \right\|_\infty < \infty$, *for all* $\alpha_i \in \mathbb{Z}^+$, $i = 1, ..., N : |\overline{\alpha}| := \sum_{i=1}^{N} \alpha_i = m$. *Let* λ_{ξ_n} *be the Borel prob-*

ability measure on $\mathbb{R}^N$, see (24.26), where $\xi_n \in (0,1]$, $n \in \mathbb{N}$. Here $\alpha, r \in \mathbb{N}$, $\beta > \frac{m+r+1}{2\alpha}$, and $w_{\xi_n}(\overline{\alpha})$ as in (24.30), and $\overline{c}_{\overline{\alpha},n} := \overline{c}_{\overline{\alpha},n,\widetilde{j}}$ as in (24.48). Then

(i)

$$\overline{E}_{r,n}^{[m]}(x) := \left| U_{r,n}^{[m]}(f;x) - f(x) - \sum_{\widetilde{j}=1}^{m} \delta_{\widetilde{j},r}^{[m]} \left(\sum_{\substack{\alpha_1,\ldots,\alpha_N \geq 0; \\ |\overline{\alpha}|=\widetilde{j}}} \frac{\overline{c}_{\overline{\alpha},n,\widetilde{j}} f_{\overline{\alpha}}(x)}{\prod_{i=1}^{N} \alpha_i!} \right) \right|$$

$$\leq \sum_{\substack{\alpha_1,\ldots,\alpha_N \geq 0; \\ |\overline{\alpha}|=m}} \frac{(\omega_r(f_{\overline{\alpha}}, \xi_n))}{\left(\prod_{i=1}^{N} \alpha_i!\right)} w_{\xi_n}(\overline{\alpha}), \quad (24.56)$$

$\forall\, x \in \mathbb{R}^N$.

(ii)

$$\left\| \overline{E}_{r,n}^{[m]} \right\|_\infty \leq R.H.S.(24.56). \quad (24.57)$$

Given that $\xi_n \to 0$, as $n \to +\infty$, we have that $w_{\xi_n}(\overline{\alpha}) \to 0$ and are uniformly bounded, and then we derive that $\left\| \overline{E}_{r,n}^{[m]} \right\|_\infty \to 0$ with rates.

(iii) It holds also that

$$\left\| U_{r,n}^{[m]}(f) - f \right\|_\infty \leq \sum_{\widetilde{j}=1}^{m} \left| \delta_{\widetilde{j},r}^{[m]} \right| \left(\sum_{\substack{\alpha_1,\ldots,\alpha_N \geq 0; \\ |\overline{\alpha}|=\widetilde{j}}} \frac{\left|\overline{c}_{\overline{\alpha},n,\widetilde{j}}\right| \left\| f_{\overline{\alpha}} \right\|_\infty}{\prod_{i=1}^{N} \alpha_i!} \right) + R.H.S.(24.56), \quad (24.58)$$

given that $\|f_{\overline{\alpha}}\|_\infty < +\infty$, for all $\overline{\alpha} : |\overline{\alpha}| = \widetilde{j}$, $\widetilde{j} = 1, \ldots, m$. Furthermore, as $\xi_n \to 0$ when $n \to +\infty$, we have that $\overline{c}_{\overline{\alpha},n,\widetilde{j}} \to 0$ and $w_{\xi_n}(\overline{\alpha}) \to 0$, and both are uniformly bounded, and we conclude that

$$\left\| U_{r,n}^{[m]}(f) - f \right\|_\infty \to 0 \quad (24.59)$$

with rates.

Proof Mainly by applying Theorem 24.14. By Theorem 24.8 we get that $w_{\xi_n}(\overline{\alpha}) \to 0$ and $w_{\xi_n}(\overline{\alpha})$ are uniformly bounded. By Theorem 24.12 and Remark 24.13 we get $\overline{c}_{\overline{\alpha},n} \to 0$ and $\overline{c}_{\overline{\alpha},n}$ are uniformly bounded. ∎

We mention

Theorem 24.16 ([3], p. 14) *Let $f \in C_B(\mathbb{R}^N)$, uniformly continuous, $N \geq 1$, $\xi_n \in (0,1]$. Then*

$$\left\|\theta_{r,n}^{[0]}f-f\right\|_{\infty}\leq\left(\int_{\mathbb{R}^N}\left(1+\frac{\|s\|_2}{\xi_n}\right)^r d\mu_{\xi_n}(s)\right)\omega_r(f,\xi_n), \tag{24.60}$$

under the assumption

$$\Phi_{\xi_n}:=\int_{\mathbb{R}^N}\left(1+\frac{\|s\|_2}{\xi_n}\right)^r d\mu_{\xi_n}(s)<\infty. \tag{24.61}$$

As $n\to\infty$ *and* $\xi_n\to 0$, *given that* Φ_{ξ_n} *are uniformly bounded, we derive*

$$\left\|\theta_{r,n}^{[0]}f-f\right\|_{\infty}\to 0 \tag{24.62}$$

with rates.

We give

Theorem 24.17 *Let* $f\in C_B\left(\mathbb{R}^N\right)$, *uniformly continuous,* $N,\alpha,r\in\mathbb{N}$, $\beta>\frac{r+1}{2\alpha}$, $\xi_n\in(0,1]$, $n\in\mathbb{N}$. *Then*

$$\left\|U_{r,n}^{[0]}f-f\right\|_{\infty}\leq\left(\frac{2\alpha\Gamma(\beta)}{\Gamma\left(\frac{1}{2\alpha}\right)\Gamma\left(\beta-\frac{1}{2\alpha}\right)}\right)^N$$

$$\left[(1+N)^r+2^r\left(\frac{\Gamma\left(\frac{r+1}{2\alpha}\right)\Gamma\left(\beta-\left(\frac{r+1}{2\alpha}\right)\right)}{2\alpha\Gamma(\beta)}\right)^N\right]\xi_n^{2\alpha\beta(N-1)}\omega_r(f,\xi_n). \tag{24.63}$$

As $n\to\infty$ *and* $\xi_n\to 0$, *we derive*

$$\left\|U_{r,n}^{[0]}f-f\right\|_{\infty}\to 0 \tag{24.64}$$

with rates.

Proof By Theorems 24.9 and 24.16. ■

24.3.2 L_p Approximation for $U_{r,n}^{[m]}$

We need

Definition 24.18 ([5, 10]) We call

$$\Delta_u^r f(x):=\Delta_{u_1,u_2,\dots,u_N}^r f(x_1,\dots,x_N):= \tag{24.65}$$

$$\sum_{j=0}^{r} (-1)^{r-j} \binom{r}{j} f\left(x_1 + ju_1, x_2 + ju_2, ..., x_N + ju_N\right).$$

Let $p \geq 1$, the modulus of smoothness of order r is given by

$$\omega_r\left(f; h\right)_p := \sup_{\|u\|_2 \leq h} \left\|\Delta_u^r\left(f\right)\right\|_p, \tag{24.66}$$

$h > 0$.

We will apply

Theorem 24.19 ([3], p. 24) *Let $f \in C^m\left(\mathbb{R}^N\right)$, $m \in \mathbb{N}$, $N \geq 1$, with $f_{\overline{\alpha}} \in L_p\left(\mathbb{R}^N\right)$, $|\overline{\alpha}| = m$, $x \in \mathbb{R}^N$. Let $p, q > 1 : \frac{1}{p} + \frac{1}{q} = 1$. Here, μ_{ξ_n} is a Borel probability measure on $\mathbb{R}^N$ for $\xi_n > 0$, $(\xi_n)_{n\in\mathbb{N}}$ bounded sequence. Assume for all $\overline{\alpha} := (\alpha_1, ..., \alpha_N)$, $\alpha_i \in \mathbb{Z}^+$, $i = 1, ..., N$, $|\overline{\alpha}| := \sum_{i=1}^{N} \alpha_i = m$ that we have*

$$\int_{\mathbb{R}^N} \left(\left(\prod_{i=1}^{N} |s_i|^{\alpha_i}\right)\left(1 + \frac{\|s\|_2}{\xi_n}\right)^r\right)^p d\mu_{\xi_n}(s) < \infty. \tag{24.67}$$

For $\widetilde{j} = 1, ..., m$, and $\overline{\alpha} := (\alpha_1, ..., \alpha_N)$, $\alpha_i \in \mathbb{Z}^+$, $i = 1, ..., N$, $|\overline{\alpha}| := \sum_{j=1}^{N} \alpha_j = \widetilde{j}$, call

$$c_{\overline{\alpha},n,\widetilde{j}} := \int_{\mathbb{R}^N} \prod_{i=1}^{N} s_i^{\alpha_i} d\mu_{\xi_n}(s). \tag{24.68}$$

Then

$$\left\|E_{r,n}^{[m]}\right\|_p = \left\|\theta_{r,n}^{[m]}\left(f; x\right) - f\left(x\right) - \sum_{\widetilde{j}=1}^{m} \delta_{\widetilde{j},r}^{[m]} \left(\sum_{|\overline{\alpha}|=\widetilde{j}} \frac{c_{\overline{\alpha},n,\widetilde{j}} f_{\overline{\alpha}}\left(x\right)}{\left(\prod_{i=1}^{N} \alpha_i!\right)}\right)\right\|_{p,x} \tag{24.69}$$

$$\leq \left(\frac{m}{\left(q\left(m-1\right)+1\right)^{\frac{1}{q}}}\right)\left(\sum_{|\overline{\alpha}|=m} \frac{1}{\prod_{i=1}^{N} \alpha_i!}\right)$$

$$\left[\int_{\mathbb{R}^N} \left[\left(\prod_{i=1}^{N} |s_i|^{\alpha_i}\right)\left(1 + \frac{\|s\|_2}{\xi_n}\right)^r\right]^p d\mu_{\xi_n}(s)\right]^{\frac{1}{p}} \omega_r\left(f_{\overline{\alpha}}, \xi_n\right)_p.$$

As $n \to \infty$ and $\xi_n \to 0$, by (24.69) we obtain that $\left\| E_{r,n}^{[m]} \right\| \to 0$ with rates.

One also finds by (24.69) that

$$\left\| \theta_{r,n}^{[m]} (f; x) - f(x) \right\|_{p,x} \leq \sum_{\widetilde{j}=1}^{m} \left| \delta_{\widetilde{j},r}^{[m]} \right| \left(\sum_{|\overline{\alpha}|=\widetilde{j}} \frac{\left| c_{\overline{\alpha},n,\widetilde{j}} \right|}{\prod_{i=1}^{N} \alpha_i!} \| f_{\overline{\alpha}} \|_p \right) + R.H.S.(24.69),$$

given that $\| f_{\overline{\alpha}} \|_p < \infty$, $|\overline{\alpha}| = \widetilde{j}$, $\widetilde{j} = 1, ..., m$.

Assuming that $c_{\overline{\alpha},n,\widetilde{j}} \to 0$, $\xi_n \to 0$, as $n \to \infty$, we get $\left\| \theta_{r,n}^{[m]} (f) - f \right\|_p \to 0$, that is $\theta_{r,n}^{[m]} \to I$ the unit operator, in L_p norm, with rates.

We present

Theorem 24.20 *Let $f \in C^m \left(\mathbb{R}^N\right)$, $m \in \mathbb{N}$, $N \in \mathbb{N} - \{1\}$, with $f_{\overline{\alpha}} \in L_p \left(\mathbb{R}^N\right)$, $|\overline{\alpha}| = m$, $x \in \mathbb{R}^N$. Let $p, q > 1 : \frac{1}{p} + \frac{1}{q} = 1$. Here λ_{ξ_n} is a Borel probability measure on $\mathbb{R}^N$ as in (24.26), for $\xi_n \in (0, 1]$, $n \in \mathbb{N}$. Let also $r, \alpha \in \mathbb{N}$ and $\beta > \frac{(m+r)p+1}{2\alpha}$; $\alpha_j \in \mathbb{Z}^+$, $j = 1, ..., N : |\overline{\alpha}| := \sum_{j=1}^{N} \alpha_j = m$. Here $\overline{c}_{\overline{\alpha},n} := \overline{c}_{\overline{\alpha},n,\widetilde{j}}$ as in (24.48), where $\widetilde{j} = 1, ..., m$, and $\overline{\alpha} := (\alpha_1, ..., \alpha_N)$, $\alpha_i \in \mathbb{Z}^+$, $i = 1, ..., N : |\overline{\alpha}| := \sum_{i=1}^{N} \alpha_i = \widetilde{j}$. Then*

$$\left\| \overline{E}_{r,n}^{[m]} \right\|_p = \left\| U_{r,n}^{[m]} (f; x) - f(x) - \sum_{\widetilde{j}=1}^{m} \delta_{\widetilde{j},r}^{[m]} \left(\sum_{|\overline{\alpha}|=\widetilde{j}} \frac{\overline{c}_{\overline{\alpha},n,\widetilde{j}} f_{\overline{\alpha}}(x)}{\left(\prod_{i=1}^{N} \alpha_i! \right)} \right) \right\|_{p,x}$$

$$\leq \left(\frac{m}{(q(m-1)+1)^{\frac{1}{q}}} \right) \left(\frac{2\alpha\Gamma(\beta)}{\Gamma\left(\frac{1}{2\alpha}\right)\Gamma\left(\beta - \frac{1}{2\alpha}\right)} \right)^{\frac{N}{p}}$$

$$\left[(1+N)^{rp} + 2^{rp} \left(\frac{\Gamma\left(\frac{(m+r)p+1}{2\alpha}\right)\Gamma\left(\beta - \left(\frac{(m+r)p+1}{2\alpha}\right)\right)}{2\alpha\Gamma(\beta)} \right)^N \right]^{\frac{1}{p}}$$

$$\left(\sum_{|\overline{\alpha}|=m} \frac{1}{\prod_{i=1}^{N} \alpha_i!} \right) \xi_n^{\left(\frac{2\alpha\beta(N-1)}{p} + m\right)} \omega_r (f_{\overline{\alpha}}, \xi_n)_p . \tag{24.70}$$

As $n \to \infty$ and $\xi_n \to 0$, by (24.70) we obtain that $\left\| \overline{E}_{r,n}^{[m]} \right\| \to 0$ with rates.

One also finds by (24.70) that

$$\left\| U_{r,n}^{[m]}(f;x) - f(x) \right\|_{p,x} \leq \sum_{\widetilde{j}=1}^{m} \left| \delta_{\widetilde{j},r}^{[m]} \right| \left(\sum_{|\overline{\alpha}|=\widetilde{j}} \frac{\left| \overline{c}_{\overline{\alpha},n,\widetilde{j}} \right|}{\prod_{i=1}^{N} \alpha_i!} \| f_{\overline{\alpha}} \|_p \right) + R.H.S.(24.70), \tag{24.71}$$

given that $\| f_{\overline{\alpha}} \|_p < \infty$, $|\overline{\alpha}| = \widetilde{j}$, $\widetilde{j} = 1, ..., m$.

Assuming that $\xi_n \to 0$, *as* $n \to \infty$, *we get* $\left\| U_{r,n}^{[m]}(f) - f \right\|_p \to 0$, *that is* $U_{r,n}^{[m]} \to I$ *the unit operator, in* L_p *norm, with rates.*

Proof By Theorem 24.19. From Theorem 24.10 we get that $\overline{\theta}_{\xi_n}(\overline{\alpha})$ is uniformly bounded, see (24.39) and $\overline{\theta}_{\xi_n}(\overline{\alpha}) \to 0$, as $\xi_n \to 0$, when $n \to \infty$. Also by Theorem 24.12 and Remark 24.13 we get that $\overline{c}_{\overline{\alpha},n} := \overline{c}_{\overline{\alpha},n,\widetilde{j}}$ are uniformly bounded and $\overline{c}_{\overline{\alpha},n} \to 0$, as $\xi_n \to 0$, when $n \to \infty$. ■

We continue with an application of

Theorem 24.21 ([3], p. 26) *Let* $f \in \left(C\left(\mathbb{R}^N\right) \cap L_p\left(\mathbb{R}^N\right)\right)$; $N \geq 1$; $p, q > 1 : \frac{1}{p} + \frac{1}{q} = 1$. *Assume* μ_{ξ_n} *probability Borel measure on* $\mathbb{R}^N$, $(\xi_n)_{n\in\mathbb{N}} > 0$ *and bounded. Also suppose*

$$\int_{\mathbb{R}^N} \left(1 + \frac{\|s\|_2}{\xi_n}\right)^{rp} d\mu_{\xi_n}(s) < \infty. \tag{24.72}$$

Then

$$\left\| \theta_{r,n}^{[0]}(f) - f \right\|_p \leq \tag{24.73}$$

$$\left(\int_{\mathbb{R}^N} \left(1 + \frac{\|s\|_2}{\xi_n}\right)^{rp} d\mu_{\xi_n}(s) \right)^{\frac{1}{p}} \omega_r(f, \xi_n)_p .$$

As $\xi_n \to 0$, *when* $n \to \infty$, *we derive* $\left\| \theta_{r,n}^{[0]}(f) - f \right\|_p \to 0$, *i.e.* $\theta_{r,n}^{[0]} \to I$, *the unit operator, in* L_p *norm.*

We give

Theorem 24.22 *Let* $f \in \left(C\left(\mathbb{R}^N\right) \cap L_p\left(\mathbb{R}^N\right)\right)$; $\alpha, N, r \in \mathbb{N}$; $p, q > 1 : \frac{1}{p} + \frac{1}{q} = 1$; $\beta > \frac{rp+1}{2\alpha}$, $\xi_n \in (0, 1]$, $n \in \mathbb{N}$. *Then*

$$\left\| U_{r,n}^{[0]}(f) - f \right\|_p \leq \left(\frac{2\alpha\Gamma(\beta)}{\Gamma\left(\frac{1}{2\alpha}\right)\Gamma\left(\beta - \frac{1}{2\alpha}\right)} \right)^{\frac{N}{p}}$$

$$\left[(1+N)^{rp}+2^{rp}\left(\frac{\Gamma\left(\frac{rp+1}{2\alpha}\right)\Gamma\left(\beta-\left(\frac{rp+1}{2\alpha}\right)\right)}{2\alpha\Gamma(\beta)}\right)^{N}\right]^{\frac{1}{p}}\xi_n^{\frac{2\alpha\beta(N-1)}{p}}\omega_r(f,\xi_n)_p. \tag{24.74}$$

As $\xi_n \to 0$, *when* $n \to \infty$, *we derive* $\left\|U_{r,n}^{[0]}(f)-f\right\|_p \to 0$, *i.e.* $U_{r,n}^{[0]} \to I$, *the unit operator, in* L_p *norm.*

Proof By Theorems 24.11, 24.21. ∎

We mention

Theorem 24.23 ([3], p. 27) *Let* $f \in \left(C\left(\mathbb{R}^N\right) \cap L_1\left(\mathbb{R}^N\right)\right)$; $N \geq 1$. *Assume* μ_{ξ_n} *probability Borel measure on* $\mathbb{R}^N$, $(\xi_n)_{n\in\mathbb{N}} > 0$ *and bounded. Also suppose*

$$\int_{\mathbb{R}^N}\left(1+\frac{\|s\|_2}{\xi_n}\right)^r d\mu_{\xi_n}(s) < \infty. \tag{24.75}$$

Then

$$\left\|\theta_{r,n}^{[0]}(f)-f\right\|_1 \leq \tag{24.76}$$

$$\left(\int_{\mathbb{R}^N}\left(1+\frac{\|s\|_2}{\xi_n}\right)^r d\mu_{\xi_n}(s)\right)\omega_r(f,\xi_n)_1.$$

As $\xi_n \to 0$, *we get* $\theta_{r,n}^{[0]} \to I$, *in* L_1 *norm.*

We give

Theorem 24.24 *Let* $f \in \left(C\left(\mathbb{R}^N\right) \cap L_1\left(\mathbb{R}^N\right)\right)$; $r, \alpha, N \in \mathbb{N}$, $\beta > \frac{r+1}{2\alpha}$, $\xi_n \in (0, 1]$, $n \in \mathbb{N}$. *Then*

$$\left\|U_{r,n}^{[0]}(f)-f\right\|_1 \leq \left(\frac{2\alpha\Gamma(\beta)}{\Gamma\left(\frac{1}{2\alpha}\right)\Gamma\left(\beta-\frac{1}{2\alpha}\right)}\right)^N$$

$$\left[(1+N)^r+2^r\left(\frac{\Gamma\left(\frac{r+1}{2\alpha}\right)\Gamma\left(\beta-\left(\frac{r+1}{2\alpha}\right)\right)}{2\alpha\Gamma(\beta)}\right)^N\right]\xi_n^{2\alpha\beta(N-1)}\omega_r(f,\xi_n)_1. \tag{24.77}$$

As $\xi_n \to 0$, *we get* $U_{r,n}^{[0]} \to I$, *in* L_1 *norm.*

Proof By Theorems 24.9, 24.23. ∎

We mention

Theorem 24.25 ([3], p. 29) *Let* $f \in C^m\left(\mathbb{R}^N\right)$, $m, N \in \mathbb{N}$, *with* $f_{\overline{\alpha}} \in L_1\left(\mathbb{R}^N\right)$, $|\overline{\alpha}| = m$, $x \in \mathbb{R}^N$. *Here,* μ_{ξ_n} *is a Borel probability measure on* $\mathbb{R}^N$ *for* $\xi_n > 0$, $(\xi_n)_{n\in\mathbb{N}}$ *is a bounded sequence. Suppose for all* $\overline{\alpha} := (\alpha_1, ..., \alpha_N)$, $\alpha_i \in \mathbb{Z}^+$, $i = 1, ..., N$, $|\overline{\alpha}| := \sum_{i=1}^{N}\alpha_i = m$ *that we have*

$$\int_{\mathbb{R}^N}\left(\left(\prod_{i=1}^{N}|s_i|^{\alpha_i}\right)\left(1+\frac{\|s\|_2}{\xi_n}\right)^r\right)d\mu_{\xi_n}(s)<\infty. \tag{24.78}$$

For $\widetilde{j}=1,...,m,$ *and* $\overline{\alpha}:=(\alpha_1,...,\alpha_N)$, $\alpha_i\in\mathbb{Z}^+$, $i=1,...,N$, $|\overline{\alpha}|:=\sum_{i=1}^{N}\alpha_i=\widetilde{j}$, *call*

$$c_{\overline{\alpha},n,\widetilde{j}}:=\int_{\mathbb{R}^N}\prod_{i=1}^{N}s_i^{\alpha_i}d\mu_{\xi_n}(s). \tag{24.79}$$

Then

$$\left\|E_{r,n}^{[m]}\right\|_1=\left\|\theta_{r,n}^{[m]}(f;x)-f(x)-\sum_{\widetilde{j}=1}^{m}\delta_{\widetilde{j},r}^{[m]}\left(\sum_{|\overline{\alpha}|=\widetilde{j}}\frac{c_{\overline{\alpha},n,\widetilde{j}}f_{\overline{\alpha}}(x)}{\prod_{i=1}^{N}\alpha_i!}\right)\right\|_{1,x} \tag{24.80}$$

$$\leq\sum_{|\overline{\alpha}|=m}\left(\frac{1}{\prod_{i=1}^{N}\alpha_i!}\right)\omega_r(f_{\overline{\alpha}},\xi_n)_1\int_{\mathbb{R}^N}\left(\prod_{i=1}^{N}|s_i|^{\alpha_i}\right)\left(1+\frac{\|s\|_2}{\xi_n}\right)^r d\mu_{\xi_n}(s).$$

As $\xi_n\to 0$, *we get* $\left\|E_{r,n}^{[m]}\right\|_1\to 0$ *with rates.*

From (24.80) we get

$$\left\|\theta_{r,n}^{[m]}f-f\right\|_1\leq\sum_{\widetilde{j}=1}^{m}\left|\delta_{\widetilde{j},r}^{[m]}\right|\left(\sum_{|\overline{\alpha}|=\widetilde{j}}\frac{\left|c_{\overline{\alpha},n,\widetilde{j}}\right|}{\prod_{i=1}^{N}\alpha_i!}\|f_{\overline{\alpha}}\|_1\right)+R.H.S.(24.80), \tag{24.81}$$

given that $\|f_{\overline{\alpha}}\|_1<\infty$, $|\overline{\alpha}|=\widetilde{j}$, $\widetilde{j}=1,...,m$.

As $n\to\infty$, *assuming* $\xi_n\to 0$ *and* $c_{\overline{\alpha},n,\widetilde{j}}\to 0$, *we obtain* $\left\|\theta_{r,n}^{[m]}(f)-f\right\|_1\to 0$, *that is* $\theta_{r,n}^{[m]}\to I$ *in* L_1 *norm, with rates.*

We give

Theorem 24.26 *Let* $f\in C^m(\mathbb{R}^N)$, $m,r\in\mathbb{N}$, $N\in\mathbb{N}-\{1\}$, *with* $f_{\overline{\alpha}}\in L_1(\mathbb{R}^N)$, *where* $\alpha_j\in\mathbb{Z}^+$, $j=1,...,N:|\overline{\alpha}|:=\sum_{j=1}^{N}\alpha_j=m$, $x\in\mathbb{R}^N$, $\xi_n\in(0,1]$, $n\in\mathbb{N}$, *and* $\alpha\in\mathbb{N}$ *such that* $\beta>\frac{m+r+1}{2\alpha}$. *Here* $\overline{c}_{\overline{\alpha},n}:=\overline{c}_{\overline{\alpha},n,\widetilde{j}}$ *as in (24.48), where* $\widetilde{j}=1,...,m$, *and* $\overline{\alpha}:=(\alpha_1,...,\alpha_N)$, $\alpha_i\in\mathbb{Z}^+$, $i=1,...,N:|\overline{\alpha}|:=\sum_{i=1}^{N}\alpha_i=\widetilde{j}$. *Also here* λ_{ξ_n} *is the Borel probability measure on* $\mathbb{R}^N$, *see (24.26). Then*

$$\left\| \overline{E}_{r,n}^{[m]} \right\|_1 = \left\| U_{r,n}^{[m]} (f; x) - f(x) - \sum_{\widetilde{j}=1}^{m} \delta_{\widetilde{j},r}^{[m]} \left(\sum_{|\overline{\alpha}|=\widetilde{j}} \frac{\overline{c}_{\overline{\alpha},n,\widetilde{j}} f_{\overline{\alpha}}(x)}{\prod_{i=1}^{N} \alpha_i!} \right) \right\|_{1,x} \tag{24.82}$$

$$\leq \left(\sum_{|\alpha|=m} \left(\frac{1}{\prod_{i=1}^{N} \alpha_i!} \right) \omega_r (f_{\overline{\alpha}}, \xi_n)_1 \right) \xi_n^{2\alpha\beta(N-1)+m} \left(\frac{2\alpha\Gamma(\beta)}{\Gamma\left(\frac{1}{2\alpha}\right)\Gamma\left(\beta - \frac{1}{2\alpha}\right)} \right)^N$$

$$\left[(1+N)^r + 2^r \left(\frac{\Gamma\left(\frac{m+r+1}{2\alpha}\right)\Gamma\left(\beta - \left(\frac{m+r+1}{2\alpha}\right)\right)}{2\alpha\Gamma(\beta)} \right)^N \right].$$

As $\xi_n \to 0$, *we get* $\left\| \overline{E}_{r,n}^{[m]} \right\|_1 \to 0$ *with rates.*

From (24.82) we get

$$\left\| U_{r,n}^{[m]} f - f \right\|_1 \leq \sum_{\widetilde{j}=1}^{m} \left| \delta_{\widetilde{j},r}^{[m]} \right| \left(\sum_{|\overline{\alpha}|=\widetilde{j}} \frac{\left| \overline{c}_{\overline{\alpha},n,\widetilde{j}} \right|}{\prod_{i=1}^{N} \alpha_i!} \| f_{\overline{\alpha}} \|_1 \right) + R.H.S.(24.82), \tag{24.83}$$

given that $\| f_{\overline{\alpha}} \|_1 < \infty$, $|\alpha| = \widetilde{j}$, $\widetilde{j} = 1, ..., m$.

As $n \to \infty$, *assuming* $\xi_n \to 0$, *we get* $\overline{c}_{\overline{\alpha},n,\widetilde{j}} \to 0$ *and* $\left\| U_{r,n}^{[m]} (f) - f \right\|_1 \to 0$, *that is* $U_{r,n}^{[m]} \to I$ *in* L_1 *norm, with rates.*

Proof By Theorem 24.25, also by Theorem 24.8, see (24.30), Theorem 24.12 and Remark 24.13. ■

24.3.3 Global Smoothness Preservation and Simultaneous Approximation of $U_{r,n}^{[m]}$

We need

Definition 24.27 ([3], p. 34) Let $f \in C\left(\mathbb{R}^N\right)$, $N \geq 1$, $m \in \mathbb{N}$, the mth modulus of smoothness for $1 \leq p \leq \infty$, is given by

$$\omega_m (f; h)_p := \sup_{\|t\|_2 \leq h} \left\| \Delta_t^m (f) \right\|_{p,x}, \tag{24.84}$$

$h > 0$, where

$$\Delta_t^m f(x) := \sum_{j=0}^{m} (-1)^{m-j} \binom{m}{j} f(x + jt). \quad (24.85)$$

Denote

$$\omega_m(f; h)_\infty = \omega_m(f, h). \quad (24.86)$$

Above, $x, t \in \mathbb{R}^N$.

We present the related global smoothness preservation result

Theorem 24.28 *We assume $U_{r,n}^{[\widetilde{m}]}(f; x) \in \mathbb{R}$, $\widetilde{m} \in \mathbb{Z}_+$, $\forall\, x \in \mathbb{R}$. Let $h > 0$, $f \in C(\mathbb{R}^N)$, $N \geq 1$.*

(i) Assume $\omega_m(f, h) < \infty$. Then

$$\omega_m\left(U_{r,n}^{[\widetilde{m}]} f, h\right) \leq \left(\sum_{\widetilde{j}=0}^{r} \left|\alpha_{\widetilde{j},r}^{[\widetilde{m}]}\right|\right) \omega_m(f, h). \quad (24.87)$$

(ii) Assume $f \in \left(C(\mathbb{R}^N) \cap L_1(\mathbb{R}^N)\right)$. Then

$$\omega_m\left(U_{r,n}^{[\widetilde{m}]} f, h\right)_1 \leq \left(\sum_{\widetilde{j}=0}^{r} \left|\alpha_{\widetilde{j},r}^{[\widetilde{m}]}\right|\right) \omega_m(f, h)_1. \quad (24.88)$$

(iii) Assume $f \in \left(C(\mathbb{R}^N) \cap L_p(\mathbb{R}^N)\right)$, $p > 1$. Then

$$\omega_m\left(U_{r,n}^{[\widetilde{m}]} f, h\right)_p \leq \left(\sum_{\widetilde{j}=0}^{r} \left|\alpha_{\widetilde{j},r}^{[\widetilde{m}]}\right|\right) \omega_m(f, h)_p. \quad (24.89)$$

Proof Direct application of [3] Theorem 3.2, p. 35. ∎

We make

Remark 24.29 Let $r = 1$, $\widetilde{m} \in \mathbb{Z}_+$, then $\alpha_{0,1}^{[\widetilde{m}]} = 0$, $\alpha_{1,1}^{[\widetilde{m}]} = 1$. Hence

$$U_{1,n}^{[\widetilde{m}]}(f; x) = W_n^N \int_{\mathbb{R}^N} f(x + s) \prod_{i=1}^{N} \frac{1}{\left(s_i^{2\alpha} + \xi_n^{2\alpha}\right)^\beta} ds_1 \ldots ds_N =: U_n(f; x). \quad (24.90)$$

By Theorem 24.28, we get

Theorem 24.30 *We suppose $U_n(f; x) \in \mathbb{R}$, $\forall\, x \in \mathbb{R}$. Let $h > 0$, $f \in C(\mathbb{R}^N)$, $N \geq 1$.*

(i) Assume $\omega_m(f, h) < \infty$. Then

$$\omega_m\left(U_n f, h\right) \leq \omega_m\left(f, h\right). \tag{24.91}$$

(ii) Assume $f \in \left(C\left(\mathbb{R}^N\right) \cap L_1\left(\mathbb{R}^N\right)\right)$. *Then*

$$\omega_m\left(U_n f, h\right)_1 \leq \omega_m\left(f, h\right)_1. \tag{24.92}$$

(iii) Assume $f \in \left(C\left(\mathbb{R}^N\right) \cap L_p\left(\mathbb{R}^N\right)\right)$, $p > 1$. *Then*

$$\omega_m\left(U_n f, h\right)_p \leq \omega_m\left(f, h\right)_p. \tag{24.93}$$

Next, we get an optimality result

Proposition 24.31 *Above inequality (24.91):*

$$\omega_m\left(U_n f, h\right) \leq \omega_m\left(f, h\right)$$

is sharp, namely it is attained by any

$$f_j^*\left(x\right) = x_j^m,\ j = 1, ..., N,\ x = \left(x_1, ..., x_j, ..., x_N\right) \in \mathbb{R}^N. \tag{24.94}$$

Proof Apply Proposition 3.5, p. 38, of [3]. ■

We need

Theorem 24.32 ([3], p. 39) *Let* $f \in C^l\left(\mathbb{R}^N\right)$, $l, N \in \mathbb{N}$. *Here,* μ_{ξ_n} *is a Borel probability measure on* $\mathbb{R}^N$, $\xi_n > 0$, $(\xi_n)_{n\in\mathbb{N}}$ *a bounded sequence. Let* $\beta := (\beta_1, ..., \beta_N)$, $\beta_i \in \mathbb{Z}^+$, $i = 1, ..., N$; $|\beta| := \sum_{i=1}^{N} \beta_i = l$. *Here* $f\ (x + sj)$, $x, s \in \mathbb{R}^N$, *is* μ_{ξ_n}*-integrable w.r.t.* s, *for* $j = 1, ..., r$. *There exist* μ_{ξ_n}*-integrable functions* $h_{i_1,j}$, $h_{\beta_1,i_2,j}$, $h_{\beta_1,\beta_2,i_3,j}$, ..., $h_{\beta_1,\beta_2,...,\beta_{N-1},i_N,j} \geq 0$ $(j = 1, ..., r)$ *on* $\mathbb{R}^N$ *such that*

$$\left|\frac{\partial^{i_1} f\left(x + sj\right)}{\partial x_1^{i_1}}\right| \leq h_{i_1,j}\left(s\right),\ \ i_1 = 1, ..., \beta_1, \tag{24.95}$$

$$\left|\frac{\partial^{\beta_1 + i_2} f\left(x + sj\right)}{\partial x_2^{i_2} \partial x_1^{\beta_1}}\right| \leq h_{\beta_1,i_2,j}\left(s\right),\ \ i_2 = 1, ..., \beta_2,$$

$$\vdots$$

$$\left|\frac{\partial^{\beta_1+\beta_2+...+\beta_{N-1}+i_N} f\left(x + sj\right)}{\partial x_N^{i_N} \partial x_{N-1}^{\beta_{N-1}} ... \partial x_2^{\beta_2} \partial x_1^{\beta_1}}\right| \leq h_{\beta_1,\beta_2,...,\beta_{N-1},i_N,j}\left(s\right),\ \ i_N = 1, ..., \beta_N,$$

$\forall\, x, s \in \mathbb{R}^N$.

Then, both of the next exist and

$$\left(\theta_{r,n}^{[\widetilde{m}]}(f;x)\right)_{\beta}=\theta_{r,n}^{[\widetilde{m}]}\left(f_{\beta};x\right),\ \ \widetilde{m}\in\mathbb{Z}_{+}. \tag{24.96}$$

In particular it holds

$$\left(U_{r,n}^{[\widetilde{m}]}(f;x)\right)_{\beta}=U_{r,n}^{[\widetilde{m}]}\left(f_{\beta};x\right), \tag{24.97}$$

when

$$d\mu_{\xi_n}=d\lambda_{\xi_n}(s),\ s\in\mathbb{R}^N,$$

see (24.26).

Corollary 24.33 (by Theorem 24.32, $r=1$) *It holds*

$$\left(U_n(f;x)\right)_{\beta}=U_n\left(f_{\beta};x\right). \tag{24.98}$$

We present simultaneous global smoothness results.

Theorem 24.34 *Let $h>0$ and the assumptions of Theorem 24.32 are valid for $d\mu_{\xi_n}=d\lambda_{\xi_n}$. Here $\gamma=0,\beta$ ($0=(0,...,0)$), $\widetilde{m}\in\mathbb{Z}_{+}$.*

(i) Assume $\omega_m\left(f_{\gamma},h\right)<\infty$. Then

$$\omega_m\left(\left(U_{r,n}^{[\widetilde{m}]}(f)\right)_{\gamma},h\right)\leq\left(\sum_{\widetilde{j}=0}^{r}\left|\alpha_{\widetilde{j},r}^{[\widetilde{m}]}\right|\right)\omega_m\left(f_{\gamma},h\right). \tag{24.99}$$

(ii) Additionally suppose $f_{\gamma}\in L_1\left(\mathbb{R}^N\right)$. Then

$$\omega_m\left(\left(U_{r,n}^{[\widetilde{m}]}(f)\right)_{\gamma},h\right)_1\leq\left(\sum_{\widetilde{j}=0}^{r}\left|\alpha_{\widetilde{j},r}^{[\widetilde{m}]}\right|\right)\omega_m\left(f_{\gamma},h\right)_1. \tag{24.100}$$

(iii) Additionally suppose $f_{\gamma}\in L_p\left(\mathbb{R}^N\right)$, $p>1$. Then

$$\omega_m\left(\left(U_{r,n}^{[\widetilde{m}]}(f)\right)_{\gamma},h\right)_p\leq\left(\sum_{\widetilde{j}=0}^{r}\left|\alpha_{\widetilde{j},r}^{[\widetilde{m}]}\right|\right)\omega_m\left(f_{\gamma},h\right)_p. \tag{24.101}$$

It follows

Corollary 24.35 (to Theorem 24.34) *Let $h>0$, $r=1$ and $\gamma=0,\beta$.*

(i) Assume $\omega_m\left(f_{\gamma},h\right)<\infty$. Then

$$\omega_m\left(\left(U_n(f)\right)_{\gamma},h\right)\leq\omega_m\left(f_{\gamma},h\right). \tag{24.102}$$

(ii) Additionally suppose $f_{\gamma}\in L_1\left(\mathbb{R}^N\right)$. Then

$$\omega_m\left(\left(U_n(f)\right)_\gamma, h\right)_1 \leq \omega_m\left(f_\gamma, h\right)_1. \tag{24.103}$$

(iii) Additionally suppose $f_\gamma \in L_p\left(\mathbb{R}^N\right)$, $p > 1$. *Then*

$$\omega_m\left(\left(U_n(f)\right)_\gamma, h\right)_p \leq \omega_m\left(f_\gamma, h\right)_p. \tag{24.104}$$

Next comes multi-simultaneous approximation. We give

Theorem 24.36 *Let* $f \in C^{m+l}\left(\mathbb{R}^N\right)$, $m, l \in \mathbb{N}$, $N \geq 2$. *The assumptions of Theorem 24.32 are valid for* $d\mu_{\xi_n} = d\lambda_{\xi_n}$. *Call* $\gamma = 0, \beta$. *Assume* $\left\|f_{\gamma+\overline{\alpha}}\right\|_\infty < \infty$, *and let* $\xi_n \in (0,1]$, $n \in \mathbb{N}$. *Here* $\alpha, r \in \mathbb{N}$, $\beta > \frac{m+r+1}{2\alpha}$, *and* $w_{\xi_n}(\overline{\alpha})$ *as in (24.30), and* $\overline{c}_{\overline{\alpha},n} := \overline{c}_{\overline{\alpha},n,\widetilde{j}}$ *as in (24.48). Then*

$$\left\|\left(U_{r,n}^{[m]}(f;\cdot)\right)_\gamma - f_\gamma(\cdot) - \sum_{\widetilde{j}=1}^{m} \delta_{\widetilde{j},r}^{[m]}\left(\sum_{\substack{\alpha_1,\ldots,\alpha_N \geq 0; \\ |\overline{\alpha}| = \widetilde{j}}} \frac{\overline{c}_{\overline{\alpha},n,\widetilde{j}}\, f_{\gamma+\overline{\alpha}}(\cdot)}{\prod_{i=1}^{N} \alpha_i!}\right)\right\|_\infty \tag{24.105}$$

$$\leq \sum_{\substack{\alpha_1,\ldots,\alpha_N \geq 0; \\ |\overline{\alpha}| = m}} \frac{\left(\omega_r\left(f_{\gamma+\overline{\alpha}}, \xi_n\right)\right)}{\left(\prod_{i=1}^{N} \alpha_i!\right)} w_{\xi_n}(\overline{\alpha}).$$

Proof Based on Theorems 24.15, 24.32. ∎

We continue with

Theorem 24.37 *Let* $f \in C_B^l\left(\mathbb{R}^N\right)$, $l, N \in \mathbb{N}$ *(functions l-times continuously differentiable and bounded). The assumptions of Theorem 24.32 are valid for* $d\mu_{\xi_n} = d\lambda_{\xi_n}$. *Call* $\gamma = 0, \beta$, $\xi_n \in (0,1]$, $n \in \mathbb{N}$. *Let also* $\alpha, r \in \mathbb{N}$, $\beta > \frac{r+1}{2\alpha}$. *Then*

$$\left\|\left(U_{r,n}^{[0]} f\right)_\gamma - f_\gamma\right\|_\infty \leq \left(\frac{2\alpha\Gamma(\beta)}{\Gamma\left(\frac{1}{2\alpha}\right)\Gamma\left(\beta - \frac{1}{2\alpha}\right)}\right)^N$$

$$\left[(1+N)^r + 2^r\left(\frac{\Gamma\left(\frac{r+1}{2\alpha}\right)\Gamma\left(\beta - \left(\frac{r+1}{2\alpha}\right)\right)}{2\alpha\Gamma(\beta)}\right)^N\right] \xi_n^{2\alpha\beta(N-1)} \omega_r\left(f_\gamma, \xi_n\right). \tag{24.106}$$

If $N \geq 2$ *or* f_γ *is uniformly continuous or both, and* $\xi_n \to 0$, *as* $n \to \infty$, *then* $\left(U_{r,n}^{[0]} f\right)_\gamma \to f_\gamma$ *uniformly.*

Proof By Theorems 24.17, 24.32. ∎

We present

Theorem 24.38 *Let $f \in C^{m+l}\left(\mathbb{R}^N\right)$, $m, l \in \mathbb{N}$, $N \in \mathbb{N} - \{1\}$. The assumptions of Theorem 24.32 are valid for $d\mu_{\xi_n} = d\lambda_{\xi_n}$, $\xi_n \in (0,1]$, $n \in \mathbb{N}$. Call $\gamma = 0, \beta$. Let $f_{(\gamma+\overline{\alpha})} \in L_p\left(\mathbb{R}^N\right)$, $|\overline{\alpha}| = m$, $x \in \mathbb{R}^N$, and $p, q > 1 : \frac{1}{p} + \frac{1}{q} = 1$. Let also $r, \alpha \in \mathbb{N}$ and $\beta > \frac{(m+r)\beta+1}{2\alpha}$; $\alpha_j \in \mathbb{Z}^+$, $j = 1, ..., N$, $|\overline{\alpha}| := \sum_{j=1}^{N} \alpha_j = m$. Here $\overline{c}_{\overline{\alpha},n} := \overline{c}_{\overline{\alpha},n,\widetilde{j}}$ as in (24.48), where $\widetilde{j} = 1, ..., m$, and $\overline{\alpha} := (\alpha_1, ..., \alpha_N)$, $\alpha_i \in \mathbb{Z}^+$, $i = 1, ..., N$, $|\overline{\alpha}| := \sum_{i=1}^{N} \alpha_i = \widetilde{j}$. Then*

$$\left\| \left(U_{r,n}^{[m]}(f;x)\right)_\gamma - f_\gamma(x) - \sum_{\widetilde{j}=1}^{m} \delta_{\widetilde{j},r}^{[m]} \left(\sum_{|\overline{\alpha}|=\widetilde{j}} \frac{\overline{c}_{\overline{\alpha},n,\widetilde{j}} f_{\gamma+\overline{\alpha}}(x)}{\prod_{i=1}^{N} \alpha_i!} \right) \right\|_{p,x}$$

$$\leq \left(\frac{m}{(q(m-1)+1)^{\frac{1}{q}}} \right) \left(\frac{2\alpha\Gamma(\beta)}{\Gamma\left(\frac{1}{2\alpha}\right)\Gamma\left(\beta - \frac{1}{2\alpha}\right)} \right)^{\frac{N}{p}}$$

$$\left[(1+N)^{rp} + 2^{rp} \left(\frac{\Gamma\left(\frac{(m+r)p+1}{2\alpha}\right)\Gamma\left(\beta - \left(\frac{(m+r)p+1}{2\alpha}\right)\right)}{2\alpha\Gamma(\beta)} \right)^{N} \right]^{\frac{1}{p}}$$

$$\left(\sum_{|\overline{\alpha}|=m} \frac{1}{\prod_{i=1}^{N} \alpha_i!} \right) \xi_n^{\left(\frac{2\alpha\beta(N-1)}{p}+m\right)} \omega_r\left(f_{\gamma+\overline{\alpha}}, \xi_n\right)_p. \tag{24.107}$$

Proof Theorems 24.20 and 24.32. ∎

We continue with

Theorem 24.39 *Let $f \in C^l\left(\mathbb{R}^N\right)$, $l, N \in \mathbb{N}$. The assumptions of Theorem 24.32 are valid for $d\mu_{\xi_n} = d\lambda_{\xi_n}$, $\xi_n \in (0,1]$, $n \in \mathbb{N}$. Call $\gamma = 0, \beta$. Let $f_\gamma \in L_p\left(\mathbb{R}^N\right)$ and $p, q > 1 : \frac{1}{p} + \frac{1}{q} = 1$. Here $\alpha, r \in \mathbb{N}$, $\beta > \frac{rp+1}{2\alpha}$. Then*

$$\left\| \left(U_{r,n}^{[0]}(f)\right)_\gamma - f_\gamma \right\|_p \leq \left(\frac{2\alpha\Gamma(\beta)}{\Gamma\left(\frac{1}{2\alpha}\right)\Gamma\left(\beta - \frac{1}{2\alpha}\right)} \right)^{\frac{N}{p}}$$

$$\left[(1+N)^{rp}+2^{rp}\left(\frac{\Gamma\left(\frac{rp+1}{2\alpha}\right)\Gamma\left(\beta-\left(\frac{rp+1}{2\alpha}\right)\right)}{2\alpha\Gamma(\beta)}\right)^{N}\right]^{\frac{1}{p}}\xi_n^{\frac{2\alpha\beta(N-1)}{p}}\omega_r\left(f_\gamma,\xi_n\right)_p. \tag{24.108}$$

As $n\to+\infty$ *and* $\xi_n\to0$, *then* $\left(U_{r,n}^{[0]}(f)\right)_\gamma\overset{\|\cdot\|_p}{\to}f_\gamma$.

Proof By Theorems 24.22 and 24.32. ■

We continue with

Theorem 24.40 *Let* $f\in C^l\left(\mathbb{R}^N\right)$, $l,N\in\mathbb{N}$. *The assumptions of Theorem 24.32 are valid for* $d\mu_{\xi_n}=d\lambda_{\xi_n}$, $\xi_n\in(0,1]$, $n\in\mathbb{N}$. *Call* $\gamma=0,\beta$. *Let* $f_\gamma\in L_1\left(\mathbb{R}^N\right)$ *and* $\alpha,r\in\mathbb{N}$, $\beta>\frac{r+1}{2\alpha}$. *Then*

$$\left\|\left(U_{r,n}^{[0]}(f)\right)_\gamma-f_\gamma\right\|_1\le\left(\frac{2\alpha\Gamma(\beta)}{\Gamma\left(\frac{1}{2\alpha}\right)\Gamma\left(\beta-\frac{1}{2\alpha}\right)}\right)^N$$

$$\left[(1+N)^{r}+2^{r}\left(\frac{\Gamma\left(\frac{r+1}{2\alpha}\right)\Gamma\left(\beta-\left(\frac{r+1}{2\alpha}\right)\right)}{2\alpha\Gamma(\beta)}\right)^{N}\right]\xi_n^{2\alpha\beta(N-1)}\omega_r\left(f_\gamma,\xi_n\right)_1. \tag{24.109}$$

As $n\to+\infty$ *and* $\xi_n\to0$, *then* $\left(U_{r,n}^{[0]}(f)\right)_\gamma\overset{\|\cdot\|_1}{\to}f_\gamma$.

Proof By Theorems 24.24, 24.32. ■

We continue with

Theorem 24.41 *Let* $f\in C^{m+l}\left(\mathbb{R}^N\right)$, $r,m,l\in\mathbb{N}$, $N\in\mathbb{N}-\{1\}$. *The assumptions of Theorem 24.32 are valid for* $d\mu_{\xi_n}=d\lambda_{\xi_n}$, $\xi_n\in(0,1]$, $n\in\mathbb{N}$. *Call* $\gamma=0,\beta$. *Let* $f_{(\gamma+\overline{\alpha})}\in L_1\left(\mathbb{R}^N\right)$, $|\overline{\alpha}|=m$, $x\in\mathbb{R}^N$. *Here* $\alpha\in\mathbb{N}:\beta>\frac{m+r+1}{2\alpha}$. *Here* $\overline{c}_{\overline{\alpha},n}:=\overline{c}_{\overline{\alpha},n,\widetilde{j}}$ *as in (24.48), where* $\widetilde{j}=1,\ldots,m$, *and* $\overline{\alpha}:=(\alpha_1,\ldots,\alpha_N)$, $\alpha_i\in\mathbb{Z}^+$, $i=1,\ldots,N$, $|\overline{\alpha}|:=\sum_{i=1}^N\alpha_i=\widetilde{j}$. *Then*

$$\left\|\left(U_{r,n}^{[m]}(f;x)\right)_\gamma-f_\gamma(x)-\sum_{\widetilde{j}=1}^m\delta_{\widetilde{j},r}^{[m]}\left(\sum_{|\overline{\alpha}|=\widetilde{j}}\frac{\overline{c}_{\overline{\alpha},n,\widetilde{j}}f_{\gamma+\overline{\alpha}}(x)}{\prod_{i=1}^N\alpha_i!}\right)\right\|_{1,x}$$

$$\le\left(\sum_{|\overline{\alpha}|=m}\frac{1}{\prod_{i=1}^N\alpha_i!}\right)\omega_r\left(f_{\gamma+\overline{\alpha}},\xi_n\right)_1\xi_n^{2\alpha\beta(N-1)+m}\left(\frac{2\alpha\Gamma(\beta)}{\Gamma\left(\frac{1}{2\alpha}\right)\Gamma\left(\beta-\frac{1}{2\alpha}\right)}\right)^N$$

$$\left[(1+N)^r + 2^r \left(\frac{\Gamma\left(\frac{m+r+1}{2\alpha}\right)\Gamma\left(\beta - \left(\frac{m+r+1}{2\alpha}\right)\right)}{2\alpha\Gamma(\beta)} \right)^N \right]. \tag{24.110}$$

Proof By Theorems 24.26, 24.32. ■

24.3.4 Voronovskaya Asymptotic Expansions for $U_{r,n}^{[m]}$

We will apply

Theorem 24.42 ([3], p. 53) *Let $f \in C^m\left(\mathbb{R}^N\right)$, $m, N \in \mathbb{N}$, with all $\|f_\alpha\|_\infty \leq M$, $M > 0$, all $\alpha : |\alpha| = m$. Let $\xi_n > 0$, $(\xi_n)_{n\in\mathbb{N}}$ bounded sequence, μ_{ξ_n} probability Borel measures on $\mathbb{R}^N$.*

Call $c_{\alpha,n,\widetilde{j}} := \int_{\mathbb{R}^N} \left(\prod_{i=1}^N s_i^{\alpha_i} \right) d\mu_{\xi_n}(s)$, all $|\alpha| = \widetilde{j} = 1, ..., m-1$. Suppose $\xi_n^{-m} \int_{\mathbb{R}^N} \left(\prod_{i=1}^N |s_i|^{\alpha_i} \right) d\mu_{\xi_n}(s) \leq \rho$, all $\alpha : |\alpha| = m$, $\rho > 0$, for any such $(\xi_n)_{n\in\mathbb{N}}$. Also $0 < \gamma \leq 1$, $x \in \mathbb{R}^N$. Then

$$\theta_{r,n}^{[m]}(f;x) - f(x) = \sum_{\widetilde{j}=1}^{m-1} \delta_{\widetilde{j},r}^{[m]} \left(\sum_{|\alpha|=\widetilde{j}} \frac{c_{\alpha,n,\widetilde{j}} f_\alpha(x)}{\left(\prod_{i=1}^N \alpha_i!\right)} \right) + 0\left(\xi_n^{m-\gamma}\right). \tag{24.111}$$

When $m = 1$, the sum collapses.

Above we assume $\theta_{r,n}^{[m]}(f;x) \in \mathbb{R}$, $\forall\, x \in \mathbb{R}^N$.

We give

Theorem 24.43 *Let $r, m, \alpha, N \in \mathbb{N}$, $\beta > \frac{m+1}{2\alpha}$, $\xi_n \in (0,1]$, $n \in \mathbb{N}$. Also $\alpha_j \in \mathbb{Z}^+$, $j = 1, ..., N : |\overline{\alpha}| := \sum_{j=1}^N \alpha_j = m$. Here $f \in C^m\left(\mathbb{R}^N\right)$, with all $\|f_{\overline{\alpha}}\|_\infty \leq M$, $M > 0$, for all $\overline{\alpha} : |\overline{\alpha}| = m$; and $d\mu_{\xi_n}(s) = d\lambda_{\xi_n}(s)$, as in (24.26), $\forall\, s \in \mathbb{R}^N$. Assume $U_{r,n}^{[m]}(f;x) \in \mathbb{R}$, $\forall\, x \in \mathbb{R}^N$. Here $\overline{c}_{\overline{\alpha},n,\widetilde{j}}$ as in (24.48), all $|\overline{\alpha}| = \widetilde{j} = 1, ..., m-1$. Let $0 < \gamma \leq 1$, $x \in \mathbb{R}^N$. Then*

$$U_{r,n}^{[m]}(f;x) - f(x) = \sum_{\widetilde{j}=1}^{m-1} \delta_{\widetilde{j},r}^{[m]} \left(\sum_{|\overline{\alpha}|=\widetilde{j}} \frac{\overline{c}_{\overline{\alpha},n,\widetilde{j}} f_{\overline{\alpha}}(x)}{\left(\prod_{i=1}^N \alpha_i!\right)} \right) + 0\left(\xi_n^{m-\gamma}\right). \tag{24.112}$$

When $m = 1$, the sum collapses.

Proof By Theorems 24.12, 24.42. Here $\rho = \varphi$, see (24.45). ■

We give

Corollary 24.44 (to Theorem 24.43) *Let $f \in C^1(\mathbb{R}^N)$, $N \geq 1$, with all $\left\| \frac{\partial f}{\partial x_i} \right\|_\infty \leq M$, $M > 0$, $i = 1, ..., N$. Let $0 < \gamma \leq 1$. Assume $U_{r,n}^{[1]}(f; x) \in \mathbb{R}$, $\forall\, x \in \mathbb{R}^N$. Here $r, \alpha \in \mathbb{N}$ and $\beta > \frac{1}{\alpha}$, $\xi_n \in (0, 1]$, $n \in \mathbb{N}$. Then*

$$U_{r,n}^{[1]}(f; x) - f(x) = 0\left(\xi_n^{1-\gamma}\right). \tag{24.113}$$

Proof By Theorems 24.12, 24.42. Here it is $\rho = \left(\frac{2\alpha\Gamma(\beta)+\Gamma\left(\frac{1}{\alpha}\right)\Gamma\left(\beta-\frac{1}{\alpha}\right)}{\Gamma\left(\frac{1}{2\alpha}\right)\Gamma\left(\beta-\frac{1}{2\alpha}\right)} \right)^N$. ■

We continue with

Corollary 24.45 (to Theorem 24.43) *Let $f \in C^2(\mathbb{R}^2)$, with all $\left\| \frac{\partial^2 f}{\partial x_1^2} \right\|_\infty$, $\left\| \frac{\partial^2 f}{\partial x_2^2} \right\|_\infty$, $\left\| \frac{\partial^2 f}{\partial x_1 \partial x_2} \right\|_\infty \leq M$, $M > 0$, $\xi_n \in (0, 1]$, $n \in \mathbb{N}$. Call*

$$c_1 = \int_{\mathbb{R}^2} s_1 d\lambda_{\xi_n}^*(s), \quad c_2 = \int_{\mathbb{R}^2} s_2 d\lambda_{\xi_n}^*(s), \tag{24.114}$$

where

$$d\lambda_{\xi_n}^* = W_n^2 \prod_{i=1}^{2} \frac{1}{\left(s_i^{2\alpha} + \xi_n^{2\alpha}\right)^\beta} ds_1 ds_2, \; s \in \mathbb{R}^2.$$

Let $0 < \gamma \leq 1$ and assume $U_{r,n}^{[2]}(f; x) \in \mathbb{R}$, $\forall\, x \in \mathbb{R}^2$. Here $r, \alpha \in \mathbb{N}$ and $\beta > \frac{3}{2\alpha}$. Then

$$U_{r,n}^{[2]}(f; x) - f(x) = \left(\sum_{j=1}^{r} \alpha_{j,r}^{[2]} j \right) \left(c_1 \frac{\partial f}{\partial x_1}(x) + c_2 \frac{\partial f}{\partial x_2}(x) \right) + 0\left(\xi_n^{2-\gamma}\right). \tag{24.115}$$

Proof By Theorems 24.12, 24.42. Here it is $\rho = \left[\frac{2\alpha\Gamma(\beta)+\Gamma\left(\frac{3}{2\alpha}\right)\Gamma\left(\beta-\frac{3}{2\alpha}\right)}{\Gamma\left(\frac{1}{2\alpha}\right)\Gamma\left(\beta-\frac{1}{2\alpha}\right)} \right]^2$. ■

We also give

Theorem 24.46 *Let $f \in C^{m+l}(\mathbb{R}^N)$, $m, l, N \in \mathbb{N}$. Assumptions of Theorem 24.32 are valid for $d\lambda_{\xi_n}(s)$, $s \in \mathbb{R}^N$, $\xi_n \in (0, 1]$, $n \in \mathbb{N}$. Call $\gamma = 0, \beta$. Suppose $\left\| f_{\gamma+\overline{\alpha}} \right\|_\infty \leq M$, $M > 0$, for all $\overline{\alpha} : |\overline{\alpha}| = m$. Here $\overline{c}_{\overline{\alpha},n,\widetilde{j}}$ is as in (24.48), all $|\overline{\alpha}| = \widetilde{j} = 1, ..., m-1$; $0 < \gamma \leq 1$. Assume $U_{r,n}^{[m]}\left(f_\gamma; x\right) \in \mathbb{R}$, $\forall\, x \in \mathbb{R}^N$. Let also $r, \alpha \in \mathbb{N}$ and $\beta > \frac{m+1}{2\alpha}$. Then*

$$\left(U_{r,n}^{[m]}(f;x)\right)_{\gamma} - f_{\gamma}(x) = \sum_{\widetilde{j}=1}^{m-1} \delta_{\widetilde{j},r}^{[m]} \left(\sum_{|\overline{\alpha}|=\widetilde{j}} \frac{\overline{c}_{\overline{\alpha},n,\widetilde{j}} f_{\gamma+\overline{\alpha}}(x)}{\left(\prod_{i=1}^{N} \alpha_i!\right)} \right) + 0\left(\xi_n^{m-\gamma}\right). \tag{24.116}$$

When $m = 1$, the sum collapses.

Proof Use of Theorem 24.12 and Theorem 4.6, p. 54 of [3]. Here it is $\rho = \varphi$, see (24.45). ■

24.3.5 Simultaneous Approximation by Multivariate Complex $U_{r,n}^{[m]}$

We make

Remark 24.47 We consider here complex valued Borel measurable functions $f : \mathbb{R}^N \to \mathbb{C}$ such that $f = f_1 + if_2$, $i = \sqrt{-1}$, where $f_1, f_2 : \mathbb{R}^N \to \mathbb{R}$ are implied to be real valued Borel measurable functions.

We define the multivariate complex Poisson–Cauchy singular operators

$$U_{r,n}^{[m]}(f;x) := U_{r,n}^{[m]}(f_1;x) + iU_{r,n}^{[m]}(f_2;x), \; x \in \mathbb{R}^N. \tag{24.117}$$

We assume that $U_{r,n}^{[m]}(f_j;x) \in \mathbb{R}$, $\forall\, x \in \mathbb{R}^N$, $j = 1, 2$.

One notices easily that

$$\left|U_{r,n}^{[m]}(f;x) - f(x)\right| \le \left|U_{r,n}^{[m]}(f_1;x) - f_1(x)\right| + \left|U_{r,n}^{[m]}(f_2;x) - f_2(x)\right| \tag{24.118}$$

also

$$\left\|U_{r,n}^{[m]}(f;x) - f(x)\right\|_{\infty,x} \le \left\|U_{r,n}^{[m]}(f_1;x) - f_1(x)\right\|_{\infty,x} + \left\|U_{r,n}^{[m]}(f_2;x) - f_2(x)\right\|_{\infty,x} \tag{24.119}$$

and

$$\left\|U_{r,n}^{[m]}(f) - f\right\|_p \le \left\|U_{r,n}^{[m]}(f_1) - f_1\right\|_p + \left\|U_{r,n}^{[m]}(f_2) - f_2\right\|_p, \; p \ge 1. \tag{24.120}$$

Furthermore, it holds

$$f_{\overline{\alpha}}(x) = f_{1,\overline{\alpha}}(x) + if_{2,\overline{\alpha}}(x), \tag{24.121}$$

where $\overline{\alpha}$ denotes a partial derivative of any order and arrangement.

We give

Theorem 24.48 *Let $f : \mathbb{R}^N \to \mathbb{C}$, $N \ge 2$, such that $f = f_1 + if_2$, $j = 1, 2$. Here $m \in \mathbb{N}$, $f_j \in C^m\left(\mathbb{R}^N\right)$, $x \in \mathbb{R}^N$. Assume $\left\|\frac{\partial^m f_j(\cdot,\cdot,\ldots,\cdot)}{\partial x_1^{\alpha_1}\ldots\partial x_N^{\alpha_N}}\right\|_{\infty} < \infty$, for all $\alpha_i \in \mathbb{Z}^+$, $i =$*

$1, ..., N : |\overline{\alpha}| := \sum_{i=1}^{N} \alpha_i = m$. *Let* λ_{ξ_n} *be the Borel probability measure on* $\mathbb{R}^N$*, see (24.26), where* $\xi_n \in (0, 1]$, $n \in \mathbb{N}$. *Here* $\alpha, r \in \mathbb{N}$, $\beta > \frac{m+r+1}{2\alpha}$, *and* $w_{\xi_n}(\overline{\alpha})$ *as in (24.30), and* $\overline{c}_{\overline{\alpha},n} := \overline{c}_{\overline{\alpha},n,\widetilde{j}}$ *as in (24.48). Then*

$$\left\| U_{r,n}^{[m]}(f; x) - f(x) - \sum_{\widetilde{j}=1}^{m} \delta_{\widetilde{j},r}^{[m]} \left(\sum_{\substack{\alpha_1,\dots,\alpha_N \geq 0; \\ |\overline{\alpha}| = \widetilde{j}}} \frac{\overline{c}_{\overline{\alpha},n,\widetilde{j}} f_{\overline{\alpha}}(x)}{\prod_{i=1}^{N} \alpha_i!} \right) \right\|_{\infty,x}$$

$$\leq \sum_{\substack{\alpha_1,\dots,\alpha_N \geq 0; \\ |\overline{\alpha}| = m}} \frac{\left(\omega_r\left(f_{1,\overline{\alpha}}, \xi_n\right) + \omega_r\left(f_{2,\overline{\alpha}}, \xi_n\right)\right)}{\left(\prod_{i=1}^{N} \alpha_i!\right)} w_{\xi_n}(\overline{\alpha}). \quad (24.122)$$

Proof By Theorem 24.15. ∎

We proceed with

Theorem 24.49 *Let* $f : \mathbb{R}^N \to \mathbb{C} : f = f_1 + i f_2$, $N \in \mathbb{N}$, $j = 1, 2$. *Here* $f_j \in C_B\left(\mathbb{R}^N\right)$ *uniformly continuous,* $\alpha, r \in \mathbb{N}$, $\beta > \frac{r+1}{2\alpha}$, $\xi_n \in (0, 1]$, $n \in \mathbb{N}$. *Then*

$$\left\| U_{r,n}^{[0]} f - f \right\|_{\infty} \leq \left(\frac{2\alpha\Gamma(\beta)}{\Gamma\left(\frac{1}{2\alpha}\right)\Gamma\left(\beta - \frac{1}{2\alpha}\right)} \right)^{N} \quad (24.123)$$

$$\left[(1+N)^r + 2^r \left(\frac{\Gamma\left(\frac{r+1}{2\alpha}\right)\Gamma\left(\beta - \left(\frac{r+1}{2\alpha}\right)\right)}{2\alpha\Gamma(\beta)} \right)^{N} \right] \xi_n^{2\alpha\beta(N-1)} \left(\omega_r\left(f_1, \xi_n\right) + \omega_r\left(f_2, \xi_n\right)\right),$$

As $n \to \infty$ *and* $\xi_n \to 0$, *we derive*

$$\left\| U_{r,n}^{[0]} f - f \right\|_{\infty} \to 0 \quad (24.124)$$

with rates.

Proof By Theorem 24.17. ∎

Next comes multi-simultaneous approximation.

Theorem 24.50 *Let* $f : \mathbb{R}^N \to \mathbb{C} : f = f_1 + i f_2$, $j = 1, 2$. *Here* $f_j \in C^{m+l}\left(\mathbb{R}^N\right)$, $m, l \in \mathbb{N}$, $N \geq 2$. *The assumptions of Theorem 24.32 are valid for* f_j *and* $d\mu_{\xi_n} = d\lambda_{\xi_n}$. *Call* $\gamma = 0, \beta$. *Assume* $\left\| f_{j,\gamma+\overline{\alpha}} \right\|_{\infty} < \infty$, *and let* $\xi_n \in (0, 1]$, $n \in \mathbb{N}$. *Here* $\alpha, r \in \mathbb{N}$, $\beta > \frac{m+r+1}{2\alpha}$, *and* $w_{\xi_n}(\overline{\alpha})$ *as in (24.30), and* $\overline{c}_{\overline{\alpha},n} := \overline{c}_{\overline{\alpha},n,\widetilde{j}}$ *as in (24.48). Then*

$$\left\| \left(U_{r,n}^{[m]}(f;\cdot)\right)_{\gamma} - f_{\gamma}(\cdot) - \sum_{\widetilde{j}=1}^{m} \delta_{\widetilde{j},r}^{[m]} \left(\sum_{\substack{\alpha_1,\dots,\alpha_N \geq 0; \\ |\overline{\alpha}|=\widetilde{j}}} \frac{\overline{c}_{\overline{\alpha},n,\widetilde{j}} f_{\gamma+\overline{\alpha}}(\cdot)}{\prod_{i=1}^{N} \alpha_i!} \right) \right\|_{\infty} \tag{24.125}$$

$$\leq \sum_{\substack{\alpha_1,\dots,\alpha_N \geq 0; \\ |\overline{\alpha}|=m}} \frac{\left(\omega_r\left(f_{1,\gamma+\overline{\alpha}},\xi_n\right) + \omega_r\left(f_{2,\gamma+\overline{\alpha}},\xi_n\right)\right)}{\left(\prod_{i=1}^{N} \alpha_i!\right)} w_{\xi_n}(\overline{\alpha}).$$

Proof Based on Theorems 24.32, 24.36. ∎

We continue with

Theorem 24.51 *Let $f : \mathbb{R}^N \to \mathbb{C} : f = f_1 + if_2$, $j = 1, 2$. Here $f_j \in C_B^l\left(\mathbb{R}^N\right)$, $l, N \in \mathbb{N}$ (functions l-times continuously differentiable and bounded). The assumptions of Theorem 24.32 are valid for f_j and $d\mu_{\xi_n} = d\lambda_{\xi_n}$. Call $\gamma = 0, \beta, \xi_n \in (0, 1]$, $n \in \mathbb{N}$. Let also $\alpha, r \in \mathbb{N}$, $\beta > \frac{r+1}{2\alpha}$. Then*

$$\left\| \left(U_{r,n}^{[0]} f\right)_{\gamma} - f_{\gamma} \right\|_{\infty} \leq \left(\frac{2\alpha\Gamma(\beta)}{\Gamma\left(\frac{1}{2\alpha}\right)\Gamma\left(\beta - \frac{1}{2\alpha}\right)} \right)^N$$

$$\left[(1+N)^r + 2^r \left(\frac{\Gamma\left(\frac{r+1}{2\alpha}\right)\Gamma\left(\beta - \left(\frac{r+1}{2\alpha}\right)\right)}{2\alpha\Gamma(\beta)} \right)^N \right] \xi_n^{2\alpha\beta(N-1)} \left(\omega_r\left(f_{1,\gamma},\xi_n\right) + \omega_r\left(f_{2,\gamma},\xi_n\right)\right). \tag{24.126}$$

If $N \geq 2$ or $f_{j,\gamma}$ is uniformly continuous ($j = 1, 2$), or both, and $\xi_n \to 0$, as $n \to \infty$, then $\left(U_{r,n}^{[0]} f\right)_{\gamma} \to f_{\gamma}$ uniformly.

Proof By Theorems 24.32 and 24.37. ∎

We proceed with L_p approximations

Theorem 24.52 *Let $f : \mathbb{R}^N \to \mathbb{C} : f = f_1 + if_2$, $j = 1, 2$. Here $f_j \in C^m\left(\mathbb{R}^N\right)$, $m \in \mathbb{N}$, $N \in \mathbb{N} - \{1\}$, with $f_{j,\overline{\alpha}} \in L_p\left(\mathbb{R}^N\right)$, $|\overline{\alpha}| = m$, $x \in \mathbb{R}^N$. Let $p, q > 1 : \frac{1}{p} + \frac{1}{q} = 1$. Here λ_{ξ_n} is a Borel probability measure on $\mathbb{R}^N$ as in (24.26), for $\xi_n \in (0, 1]$, $n \in \mathbb{N}$. Let also $r, \alpha \in \mathbb{N}$ and $\beta > \frac{(m+r)p+1}{2\alpha}$; $\alpha_i \in \mathbb{Z}^+$, $i = 1, \dots, N : |\overline{\alpha}| := \sum_{i=1}^{N} \alpha_i = m$. Here $\overline{c}_{\overline{\alpha},n} := \overline{c}_{\overline{\alpha},n,\widetilde{j}}$ as in (24.48), where $\widetilde{j} = 1, \dots, m$, and $\overline{\alpha} := (\alpha_1, \dots, \alpha_N)$, $\alpha_i \in \mathbb{Z}^+$, $i = 1, \dots, N$, $|\overline{\alpha}| := \sum_{i=1}^{N} \alpha_i = \widetilde{j}$. Then*

$$\left\| U_{r,n}^{[m]}(f;x) - f(x) - \sum_{\widetilde{j}=1}^{m} \delta_{\widetilde{j},r}^{[m]} \left(\sum_{|\overline{\alpha}|=\widetilde{j}} \frac{\overline{c}_{\overline{\alpha},n,\widetilde{j}} f_{\overline{\alpha}}(x)}{\left(\prod_{i=1}^{N} \alpha_i!\right)} \right) \right\|_{p,x}$$

$$\leq \left(\frac{m}{(q(m-1)+1)^{\frac{1}{q}}} \right) \left(\frac{2\alpha\Gamma(\beta)}{\Gamma\left(\frac{1}{2\alpha}\right)\Gamma\left(\beta - \frac{1}{2\alpha}\right)} \right)^{\frac{N}{p}}$$

$$\left[(1+N)^{rp} + 2^{rp} \left(\frac{\Gamma\left(\frac{(m+r)p+1}{2\alpha}\right)\Gamma\left(\beta - \left(\frac{(m+r)p+1}{2\alpha}\right)\right)}{2\alpha\Gamma(\beta)} \right)^{N} \right]^{\frac{1}{p}}$$

$$\left(\sum_{|\overline{\alpha}|=m} \frac{1}{\prod_{i=1}^{N} \alpha_i!} \right) \xi_n^{\left(\frac{2\alpha\beta(N-1)}{p}+m\right)} \left[\omega_r \left(f_{1,\overline{\alpha}}, \xi_n\right)_p + \omega_r \left(f_{2,\overline{\alpha}}, \xi_n\right)_p \right]. \tag{24.127}$$

Proof By Theorem 24.20. ■

We continue with

Theorem 24.53 *Let* $f : \mathbb{R}^N \to \mathbb{C} : f = f_1 + if_2$, $j = 1, 2$. *Here* $f_j \in \left(C\left(\mathbb{R}^N\right) \cap L_p\left(\mathbb{R}^N\right)\right)$; $\alpha, N, r \in \mathbb{N}$; $p, q > 1 : \frac{1}{p} + \frac{1}{q} = 1$; $\beta > \frac{rp+1}{2\alpha}$, $\xi_n \in (0, 1]$, $n \in \mathbb{N}$. *Then*

$$\left\| U_{r,n}^{[0]}(f) - f \right\|_p \leq \left(\frac{2\alpha\Gamma(\beta)}{\Gamma\left(\frac{1}{2\alpha}\right)\Gamma\left(\beta - \frac{1}{2\alpha}\right)} \right)^{\frac{N}{p}}$$

$$\left[(1+N)^{rp} + 2^{rp} \left(\frac{\Gamma\left(\frac{rp+1}{2\alpha}\right)\Gamma\left(\beta - \left(\frac{rp+1}{2\alpha}\right)\right)}{2\alpha\Gamma(\beta)} \right)^{N} \right]^{\frac{1}{p}}$$

$$\xi_n^{\frac{2\alpha\beta(N-1)}{p}} \left[\omega_r (f_1, \xi_n)_p + \omega_r (f_2, \xi_n)_p \right]. \tag{24.128}$$

As $\xi_n \to 0$, *when* $n \to \infty$, *we derive* $\left\| U_{r,n}^{[0]} f - f \right\|_p \to 0$, *i.e.* $U_{r,n}^{[0]} \to I$, *the unit operator, in* L_p *norm.*

Proof By Theorem 24.22. ■

We also give

Theorem 24.54 *Let* $f : \mathbb{R}^N \to \mathbb{C} : f = f_1 + if_2$, $j = 1, 2$. *Here* $f_j \in \left(C\left(\mathbb{R}^N\right) \cap L_1\left(\mathbb{R}^N\right)\right)$; $r, \alpha, N \in \mathbb{N}$, $\beta > \frac{r+1}{2\alpha}$, $\xi_n \in (0, 1]$, $n \in \mathbb{N}$. *Then*

$$\left\| U_{r,n}^{[0]}(f) - f \right\|_1 \leq \left(\frac{2\alpha\Gamma(\beta)}{\Gamma\left(\frac{1}{2\alpha}\right)\Gamma\left(\beta - \frac{1}{2\alpha}\right)} \right)^{N}$$

$$\left[(1+N)^r + 2^r \left(\frac{\Gamma\left(\frac{r+1}{2\alpha}\right)\Gamma\left(\beta - \left(\frac{r+1}{2\alpha}\right)\right)}{2\alpha\Gamma(\beta)}\right)^N\right] \xi_n^{2\alpha\beta(N-1)} \left(\omega_r(f_1,\xi_n)_1 + \omega_r(f_2,\xi_n)_1\right). \tag{24.129}$$

As $\xi_n \to 0$, we get $U_{r,n}^{[0]} \to I$, in L_1 norm.

Proof By Theorem 24.24. ∎

We further present

Theorem 24.55 *Let $f : \mathbb{R}^N \to \mathbb{C} : f = f_1 + if_2$, $j = 1, 2$. Here $f_j \in C^m\left(\mathbb{R}^N\right)$, $m, r \in \mathbb{N}$, $N \in \mathbb{N} - \{1\}$, with $f_{j,\overline{\alpha}} \in L_1\left(\mathbb{R}^N\right)$, where $\alpha_i \in \mathbb{Z}^+$, $i = 1, ..., N$, $|\overline{\alpha}| := \sum_{i=1}^{N} \alpha_i = m$, $x \in \mathbb{R}^N$, $\xi_n \in (0, 1]$, $n \in \mathbb{N}$ and $\alpha \in \mathbb{N}$ such that $\beta > \frac{m+r+1}{2\alpha}$. Here $\overline{c}_{\overline{\alpha},n} := \overline{c}_{\overline{\alpha},n,\widetilde{j}}$ as in (24.48), where $\widetilde{j} = 1, ..., m$, and $\overline{\alpha} := (\alpha_1, ..., \alpha_N)$, $\alpha_i \in \mathbb{Z}^+$, $i = 1, ..., N$, $|\overline{\alpha}| := \sum_{i=1}^{N} \alpha_i = \widetilde{j}$. Also here λ_{ξ_n} is the Borel probability measure on $\mathbb{R}^N$, see (24.26). Then*

$$\left\| U_{r,n}^{[m]}(f;x) - f(x) - \sum_{\widetilde{j}=1}^{m} \delta_{\widetilde{j},r}^{[m]} \left(\sum_{|\overline{\alpha}|=\widetilde{j}} \frac{\overline{c}_{\overline{\alpha},n,\widetilde{j}} f_{\overline{\alpha}}(x)}{\prod_{i=1}^{N} \alpha_i!} \right) \right\|_{1,x} \tag{24.130}$$

$$\leq \left\{ \sum_{|\overline{\alpha}|=m} \left(\frac{1}{\prod_{i=1}^{N} \alpha_i!} \right) \left[\omega_r\left(f_{1,\overline{\alpha}},\xi_n\right)_1 + \omega_r\left(f_{2,\overline{\alpha}},\xi_n\right)_1 \right] \right\} \xi_n^{2\alpha\beta(N-1)+m}$$

$$\left(\frac{2\alpha\Gamma(\beta)}{\Gamma\left(\frac{1}{2\alpha}\right)\Gamma\left(\beta - \frac{1}{2\alpha}\right)} \right)^N \left[(1+N)^r + 2^r \left(\frac{\Gamma\left(\frac{m+r+1}{2\alpha}\right)\Gamma\left(\beta - \left(\frac{m+r+1}{2\alpha}\right)\right)}{2\alpha\Gamma(\beta)} \right)^N \right].$$

Proof By Theorem 24.26. ∎

We continue with simultaneous L_p approximations.

Theorem 24.56 *Let $f : \mathbb{R}^N \to \mathbb{C} : f = f_1 + if_2$, $j = 1, 2$. Here $f_j \in C^{m+l}\left(\mathbb{R}^N\right)$, $m, l \in \mathbb{N}$, $N \in \mathbb{N} - \{1\}$. The assumptions of Theorem 24.32 are valid for $d\mu_{\xi_n} = d\lambda_{\xi_n}$, $\xi_n \in (0, 1]$, $n \in \mathbb{N}$ and f_j. Call $\gamma = 0, \beta$. Let $f_{j,(\gamma+\overline{\alpha})} \in L_p\left(\mathbb{R}^N\right)$, $|\overline{\alpha}| = m$, $x \in \mathbb{R}^N$, and $p, q > 1 : \frac{1}{p} + \frac{1}{q} = 1$. Let also $r, \alpha \in \mathbb{N}$ and $\beta > \frac{(m+r)p+1}{2\alpha}$; $\alpha_i \in \mathbb{Z}^+$, $i = 1, ..., N : |\overline{\alpha}| := \sum_{i=1}^{N} \alpha_i = m$. Here $\overline{c}_{\overline{\alpha},n} := \overline{c}_{\overline{\alpha},n,\widetilde{j}}$ as in (24.48), where $\widetilde{j} = 1, ..., m$, and $\overline{\alpha} := (\alpha_1, ..., \alpha_N)$, $\alpha_i \in \mathbb{Z}^+$, $i = 1, ..., N$, $|\overline{\alpha}| := \sum_{i=1}^{N} \alpha_i = \widetilde{j}$. Then*

$$\left\| \left(U_{r,n}^{[m]}(f;x)\right)_{\gamma} - f_{\gamma}(x) - \sum_{\widetilde{j}=1}^{m} \delta_{\widetilde{j},r}^{[m]} \left(\sum_{|\overline{\alpha}|=\widetilde{j}} \frac{\overline{c}_{\overline{\alpha},n,\widetilde{j}} f_{\gamma+\overline{\alpha}}(x)}{\prod_{i=1}^{N} \alpha_i!} \right) \right\|_{p,x}$$

$$\leq \left(\frac{m}{(q(m-1)+1)^{\frac{1}{q}}} \right) \left(\frac{2\alpha\Gamma(\beta)}{\Gamma\left(\frac{1}{2\alpha}\right)\Gamma\left(\beta-\frac{1}{2\alpha}\right)} \right)^{\frac{N}{p}} \tag{24.131}$$

$$\left[(1+N)^{rp} + 2^{rp} \left(\frac{\Gamma\left(\frac{(m+r)p+1}{2\alpha}\right)\Gamma\left(\beta-\left(\frac{(m+r)p+1}{2\alpha}\right)\right)}{2\alpha\Gamma(\beta)} \right)^{N} \right]^{\frac{1}{p}}$$

$$\left(\sum_{|\overline{\alpha}|=m} \frac{1}{\prod_{i=1}^{N} \alpha_i!} \right) \xi_n^{\left(\frac{2\alpha\beta(N-1)}{p}+m\right)} \left[\omega_r\left(f_{1,\gamma+\overline{\alpha}}, \xi_n\right)_p + \omega_r\left(f_{2,\gamma+\overline{\alpha}}, \xi_n\right)_p \right].$$

Proof By Theorems 24.32 and 24.38. ∎

We give also

Theorem 24.57 *Let $f : \mathbb{R}^N \to \mathbb{C} : f = f_1 + if_2$, $j = 1, 2$. Here $f_j \in C^l\left(\mathbb{R}^N\right)$, $l, N \in \mathbb{N}$. The assumptions of Theorem 24.32 are valid for $d\mu_{\xi_n} = d\lambda_{\xi_n}$, $\xi_n \in (0, 1]$, $n \in \mathbb{N}$ and f_j. Call $\gamma = 0, \beta$. Let $f_{j,\gamma} \in L_p\left(\mathbb{R}^N\right)$ and $p, q > 1 : \frac{1}{p} + \frac{1}{q} = 1$. Here $\alpha, r \in \mathbb{N}$, $\beta > \frac{rp+1}{2\alpha}$. Then*

$$\left\| \left(U_{r,n}^{[0]}(f)\right)_{\gamma} - f_{\gamma} \right\|_p \leq \left(\frac{2\alpha\Gamma(\beta)}{\Gamma\left(\frac{1}{2\alpha}\right)\Gamma\left(\beta-\frac{1}{2\alpha}\right)} \right)^{\frac{N}{p}}$$

$$\left[(1+N)^{rp} + 2^{rp} \left(\frac{\Gamma\left(\frac{rp+1}{2\alpha}\right)\Gamma\left(\beta-\left(\frac{rp+1}{2\alpha}\right)\right)}{2\alpha\Gamma(\beta)} \right)^{N} \right]^{\frac{1}{p}}$$

$$\xi_n^{\frac{2\alpha\beta(N-1)}{p}} \left[\omega_r\left(f_{1,\gamma}, \xi_n\right)_p + \omega_r\left(f_{2,\gamma}, \xi_n\right)_p \right]. \tag{24.132}$$

As $n \to +\infty$ and $\xi_n \to 0$, then $\left(U_{r,n}^{[0]} f\right)_{\gamma} \overset{\|\cdot\|_p}{\to} f_{\gamma}$.

Proof By Theorems 24.32 and 24.39. ∎

We continue with

Theorem 24.58 *Let $f : \mathbb{R}^N \to \mathbb{C} : f = f_1 + i f_2$, $j = 1, 2$. Here $f_j \in C^l\left(\mathbb{R}^N\right)$, $l, N \in \mathbb{N}$. The assumptions of Theorem 24.32 are valid for $d\mu_{\xi_n} = d\lambda_{\xi_n}$, $\xi_n \in (0, 1]$, $n \in \mathbb{N}$ and f_j. Call $\gamma = 0, \beta$. Let $f_{j,\gamma} \in L_1\left(\mathbb{R}^N\right)$ and $\alpha, r \in \mathbb{N}$, $\beta > \frac{r+1}{2\alpha}$. Then*

$$\left\| \left(U_{r,n}^{[0]}(f)\right)_\gamma - f_\gamma \right\|_1 \leq \left(\frac{2\alpha\Gamma(\beta)}{\Gamma\left(\frac{1}{2\alpha}\right)\Gamma\left(\beta - \frac{1}{2\alpha}\right)} \right)^N$$

$$\left[(1+N)^r + 2^r \left(\frac{\Gamma\left(\frac{r+1}{2\alpha}\right)\Gamma\left(\beta - \left(\frac{r+1}{2\alpha}\right)\right)}{2\alpha\Gamma(\beta)} \right)^N \right]$$

$$\xi_n^{2\alpha\beta(N-1)} \left[\omega_r\left(f_{1,\gamma}, \xi_n\right)_1 + \omega_r\left(f_{2,\gamma}, \xi_n\right)_1 \right]. \tag{24.133}$$

As $n \to +\infty$ and $\xi_n \to 0$, then $\left(U_{r,n}^{[0]}(f)\right)_\gamma \overset{\|\cdot\|_1}{\to} f_\gamma$.

Proof By Theorems 24.32, 24.40. ∎

We finish with

Theorem 24.59 *Let $f : \mathbb{R}^N \to \mathbb{C} : f = f_1 + i f_2$, $j = 1, 2$. Here $f_j \in C^{m+l}\left(\mathbb{R}^N\right)$, $r, m, l \in \mathbb{N}$, $N \in \mathbb{N} - \{1\}$. The assumptions of Theorem 24.32 are valid for $d\mu_{\xi_n} = d\lambda_{\xi_n}$, $\xi_n \in (0, 1]$, $n \in \mathbb{N}$ and f_j. Call $\gamma = 0, \beta$. Let $f_{j,(\gamma+\overline{\alpha})} \in L_1\left(\mathbb{R}^N\right)$, $|\overline{\alpha}| = m$, $x \in \mathbb{R}^N$. Here $\alpha \in \mathbb{N} : \beta > \frac{m+r+1}{2\alpha}$. Here $\overline{c}_{\overline{\alpha},n} := \overline{c}_{\overline{\alpha},n,\widetilde{j}}$ as in (24.48), where $\widetilde{j} = 1, ..., m$, and $\overline{\alpha} := (\alpha_1, ..., \alpha_N)$, $\alpha_i \in \mathbb{Z}^+$, $i = 1, ..., N$, $|\overline{\alpha}| := \sum_{i=1}^N \alpha_i = \widetilde{j}$. Then*

$$\left\| \left(U_{r,n}^{[m]}(f; x)\right)_\gamma - f_\gamma(x) - \sum_{\widetilde{j}=1}^m \delta_{\widetilde{j},r}^{[m]} \left(\sum_{|\overline{\alpha}|=\widetilde{j}} \frac{\overline{c}_{\overline{\alpha},n,\widetilde{j}} f_{\gamma+\overline{\alpha}}(x)}{\prod_{i=1}^N \alpha_i!} \right) \right\|_{1,x}$$

$$\leq \left(\sum_{|\overline{\alpha}|=m} \left(\frac{1}{\prod_{i=1}^N \alpha_i!} \right) \left[\omega_r\left(f_{1,\gamma+\overline{\alpha}}, \xi_n\right)_1 + \omega_r\left(f_{2,\gamma+\overline{\alpha}}, \xi_n\right)_1 \right] \right)$$

$$\xi_n^{2\alpha\beta(N-1)+m} \left(\frac{2\alpha\Gamma(\beta)}{\Gamma\left(\frac{1}{2\alpha}\right)\Gamma\left(\beta - \frac{1}{2\alpha}\right)} \right)^N$$

$$\left[(1+N)^r + 2^r \left(\frac{\Gamma\left(\frac{m+r+1}{2\alpha}\right)\Gamma\left(\beta - \left(\frac{m+r+1}{2\alpha}\right)\right)}{2\alpha\Gamma(\beta)} \right)^N \right]. \tag{24.134}$$

Proof By Theorems 24.32, 24.41. ■

References

1. Anastassiou, G.A.: Rate of convergence of non-positive linear convolution type operators. A sharp inequality. J. Math. Anal. Appl. **142**, 441–451 (1989)
2. Anastassiou, G.A.: Moments in Probability and Approximation Theory, Pitman Research Notes in Math, vol. 287. Longman Scientific & Technical, Harlow (1993)
3. Anastassiou, G.A.: Approximation by Multivariate Singular Integrals. Springer, New York (2011)
4. Anastassiou, G.A.: Approximation by Multivariate Generalized Poisson-Cauchy type Singular Integral operators submitted (2020)
5. Anastassiou, G., Gal, S.: Approximation Theory. Birkhaüser, Boston (2000)
6. Anastassiou, G.A., Mezei, R.A.: Global Smoothness and uniform convergence of smooth Poisson-Cauchy type singular operators. Appl. Math. Comput. **215**, 1718–1731 (2009)
7. Anastassiou, G., Mezei, R.: Uniform convergence with rates of smooth Poisson-Cauchy type singular integral operators. Math. Comput. Modell. **50**, 1553–1570 (2009)
8. Anastassiou, G.A., Mezei, R.A.: A Voronovskaya type theorem for Poisson-Cauchy type singular operators. J. Math. Anal. Appl. **336**, 525–529 (2010)
9. Anastassiou, G.A., Mezei, R.A.: L_p convergence with rates of smooth Poisson-Cauchy Type Singular Operators. Int. J. Math. Sci. (Korea) **11**(1–2), 1–26 (2012)
10. DeVore, R.A., Lorentz, G.G.: Constructive Approximation, vol. 303. Springer, Berlin (1993)
11. Zwillinger, D.: CRC standard Mathematical Tables and Formulae, 30th edn. Chapman & Hall/CRC, Boca Raton (1995)

Chapter 25
High Order Approximation with Multivariate Generalized Trigonometric Type Singular Integral Operators

This research and survey chapter deals exclusively with the chapter of the approximation of generalized multivariate Trigonometric type singular integrals to the identity-unit operator. Here we study quantitatively most of their approximation properties. These operators are not in general positive linear operators. In particular we study the rate of convergence of these integral operators to the unit operator, as well as the related simultaneous approximation. These are given via Jackson type inequalities and by the use of multivariate high order modulus of smoothness of the high order partial derivatives of the involved function. Also we study the global smoothness preservation properties of these integral operators. These multivariate inequalities are nearly sharp and in one case the inequality is attained, that is sharp. Furthermore we give asymptotic expansions of Voronovskaya type for the error of approximation. The above properties are studied with respect to L_p norm, $1 \leq p \leq \infty$. It follows [4].

25.1 Introduction

We start with our motivation for this chapter. The following come from [6].

For $r \in \mathbb{N}$ and $n \in \mathbb{Z}_+$ we set

$$\alpha_j = \begin{cases} (-1)^{r-j} \binom{r}{j} j^{-n}, & j = 1, ..., r, \\ 1 - \sum_{j=1}^{r} (-1)^{r-j} \binom{r}{j} j^{-n}, & j = 0, \end{cases} \tag{25.1}$$

that is $\sum_{j=0}^{r} \alpha_j = 1$. Here it is $\xi \in (0, 1]$.

G. A. Anastassiou, *Constructive Fractional Analysis with Applications*,
Studies in Systems, Decision and Control 362,
https://doi.org/10.1007/978-3-030-71481-9_25

Let $f \in C^n(\mathbb{R})$, $n \in \mathbb{Z}_+$, and $f^{(n)} \in L_p(\mathbb{R})$, $1 \leq p < \infty$, $\beta \in \mathbb{N}$, we define for $x \in \mathbb{R}$, the trigonometric integral

$$T_{r,\xi}(f; x) := \frac{1}{W} \int_{-\infty}^{\infty} \left(\sum_{j=0}^{r} \alpha_j f(x + jt) \right) \left(\frac{\sin\left(\frac{t}{\xi}\right)}{t} \right)^{2\beta} dt, \tag{25.2}$$

where

$$W = \int_{-\infty}^{\infty} \left(\frac{\sin\left(\frac{t}{\xi}\right)}{t} \right)^{2\beta} dt = 2\xi^{1-2\beta} \int_{0}^{\infty} \left(\frac{\sin t}{t} \right)^{2\beta} dt \overset{(25.6.)}{=}$$

$$2\xi^{1-2\beta} \pi (-1)^{\beta} \beta \sum_{k=1}^{\beta} (-1)^k \frac{k^{2\beta-1}}{(\beta-k)!(\beta+k)!}. \tag{25.3}$$

$T_{r,\xi}$ operators are not positive operators, see [8].

We mention:

Let p and m be integers with $1 \leq p \leq m$. We define the integral

$$I(m, p) := \int_{-\infty}^{\infty} \frac{(\sin x)^{2m}}{x^{2p}} dx = 2 \int_{0}^{\infty} \frac{(\sin x)^{2m}}{x^{2p}} dx. \tag{25.4}$$

That is an (absolutely) convergent integral.

According to [10], page 210, item 1033, we obtain

$$I(m, p) = \pi \frac{(-1)^p (2m)!}{4^{m-p}(2p-1)!} \sum_{k=1}^{m} (-1)^k \frac{k^{2p-1}}{(m-k)!(m+k)!}. \tag{25.5}$$

In particular, for $p = m$ the above formula becomes

$$\int_{0}^{\infty} \frac{(\sin x)^{2m}}{x^{2m}} dx = \pi (-1)^m m \sum_{k=1}^{m} (-1)^k \frac{k^{2m-1}}{(m-k)!(m+k)!}. \tag{25.6}$$

We need the rth L_p-modulus of smoothness

$$\omega_r\left(f^{(n)}, h\right)_p := \sup_{|t| \leq h} \left\| \Delta_t^r f^{(n)}(x) \right\|_{p,x}, \quad h > 0, \tag{25.7}$$

where

$$\Delta_t^r f^{(n)}(x) := \sum_{j=0}^{r} (-1)^{r-j} \binom{r}{j} f^{(n)}(x + jt), \tag{25.8}$$

see [9], p. 44. Here we have that $\omega_r\left(f^{(n)}, h\right)_p < \infty$, $h > 0$.

We need to introduce

$$\delta_k := \sum_{j=1}^{r} \alpha_j j^k, \; k = 1, ..., n \in \mathbb{N}. \tag{25.9}$$

Call

$$\tau(w, x) := \sum_{j=0}^{r} \alpha_j j^n f^{(n)}(x + jw) - \delta_n f^{(n)}(x). \tag{25.10}$$

Notice also that

$$-\sum_{j=1}^{r} (-1)^{r-j} \binom{r}{j} = (-1)^r \binom{r}{0}.$$

According to [2], p. 306, [1], we get

$$\tau(w, x) = \Delta_w^r f^{(n)}(x). \tag{25.11}$$

Thus

$$\|\tau(w, x)\|_{p,x} \le \omega_r \left(f^{(n)}, |w|\right)_p, \; w \in \mathbb{R}. \tag{25.12}$$

Using Taylor's formula one has

$$\sum_{j=0}^{r} \alpha_j [f(x + jt) - f(x)] = \sum_{k=1}^{n} \frac{f^{(k)}(x)}{k!} \delta_k t^k + R_n(0, t, x), \tag{25.13}$$

where

$$R_n(0, t, x) := \int_0^t \frac{(t-w)^{n-1}}{(n-1)!} \tau(w, x)\, dw, \; n \in \mathbb{N}. \tag{25.14}$$

Assume

$$c_{k,\xi} := \int_{-\infty}^{\infty} t^k d\mu_\xi(t) \in \mathbb{R}, \; k = 1, ..., n, \tag{25.15}$$

where

$$d\mu_\xi(t) := \frac{1}{W} \left(\frac{\sin\left(\frac{t}{\xi}\right)}{t} \right)^{2\beta} dt, \; \forall t \in \mathbb{R}.$$

Using the above terminology we derive

$$\Delta(x) := T_{r,\xi}(f; x) - f(x) - \sum_{k=1}^{n} \frac{f^{(k)}(x)}{k!} \delta_k c_{k,\xi} = R_n^*(x), \tag{25.16}$$

where

$$R_n^*(x) := \int_{-\infty}^{\infty} R_n(0, t, x)\, d\mu_\xi(t), \quad n \in \mathbb{N}. \tag{25.17}$$

Let $\lceil\cdot\rceil$ denote the ceiling of a real number. We mention

Theorem 25.1 ([6]) *Let $p, q > 1$ such that $\frac{1}{p} + \frac{1}{q} = 1$, $n, \beta \in \mathbb{N}$, $\beta > \frac{\lceil rp\rceil + np + 1}{2}$ and the rest as above. Then*

$$\|\Delta(x)\|_p \leq \frac{1}{((n-1)!)\,(q(n-1)+1)^{\frac{1}{q}}\,(rp+1)^{\frac{1}{p}}} \tag{25.18}$$

$$\left[\frac{1}{\int_0^\infty \left(\frac{\sin t}{t}\right)^{2\beta} dt} \sum_{j=1}^{\lceil rp\rceil+1} \int_0^\infty \left[t^{np-1+j}\left(\frac{\sin t}{t}\right)^{2\beta}\right] dt\right]^{\frac{1}{p}} \xi^n \omega_r\left(f^{(n)}, \xi\right)_p.$$

Moreover, as $\xi \to 0$ we get that $\|\Delta(x)\|_p \to 0$.

The counterpart of Theorem 25.1 follows, case of $p = 1$.

Theorem 25.2 ([6]) *Let $f \in C^n(\mathbb{R})$ and $f^{(n)} \in L_1(\mathbb{R})$, $n \in \mathbb{N}$, $\beta \in \mathbb{N}$, $\beta > \frac{r+1+n}{2}$. Then*

$$\|\Delta(x)\|_1 \leq \frac{1}{(r+1)(n-1)!\left[\int_0^\infty \left(\frac{\sin t}{t}\right)^{2\beta} dt\right]} \tag{25.19}$$

$$\sum_{j=1}^{r+1}\left(\int_0^\infty \left[t^{n-1+j}\left(\frac{\sin t}{t}\right)^{2\beta}\right] dt\right) \xi^n \omega_r\left(f^{(n)}, \xi\right)_1.$$

Hence as $\xi \to 0$ we obtain $\|\Delta(x)\|_1 \to 0$.

The case $n = 0$ is mentioned next.

Proposition 25.3 ([6]) *Let $p, q > 1$ such that $\frac{1}{p} + \frac{1}{q} = 1$, $\beta \in \mathbb{N}$, $\beta > \frac{\lceil rp\rceil + 1}{2}$ and the rest as above. Then*

$$\left\|T_{r,\xi}(f) - f\right\|_p \leq \omega_r(f, \xi)_p \left(\frac{1}{\left[\int_0^\infty \left(\frac{\sin t}{t}\right)^{2\beta} dt\right]} \sum_{j=0}^{\lceil rp\rceil}\left[\int_0^\infty t^j \left(\frac{\sin t}{t}\right)^{2\beta} dt\right]\right)^{\frac{1}{p}}. \tag{25.20}$$

Also as $\xi \to 0$ we obtain $T_{r,\xi} \to$ unit operator I in the L_p norm, $p > 1$.

We also give

Proposition 25.4 ([6]) *For $\beta \in \mathbb{N}$, $\beta > \frac{r+1}{2}$, we have*

$$\left\|T_{r,\xi}(f)-f\right\|_1 \leq \frac{\omega_r(f,\xi)_1}{\left[\int_0^{\infty}\left(\frac{\sin t}{t}\right)^{2\beta}dt\right]}\sum_{j=0}^{r}\left[\int_0^{\infty}t^j\left(\frac{\sin t}{t}\right)^{2\beta}dt\right]. \tag{25.21}$$

Moreover as $\xi \to 0$ we get that $T_{r,\xi} \to I$ in the L_1 norm.

We also mention:
Case $\beta = 2$.

Corollary 25.5 ([6]) *Let $f \in C^1(\mathbb{R})$ and $f' \in L_1(\mathbb{R})$. Then*

$$\left\|T_{1,\xi}(f;x)-f(x)\right\|_1 \leq \frac{3}{2\pi}\left(\ln 2+\frac{\pi}{4}\right)\xi\omega_1\left(f',\xi\right)_1. \tag{25.22}$$

Corollary 25.6 ([6]) *Let $f \in C^1(\mathbb{R})$ and $f' \in L_1(\mathbb{R})$. Then*

$$\left\|T_{2,\xi}(f;x)-f(x)\right\|_1 \leq \left(\frac{40}{33\pi}\ln\left(\frac{32^{\frac{27}{16}}}{4}\right)+\frac{5}{33}+\frac{5}{22\pi}\ln\frac{256}{27}\right)\xi\omega_2\left(f',\xi\right)_1. \tag{25.23}$$

Corollary 25.7 ([6]) *It holds*

$$\left\|T_{1,\xi}(f)-f\right\|_4 \leq \omega_1(f,\xi)_4\sqrt[4]{\frac{40}{11\pi}\ln\left(\frac{3^{\frac{27}{16}}}{4}\right)+\frac{15}{22\pi}\ln\frac{256}{27}+\frac{47}{22}}. \tag{25.24}$$

Also as $\xi \to 0$ we obtain $T_{1,\xi} \to$ unit operator I in the L_4 norm.

Corollary 25.8 ([6]) *We have*

$$\left\|T_{6,\xi}(f)-f\right\|_1 \leq \omega_6(f,\xi)_1\left(\frac{630}{151\pi}\ln\frac{2^{\frac{251}{60}}}{3^{\frac{9}{5}}}+\frac{5671}{2416}\right). \tag{25.25}$$

Moreover as $\xi \to 0$ we get that $T_{6,\xi} \to I$ in the L_1 norm.

We will use a lot the following:

Remark 25.9 ([7]) Let $j, m \in \mathbb{Z}$, $m \geq 1$ such that $0 \leq j < 2m-1$. The integral

$$\int_{-\infty}^{\infty}x^j\left(\frac{\sin x}{x}\right)^{2m}dx = \begin{cases} 2\int_0^{\infty}x^j\left(\frac{\sin x}{x}\right)^{2m}dx, & \text{if } j \text{ is even} \\ 0, & \text{if } j \text{ is odd}\end{cases}, \tag{25.26}$$

is an (absolutely) convergent integral.

According to [10], page 210, item 1033, we obtain
case 1: j is even, $j < 2m-1$

$$\int_0^\infty x^j \left(\frac{\sin x}{x}\right)^{2m} dx = \frac{\pi (-1)^{\frac{2m-j}{2}} (2m)!}{2^{j+1} (2m-j-1)!} \sum_{k=1}^{m} (-1)^k \frac{k^{2m-j-1}}{(m-k)! (m+k)!}, \tag{25.27}$$

and

case 2: j is odd, $j < 2m-1$

$$\int_0^\infty x^j \left(\frac{\sin x}{x}\right)^{2m} dx = \frac{(-1)^{\frac{j-1}{2}} (2m)!}{2^{j} (2m-j-1)!} \sum_{k=1}^{m} (-1)^{m-k} \frac{k^{2m-j-1} [\ln (2k)]}{(m-k)! (m+k)!}. \tag{25.28}$$

In particular, for $j = 0$ the formula (25.27) becomes

$$\int_0^\infty \left(\frac{\sin x}{x}\right)^{2m} dx = \pi (-1)^m m \sum_{k=1}^{m} (-1)^k \frac{k^{2m-1}}{(m-k)! (m+k)!}. \tag{25.29}$$

In this chapter we study the approximation properties of general multivariate smooth Trigonometric singular integral operators:

$$T_{r,n}^{[m]} (f; x_1, ..., x_N) :=$$

$$\lambda_n^{-N} \sum_{j=0}^{r} \alpha_{j,r}^{[m]} \int_{\mathbb{R}^N} f (x_1 + s_1 j, ..., x_N + s_N j) \prod_{i=1}^{N} \left(\frac{\sin\left(\frac{s_i}{\xi_n}\right)}{s_i}\right)^{2\beta} ds_1 ... ds_N, \tag{25.30}$$

with $\beta \in N$, and

$$\lambda_n := 2\xi_n^{1-2\beta} \pi (-1)^\beta \beta \sum_{k=1}^{\beta} (-1)^k \frac{k^{2\beta-1}}{(\beta-k)! (\beta+k)!}, \tag{25.31}$$

see [8, 10], p. 210, item 1033.

Notice that

$$\lambda_n^{-N} \int_{\mathbb{R}^N} \prod_{i=1}^{N} \left(\frac{\sin\left(\frac{s_i}{\xi_n}\right)}{s_i}\right)^{2\beta} ds_1 ... ds_N = 1, \tag{25.32}$$

see also [8, 10], p. 210, item 1033, and [3], p. 16.

We call

$$\gamma := 2\pi (-1)^\beta \beta \sum_{k=1}^{\beta} (-1)^k \frac{k^{2\beta-1}}{(\beta-k)! (\beta+k)!}, \tag{25.33}$$

that is

$$\lambda_n = \gamma \xi_n^{1-2\beta}. \tag{25.34}$$

Here $r \in \mathbb{N}$, $m \in Z_{+}$, and

$$\alpha_{j,r}^{[m]} := \begin{cases} (-1)^{r-j} \binom{r}{j} j^{-m}, & \text{if } j = 1, 2, ..., r, \\ 1 - \sum_{i=1}^{r} (-1)^{r-i} \binom{r}{i} i^{-m}, & \text{if } j = 0, \end{cases} \tag{25.35}$$

and

$$\delta_{k,r}^{[m]} := \sum_{j=1}^{r} \alpha_{j,r}^{[m]} j^k, \quad k = 1, 2, ..., m \in \mathbb{N}. \tag{25.36}$$

See that $\sum_{j=0}^{r} \alpha_{j,r}^{[m]} = 1$.

Also here $\xi_n \in (0, 1]$, $n \in \mathbb{N}$, and $f : \mathbb{R}^N \to \mathbb{R}$ is a Borel measurable function. The above operator $T_{r,n}^{[m]}$ is a special case of a more general operator $\theta_{r,n}^{[m]}$ studied in general in [3] by the author.

We mention next about $\theta_{r,n}^{[m]}$.

Let μ_{ξ_n} be a probability Borel measure on $\mathbb{R}^N$, $N \geq 1$.

We define the multiple smooth singular integral operators

$$\theta_{r,n}^{[m]} (f; x_1, ..., x_N) := \sum_{j=0}^{r} \alpha_{j.r}^{[m]} \int_{\mathbb{R}^N} f (x_1 + s_1 j, x_2 + s_2 j, ..., x_N + s_N j) \, d\mu_{\xi_n} (s), \tag{25.37}$$

where $s := (s_1, ..., s_N)$, $x := (x_1, ..., x_N) \in \mathbb{R}^N$.

The operators $\theta_{r,n}^{[m]}$ are not in general positive. For example, consider the function $\varphi (u_1, ..., u_N) = \sum_{i=1}^{N} u_i^2$ and also take $r = 2$, $m = 3$; $x_i = 0$, $i = 1, ..., N$. See that $\varphi \geq 0$, however

$$\theta_{2,n}^{[3]} (\varphi; 0, 0, ..., 0) = \left(\sum_{j=1}^{2} j^2 \alpha_{j,2}^{[3]} \right) \int_{\mathbb{R}^N} \left(\sum_{i=1}^{N} s_i^2 \right) d\mu_{\xi_n} (s) =$$

$$\left(\alpha_{1,2}^{[3]} + 4\alpha_{2,2}^{[3]} \right) \int_{\mathbb{R}^N} \left(\sum_{i=1}^{N} s_i^2 \right) d\mu_{\xi_n} (s) = \left(-2 + \frac{1}{2} \right) \int_{\mathbb{R}^N} \left(\sum_{i=1}^{N} s_i^2 \right) d\mu_{\xi_n} (s) < 0. \tag{25.38}$$

assuming that $\int_{\mathbb{R}^N} \left(\sum_{i=1}^{N} s_i^2 \right) d\mu_{\xi_n} (s) < \infty$.

Clearly in the case of $T_{r,n}^{[m]}$ we have that

$$d\mu_{\xi_n} (s) = \lambda_n^{-N} \prod_{i=1}^{N} \left(\frac{\sin \left(\frac{s_i}{\xi_n} \right)}{s_i} \right)^{2\beta} \prod_{i=1}^{N} ds_i =: d\varphi_{\xi_n} (s), \quad s \in \mathbb{R}^N. \tag{25.39}$$

Lemma 25.10 *The operator $\theta_{r,n}^{[m]}$ preserves the constant functions in N variables.*

We need the following definition.

Definition 25.11 Let $f \in C_B\left(\mathbb{R}^N\right)$, the space of all bounded and continuous functions or uniformly continuous on $\mathbb{R}^N$. Then, the rth multivariate modulus of smoothness of f is given by (see, e.g. [5])

$$\omega_r(f; h) := \sup_{\sqrt{u_1^2+\ldots+u_N^2} \leq h} \left\| \Delta_{u_1,u_2,\ldots,u_N}^r (f) \right\|_\infty < \infty, \; h > 0, \tag{25.40}$$

where $\|\cdot\|_\infty$ is the sup-norm and

$$\Delta_u^r f(x) := \Delta_{u_1,u_2,\ldots,u_N}^r f(x_1, \ldots, x_N) =$$

$$\sum_{j=0}^{r} (-1)^{r-j} \binom{r}{j} f(x_1 + ju_1, x_2 + ju_2, \ldots, x_N + ju_N). \tag{25.41}$$

Let $m \in \mathbb{N}$ and let $f \in C^m\left(\mathbb{R}^N\right)$.

Suppose that all partial derivatives of f of order m are bounded, i.e.

$$\left\| \frac{\partial^m f(\cdot, \cdot, \ldots, \cdot)}{\partial x_1^{\alpha_1} \ldots \partial x_N^{\alpha_N}} \right\|_\infty < \infty, \tag{25.42}$$

for all $\alpha_j \in \mathbb{Z}^+$, $j = 1, \ldots, N$; $\sum_{j=1}^{N} \alpha_j = m$.

In this chapter we apply the general theory developed in [3] about $\theta_{r,n}^{[m]}$ to the operators $T_{r,n}^{[m]}$, so we can obtain computationally specific results and show that the general theory has applications and it is a valid theory.

So for the very important in various branches of mathematics operators $T_{r,n}^{[m]}$ we prove the very essential properties of uniform approximation, L_p approximation, global smoothness preservation and simultaneously approximation, Voronovskaya asymptotic expansions and complex simultaneous approximation.

25.2 Auxiliary Essential Results

We will use

Lemma 25.12 *Let $N \in \mathbb{N}$, $r > 0$, $z_i \in \mathbb{R}_+$, $i = 1, \ldots, N$. Then*

$$\left(1 + \sum_{i=1}^{N} z_i\right)^r \leq \prod_{i=1}^{N} (1 + z_i)^r. \tag{25.43}$$

Proof We have

$$\left(1+\sum_{i=1}^{N} z_i\right)^r \leq \left(N+\sum_{i=1}^{N} z_i\right)^r = [(1+z_1)+(1+z_2)+...+(1+z_N)]^r =$$

$$\left(\sum_{i=1}^{N}(1+z_i)\right)^r \leq \prod_{i=1}^{N}(1+z_i)^r, \text{ by } 1+z_i \geq 1, i=1,...,N.$$

■

We give

Theorem 25.13 *Let $r, N, \beta \in \mathbb{N}$, $\alpha_j \in \mathbb{Z}^+$, $j = 1, ..., N : |\overline{\alpha}| := \sum_{j=1}^{N} \alpha_j = m \in \mathbb{N}$, $\xi_n \in (0, 1]$, $n \in \mathbb{N}$. Here we take $\beta > \frac{r+m+1}{2}$, and γ, λ_n are as in (25.33) and (25.34), respectively. Also we take $\lambda = 0, 1, ..., r$. When λ is even we define*

$$\psi_{1\lambda} := \frac{\pi(-1)^{\frac{2\beta-\lambda}{2}}(2\beta)!}{2^{\lambda+1}(2\beta-\lambda-1)!}\left(\sum_{k=1}^{\beta}(-1)^k\frac{k^{2\beta-\lambda-1}}{(\beta-k)!(\beta+k)!}\right), \tag{25.44}$$

and when λ is odd we define

$$\psi_{2\lambda} := \frac{(-1)^{\frac{\lambda-1}{2}}(2\beta)!}{2^{\lambda}(2\beta-\lambda-1)!}\left(\sum_{k=1}^{\beta}(-1)^{\beta-k}\frac{k^{2\beta-\lambda-1}[\ln(2k)]}{(\beta-k)!(\beta+k)!}\right), \tag{25.45}$$

and we set

$$\psi_{\lambda} := \begin{cases} \psi_{1\lambda}, & \text{if } \lambda \text{ is even,} \\ \psi_{2\lambda}, & \text{if } \lambda \text{ is odd.} \end{cases} \tag{25.46}$$

Similarly, it is defined $\psi_{\lambda+m}$, just set in (25.44), (25.45), (25.46), $\lambda+m$ in place of λ. Then

$$A_{\xi_n}(\overline{\alpha}) := \lambda_n^{-N}\int_{\mathbb{R}^N}\left(\prod_{i=1}^{N}|s_i|^{\alpha_i}\right)\left(1+\frac{\|s\|_2}{\xi_n}\right)^r\prod_{i=1}^{N}\left(\frac{\sin\left(\frac{s_i}{\xi_n}\right)}{s_i}\right)^{2\beta}\prod_{i=1}^{N}ds_i \leq$$

$$\xi_n^{2\beta(N-1)+m}2^N\gamma^{-N}\left\{\sum_{\lambda=0}^{r}\binom{r}{\lambda}[\psi_\lambda+\psi_{\lambda+m}]\right\}^N \leq$$

$$2^N\gamma^{-N}\left\{\sum_{\lambda=0}^{r}\binom{r}{\lambda}[\psi_\lambda+\psi_{\lambda+m}]\right\}^N < +\infty, \tag{25.47}$$

uniformly bounded, and convergent to zero as $\xi_n \to 0$, *when* $n \to +\infty$.

Proof We estimate

$$A_{\xi_n}(\overline{\alpha}) = \lambda_n^{-N} \int_{\mathbb{R}^N} \left(\prod_{i=1}^{N} |s_i|^{\alpha_i} \right) \left(1 + \frac{\|s\|_2}{\xi_n} \right)^r \prod_{i=1}^{N} \left(\frac{\sin\left(\frac{s_i}{\xi_n}\right)}{s_i} \right)^{2\beta} \prod_{i=1}^{N} ds_i =$$

$$2^N \lambda_n^{-N} \int_{\mathbb{R}_+^N} \left(\prod_{i=1}^{N} s_i^{\alpha_i} \right) \left(1 + \frac{\|s\|_2}{\xi_n} \right)^r \prod_{i=1}^{N} \left(\frac{\sin\left(\frac{s_i}{\xi_n}\right)}{s_i} \right)^{2\beta} \prod_{i=1}^{N} ds_i \leq \tag{25.48}$$

$$2^N \lambda_n^{-N} \int_{\mathbb{R}_+^N} \left(\prod_{i=1}^{N} s_i^{\alpha_i} \right) \left(1 + \sum_{i=1}^{N} \left(\frac{s_i}{\xi_n} \right) \right)^r \prod_{i=1}^{N} \left(\frac{\sin\left(\frac{s_i}{\xi_n}\right)}{s_i} \right)^{2\beta} \prod_{i=1}^{N} ds_i =$$

$$\xi_n^{2\beta(N-1)+m} 2^N \gamma^{-N} \int_{\mathbb{R}_+^N} \left(\prod_{i=1}^{N} z_i^{\alpha_i} \right) \left(1 + \sum_{i=1}^{N} z_i \right)^r \prod_{i=1}^{N} \left(\frac{\sin z_i}{z_i} \right)^{2\beta} \prod_{i=1}^{N} dz_i \overset{(25.43)}{\leq}$$

$$\xi_n^{2\beta(N-1)+m} 2^N \gamma^{-N} \int_{\mathbb{R}_+^N} \left(\prod_{i=1}^{N} z_i^{\alpha_i} \right) \left(\prod_{i=1}^{N} (1 + z_i)^r \right) \prod_{i=1}^{N} \left(\frac{\sin z_i}{z_i} \right)^{2\beta} \prod_{i=1}^{N} dz_i =$$

$$\xi_n^{2\beta(N-1)+m} 2^N \gamma^{-N} \prod_{i=1}^{N} \left(\int_0^{\infty} z^{\alpha_i} (1+z)^r \left(\frac{\sin z}{z} \right)^{2\beta} dz \right) =$$

$$\xi_n^{2\beta(N-1)+m} 2^N \gamma^{-N} \prod_{i=1}^{N} \left[\sum_{\lambda=0}^{r} \binom{r}{\lambda} \left(\int_0^{\infty} z^{\lambda+\alpha_i} \left(\frac{\sin z}{z} \right)^{2\beta} dz \right) \right] =$$

$$\xi_n^{2\beta(N-1)+m} 2^N \gamma^{-N} \prod_{i=1}^{N} \left\{ \sum_{\lambda=0}^{r} \binom{r}{\lambda} \left[\int_0^{1} z^{\lambda+\alpha_i} \left(\frac{\sin z}{z} \right)^{2\beta} dz + \right.\right.$$

$$\left.\left. \int_1^{\infty} z^{\lambda+\alpha_i} \left(\frac{\sin z}{z} \right)^{2\beta} dz \right] \right\} \leq \tag{25.49}$$

$$\xi_n^{2\beta(N-1)+m} 2^N \gamma^{-N} \left\{ \sum_{\lambda=0}^{r} \binom{r}{\lambda} \left[\int_0^{1} z^{\lambda} \left(\frac{\sin z}{z} \right)^{2\beta} dz + \right.\right.$$

$$\left.\left. \int_1^{\infty} z^{\lambda+m} \left(\frac{\sin z}{z} \right)^{2\beta} dz \right] \right\}^N \leq$$

$$\xi_n^{2\beta(N-1)+m} 2^N \gamma^{-N} \left\{ \sum_{\lambda=0}^{r} \binom{r}{\lambda} \left[\int_0^\infty z^\lambda \left(\frac{\sin z}{z} \right)^{2\beta} dz + \right.\right.$$

$$\left.\left. \int_0^\infty z^{\lambda+m} \left(\frac{\sin z}{z} \right)^{2\beta} dz \right] \right\}^N =: I. \quad (25.50)$$

Based on [10], p. 210, item 1033 and [7], see (25.27), (25.28), and by assuming $\mathbb{N} \ni \beta > \frac{r+m+1}{2}$, i.e. $\lambda < \lambda + m < 2\beta - 1$, for all $\lambda = 0, 1, ..., r$, we have the following calculations:

Let λ be even, then

$$\int_0^\infty z^\lambda \left(\frac{\sin z}{z} \right)^{2\beta} dz = \frac{\pi \left(-1\right)^{\frac{2\beta-\lambda}{2}} (2\beta)!}{2^{\lambda+1} (2\beta - \lambda - 1)!} \sum_{k=1}^{\beta} (-1)^k \frac{k^{2\beta-\lambda-1}}{(\beta - k)! \, (\beta + k)!} = \psi_{1\lambda}. \quad (25.51)$$

Let λ be odd, then

$$\int_0^\infty z^\lambda \left(\frac{\sin z}{z} \right)^{2\beta} dz = \frac{\left(-1\right)^{\frac{\lambda-1}{2}} (2\beta)!}{2^{\lambda} (2\beta - \lambda - 1)!} \sum_{k=1}^{\beta} (-1)^{\beta-k} \frac{k^{2\beta-\lambda-1} \left[\ln (2k)\right]}{(\beta - k)! \, (\beta + k)!} = \psi_{2\lambda}. \quad (25.52)$$

Therefore

$$\int_0^\infty z^\lambda \left(\frac{\sin z}{z} \right)^{2\beta} dz = \psi_\lambda = \begin{cases} \psi_{1\lambda}, & \text{when } \lambda \text{ is even,} \\ \psi_{2\lambda}, & \text{when } \lambda \text{ is odd.} \end{cases} \quad (25.53)$$

Similarly, for $\lambda + m$ being even, we get

$$\int_0^\infty z^{\lambda+m} \left(\frac{\sin z}{z} \right)^{2\beta} dz = \frac{\pi \left(-1\right)^{\frac{2\beta-\lambda-m}{2}} (2\beta)!}{2^{\lambda+m+1} (2\beta - \lambda - m - 1)!} \quad (25.54)$$

$$\sum_{k=1}^{\beta} (-1)^k \frac{k^{2\beta-\lambda-m-1}}{(\beta - k)! \, (\beta + k)!} = \psi_{1(\lambda+m)}.$$

And when $\lambda + m$ is odd we get that

$$\int_0^\infty z^{\lambda+m} \left(\frac{\sin z}{z} \right)^{2\beta} dz = \frac{\left(-1\right)^{\frac{\lambda+m-1}{2}} (2\beta)!}{2^{\lambda+m} (2\beta - \lambda - m - 1)!} \quad (25.55)$$

$$\sum_{k=1}^{\beta} (-1)^{\beta-k} \frac{k^{2\beta-\lambda-m-1} \left[\ln (2k)\right]}{(\beta - k)! \, (\beta + k)!} = \psi_{2(\lambda+m)}.$$

Therefore, it holds

$$\int_0^{\infty} z^{\lambda+m} \left(\frac{\sin z}{z}\right)^{2\beta} dz = \psi_{\lambda+m} = \begin{cases} \psi_{1(\lambda+m)}, & \text{when } \lambda+m \text{ is even,} \\ \psi_{2(\lambda+m)}, & \text{when } \lambda+m \text{ is odd.} \end{cases} \quad (25.56)$$

That is

$$I = \xi_n^{2\beta(N-1)+m} 2^N \gamma^{-N} \left\{ \sum_{\lambda=0}^{r} \binom{r}{\lambda} [\psi_\lambda + \psi_{\lambda+m}] \right\}^N \leq$$

$$2^N \gamma^{-N} \left\{ \sum_{\lambda=0}^{r} \binom{r}{\lambda} [\psi_\lambda + \psi_{\lambda+m}] \right\}^N < +\infty. \quad (25.57)$$

I.e. $A_{\xi_n}(\overline{\alpha})$ is uniformly bounded. The theorem is proved. ∎

We continue with

Theorem 25.14 *Let* $r, n \in \mathbb{N}$, $\xi_n \in (0, 1]$, $\beta \in \mathbb{N} : \beta > \frac{r+1}{2}$, $N \in \mathbb{N} - \{1\}$. *Here* γ, λ_n *are as in (25.33) and (25.34), respectively, and* ψ_λ *is defined by (25.44), (25.45) and (25.46),* $\lambda = 0, 1, \ldots, r$. *Then*

$$B_{\xi_n} := \lambda_n^{-N} \int_{\mathbb{R}^N} \left(1 + \frac{\|s\|_2}{\xi_n}\right)^r \prod_{i=1}^{N} \left(\frac{\sin\left(\frac{s_i}{\xi_n}\right)}{s_i}\right)^{2\beta} \prod_{i=1}^{N} ds_i \leq$$

$$\xi_n^{2\beta(N-1)} 2^N \gamma^{-N} \left[\sum_{\lambda=0}^{r} \binom{r}{\lambda} \psi_\lambda \right]^N \leq$$

$$2^N \gamma^{-N} \left[\sum_{\lambda=0}^{r} \binom{r}{\lambda} \psi_\lambda \right]^N < +\infty, \quad (25.58)$$

uniformly bounded, and convergent to zero as $\xi_n \to 0$, *when* $n \to +\infty$.

Proof We estimate

$$B_{\xi_n} = \lambda_n^{-N} \int_{\mathbb{R}^N} \left(1 + \frac{\|s\|_2}{\xi_n}\right)^r \prod_{i=1}^{N} \left(\frac{\sin\left(\frac{s_i}{\xi_n}\right)}{s_i}\right)^{2\beta} \prod_{i=1}^{N} ds_i =$$

$$2^N \lambda_n^{-N} \int_{\mathbb{R}_+^N} \left(1 + \frac{\|s\|_2}{\xi_n}\right)^r \prod_{i=1}^{N} \left(\frac{\sin\left(\frac{s_i}{\xi_n}\right)}{s_i}\right)^{2\beta} \prod_{i=1}^{N} ds_i \leq \quad (25.59)$$

$$2^N\lambda_n^{-N}\int_{\mathbb{R}_+^N}\left(1+\sum_{i=1}^{N}\left(\frac{s_i}{\xi_n}\right)\right)^r\prod_{i=1}^{N}\left(\frac{\sin\left(\frac{s_i}{\xi_n}\right)}{s_i}\right)^{2\beta}\prod_{i=1}^{N}ds_i =$$

$$\xi_n^{2\beta(N-1)}2^N\gamma^{-N}\int_{\mathbb{R}_+^N}\left(1+\sum_{i=1}^{N}z_i\right)^r\prod_{i=1}^{N}\left(\frac{\sin z_i}{z_i}\right)^{2\beta}\prod_{i=1}^{N}dz_i \overset{(25.43)}{\leq}$$

$$\xi_n^{2\beta(N-1)}2^N\gamma^{-N}\int_{\mathbb{R}_+^N}\left(\prod_{i=1}^{N}(1+z_i)^r\right)\prod_{i=1}^{N}\left(\frac{\sin z_i}{z_i}\right)^{2\beta}\prod_{i=1}^{N}dz_i =$$

$$\xi_n^{2\beta(N-1)}2^N\gamma^{-N}\left(\int_0^{\infty}(1+z)^r\left(\frac{\sin z}{z}\right)^{2\beta}dz\right)^N =$$

$$\xi_n^{2\beta(N-1)}2^N\gamma^{-N}\left[\sum_{\lambda=0}^{r}\binom{r}{\lambda}\left(\int_0^{\infty}z^{\lambda}\left(\frac{\sin z}{z}\right)^{2\beta}dz\right)\right]^N \overset{(25.53)}{=}$$

$$\xi_n^{2\beta(N-1)}2^N\gamma^{-N}\left[\sum_{\lambda=0}^{r}\binom{r}{\lambda}\psi_\lambda\right]^N \leq \tag{25.60}$$

$$2^N\gamma^{-N}\left[\sum_{\lambda=0}^{r}\binom{r}{\lambda}\psi_\lambda\right]^N < +\infty,$$

under $\beta > \frac{r+1}{2}$. The theorem is proved. ■

We also give

Theorem 25.15 *Let $p > 1$; $r, \beta, N \in \mathbb{N}$, $\alpha_j \in \mathbb{Z}^+$, $j = 1, \ldots, N$: $|\overline{\alpha}| := \sum_{j=1}^{N}\alpha_j = m \in \mathbb{N}$, $\xi_n \in (0, 1]$, $n \in \mathbb{N}$. Here we take $\beta > \frac{\lceil rp\rceil+m+1}{2}$, and γ, λ_n are as in (25.33) and (25.34), respectively, and λ runs as $\lambda = 0, 1, \ldots, \lceil rp\rceil$. Furthermore ψ_λ is defined as in (25.44), (25.45) and (25.46). Similarly, it is defined $\psi_{\lambda+mp}$, just set in (25.44), (25.45), (25.46), $(\lambda + mp)$ in place of λ. Then*

$$C_{\xi_n}(\overline{\alpha}) := \lambda_n^{-N}\int_{\mathbb{R}^N}\left(\left(\prod_{i=1}^{N}|s_i|^{\alpha_i}\right)\left(1+\frac{\|s\|_2}{\xi_n}\right)^r\right)^p\prod_{i=1}^{N}\left(\frac{\sin\left(\frac{s_i}{\xi_n}\right)}{s_i}\right)^{2\beta}\prod_{i=1}^{N}ds_i \leq \tag{25.61}$$

$$\xi_n^{2\beta(N-1)+mp}2^N\gamma^{-N}\left\{\sum_{\lambda=0}^{\lceil rp\rceil}\binom{\lceil rp\rceil}{\lambda}\left[\psi_\lambda+\psi_{\lambda+mp}\right]\right\}^N \leq$$

$$2^N \gamma^{-N} \left\{ \sum_{\lambda=0}^{\lceil rp \rceil} \binom{\lceil rp \rceil}{\lambda} \left[\psi_\lambda + \psi_{\lambda+mp} \right] \right\}^N < +\infty,$$

uniformly bounded, and convergent to zero as $\xi_n \to 0$*, when* $n \to +\infty$.
Above $\lceil \cdot \rceil$ *is the ceiling of the number.*

Proof We estimate

$$C_{\xi_n}(\overline{\alpha}) = \lambda_n^{-N} \int_{\mathbb{R}^N} \left(\left(\prod_{i=1}^N |s_i|^{\alpha_i} \right) \left(1 + \frac{\|s\|_2}{\xi_n} \right)^r \right)^p \prod_{i=1}^N \left(\frac{\sin\left(\frac{s_i}{\xi_n}\right)}{s_i} \right)^{2\beta} \prod_{i=1}^N ds_i =$$

$$2^N \lambda_n^{-N} \int_{\mathbb{R}_+^N} \left(\left(\prod_{i=1}^N s_i^{\alpha_i} \right) \left(1 + \frac{\|s\|_2}{\xi_n} \right)^r \right)^p \prod_{i=1}^N \left(\frac{\sin\left(\frac{s_i}{\xi_n}\right)}{s_i} \right)^{2\beta} \prod_{i=1}^N ds_i \le$$

$$2^N \lambda_n^{-N} \int_{\mathbb{R}_+^N} \left(\prod_{i=1}^N s_i^{\alpha_i p} \right) \left(1 + \sum_{i=1}^N \left(\frac{s_i}{\xi_n} \right) \right)^{rp} \prod_{i=1}^N \left(\frac{\sin\left(\frac{s_i}{\xi_n}\right)}{s_i} \right)^{2\beta} \prod_{i=1}^N ds_i = \tag{25.62}$$

$$2^N \gamma^{-N} \xi_n^{2\beta(N-1)+mp} \int_{\mathbb{R}_+^N} \left(\prod_{i=1}^N z_i^{\alpha_i p} \right) \left(1 + \sum_{i=1}^N z_i \right)^{rp} \prod_{i=1}^N \left(\frac{\sin z_i}{z_i} \right)^{2\beta} \prod_{i=1}^N dz_i \overset{(25.43)}{\le}$$

$$\xi_n^{2\beta(N-1)+mp} 2^N \gamma^{-N} \int_{\mathbb{R}_+^N} \left(\prod_{i=1}^N z_i^{\alpha_i p} \right) \left(\prod_{i=1}^N (1+z_i)^{rp} \right) \prod_{i=1}^N \left(\frac{\sin z_i}{z_i} \right)^{2\beta} \prod_{i=1}^N dz_i =$$

$$\xi_n^{2\beta(N-1)+mp} 2^N \gamma^{-N} \prod_{i=1}^N \left(\int_0^\infty (1+z)^{rp} z^{\alpha_i p} \left(\frac{\sin z}{z} \right)^{2\beta} dz \right) \le$$

$$\xi_n^{2\beta(N-1)+mp} 2^N \gamma^{-N} \prod_{i=1}^N \left(\int_0^\infty (1+z)^{\lceil rp \rceil} z^{\alpha_i p} \left(\frac{\sin z}{z} \right)^{2\beta} dz \right) = \tag{25.63}$$

$$\xi_n^{2\beta(N-1)+mp} 2^N \gamma^{-N} \prod_{i=1}^{N} \left\{ \sum_{\lambda=0}^{\lceil rp \rceil} \binom{\lceil rp \rceil}{\lambda} \left(\int_0^{\infty} z^{\lambda+\alpha_i p} \left(\frac{\sin z}{z} \right)^{2\beta} dz \right) \right\} =$$

$$\xi_n^{2\beta(N-1)+mp} 2^N \gamma^{-N} \prod_{i=1}^{N} \left\{ \sum_{\lambda=0}^{\lceil rp \rceil} \binom{\lceil rp \rceil}{\lambda} \left[\int_0^{1} z^{\lambda+\alpha_i p} \left(\frac{\sin z}{z} \right)^{2\beta} dz + \right. \right.$$

$$\left. \left. \int_1^{\infty} z^{\lambda+\alpha_i p} \left(\frac{\sin z}{z} \right)^{2\beta} dz \right] \right\} \leq$$

$$\xi_n^{2\beta(N-1)+mp} 2^N \gamma^{-N} \left\{ \sum_{\lambda=0}^{\lceil rp \rceil} \binom{\lceil rp \rceil}{\lambda} \left[\int_0^{1} z^{\lambda} \left(\frac{\sin z}{z} \right)^{2\beta} dz + \right. \right.$$

$$\left. \left. \int_1^{\infty} z^{\lambda+mp} \left(\frac{\sin z}{z} \right)^{2\beta} dz \right] \right\}^N \leq$$

$$\xi_n^{2\beta(N-1)+mp} 2^N \gamma^{-N} \left\{ \sum_{\lambda=0}^{\lceil rp \rceil} \binom{\lceil rp \rceil}{\lambda} \left[\int_0^{\infty} z^{\lambda} \left(\frac{\sin z}{z} \right)^{2\beta} dz + \right. \right.$$

$$\left. \left. \int_0^{\infty} z^{\lambda+mp} \left(\frac{\sin z}{z} \right)^{2\beta} dz \right] \right\}^N \overset{((25.53),(25.56))}{=}$$

$$\xi_n^{2\beta(N-1)+mp} 2^N \gamma^{-N} \left\{ \sum_{\lambda=0}^{\lceil rp \rceil} \binom{\lceil rp \rceil}{\lambda} \left[\psi_\lambda + \psi_{\lambda+mp} \right] \right\}^N \leq \tag{25.64}$$

$$2^N \gamma^{-N} \left\{ \sum_{\lambda=0}^{\lceil rp \rceil} \binom{\lceil rp \rceil}{\lambda} \left[\psi_\lambda + \psi_{\lambda+mp} \right] \right\}^N < +\infty.$$

I.e. $C_{\xi_n}(\overline{\alpha})$ is uniformly bounded.

Above we assumed that $\mathbb{N} \ni \beta > \frac{\lceil rp \rceil + m + 1}{2}$, i.e. $\lambda < \lambda + m < 2\beta - 1$, for all $\lambda = 0, 1, ..., \lceil rp \rceil$.

The theorem is proved. ∎

We also present

Theorem 25.16 *Let $p > 1$; $r, \beta \in \mathbb{N}$, $N \in \mathbb{N} - \{1\}$, $\xi_n \in (0, 1]$, $n \in \mathbb{N}$. Here we take $\beta > \frac{\lceil rp \rceil + 1}{2}$, and γ, λ_n are as in (25.33) and (25.34), respectively, and λ runs as $\lambda = 0, 1, ..., \lceil rp \rceil$. Furthermore ψ_λ is defined as in (25.44), (25.45) and (25.46). Then*

$$D_{\xi_n} := \lambda_n^{-N} \int_{\mathbb{R}^N} \left(1 + \frac{\|s\|_2}{\xi_n}\right)^{rp} \prod_{i=1}^{N} \left(\frac{\sin\left(\frac{s_i}{\xi_n}\right)}{s_i}\right)^{2\beta} \prod_{i=1}^{N} ds_i \leq$$

$$\xi_n^{2\beta(N-1)} 2^N \gamma^{-N} \left[\sum_{\lambda=0}^{\lceil rp \rceil} \binom{\lceil rp \rceil}{\lambda} \psi_\lambda\right]^N \leq \tag{25.65}$$

$$2^N \gamma^{-N} \left[\sum_{\lambda=0}^{\lceil rp \rceil} \binom{\lceil rp \rceil}{\lambda} \psi_\lambda\right]^N < +\infty,$$

uniformly bounded, and convergent to zero as $\xi_n \to 0$*, when* $n \to +\infty$.

Proof We estimate

$$D_{\xi_n} = \lambda_n^{-N} \int_{\mathbb{R}^N} \left(1 + \frac{\|s\|_2}{\xi_n}\right)^{rp} \prod_{i=1}^{N} \left(\frac{\sin\left(\frac{s_i}{\xi_n}\right)}{s_i}\right)^{2\beta} \prod_{i=1}^{N} ds_i = \tag{25.66}$$

$$2^N \lambda_n^{-N} \int_{\mathbb{R}_+^N} \left(1 + \frac{\|s\|_2}{\xi_n}\right)^{rp} \prod_{i=1}^{N} \left(\frac{\sin\left(\frac{s_i}{\xi_n}\right)}{s_i}\right)^{2\beta} \prod_{i=1}^{N} ds_i \leq$$

$$2^N \lambda_n^{-N} \int_{\mathbb{R}_+^N} \left(1 + \sum_{i=1}^{N} \left(\frac{s_i}{\xi_n}\right)\right)^{rp} \prod_{i=1}^{N} \left(\frac{\sin\left(\frac{s_i}{\xi_n}\right)}{s_i}\right)^{2\beta} \prod_{i=1}^{N} ds_i =$$

$$2^N \gamma^{-N} \xi_n^{2\beta(N-1)} \int_{\mathbb{R}_+^N} \left(1 + \sum_{i=1}^{N} z_i\right)^{rp} \prod_{i=1}^{N} \left(\frac{\sin z_i}{z_i}\right)^{2\beta} \prod_{i=1}^{N} dz_i \overset{(25.43)}{\leq}$$

$$\xi_n^{2\beta(N-1)} 2^N \gamma^{-N} \int_{\mathbb{R}_+^N} \left(\prod_{i=1}^{N} (1 + z_i)^{rp}\right) \prod_{i=1}^{N} \left(\frac{\sin z_i}{z_i}\right)^{2\beta} \prod_{i=1}^{N} dz_i =$$

$$\xi_n^{2\beta(N-1)} 2^N \gamma^{-N} \left(\int_0^\infty (1 + z)^{rp} \left(\frac{\sin z}{z}\right)^{2\beta} dz\right)^N \leq$$

$$\xi_n^{2\beta(N-1)} 2^N \gamma^{-N} \left(\int_0^\infty (1 + z)^{\lceil rp \rceil} \left(\frac{\sin z}{z}\right)^{2\beta} dz\right)^N =$$

$$\xi_n^{2\beta(N-1)} 2^N \gamma^{-N} \left[\sum_{\lambda=0}^{\lceil rp \rceil} \binom{\lceil rp \rceil}{\lambda} \left(\int_0^{\infty} z^{\lambda} \left(\frac{\sin z}{z} \right)^{2\beta} dz \right) \right]^N \overset{(25.53)}{=} \tag{25.67}$$

$$\xi_n^{2\beta(N-1)} 2^N \gamma^{-N} \left[\sum_{\lambda=0}^{\lceil rp \rceil} \binom{\lceil rp \rceil}{\lambda} \psi_{\lambda} \right]^N \leq$$

$$2^N \gamma^{-N} \left[\sum_{\lambda=0}^{\lceil rp \rceil} \binom{\lceil rp \rceil}{\lambda} \psi_{\lambda} \right]^N < +\infty,$$

under $\beta > \frac{\lceil rp \rceil + 1}{2}$. The theorem is proved. ■

We proceed it

Theorem 25.17 *Let $n, N \in \mathbb{N}$, $\xi_n \in (0, 1]$, $\alpha_j \in \mathbb{Z}^+$, $j = 1, ..., N : |\overline{\alpha}| := \sum_{j=1}^{N} \alpha_j = m \in \mathbb{N}$. Here $\beta \in \mathbb{N} : \beta > \frac{m+1}{2}$, and γ, λ_n are as in (25.33) and (25.34), respectively. Furthermore ψ_{α_i} is defined as in (25.44), (25.45) and (25.46), just replace λ by α_i, $i = 1, ..., N$. Then*

$$F_{\xi_n}(\overline{\alpha}) := \xi_n^{-m} \lambda_n^{-N} \int_{\mathbb{R}^N} \left(\prod_{i=1}^{N} |s_i|^{\alpha_i} \right) \prod_{i=1}^{N} \left(\frac{\sin\left(\frac{s_i}{\xi_n}\right)}{s_i} \right)^{2\beta} \prod_{i=1}^{N} ds_i \leq$$

$$\xi_n^{2\beta(N-1)} 2^N \gamma^{-N} \max_{|\alpha|=m} \left(\prod_{i=1}^{N} \psi_{\alpha_i} \right) \leq \tag{25.68}$$

$$2^N \gamma^{-N} \max_{|\alpha|=m} \left(\prod_{i=1}^{N} \psi_{\alpha_i} \right) =: \varphi < +\infty.$$

Proof We estimate

$$F_{\xi_n}(\overline{\alpha}) = \xi_n^{-m} \lambda_n^{-N} \int_{\mathbb{R}^N} \left(\prod_{i=1}^{N} |s_i|^{\alpha_i} \right) \prod_{i=1}^{N} \left(\frac{\sin\left(\frac{s_i}{\xi_n}\right)}{s_i} \right)^{2\beta} \prod_{i=1}^{N} ds_i =$$

$$\xi_n^{-m} \lambda_n^{-N} 2^N \int_{\mathbb{R}_+^N} \left(\prod_{i=1}^{N} s_i^{\alpha_i} \right) \prod_{i=1}^{N} \left(\frac{\sin\left(\frac{s_i}{\xi_n}\right)}{s_i} \right)^{2\beta} \prod_{i=1}^{N} ds_i =$$

$$\xi_n^{2\beta(N-1)} 2^N \gamma^{-N} \int_{\mathbb{R}_+^N} \left(\prod_{i=1}^N z_i^{\alpha_i} \right) \prod_{i=1}^N \left(\frac{\sin z_i}{z_i} \right)^{2\beta} \prod_{i=1}^N dz_i = \tag{25.69}$$

$$\xi_n^{2\beta(N-1)} 2^N \gamma^{-N} \prod_{i=1}^N \left(\int_0^\infty z^{\alpha_i} \left(\frac{\sin z}{z} \right)^{2\beta} dz \right) \overset{(25.53)}{=}$$

$$\xi_n^{2\beta(N-1)} 2^N \gamma^{-N} \left(\prod_{i=1}^N \psi_{\alpha_i} \right) \leq \xi_n^{2\beta(N-1)} 2^N \gamma^{-N} \max_{|\alpha|=m} \left(\prod_{i=1}^N \psi_{\alpha_i} \right) \leq$$

$$2^N \gamma^{-N} \max_{|\alpha|=m} \left(\prod_{i=1}^N \psi_{\alpha_i} \right) < +\infty,$$

under $\beta > \frac{m+1}{2}$, i.e. $\alpha_i < 2\beta - 1$, $i = 1, ..., N$. The theorem is proved. ■

We make

Remark 25.18 All as in Theorem 25.17. We denote

$$\overline{c}_{\overline{\alpha},n} := \overline{c}_{\overline{\alpha},n,\widetilde{j}} := \lambda_n^{-N} \int_{\mathbb{R}^N} \left(\prod_{i=1}^N s_i^{\alpha_i} \right) \prod_{i=1}^N \left(\frac{\sin\left(\frac{s_i}{\xi_n}\right)}{s_i} \right)^{2\beta} \prod_{i=1}^N ds_i, \tag{25.70}$$

where $\widetilde{j} = 1, ..., m$, and $\overline{\alpha} := (\alpha_1, ..., \alpha_N)$, $\alpha_i \in \mathbb{Z}^+$, $i = 1, ..., N$, $|\overline{\alpha}| := \sum_{j=1}^N \alpha_j = \widetilde{j}$.

By (25.68) we obtain

$$\left|\overline{c}_{\overline{\alpha},n}\right| = \left|\overline{c}_{\overline{\alpha},n,\widetilde{j}}\right| \leq \varphi \xi_n^m \leq \varphi. \tag{25.71}$$

25.3 Main Results for $T_{r,n}^{[m]}$

25.3.1 Uniform Approximation

We start with an application to $T_{r,n}^{[m]}$ of the following theorem.

Theorem 25.19 ([3], p. 11) *Let $m \in \mathbb{N}$, $f \in C^m\left(\mathbb{R}^N\right)$, $N \geq 1$, $x \in \mathbb{R}^N$. Assume $\left\| \frac{\partial^m f(\cdot,\cdot,\ldots,\cdot)}{\partial x_1^{\alpha_1} \ldots \partial x_N^{\alpha_N}} \right\|_\infty < \infty$, for all $\alpha_j \in \mathbb{Z}^+$, $j = 1, ..., N : |\overline{\alpha}| := \sum_{i=1}^N \alpha_i = m$. Let μ_{ξ_n} be a Borel probability measure on $\mathbb{R}^N$, for $\xi_n > 0$, $(\xi_n)_{n \in \mathbb{N}}$ bounded sequence.*

Suppose that for all $\overline{\alpha} := (\alpha_1, ..., \alpha_N)$, $\alpha_i \in \mathbb{Z}^+$, $i = 1, ..., N$, $|\overline{\alpha}| := \sum_{i=1}^{N} \alpha_i = m$ *we have that*

$$u_{\xi_n}(\overline{\alpha}) := \int_{\mathbb{R}^N} \left(\prod_{i=1}^{N} |s_i|^{\alpha_i} \right) \left(1 + \frac{\|s\|_2}{\xi_n} \right)^r d\mu_{\xi_n}(s) < \infty. \tag{25.72}$$

For $\widetilde{j} = 1, ..., m$, *and* $\overline{\alpha} := (\alpha_1, ..., \alpha_N)$, $\alpha_i \in \mathbb{Z}^+$, $i = 1, ..., N$, $|\overline{\alpha}| := \sum_{j=1}^{N} \alpha_j = \widetilde{j}$, *call*

$$c_{\overline{\alpha},n} := c_{\overline{\alpha},n,\widetilde{j}} := \int_{\mathbb{R}^N} \prod_{i=1}^{N} s_i^{\alpha_i} d\mu_{\xi_n}(s_1, ..., s_N). \tag{25.73}$$

Then

(i)

$$E_{r,n}^{[m]}(x) := \left| \theta_{r,n}^{[m]}(f; x) - f(x) - \sum_{\widetilde{j}=1}^{m} \delta_{\widetilde{j},r}^{[m]} \left(\sum_{\substack{\alpha_1, ..., \alpha_N \geq 0; \\ |\overline{\alpha}| = \widetilde{j}}} \frac{c_{\overline{\alpha},n,\widetilde{j}} f_{\overline{\alpha}}(x)}{\prod_{i=1}^{N} \alpha_i!} \right) \right|$$

$$\leq \sum_{\substack{\alpha_1, ..., \alpha_N \geq 0; \\ |\overline{\alpha}| = m}} \frac{(\omega_r(f_{\overline{\alpha}}, \xi_n))}{\left(\prod_{i=1}^{N} \alpha_i \right)} \left(\int_{\mathbb{R}^N} \left(\prod_{i=1}^{N} |s_i|^{\alpha_i} \right) \left(1 + \frac{\|s\|_2}{\xi_n} \right)^r d\mu_{\xi_n}(s) \right). \tag{25.74}$$

$\forall\, x \in \mathbb{R}^N$.

(ii)

$$\left\| E_{r,n}^{[m]} \right\|_\infty \leq R.H.S.(25.74). \tag{25.75}$$

Given that $\xi_n \to 0$, *as* $n \to \infty$, *and* u_{ξ_n} *is uniformly bounded, then we derive that* $\left\| E_{r,n}^{[m]} \right\| \to 0$ *with rates.*

(iii) It holds also that

$$\left\| \theta_{r,n}^{[m]}(f) - f \right\|_\infty \leq \sum_{\widetilde{j}=1}^{m} \delta_{\widetilde{j},r}^{[m]} \left(\sum_{\substack{\alpha_1, ..., \alpha_N \geq 0; \\ |\overline{\alpha}| = \widetilde{j}}} \frac{\left| c_{\overline{\alpha},n,\widetilde{j}} \right| \|f_{\overline{\alpha}}\|_\infty}{\prod_{i=1}^{N} \alpha_i!} \right) + R.H.S.(25.74), \tag{25.76}$$

given that $\|f_{\overline{\alpha}}\|_\infty < \infty$, *for all* $\overline{\alpha} : |\overline{\alpha}| = \widetilde{j}$, $\widetilde{j} = 1, ..., m$. *Furthermore, as* $\xi_n \to 0$ *when* $n \to \infty$, *assuming that* $c_{\overline{\alpha},n,\widetilde{j}} \to 0$, *while* u_{ξ_n} *is uniformly bounded, we conclude that*

$$\left\| \theta_{r,n}^{[m]}(f) - f \right\|_\infty \to 0 \tag{25.77}$$

with rates.

A uniform approximation result for $T_{r,n}^{[m]}$ follows:

Theorem 25.20 *Let $r, N, \beta, m \in \mathbb{N}$, $f \in C^m\left(\mathbb{R}^N\right)$, $x \in \mathbb{R}^N$. Assume $\left\| \frac{\partial^m f(\cdot,\cdot,\ldots,\cdot)}{\partial x_1^{\alpha_1} \ldots \partial x_N^{\alpha_N}} \right\|_\infty < \infty$, for all $\alpha_i \in \mathbb{Z}^+$, $i = 1, \ldots, N : |\overline{\alpha}| := \sum_{i=1}^N \alpha_i = m$. Let φ_{ξ_n} be the Borel probability measure on $\mathbb{R}^N$, see (25.39), where $\xi_n \in (0, 1]$, $n \in \mathbb{N}$. Here $\beta > \frac{m+r+1}{2}$, and $A_{\xi_n}(\overline{\alpha})$ as in (25.47), and $\overline{c}_{\overline{\alpha},n} := \overline{c}_{\overline{\alpha},n,\widetilde{j}}$ as in (25.70). Then*

(i)

$$\overline{E}_{r,n}^{[m]}(x) := \left| T_{r,n}^{[m]}(f; x) - f(x) - \sum_{\widetilde{j}=1}^{m} \delta_{\widetilde{j},r}^{[m]} \left(\sum_{\substack{\alpha_1,\ldots,\alpha_N \geq 0; \\ |\overline{\alpha}| = \widetilde{j}}} \frac{\overline{c}_{\overline{\alpha},n,\widetilde{j}} f_{\overline{\alpha}}(x)}{\prod_{i=1}^{N} \alpha_i!} \right) \right|$$

$$\leq \sum_{\substack{\alpha_1,\ldots,\alpha_N \geq 0; \\ |\overline{\alpha}| = m}} \frac{(\omega_r(f_{\overline{\alpha}}, \xi_n))}{\left(\prod_{i=1}^{N} \alpha_i!\right)} A_{\xi_n}(\overline{\alpha}), \tag{25.78}$$

$\forall\, x \in \mathbb{R}^N$.

(ii)

$$\left\| \overline{E}_{r,n}^{[m]} \right\|_\infty \leq R.H.S.(25.78). \tag{25.79}$$

Given that $\xi_n \to 0$, as $n \to +\infty$, we have that $A_{\xi_n}(\overline{\alpha}) \to 0$ and are uniformly bounded, and then we derive that $\left\| \overline{E}_{r,n}^{[m]} \right\|_\infty \to 0$ with rates.

(iii) It holds also that

$$\left\| T_{r,n}^{[m]}(f) - f \right\|_\infty \leq \sum_{\widetilde{j}=1}^{m} \left| \delta_{\widetilde{j},r}^{[m]} \right| \left(\sum_{\substack{\alpha_1,\ldots,\alpha_N \geq 0; \\ |\overline{\alpha}| = \widetilde{j}}} \frac{\left| \overline{c}_{\overline{\alpha},n,\widetilde{j}} \right| \left\| f_{\overline{\alpha}} \right\|_\infty}{\prod_{i=1}^{N} \alpha_i!} \right) + R.H.S.(25.78), \tag{25.80}$$

given that $\| f_{\overline{\alpha}} \|_\infty < +\infty$, for all $\overline{\alpha} : |\overline{\alpha}| = \widetilde{j}$, $\widetilde{j} = 1, \ldots, m$. Furthermore, as $\xi_n \to 0$ when $n \to +\infty$, we have that $\overline{c}_{\overline{\alpha},n,\widetilde{j}} \to 0$ and $A_{\xi_n}(\overline{\alpha}) \to 0$, and both are uniformly bounded, and we conclude that

$$\left\| T_{r,n}^{[m]}(f) - f \right\|_\infty \to 0 \tag{25.81}$$

with rates.

Proof Mainly by applying Theorem 25.19. By Theorem 25.13 we get that $A_{\xi_n}(\overline{\alpha}) \to 0$ and $A_{\xi_n}(\overline{\alpha})$ are uniformly bounded. By Theorem 25.17 and Remark 25.18 we get $\overline{c}_{\overline{\alpha},n} \to 0$ and $\overline{c}_{\overline{\alpha},n}$ are uniformly bounded. ■

We mention

Theorem 25.21 ([3], p. 14) *Let $f \in C_B(\mathbb{R}^N)$, uniformly continuous, $N \geq 1$, $\xi_n \in (0,1]$. Then*

$$\left\| \theta_{r,n}^{[0]} f - f \right\|_\infty \leq \left(\int_{\mathbb{R}^N} \left(1 + \frac{\|s\|_2}{\xi_n} \right)^r d\mu_{\xi_n}(s) \right) \omega_r(f, \xi_n), \tag{25.82}$$

under the assumption

$$\Phi_{\xi_n} := \int_{\mathbb{R}^N} \left(1 + \frac{\|s\|_2}{\xi_n} \right)^r d\mu_{\xi_n}(s) < \infty. \tag{25.83}$$

As $n \to \infty$ and $\xi_n \to 0$, given that Φ_{ξ_n} are uniformly bounded, we derive

$$\left\| \theta_{r,n}^{[0]} f - f \right\|_\infty \to 0 \tag{25.84}$$

with rates.

We give

Theorem 25.22 *Let $f \in C_B(\mathbb{R}^N)$, uniformly continuous, $\beta, r \in \mathbb{N}$, $N \in \mathbb{N} - \{1\}$, $\beta > \frac{r+1}{2}$, $\xi_n \in (0,1]$, $n \in \mathbb{N}$. Then*

$$\left\| T_{r,n}^{[0]} f - f \right\|_\infty \leq 2^N \gamma^{-N} \left[\sum_{\lambda=0}^{r} \binom{r}{\lambda} \psi_\lambda \right]^N \xi_n^{2\beta(N-1)} \omega_r(f, \xi_n). \tag{25.85}$$

As $n \to \infty$ and $\xi_n \to 0$, we derive

$$\left\| T_{r,n}^{[0]} f - f \right\|_\infty \to 0 \tag{25.86}$$

with rates.

Proof By Theorems 25.14 and 25.21. ■

25.3.2 L_p Approximation for $T_{r,n}^{[m]}$

We need

Definition 25.23 ([5, 9]) We call

$$\Delta_u^r f(x) := \Delta_{u_1,u_2,\ldots,u_N}^r f(x_1, \ldots, x_N) := \tag{25.87}$$

$$\sum_{j=0}^{r} (-1)^{r-j} \binom{r}{j} f(x_1 + ju_1, x_2 + ju_2, \ldots, x_N + ju_N).$$

Let $p \geq 1$, the modulus of smoothness of order r is given by

$$\omega_r(f; h)_p := \sup_{\|u\|_2 \leq h} \left\| \Delta_u^r(f) \right\|_p, \tag{25.88}$$

$h > 0$.

We will apply

Theorem 25.24 ([3], p. 24) *Let $f \in C^m(\mathbb{R}^N)$, $m \in \mathbb{N}$, $N \geq 1$, with $f_{\overline{\alpha}} \in L_p(\mathbb{R}^N)$, $|\overline{\alpha}| = m$, $x \in \mathbb{R}^N$. Let $p, q > 1 : \frac{1}{p} + \frac{1}{q} = 1$. Here, μ_{ξ_n} is a Borel probability measure on $\mathbb{R}^N$ for $\xi_n > 0$, $(\xi_n)_{n \in \mathbb{N}}$ bounded sequence. Assume for all $\overline{\alpha} := (\alpha_1, \ldots, \alpha_N)$, $\alpha_i \in \mathbb{Z}^+$, $i = 1, \ldots, N$, $|\overline{\alpha}| := \sum_{i=1}^{N} \alpha_i = m$ that we have*

$$\int_{\mathbb{R}^N} \left(\left(\prod_{i=1}^{N} |s_i|^{\alpha_i} \right) \left(1 + \frac{\|s\|_2}{\xi_n} \right)^r \right)^p d\mu_{\xi_n}(s) < \infty. \tag{25.89}$$

For $\widetilde{j} = 1, \ldots, m$, and $\overline{\alpha} := (\alpha_1, \ldots, \alpha_N)$, $\alpha_i \in \mathbb{Z}^+$, $i = 1, \ldots, N$, $|\overline{\alpha}| := \sum_{j=1}^{N} \alpha_j = \widetilde{j}$, call

$$c_{\overline{\alpha}, n, \widetilde{j}} := \int_{\mathbb{R}^N} \prod_{i=1}^{N} s_i^{\alpha_i} d\mu_{\xi_n}(s). \tag{25.90}$$

Then

$$\left\| E_{r,n}^{[m]} \right\|_p = \left\| \theta_{r,n}^{[m]}(f; x) - f(x) - \sum_{\widetilde{j}=1}^{m} \delta_{\widetilde{j},r}^{[m]} \left(\sum_{|\overline{\alpha}| = \widetilde{j}} \frac{c_{\overline{\alpha}, n, \widetilde{j}} f_{\overline{\alpha}}(x)}{\left(\prod_{i=1}^{N} \alpha_i! \right)} \right) \right\|_{p,x} \tag{25.91}$$

$$\leq \left(\frac{m}{(q(m-1)+1)^{\frac{1}{q}}} \right) \left(\sum_{|\overline{\alpha}| = m} \frac{1}{\prod_{i=1}^{N} \alpha_i!} \right)$$

$$\left[\int_{\mathbb{R}^N}\left[\left(\prod_{i=1}^{N}|s_i|^{\alpha_i}\right)\left(1+\frac{\|s\|_2}{\xi_n}\right)^r\right]^p d\mu_{\xi_n}(s)\right]^{\frac{1}{p}}\omega_r\left(f_{\overline{\alpha}},\xi_n\right)_p.$$

As $n \to \infty$ and $\xi_n \to 0$, by (25.91) we obtain that $\left\|E_{r,n}^{[m]}\right\| \to 0$ with rates.

One also finds by (25.91) that

$$\left\|\theta_{r,n}^{[m]}(f;x)-f(x)\right\|_{p,x} \le \sum_{\widetilde{j}=1}^{m}\left|\delta_{\widetilde{j},r}^{[m]}\right|\left(\sum_{|\overline{\alpha}|=\widetilde{j}}\frac{\left|c_{\overline{\alpha},n,\widetilde{j}}\right|}{\prod_{i=1}^{N}\alpha_i!}\|f_{\overline{\alpha}}\|_p\right)+R.H.S.(25.91.), \tag{25.92}$$

given that $\|f_{\overline{\alpha}}\|_p < \infty$, $|\overline{\alpha}| = \widetilde{j}$, $\widetilde{j} = 1, ..., m$.

Assuming that $c_{\overline{\alpha},n,\widetilde{j}} \to 0$, $\xi_n \to 0$, as $n \to \infty$, we get $\left\|\theta_{r,n}^{[m]}(f) - f\right\|_p \to 0$, that is $\theta_{r,n}^{[m]} \to I$ the unit operator, in L_p norm, with rates.

We present

Theorem 25.25 *Let $f \in C^m\left(\mathbb{R}^N\right)$, $r, \beta, N, m \in \mathbb{N}$, with $f_{\overline{\alpha}} \in L_p\left(\mathbb{R}^N\right)$, $|\overline{\alpha}| = m$, $x \in \mathbb{R}^N$. Let $p, q > 1 : \frac{1}{p}+\frac{1}{q} = 1$. Here φ_{ξ_n} is a Borel probability measure on $\mathbb{R}^N$ as in (25.39), for $\xi_n \in (0, 1]$, $n \in \mathbb{N}$. Let $\beta > \frac{\lceil rp\rceil+m+1}{2}$; $\alpha_j \in \mathbb{Z}^+$, $j = 1, ..., N$: $|\overline{\alpha}| := \sum_{j=1}^{N}\alpha_j = m$. Here $\overline{c}_{\overline{\alpha},n} := \overline{c}_{\overline{\alpha},n,\widetilde{j}}$ as in (25.70), where $\widetilde{j} = 1, ..., m$, and $\overline{\alpha} := (\alpha_1, ..., \alpha_N)$, $\alpha_i \in \mathbb{Z}^+$, $i = 1, ..., N : |\overline{\alpha}| := \sum_{i=1}^{N}\alpha_i = \widetilde{j}$. Then*

$$\left\|\overline{E}_{r,n}^{[m]}\right\|_p = \left\|T_{r,n}^{[m]}(f;x)-f(x)-\sum_{\widetilde{j}=1}^{m}\delta_{\widetilde{j},r}^{[m]}\left(\sum_{|\overline{\alpha}|=\widetilde{j}}\frac{\overline{c}_{\overline{\alpha},n,\widetilde{j}}f_{\overline{\alpha}}(x)}{\left(\prod_{i=1}^{N}\alpha_i!\right)}\right)\right\|_{p,x}$$

$$\le\left(\frac{m}{(q(m-1)+1)^{\frac{1}{q}}}\right)2^{\frac{N}{p}}\gamma^{-\frac{N}{p}}\left\{\sum_{\lambda=0}^{\lceil rp\rceil}\binom{\lceil rp\rceil}{\lambda}\left(\psi_\lambda+\psi_{\lambda+mp}\right)\right\}^{\frac{N}{p}}$$

$$\left(\sum_{|\overline{\alpha}|=m}\frac{1}{\prod_{i=1}^{N}\alpha_i!}\right)\xi_n^{\left(\frac{2\beta(N-1)}{p}+m\right)}\omega_r\left(f_{\overline{\alpha}},\xi_n\right)_p. \tag{25.93}$$

As $n \to \infty$ and $\xi_n \to 0$, by (25.93) we obtain that $\left\|\overline{E}_{r,n}^{[m]}\right\| \to 0$ with rates.

One also finds by (25.93) that

$$\left\| T_{r,n}^{[m]}(f;x) - f(x) \right\|_{p,x} \leq \sum_{\widetilde{j}=1}^{m} \left| \delta_{\widetilde{j},r}^{[m]} \right| \left(\sum_{|\overline{\alpha}|=\widetilde{j}} \frac{\left| \overline{c}_{\overline{\alpha},n,\widetilde{j}} \right|}{\prod_{i=1}^{N} \alpha_i!} \| f_{\overline{\alpha}} \|_p \right) + R.H.S.(25.93.), \tag{25.94}$$

given that $\|f_{\overline{\alpha}}\|_p < \infty$, $|\overline{\alpha}| = \widetilde{j}$, $\widetilde{j} = 1, ..., m$.

Assuming that $\xi_n \to 0$, *as* $n \to \infty$, *we get* $\left\| T_{r,n}^{[m]}(f) - f \right\|_p \to 0$, *that is* $T_{r,n}^{[m]} \to I$ *the unit operator, in* L_p *norm, with rates.*

Proof By Theorem 25.24. From Theorem 25.15 we get that $C_{\xi_n}(\overline{\alpha})$ is uniformly bounded, see (25.61) and $C_{\xi_n}(\overline{\alpha}) \to 0$, as $\xi_n \to 0$, when $n \to \infty$. Also by Theorem 25.17 and Remark 25.18 we get that $\overline{c}_{\overline{\alpha},n} := \overline{c}_{\overline{\alpha},n,\widetilde{j}}$ are uniformly bounded and $\overline{c}_{\overline{\alpha},n} \to 0$, as $\xi_n \to 0$, when $n \to \infty$. ■

We continue with an application of

Theorem 25.26 ([3], p. 26) *Let* $f \in \left(C\left(\mathbb{R}^N\right) \cap L_p\left(\mathbb{R}^N\right)\right)$; $N \geq 1$; $p, q > 1 : \frac{1}{p} + \frac{1}{q} = 1$. *Assume* μ_{ξ_n} *probability Borel measure on* $\mathbb{R}^N$, $(\xi_n)_{n\in\mathbb{N}} > 0$ *and bounded. Also suppose*

$$\int_{\mathbb{R}^N} \left(1 + \frac{\|s\|_2}{\xi_n}\right)^{rp} d\mu_{\xi_n}(s) < \infty. \tag{25.95}$$

Then

$$\left\| \theta_{r,n}^{[0]}(f) - f \right\|_p \leq \tag{25.96}$$

$$\left(\int_{\mathbb{R}^N} \left(1 + \frac{\|s\|_2}{\xi_n}\right)^{rp} d\mu_{\xi_n}(s) \right)^{\frac{1}{p}} \omega_r(f, \xi_n)_p .$$

As $\xi_n \to 0$, *when* $n \to \infty$, *we derive* $\left\| \theta_{r,n}^{[0]}(f) - f \right\|_p \to 0$, *i.e.* $\theta_{r,n}^{[0]} \to I$, *the unit operator, in* L_p *norm.*

We give

Theorem 25.27 *Let* $f \in \left(C\left(\mathbb{R}^N\right) \cap L_p\left(\mathbb{R}^N\right)\right)$; $N \in \mathbb{N} - \{1\}$, $\beta, r \in \mathbb{N}$; $p, q > 1 : \frac{1}{p} + \frac{1}{q} = 1$; $\beta > \frac{\lceil rp \rceil + 1}{2}$, $\xi_n \in (0, 1]$, $n \in \mathbb{N}$. *Then*

$$\left\| T_{r,n}^{[0]}(f) - f \right\|_p \leq 2^{\frac{N}{p}} \gamma^{-\frac{N}{p}} \left[\sum_{\lambda=0}^{\lceil rp \rceil} \binom{\lceil rp \rceil}{\lambda} \psi_\lambda \right]^{\frac{N}{p}} \xi_n^{\frac{2\beta(N-1)}{p}} \omega_r(f, \xi_n)_p . \tag{25.97}$$

As $\xi_n \to 0$, *when* $n \to \infty$, *we derive* $\left\| T_{r,n}^{[0]}(f) - f \right\|_p \to 0$, *i.e.* $T_{r,n}^{[0]} \to I$, *the unit operator, in* L_p *norm.*

Proof By Theorems 25.26 and 25.16. ■

We mention

Theorem 25.28 ([3], p. 27) *Let $f \in \left(C\left(\mathbb{R}^N\right) \cap L_1\left(\mathbb{R}^N\right)\right)$; $N \geq 1$. Assume μ_{ξ_n} probability Borel measure on $\mathbb{R}^N$, $(\xi_n)_{n\in\mathbb{N}} > 0$ and bounded. Also suppose*

$$\int_{\mathbb{R}^N} \left(1 + \frac{\|s\|_2}{\xi_n}\right)^r d\mu_{\xi_n}(s) < \infty. \tag{25.98}$$

Then

$$\left\|\theta_{r,n}^{[0]}(f) - f\right\|_1 \leq \left(\int_{\mathbb{R}^N} \left(1 + \frac{\|s\|_2}{\xi_n}\right)^r d\mu_{\xi_n}(s)\right) \omega_r(f, \xi_n)_1. \tag{25.99}$$

As $\xi_n \to 0$, we get $\theta_{r,n}^{[0]} \to I$, in L_1 norm.

We give

Theorem 25.29 *Let $f \in \left(C\left(\mathbb{R}^N\right) \cap L_1\left(\mathbb{R}^N\right)\right)$; $N \in \mathbb{N} - \{1\}$, $r, \beta \in \mathbb{N}$, $\beta > \frac{r+1}{2}$, $\xi_n \in (0, 1]$, $n \in \mathbb{N}$. Then*

$$\left\|T_{r,n}^{[0]}(f) - f\right\|_1 \leq 2^N \gamma^{-N} \left[\sum_{\lambda=0}^{r} \binom{r}{\lambda} \psi_\lambda\right]^N \xi_n^{2\beta(N-1)} \omega_r(f, \xi_n)_1. \tag{25.100}$$

As $\xi_n \to 0$, we get $T_{r,n}^{[0]} \to I$, in L_1 norm.

Proof By Theorems 25.14 and 25.28. ∎

We mention

Theorem 25.30 ([3], p. 29) *Let $f \in C^m\left(\mathbb{R}^N\right)$, $m, N \in \mathbb{N}$, with $f_{\overline{\alpha}} \in L_1\left(\mathbb{R}^N\right)$, $|\overline{\alpha}| = m$, $x \in \mathbb{R}^N$. Here, μ_{ξ_n} is a Borel probability measure on $\mathbb{R}^N$ for $\xi_n > 0$, $(\xi_n)_{n\in\mathbb{N}}$ is a bounded sequence. Suppose for all $\overline{\alpha} := (\alpha_1, ..., \alpha_N)$, $\alpha_i \in \mathbb{Z}^+$, $i = 1, ..., N$, $|\overline{\alpha}| := \sum_{i=1}^{N} \alpha_i = m$ that we have*

$$\int_{\mathbb{R}^N} \left(\left(\prod_{i=1}^{N} |s_i|^{\alpha_i}\right)\left(1 + \frac{\|s\|_2}{\xi_n}\right)^r\right) d\mu_{\xi_n}(s) < \infty. \tag{25.101}$$

For $\widetilde{j} = 1, ..., m$, and $\overline{\alpha} := (\alpha_1, ..., \alpha_N)$, $\alpha_i \in \mathbb{Z}^+$, $i = 1, ..., N$, $|\overline{\alpha}| := \sum_{i=1}^{N} \alpha_i = \widetilde{j}$, call

$$c_{\overline{\alpha}, n, \widetilde{j}} := \int_{\mathbb{R}^N} \prod_{i=1}^{N} s_i^{\alpha_i} d\mu_{\xi_n}(s). \tag{25.102}$$

Then

$$\left\| E_{r,n}^{[m]} \right\|_1 = \left\| \theta_{r,n}^{[m]}(f;x) - f(x) - \sum_{\widetilde{j}=1}^{m} \delta_{\widetilde{j},r}^{[m]} \left(\sum_{|\overline{\alpha}|=\widetilde{j}} \frac{c_{\overline{\alpha},n,\widetilde{j}} f_{\overline{\alpha}}(x)}{\prod_{i=1}^{N} \alpha_i!} \right) \right\|_{1,x} \tag{25.103}$$

$$\leq \sum_{|\overline{\alpha}|=m} \left(\frac{1}{\prod_{i=1}^{N} \alpha_i!} \right) \omega_r(f_{\overline{\alpha}}, \xi_n)_1 \int_{\mathbb{R}^N} \left(\prod_{i=1}^{N} |s_i|^{\alpha_i} \right) \left(1 + \frac{\|s\|_2}{\xi_n} \right)^r d\mu_{\xi_n}(s).$$

As $\xi_n \to 0$*, we get* $\left\| E_{r,n}^{[m]} \right\|_1 \to 0$ *with rates.*

From (25.103) we get

$$\left\| \theta_{r,n}^{[m]} f - f \right\|_1 \leq \sum_{\widetilde{j}=1}^{m} \left| \delta_{\widetilde{j},r}^{[m]} \right| \left(\sum_{|\overline{\alpha}|=\widetilde{j}} \frac{\left| c_{\overline{\alpha},n,\widetilde{j}} \right|}{\prod_{i=1}^{N} \alpha_i!} \| f_{\overline{\alpha}} \|_1 \right) + R.H.S.(25.103.), \tag{25.104}$$

given that $\|f_{\overline{\alpha}}\|_1 < \infty$, $|\overline{\alpha}| = \widetilde{j}$, $\widetilde{j} = 1, ..., m$.

As $n \to \infty$*, assuming* $\xi_n \to 0$ *and* $c_{\overline{\alpha},n,\widetilde{j}} \to 0$*, we obtain* $\left\| \theta_{r,n}^{[m]}(f) - f \right\|_1 \to 0$*, that is* $\theta_{r,n}^{[m]} \to I$ *in* L_1 *norm, with rates.*

We give

Theorem 25.31 *Let* $f \in C^m(\mathbb{R}^N)$*,* $r, N, \beta, m \in \mathbb{N}$*, with* $f_{\overline{\alpha}} \in L_1(\mathbb{R}^N)$*, where* $\alpha_j \in \mathbb{Z}^+$*,* $j = 1, ..., N : |\overline{\alpha}| := \sum_{j=1}^{N} \alpha_j = m$*,* $x \in \mathbb{R}^N$*,* $\xi_n \in (0, 1]$*,* $n \in \mathbb{N}$*, and* $\beta > \frac{m+r+1}{2}$*. Here* $\overline{c}_{\overline{\alpha},n} := \overline{c}_{\overline{\alpha},n,\widetilde{j}}$ *as in (25.70), where* $\widetilde{j} = 1, ..., m$*, and* $\overline{\alpha} := (\alpha_1, ..., \alpha_N)$*,* $\alpha_i \in \mathbb{Z}^+$*,* $i = 1, ..., N : |\overline{\alpha}| := \sum_{i=1}^{N} \alpha_i = \widetilde{j}$*. Also here* φ_{ξ_n} *is the Borel probability measure on* $\mathbb{R}^N$*, see (25.39). Then*

$$\left\| \overline{E}_{r,n}^{[m]} \right\|_1 = \left\| T_{r,n}^{[m]}(f;x) - f(x) - \sum_{\widetilde{j}=1}^{m} \delta_{\widetilde{j},r}^{[m]} \left(\sum_{|\overline{\alpha}|=\widetilde{j}} \frac{\overline{c}_{\overline{\alpha},n,\widetilde{j}} f_{\overline{\alpha}}(x)}{\prod_{i=1}^{N} \alpha_i!} \right) \right\|_{1,x} \tag{25.105}$$

$$\leq \left(\sum_{|\alpha|=m} \left(\frac{1}{\prod_{i=1}^{N} \alpha_i!} \right) \omega_r(f_{\overline{\alpha}}, \xi_n)_1 \right) \xi_n^{2\beta(N-1)+m}$$

$$2^{N}\gamma^{-N}\left\{\sum_{\lambda=0}^{r}\binom{r}{\lambda}\left[\psi_{\lambda}+\psi_{\lambda+m}\right]\right\}^{N}.$$

As $\xi_n \to 0$, we get $\left\|\overline{E}_{r,n}^{[m]}\right\|_1 \to 0$ with rates.

From (25.105) we get

$$\left\|T_{r,n}^{[m]}f-f\right\|_1 \le \sum_{\widetilde{j}=1}^{m}\left|\delta_{\widetilde{j},r}^{[m]}\right|\left(\sum_{|\overline{\alpha}|=\widetilde{j}}\frac{\left|\overline{c}_{\overline{\alpha},n,\widetilde{j}}\right|}{\prod_{i=1}^{N}\alpha_i!}\left\|f_{\overline{\alpha}}\right\|_1\right)+R.H.S.(25.105.), \tag{25.106}$$

given that $\|f_{\overline{\alpha}}\|_1 < \infty$, $|\alpha| = \widetilde{j}$, $\widetilde{j} = 1, ..., m$.

As $n \to \infty$, assuming $\xi_n \to 0$, we get $\overline{c}_{\overline{\alpha},n,\widetilde{j}} \to 0$ and $\left\|T_{r,n}^{[m]}(f)-f\right\|_1 \to 0$, that is $T_{r,n}^{[m]} \to I$ in L_1 norm, with rates.

Proof By Theorem 25.30, also by Theorem 25.13, see (25.47) and by Theorem 25.17 and Remark 25.18. ■

25.3.3 Global Smoothness Preservation and Simultaneous Approximation Of $T_{r,n}^{[m]}$

We need

Definition 25.32 ([3], *p. 34*) Let $f \in C\left(\mathbb{R}^N\right)$, $N \ge 1$, $m \in \mathbb{N}$, the mth modulus of smoothness for $1 \le p \le \infty$, is given by

$$\omega_m(f;h)_p := \sup_{\|t\|_2 \le h}\left\|\Delta_t^m(f)\right\|_{p,x}, \tag{25.107}$$

$h > 0$, where

$$\Delta_t^m f(x) := \sum_{j=0}^{m}(-1)^{m-j}\binom{m}{j}f(x+jt). \tag{25.108}$$

Denote

$$\omega_m(f;h)_\infty = \omega_m(f,h). \tag{25.109}$$

Above, $x, t \in \mathbb{R}^N$.

We present the related global smoothness preservation result

Theorem 25.33 *We assume $T_{r,n}^{[\widetilde{m}]}(f;x) \in \mathbb{R}$, $\widetilde{m} \in \mathbb{Z}_+$, $\forall\ x \in \mathbb{R}$. Let $h > 0$, $f \in C\left(\mathbb{R}^N\right)$, $N \ge 1$.*

(i) Assume $\omega_m(f,h) < \infty$. *Then*

$$\omega_m\left(T_{r,n}^{[\widetilde{m}]} f, h\right) \leq \left(\sum_{\widetilde{j}=0}^{r} \left|\alpha_{\widetilde{j},r}^{[\widetilde{m}]}\right|\right) \omega_m(f,h). \tag{25.110}$$

(ii) Assume $f \in \left(C\left(\mathbb{R}^N\right) \cap L_1\left(\mathbb{R}^N\right)\right)$. *Then*

$$\omega_m\left(T_{r,n}^{[\widetilde{m}]} f, h\right)_1 \leq \left(\sum_{\widetilde{j}=0}^{r} \left|\alpha_{\widetilde{j},r}^{[\widetilde{m}]}\right|\right) \omega_m(f,h)_1. \tag{25.111}$$

(iii) Assume $f \in \left(C\left(\mathbb{R}^N\right) \cap L_p\left(\mathbb{R}^N\right)\right)$, $p > 1$. *Then*

$$\omega_m\left(T_{r,n}^{[\widetilde{m}]} f, h\right)_p \leq \left(\sum_{\widetilde{j}=0}^{r} \left|\alpha_{\widetilde{j},r}^{[\widetilde{m}]}\right|\right) \omega_m(f,h)_p. \tag{25.112}$$

Proof Direct application of ([3]) Theorem 3.2, p. 35. ∎

We make

Remark 25.34 Let $r = 1$, $\widetilde{m} \in \mathbb{Z}_+$, then $\alpha_{0,1}^{[\widetilde{m}]} = 0$, $\alpha_{1,1}^{[\widetilde{m}]} = 1$. Hence

$$T_{1,n}^{[\widetilde{m}]}(f;x) = \lambda_n^{-N} \int_{\mathbb{R}^N} f(x+s) \prod_{i=1}^{N} \left(\frac{\sin\left(\frac{s_i}{\xi_n}\right)}{s_i}\right)^{2\beta} ds_1 \dots ds_N =: T_n(f;x). \tag{25.113}$$

By Theorem 25.33, we get

Theorem 25.35 *We suppose* $T_n(f;x) \in \mathbb{R}$, $\forall\, x \in \mathbb{R}$. *Let* $h > 0$, $f \in C\left(\mathbb{R}^N\right)$, $N \geq 1$.

(i) Assume $\omega_m(f,h) < \infty$. *Then*

$$\omega_m(T_n f, h) \leq \omega_m(f,h). \tag{25.114}$$

(ii) Assume $f \in \left(C\left(\mathbb{R}^N\right) \cap L_1\left(\mathbb{R}^N\right)\right)$. *Then*

$$\omega_m(T_n f, h)_1 \leq \omega_m(f,h)_1. \tag{25.115}$$

(iii) Assume $f \in \left(C\left(\mathbb{R}^N\right) \cap L_p\left(\mathbb{R}^N\right)\right)$, $p > 1$. *Then*

$$\omega_m(T_n f, h)_p \leq \omega_m(f,h)_p. \tag{25.116}$$

Next, we get an optimality result

Proposition 25.36 *Above inequality (25.114):*

$$\omega_m\left(T_n f, h\right) \leq \omega_m\left(f, h\right)$$

is sharp, namely it is attained by any

$$f_j^*\left(x\right) = x_j^m,\ j = 1, ..., N,\ x = \left(x_1, ..., x_j, ..., x_N\right) \in \mathbb{R}^N. \tag{25.117}$$

Proof Apply Proposition 3.5, p. 38, of [3]. ∎

We need

Theorem 25.37 ([3], p. 39) *Let $f \in C^l\left(\mathbb{R}^N\right)$, $l, N \in \mathbb{N}$. Here, μ_{ξ_n} is a Borel probability measure on $\mathbb{R}^N$, $\xi_n > 0$, $(\xi_n)_{n \in \mathbb{N}}$ a bounded sequence. Let $\overline{\beta} := (\beta_1, ..., \beta_N)$, $\beta_i \in \mathbb{Z}^+$, $i = 1, ..., N$; $\left|\overline{\beta}\right| := \sum_{i=1}^{N} \beta_i = l$. Here $f\left(x + sj\right)$, $x, s \in \mathbb{R}^N$, is μ_{ξ_n}-integrable wrt s, for $j = 1, ..., r$. There exist μ_{ξ_n}-integrable functions $h_{i_1,j}$, $h_{\beta_1,i_2,j}$, $h_{\beta_1,\beta_2,i_3,j}$, ..., $h_{\beta_1,\beta_2,...,\beta_{N-1},i_N,j} \geq 0$ $(j = 1, ..., r)$ on $\mathbb{R}^N$ such that*

$$\left|\frac{\partial^{i_1} f\left(x + sj\right)}{\partial x_1^{i_1}}\right| \leq h_{i_1,j}\left(s\right),\ \ i_1 = 1, ..., \beta_1, \tag{25.118}$$

$$\left|\frac{\partial^{\beta_1 + i_2} f\left(x + sj\right)}{\partial x_2^{i_2} \partial x_1^{\beta_1}}\right| \leq h_{\beta_1,i_2,j}\left(s\right),\ \ i_2 = 1, ..., \beta_2,$$

$$\vdots$$

$$\left|\frac{\partial^{\beta_1 + \beta_2 + ... + \beta_{N-1} + i_N} f\left(x + sj\right)}{\partial x_N^{i_N} \partial x_{N-1}^{\beta_{N-1}} ... \partial x_2^{\beta_2} \partial x_1^{\beta_1}}\right| \leq h_{\beta_1,\beta_2,...,\beta_{N-1},i_N,j}\left(s\right),\ \ i_N = 1, ..., \beta_N,$$

$\forall\, x, s \in \mathbb{R}^N$.

Then, both of the next exist and

$$\left(\theta_{r,n}^{[\widetilde{m}]}\left(f; x\right)\right)_{\overline{\beta}} = \theta_{r,n}^{[\widetilde{m}]}\left(f_{\overline{\beta}}; x\right),\ \ \widetilde{m} \in \mathbb{Z}_+. \tag{25.119}$$

In particular it holds

$$\left(T_{r,n}^{[\widetilde{m}]}\left(f; x\right)\right)_{\overline{\beta}} = T_{r,n}^{[\widetilde{m}]}\left(f_{\overline{\beta}}; x\right), \tag{25.120}$$

when

$$d\mu_{\xi_n} = d\varphi_{\xi_n}\left(s\right),\ \ s \in \mathbb{R}^N,$$

see (25.39).

Corollary 25.38 (by Theorem 25.37, $r = 1$) *It holds*

$$\left(T_n\left(f;x\right)\right)_{\overline{\beta}} = T_n\left(f_{\overline{\beta}};x\right). \tag{25.121}$$

We present simultaneous global smoothness results.

Theorem 25.39 *Let $h > 0$ and the assumptions of Theorem 25.37 are valid for $d\mu_{\xi_n} = d\varphi_{\xi_n}$. Here $\overline{\gamma} = 0, \overline{\beta}$ $(0 = (0, ..., 0))$, $\widetilde{m} \in \mathbb{Z}_+$.*
(i) Assume $\omega_m\left(f_{\overline{\gamma}}, h\right) < \infty$. Then

$$\omega_m\left(\left(T_{r,n}^{[\widetilde{m}]}\left(f\right)\right)_{\overline{\gamma}}, h\right) \leq \left(\sum_{\widetilde{j}=0}^{r}\left|\alpha_{\widetilde{j},r}^{[\widetilde{m}]}\right|\right)\omega_m\left(f_{\overline{\gamma}}, h\right). \tag{25.122}$$

(ii) Additionally suppose $f_{\overline{\gamma}} \in L_1\left(\mathbb{R}^N\right)$. Then

$$\omega_m\left(\left(T_{r,n}^{[\widetilde{m}]}\left(f\right)\right)_{\overline{\gamma}}, h\right)_1 \leq \left(\sum_{\widetilde{j}=0}^{r}\left|\alpha_{\widetilde{j},r}^{[\widetilde{m}]}\right|\right)\omega_m\left(f_{\overline{\gamma}}, h\right)_1. \tag{25.123}$$

(iii) Additionally suppose $f_{\overline{\gamma}} \in L_p\left(\mathbb{R}^N\right)$, $p > 1$. Then

$$\omega_m\left(\left(T_{r,n}^{[\widetilde{m}]}\left(f\right)\right)_{\overline{\gamma}}, h\right)_p \leq \left(\sum_{\widetilde{j}=0}^{r}\left|\alpha_{\widetilde{j},r}^{[\widetilde{m}]}\right|\right)\omega_m\left(f_{\overline{\gamma}}, h\right)_p. \tag{25.124}$$

It follows

Corollary 25.40 (to Theorem 25.39) *Let $h > 0$, $r = 1$ and $\overline{\gamma} = 0, \overline{\beta}$.*
(i) Assume $\omega_m\left(f_{\overline{\gamma}}, h\right) < \infty$. Then

$$\omega_m\left(\left(T_n\left(f\right)\right)_{\overline{\gamma}}, h\right) \leq \omega_m\left(f_{\overline{\gamma}}, h\right). \tag{25.125}$$

(ii) Additionally suppose $f_{\overline{\gamma}} \in L_1\left(\mathbb{R}^N\right)$. Then

$$\omega_m\left(\left(T_n\left(f\right)\right)_{\overline{\gamma}}, h\right)_1 \leq \omega_m\left(f_{\overline{\gamma}}, h\right)_1. \tag{25.126}$$

(iii) Additionally suppose $f_{\overline{\gamma}} \in L_p\left(\mathbb{R}^N\right)$, $p > 1$. Then

$$\omega_m\left(\left(T_n\left(f\right)\right)_{\overline{\gamma}}, h\right)_p \leq \omega_m\left(f_{\overline{\gamma}}, h\right)_p. \tag{25.127}$$

Next comes multi-simultaneous approximation. We give

Theorem 25.41 *Let $f \in C^{m+l}\left(\mathbb{R}^N\right)$, $m, l, N \in \mathbb{N}$. The assumptions of Theorem 25.37 are valid for $d\mu_{\xi_n} = d\varphi_{\xi_n}$. Call $\overline{\gamma} = 0, \overline{\beta}$. Assume $\left\|f_{\overline{\gamma}+\overline{\alpha}}\right\|_\infty < \infty$,* *and let* $\xi_n \in (0,1]$, $n \in \mathbb{N}$. *Here* $\beta, r \in \mathbb{N}$, $\beta > \frac{m+r+1}{2}$, *and* $A_{\xi_n}(\overline{\alpha})$ *as in (25.47), and* $\overline{c}_{\overline{\alpha},n} := \overline{c}_{\overline{\alpha},n,\widetilde{j}}$ *as in (25.70). Then*

$$\left\| \left(T_{r,n}^{[m]}(f;\cdot)\right)_{\overline{\gamma}} - f_{\overline{\gamma}}(\cdot) - \sum_{\widetilde{j}=1}^{m} \delta_{\widetilde{j},r}^{[m]} \left(\sum_{\substack{\alpha_1,\ldots,\alpha_N \geq 0; \\ |\overline{\alpha}|=\widetilde{j}}} \frac{\overline{c}_{\overline{\alpha},n,\widetilde{j}} f_{\overline{\gamma}+\overline{\alpha}}(\cdot)}{\prod_{i=1}^{N} \alpha_i!} \right) \right\|_{\infty} \tag{25.128}$$

$$\leq \sum_{\substack{\alpha_1,\ldots,\alpha_N \geq 0; \\ |\overline{\alpha}|=m}} \frac{\left(\omega_r\left(f_{\overline{\gamma}+\overline{\alpha}}, \xi_n\right)\right)}{\left(\prod_{i=1}^{N} \alpha_i!\right)} A_{\xi_n}(\overline{\alpha}).$$

Proof Based on Theorems 25.20 and 25.37. ∎

We continue with

Theorem 25.42 *Let* $f \in C_B^l\left(\mathbb{R}^N\right)$, $r, l, \beta \in \mathbb{N}$ *(functions l-times continuously differentiable and bounded),* $N \in \mathbb{N} - \{1\}$. *The assumptions of Theorem 25.37 are valid for* $d\mu_{\xi_n} = d\varphi_{\xi_n}$. *Call* $\overline{\gamma} = 0, \overline{\beta}$, $\xi_n \in (0,1]$, $n \in \mathbb{N}$. *Let also* $\beta > \frac{r+1}{2}$. *Then*

$$\left\| \left(T_{r,n}^{[0]} f\right)_{\overline{\gamma}} - f_{\overline{\gamma}} \right\|_{\infty} \leq 2^N \gamma^{-N} \left[\sum_{\lambda=0}^{r} \binom{r}{\lambda} \psi_\lambda \right]^N \xi_n^{2\beta(N-1)} \omega_r\left(f_{\overline{\gamma}}, \xi_n\right). \tag{25.129}$$

If $\xi_n \to 0$, *as* $n \to \infty$, *then* $\left(T_{r,n}^{[0]} f\right)_{\overline{\gamma}} \to f_{\overline{\gamma}}$ *uniformly.*

Proof By Theorems 25.22 and 25.37. ∎

We present

Theorem 25.43 *Let* $f \in C^{m+l}\left(\mathbb{R}^N\right)$, $r, \beta, N, m, l \in \mathbb{N}$. *The assumptions of Theorem 25.37 are valid for* $d\mu_{\xi_n} = d\varphi_{\xi_n}$, $\xi_n \in (0,1]$, $n \in \mathbb{N}$. *Call* $\overline{\gamma} = 0, \overline{\beta}$. *Let* $f_{(\overline{\gamma}+\overline{\alpha})} \in L_p\left(\mathbb{R}^N\right)$, $|\overline{\alpha}| = m$, $x \in \mathbb{R}^N$, *and* $p, q > 1 : \frac{1}{p} + \frac{1}{q} = 1$. *Let also* $\beta > \frac{\lceil rp \rceil + m + 1}{2}$; $\alpha_j \in \mathbb{Z}^+$, $j = 1, \ldots, N$, $|\overline{\alpha}| := \sum_{j=1}^{N} \alpha_j = m$. *Here* $\overline{c}_{\overline{\alpha},n} := \overline{c}_{\overline{\alpha},n,\widetilde{j}}$ *as in (25.70), where* $\widetilde{j} = 1, \ldots, m$, *and* $\overline{\alpha} := (\alpha_1, \ldots, \alpha_N)$, $\alpha_i \in \mathbb{Z}^+$, $i = 1, \ldots, N$, $|\overline{\alpha}| := \sum_{i=1}^{N} \alpha_i = \widetilde{j}$. *Then*

$$\left\| \left(T_{r,n}^{[m]}(f;x)\right)_{\overline{\gamma}} - f_{\overline{\gamma}}(x) - \sum_{\widetilde{j}=1}^{m} \delta_{\widetilde{j},r}^{[m]} \left(\sum_{|\overline{\alpha}|=\widetilde{j}} \frac{\overline{c}_{\overline{\alpha},n,\widetilde{j}} f_{\overline{\gamma}+\overline{\alpha}}(x)}{\prod_{i=1}^{N} \alpha_i!} \right) \right\|_{p,x}$$

$$\leq \left(\frac{m}{(q(m-1)+1)^{\frac{1}{q}}} \right) 2^{\frac{N}{p}} \gamma^{-\frac{N}{p}} \left\{ \sum_{\lambda=0}^{\lceil rp \rceil} \binom{\lceil rp \rceil}{\lambda} (\psi_\lambda + \psi_{\lambda+mp}) \right\}^{\frac{N}{p}}$$

$$\left(\sum_{|\overline{\alpha}|=m} \frac{1}{\prod_{i=1}^{N} \alpha_i!} \right) \xi_n^{\left(\frac{2\beta(N-1)}{p} + m \right)} \omega_r \left(f_{\overline{\gamma}+\overline{\alpha}}, \xi_n \right)_p . \tag{25.130}$$

Proof Theorems 25.25 and 25.37. ■

We continue with

Theorem 25.44 *Let $f \in C^l(\mathbb{R}^N)$, $\beta, r, l \in \mathbb{N}$, $N \in \mathbb{N} - \{1\}$. The assumptions of Theorem 25.37 are valid for $d\mu_{\xi_n} = d\varphi_{\xi_n}$, $\xi_n \in (0,1]$, $n \in \mathbb{N}$. Call $\overline{\gamma} = 0, \overline{\beta}$. Let $f_{\overline{\gamma}} \in L_p(\mathbb{R}^N)$ and $p, q > 1 : \frac{1}{p} + \frac{1}{q} = 1$. Here $\beta > \frac{\lceil rp \rceil + 1}{2}$. Then*

$$\left\| \left(T_{r,n}^{[0]}(f) \right)_{\overline{\gamma}} - f_{\overline{\gamma}} \right\|_p \leq$$

$$2^{\frac{N}{p}} \gamma^{-\frac{N}{p}} \left[\sum_{\lambda=0}^{\lceil rp \rceil} \binom{\lceil rp \rceil}{\lambda} \psi_\lambda \right]^{\frac{N}{p}} \xi_n^{\frac{2\beta(N-1)}{p}} \omega_r \left(f_{\overline{\gamma}}, \xi_n \right)_p . \tag{25.131}$$

As $n \to +\infty$ and $\xi_n \to 0$, then $\left(T_{r,n}^{[0]}(f) \right)_{\overline{\gamma}} \overset{\|\cdot\|_p}{\rightarrow} f_{\overline{\gamma}}$.

Proof By Theorems 25.27 and 25.37. ■

We continue with

Theorem 25.45 *Let $f \in C^l(\mathbb{R}^N)$, $l \in \mathbb{N}$, $N \in \mathbb{N} - \{1\}$. The assumptions of Theorem 25.37 are valid for $d\mu_{\xi_n} = d\varphi_{\xi_n}$, $\xi_n \in (0,1]$, $n \in \mathbb{N}$. Call $\overline{\gamma} = 0, \overline{\beta}$. Let $f_{\overline{\gamma}} \in L_1(\mathbb{R}^N)$ and $\beta, r \in \mathbb{N}$, $\beta > \frac{r+1}{2}$. Then*

$$\left\| \left(T_{r,n}^{[0]}(f) \right)_{\overline{\gamma}} - f_{\overline{\gamma}} \right\|_1 \leq 2^N \gamma^{-N} \left[\sum_{\lambda=0}^{r} \binom{r}{\lambda} \psi_\lambda \right]^N \xi_n^{2\beta(N-1)} \omega_r \left(f_{\overline{\gamma}}, \xi_n \right)_1 . \tag{25.132}$$

As $n \to +\infty$ and $\xi_n \to 0$, then $\left(T_{r,n}^{[0]}(f) \right)_{\overline{\gamma}} \overset{\|\cdot\|_1}{\rightarrow} f_{\overline{\gamma}}$.

Proof By Theorems 25.29 and 25.37. ■

We continue with

Theorem 25.46 *Let $f \in C^{m+l}\left(\mathbb{R}^N\right)$, $r, N, \beta, m, l \in \mathbb{N}$. The assumptions of Theorem 25.37 are valid for $d\mu_{\xi_n} = d\varphi_{\xi_n}$, $\xi_n \in (0, 1]$, $n \in \mathbb{N}$. Call $\overline{\gamma} = 0, \overline{\beta}$. Let $f_{(\overline{\gamma}+\overline{\alpha})} \in L_1\left(\mathbb{R}^N\right)$, $|\overline{\alpha}| = m$, $x \in \mathbb{R}^N$, $\beta > \frac{m+r+1}{2}$. Here $\overline{c}_{\overline{\alpha},n} := \overline{c}_{\overline{\alpha},n,\widetilde{j}}$ as in (25.70), where $\widetilde{j} = 1, ..., m$, and $\overline{\alpha} := (\alpha_1, ..., \alpha_N)$, $\alpha_i \in \mathbb{Z}^+$, $i = 1, ..., N$, $|\overline{\alpha}| := \sum_{i=1}^{N} \alpha_i = \widetilde{j}$. Then*

$$\left\| \left(T_{r,n}^{[m]}(f; x)\right)_{\overline{\gamma}} - f_{\overline{\gamma}}(x) - \sum_{\widetilde{j}=1}^{m} \delta_{\widetilde{j},r}^{[m]} \left(\sum_{|\overline{\alpha}|=\widetilde{j}} \frac{\overline{c}_{\overline{\alpha},n,\widetilde{j}} f_{\overline{\gamma}+\overline{\alpha}}(x)}{\prod_{i=1}^{N} \alpha_i!} \right) \right\|_{1,x} \leq$$

$$\left(\sum_{|\overline{\alpha}|=m} \frac{1}{\prod_{i=1}^{N} \alpha_i!} \right) \omega_r\left(f_{\overline{\gamma}+\overline{\alpha}}, \xi_n\right)_1 \xi_n^{2\beta(N-1)+m} 2^N \gamma^{-N} \left\{ \sum_{\lambda=0}^{r} \binom{r}{\lambda} \left[\psi_\lambda + \psi_{\lambda+m}\right] \right\}^N. \tag{25.133}$$

Proof By Theorems 25.31 and 25.37. ∎

25.3.4 Voronovskaya Asymptotic Expansions for $T_{r,n}^{[m]}$

We will apply

Theorem 25.47 ([3], p. 53) *Let $f \in C^m\left(\mathbb{R}^N\right)$, $m, N \in \mathbb{N}$, with all $\|f_{\overline{\alpha}}\|_\infty \leq M$, $M > 0$, all $\overline{\alpha} : |\overline{\alpha}| = m$. Let $\xi_n > 0$, $(\xi_n)_{n\in\mathbb{N}}$ bounded sequence, μ_{ξ_n} probability Borel measures on $\mathbb{R}^N$.*

Call $c_{\overline{\alpha},n,\widetilde{j}} := \int_{\mathbb{R}^N} \left(\prod_{i=1}^{N} s_i^{\alpha_i}\right) d\mu_{\xi_n}(s)$, all $|\overline{\alpha}| = \widetilde{j} = 1, ..., m-1$. Suppose $\xi_n^{-m} \int_{\mathbb{R}^N} \left(\prod_{i=1}^{N} |s_i|^{\alpha_i}\right) d\mu_{\xi_n}(s) \leq \rho$, all $\overline{\alpha} : |\overline{\alpha}| = m$, $\rho > 0$, for any such $(\xi_n)_{n\in\mathbb{N}}$. Also $0 < \gamma^* \leq 1$, $x \in \mathbb{R}^N$. Then

$$\theta_{r,n}^{[m]}(f; x) - f(x) = \sum_{\widetilde{j}=1}^{m-1} \delta_{\widetilde{j},r}^{[m]} \left(\sum_{|\overline{\alpha}|=\widetilde{j}} \frac{c_{\overline{\alpha},n,\widetilde{j}} f_{\overline{\alpha}}(x)}{\left(\prod_{i=1}^{N} \alpha_i!\right)} \right) + 0\left(\xi_n^{m-\gamma^*}\right). \tag{25.134}$$

When $m = 1$, the sum collapses.

Above we assume $\theta_{r,n}^{[m]}(f; x) \in \mathbb{R}$, $\forall\, x \in \mathbb{R}^N$.

We give

Theorem 25.48 *Let* $r, m, \beta, N \in \mathbb{N}$, $\beta > \frac{m+1}{2}$, $\xi_n \in (0,1]$, $n \in \mathbb{N}$. *Also* $\alpha_j \in \mathbb{Z}^+$, $j = 1, ..., N : |\overline{\alpha}| := \sum_{j=1}^{N} \alpha_j = m$. *Here* $f \in C^m(\mathbb{R}^N)$, *with all* $\|f_{\overline{\alpha}}\|_\infty \leq M$, $M > 0$, *for all* $\overline{\alpha} : |\overline{\alpha}| = m$; *and* $d\mu_{\xi_n}(s) = d\varphi_{\xi_n}(s)$, *as in (25.39),* $\forall\, \underline{s} \in \mathbb{R}^N$. *Assume* $T_{r,n}^{[m]}(f; x) \in \mathbb{R}$, $\forall\, x \in \mathbb{R}^N$. *Here* $\overline{c}_{\overline{\alpha},n,\widetilde{j}}$ *as in (25.70), all* $|\overline{\alpha}| = \widetilde{j} = 1, ..., m-1$. *Let* $0 < \gamma^* \leq 1$, $x \in \mathbb{R}^N$. *Then*

$$T_{r,n}^{[m]}(f; x) - f(x) = \sum_{\widetilde{j}=1}^{m-1} \delta_{\widetilde{j},r}^{[m]} \left(\sum_{|\overline{\alpha}|=\widetilde{j}} \frac{\overline{c}_{\overline{\alpha},n,\widetilde{j}} f_{\overline{\alpha}}(x)}{\left(\prod_{i=1}^{N} \alpha_i!\right)} \right) + 0\left(\xi_n^{m-\gamma^*}\right). \qquad (25.135)$$

When $m = 1$, *the sum collapses.*

Proof By Theorems 25.17 and 25.47. Here $\rho = \varphi$, see (25.68). ■

We give

Corollary 25.49 (to Theorem *25.48*) *Let* $f \in C^1(\mathbb{R}^N)$, $N \in \mathbb{N}$, *with all* $\left\|\frac{\partial f}{\partial x_i}\right\|_\infty \leq M$, $M > 0$, $i = 1, ..., N$. *Let* $0 < \gamma^* \leq 1$. *Assume* $T_{r,n}^{[1]}(f; x) \in \mathbb{R}$, $\forall\, x \in \mathbb{R}^N$. *Here* $r \in \mathbb{N}$ *and* $\beta \in \mathbb{N} - \{1\}$, $\xi_n \in (0,1]$, $n \in \mathbb{N}$. *Then*

$$T_{r,n}^{[1]}(f; x) - f(x) = 0\left(\xi_n^{1-\gamma^*}\right). \qquad (25.136)$$

Proof By Theorems 25.17 and 25.48. Here it is $\rho = \varphi$, apply (25.68) for $m = 1$. ■

We continue with

Corollary 25.50 (to Theorem *25.48*) *Let* $f \in C^2(\mathbb{R}^2)$, *with all* $\left\|\frac{\partial^2 f}{\partial x_1^2}\right\|_\infty$, $\left\|\frac{\partial^2 f}{\partial x_2^2}\right\|_\infty$, $\left\|\frac{\partial^2 f}{\partial x_1 \partial x_2}\right\|_\infty \leq M$, $M > 0$, $\xi_n \in (0,1]$, $n \in \mathbb{N}$. *Call*

$$c_1 = \int_{\mathbb{R}^2} s_1 d\varphi_{\xi_n}^*(s), \ c_2 = \int_{\mathbb{R}^2} s_2 d\varphi_{\xi_n}^*(s), \qquad (25.137)$$

where

$$d\varphi_{\xi_n}^* = \lambda_n^{-2} \prod_{i=1}^{2} \left(\frac{\sin\left(\frac{s_i}{\xi_n}\right)}{s_i} \right)^{2\beta} ds_1 ds_2, \ s = (s_1, s_2) \in \mathbb{R}^2.$$

Let $0 < \gamma^* \leq 1$ *and assume* $T_{r,n}^{[2]}(f; x) \in \mathbb{R}$, $\forall\, x \in \mathbb{R}^2$. *Here* $r, \beta \in \mathbb{N}$ *and* $\beta > \frac{3}{2}$. *Then*

$$T_{r,n}^{[2]}(f; x) - f(x) = \left(\sum_{j=1}^{r} \alpha_{j,r}^{[2]} j \right) \left(c_1 \frac{\partial f}{\partial x_1}(x) + c_2 \frac{\partial f}{\partial x_2}(x) \right) + 0\left(\xi_n^{2-\gamma^*}\right). \qquad (25.138)$$

Proof By Theorems 25.17 and 25.48. ■

We also give

Theorem 25.51 *Let* $f \in C^{m+l}(\mathbb{R}^N)$, $m, l, N \in \mathbb{N}$. *Assumptions of Theorem 25.37 are valid for* $d\varphi_{\xi_n}(s)$, $s \in \mathbb{R}^N$, $\xi_n \in (0,1]$, $n \in \mathbb{N}$. *Call* $\overline{\gamma} = 0, \overline{\beta}$. *Suppose* $\|f_{\overline{\gamma}+\overline{\alpha}}\|_\infty \le M$, $M > 0$, *for all* $\overline{\alpha}: |\overline{\alpha}| = m$. *Here* $\overline{c}_{\overline{\alpha},n,\widetilde{j}}$ *is as in (25.70), all* $|\overline{\alpha}| = \widetilde{j} = 1, ..., m-1$; $0 < \gamma^* \le 1$. *Assume* $T_{r,n}^{[m]}(f_{\overline{\gamma}}; x) \in \mathbb{R}$, $\forall\, x \in \mathbb{R}^N$. *Let also* $r, \beta \in \mathbb{N}$ *and* $\beta > \frac{m+1}{2}$. *Then*

$$\left(T_{r,n}^{[m]}(f;x)\right)_{\overline{\gamma}} - f_{\overline{\gamma}}(x) = \sum_{\widetilde{j}=1}^{m-1} \delta_{\widetilde{j},r}^{[m]} \left(\sum_{|\overline{\alpha}|=\widetilde{j}} \frac{\overline{c}_{\overline{\alpha},n,\widetilde{j}}\, f_{\overline{\gamma}+\overline{\alpha}}(x)}{\left(\prod_{i=1}^{N} \alpha_i!\right)} \right) + 0\left(\xi_n^{m-\gamma^*}\right). \tag{25.139}$$

When $m = 1$, *the sum collapses.*

Proof Use of Theorem 25.17 and Theorem 4.6, p. 54 of [3]. Here it is $\rho = \varphi$, see (25.68). ■

25.3.5 Simultaneous Approximation by Multivariate Complex $T_{r,n}^{[m]}$

We make

Remark 25.52 We consider here complex valued Borel measurable functions $f : \mathbb{R}^N \to \mathbb{C}$ such that $f = f_1 + i f_2$, $i = \sqrt{-1}$, where $f_1, f_2 : \mathbb{R}^N \to \mathbb{R}$ are implied to be real valued Borel measurable functions.

We define the multivariate complex Trigonometric singular operators

$$T_{r,n}^{[m]}(f;x) := T_{r,n}^{[m]}(f_1;x) + i T_{r,n}^{[m]}(f_2;x), \; x \in \mathbb{R}^N. \tag{25.140}$$

We assume that $T_{r,n}^{[m]}(f_j;x) \in \mathbb{R}$, $\forall\, x \in \mathbb{R}^N$, $j = 1, 2$.

One notices easily that

$$\left|T_{r,n}^{[m]}(f;x) - f(x)\right| \le \left|T_{r,n}^{[m]}(f_1;x) - f_1(x)\right| + \left|T_{r,n}^{[m]}(f_2;x) - f_2(x)\right| \tag{25.141}$$

also

$$\left\|T_{r,n}^{[m]}(f;x) - f(x)\right\|_{\infty,x} \le \left\|T_{r,n}^{[m]}(f_1;x) - f_1(x)\right\|_{\infty,x} + \left\|T_{r,n}^{[m]}(f_2;x) - f_2(x)\right\|_{\infty,x} \tag{25.142}$$

and

$$\left\| T_{r,n}^{[m]}(f) - f \right\|_p \leq \left\| T_{r,n}^{[m]}(f_1) - f_1 \right\|_p + \left\| T_{r,n}^{[m]}(f_2) - f_2 \right\|_p, \ p \geq 1. \quad (25.143)$$

Furthermore, it holds

$$f_{\overline{\alpha}}(x) = f_{1,\overline{\alpha}}(x) + i f_{2,\overline{\alpha}}(x), \quad (25.144)$$

where $\overline{\alpha}$ denotes a partial derivative of any order and arrangement.

We give

Theorem 25.53 *Let $f : \mathbb{R}^N \to \mathbb{C}$, such that $f = f_1 + i f_2$, $j = 1, 2$. Here $r, N, \beta, m \in \mathbb{N}$, $f_j \in C^m(\mathbb{R}^N)$, $x \in \mathbb{R}^N$. Assume $\left\| \frac{\partial^m f_j(\cdot, \cdot, \ldots, \cdot)}{\partial x_1^{\alpha_1} \ldots \partial x_N^{\alpha_N}} \right\|_\infty < \infty$, for all $\alpha_i \in \mathbb{Z}^+$, $i = 1, \ldots, N : |\overline{\alpha}| := \sum_{i=1}^N \alpha_i = m$. Let φ_{ξ_n} be the Borel probability measure on $\mathbb{R}^N$, see (25.39), where $\xi_n \in (0, 1]$, $n \in \mathbb{N}$. Here $\beta > \frac{m+r+1}{2}$, and $A_{\xi_n}(\overline{\alpha})$ as in (25.47), and $\overline{c}_{\overline{\alpha},n} := \overline{c}_{\overline{\alpha},n,\widetilde{j}}$ as in (25.70). Then*

$$\left\| T_{r,n}^{[m]}(f; x) - f(x) - \sum_{\widetilde{j}=1}^{m} \delta_{\widetilde{j},r}^{[m]} \left(\sum_{\substack{\alpha_1, \ldots, \alpha_N \geq 0; \\ |\overline{\alpha}| = \widetilde{j}}} \frac{\overline{c}_{\overline{\alpha},n,\widetilde{j}} f_{\overline{\alpha}}(x)}{\prod_{i=1}^N \alpha_i!} \right) \right\|_{\infty, x}$$

$$\leq \sum_{\substack{\alpha_1, \ldots, \alpha_N \geq 0; \\ |\overline{\alpha}| = m}} \frac{\left(\omega_r\left(f_{1,\overline{\alpha}}, \xi_n\right) + \omega_r\left(f_{2,\overline{\alpha}}, \xi_n\right) \right)}{\left(\prod_{i=1}^N \alpha_i! \right)} A_{\xi_n}(\overline{\alpha}). \quad (25.145)$$

Proof By Theorem 25.20. ∎

We proceed with

Theorem 25.54 *Let $f : \mathbb{R}^N \to \mathbb{C} : f = f_1 + i f_2$, $N \in \mathbb{N} - \{1\}$, $j = 1, 2$. Here $f_j \in C_B(\mathbb{R}^N)$ uniformly continuous, $\beta, r \in \mathbb{N}$, $\beta > \frac{r+1}{2}$, $\xi_n \in (0, 1]$, $n \in \mathbb{N}$. Then*

$$\left\| T_{r,n}^{[0]} f - f \right\|_\infty \leq 2^N \gamma^{-N} \left[\sum_{\lambda=0}^{r} \binom{r}{\lambda} \psi_\lambda \right]^N \xi_n^{2\beta(N-1)} \quad (25.146)$$

$$\left(\omega_r(f_1, \xi_n) + \omega_r(f_2, \xi_n) \right),$$

As $n \to \infty$ and $\xi_n \to 0$, we derive

$$\left\| T_{r,n}^{[0]} f - f \right\|_\infty \to 0 \quad (25.147)$$

with rates.

Proof By Theorem 25.22. ∎

Next comes multi-simultaneous approximation.

Theorem 25.55 *Let $f : \mathbb{R}^N \to \mathbb{C} : f = f_1 + if_2$, $j = 1, 2$. Here $f_j \in C^{m+l}(\mathbb{R}^N)$, $N, m, l \in \mathbb{N}$. The assumptions of Theorem 25.37 are valid for f_j and $d\mu_{\xi_n} = d\varphi_{\xi_n}$. Call $\overline{\gamma} = 0, \overline{\beta}$. Assume $\|f_{j,\overline{\gamma}+\overline{\alpha}}\|_\infty < \infty$, and let $\xi_n \in (0, 1]$, $n \in \mathbb{N}$. Here $\beta, r \in \mathbb{N}$, $\beta > \frac{m+r+1}{2}$, and $A_{\xi_n}(\overline{\alpha})$ as in (25.47), and $\overline{c}_{\overline{\alpha},n} := \overline{c}_{\overline{\alpha},n,\widetilde{j}}$ as in (25.70). Then*

$$\left\| \left(T_{r,n}^{[m]}(f;\cdot)\right)_{\overline{\gamma}} - f_{\overline{\gamma}}(\cdot) - \sum_{\widetilde{j}=1}^{m} \delta_{\widetilde{j},r}^{[m]} \left(\sum_{\substack{\alpha_1,\ldots,\alpha_N \geq 0; \\ |\overline{\alpha}| = \widetilde{j}}} \frac{\overline{c}_{\overline{\alpha},n,\widetilde{j}} f_{\overline{\gamma}+\overline{\alpha}}(\cdot)}{\prod_{i=1}^{N} \alpha_i!} \right) \right\|_\infty \tag{25.148}$$

$$\leq \sum_{\substack{\alpha_1,\ldots,\alpha_N \geq 0; \\ |\overline{\alpha}| = m}} \frac{\left(\omega_r\left(f_{1,\overline{\gamma}+\overline{\alpha}}, \xi_n\right) + \omega_r\left(f_{2,\overline{\gamma}+\overline{\alpha}}, \xi_n\right)\right)}{\left(\prod_{i=1}^{N} \alpha_i!\right)} A_{\xi_n}(\overline{\alpha}).$$

Proof Based on Theorems 25.37 and 25.41. ■

We continue with

Theorem 25.56 *Let $f : \mathbb{R}^N \to \mathbb{C} : f = f_1 + if_2$, $j = 1, 2$. Here $f_j \in C_B^l(\mathbb{R}^N)$, $N \in \mathbb{N} - \{1\}$, $l \in \mathbb{N}$ (functions l-times continuously differentiable and bounded). The assumptions of Theorem 25.37 are valid for f_j and $d\mu_{\xi_n} = d\varphi_{\xi_n}$. Call $\overline{\gamma} = 0, \overline{\beta}$, $\xi_n \in (0, 1]$, $n \in \mathbb{N}$. Let also $\beta, r \in \mathbb{N}$, $\beta > \frac{r+1}{2}$. Then*

$$\left\| \left(T_{r,n}^{[0]} f\right)_{\overline{\gamma}} - f_{\overline{\gamma}} \right\|_\infty \leq 2^N \gamma^{-N} \left[\sum_{\lambda=0}^{r} \binom{r}{\lambda} \psi_\lambda \right]^N$$

$$\xi_n^{2\beta(N-1)} \left(\omega_r\left(f_{1,\overline{\gamma}}, \xi_n\right) + \omega_r\left(f_{2,\overline{\gamma}}, \xi_n\right)\right). \tag{25.149}$$

If $\xi_n \to 0$, as $n \to \infty$, then $\left(T_{r,n}^{[0]} f\right)_{\overline{\gamma}} \to f_{\overline{\gamma}}$ uniformly.

Proof By Theorems 25.42 and 25.37. ■

We proceed with L_p approximations

Theorem 25.57 *Let $f : \mathbb{R}^N \to \mathbb{C} : f = f_1 + if_2$, $j = 1, 2$. Here $f_j \in C^m(\mathbb{R}^N)$, $r, \beta, N, m \in \mathbb{N}$, with $f_{j,\overline{\alpha}} \in L_p(\mathbb{R}^N)$, $|\overline{\alpha}| = m$, $x \in \mathbb{R}^N$. Let $p, q > 1 : \frac{1}{p} + \frac{1}{q} = 1$. Here φ_{ξ_n} is a Borel probability measure on $\mathbb{R}^N$ as in (25.39), for $\xi_n \in (0, 1]$, $n \in \mathbb{N}$. Let $\beta > \frac{[rp]+m+1}{2}$; $\alpha_i \in \mathbb{Z}^+$, $i = 1, \ldots, N : |\overline{\alpha}| := \sum_{i=1}^{N} \alpha_i = m$. Here $\overline{c}_{\overline{\alpha},n} := \overline{c}_{\overline{\alpha},n,\widetilde{j}}$ as in (25.70), where $\widetilde{j} = 1, \ldots, m$, and $\overline{\alpha} := (\alpha_1, \ldots, \alpha_N)$, $\alpha_i \in \mathbb{Z}^+$, $i = 1, \ldots, N$, $|\overline{\alpha}| := \sum_{i=1}^{N} \alpha_i = \widetilde{j}$. Then*

$$\left\| T_{r,n}^{[m]}(f;x) - f(x) - \sum_{\widetilde{j}=1}^{m} \delta_{\widetilde{j},r}^{[m]} \left(\sum_{|\widetilde{\alpha}|=\widetilde{j}} \frac{\overline{c}_{\widetilde{\alpha},n,\widetilde{j}} f_{\widetilde{\alpha}}(x)}{\left(\prod_{i=1}^{N} \alpha_i!\right)} \right) \right\|_{p,x}$$

$$\leq \left(\frac{m}{(q(m-1)+1)^{\frac{1}{q}}} \right) 2^{\frac{N}{p}} \gamma^{-\frac{N}{p}} \left\{ \sum_{\lambda=0}^{\lceil rp \rceil} \binom{\lceil rp \rceil}{\lambda} (\psi_\lambda + \psi_{\lambda+mp}) \right\}^{\frac{N}{p}}$$

$$\left(\sum_{|\widetilde{\alpha}|=m} \frac{1}{\prod_{i=1}^{N} \alpha_i!} \right) \xi_n^{\left(\frac{2\beta(N-1)}{p}+m\right)} \left[\omega_r\left(f_{1,\widetilde{\alpha}}, \xi_n\right)_p + \omega_r\left(f_{2,\widetilde{\alpha}}, \xi_n\right)_p \right]. \tag{25.150}$$

Proof By Theorem 25.25. ■

We continue with

Theorem 25.58 *Let* $f : \mathbb{R}^N \to \mathbb{C} : f = f_1 + i f_2$, $j = 1, 2$. *Here* $f_j \in \left(C\left(\mathbb{R}^N\right) \cap L_p\left(\mathbb{R}^N\right)\right)$; $N \in \mathbb{N} - \{1\}$, $\beta, r \in \mathbb{N}$; $p, q > 1 : \frac{1}{p} + \frac{1}{q} = 1$; $\beta > \frac{\lceil rp \rceil + 1}{2}$, $\xi_n \in (0, 1]$, $n \in \mathbb{N}$. *Then*

$$\left\| T_{r,n}^{[0]}(f) - f \right\|_p \leq 2^{\frac{N}{p}} \gamma^{-\frac{N}{p}} \left[\sum_{\lambda=0}^{\lceil rp \rceil} \binom{\lceil rp \rceil}{\lambda} \psi_\lambda \right]^{\frac{N}{p}}$$

$$\xi_n^{\frac{2\beta(N-1)}{p}} \left[\omega_r(f_1, \xi_n)_p + \omega_r(f_2, \xi_n)_p \right]. \tag{25.151}$$

As $\xi_n \to 0$, *when* $n \to \infty$, *we derive* $\left\| T_{r,n}^{[0]} f - f \right\|_p \to 0$, *i.e.* $T_{r,n}^{[0]} \to I$, *the unit operator, in* L_p *norm.*

Proof By Theorem 25.27. ■

We also give

Theorem 25.59 *Let* $f : \mathbb{R}^N \to \mathbb{C} : f = f_1 + i f_2$, $j = 1, 2$. *Here* $f_j \in \left(C\left(\mathbb{R}^N\right) \cap L_1\left(\mathbb{R}^N\right)\right)$; $N \in \mathbb{N} - \{1\}$, $r, \beta \in \mathbb{N}$, $\beta > \frac{r+1}{2}$, $\xi_n \in (0, 1]$, $n \in \mathbb{N}$. *Then*

$$\left\| T_{r,n}^{[0]}(f) - f \right\|_1 \leq 2^N \gamma^{-N} \left[\sum_{\lambda=0}^{r} \binom{r}{\lambda} \psi_\lambda \right]^N$$

$$\xi_n^{2\beta(N-1)} \left(\omega_r(f_1, \xi_n)_1 + \omega_r(f_2, \xi_n)_1 \right). \tag{25.152}$$

As $\xi_n \to 0$, *we get* $T_{r,n}^{[0]} \to I$, *in* L_1 *norm.*

Proof By Theorem 25.29. ∎

We further present

Theorem 25.60 *Let* $f : \mathbb{R}^N \to \mathbb{C} : f = f_1 + if_2$, $j = 1, 2$. *Here* $f_j \in C^m(\mathbb{R}^N)$, $N, \beta, m, r \in \mathbb{N}$, *with* $f_{j,\overline{\alpha}} \in L_1(\mathbb{R}^N)$, *where* $\alpha_i \in \mathbb{Z}^+$, $i = 1, ..., N$, $|\overline{\alpha}| := \sum_{i=1}^{N} \alpha_i = m$, $x \in \mathbb{R}^N$, $\xi_n \in (0, 1]$, $n \in \mathbb{N}$ *and* $\beta > \frac{m+r+1}{2}$. *Here* $\overline{c}_{\overline{\alpha},n} := \overline{c}_{\overline{\alpha},n,\widetilde{j}}$ *as in (25.70), where* $\widetilde{j} = 1, ..., m$, *and* $\overline{\alpha} := (\alpha_1, ..., \alpha_N)$, $\alpha_i \in \mathbb{Z}^+$, $i = 1, ..., N$, $|\overline{\alpha}| := \sum_{i=1}^{N} \alpha_i = \widetilde{j}$. *Also here* φ_{ξ_n} *is the Borel probability measure on* $\mathbb{R}^N$, *see (25.39). Then*

$$\left\| T_{r,n}^{[m]}(f; x) - f(x) - \sum_{\widetilde{j}=1}^{m} \delta_{\widetilde{j},r}^{[m]} \left(\sum_{|\overline{\alpha}|=\widetilde{j}} \frac{\overline{c}_{\overline{\alpha},n,\widetilde{j}} f_{\overline{\alpha}}(x)}{\prod_{i=1}^{N} \alpha_i!} \right) \right\|_{1,x} \tag{25.153}$$

$$\leq \left\{ \sum_{|\overline{\alpha}|=m} \left(\frac{1}{\prod_{i=1}^{N} \alpha_i!} \right) \left[\omega_r\left(f_{1,\overline{\alpha}}, \xi_n\right)_1 + \omega_r\left(f_{2,\overline{\alpha}}, \xi_n\right)_1 \right] \right\} \xi_n^{2\beta(N-1)+m}$$

$$2^N \gamma^{-N} \left\{ \sum_{\lambda=0}^{r} \binom{r}{\lambda} \left[\psi_\lambda + \psi_{\lambda+m} \right] \right\}^N.$$

Proof By Theorem 25.31. ∎

We continue with simultaneous L_p approximations.

Theorem 25.61 *Let* $f : \mathbb{R}^N \to \mathbb{C} : f = f_1 + if_2$, $j = 1, 2$. *Here* $f_j \in C^{m+l}(\mathbb{R}^N)$, $r, \beta, N, m, l \in \mathbb{N}$. *The assumptions of Theorem 25.37 are valid for* $d\mu_{\xi_n} = d\varphi_{\xi_n}$, $\xi_n \in (0, 1]$, $n \in \mathbb{N}$ *and* f_j. *Call* $\overline{\gamma} = 0, \overline{\beta}$. *Let* $f_{j,(\overline{\gamma}+\overline{\alpha})} \in L_p(\mathbb{R}^N)$, $|\overline{\alpha}| = m$, $x \in \mathbb{R}^N$, *and* $p, q > 1 : \frac{1}{p} + \frac{1}{q} = 1$. *Let* $\beta > \frac{\lceil rp \rceil + m + 1}{2}$; $\alpha_i \in \mathbb{Z}^+$, $i = 1, ..., N : |\overline{\alpha}| := \sum_{i=1}^{N} \alpha_i = m$. *Here* $\overline{c}_{\overline{\alpha},n} := \overline{c}_{\overline{\alpha},n,\widetilde{j}}$ *as in (25.70), where* $\widetilde{j} = 1, ..., m$, *and* $\overline{\alpha} := (\alpha_1, ..., \alpha_N)$, $\alpha_i \in \mathbb{Z}^+$, $i = 1, ..., N$, $|\overline{\alpha}| := \sum_{i=1}^{N} \alpha_i = \widetilde{j}$. *Then*

$$\left\| \left(T_{r,n}^{[m]}(f; x) \right)_{\overline{\gamma}} - f_{\overline{\gamma}}(x) - \sum_{\widetilde{j}=1}^{m} \delta_{\widetilde{j},r}^{[m]} \left(\sum_{|\overline{\alpha}|=\widetilde{j}} \frac{\overline{c}_{\overline{\alpha},n,\widetilde{j}} f_{\overline{\gamma}+\overline{\alpha}}(x)}{\prod_{i=1}^{N} \alpha_i!} \right) \right\|_{p,x}$$

$$\leq \left(\frac{m}{(q(m-1)+1)^{\frac{1}{q}}} \right) 2^{\frac{N}{p}} \gamma^{-\frac{N}{p}} \left\{ \sum_{\lambda=0}^{\lceil rp \rceil} \binom{\lceil rp \rceil}{\lambda} (\psi_\lambda + \psi_{\lambda+mp}) \right\}^{\frac{N}{p}} \tag{25.154}$$

$$\left(\sum_{|\overline{\alpha}|=m} \frac{1}{\prod_{i=1}^{N} \alpha_i!} \right) \xi_n^{\left(\frac{2\beta(N-1)}{p}+m\right)} \left[\omega_r \left(f_{1,\overline{\gamma}+\overline{\alpha}}, \xi_n\right)_p + \omega_r \left(f_{2,\overline{\gamma}+\overline{\alpha}}, \xi_n\right)_p \right].$$

Proof By Theorems 25.37 and 25.43. ■

We give also

Theorem 25.62 *Let $f : \mathbb{R}^N \to \mathbb{C} : f = f_1 + i f_2$, $j = 1, 2$. Here $f_j \in C^l(\mathbb{R}^N)$, $N \in \mathbb{N} - \{1\}$, $l \in \mathbb{N}$. The assumptions of Theorem 25.37 are valid for $d\mu_{\xi_n} = d\varphi_{\xi_n}$, $\xi_n \in (0, 1]$, $n \in \mathbb{N}$ and f_j. Call $\overline{\gamma} = 0, \overline{\beta}$. Let $f_{j,\overline{\gamma}} \in L_p(\mathbb{R}^N)$ and $p, q > 1 : \frac{1}{p} + \frac{1}{q} = 1$. Here $\beta, r \in \mathbb{N}$, $\beta > \frac{\lceil rp \rceil + 1}{2}$. Then*

$$\left\| \left(T_{r,n}^{[0]}(f)\right)_{\overline{\gamma}} - f_{\overline{\gamma}} \right\|_p \leq 2^{\frac{N}{p}} \gamma^{-\frac{N}{p}} \left[\sum_{\lambda=0}^{\lceil rp \rceil} \binom{\lceil rp \rceil}{\lambda} \psi_\lambda \right]^{\frac{N}{p}}$$

$$\xi_n^{\frac{2\beta(N-1)}{p}} \left[\omega_r \left(f_{1,\overline{\gamma}}, \xi_n\right)_p + \omega_r \left(f_{2,\overline{\gamma}}, \xi_n\right)_p \right]. \tag{25.155}$$

As $n \to +\infty$ and $\xi_n \to 0$, then $\left(T_{r,n}^{[0]} f\right)_{\overline{\gamma}} \overset{\|\cdot\|_p}{\to} f_{\overline{\gamma}}$.

Proof By Theorems 25.37 and 25.44. ■

We continue with

Theorem 25.63 *Let $f : \mathbb{R}^N \to \mathbb{C} : f = f_1 + i f_2$, $j = 1, 2$. Here $f_j \in C^l(\mathbb{R}^N)$, $N \in \mathbb{N} - \{1\}$, $l \in \mathbb{N}$. The assumptions of Theorem 25.37 are valid for $d\mu_{\xi_n} = d\varphi_{\xi_n}$, $\xi_n \in (0, 1]$, $n \in \mathbb{N}$ and f_j. Call $\overline{\gamma} = 0, \overline{\beta}$. Let $f_{j,\overline{\gamma}} \in L_1(\mathbb{R}^N)$ and $\beta, r \in \mathbb{N}$, $\beta > \frac{r+1}{2}$. Then*

$$\left\| \left(T_{r,n}^{[0]}(f)\right)_{\overline{\gamma}} - f_{\overline{\gamma}} \right\|_1 \leq 2^N \gamma^{-N} \left[\sum_{\lambda=0}^{r} \binom{r}{\lambda} \psi_\lambda \right]^N$$

$$\xi_n^{2\beta(N-1)} \left[\omega_r \left(f_{1,\overline{\gamma}}, \xi_n\right)_1 + \omega_r \left(f_{2,\overline{\gamma}}, \xi_n\right)_1 \right]. \tag{25.156}$$

As $n \to +\infty$ and $\xi_n \to 0$, then $\left(T_{r,n}^{[0]}(f)\right)_{\overline{\gamma}} \overset{\|\cdot\|_1}{\to} f_{\overline{\gamma}}$.

Proof By Theorems 25.37 and 25.45. ■

We finish with

Theorem 25.64 *Let* $f : \mathbb{R}^N \to \mathbb{C} : f = f_1 + if_2$, $j = 1, 2$. *Here* $f_j \in C^{m+l}\left(\mathbb{R}^N\right)$, $N, \beta.r, m, l \in \mathbb{N}$. *The assumptions of Theorem 25.37 are valid for* $d\mu_{\xi_n} = d\varphi_{\xi_n}$, $\xi_n \in (0, 1]$, $n \in \mathbb{N}$ *and* f_j. *Call* $\overline{\gamma} = 0, \overline{\beta}$. *Let* $f_{j,(\overline{\gamma}+\overline{\alpha})} \in L_1\left(\mathbb{R}^N\right)$, $|\overline{\alpha}| = m$, $x \in \mathbb{R}^N$. *Here* $\beta > \frac{m+r+1}{2}$ *and* $\overline{c}_{\overline{\alpha},n} := \overline{c}_{\overline{\alpha},n,\widetilde{j}}$ *as in (25.70), where* $\widetilde{j} = 1, ..., m$, *and* $\overline{\alpha} := (\alpha_1, ..., \alpha_N)$, $\alpha_i \in \mathbb{Z}^+$, $i = 1, ..., N$, $|\overline{\alpha}| := \sum_{i=1}^{N} \alpha_i = \widetilde{j}$. *Then*

$$\left\| \left(T_{r,n}^{[m]}(f;x)\right)_{\overline{\gamma}} - f_{\overline{\gamma}}(x) - \sum_{\widetilde{j}=1}^{m} \delta_{\widetilde{j},r}^{[m]} \left(\sum_{|\overline{\alpha}|=\widetilde{j}} \frac{\overline{c}_{\overline{\alpha},n,\widetilde{j}} f_{\overline{\gamma}+\overline{\alpha}}(x)}{\prod_{i=1}^{N} \alpha_i!} \right) \right\|_{1,x}$$

$$\leq \left(\sum_{|\overline{\alpha}|=m} \left(\frac{1}{\prod_{i=1}^{N} \alpha_i!} \right) \left[\omega_r\left(f_{1,\overline{\gamma}+\overline{\alpha}}, \xi_n\right)_1 + \omega_r\left(f_{2,\overline{\gamma}+\overline{\alpha}}, \xi_n\right)_1 \right] \right)$$

$$\xi_n^{2\beta(N-1)+m} 2^N \gamma^{-N} \left\{ \sum_{\lambda=0}^{r} \binom{r}{\lambda} \left[\psi_\lambda + \psi_{\lambda+m} \right] \right\}^N . \tag{25.157}$$

Proof By Theorems 25.37 and 25.46. ∎

References

1. Anastassiou, G.A.: Rate of convergence of non-positive linear convolution type operators. A sharp inequality. J. Math. Anal. and Appl. **142**, 441–451 (1989)
2. Anastassiou, G.A.: Moments in Probability and Approximation Theory. Pitman Research Notes in Mathematics, vol. 287. Longman Scientific & Technical, Harlow (1993)
3. Anastassiou, G.A.: Approximation by Multivariate Singular Integrals. Springer, New York (2011)
4. Anastassiou, G.A.: Approximations by multivariate generalized trigonometric type singular integral operators. Submitted (2020)
5. Anastassiou, G., Gal, S.: Approximation Theory. Birkhaüser, Boston (2000)
6. Anastassiou, G.A., Mezei, R.A.: L_p convergence with rates of general singular integral operators. J. Comput. Anal. Appl. **14**(6), 1067–1083 (2012)
7. Anastassiou, G.A., Mezei, R.A.: Convergence of complex general singular integral operators. J. Concr. Appl. Math. **10**(3–4), 259–283 (2012)
8. Anastassiou, G.A., Mezei, R.A.: Uniform convergence with rates of general singular operators. CUBO **15**(2), 1–19 (2013)
9. DeVore, R.A., Lorentz, G.G.: Constructive Approximation, vol. 303. Springer, Berlin (1993)
10. Edwards, J.: A Treatise on the Integral Calculus, vol. II. Chelsea, New York (1954)

Chapter 26
Concluding Remarks

In a 1695 letter to L'Hospital, Leibniz asked, "Can we generalize ordinary derivatives to ones of arbitrary order?" L'Hospital replied to Leibniz with another question, "What if the order is 1/2?" Then Leibniz, in a letter dated September 30, 1695 replied, "It will lead to a paradox, from which one day many useful consequences will be drawn." That was the beginning of fractional calculus. The subject has been ongoing for more than 300 years now. Many famous mathematicians contributed to the topic over the years such as Liouville, Euler, Laplace, Lagrange, Riemann, Weyl, Fourier, Abel, Lacroix, Grunwald, and Letnikov.

The first attempt to give a rigorous definition of fractional derivatives and study the subject in depth was by Liouville during 1832–1855. The concept grew out of his earlier work in electromagnetism. Fractional derivatives describe solutions of fractional integral equations, many times arising from physics, as done by Abel in 1823 to solve the brachistochrone problem.

Fractional calculus has a long history but, from the applicative point of view, it fell into oblivion for hundreds of years because of lack of applications to other sciences such as physics and engineering. This oblivion was also due to its complexity and lack of physical and geometric connection. Fortunately, that was then as things have changed dramatically. Fractional calculus has developed a lot in the last fifty years and especially since 1974 when the first international conference in the field took place, organized by B. Ross, in New Haven, Connecticut, USA. Now we frequently see such conferences around the world. There exist a lot of related research activities resulting in many interesting articles and books. AMS/Mathscinet lists over 50000 fractional publications during 2020. We now see many important applications: in acoustic wave propagation in inhomogeneous porous material, diffusive transport, fluid flow, dynamical processes in self-similar structures, dynamics of earthquakes, optics, geology, viscoelastic materials, biosciences, bioengineering, medicine, economics, probability and statistics, astrophysics, chemical engineering, physics, splines, tomography, fluid mechanics, electromagnetic waves, non-linear control, signal processing, control of electronic power, converters, chaotic dynam-

G. A. Anastassiou, *Constructive Fractional Analysis with Applications*,
Studies in Systems, Decision and Control 362,
https://doi.org/10.1007/978-3-030-71481-9_26

ics, polymer science, proteins, polymer physics, electrochemistry, statistical physics, rheology, thermodynamics, neural networks, and many others. By now almost all fields of research in science and engineering use fractional calculus to better describe them. The current mathematical theory of fractional calculus seems to lag behind the needs for mathematical modeling of all the above applications. Therefore there exists a lot of room for mathematical theoretical expansion on the subject. The recent development of fractal theory has shown great connections to fractional calculus. In general, fractional calculus currently provides the best description for fractals and chaotic situations. All in all fractional calculus provides the best expression for the complexity of our modern science. In recent years researchers came up with more efficient and practical fractional derivatives, e.g. M. Caputo (1967) inspired by studies in Geophysics, J.A. Canavati (1987) and by many the new method of fractional variable order (2019). The author was the first to study fractional inequalities and fractional approximation theory by publishing numerous of articles and monographs. In this monograph the author continuous his studies by employing his constructive and computational fractional calculus methods (2020).

He applies his generalized fractional differentiation techniques of Riemann-Liouville, Caputo and Canavati types and of the very new fractional variable order to various kinds of inequalities such as of Opial, Hardy, Hilbert-Pachpatte and over the spherical shell. He continues with E. R. Love left and right sides fractional integral inequalities. The fractional Landau inequalities, of left and right sides, univariate and multivariate, including ones for Semigroups occupied fourteen chapters of this monograph. These were developed to all possible directions and right side multivariate fractional Taylor formulae were proven for that purpose. He continued with several of the very new Gronwall fractional inequalities of variable order. The last part of book was about constructive approximation theory, it included: ordinary and fractional approximations by positive sublinear operators, and high order approximation by multivariate generalized Picard, Gauss-Weierstrass, Poisson-Cauchy and Trigonometric singular integrals which are not positive operators. Definitely what is developed in this monograph can have a great impact and can lead to more interesting avenues of research.